The Flora of Canada
Part 4 – Dicotyledoneae (Loasaceae to Compositae)

National Museum of Natural Sciences
Publications in Botany, No. 7(4)

Published by the
National Museums of Canada

Staff editor
Bonnie Livingstone

Musée national des Sciences naturelles
Publications de Botanique, n⁰ 7(4)

Publié par
Les musées nationaux du Canada

The Flora of Canada

Part 4 – Dicotyledoneae (Loasaceae to Compositae)

H.J. Scoggan

LOASACEAE (Loasa Family)

MENTZELIA L. [5383] Blazing Star

Annual, biennial, or perennial herbs with mostly alternate, sessile to slender-petioled, entire to pinnatifid, exstipulate, brittle leaves (floral leaves sometimes opposite), these very adhesive by the barbed (glochidiate) pubescence. Flowers perfect, regular, epigynous, solitary at the ends of the branches or irregularly cymose. Calyx 5-lobed. Petals 5 or sometimes apparently 10 (but the inner 5 then more or less staminodial and narrower), cream-colour to yellow, distinct. Stamens very numerous. Ovary inferior. Fruit a capsule.

1 Petals 5, yellow, to 6 mm long; calyx-lobes to 4 mm long; stamens 15–35; seeds prismatic, not flattened; flowers in irregular leafy cymes; leaves to 1 dm long, the upper ones sessile, the lower ones subsessile or short-petioled; annuals to 4 dm tall; (s B.C.).

  2 Cymes congested, their bracts mostly broadly lanceolate to ovate; capsules linear, to 3 cm long, their seeds 1-rowed throughout, grooved on the vertical margins, very obscurely pebbled; leaves narrowly to broadly lanceolate, entire to sinuate-lobed or even subpinnatifid; (s ?B.C.) .... [*M. dispersa*]

  2 Cymes not congested, their bracts narrowly to broadly lanceolate; capsules linear-clavate, to 2.5 cm long, their basal seeds usually 1-rowed and often grooved on the angles, the upper seeds usually irregularly disposed and not grooved, all of the seeds rather prominently pebbled; leaves various, the basal ones usually linear and entire to shallowly few-lobed, the cauline ones linear to lanceolate, from subentire to laciniate into linear lobes; (s B.C.) .... *M. albicaulis*

1 Petals to 8 cm long, the 5 true petals alternating with 5 more or less petaloid staminodia; calyx-lobes to 4 cm long; stamens very numerous; seeds flattened; flowers solitary at the ends of the branches (and often in the top 1 or 2 leaf-axils in *M. laevicaulis*); leaves to about 1.5 dm long; biennials or perennials to about 1 m tall, from a deep taproot.

  3 Petals creamy or pale yellow, apparently 10 (but the inner 5 staminodial and not quite as broad as the others); floral bracts adherent to the ovary; seeds thin-margined but not winged; leaves fleshy, lanceolate, sharply sinuate-pinnatifid, the lower ones petioled, the upper ones sessile and often somewhat clasping; stems usually solitary (occasionally 2 or 3), simple below, branching above; (s Alta. to sw Man.) .... *M. decapetala*

  3 Petals lemon-yellow, 5 (the 5 outer stamens sometimes flattened and somewhat petaloid but much narrower than the true petals); floral bracts not adherent to the ovary; seeds distinctly wing-margined; lower leaves petioled, oblanceolate, deeply sinuate-pinnatifid and somewhat runcinate-lobed, the upper ones sessile, oblong to ovate-oblong, less deeply lobed; stem usually solitary and branched above but also often branched near the base; (s B.C.) .... *M. laevicaulis*

M. albicaulis Dougl.
/t/W/ (T) Dry (usually sandy) valleys and foothills from s B.C. (Fraser–Thompson Valley N to Lillooet and Cache Creek and in the Dry Interior s to Keremeos, about 25 mi sw of Penticton) and Mont. to s Calif. and N.Mex. [*Acrolasia* Rydb.; *Bartonia* Dougl.; *A. gracilis* and *A. ctenophora* Rydb.; *M. cten.* and *M. tweedyi* Rydb.; *M. ?gracilenta* T. & G.].

M. decapetala (Pursh) Urban & Gilg   Stickleaf, Gumbo-Lily, Evening Star
/T/WW/ (Hp) Dry prairies, plains, and lower montane slopes from s Alta. (West Butte; Fort Macleod; Belly R.; Lethbridge; Medicine Hat), s Sask. (Cypress Hills, Eastend, Empress, and Roche Percée; Breitung 1957*a*), and sw Man. (Boissevain, about 40 mi s of Brandon; WIN) to Nev., Mexico, Tex., and Okla. [*Bartonia* Pursh; *Nuttallia* Greene; *B. (M.) ornata* Pursh].

[*M. dispersa* Wats.]
[Collections in CAN from s B.C. (N to Lillooet) have been placed here but might apparently

equally well be referred to *M. albicaulis*. The species ranges in the w U.S.A. from Wash. and Mont. to s Calif. and Colo. (*Acrolasia* Davidson; *M. albicaulis* var. *integrifolia* Wats.).]

M. laevicaulis (Dougl.) T. & G.
/t/W/   (Hs)   Dry valleys and lower montane slopes from s B.C. (Dry Interior N to near Lillooet, E to Keremeos, about 25 mi sw of Penticton) and Mont. to Calif., Utah, and Wyo. [*Bartonia* Dougl.; *Nuttallia* Greene].

   The B.C. plant is referable to var. *parviflora* (Dougl.) Hitchc. (*Bartonia parv.* Dougl.; petals to 4 cm long rather than to 8 cm, capsules usually less than 2 cm long rather than to 3.5 cm).

CACTACEAE (Cactus Family)

Stems very fleshy, ovoid to globose (*Coryphantha*) or consisting of subterete or flattened jointed segments (*Opuntia*), green, leafless but commonly with long slender sharp spines subtended by a cluster of woolly hairs or small bristles. Flowers large and showy, regular, perfect, epigynous, sessile (or the base of the ovary prolonged). Sepals, petals, and stamens each numerous, distinct. Ovary inferior. Fruit a dry, pulpy, or juicy berry.

1 Stem a single ovoid to globose body covered with spine-bearing tubercles; flowers greenish white or purple, arising from the base of young tubercles near the summit of the stem, they and the spines subtended by a cluster of woolly hairs . . . . *Coryphantha*
1 Stem jointed, the segments subterete or distinctly flattened, the large yellow flowers borne along the margins of the newer segments; spines subtended by spinose barbed bristles . . . . *Opuntia*

CORYPHANTHA (Engelm.) Lemaire [5411]

1 Flowers greenish white, barely reddish-tinged; fruit reddish, subglobose, less than 1 cm long; main spines solitary, greyish; (?Man.) . . . . [*C. missouriensis*]
1 Flowers reddish purple; fruit greenish, oblong, 1 or 2 cm long; main spines in clusters of 3 or more, one of them deflexed, the others ascending, reddish brown; (s Alta. to sw Man.) . . . . *C. vivipara*

[C. missouriensis (Sweet) Britt. & Rose]
[Reports of this species of the w U.S.A. (N to Idaho and Mont.) from Man. by Burman (1909), Jackson et al. (1922), Rydberg (1932), and Hitchcock et al. (1961) require confirmation. (*Mamillaria* Sweet; *Cactus* Ktze.; *Neomamillaria* Britt. & Rose; not *Opuntia missouriensis* DC.).]

C. vivipara (Nutt.) Britt. & Brown
/T/WW/ (Ch (succulent)) Dry sandy prairies and hillsides from s Alta. (Fort Macleod; Medicine Hat), Sask. (N to near Saskatoon), and sw Man. (Lauder; Grande Clarière; Oak Lake; Virden; Spruce Woods Forest Reserve) to Oreg., Colo., Kans., and w Minn. [*Cactus* Nutt.; *Mamillaria* Haw.; *Neomamillaria* Britt. & Rose].

OPUNTIA Mill. [5417] Prickly Pear, Indian Fig

1 Segments of stem not greatly flattened, readily detached, usually less than 5 cm long, their spines about 5 in a cluster, to about 2 cm long; fruit dry, spiny, to 2 cm long; (B.C. to Ont.) . . . . *O. fragilis*
1 Segments of stem conspicuously flattened, not easily detached.
  2 Fruit juicy, red or red-purple, not spiny, to 5 cm long; segments of stem to 2.5 dm long, the spines wanting or solitary and up to 5 cm long; (s Ont.) . . . . *O. compressa*
  2 Fruit dry, spiny, about 2 cm long; segments of stem to 1.5 dm long, the spines in clusters of at least 5; (B.C. to Sask.) . . . . *O. polyacantha*

O. compressa (Salisb.) Macbr.
/t/EE/ (Ch (succulent)) Dry sands and rocks from s Minn. to s Ont. (Pelee Point and Pelee Is., Essex Co.; a report from Long Point, Norfolk Co., noted by John Macoun 1883) and Mass., s to Okla., Mo., Miss., Ala., and Ga. [*Cactus* Salisb.; *O. humifusa* Raf.; *O. opuntia* Karst.; *O. rafinesquii* Engelm.; *O. vulgaris* of auth., not Mill.]. MAP: Benson 1962: fig. 3-16, p. 74.

O. fragilis (Nutt.) Haw.
/T/WW/ (Ch (succulent)) Dry prairies, sand-hills, and rocks from B.C. (N to Taylor Flats, in the Peace R. system at 56°08′N) to Alta. (N to the Peace R. system at 56°12′N), s Sask. (N to Swift Current), s Man. (Spruce Woods Forest Reserve SE of Brandon; Whiteshell Forest Reserve and the Lake of the Woods region E of Winnipeg), and Ont. (islands and shores of Rainy L. and Lake of the Woods; reported from near Kaladar, Lennox & Addington Co., by Roland Beschel, Blue Bill 14:11.

1967), s to N Calif., Tex., Kans., Iowa, and Ill. [*Cactus* Nutt.; *O. missouriensis sensu* John Macoun 1883, as to the Peace River, Alta., plant, not DC., and *O. polyacantha sensu* Raup 1934, not Haw., the relevant collections in CAN].

O. polyacantha Haw.
/T/WW/ (Ch (succulent)) Dry prairies, sand-hills, and rocks from s B.C. (Saltspring Is.; Dry Interior N to Kamloops and Kelowna), s Alta. (along the Belly, Milk, and Red Deer rivers), and s Sask. (N to Saskatoon) to Oreg., Ariz., Tex., and Mo. [*O. missouriensis* DC., not *Mamillaria (Coryphantha) miss.* Sweet; *Cactus ferox* Nutt.].

THYMELAEACEAE (Mezereum Family)

Shrubs to 1 or 2 m tall, with simple, alternate, entire, stipulate, oblanceolate to oblong-obovate leaves to about 8 cm long. Flowers regular, perfect, perigynous, about 1 cm long, sessile or subsessile in lateral clusters, preceding the leaves in early spring. Calyx coloured and petaloid, 4-lobed or the lobes obsolete. Petals none. Stamens 8. Ovary partially inferior. Fruit a 1-seeded berry-like drupe.

1 Calyx rose-purple, pubescent, with 4 somewhat spreading lobes; stamens and style included, the anthers nearly sessile at the top of the calyx-tube; drupe red; leaves oblanceolate, tapering to the subpetiolar base; (introd.) .... *Daphne*
1 Calyx light yellow, glabrous, its lobes obsolete; stamens and style long-exserted, the anther-filaments elongate; drupe green, yellowish, or red; leaves oval-obovate, usually rounded at the very short-petioled base; twigs jointed; bark fibrous and remarkably strong; (Ont. to N.S.) .... *Dirca*

DAPHNE L.   [5455]   Mezereum

D. mezereum L.   Daphne.   Bois gentil or Bois joli
Eurasian; a garden-escape to roadsides, thickets, and limestone quarries in N. America, as in Ont. (N to the Ottawa dist.), sw Que. (N to the Montreal dist.), Nfld. (Boivin 1966b), N.B. (Fowler 1885; near Fredericton), P.E.I., and N.S.

DIRCA L.   [5458]

D. palustris L.   Leatherwood, Wicopy.   Bois de plomb
/T/EE/   (N)   Rich deciduous or mixed woods from Minn. to Ont. (N to near Thunder Bay and Haileybury, 47°27'N), Que. (N to L. Nominingue, Labelle Co., and Montmorency Falls, E of Quebec City), N.B. (York and Madawaska counties; not known from P.E.I.), and N.S. (gypsum quarries in Hants Co.), s to La. and N Fla. MAPS: *Atlas of Canada* 1957: map 13, sheet 38; Stebbins 1942: fig. 1, p. 246; the s Ont. distribution is shown in a map by Soper and Heimburger 1961:25.

ELAEAGNACEAE (Oleaster Family)

Shrubs or small trees with silvery- or rusty-scurfy, entire, exstipulate, short-petioled, opposite or alternate leaves. Flowers perfect or unisexual, regular, perigynous, in small clusters in the leaf-axils. Calyx-lobes usually 4. Petals none. Stamens 2, 4, or 8. Ovary apparently inferior (but not actually adnate to the calyx-tube). Fruit pulpy and drupe-like.

1 Leaves opposite, densely rusty- or silvery-scurfy beneath; flowers unisexual; sepals 4;
  stamens 8; fruit fleshy . . . . *Shepherdia*
1 Leaves alternate, densely silvery-scurfy at least beneath; sepals and stamens each 2 or 4.
  2 Sepals and stamens each 4; flowers perfect or polygamous (mixed perfect and
    unisexual); hypanthium-tube elongated; fruit silvery-scurfy, dry and mealy; leaves
    lanceolate to ovate or obovate . . . . *Elaeagnus*
  2 Sepals and stamens each 2; flowers unisexual; hypanthium-tube short; fruit orange;
    leaves linear-lanceolate, to 8 cm long, silvery-scurfy on both sides or becoming
    subglabrous and dull green above; (introd. in Alta.) . . . . *Hippophaë*

ELAEAGNUS L. [5472] Oleaster. Chalef

1 Leaves lanceolate, commonly over 3 times as long as broad, silvery-scurfy beneath,
  greener above, flattish-margined; young twigs silvery-scurfy; style surrounded or
  covered by a disc; plant to 7 m tall, sometimes thorny; (introd.) . . . . *E. angustifolia*
1 Leaves ovate- to obovate-oblong, commonly less than 3 times as long as broad,
  silvery-scurfy on both sides, strongly wavy-margined; young twigs rusty-scurfy; disc
  none; plant to 4 m tall, not thorny; (B.C. to Que.) . . . . *E. commutata*

E. angustifolia L. Russian Olive, Oleaster
Eurasian; spreading from cult. in N. America, as in B.C. (Boivin 1966*b*), Alta. (Moss 1959), Man. (Boivin 1966*b*), and s Ont. (York, Middlesex, and Lincoln counties).

E. commutata Bernh. Silverberry. Bois d'argent or Chalef
/ST/(X)/ (N (Mc)) Prairies, dry fields, gravel ridges, and shores from N-cent. Alaska–Yukon (N to ca. 67°30′N) and w-cent. Dist. Mackenzie (N to Norman Wells, ca. 65°N) to B.C.–Alta., Sask. (N to Prince Albert), Man. (N to the Nelson R. about 20 mi sw of York Factory, Hudson Bay; CAN), and northernmost Ont., s to Idaho, Utah, S.Dak., Minn., cent. Ont. (s to the NW shore of L. Superior near Thunder Bay, the Blackwater R., L. Nipigon, and the Albany and Moose rivers sw of James Bay at ca. 51°N), and sw-cent. Que. (SE James Bay at ca. 52°30′N s to the Harricanaw R. at ca. 50°10′N; near L. Timiskaming at ca. 47°25′N, 79°30′W; *see* the James Bay watershed map by Dutilly, Lepage, and Duman 1958: fig. 12, p. 136); isolated stations in E Que. (Quebec City dist.; St-Augustin, Portneuf Co.; Bic, Rimouski Co.; Gaspé Pen. at Mont St-Pierre and Percé and along the Métis and Bonaventure rivers). [*E. argentea* Nutt., not Moench; *E. veteris-castelli* Lepage]. MAP: Hultén 1968*b*:684.

HIPPOPHAË L. [5470]

H. rhamnoides L. Sea Buckthorn
Eurasian; reported as introd. in Alta. by Moss (1959; "Sometimes found as an escape and well adapted for growth on sandy soil and steep slopes.").

SHEPHERDIA Nutt. [5471]

1 Leaves wedge-oblong, densely silvery-scurfy beneath; young twigs silvery-scurfy; older
  branches commonly spine-tipped; fruit scarlet; shrub or small tree to about 6 m tall; (Alta.
  to Man.) . . . . *S. argentea*
1 Leaves ovate-lanceolate to ovate, densely rusty-scurfy beneath; young twigs rusty-
  scurfy; fruit yellowish red; unarmed shrub usually less than 2 m tall; (transcontinental) . . . .
  . . . . *S. canadensis*

S. argentea (Pursh) Nutt.   Buffalo-berry
/T/WW/   (Mc)   Open woods, thickets, rocks, and shores from s Alta. (N to Medicine Hat), s Sask. (N to near Saskatoon), and s Man. (N to Lac du Bonnet, about 50 mi NE of Winnipeg) to s Calif., N.Mex., Kans., and Iowa. [*Hippophaë* Pursh; *Lepargyraea* Greene; reports from B.C. probably refer to *Elaeagnus commutata; see* Boivin 1967a]. The Canadian area in the MAPS by Preston (1961:328; 1947:252) is completely erroneous, apparently applicable to *S. canadensis* but incomplete northwards for that species.

S. canadensis (L.) Nutt.   Soapberry
/ST/X/   (N (Mc))   Open woods, thickets, rocks, and shores from N Alaska–Yukon–Dist. Mackenzie (N to ca. 69°N) to Great Slave L., L. Athabasca (Alta. and Sask.), Man. (N to Churchill), northernmost Ont., Que. (N to s Ungava Bay, L. St. John, and the Côte-Nord), Nfld., N.B., and N.S. (Hants, Inverness, and Victoria counties; not known from P.E.I.), s to Oreg., N.Mex., S.Dak., Minn., Ohio, and New Eng. [*Hippophaë* L.; *Elaeagnus* Nels.; *Lepargyraea* Greene]. MAPS: Hultén 1968b:684; Raup 1947: pl. 30.

LYTHRACEAE (Loosestrife Family)

Herbs with chiefly opposite or whorled, entire, exstipulate leaves and 4-angled stems. Flowers perfect, perigynous, regular (somewhat irregular in *Cuphea*). Calyx-lobes and petals each 5, 6, or 7, the stamens as many or twice as many (then in 2 sets of unequal length), inserted at the throat of the calyx-tube below the petals, the calyx with a small appendage in each sinus. Ovary partially inferior. Fruit a many-seeded capsule.

1 Stems prostrate, rooting at the nodes; leaves obovate-spatulate, 1 or 2 cm long; flowers about 1 mm long, subsessile, solitary in the leaf-axils, the petals early deciduous; capsule subglobose, about 1.5 mm long; plant glabrous .... [*Peplis*]
1 Stems erect or ascending; flowers and capsules larger.
  2 Calyx (and capsule) globose or nearly so; petals purplish; plants of wet or muddy habitats.
    3 Flowers few and crowded in nearly or quite sessile axillary cymes; petals 4, small (the blade about 1 mm long), deciduous or sometimes wanting; calyx 4-angled and 4-toothed; leaves opposite, oblong-lanceolate to oblong, cordate-clasping at base, to 4 cm long; annual to about 4 dm tall; (s B.C.) .... *Ammannia*
    3 Flowers long-pedicelled in clusters in the upper axils; petals 5, about 12 mm long; calyx 5–7-angled and with 5–7 teeth; leaves opposite or whorled, nearly sessile, lanceolate, to 1.5 dm long; stems to 2.5 dm long, the arching submersed base spongy-thickened; perennial; (Ont., sw Que., and N.S.) .... *Decodon*
  2 Calyx cylindric; petals usually 6, purple or red-purple; stems softer.
    4 Calyx saccate on one side at base; 2 petals larger than the other 4; flowers solitary in the leaf-axils or in short racemes; leaves lanceolate to ovate-lanceolate, long-petioled, opposite; very viscid-hairy branching annual; (introd.) .... [*Cuphea*]
    4 Calyx regular at base; petals equal; leaves sessile; plants perennial .... *Lythrum*

AMMANNIA L. [5474]

A. coccinea Rottb.
/t/X/ (T) Wet shores and muddy (often alkaline) places from s B.C. (Osoyoos L., near the U.S.A. boundary about 30 mi s of Penticton, where "apparently thriving in alkali encrustations" along muddy edges of the lake; Eastham 1947) and Wash. to Nebr., Minn., Ill., and Ohio, s to Tex., Mexico, and Fla.; tropical America.

[CUPHEA P. Br.] [5478]

[*C. petiolata* (L.) Koehne] Clammy Cuphea, Blue Waxweed
[Native in the E U.S.A. (Iowa to Ind., Ohio, and New Eng., s to La. and Ga.; introd. elsewhere, as in s Ont. (John Macoun 1890; *C. viscosissima*, "In cultivated fields between Hamilton and St. Catherines, Ont."), where apparently not taken since that time and scarcely established. (*Lythrum* L.; *C. viscosissima* Jacq.).]

DECODON Gmel. [5488] Swamp Loosestrife

D. verticillatus (L.) Ell. Water-willow, Water-oleander
/T/EE/ (Ch) Swamps and shallow pools from Minn. to Wisc., Ont. (N to near Chalk River, Renfrew Co.; DAO; *see* s Ont. map by Soper 1962: fig. 19, p. 31), sw Que. (N to Pontiac and Gatineau counties and the Montreal dist.; *see* s Que. map by Robert Joyal, Nat. can. (Que.) 97(5): map D, fig. 2, p. 564. 1970), and N.S. (Shelburne and Digby counties; not known from N.B. or P.E.I.), s to La. and Fla. [*Lythrum* L.; *Nesaea* HBK.].

Some of our material is referable to var. *laevigatus* T. & G. (plant nearly or quite glabrous rather than soft-pubescent).

LYTHRUM L.   [5476]   Loosestrife

1   Median and upper leaves alternate, the flowers solitary in their axils; petals purple, about
5 mm long; stamens 5, 6, or 7; leaves to about 5 cm long; plant glabrous; (?B.C. and
s Ont.) .... *L. alatum*
1   Leaves mostly opposite or whorled (or the uppermost ones and the floral bracts alternate); flowers in whorls in a terminal interrupted leafy spike; petals red-purple, to 1 cm
long; stamens usually 12; (introd.) .... *L. salicaria*

L. alatum Pursh
/t/EE/   (Hpr)   Swamps, meadows, prairies, and ditches from S.Dak. to s Ont. (Essex, Kent,
Lambton, Middlesex, Norfolk, Waterloo, and York counties) and N.Y., s to Tex., La., and Ga.
Reports from B.C. require confirmation. It may have been introd. there.

L. salicaria L.   Spiked Loosestrife.   Salicaire or Bouquet violet
Eurasian; locally abundant in N. America (ranges of Canadian taxa outlined below) along shores
and in wet meadows and river-floodplains. MAP and synonymy: *see* below.
1   Plant essentially glabrous; spike slender and loosely flowered; [a nursery-escape becoming natzd. near Otterburne, about 30 mi s of Winnipeg, Man.; Löve and Bernard
1959] .... var. *gracilior* Turcz.
1   Plant more or less pubescent; spike relatively thick and compact.
2   Calyx and bracts white-tomentose; [var. *pubescens* Pursh; *L. tomentosum* Mill.;
B.C. (Vancouver Is.), Man. (N to Neepawa, about 35 mi NE of Brandon), Ont. (N to
Kapuskasing, 49°24′N), and Que. (N to Anticosti Is. and the Gaspé Pen.)] ....
.... var. *tomentosum* (Mill.) DC.
2   Calyx and bracts greenish, somewhat pubescent; [s B.C. (Vancouver Is.; Vancouver;
Cloverdale; Chilliwack), s Alta. (Calgary), Man. (N to Neepawa), Ont. (N to Lanark
and Carleton counties), Que. (recorded N to near Montreal but probably extending
considerably farther northeastwards along the St. Lawrence R.), Nfld., N.B., P.E.I.,
and N.S.; MAP (aggregate species): Meusel 1943: fig. 35b] .... var. *salicaria*

[PEPLIS L.]   [5475]

[P. portula L.]   Water-Purslane
[Reports of this European species from Labrador by Schrank (1818), E. Meyer (1830), and
Schlechtendal (1836) undoubtedly refer to some other species.]

NYSSACEAE (Sour Gum Family)

NYSSA L. [6151] Tupelo, Sour Gum

Tree to over 30 m tall with alternate, commonly entire, exstipulate, elliptic to obovate leaves to about 1.5 dm long. Flowers unisexual, the staminate ones borne in peduncled umbels or umbel-like racemes, with commonly about 10 stamens. Pistillate flowers sessile in clusters of up to 8 at the end of a peduncle. Calyx-lobes 5, minute. Petals 5, very small and fleshy, or none. Ovary inferior. Fruit a dark-blue or black drupe. (S Ont.).

N. sylvatica Marsh.   Black Gum
/t/EE/   (Ms)   Woods, swamps, damp sands, and shores from SE Minn. to Wisc., S Ont. (all the counties bordering L. Erie N to Lambton, Middlesex, and Lincoln counties; *see* S Ont. map by Fox and Soper 1953: map 25, p. 27), and Maine, S to Mexico, Tex., and Fla. MAPS: Little 1971: map 144-N; Hosie 1969:290; Fowells 1965:278; Canada Department of Northern Affairs and Natural Resources 1956:266; Preston 1961:326; Hough 1947:363; Munns 1938: map 157, p. 161; Braun 1935: fig. 1, p. 352; combine the maps by M.L. Fernald, Rhodora 37(444): map 1, p. 435, and map 4 (var. *caroliniana*), p. 436. 1935.

Fernald's map 1 indicates a station for the typical form in S Ont., apparently in Norfolk Co. However, most of our material is referable to var. *caroliniana* (Poir.) Fern. (*N. car.* Poir.; *N. multiflora sensu* John Macoun 1884, not Wang.; leaves of the fertile shoots relatively thin and broad, papillate beneath rather than glabrous or glabrate, tapering to the acuminate apex rather than often abruptly short-acuminate).

MELASTOMATACEAE (Melastoma Family)

RHEXIA L.   [5664]   Deergrass, Meadow-beauty

Perennial herb with opposite minutely dentate ovate-lanceolate to ovate sessile leaves, the narrowly 4-winged stem to about 1 m tall, from tuberous-thickened roots. Flowers showy, purple, perfect, regular, perigynous, to 2 cm long, in terminal cymes. Calyx-lobes and petals each 4. Stamens 8, inserted with the petals at the summit of the calyx-tube, the anthers opening by apical pores. Ovary apparently inferior (but not adnate to the calyx-tube). Fruit a 4-locular capsule.

R. virginica L.   Common Meadow-beauty
/T/EE/   (Grt)   Peats, wet sands, and gravels from s Ont. (N to the Georgian Bay dist., L. Huron, and Algonquin Provincial Park, about 150 mi E of Ottawa; *see* s Ont. map by Soper 1956: map 10, p. 84) and N.S. (Yarmouth, Shelburne, Queens, Annapolis, and Lunenburg counties) to Mo., Tenn., Ala., and Ga.; introd. in sw B.C. (Lulu Is., where presumably introd. with blueberry plants imported from the East; Herb. V).

ONAGRACEAE (Evening-Primrose Family)

Herbs with simple, opposite or alternate, entire to deeply pinnatifid, exstipulate leaves. Flowers perfect, regular or nearly so, epigynous. Sepals and petals each 2 or 4. Stamens 2, 4, or 8. Style solitary, slender, the stigma 2–4-lobed or capitate. Ovary inferior. Fruit few-seeded and indehiscent or a many-seeded dehiscent capsule.

1   Sepals, petals, and stamens each 2; fruits obovoid or pear-shaped, usually slightly
    compressed, bristly with hooked hairs, to 5 mm long, on reflexed pedicels, indehiscent;
    leaves ovate, rounded or cordate at base, long-petioled, opposite .... Circaea
1   Sepals and petals each 4 (the petals sometimes wanting in *Ludwigia*); stamens 4 or 8;
    fruit not bristly; leaves mostly narrower in outline, sessile or short-petioled.
    2   Fruit indehiscent, obtusely 4-angled, 6 or 7 mm long, with at most 4 seeds; petals
        white to pink or red, to 5 or 6 mm long, slender-clawed, ephemeral; stamens 8; stigma
        shortly 4-lobed; stamens 8; stem-leaves alternate, sessile .... Gaura
    2   Fruit a dehiscent many-seeded capsule opening lengthwise by valves, usually
        loculicidal.
        3   Seeds with a tuft of silky hairs (a coma) at summit (except in *E. glandulosum* var.
            ecomosum); capsule slender, its separated valves recurving; stamens 8; petals
            white, pink, or purple (yellow only in *E. luteum*); leaves opposite or alternate or
            both .... Epilobium
        3   Seeds naked, lacking a coma; leaves nearly always alternate.
            4   Capsule less than twice as long as broad, 4-angled, many-seeded, less than
                1 cm long; sepals persistent; hypanthium not prolonged above the ovary;
                stamens 4; stems to over 1 m long, erect to depressed, creeping, or
                floating; perennials of aquatic, marshy, or muddy habitats .... Ludwigia
            4   Capsule usually several times as long as broad, seldom 4-angled, often over 1
                cm long; hypanthium usually prolonged as a tube above the ovary (and
                capsule), but usually deciduous as the fruit matures; stamens 8 (but 4 of the
                anthers reduced and non-functional in *Clarkia pulchella*).
                5   Ovary (and capsule) 2-locular, the capsule dehiscing by 4 valves, linear to
                    linear-clavate, its seeds 1-rowed in each locule; petals white to pinkish, at
                    most 1 mm long; stigma capitate; leaves linear or linear-spatulate, entire;
                    annuals; (s B.C. and sw Alta.) .... Gayophytum
                5   Ovary (and capsule) 4-locular; petals usually conspicuous.
                    6   Petals mostly yellow, sometimes white (frequently aging reddish or
                      purplish); anthers usually versatile (attached near the middle to the
                      filaments); annuals or perennials .... Oenothera
                  6   Petals pink to lavender, rose-purple, or purple; anthers erect, attached
                    near the base; leaves linear to lanceolate (those subtending the flowers
                    often broader), entire or denticulate.
                    7   Flowers axillary, sessile or subsessile; calyx-lobes erect; (B.C. to
                      sw Sask.) .... Boisduvalia
                    7   Flowers pedicelled or, if sessile, the calyx-lobes either reflexed or
                    connate and turned to one side; (s B.C.) .... Clarkia

BOISDUVALIA Spach   [5798]

1   Petals to over 8 mm long; capsules slenderly fusiform, straight, very short-beaked, to 1
    cm long, the internal partitioning septa completely free from the valves, almost entirely
    adherent to the seed-bearing placentae, these persisting as a central core until the seeds
    (usually less than 6 in each locule) are shed; flowers crowded but the inflorescence
    elongating in fruit; leaves to 5 cm long, lanceolate to ovate, entire or remotely denticulate;
    plant usually densely ashy-strigose to softly pilose and also often glandular; (s B.C. and
    s Alta.) .... B. densiflora

1 Petals at most 4 mm long; capsules narrower in outline, their internal septa adherent to the valves, the seed-bearing placentae usually disintegrating as the seeds (usually 6 or more in each locule) are shed.

    2 Plant villous-pilose, usually ashy; flowers usually not crowded; leaves to 4 cm long, linear to narrowly lanceolate, entire or remotely denticulate; capsules to about 1 cm long, short-beaked, usually somewhat curved and constricted between the seeds, these at most 8 in each locule; (sw B.C.) . . . . *B. stricta*

    2 Plant strigose to glabrate, greenish; flowers usually crowded; leaves less than 2 cm long, lanceolate to narrowly ovate, denticulate; capsules averaging about 7 mm long, pointed but not beaked, nearly straight, with often more than 8 seeds in each locule; (B.C. to sw Sask.) . . . . *B. glabella*

B. densiflora (Lindl.) Wats.
/T/W/ (T) Moist ground from sw B.C. (Vancouver Is., where first taken by Dawson in 1876; CAN) and s Alta. (Lethbridge; CAN) to Baja Calif., Nev., Idaho, and Mont. [*Oenothera* Lindl.; *O. (B.) salicina* Nutt.; *B. douglasii* Spach]. MAP: P.H. Raven and D.M. Moore, Brittonia 17(3): fig. 14, p. 248. 1965.

B. glabella (Nutt.) Walp.
/T/WW/ (T) Moist ground from sw B.C. (Douglas L., Vancouver Is.), s Alta. (Milk River and near Medicine Hat; CAN), and Sask. (Cypress Hills; Mortlach; Bracken) to s Calif., Utah, and S.Dak. [*Oenothera* Nutt.]. MAP: Raven and Moore, loc. cit., fig. 15, p. 252.

B. stricta (Gray) Greene
/t/W/ (T) Moist ground from sw B.C. (Douglas L., Vancouver Is.; Herb. V) to cent. Calif. and Idaho. [*Gayophytum* Gray]. MAP: Raven and Moore, loc. cit., fig. 12, p. 248.

CIRCAEA L. [5828] Enchanter's Nightshade

1 Fruit longitudinally furrowed, to 5 mm long, equally 2-locular and 2-seeded; calyx-lobes to 2.5 mm broad; leaves usually more than twice as long as broad, firm, dark green above, rounded or at most subcordate at base; stem firm; (Man. to N.S.) . . . . *C. quadrisulcata*

1 Fruit not furrowed, at most about 3 mm long; calyx-lobes less than 2 mm broad; leaves usually less than twice as long as broad, pale green and more or less flaccid, cordate or subcordate at base; stem weak.

    2 Fruit 1-locular and 1-seeded; anthers less than 0.5 mm long; leaves mostly not over 5 cm long; stem usually less than 3 dm tall; (transcontinental) . . . . *C. alpina*

    2 Fruit unequally 3-locular but often only 1-seeded; anthers at least 0.5 mm long; leaves to over 8 cm long; stem to about 4.5 dm tall; (Que. to N.S.) . . . . *C. canadensis*

C. alpina L.
/ST/X/EA/ (Grh) Cool moist woods and clearings from the E Aleutian Is., cent. Alaska, and sw Dist. Mackenzie (not known from the Yukon but very close to it along the Alaska–B.C. boundary) to B.C.–Alta., Sask. (N to Windrum L. at ca. 56°N), Man. (N to Wekusko L., about 90 mi NE of The Pas), Ont. (N to Sandy L. at ca. 53°N), Que. (N to the E James Bay watershed at ca. 54°N, L. St. John, and the Côte-Nord), Labrador (N to the Hamilton R. basin), Nfld., N.B., P.E.I., and N.S., s to s Calif., Utah, Colo., S.Dak., Mich., Tenn., and Ga.; Eurasia. [Incl. var. *pacifica* (Aschers. & Magnus) Jones (*C. pac.* A. & M.), the leaves less deeply toothed and less deeply cordate at base than in the typical form]. MAP: Hultén 1968b:693.

C. canadensis Hill
/T/E/ (Grh) Rich or alluvial woods from Que. (N to L. St. John and the Gaspé Pen.; reports from Man. and Ont. require confirmation), N.B., and N.S. to W.Va. and Va. [*C. intermedia* of Canadian reports, not Ehrh.]. MAP: Hultén 1958: map 58, p. 77.

    Concerning the probable origin of this taxon through hybridization between *C. alpina* and *C. quadrisulcata* (var. ?*canadensis*), see T.S. Cooperrider (Rhodora 64(757):63–67. 1962).

C. quadrisulcata (Maxim.) Franch. & Sav.

/T/EE/A/ (Grh) Rich woods, thickets, and ravines from SE ?Man. (Lowe 1943; near the Ont. boundary) to Ont. (N to the Ottawa dist.), Que. (N to the S Gaspé Pen.; *see* Que. map by Doyon and Lavoie 1966: fig. 25 (*C. lut.* var. *can.*), p. 821), ?Nfld. (Boivin 1966*b*; not known from P.E.I.), N.B., and N.S., S to Okla., Mo., Tenn., and Ga.; Asia. [*C. lutetiana* var. *quad.* Maxim.]. MAP: Hultén 1958: map 57, p. 77.

The N. American plant is referable to var. *canadensis* (L.) Hara (*C. lut.* var. *can.* L.; *C. latifolia* Hill; petals white rather than roseate; sepals mostly greenish, sparingly glandular-pilose or glabrate, rather than brownish red and copiously glandular-pilose).

CLARKIA Pursh [5799]

(Ref.: Lewis and Lewis 1955)
1 Petals distinctly clawed, the claw usually with a pair of opposite short blunt teeth near base; flowers slightly irregular, not closing at night; flower-buds nodding; hypanthium to 3 mm long; plants finely strigose; (S B.C.).
   2 Petals 3-lobed to about the middle, the central lobe the broadest; fertile stamens 4, the anthers coiling after dehiscence, the other 4 stamens reduced and sterile; stigmas white; capsule about 2 cm long; leaves alternate, linear-lanceolate or spatulate, entire or denticulate, to 7 cm long and 1 cm broad; (S B.C.) . . . . *C. pulchella*
   2 Petals with a narrow entire rhomboidal blade; fertile stamens 8; stigma-lobes white or purple; capsule to over 3 cm long; leaves mostly subopposite, lanceolate to elliptic, to 7 cm long and 2 cm broad; (S ?B.C.) . . . . [*C. rhomboidea*]
1 Petals neither clawed nor lobed; flowers regular, tending to close at night.
   3 Calyx-lobes usually distinct and sharply reflexed (sometimes partly united); petals to 1.5 cm long, with or without a carmine or purplish central spot; hypanthium to 7 mm long; stigma-lobes about 1 mm long, purplish; capsule nearly or quite sessile, to 2.5 cm long and rather uniformly 2 or 3 mm thick, often becoming somewhat 4-angled; leaves to 5 cm long and 7 mm broad; (introd. in SW B.C.) . . . . [*C. quadrivulnera*]
   3 Calyx-lobes usually united and turned to one side under the flower; stigma-lobes creamy or yellow; capsules on pedicels to 1 cm long, linear to somewhat clavate, to 5 cm long; leaves to 7 cm long and 6 mm broad.
      4 Flower-buds and tip of inflorescence reflexed, becoming erect only as flowering progresses; hypanthium to 3 mm long; petals to 2 cm long, usually not spotted; style shorter than the stamens, the stigma-lobes creamy, to 1.5 mm long; capsules attenuate to a distinct slender beak; (SW ?B.C.) . . . . [*C. gracilis*]
      4 Flower-buds and tip of inflorescence erect; hypanthium to 1 cm long; petals to 4 cm long, usually carmine-spotted at the centre; style usually surpassing the stamens, the stigma-lobes yellow, to over 5 mm long; capsules beakless or with a beak to several mm long; (SW B.C.) . . . . *C. amoena*

C. amoena (Lehm.) Nels. & Macbr.

/t/W/ (T) Valleys and lower montane slopes from SW B.C. (Mt. Finlayson, Vancouver Is.; Mayne Is.; CAN) to cent. Calif. The species is reported as introd. in Que. by C. Rousseau (Nat. can. (Que.) 98(4):715. 1971; Ste-Foy, near Quebec City, as *Godetia amoena*). MAP and synonymy: *see* below.
1 Stigma linear, usually well over 2 mm long and surpassing the stamens; petals mostly over 2 cm long.
   2 Plant often sprawling, usually less than 1 m tall; [*Oenothera* Lehm.; *Godetia* Don; *G. epilobioides* Wats.; Calif. only, B.C. reports referring to the following varieties; MAP: Lewis and Lewis 1955: fig. 2, p. 266] . . . . [var. *amoena*]
   2 Plant erect, to over 2 m tall; [*Oenothera lindleyi* Dougl.; Vancouver Is.; MAP: on the above-noted map] . . . . var. *lindleyi* (Dougl.) Hitchc.
1 Stigma oval, usually less than 2 mm long and not surpassing the stamens; petals mostly less than 2 cm long; [*Godetia caurina* Abrams; *G. epilobioides* of B.C. reports, not Wats.; Vancouver Is.; MAP: on the above-noted map] . . . . var. *caurina* (Abrams) Hitchc.

[C. gracilis (Piper) Nels. & Macbr.]
[A report of this species of the W U.S.A. (S Oreg. and N Calif.) from Vancouver Is., B.C., by Piper (*see* Eastham 1947) requires confirmation. The MAP by Lewis and Lewis (1955: fig. 6, p. 281) indicates no Canadian stations. (*Godetia* Piper).]

C. pulchella Pursh
/t/W/   (T)   Valleys and shores from S B.C. (Vancouver Is. to Creston, S of 50°N) to Oreg., Idaho, and Mont. [*Oenothera* Levl.]. MAP: Lewis and Lewis 1955: fig. 27, p. 357.

[C. quadrivulnera (Dougl.) Nels. & Macbr.]
[Native in the W U.S.A. (Wash. to Baja Calif.); known in Canada only from SW B.C. (Nanaimo, Vancouver Is., where taken by John Macoun on wharf-ballast in 1887 but not noted since that date and thus scarcely established; CAN). (*Oenothera* Dougl.; *Godetia* Spach; *G. hispidula sensu* Macoun 1890, not Wats., the relevant collection being the above-noted one in CAN). The MAP by Lewis and Lewis (1955: fig. 13, p. 302) indicates the northernmost stations as being along the Juan de Fuca Strait in N Wash., just S of the S tip of Vancouver Is.]

[C. rhomboidea Dougl.]
[This species of the W U.S.A. (Wash. to S Calif., Utah, and Ariz.) is noted by Eastham (1947) as having been collected near the B.C. boundary and the MAP by Lewis and Lewis (1955: fig. 25, p. 349) indicates stations either in or very close to S B.C. The authors, however, do not cite any B.C. localities and its occurrence there requires confirmation. (*Oenothera* Levl.; *Phaeostoma* Nels.).]

EPILOBIUM L.   [5795]   Willow-herb

1   Stigma deeply 4-lobed, the lobes usually at least about 2 mm long; capsules to 8 cm long; leaves subsessile.
　　2   Petals yellow, shallowly obcordate, to 18 mm long; flowers 2–10 in the axils of the somewhat reduced upper leaves, the ovaries and usually the pedicels glandular-puberulent; corolla regular; stamens and style erect or ascending, the style usually considerably surpassing the petals and stamens; stigma-lobes about 2 mm long; free part of hypanthium 1 or 2 mm long; leaves mostly opposite, narrowly to broadly lanceolate, rounded to acuminate at apex, glandular-toothed, glabrous except for the usually puberulent margins; (B.C. and Alta.) .... *E. luteum*
　　2   Petals roseate to purplish; flowers in racemes.
　　　　3   Petals obcordate, purple, to 2 cm long, the corolla regular; stamens and style erect or ascending, the former in 2 or more series; free part of hypanthium about 2 mm long; leaves lanceolate to oblong, partly clasping at base, soft and hairy, sharply serrulate; plant long-villous with spreading hairs, from stout rope-like rhizomes; (introd.) .... *E. hirsutum*
　　　　3   Petals rounded at summit, roseate to purplish, to 3 cm long, the corolla slightly irregular; stamens and style successively declined, the former in a single series; free hypanthium lacking (the calyx cleft to the top of the ovary); plants glabrous or finely strigillose-puberulent; (transcontinental).
　　　　　　4   Stems depressed and densely matted, upwardly arching, commonly several from a caespitose rootstock; leaves mostly opposite, thick and fleshy, whitish or strongly glaucous, not veiny, less than 8 cm long; racemes usually not more than 15-flowered, their bracts not much reduced; style shorter than the stamens, glabrous; stigma-lobes short and thick .... *E. latifolium*
　　　　　　4   Stems erect, usually solitary, to about 2 m tall, from rhizome-like roots; leaves mostly alternate, relatively thin, green above, pale and reticulate-veiny beneath, commonly over 8 cm long; racemes usually at least 15-flowered, their bracts greatly reduced; style surpassing the stamens, hairy at base; stigma-lobes slender, soon recurved .... *E. angustifolium*
1   Stigma entire or only obscurely 4-lobed (the lobes coalescing); petals 2-lobed at summit, at most 1 cm long, whitish, pink, roseate, purplish, or violet, the corolla regular; stamens and style erect or ascending, the former in 2 or more series; hypanthium more or less

prolonged between the summit of the ovary and the base of the calyx; flowers corymbed, panicled, or few in the upper leaf-axils; plants glabrous or short-pubescent.

5 Plants annual, with a taproot, the pale dry epidermal cortex soon exfoliating from the lower part of the stem; leaves short-petioled, entire to somewhat denticulate; capsules linear-clavate, to 2.5 cm long; seeds beakless, their pale-yellow comas soon deciduous; petals white to roseate.

  6 Leaves chiefly opposite (at least below), the blades mostly oblong-lanceolate, to 2.5 cm long; petals seldom over 4 mm long; free part of hypanthium scarcely 1 mm long; seeds less than 1 mm long; stems soft-pubescent, mostly less than 4 dm tall; (B.C.) . . . . *E. minutum*

  6 Leaves chiefly alternate (except the uppermost), linear to narrowly lanceolate, those of the main stem to 7 cm long but those of the axillary clusters and branches usually much shorter; free part of hypanthium 2–12 mm long; seeds at least 1 mm long; plant glabrous except sometimes in the inflorescence, to 1(2) m tall; (B.C. to Man; introd. eastwards) . . . . *E. paniculatum*

5 Perennials (or biennials), often with bulblet-like offsets (turions), stolons, or autumnal rosettes, the epidermis of the stem usually not exfoliating; seeds often tapering above to a short collar or beak, their comas more persistent.

  7 Stem-leaves linear, lanceolate, or narrowly oblong, often revolute-margined, mostly entire or merely undulate, without decurrent lines running from their bases down the terete stem.

    8 Plant with short horizontally spreading pubescence; (Ont. to N.S.) . . . . *E. strictum*

    8 Plant glabrous or with fine incurved pubescence; (transcontinental) . . . . . . *E. palustre*

  7 Stem-leaves lanceolate to elliptic or ovate, flat, with lines running down from their bases along the usually distinctly 4-angled stem.

    9 Leaves entire or merely denticulate, to about 5 cm long; stems tufted or matted, from slender creeping basal offshoots, often sending out slender elongate stolons but not producing turions.

      10 Plant glabrous throughout or minutely pubescent in the inflorescence or on the ovary, usually distinctly glaucous, to 5 dm tall; seeds minutely papillate, with a white coma; (s B.C. and sw Alta.) . . . . *E. glaberrimum*

      10 Plant usually pubescent and not glaucous, the stem seldom over 3 dm tall, usually glabrous toward base and crisp-puberulent in lines above; seeds smooth or inconspicuously papillate, the coma dingy-white; (transcontinental) . . . . *E. alpinum*

    9 Leaves mostly distinctly toothed; stems usually solitary, lacking elongate basal offshoots, finally usually bearing sessile or short-stalked leafy rosettes or fleshy and scaly bulb-like offsets (turions).

      11 Stem-leaves grey-green, distinctly petioled, rugose-veiny; seeds blackish, with a cinnamon-coloured coma; (Ont. to N.S.) . . . . *E. coloratum*

      11 Stem-leaves not greyish, scarcely rugose-veiny; seeds tawny, with a white or whitish coma; (transcontinental).

        12 Turions (bulblet-like offsets) terminating the slender rhizomes . . . . *E. glandulosum*

        12 Turions wanting . . . . *E. watsonii*

E. alpinum L.

/aST/D/GEA/ (Hpr) Moist banks, rocky slopes, and meadows (often above timberline), the aggregate species from the Aleutian Is., cent. Alaska–Yukon, and Great Bear L. through B.C. and the mts. of sw Alta. to s Calif. and Colo., with a disjunct area in the East from northernmost Ungava–Labrador to E Que. (Côte-Nord, Anticosti Is., and Gaspé Pen.; not known from the Maritime Provinces), Nfld., and the mts. of N.H. and Maine; w and E Greenland N to ca. 70°N; Iceland; N Eurasia. MAPS and synonymy: *see* below.

1 Stems erect, glabrous or sparingly glandular above, not puberulent in lines; leaves sessile, linear to narrowly oblong-lanceolate, usually nearly erect, mostly rather crowded on the lower part of the stem; seeds smooth; [*E. oregonense* Haussk. and its var. *grac.* Trel.; Vancouver Is.] . . . . var. *gracillimum* (Trel.) Hitchc.

1　Stems usually decumbent at base, generally crisp-puberulent in lines decurrent from the
　　leaf-bases; leaves mostly short-petioled, lanceolate to ovate or ovate-oblong, spreading,
　　generally rather uniformly distributed along the stem.
　　2　Petals white or creamy to pale pinkish, to 5 mm long; seeds smooth; leaves to 5 cm
　　　　long; stems usually not matted, to about 3 dm tall; [*E. lactiflorum* Haussk.; trans-
　　　　continental; MAPS (*E. lact.*): Hultén 1968*b*:692, and 1958: map 176, p. 195; Raup
　　　　1947: pl. 31] . . . . var. *lactiflorum* (Haussk.) Hitchc.
　　2　Petals usually either pink or over 5 mm long; seeds often papillate; leaves often less
　　　　than 2 cm long; stems usually matted.
　　　　3　Seeds smooth; capsules linear, about 1 mm thick; petals to 6 mm long; leaves
　　　　　　mostly ovate, 1 or 2 cm long; stems mostly less than 1.5 dm tall; [*E. anagalli-
　　　　　　difolium* Lam.; incl. *E. behringianum* and *E. pseudoscaposum* Haussk.; *E. roseum*
　　　　　　of Alaskan reports in part, not Schreb.; transcontinental; MAPS (*E. anag.*): Hultén
　　　　　　1958: map 221, p. 241; Porsild 1957: map 259, p. 193; Raup 1947: pl. 31; combine
　　　　　　the maps by Hultén 1968*b*:689 (*E. anag.*) and 691 (*E. behr.*)] . . . . var. *alpinum*
　　　　3　Seeds papillate.
　　　　　　4　Capsules subclavate, to 2 mm thick above the middle; petals to 6 mm long;
　　　　　　　　leaves mostly ovate, 1 or 2 cm long; stems mostly less than 1.5 dm tall;
　　　　　　　　[*E. clavatum* Trel., the type, as first collection cited, from along the Kicking
　　　　　　　　Horse R., B.C.; mts. of B.C.–Alta.] . . . . var. *clavatum* (Trel.) Hitchc.
　　　　　　4　Capsules linear, about 1 mm thick; leaves lanceolate to ovate, to 5 cm long;
　　　　　　　　stems to about 3 dm tall; [*E. nutans* Hornem., not Schmidt; *E. alsinifolium* Vill.;
　　　　　　　　E. bongardii and *E. sertulatum* Haussk.; *E. hornemannii* Rchb.; *E. treleasi-
　　　　　　　　anum* Levl.; *E. origanifolium sensu* John Macoun 1883, at least in large part,
　　　　　　　　not Lam.; transcontinental; MAPS: combine the maps by Hultén 1968*b*:692 (*E.
　　　　　　　　horn.*) and 693 (*E. sert.*)] . . . . var. *nutans* (Hornem.) Hook.

E. angustifolium L.　Fireweed, Great Willow-herb.　Bouquets rouges
/aST/X/GEA/　(Hp (Gr))　Open woods, recent clearings (particularly burns), fields, and river
gravels, the aggregate species from the Aleutian Is. and coasts of Alaska–Yukon–Dist.
Mackenzie–Dist. Keewatin (floral emblem of the Yukon) to Baffin Is. (N to near the Arctic Circle) and
northernmost Ungava–Labrador, S to Calif., Ariz., N.Mex., S.Dak., Ohio, and N.C.; W Greenland N to
ca. 70°N, E Greenland N to ca. 66°N; Iceland; Eurasia. MAPS and synonymy: *see* below.
1　Leaves rarely over 1.5 dm long and at most 3 cm broad, their midribs beneath always
　　glabrous; (pollen-grains usually 3-pored and less than 85 microns in diameter) . . . .
　　. . . . . ssp. *angustifolium*
　　2　Petals roseate to purplish; [*Chamaenerion* Scop.; *C. exaltatum* Rydb.; *E. intermedium*
　　　　Wormskj. (*E. ang.* var. *int.* (Wormskj.) Fern.; *E. (C.) spicatum* Lam.; *E. pauciflorum*
　　　　Schrank; transcontinental; MAPS: Hultén 1968*b*:686; Porsild 1957: map 256, p. 192
　　　　(aggregate species); Raup 1947: pl. 31 (aggregate species); Theodore Mosquin,
　　　　Brittonia 18(2): fig. 1 (aggregate species), p. 168, and fig. 4 (ssp. *ang.*; solid dots),
　　　　p. 176. 1966] . . . . f. *angustifolium*
　　2　Petals white or whitish; [essentially transcontinental; one or both of the following forms
　　　　may eventually be referred to ssp. *circumvagum*].
　　　　3　Sepals white or whitish; [*E. spicatum* var. *alb.* Dum.] . . . . . . f. *albiflorum* (Dum.) Haussk.
　　　　3　Sepals red; [*Cham. ang.* var. *spect.* Simmons] . . . . f. *spectabile* (Simmons) Fern.
1　Leaves to over 2 dm long and as much as 4 cm broad, their midribs beneath glabrous to
　　very pubescent; (pollen usually a mixture of 3-pored and 4-pored grains, commonly over
　　85 microns in diameter); [vars. *macrophyllum* (Haussk.) Fern. (*E. ang.* f. *mac.* Haussk.)
　　and *platyphyllum* (Daniels) Fern. (*Cham. ang. plat.* Daniels); *C. (E.) ang.* var. *canes-
　　cens* Wood; *E. danielsii* D. Löve; transcontinental; MAPS: Mosquin, loc. cit., fig. 4 (open
　　rings), p. 176; Hultén 1968*b*:686 (ssp. *macrophyllum*)] . . . . ssp. *circumvagum* Mosquin

E. coloratum Biehler
/T/EE/　(Hp)　Low ground and springy slopes from S.Dak. to Minn., Ont. (N to the Ottawa dist.;
Gillett 1958; reported N to the Kaministikwia R. W of Thunder Bay by John Macoun, Can. J., n.s.
15(94, 95).1877), Que. (N to near Ste-Anne-de-la-Pocatière, Kamouraska Co.; DAO; reported N to

Cacouna, Kamouraska Co., by Penhallow 1891), N.B. (Kent and Victoria counties), and N.S. (Digby, Kings, and Colchester counties; reports from P.E.I. require confirmation), s to Kans., Ark., Tenn., Ala., and Ga.

E. glaberrimum Barbey
/T/W/ (Hp) Moist valleys and slopes from s B.C. (Skagit and Chilliwack valleys; CAN; reported from Rossland by Eastham 1947) and sw Alta. (Waterton Lakes; Breitung 1957b) to s Calif. and Utah. [E. affine var. fastigiatum Nutt.; E. platyphyllum Rydb.].

E. glandulosum Lehm.
/sT/X/eA/ (Hpr) Moist places at low to moderate elevations, the aggregate species from the Aleutian Is. and s Alaska–Yukon to L. Athabasca (Alta. and Sask.), Man. (N to about 175 mi NE of The Pas at ca. 55°30′N, 97°30′W; GH, detd. Fernald, as var. macounii), Ont. (N to the N shore of L. Superior), Que. (N to the ?Côte-Nord and Gaspé Pen.), Nfld., and N.S. (Antigonish and Inverness counties; not known from N.B. or P.E.I.), s to s Calif., Colo., Minn., N.Y., and Vt.; E Asia; introd. in Europe (Hultén 1958). (Through past confusion with E. watsonii (incl. E. gl. var. adenocaulon), the above statement of range should be considered as merely tentative). MAPS and synonymy: see below.

1   Seed-coma rudimentary or wanting; seeds closely covered with prominent ridges of pale
    papillae; [E. ecomosum (Fassett) Fern.; E. ciliatum var. ecom. (Fassett) Boivin; fresh to
    brackish tidal flats of the St. Lawrence R., Que., in Portneuf, Quebec, Lotbinière, Lévis,
    Bellechasse (type from St-Vallier), and Montmagny counties] .... var. ecomosum Fassett
1   Seed-coma well developed; seeds mostly less prominently papillate.
    2   Median stem-leaves distinctly cordate-based; [E. boreale of Nfld. reports, not
      Haussk.; transcontinental; type from Tabletop Mt., Gaspé Pen., E Que.] ....
      .... var. cardiophyllum Fern.
    2   Stem-leaves narrowed to base or barely subcordate.
      3   Petals at most 5 mm long.
        4   Capsules glandular-pilose; leaves sessile or petioled; stems not pubescent
          in lines; [E. delicatum Trel. and its var. tenue Trel.; E. brevistylum Barbey;
          E. pringleanum Haussk.; Alaska–B.C.] .... var. tenue (Trel.) Hitchc.
        4   Capsules minutely strigose, nonglandular; leaves usually petioled; stems
          pubescent in lines; [E. halleanum Haussk.; E. leptocarpum Haussk. and its
          var. macounii Trel. (type from L. Athabasca); transcontinental; MAP (E. lept.):
          Hultén 1968b:690] .... var. macounii (Trel.) Hitchc.
      3   Petals mostly over 6 mm long; capsules glandular-puberulent; leaves sessile;
        stems glabrous below but crisp-puberulent above (usually in lines decurrent from
        the leaf-bases); [E. drummondii and E. saximontanum Haussk.; E. ?affine Bong.;
        incl. the glabrous extremes, E. ?scalare and E. ?steckerianum Fern.; transconti-
        nental; MAP: Hultén 1968b:690] .... var. glandulosum

E. hirsutum L.   Great Hairy Willow-herb
Eurasian; introd. in meadows, roadside thickets, and waste places in N. America, as in Ont. (N to the Ottawa dist.), sw Que. (N to the Montreal dist.), and N.S. (Yarmouth and Halifax counties).

E. latifolium L.   River-beauty
/AST/X/GEA/ (Hp (Ch)) River-gravels, streambanks, and damp slopes at low to alpine elevations from the Aleutian Is. and coasts of Alaska–Yukon–Dist. Mackenzie–Dist. Keewatin to northernmost Ellesmere Is. and northernmost Ungava–Labrador, s to Oreg., Idaho, Colo., S.Dak., N Man. (s to the Nelson R. at Gillam Is., about 165 mi s of Churchill; CAN; not known from Sask.), N Ont. (coasts of Hudson Bay–James Bay s to ca. 53°N), Que. (s to E James Bay at ca. 53°N, Bic Mt., Rimouski Co., the Côte-Nord, and Gaspé Pen.; not known from the Maritime Provinces), and Nfld.; circumgreenlandic; Iceland; Spitsbergen; Eurasia. [Chamaenerion Spach; C. subdentatum Rydb.; incl. f. angustifolium, f. arcticum, and f. longifolium Haussk., based chiefly upon minor variations in the shape and size of the leaves]. MAPS: Hultén 1968b:687; Porsild 1957: map 257, p. 193; Raup 1947: pl. 31; Fernald 1929: map 18, p. 1496, and 1925: map 64, p. 337.

Forma *leucanthum* (Ulke) Fern. (f. *?albiflorum* Nath.; sepals and petals white or whitish rather than roseate or purplish) is known from s-cent. Alaska, B.C. (type of ssp. *leuc.* Ulke from Horsethief Creek in the Purcell Range), and N Que. (Chimo, Ungava Bay). Forma *munzii* Lepage (petals white or whitish but the sepals roseate or purplish as in the typical form) occurs nearly throughout the range.

E. luteum Pursh

/sT/W/ (Hp) Moist places at low to fairly high elevations from the Aleutian Is. and s Alaska (*see* Hultén 1947: map 876, p. 1197) through B.C. and sw Alta. (Athabasca R. near Jasper; CAN) to Oreg. MAP: Hultén 1968b:687.

E. minutum LIndl.

/T/W/ (T) Usually on gravelly or dry soil at low to fairly high elevations from B.C. (N to Queen Charlotte Is.; DAO; reported from L. Athabasca, Alta., by John Macoun 1883, where perhaps introd. at Fort Chipewyan) and Mont. to Calif. [*E. pubescens sensu* Macoun 1883, not Roth, the relevant collection in CAN].

E. palustre L.

/aST/X/GEA/ (Hpr) Damp or wet places, the aggregate species from N Alaska–Yukon and the coasts of Dist. Mackenzie–Dist. Keewatin to Southampton Is., N Ungava (Akpatok Is., Ungava Bay), Labrador (N to Hebron, 58°13′N), Nfld., N.B., P.E.I., and N.S., s to Oreg., Colo., S.Dak., Minn., N.Y., and Mass.; w Greenland N to ca. 70°N, E Greenland N to ca. 77°N; Iceland; Eurasia. MAPS and synonymy (together with a distinguishing key to four closely related species (?"microspecies")): *see* below.

1   Median and upper leaves elliptic-oblong, obtuse or rounded at apex, at most about 2 cm
      long; calyx-lobes round-tipped; seeds with a long slender neck (beak); [s Nfld., the type
      from Port-aux-Basques] . . . . *E. pylaieanum* Fern.
1   Median and upper leaves linear to lanceolate, to 1 dm long; calyx-lobes acute or
      subacute; seeds with a short thick neck.
      2   Leaves closely and evenly pubescent above; tips of stem and pedicels before
            flowering erect or arching.
            3   Calyx less than 5 mm long; petals less than 7 mm long; seeds with an evident
                  collar or neck; plant nonstoloniferous; [incl. f. *umbrosum* (Haussk.) Fern.;
                  transcontinental; MAPS: Hultén 1968b:688; Porsild 1966: map 111, p. 80] . . . .
                  . . . . *E. leptophyllum* Raf.
            3   Calyx to 7 mm long; petals to 1 cm long; seeds with a less evident collar; plant
                  finally stoloniferous . . . . *E. nesophilum* Fern.
                  4   Leaves linear to lanceolate, attenuate; fruiting pedicels to 4 cm long; [incl. var.
                        lupulinum Hodgdon & Pike; *E. densum* var. *neso.* Fern., the type from
                        Grindstone, Magdalen Is., E Que.; E Que., Nfld., and N.S.] . . . . var. *nesophilum*
                  4   Leaves oblong-lanceolate, blunt or merely acutish; longest fruiting pedicels
                        at most about 1.5 cm long; [*E. molle* var. *sab.* Fern., the type from Sable Is.,
                        N.S.] . . . . var. *sabulonense* Fern.
      2   Leaves glabrous above or nearly so; tips of stem and pedicels before flowering
            nodding.
            5   Calyx glabrous; petals whitish; leaves with flat, ciliate, finely serrulate margins,
                  broadly rounded at apex; [incl. var. *arcticum* (Sam.) Polunin (*E. arcticum* Sam.);
                  transcontinental; MAPS: Porsild 1957: map 258 (*E. arct.*), p. 193; Hultén 1958: map
                  10 (*E. arct.*), p. 29, and 1968b:689; Raup 1947: pl. 31] . . . . *E. davuricum* Fisch.
            5   Calyx sparingly pubescent; petals white, pink, lilac, or violet; leaves with entire to
                  shallowly undulate, often revolute margins . . . . *E. palustre*
                  6   Principal leaves lanceolate, to 1.5 cm broad; fruiting pedicels mostly surpassed
                        by the subtending leaves.
                        7   Stem simple, or branching above; [incl. var. *albiflora* Lehm.; transconti-
                              nental; MAPS (aggregate species): Porsild 1957 (1964 revision): map 340,
                              p. 203; Hultén 1968b:688] . . . . var. *palustre*

7 Stem branching from the lowest internodes, the strongly ascending floriferous branches nearly equalling the main stem; [E James Bay, Que., to Labrador (N to Turnavik, ca. 55°15′N); type from Blanc Sablon, Côte-Nord, E Que.] . . . . var. *longirameum* Fern. & Wieg.

6 Principal leaves linear to narrowly oblong, less than 1 cm broad; fruiting pedicels often or mostly equalling or overtopping the subtending leaves.

8 Leaves to about 7 cm long and 8 mm broad, flaccid; stem to 8 dm tall, simple or with axillary fascicles or many usually erect branches; [var. *mandjuricum* of Canadian reports, not Haussk.; *E. wyomingense* in part of reports from Nfld. and E Que., not Nels.; transcontinental] . . . .
. . . . var. *grammadophyllum* Haussk.

8 Leaves at most about 3.5 cm long and 5 mm broad, strongly ascending; stem less than 6 dm tall, rarely with axillary fascicles.

9 Pairs of lower and median leaves strongly overlapping; stem to about 2 dm tall; [W James Bay, Ont., to Labrador (N to Hebron, 58°13′N; type from Nain), and Nfld.] . . . . var. *labradoricum* Haussk.

9 Pairs of lower and median leaves scarcely overlapping.

10 Stem to 5.5 dm tall, its upper and median leaves mostly longer than the internodes; petals white or pink; capsules to 6 cm long; [var. *monticola* of Canadian reports, not Haussk.; *E. oliganthum* Michx., the type from L. Mistassini, Que.; *E. lineare* var. *olig.* (Michx.) Trel.; transcontinental] . . . . var. *oliganthum* (Michx.) Fern.

10 Stem rarely over 3 dm tall, its upper and median leaves mostly shorter than the internodes; petals deep lilac or purple; capsules rarely as much as 4 cm long; [*E. wyomingense* in part of reports from E Que. and Nfld., not Nels.; essentially transcontinental, the type from Labrador; MAP (E Canada): Dutilly, Lepage, and Duman 1953: fig. 16, p. 80] . . . . var. *lapponicum* Wahl.

E. paniculatum Nutt.
/T/WW/ (T) Open woods, clearings, prairies, and rocky slopes from B.C. (N to Kamloops; CAN), S Alta. (N to Medicine Hat), S Sask. (N to Saskatoon), and S Man. (near Virden and Winnipeg) to S Calif., N.Mex., and S.Dak.; introd. eastwards, as in Ont. (Bruce, Frontenac, and Renfrew counties) and SW Que. (Pontiac Co.; MT).

The Sask. plant is referred to f. *adenocladon* Haussk. by Breitung (1957a; capsules and pedicels glandular-puberulent rather than glabrous). Other material from our area is partly referable to f. *subulatum* Haussk. (petals less than 6 mm long rather than to 8 mm, the fruiting pedicels to 2 cm long rather than at most about 5 mm long).

E. strictum Muhl.
/sT/EE/ (Hpr) Damp or wet places from Minn. to Ont. (N to Moosonee, S of James Bay at 51°17′N), Que. (N to E James Bay at ca. 52°15′N and the Gaspé Pen.), N.B., P.E.I., and N.S., S to Ill., Ohio, and Va. [*E. densum* Raf.; *E. molle* Torr., not Lam.].

E. watsonii Barbey
/ST/X/ (Hp) Moist or wet places from N-cent. Alaska–Yukon to Great Bear L., Great Slave L., L. Athabasca (Alta. and Sask.), Man. (N to Churchill), Ont. (N to W Hudson Bay at ca. 56°50′N), Que. (N to the George R., Ungava Bay, at ca. 58°30′N), Labrador (N to the type locality of *E. steckerianum* at Ramah, 58°54′N), Nfld., N.B., P.E.I., and N.S., S to S Calif., N.Mex., Mo., Ohio, and Md.; introd. in Europe (Hultén 1958; *E. aden.*). [*E. franciscanum* Barbey]. MAP: the map for *E. adenocaulon* by Hultén (1968b:691) applies here.

The plant of our area is largely or wholly referable to var. *occidentale* (Trel.) Hitchc. (*E. adenocaulon* var. *occ.* Trel.; *E. occ.* (Trel.) Rydb.; *E. aden.* and *E. americanum* Haussk.; *E. glandulosum* vars. *aden.* (Haussk.) Fern., *brionense* Fern., and *occ.* and *perplexans* (Trel.) Fern.; *E. ciliatum* Raf.; *E. perplexans* (Trel.) Nels.; *E. ?scalare* and *E. ?steckerianum* Fern.; petals commonly less than 6 mm long, often pale, rather than to 1 cm long and deep purplish).

GAURA L.   [5819]   Gaura

1   Stem-leaves lanceolate to oblong-lanceolate, remotely sinuate-denticulate, to over 1 dm
    long; rosette-leaves oblanceolate, to about 3 dm long; petals white, becoming pink or
    reddish; capsule finely pubescent with spreading hairs, obtusely angled to base; villous
    and downy winter-annual or biennial; (s Ont. and sw Que.) . . . . *G. biennis*
1   Leaves linear to lanceolate or oblong, entire or commonly with up to 4 sinuate teeth on
    each margin, usually less than 3 cm long; petals white to brick-red, becoming scarlet;
    capsule canescent with minute incurved hairs, terete at base; canescent or glabrate
    perennial; (B.C. to s Man.; introd. in s Ont.) . . . . *G. coccinea*

G. biennis L.   Biennial Gaura
/t/EE/   (Hs (bien. or T))   Damp shores and meadows from Minn. to s Ont. (Essex, Lambton,
Lincoln, Welland, and York counties; reported from the Montreal dist., Que., by John Macoun 1883,
where perhaps introd., as in New Eng.), s to Mo., Tenn., N.C., and Va.

G. coccinea Nutt.   Scarlet Gaura
/T/WW/   (Hp)   Dry prairies, roadsides, and waste places from SE B.C. (a large roadside patch at
Windermere, Columbia Valley, where perhaps introd.; dry cliffs, Crowsnest Pass, on the B.C.–Alta.
boundary; CAN) to Alta. (N to Edmonton), Sask. (N to Saskatoon), and s Man. (N to Millwood, about
85 mi NW of Brandon), s to Calif., Mexico, Tex., and Mo.; introd. along roadsides and railways
eastwards, as in s Ont. (Sturgeon Bay., L. Huron, Simcoe Co., and Sharbot L., Frontenac Co.;
TRT) and N.Y. [Incl. *G. glabra* and *G. marginata* Lehm.].

GAYOPHYTUM Juss.   [5818]   Groundsmoke

1   Capsules sessile or subsessile, not at all constricted between the seeds, to 1.5 cm long;
    petals barely 1 mm long; leaves to about 3 cm long and 2 mm broad, commonly much
    longer than the internodes; plant low and diffusely branched basally, from nearly or quite
    glabrous to greyish-puberulent; (sw Alta.) . . . . *G. humile*
1   Capsules on filiform pedicels 2–8 mm long, usually constricted between the seeds; plants
    commonly diffusely branched especially above, the internodes often longer than the
    leaves; (s B.C.).
    2   Capsules rarely over 6 mm long, shorter than the usually sharply reflexed pedicels;
        seeds about 0.5 mm long; sepals and petals usually less than 1 mm long; leaves to
        about 3.5 cm long and 2.5 mm broad; plant glabrous . . . . [*G. ramosissimum*]
    2   Capsules to 12 mm long, longer than the erect to spreading (sometimes reflexed)
        pedicels; seeds to 1.5 mm long; sepals to 3 mm long; petals to 5 mm long; leaves to
        5 cm long and 2 (or even up to 7) mm broad; plant glabrous to minutely strigose or
        densely puberulent with soft spreading hairs . . . . *G. nuttallii*

G. humile Juss.
/T/W/   (T)   Usually along the drying margins of meadows, streams, lakes, and pools at low to
moderate elevations from Wash. and sw Alta. (*G. racemosum* reported from Mt. Glendowan,
Waterton Lakes, as new to Canada by Breitung 1957*b*, this locality accepted by Harlan Lewis and
Jerzy Szweykowski, Brittonia 16(4):366. 1964, in their monograph of the genus) to s Calif. and
Colo. [Incl. *G. racemosum* T. & G.].

G. nuttallii T. & G.
/t/WW/   (T)   Open woods, sagebrush slopes, and dry margins of meadows from s B.C. (Kleena
Kleena; near Tatla L. at ca. 51°45′N; Anarchist Mt., near Osoyoos; Marysville, NE of Cranbrook;
these localities cited by Lewis and Szweykowski, loc. cit., p. 387) to Mont. and w S.Dak., s to Baja
Calif. and N.Mex. [*G. diffusum* T. & G. and its var. *parviflorum* Lewis & Szweykowski; *G.
intermedium* Rydb.; *Oenothera micrantha* Nutt., not Hornem. nor Presl].

[G. ramosissimum Nutt.]

[The report of this species of the w U.S.A. (N to Wash. and Mont.) from B.C. by Henry (1915; Penticton) requires confirmation, probably being based upon a very scrappy collection in CAN reported from Dog L., Okanagan Valley, by J.M. Macoun (1895). Another collection in CAN is the basis of the above citation of *G. nuttallii* from Anarchist Mt., near Osoyoos.]

LUDWIGIA L.   [5793]   False Loosestrife

1   Leaves opposite, lanceolate to ovate, usually less than 3 cm long, on petioles up to about the length of the blade; flowers sessile in the leaf-axils; petals minute or none; flowering-stems prostrate, creeping, or floating; (B.C.; Ont. to N.S.) . . . . *L. palustris*
1   Leaves alternate, linear-lanceolate to lanceolate, to about 1 dm long, sessile or obscurely petioled; flowering-stems erect or ascending; (s Ont.).
  2   Flowers (and capsules) on pedicels at least 3 mm long; petals yellow or reddish, about equalling the calyx-lobes; capsule sharply 4-angled or narrowly winged, opening by terminal pores . . . . *L. alternifolia*
  2   Flowers (and capsules) sessile; petals minute or none; capsule roundly 4-sided or shallowly grooved, opening by valves . . . . *L. polycarpa*

L. alternifolia L.

/t/EE/   (Hp)   Swamps, wet meadows, and pastures from Kans. to s Mich., s Ont. (Sandwich, Essex Co., where taken by John Macoun in 1892 and 1893; CAN; reported from Windsor, Essex Co., by J.M. Macoun 1894; reported from Point Edward and the delta of the St. Clair R., Lambton Co., by Dodge 1915), and Mass., s to E Tex. and Fla. [*Isnardia* DC.].

L. palustris (L.) Ell.   Water-purslane

/T/X/EA/   (Hpr)   Wet ground, shores, and shallow water, the aggregate species from s B.C. (Vancouver Is. and Chilliwack; not known from Alta.–Sask.–Man.) to Minn., Ont. (N to the Ottawa dist.), sw Que. (N to St-Maurice and Nicolet counties), N.B., and N.S. (not known from P.E.I.), s to Calif., Mexico, Tex., and Fla.; S. America; Eurasia; N Africa.

1   Hypanthium with 4 green longitudinal bands terminating well below the sinuses between the calyx-lobes; [*Isnardia* L.; Eurasia–Africa] . . . . [var. *palustris*]
1   Hypanthium with 4 green longitudinal bands reaching or nearly reaching the sinuses.
  2   Leaf-blades usually less than twice as long as broad; [*Isnardia pal.* var. *amer.* DC.; *I. nitida* Michx.; Ont. to N.B. and N.S.] . . . . var. *americana* (DC.) Fern. & Grisc.
  2   Leaf-blades usually more than twice as long as broad; [B.C., the type from Sproat L., Vancouver Is.] . . . . var. *pacifica* Fern. & Grisc.

L. polycarpa Short & Peter

/t/EE/   (Hpr)   Wet ground and shores from Minn. to s Ont. (Essex, Lambton, and Welland counties; CAN; TRT) and Ohio, and from sw Maine to Conn. MAP: Fernald 1918*b*: map 6, pl. 12.

OENOTHERA L.   [5804]   Evening-Primrose.   Onagre

1   Plants nearly or quite stemless, the mostly oblanceolate, coarsely dentate to deeply pinnatifid leaves all in a basal rosette, the flowers solitary on basal peduncles; capsules linear- to fusiform-lanceolate, sharply 4-angled or narrowly 4-winged.
  2   Stigma capitate-globose, very slightly lobed; petals permanently yellow, less than 1 cm long; free part of hypanthium (that part of the floral-tube projecting above the ovary) to 2.5 cm long; capsules to about 3 cm long; leaves deeply sinuate-pinnatifid; (B.C. to sw Sask.) . . . . *O. breviflora*
  2   Stigmas with 4 linear-cyclindric lobes; free part of hypanthium to over 1 dm long; (s Alta. to s Man.).
    3   Petals yellow, aging to purple, 1 or 2 cm long; anthers to 8 mm long; capsules distinctly wing-margined, rarely over 2 cm long; leaves to 1.5 cm broad, deeply runcinate or runcinate-pinnatifid on the lower third, the terminal lobe entire to undulate-dentate . . . . *O. flava*

3 Petals white, aging to purple, usually at least 2.5 cm long (up to 4.5 cm); anthers to 13 mm long; capsules not wing-margined, to 4 cm long; leaves to 2.5 cm broad, from nearly entire to remotely toothed, runcinate, or pinnatifid . . . . *O. caespitosa*

1 Plants leafy-stemmed; flowers usually several in bracted or leafy spikes or racemes.

    4 Stigma capitate-globose or discoid; capsules linear to linear- or fusiform-lanceolate; stamens often in 2 series of markedly unequal length.

        5 Petals to over 8 mm long, yellow, usually more than twice as long as the stamens; capsules 4-angled, to about 3 cm long; principal leaves narrowly oblanceolate; stems to over 6 dm tall.

            6 Free part of hypanthium to 1.5 cm long, flared almost immediately above the ovary; capsule straight or slightly curved; petals to 2.5 cm long; leaves remotely serrulate to serrate; semishrubby perennial, usually many-stemmed from the base; (s Alta. to s Man.) . . . . *O. serrulata*

            6 Free part of hypanthium at most 5 mm long; capsule curved to spirally coiled; petals to about 1.5 cm long; leaves subentire to denticulate; plant usually annual, with commonly several prostrate to ascending stems; (introd in sw B.C.) . . . . [*O. bistorta*]

        5 Petals to 5 mm long; stems commonly less than 3 dm tall (but up to about 5 dm); annuals.

            7 Petals white (or drying pinkish), to 5 mm long, distinctly shorter than the stamens; anthers about 1.5 mm long; free part of hypanthium to 7 mm long; capsules to 2.5 cm long, usually slightly coiled to conspicuously contorted; principal leaves lanceolate to oblanceolate, remotely serrate . . . . [*O. alyssoides*]

            7 Petals yellow (often drying greenish or reddish); free part of hypanthium at most about 2 mm long; anthers about 0.5 mm long.

                8 Capsules 4-angled, curved or contorted, to 4 cm long; leaves lanceolate to oblanceolate, subentire, to about 2 cm broad; (introd. in sw B.C.) . . . . [*O. micrantha*]

                8 Capsules terete in cross-section; leaves usually linear-oblong and less than 5 mm broad.

                    9 Capsules almost straight, less than 1 cm long; stems largely naked, the linear to linear-spatulate, entire leaves mostly located in the inflorescence; (s B.C. and s Alta.) . . . . *O. andina*

                    9 Capsules straight to curved, to 3.5 cm long; stems leafy from the base; leaves narrowly lanceolate to oblanceolate, entire to remotely denticulate; (Vancouver Is.) . . . . *O. contorta*

    4 Stigma with 4 linear-cylindric lobes; petals usually showy; stamens subequal.

        10 Capsules obclavate to ovoid or obovoid (usually broadest near or above the middle), at most about 1 cm long, usually not much more than twice as long as thick, sharply 4-angled or 4-winged; petals yellow; stamens alternately unequal; leaves entire or minutely toothed; perennials.

            11 Inflorescence at first nodding, the buds drooping, the scattered erect flowers opening singly; calyx-tube, calyx-lobes, and petals each less than 1 cm long; anthers less than 3 mm long; capsule broadest above the middle, its stipe to 4 mm long; leaves to about 6 cm long, obtuse; stem to about 6 dm tall; (se Man. to Nfld. and N.S.; introd. in B.C.) . . . . *O. perennis*

            11 Inflorescence more compact, it and the buds erect from the first, 2 or more flowers often open simultaneously; calyx-tube, calyx-lobes, and petals each to over 2 cm long; anthers usually over 4 mm long; stem to about 1 m tall.

                12 Capsule more or less stipitate-glandular, otherwise glabrous, usually broadest near the middle, sessile or short-stipitate; hypanthium sparsely to densely puberulent with minute spreading gland-tipped hairs, the tube above the ovary less than 1.5 cm long; leaves to about 1 dm long; (introd. in s Ont. and N.S.) . . . . *O. tetragona*

                12 Capsule and hypanthium pubescent with nonglandular hairs (or glabrescent in age), the capsule usually broadest above the middle; leaves to about 13 cm long.

13 Capsule thick-clavate, usually sessile or short-stipitate, spreading-villous with hairs 1 or 2 mm long; hypanthium-tube above ovary at least 1.5 cm long; free tips of sepals distinctly hirsute, to 4 mm long; anthers to 8(10) mm long; stem spreading-hirsute, especially above; (Ont. and s Que.) .... *O. pilosella*

13 Capsule oblanceolate to broadly obovoid, on a slender stipe to 1.5 cm long; hypanthium-tube above ovary less than 1.5 cm long; free tips of sepals at most about 1 mm long; anthers to 6 mm long; stem usually appressed-strigose; (?Man.) .... [*O. fruticosa*]

10 Capsules cylindric (usually somewhat broadest toward base), to over 4 cm long and usually more than 3 times as long as thick, not prominently winged; flowers commonly opening toward evening.

14 Leaves prominently sinuate-dentate to pinnatifid (lanceolate to oblong or oblanceolate in outline); petals whitish or yellow (aging reddish); free part of hypanthium to 3.5 cm long; capsules linear-cylindric, to 3 cm long, their strongly pitted seeds 2-rowed, not sharply angled; annual; (introd. in s Ont.) .... *O. laciniata*

14 Leaves entire or merely shallowly toothed (occasionally with 1 or more prominent tooth-like basal lobes).

15 Petals white (aging pinkish to reddish purple); flower-buds nodding; capsules to about 3.5 cm long and 3 mm thick at base, their 1-rowed seeds not sharply angled; leaves linear to lanceolate; stems with whitish exfoliating epidermal cortex, rhizomatous, commonly rather freely branched.

16 Hypanthium and sepals minutely glandular-pubescent; petals to 2.5 cm long, the stamens about 2/3 as long; basal leaves to over 1 dm long and 1 cm broad, strigillose; stems glabrous, to about 1 m tall; (s ?B.C. to s Man.; introd. in w Ont.) .... *O. nuttallii*

16 Hypanthium and sepals glabrous or strigose but not glandular; petals to 3 cm long, about equalled by the stamens; basal leaves mostly not over 6 cm long and 6 or 7 mm broad; plant nearly or quite glabrous, the stems to about 5 dm tall; (s B.C.) .... *O. pallida*

15 Petals yellow (often aging orange or reddish); flower-buds erect; seeds commonly 2-rowed and rather sharply angled; leaves lanceolate to oblong or oblanceolate; stem usually simple.

17 Tips of reflexed calyx-lobes neither bent nor appendaged at base, the tips in the bud closely connivent into a tube; anthers to over 1 cm long; free part of hypanthium to 5 cm long; capsules to 5 cm long; (trans-continental) .... *O. biennis*

17 Tips of reflexed calyx-lobes slightly deflected outwards from the direction of the main body and subtended by an evident auricle, the tips in the bud not connivent into a tube; anthers to about 7.5 mm long.

18 Petals linear, at most 12 mm long and 3 mm broad; free part of hypanthium to 3.5 cm long; floral-bracts deciduous, the fruiting spike naked; capsules loosely villous, to 3.5 cm long; leaves thin, minutely pilose beneath, spreading-ascending to reflexed; (s Ont. to N.S.) .... *O. cruciata*

18 Petals obovate, to 2 cm long; free part of hypanthium to 3 cm long; bracts persistent on the fruiting spike; capsules to 4 cm long; leaves firm, mostly ascending; (transcontinental) .... *O. parviflora*

[O. alyssoides H. & A.]
[The inclusion of Canada in the range of this species of the w U.S.A. (N to Oreg. and Idaho) by Jepson (1951) requires confirmation. (*Sphaerostigma* Walp.).]

O. andina Nutt.
/T/W/ (T) Dry fields and sagebrush slopes from s B.C. (s Okanagan Valley at L. Osoyoos, s

of Penticton, where taken by J.M. Macoun in 1905; CAN) and s Alta. (near Milk River and Medicine Hat, where taken by John Macoun in 1895 and 1894, respectively; CAN; the report from Sask. by Rydberg 1922, requires confirmation) to Calif., Utah, and Wyo. [*Camissonia* Raven; *Sphaerostigma* Walp.].

O. biennis L.   Common Evening-Primrose
/T/X/ (Hs (bien.))   Dry open soil, meadows, and roadsides, the aggregate species from B.C. (N to Shuswap L., about 40 mi E of Kamloops) to Alta. (N to Calgary; CAN), Sask. (N to Prince Albert), Man. (N to Gillam, about 165 mi s of Churchill; ?introd.), Ont. (N to the sw James Bay watershed at 51°15'N), Que. (N to L. St. John, the Côte-Nord, Anticosti Is., and Gaspé Pen.), Nfld., N.B., P.E.I., and N.S., s to Wash., Mont., Idaho, Tex., and N Fla.; introd. in Eurasia.
1   Flowers very pale yellow (usually aging to orange or reddish purple), the petals to 4 cm long; calyx red; anthers to 1 cm long; leaves crinkled; blades of the stem-leaves and floral bracts to over 1/2 as broad as long; [*O. hookeri* T. & G.; sw B.C.: Nanaimo, Vancouver Is.; CAN] .... var. *hookeri* (T. & G.) Boivin
1   Flowers yellow to orange; calyx often not reddish; leaves not crinkled, their blades usually less than 1/3 as broad as long.
  2   Surface of calyx-lobes, ovaries, and capsules hidden beneath the dense pubescence; leaves firm, strongly ascending.
    3   Capsule and calyx closely appressed-short-strigose; [*O. eriensis* and *O. repandodentata* Gates; SE Alta. to N.B.] .... var. *canescens* T. & G.
    3   Capsule and calyx pubescent with long, white, loosely ascending to spreading hairs; [*O. strigosa* (Rydb.) Mack. & Bush; B.C. to N.S.] .... var. *hirsutissima* Gray
  2   Surface of calyx-lobes, ovaries, and capsules clearly visible beneath the loose-villous pubescence; leaves relatively soft, spreading or loosely ascending.
    4   Floral bracts persisting on the mature fruiting spike; [*O. pycnocarpa* Atkinson & Bartlett; s Man. (Löve and Bernard 1959) and Ont. (N to Ottawa)] .... var. *pycnocarpa* (Atkinson & Bartlett) Wieg.
    4   Floral-bracts finally deciduous, the mature fruiting spike essentially naked .... var. *biennis*
      5   Petals 3–6 cm long; calyx-lobes 2.2–5 cm long; styles mostly at least 1.8 cm long.
        6   Ovaries and capsules villous, ascending; petals to 6 cm long; [*O. grandiflora* Ait.; incl. *O. erythrosepala* Borbas (*O. lamarckiana* De Vries, not Ser.); introd. in s B.C. (Vancouver; New Westminster), Ont. (N to Ottawa), Que., N.B., and N.S.] .... f. *grandiflora* (Ait.) Carpenter
        6   Ovaries and capsules glabrous or nearly so, their bases divergent from the axis of the spike; petals to 4 cm long; [*O. argillicola* Mack.; ?Que.: Boivin 1966*b*] .... f. *argillicola* (Mack.) Boivin
      5   Petals (and calyx-lobes) 1–2.5 cm long; styles mostly not over 1.5 cm long; [transcontinental]. This is a remarkably plastic, freely hybridizing species that has been the subject of many years of experimental genetical studies by the English *Oenothera* specialist, R.R. Gates. The following species and varieties (a few of them, perhaps, more closely related to *O. parviflora*) are noted from our area by R.R. Gates in two of his most recent papers (*A conspectus of the genus Oenothera in eastern North America*, Rhodora 59(697):9–17. 1957; *Taxonomy and genetics of Oenothera*, Junk, The Hague, 115 pp., 1958); in addition, *O. canovirens* Steele is reported from Berwyn, Alta., by Groh and Frankton (1949*b*): *O. ammophiloides* Gates & Catchside (type from Guysborough, Guysborough Co., N.S.; also known from Charlevoix and Bellechasse counties, Que.); var. *flecticaulis* Gates (type from Lunenburg Co., N.S.); var. *laurensis* Gates (type from Westmorland Co., N.B.); var. *parva* Gates (type from Que.). *O. apicaborta* Gates (type from Les-Piles, Champlain Co., Que.). *O. biformiflora* Gates (type from Que.); var. *cruciata* Gates (type from Que.). *O. comosa* Gates (type from Wilmot, Annapolis Co., N.S.). *O. deflexa* Gates (type from Wentworth Co., s Ont.; also known from Que.); var. *bracteata* Gates (type from Essex Co., s Ont.).

O. grandifolia Gates (type from N.S.; also known from N.B.). *O. hazelae* Gates (type from N.S.); var. *parviflora* Gates (type from N.S.; not *O. parviflora* L.); var. *subterminalis* Gates (type from N.S.). *O. insignis* Bartl. (reported from Sask. and Que.). *O. laevigata* Barti. var. *rubipunctata* Gates (type from Que.); var. *similis* Gates (type from St-Vallier, Bellechasse Co., Que.). *O. leucophylla* Gates (type from Que.). *O. magdalena* Gates (type from Magdalen Is., E Que.). *O. novae-scotiae* Gates (type from N.S.); var. *intermedia* Gates (type from N.S.); var. *distantifolia* Gates (type from N.S.); var. *serratifolia* Gates (type from N.S.). *O. perangusta* Gates (type from the Bruce Pen., S Ont.); var. *rubricalyx* Gates (type from the NW shore of L. Superior, Ont.). *O. sackvillensis* Gates (type from Sackville, Westmorland Co., N.B.); var. *albibiridis* Gates (type from Sackville, N.B.); var. *royfraseri* Gates (type from Sackville, N.B.). *O. victorinii* Gates & Catchside (type from Que.); var. *intermedia* Gates (type from Que.); var. *parviflora* Gates (type from Que.; also known from Lincoln Co., S Ont.); var. *undulata* Gates (type from Ont.; also known from Que.) . . . . f. *biennis*

[O. bistorta Nutt.]
[This Californian species is known in Canada only through a collection in CAN by John Macoun in 1893 on ballast heaps at Nanaimo, Vancouver Is., SW B.C., where apparently not established. Its identity has been confirmed by P.H. Raven under the name *Camissonia bistorta* (Nutt.) Raven.]

O. breviflora T. & G.
/T/W/ (Hr) Drier meadowlands and streambanks from B.C. (reported N to Chilco, near Vanderhoof, at ca. 53°30′N, by Eastham 1947, who also reports it from Chilcotin, NE of Lillooet), Alta. (Milk River, Hand Hill L., and Etzikom, SW of Medicine Hat; CAN), and SW Sask. (Sidewood; Breitung 1957a) to NE Calif. and Wyo. [*Taraxia* Nutt.].

O. caespitosa Nutt.   Tufted Evening-Primrose
/T/WW/ (Hr) Dry prairies, rocky slopes, and roadsides, the aggregate species from Wash. to Alta. (Milk River and Lethbridge) and S Sask. (N to Moose Jaw; reports from Man. require confirmation), S to Calif. and Colo.
1   Leaves minutely strigose or villous at least on the veins; capsules not tuberculate, usually glabrous; [*O. (Pachylophus) montana* Nutt.; ?Sask., this included in the range by Rydberg 1922] . . . . var. *montana* (Nutt.) Durand
1   Leaves nearly or quite glabrous.
    2   Capsules not at all tuberculate, to well over 2 cm long; stem evident; [*Pachylophus psamm.* Nels. & Macbr.; Cardston, SW Alta.: Boivin 1966b] . . . .
        . . . . var. *psammophila* (Nels. & Macbr.) Munz
    2   Capsules tuberculate on the angles, to about 2 cm long; plant stemless; [*Pachylophus* Raim.; S Alta.–Sask.] . . . . var. *caespitosa*

O. contorta Dougl.
/t/W/ (T) Sandy soil along the coast from SW B.C. (near Victoria, Vancouver Is., where first taken by John Macoun in 1875 and again in 1942 by Eastham, noting it as plentiful; CAN) to Baja Calif. [*Camissonia* Raven; *Sphaerostigma* Walp.; *S. filiforme* Nels.; *S. pubens* (Wats.) Rydb.; *S. strigulosum* F. & M.].

O. cruciata Nutt.
/T/EE/ (T) Dry open soil from Mich. to S Ont. (Waterloo Co.; OAC), SW Que. (Bromptonville; DAO), N.B. (Moncton; DAO), P.E.I. (Charlottetown, Brackley Point, and Fortune Bridge; D.S. Erskine 1960), and N.S. (Lunenburg Co.; E.C. Smith and J.S. Erskine, Rhodora 56(671):249. 1954; type of var. *sabulonensis* from Sable Is.), S to New Eng. and Mass. [Incl. vars. *sabulonensis* Fern. and *stenopetala* (Bickn.) Fern.].

O. flava (Nels.) Garrett   Yellow Evening-Primrose
/T/WW/ (Hr) Usually in hard-packed mud in the plains and lower foothills from Wash. to S Alta.

(Milk River, Calgary, and Hand Hills; CAN), s Sask. (N to Saskatoon), and sw ?Man. (repor *triloba* and its var. *ecristata* by Lowe 1943, require confirmation), s to Calif., Mexico, Col Nebr. [*Lavauxia* Nels.; *O. triloba* var. *ecristata* Jones]

[O. fruticosa L.]   Common Sundrops
[This species of the E U.S.A. (N to Mo. and New Eng.) is reported from SE Man. by Lowe Victoria Beach, about 55 mi NE of Winnipeg) and a collection in CAN from Birds Hill, near Wini may belong here but requires further study. Gleason (1958) includes N.S. in the range, presur on the basis of the report by Lindsay (1878; repeated by John Macoun 1883), but a rele collection in CAN from Lucyfield is referable to *O. tetragona*. The Grand Lake, N.S., citation requires clarification. Macoun's report from Montreal, Que., is based upon *O. perennis*, the rele collection in CAN. (*Kneiffia* Raim.; incl. *O. linearis* Michx.).]

O. laciniata Hill   Cutleaf Evening-Primrose
Native in the E U.S.A. (N to N.Dak. and N.J.); introd. elsewhere, as in s Ont. (field near Simc Norfolk Co., where taken by Landon in 1951; OAC). [*Raimannia* Rose].

[O. micrantha Hornem.]
[This Californian species is known in Canada only through an 1893 collection by John Macoun ballast heaps at Nanaimo, Vancouver Is., sw B.C., the identity confirmed by Raven under the nam *Camissonia micrantha* (Hornem.) Raven. However, it is a rather meagre collection and may final prove referable to *O. bistorta*, taken by Macoun at the same locality on the same day.]

O. nuttallii Sweet   White Evening-Primrose
/T/WW/   (Hp)   Dry plains and prairies (often on dunes) from s ?B.C. (the report of *O. albicaulis* N to Spences Bridge by John Macoun 1883, may be based upon *O. pallida*) to s Alta. (N to Lethbridge; CAN), s Sask. (N to Saskatoon and Asquith; CAN), and s Man. (N to St. Lazare, about 75 mi NW of Brandon; introd. in w Ont. along railway ballast at Peninsula and Heron Bay, N shore of L. Superior; CAN), s to Colo., Nebr., Minn., and Wisc. [*Anogra* Nels.; *O. albicaulis* Nutt.].

O. pallida Lindl.   White Evening-Primrose
/t/W/   (Hp)   Dry plains and prairies (often on dunes), cliffs, and talus from s B.C. (Dry Interior in the Okanagan Valley from Kelowna s to Keremeos and Osoyoos; CAN; the report of *O. albicaulis* from near Spences Bridge in the Thompson Valley by John Macoun 1883, may be referable here) to Oreg., Ariz., and New Eng. [*Anogra* Britt.].

O. parviflora L.
/T/X/   (Hs)   Gravelly shores, sands, talus, and waste places, the aggregate species from s B.C. (N to Vernon; CAN) to Alta. (N to Edson, about 120 mi w of Edmonton; CAN), s Sask. (Indian Head; CAN; not listed by Breitung 1957a), Man. (N to the Minago R. N of L. Winnipeg), Ont. (N to the James Bay watershed at ca. 52°N), Que. (N to the E James Bay watershed at ca. 52°15′N, L. St. John, and the Côte-Nord), Nfld., N.B., P.E.I., and N.S., s to Mont., Ill., and N.J.
1   Plant essentially glabrous; [*O. angustissima* Gates; Que.: Fernald *in* Gray 1950] . . . .
. . . . var. *angustissima* (Gates) Wieg.
1   Plant copiously pubescent.
    2   Pubescence consisting almost entirely of close minute appressed hairs; leaves
        narrowly to broadly lanceolate; [*O. oakesiana* Robbins]. According to Fernald *in* Gray
        (1950), this taxon ranges from Mass. to Va. It is cited from s Ont. (London, Middlesex
        Co.; Port Colborne, Welland Co.; Southampton, Bruce Co.) and from E Que. (Gaspé
        Pen.) by J.M. Macoun (1903:215; the probable basis of the listing for s Ont. by Soper
        1949), but the relevant collections in CAN from London, Port Colborne, and the Gaspé
        Pen. are apparently referable to the typical form. The Southampton collection was not
        located . . . . var. *oakesiana* (Robbins) Fern.
    2   Pubescence consisting partly of long spreading hairs with reddish pustular bases;
        leaves narrowly lanceolate; [*O. muricata* L.; *O. biennis* f. *mur.* (L.) Boivin; *O.
        angustissima* var. *quebecensis* Gates; transcontinental] . . . . f. *parviflora*

O. perennis L.

/T/EE/ (Hs) Dry to moist open places from SE Man. (Winnipeg dist. and Lake of the Woods; the report from Killarney, SW Man., by Lowe 1943, is based upon *O. serrulata,* the relevant collection in CAN) to Ont. (N to the Nipigon R. N of L. Superior), Que. (N to the Bell R. at 50°52′N, L. Mistassini, and the Côte-Nord), Nfld., N.B., P.E.I., and N.S., S to Mo., Ind., Ohio, and Ga.; introd. in S B.C. (between Princeton and Penticton; CAN; reported from New Westminster by J.M. Macoun 1898). [*Kneiffia* Pennell; *O. chrysantha* Michx.; *O. pumila* L.; *O. ?riparia* Nutt.].

Some of our material from Ont. eastwards is referable to var. *rectipilis* Blake (hairs of the stem and branches spreading rather than appressed or incurved).

O. pilosella Raf.

/T/EE/ (Hs) Open woods, moist prairies, and meadows from Ill. to Mich., Ont. (collections in OAC and TRT from Lambton, Frontenac, and Waterloo counties; collection in CAN from Kemptville, Carleton Co., as also one from Sioux Lookout, about 175 mi NW of Thunder Bay, where probably introd.), and SW Que. (Chambly, Huntingdon, and St. John counties; ?introd.), S to Ark., Ill., Ohio, and Pa.; introd. E to New Eng., Pa., and Va. [*Kneiffia* Heller; *O. fruticosa* var. *hirsuta* Nutt.; *O. pratensis* (Small) Robins.; *O. ?canadensis* Goldie].

O. serrulata Nutt.

/T/WW/ (Hp (Ch)) Prairies, dry fields, dunes, and roadsides from Mont. to S Alta. (N to Medicine Hat), Sask. (Breitung 1957a), and S Man. (N to Rossburn, about 65 mi NW of Brandon), S to Ariz., N.Mex., Tex., Mo., and Wisc.; introd. in sandy ground along a railway in S Ont. (Dodge 1915; Lambton Co.) and reported from Thunder Bay, NW shore of L. Superior, Ont., by Hartley (1970).

O. tetragona Roth

Native in the E U.S.A.; introd. elsewhere, as in SE ?Man. (the Birds Hill plant reported under *O. fruticosa* may belong here), S Ont. (Boivin 1966b), and N.S. (Hants, Digby, and Halifax counties; CAN; GH; NSAC). [*O. fruticosa sensu* Lindsay 1878, and John Macoun 1883, at least in part as to the N.S. plant, not L., a relevant collection in CAN; *O. hybrida sensu* M.L. Fernald, Rhodora 24(285):177. 1922, not Michx., relevant collections in CAN and GH].

HALORAGACEAE (Water-Milfoil Family)

Aquatic or subaquatic herbs with alternate or whorled leaves, these commonly deeply 1-pinnatifid into filiform segments (emersed leaves or bracts sometimes merely serrate). Flowers inconspicuous, regular, perfect or unisexual, epigynous, sessile, 1, 2, or 3 in the upper leaf-axils. Sepals 3 or 4. Petals 4 or none. Stamens 3, 4, or 8. Ovary inferior. Fruit nut-like, deeply 4-lobed. (Haloragidaceae).

1   Leaves whorled or alternate, the emersed ones greatly reduced; petals sometimes
    present; stamens 4 or 8; fruit separating into 4 distinct nutlets. . . . . *Myriophyllum*
1   Leaves alternate, the emersed ones not greatly reduced; petals none; stamens 3; fruit
    indehiscent, 3-locular and 3-angled . . . . *Proserpinaca*

MYRIOPHYLLUM L.   [5834]   Water-Milfoil

1   Stem appearing leafless and scape-like, the leaves wanting or reduced to minute remote
    alternate scales; bracts ovate, blunt, entire, to 4 mm long, remotely alternate;
    stamens 4; fruit truncately ovoid, smooth or nearly so, about 1 mm long; (Ont. to Nfld.
    and N.S.) . . . . *M. tenellum*
1   Stems leafy.
    2   Flowers in the axils of submersed or emersed foliage-leaves, not in a terminal
        spike-like inflorescence; stamens 4; leaves commonly alternate, subopposite, or
        whorled on the same plant.
        3   Fruit about 1 mm long, its carpels smooth or merely scabrous; leaves commonly
            about 1 cm long (to about 3.5 cm long in submersed forms), entire or pinnatifid,
            alternate (or subverticillate in submersed forms); (Ont. to N.S.) . . . . *M. humile*
        3   Fruit to 2.5 mm long, each carpel bearing 2 or 3 tuberculate dorsal ridges;
            submersed leaves commonly 2 or 3 cm long.
            4   Plant entirely submersed, the floral-leaves similar to the foliage-leaves, the
                leaves all very delicate, dissected into filiform segments; fruit to 2.5 cm long,
                each carpel with 3 vertical ridges of low, often hooked tubercles; (Ont. to N.S.) . . . . . .
                . . . . *M. farwellii*
            4   Plant with its terminal (flowering) portion emersed; floral-leaves linear to
                oblanceolate, chiefly whorled, 1 or 2 cm long, the blade to 1 mm broad, each
                side with up to 5 ascending teeth commonly 1 or 2 mm long; fruit less than 2
                mm long, each carpel with 2 vertical tuberculate ridges; (SE Sask.) . . . . . *M. pinnatum*
    2   Flowers in a terminal spike-like inflorescence, this commonly wholly or partly emersed,
        the subtending bracts entire, toothed, or pectinate.
        5   Floral bracts chiefly alternate, the upper often entire, the lower pectinate or
            pinnatifid; submersed leaves whorled, commonly about 1 cm long, with up to 7
            pairs of stiffish capillary divisions; stamens 8; fruit to 2 mm long, its carpels
            granular-roughened with minute reddish tubercles; stem usually whitened
            on drying; (transcontinental) . . . .'. . . . . . . *M. alterniflorum*
        5   Floral bracts opposite or whorled; submersed leaves with up to 11 pairs of
            capillary, flaccid or slightly stiffish segments; fruits normally smooth or minutely
            papillate.
            6   Stamens normally 8; fruits commonly about 2.5 mm long, smooth or atypically
                more or less rugose, neither dorsally ridged nor prominently beaked; floral
                bracts entire to serrate or pectinate, from shorter than the flowers (or fruits) to
                6 times as long; stems commonly whitened on drying; (transcontinental) . . . .
                . . . . *M. spicatum*
            6   Stamens 4; carpels with 2 smooth dorsal ridges; floral bracts entire or merely
                serrate, lanceolate to narrowly elliptic or linear-oblong, several or many times
                longer than the flowers (or fruits); stems not noticeably whitened on drying.
                7   Fruits rarely over 1.5 mm long, minutely papillate, prominently beaked;
                    leaves in 4's and 6's, the emersed ones and the bracts to 5 mm broad, the
                    submersed ones to 5 cm long; (Ont. and sw Que.) . . . . *M. heterophyllum*

7 Fruits about 2 mm long, nearly or quite smooth and not evidently beaked; leaves in 4's or 5's, the emersed ones and the bracts 1 or 2 mm broad, the submersed ones to 3 cm long; (s B.C.) .... *M. hippuroides*

M. alterniflorum DC.

/aST/X/GE/ (HH) Lakes, ponds, and streams from NW-cent. ?Alaska (*see* Hultén 1947:1158, and his map 881, p. 1197; not listed by Hultén 1968*b*) and Dist. Mackenzie (Eskimo Lake Basin and Great Bear L.; CAN) to ?B.C. (collections in CAN N to Prince George are tentatively referred here by Porsild; not known from Alta.), Sask. (known only from L. Athabasca; CAN), Man. (N to Reindeer L. at 57°54'N), Ont. (N to W James Bay at ca. 52°N), Que. (N to Knob Lake, 54°48'N, Anticosti Is., and the Gaspé Pen.), Nfld., N.B., and N.S. (not known from P.E.I.), s to N Minn., N Mich., N.Y., and Conn.; W Greenland N to ca. 65°N; Iceland; Europe. MAP: Hultén 1958: map 234, p. 253.

M. farwellii Morong

/T/EE/ (HH) Ponds and streams from Ont. (N to the N shore of L. Superior at Peninsula and Timmins, 48°28'N) to Que. (N to E. Abitibi and L. St. John counties, Anticosti Is., and Tabletop Mt., Gaspé Pen.), N.B. (Boivin 1966*b*; not known from P.E.I.), and N.S. (Digby, Lunenburg, Colchester, and Pictou counties), s to N Minn., N Mich., N.Y., and New Eng.

M. heterophyllum Michx.

/T/EE/ (HH) Ponds and streams from N.Dak. to Ont. (N to the Timagami Forest Reserve and the Ottawa dist.) and SW Que. (Fernald *in* Gray 1950), s to N.Mex., Tex., and Fla.

M. hippuroides Nutt.

/T/D/ (HH) Ponds and slow streams: B.C. (N to Kamloops and Revelstoke; CAN) to Calif. and Mexico; Wisc. to N.Y.

M. humile (Raf.) Morong

/T/EE/ (HH) Ponds and sandy, peaty, or muddy shores from Ont. (Boivin 1966*b*) to Que. (reported from the Richelieu R. E of Montreal by Marcel Raymond, Ann. ACFAS 8:94. 1942, and from the Laurentide Provincial Park N of Quebec City by Y. Desmarais, Nat. can. (Que.) 80(6/7):173. 1953), N.B. (Miramichi R. in Northumberland Co.; CAN; not known from P.E.I.), and N.S., s to Pa. and Md. [*Burshia* Raf.].

Some of our material is referable to phases in which the inflorescence (as well as the foliage) is completely submersed. These include f. *capillaceum* (Torr.) Fern. (leaves abnormally large, to 3.5 cm long and 3 cm broad, with up to 8 pairs of capillary divisions; N.S.) and f. *natans* (DC.) Fern. (the leaves as in the typical form, to 1 cm long and about 4 mm broad, with up to 4 pairs of short divisions; Richelieu R., Que., and N.S.).

M. pinnatum (Walt.) BSP.

/T/EE/ (HH) Peaty or muddy shores and shallow water from s Sask. (Mortlach, near Moose Jaw; Wordsworth, about 120 mi SE of Regina; CAN; DAO; Breitung 1957*a*) to Iowa, Ky., and New Eng., s to Tex. and Fla. [*M. scabratum* Michx.].

M. spicatum L.

/aST/X/GEA/ (HH) Ponds and quiet streams (often brackish or calcareous) from N-cent. Alaska–Yukon and the Mackenzie R. Delta to Great Bear L., L. Athabasca (Alta. and Sask.), Man. (N to Mosquito Point, N of Churchill), Ont. (N to Fort Severn, Hudson Bay, ca. 56°N), W Baffin Is. (an isolated station near the Arctic Circle), Que. (N to s Ungava Bay and the Côte-Nord), Labrador (N to Goose Bay, Hamilton R. basin), Nfld., N.B., P.E.I., and N.S., s to s Calif., Ariz., Kans., Minn., Ohio, and Md.; W Greenland between ca. 66° and 71°N; Iceland; Eurasia. MAPS: Hultén 1968*b*:694; A. Löve, Sven. Bot. Tidskr. 48(1): fig. 4, p. 227. 1954; B.C. Patten, Rhodora 56(670): fig. 1, p. 221. 1954.

Most of our material (particularly northwards) appears referable to var. *exalbescens* (Fern.) Jeps. (*M. exal.* Fern., the type from the York R., Gaspé Pen., E Que.; *M. magdalenense* Fern.; *M.*

verticillatum L.; bracteal leaves shorter than or about equalling the fruits, minutely serrate to entire, rather than usually surpassing the fruits and serrate to pectinate).

M. tenellum Bigel.
/T/EE/ (HH) Shallow pools and shores (chiefly acidic) from Ont. (N to the NW shore of L. Superior; TRT) to Que. (N to Charlevoix Co.; MT; reported N to Tadoussac, Saguenay Co., by A.T. Drummond, Can. Naturalist 4 (Ser. ii):265. 1869), Nfld., N.B., and N.S. (not known from P.E.I.), s to Minn., Mich., N.Y., and N.J.

PROSERPINACA L. [5835] Mermaid-weed. Proserpinie

1 Upper (bracteal) leaves finely serrate; submersed leaves (if present) deeply pinnatifid into linear or filiform divisions; fruit to 4 mm broad; (Ont. to N.S.) .... *P. palustris*
1 Leaves nearly uniform, all pinnatifid or pinnatisect; (N.S.).
  2 Bracteal (emersed) leaves pectinately pinnatifid, with up to 12 pairs of strongly ascending divisions less than 4 mm long, the rachis (median portion) to 4 mm broad ....
    .... *P. intermedia*
  2 Bracteal (emersed) leaves deeply pinnatisect into less than 10 spreading-ascending divisions up to 7.5 mm long, the rachis scarcely broader than the divisions .... *P. pectinata*

P. intermedia Mackenz.
/T/E/ (HH) Shallow water and shores: SW N.S. (Butler's L., Gavelton, Yarmouth Co.; CAN; GH); E Mass. to SE Va. (Suggesting a hybrid between *P. palustris* and *P. pectinata* but often isolated or growing with only one of these species).

P. palustris L.
/T/EE/ (HH) Shallow water and shores from Ont. (N to the Ottawa dist.) to SW Que. (N to Gatineau Co. and the Montreal dist.), N.B. (St. Patrick, Charlotte Co.; CAN), and N.S. (not known from P.E.I.), s to Tex. and Ga.; Mexico; Central America; W.I. [Incl. var. *crebra* Fern. & Grisc.].

Some of our material from s Ont. is referable to var. *amblyogona* Fern. (angles of fruit rounded or nearly obsolete rather than subacute to more or less winged).

P. pectinata Lam.
/T/EE/ (HH) Shallow water and sandy bogs from N.S. (Yarmouth, Annapolis, Shelburne, Queens, Lunenburg, and Cumberland counties; *see* N.S. map by Roland 1947: map 341, p. 471) and SW Maine s (chiefly on the Coastal Plain) to Tenn., Fla., and Tex.

HIPPURIDACEAE (Mare's-tail Family)

HIPPURIS L. [5837] Mare's-tail. Hippuride

Aquatic or subaquatic herbs with erect unbranched flowering-stems bearing whorls of entire leaves at nearly regular intervals. Flowers minute, perfect or mixed perfect and unisexual, epigynous, sessile in whorls in the axils of the upper leaves. Calyx entire. Petals none. Stamens, pistils, and styles each 1 (when present). Ovary inferior. Fruit nut-like, 1-locular, 1-seeded.

1   Leaves at most about 1 cm long and 1 mm broad, in whorls of 5–8; flowers nearly all unisexual, the staminate mostly in whorls below the pistillate (but both types often intermixed); fruit about 1 mm long; flowering-stems to about 1 dm tall, scarcely 0.5 mm thick; (Alaska–Yukon–Dist. Mackenzie–B.C.) . . . . *H. montana*
1   Leaves and flowering-stems commonly larger; flowers mostly perfect; fruits usually about 2 mm long; (transcontinental).
    2   Leaves elliptic to oblong-obovate, obtuse, to about 1.5 cm long, in whorls of 3–6 (mostly 4); stems rarely over 4 dm tall . . . . *H. tetraphylla*
    2   Leaves linear-attenuate, to about 3.5 cm long (or flaccid submersed ones to over 6 cm long), mostly in whorls of 6 or more; stems to over 8 dm tall . . . . *H. vulgaris*

H. montana Ledeb.
/ST/W/ (Hel (Grh)) Shallow streams, mossy banks, and wet or boggy meadows from the Aleutian Is. and Alaska–Yukon–w Dist. Mackenzie (N to ca. 62°N; type from Unalaska, Aleutian Is.) locally through B.C. (Queen Charlotte Is.; Vancouver Is. and adjacent islands and mainland E to Griffin L., Kamloops dist.) to NW Wash. MAPS: Hultén 1968*b*:696; Raup 1947: pl. 31.

H. tetraphylla L. f.
/aST/X/EA/ (Hel (Grh)) Saline or brackish marshes and shallow water from the coasts of Alaska–Yukon–Dist. Mackenzie (E to Coronation Gulf) to s-cent. B.C. (tidal flats at Bella Coola, E of Ocean Falls at ca. 52°20′N; CAN), then along the coasts of Hudson Bay in E Dist. Keewatin, NE Man., Ont., and Que. to s James Bay, and from northernmost Ungava–Labrador s along the Atlantic coast to E Que. (Côte-Nord, Anticosti Is., and Gaspé Pen.; reported from Ile-aux-Coudres, about 60 mi NE of Quebec City, by J. Rousseau, Ann. ACFAS 8:92. 1942, as *H. maritima*) and Nfld. (not known from the Maritime Provinces); Eurasia. [*H. lanceolata* Retz.; *H. maritima* Hellen.]. MAP: Hultén 1968*b*:695.

Forma *lacunarum* Dutilly & Lepage (leaves relatively flaccid, to 2.5 cm long rather than at most about 1.5 cm) is known from the type locality, the mouth of the Attawapiskat R., SW of James Bay, Ont., at 52°57′N.

H. vulgaris L. Mare's-tail. Queue de cheval
/AST/X/GEA/ (Hel (Grh)) Shallow pools and margins of streams and lakes from the Aleutian Is. and coasts of Alaska–Yukon–Dist. Mackenzie–Dist. Keewatin to Banks Is., Victoria Is., N Baffin Is. (an isolated station in Ellesmere Is. at ca. 80°N), and northernmost Ungava–Labrador, s to Calif., N.Mex., Nebr., Ind., and New Eng.; Chile and Argentina; nearly circumgreenlandic; Iceland; Eurasia. MAPS: Hultén 1968*b*:95; Porsild 1957: map 261, p. 193.

Forma *fluviatilis* (Coss. & Germ.) Glück, the completely submersed phase with flaccid leaves to 6 cm long, occurs throughout the range. Forma *litoralis* Lindb. f. (leaves at most 8 in a whorl, less than 2 cm long, rather than up to 12 in a whorl and to about 3.5 cm long) is reported from the James Bay shores of Ont. and Que. by Dutilly, Lepage, and Duman (1954; 1958).

ARALIACEAE (Ginseng Family)

(Ref.: L.C. Smith, N. Am. Flora 28B:3–11. 1944)

Unarmed herbs or bristly or spiny shrubs or semishrubs with alternate, whorled, or basal leaves. Flowers small, regular, perfect or unisexual, epigynous, white or green, in simple or corymbose, racemose, or panicled umbels. Calyx-lobes, petals, and stamens each 5 (or the calyx-lobes obsolete), the stamens inserted on a disk within the calyx. Styles 2 or more. Ovary inferior. Fruit a yellowish, red, or blackish drupe.

1 Leaves compound.
  2 Umbel solitary; fruit red or yellow; leaves in a single whorl at the top of the stem, subtending the terminal peduncle, 2-ternate or ternate-quinate; (Ont. eastwards) .... *Panax*
  2 Umbels 2 or more; fruit blackish; leaves alternate or basal, the principal ones 2–3-compound, their major divisions pinnate .... *Aralia*
1 Leaves simple, alternate; inflorescence a terminal raceme of usually several globose umbels; berries to 8 or 9 cm long.
  3 Stems creeping or climbing by aerial roots, to over 30 m long; leaves ovate, evergreen and coriaceous, glabrous, often with paler markings, entire to fairly deeply 3–5-lobed, to about 1 dm long; berries deep bluish-black; (introd. in sw B.C.) .... *Hedera*
  3 Stems erect, thick and rather punky, to about 3 m tall, they, the petioles, and the leaf-veins copiously armed with yellowish spines to 1 cm long; leaves roundish-cordate, deciduous, palmately 7–9-lobed and irregularly doubly serrate, to about 3.5 dm long and broad; berries bright red; (B.C. and Alta.; L. Superior, Ont.) .... *Oplopanax*

ARALIA L. [5881] Sarsaparilla

1 Umbels in a corymb, their rays arising from the same terminal point or from the upper leaf-axils.
  2 Stem leafy throughout, to about 9 dm tall, bristly at the woody base; leaves 2-pinnate or ternate-pinnate, the terminal leaflet long-stalked; (Sask. to Labrador, Nfld., and N.S.) .... *A. hispida*
  2 Stem scarcely rising above the ground, not bristly, bearing a single long-petioled ternate-pinnate leaf and a shorter naked scape with commonly about 3 (up to 7) umbels; leaflets subsessile, usually 5 on each of the 3 major divisions; (transcontinental) ..... *A. nudicaulis*
1 Umbels numerous in a large raceme or panicle.
  3 Stem and branches (and often the petioles and leaf-rachises) armed with stout thorns; leaves 2–3-pinnate; leaflets ovate, tapering or rounded at base; umbels in a large broad panicle; shrub or small tree up to 10 m tall; (introd. in s Ont.) .... *A. spinosa*
  3 Plant an unarmed herb at most about 3 m tall; leaves ternate-pinnate; leaflets cordate-ovate; umbels in a raceme-like panicle; (s Man. to N.S.) .... *A. racemosa*

A. hispida Vent. Bristly Sarsaparilla. Salsepareille
/T/EE/ (Ch) Rocky or sandy sterile soil from ?Alta. (Boivin 1966*b*; not listed by Moss 1959) to Sask. (N to Windrum L. at ca. 56°N; CAN), Man. (N to Wekusko L., about 90 mi NE of The Pas), Ont. (N to Sandy L. at ca. 53°N, 93°W), Que. (N to the Opinaca R. SE of James Bay at 52°33′N, the Côte-Nord, Anticosti Is., and Gaspé Pen.), Labrador (N to Goose Bay, 53°18′N), Nfld., N.B., P.E.I., and N.S., s to Minn., Ill., Ohio, and N.C.

A. nudicaulis L. Wild Sarsaparilla. Salsepareille
/sT/X/ (Hpr (Grh)) Moist or dry woods and clearings (ranges of Canadian taxa outlined below), s to Wash., Mont., Colo., Nebr., Mo., Tenn., and Ga.
1 At least one of the umbels sessile or subsessile; [s Man. (near Otterburne) and s Que. (the type, as first collection cited, from St-Maurice Co.)] .... f. *abortiva* Dansereau
1 All of the umbels long-peduncled.
  2 Leaves ternate, often inequilateral; [type from Longueuil, near Montreal, Que.] ....
  .... f. *depauperata* Vict.
  2 Leaves ternate-pinnate (the 3 primary divisions divided into mostly 5 distinct leaflets).

3 Flowers with 5 stamens but the carpels modified into reduced simple leaves; [Que.: type from Ile Ste-Thérèse in the Richelieu R.; also known from Montreal] . . . .
. . . . f. *virescens* Vict. & Rousseau
3 Flowers perfect; ["*Azalea*" *nud.*, orthographic error, *in* Richardson 1823; N B.C. and sw Dist. Mackenzie to L. Athabasca (Alta. and Sask.), Man. (N to Reindeer L. at 57°23′N), Ont. (N to the Fawn R. at ca. 54°40′N, 89°W), Que. (N to the Opinaca R. E of James Bay at 52°40′N, L. Mistassini, and the Côte-Nord; reports from Labrador may refer to the Côte-Nord, E Que.), and Nfld., throughout s Canada from B.C. to N.B., P.E.I., and N.S.] . . . . f. *nudicaulis*

A. racemosa L.   Spikenard.   Grande Salsepareille or Anis sauvage
/T/EE/ (Hpr)   Rich woods and thickets from s Man. (N to Victoria Beach, about 50 mi NE of Winnipeg; Gimli; Camp Morton; Waugh) to Ont. (N to the N shore of L. Superior near Thunder Bay), Que. (N to L. St. John and the Gaspé Pen. at Matapédia and near Carleton), N.B., P.E.I., and N.S., s to E Kans., Mo., Ala., and Ga.
Forma *foliosa* (Vict. & Rousseau) Scoggan (the inflorescence leafy rather than essentially leafless; type of var. *foliosa* V. & R. from L. St. John, Que.) is known from s Ont., Que., and N.S.

A. spinosa L.   Hercules'-club, Devil's-walking-stick, Angelica-tree
Native in the E U.S.A. (N to Iowa and N.J.); introd. elsewhere, as in s Ont. (Norfolk, Waterloo, Welland, and York counties). The native area is shown in MAPS by Preston (1961:322) and Hough (1947:355).

HEDERA L.   [5855]

H. helix L.   English Ivy
Eurasian; commonly cult. in N. America and escaping to open woods, as in sw B.C. (moist wooded ravines at Victoria, Vancouver Is.; Mayne Is.; CAN; V).

OPLOPANAX (T. & G.) Miq.   [5850]

O. horridus (Sm.) Miq.   Devil's-club
/sT/D/eA/ (N (Mc))   Moist woods and rocky thickets from sw Alaska–w Yukon through B.C. and Alta. (Lesser Slave L.; CAN) to Oreg., Idaho, and Mont.; a disjunct area on L. Superior (near Thunder Bay, Ont.; Isle Royale and adjacent islands, N Mich.); ssp. *japonicus* (Nakai) Hult. in Japan. [*Panax (Echinopanax; Fatsia; Ricinophyllum) horridum* Sm., the type from Nootka Sound, Vancouver Is., B.C.; *Aralia erinacea* Hook.]. MAPS: Hultén 1968b:696 *(Echinopanax)*; Fernald 1925: map 22, p. 257.

PANAX L.   [5883] Ginseng.   Ginseng

1 Leaflets long-stalked, acuminate, commonly 5, to about 1.5 dm long; styles usually 2; berries bright white, about 1 cm thick; plant to about 3 dm tall; (Ont. and Que.) . . . .
. . . . *P. quinquefolius*
1 Leaflets sessile, obtuse or subacute, 3 or 5 (sometimes 4), less than 1 dm long; styles usually 3; berries yellow, about 5 mm thick; stem usually not over 1.5 dm tall; (Ont. to N.S.)   . . . . *P. trifolius*

P. quinquefolius L.   Ginseng, Sang
/T/EE/ (Grt)   Rich moist woods from Ont. (N to the Ottawa dist.; reports from Man. require confirmation) and sw Que. (N to Cap-Tourmente, about 30 mi NE of Quebec City; Raymond 1950b) to Okla., La., Ala., and N Fla. [*Aralia* Dcne. & Planch.; *Ginseng* Wood].

P. trifolius L.   Dwarf Ginseng, Ground-nut.   Petit Ginseng
/T/EE/ (Grt)   Rich moist woods and clearings from Ont. (N to Georgian Bay, L. Huron, and the Ottawa dist.) to Que. (N to Charlesbourgh, near Quebec City; John Macoun 1883), N.B., P.E.I., and N.S., s to Nebr., Iowa, Ohio, and N Ga. [*Aralia* Dcne. & Planch.; *Ginseng* Wood].

UMBELLIFERAE (Parsley Family)

(Ref.: M.E. Mathias and Lincoln Constance, N. Am. Flora 28B:43–397. 1944–1945)
Herbs with compound (rarely simple), chiefly alternate or basal (rarely opposite) leaves on sheathing-based petioles. Stems usually hollow. Flowers small, usually regular and perfect, commonly white, greenish, or yellowish, epigynous, in usually compound umbels (these simple only in *Hydrocotyle* and *Lilaeopsis*; flowers in heads in *Eryngium*). Umbels with or without a subtending involucre of bracts. Umbellets (the small secondary umbels) with or without a subtending involucel of bractlets. Calyx-teeth, petals, and stamens each 5, or the calyx-teeth obsolete. Petals and stamens inserted on a disk crowning the inferior ovary and surrounding the usually thickened base (stylopodium) of the 2 styles. Fruit a pair of dry 1-seeded seed-like carpels (mericarps), these separating at maturity along their adjoining surfaces (the commissure) and commonly suspended from the summit of a slender prolongation of the common axis, the intervals between their ribs usually with internal longitudinal oil-tubes. (Ammiaceae).

1 Leaves all simple (sometimes deeply parted but not into separate leaflets); fruit nearly terete or flattened laterally (at right angles to the adjoining surfaces or commissure), wingless, broadly oblong to oval or orbicular in outline, to about 4 mm long; plants glabrous.
  2 Fruit covered with stout hooked prickles, the stylopodium flattened and disk-like or wanting; flowers mostly greenish-white or -yellow to yellow (sometimes purple-tinged; purplish red in *S. bipinnatifida*) .... *Sanicula*
  2 Fruit smooth or merely covered with scales.
    3 Inflorescence a dense spiny-bracted head of greenish, bluish, or purplish flowers; fruit densely covered with linear to lanceolate white scales to 1 or 2 mm long, the ribs obsolete and the stylopodium wanting; leaves spiny-toothed, green or slightly glaucous; (introd.) .... *Eryngium*
    3 Inflorescence an umbel; lateral ribs of fruit usually more or less corky-thickened.
      4 Umbel compound, with or without an involucre of bracts; flowers yellowish or purplish; fruit smooth, the stylopodium depressed-conic; leaves entire, more or less glaucous .... *Bupleurum*
      4 Umbel simple; flowers white or whitish; stylopodium of fruit depressed or obsolete; involucre a few small bracts or wanting; leaves (and peduncles) mostly arising from the rooting nodes of the creeping or floating stems.
        5 Leaves reduced to a tuft of long narrow hollow septate phyllodia; fruit nearly terete or only slightly flattened laterally; (sw B.C.; N.S.) .... *Lilaeopsis*
        5 Leaves with broad orbicular to reniform blades, crenate or shallowly to deeply lobed, sometimes peltate, long-petioled, solitary at the nodes; fruit strongly flattened laterally .... *Hydrocotyle*
1 At least some of the leaves compound; umbel compound.
  6 Leaves all or nearly all basal or low-cauline (*Cnidium* may be sought here), the more or less scapose flowering stem from a thickened taproot; (western species).
    7 Fruit tipped by a conic stylopodium, oval or oval-oblong, slightly flattened laterally, prominently ribbed, to 5 mm long and 3 mm broad; rays of umbel to 2 cm long; umbellets subcapitate, the pedicels 2 or 3 mm long; leaves mostly 1-pinnate (or the lowest ternate-pinnate), minutely scabrous on the veins and margins, otherwise glabrous; root bearing a crown of dark reddish-brown fibrous sheaths; stem to 1.5 dm tall; (Alaska–Yukon) .... *Podistera*
    7 Fruit lacking a stylopodium.
      8 Leaflets broadly elliptic to obovate (or some of them broader and deeply 3-lobed), with cartilaginous, minutely crenate-serrate margins, the whole leaf 2-ternate or ternate-pinnate, spreading, often prostrate, thick and firm, glabrous above, tomentose beneath; umbel with up to 13 woolly rays to 4.5 cm long, the involucre wanting or of a few narrow bracts; umbellets capitate (pedicels obsolete), the involucels of several well-developed lance-attenuate bractlets; petals white; fruit ovate-oblong to subglobose, somewhat com-

pressed dorsally (parallel with the commissure), glabrous or with a few long hairs toward tip, to 13 mm long, the ribs all broadly corky-winged; (s Alaska–B.C.) .... *Glehnia*

8 Leaflets mostly finely dissected into narrow segments (except in *Lomatium nudicaule*) and often more or less "fern"-like.

  9 Fruit slightly flattened laterally, to 6 mm long, scabrous-tuberculate, none of the coarse ribs prominently winged; flowers bright yellow; involucre usually wanting; involucels of a few linear-lanceolate bractlets up to 4 mm long; leaves ternate-pinnate-pinnatifid or 2-pinnate-pinnatifid; plant short-stemmed, minutely puberulent or glabrate, commonly about 1 dm tall; (s Alta. to s Man.) .... *Musineon*

  9 Fruit flattened dorsally, at least the lateral ribs winged; flowers white, yellow, or purplish.

    10 Fruit strongly flattened, the dorsal ribs filiform, the lateral ribs more or less broadly winged; bracts of involucre wanting or inconspicuous; bractlets of involucels usually numerous and conspicuous, often connate at base (sometimes wanting) .... *Lomatium*

    10 Fruit only slightly flattened, to about 12 mm long, the lateral ribs strongly winged (as also usually some of the dorsal ribs); leaflets 2-pinnate or 2-pinnate-pinnatifid; plants glabrous or nearly so .... *Cymopteris*

6 Leaves distinctly cauline (as well as often basal).

  11 Fruit bristly or prickly, or at least warty-tuberculate; petals white or greenish white (sometimes yellowish in *Daucus* and *Sanicula* or pinkish in *Daucus*).

    12 Bristles of the linear-cylindric to narrowly clavate fruit strongly appressed and directed toward apex, not hooked at tip; involucre present or none; leaves 2–3-ternate, the large ultimate divisions lanceolate to ovate, coarsely toothed or moderately lobed; plants more or less pubescent .... *Osmorhiza*

    12 Bristles of the ovoid-lanceolate to oblong or subglobose fruit spreading and usually hooked or barbed at tip.

      13 Leaves mostly palmately 3–5-foliolate (pinnately divided in *S. bipinnatifida* and *S. graveolens*), the large obovate to elliptic leaflets coarsely toothed or the lateral ones deeply 1-cleft; umbels dense or almost capitate; involucre and involucels foliaceous; plants glabrous .... *Sanicula*

      13 Leaves pinnately or ternate-pinnately parted (*Sanicula bipinnatifida* and *S. graveolens* may be sought here), the divisions of the leaflets linear or deeply serrate or lobed, the whole leaf more or less "fern"-like in appearance; umbels open; plants usually more or less pubescent.

        14 Involucre consisting of large leaf-like bracts, the involucels of several linear or pinnate bractlets; bristles of fruit in straight rows corresponding to the ribs; leaves pinnately decompound, the ultimate segments linear to lanceolate; plants more or less pubescent.

          15 Fruit somewhat dorsally compressed, the bristles barbed at tip; rays of umbel numerous; plant to about 1 m tall .... *Daucus*

          15 Fruit laterally compressed, to 7 mm long, the bristles hooked at tip, those of alternate rows larger and tending to be confluent at base; rays of umbel less than 10, very unequal, ascending, to 8 cm long, all of the flowers white; plant to about 4 dm tall, more or less spreading-hirsute throughout; (sw B.C.) .... *Caucalis*

        14 Involucre consisting of a few small simple or pinnate bracts, or wanting; bristles of fruit not in regular rows; leaflets composed of broad incised segments; (introd.).

          16 Sheaths of the upper leaves villous-ciliate; rays of umbel glabrous; fruit lance-ovoid, short-beaked, covered with sharp warty tubercles; bractlets of involucels ovate-lanceolate; leaves minutely hispid beneath; stem glabrous; (*A. neglecta*) .... *Anthriscus*

        16 Sheaths entire; rays of umbel pubescent; fruit ovoid, obscurely beaked; bractlets linear; plants hispid throughout .... *Torilis*

11 Fruit not bristly.
    17 Axils of upper leaves bearing clusters of bulblets, normal fruit rarely maturing; leaves 2–3-pinnate (the lower often ternate), the linear segments entire or remotely toothed; plant glabrous; (*C. bulbifera*; transcontinental) .... *Cicuta*
    17 Leaf-axils lacking bulblets.
       18 Larger divisions of at least the stem-leaves finely dissected into filiform, linear, or deeply incised segments, the whole leaf often more or less "fern"-like in appearance, 2–4-pinnate or ternate-pinnate.
          19 Fruit with a beak to 7 cm long, the linear body hispid or scabrous, commonly about 1 cm long; petals white, the marginal ones commonly enlarged; umbel simple or with 2 stout rays; involucre none or a single small bract; involucels consisting of foliaceous lobed bractlets; leaves 3–4-pinnate; leaflets composed entirely of linear segments; minutely hispid annual; (introd.) .... *Scandix*
          19 Fruit beakless or short-beaked, usually nearly or quite glabrous.
             20 Petals white (rarely pink or purplish).
                21 Fruits broader than long, laterally flattened, about 5 mm broad, emarginate at summit, the ribs evident but low and obtuse; umbel sessile, solitary, with 2–4 umbellets of comparatively large flowers, subtended by a single reduced foliage-leaf; involucels consisting of spatulate foliaceous bractlets; leaves 3-ternate-pinnate; leaflets composed entirely of narrowly spatulate segments; glabrous perennial from a subglobose tuber; (s Ont.) .... *Erigenia*
                21 Fruits longer than broad .... *GROUP A* (p. 1154)
             20 Petals yellow or greenish yellow; plants glabrous (or more or less puberulent or granular-scaberulous in *Lomatium*).
                22 Fruit 5 mm long or more, beakless, dorsally flattened and with more or less well-developed corky marginal wings; petals yellow; (western species) .... *Lomatium*
                22 Fruit rarely as much as 5 mm long, beakless or beaked; (introd.).
                    23 Leaves 2–3-ternate-pinnate; divisions of leaflets linear and entire to ovate and toothed or incised; involucre consisting of 2 or 3 entire or 3-lobed bracts with sheathing base; involucels consisting of about 5 lanceolate bractlets; fruit ovate, 2 or 3 mm long, laterally compressed, tipped by a cushion-like stylopodium .... [*Petroselinum*]
                    23 Leaves 3–4-pinnate; leaflets composed entirely of filiform or narrowly linear segments; bracts and bractlets usually none; fruit at least 3 mm long, the short stylopodium conic; stem more or less glaucous.
                        24 Annual; petiolar sheaths of larger leaves at most about 3 cm long; fruit somewhat compressed dorsally, the lateral ribs distinctly winged .... *Anethum*
                        24 Perennial; petiolar sheaths to about 1 dm long; fruit scarcely compressed, the prominent slender ribs merely acute .... *Foeniculum*
       18 Larger divisions of leaves entire or only moderately toothed or lobed, not incised and "fern"-like; (*Pimpinella,* with much reduced upper leaves with linear segments, is included here).
         25 Leaflets or leaf-segments entire or nearly so; plants glabrous.
            26 Petals yellow; fruit oblong, 3 or 4 mm long, somewhat compressed laterally, the low ribs wingless; bracts and bractlets none; leaves 2–3-ternate or ternate-bipinnate; leaflets lanceolate to ovate, at

most about 2.5 cm long; plant glaucous, from a taproot; (s Ont. and sw Que.) . . . . *Taenidia*

26 Petals white; involucre wanting or consisting of a few linear bracts; involucels a few linear bractlets to 4 or 5 mm long; leaflets to about 1.5 dm long; plants from fascicled tubers.

27 Fruit ellipsoid to ovoid, to 7 mm long, rather strongly compressed dorsally, the lateral ribs broadly winged, the dorsal ribs filiform; leaves 1-pinnate, the leaflets narrowly linear to oblanceolate (the broader ones coarsely toothed); (s Ont.) . . . .
. . . . [*Oxypolis*]

27 Fruit oblong-ovate to orbicular, to 4 mm long, somewhat flattened laterally, wingless, the ribs filiform; leaves 1-pinnate (the upper ones simple), the leaflets themselves occasionally pinnately cleft, filiform to linear or rarely narrowly lanceolate, all entire; (B.C. to sw Sask.) . . . . *Perideridia*

25 Leaflets or leaf-segments toothed or lobed.

28 Principal leaves with only 3 leaflets.

29 Petals white (sometimes purplish).

30 Fruit linear-oblong, often curved, pointed at both ends, glabrous, to 8 mm long including the slender-subulate stylopodium, slightly flattened laterally, with low obtuse equal ribs; umbel-rays few, very unequal; involucre none or a single small bract; involucels none or at most 3 minute bractlets; leaflets lanceolate to ovate, broadly cuneate at base, doubly serrate and often moderately lobed, to 1.5 dm long; plant glabrous; (s Man. to Que. and N.B.) . . . . *Cryptotaenia*

30 Fruit obovate, about 1 cm long, often pubescent, strongly flattened dorsally, the lateral ribs broadly winged, the dorsal ribs separated from below middle to apex by dark lines (oil-tubes); umbel regular, several-rayed, flat-topped; bracts of involucre deciduous; involucels consisting of several linear bractlets; leaflets broadly ovate to suborbicular, coarsely serrate and deeply lobed, to 4 dm long; plant woolly; (*H. lanatum*; transcontinental) . . . . *Heracleum*

29 Petals yellow (rarely purplish); fruit ovate to oblong, to 6 mm long, beakless; umbels regular; involucre none; involucels consisting of a few short bractlets; leaflets firm, simply serrate.

31 Central flower and fruit of each umbellet sessile; fruit somewhat compressed laterally, its filiform ribs wingless; plant glabrous or nearly so; (*Z. aptera;* B.C. to s Ont.; introd. eastwards) . . . . *Zizia*

31 Central flower and fruit of each umbellet pedicelled; fruit nearly terete or slightly compressed dorsally, some or all of the ribs prominently winged; (s Ont.) . . . . *Thaspium*

28 Principal leaves with more than 3 leaflets.

32 Petals yellow or greenish yellow; plants essentially glabrous (stems of *Thaspium barbinode* hairy at the nodes) . . . .
. . . . GROUP B (p. 1155)

32 Petals white or greenish white (sometimes pinkish) . . . .
. . . . GROUP C (p. 1156)

GROUP A (see p. 1153)

1 Fruit to about 2.5 cm long, oblong-linear, strongly and sharply ridged, beaked; involucre deciduous; involucels consisting of several lanceolate bractlets; leaves 2–3 pinnate; leaflets consisting of broad incised segments; finely pubescent perennial; (introd.) . . . . [*Myrrhis*]

1 Fruit rarely over 1 cm long, beakless or short-beaked.

2 Fruits flattened dorsally, at least the lateral ribs broadly winged; involucre a few bracts or none; involucels consisting of linear bractlets; petioles sheathing, dilated and scarious-margined.

  3 Bractlets of the involucels scarious and awn-tipped; lateral ribs of the fruit about as broad as the dorsal ribs or only slightly broader; (Alaska–Yukon–Dist. Mackenzie) . . . . *Cnidium*

  3 Bractlets of the involucels at most only scarious-margined, not awn-tipped; lateral ribs of the fruit broadly winged, the corky other ribs usually narrowly winged; (Alaska–B.C.; Ont. to s Labrador, Nfld., and N.S.) . . . . *Conioselinum*

2 Fruits nearly terete or somewhat flattened laterally, wingless or narrowly winged.

  4 Fruit with prominent pale-brown undulate ribs, broadly ovoid, about 3 mm long; bracts and bractlets lanceolate to ovate, conspicuous; leaves commonly ternate-tripinnate, their leaflets composed of broad incised segments; stem purple-spotted; plant glabrous; (introd.) . . . . *Conium*

  4 Fruit with straight ribs; involucre none or consisting of 1 or few subulate bracts (these numerous only in *Carum bulbocastanum*).

    5 Fruits linear or lanceolate, about 6 mm long, nearly ribless, the beak to 3 mm long; involucels consisting of narrowly to broadly lanceolate bractlets to 6 mm long; leaves commonly ternate-bipinnate; leaflets composed of broad incised segments; (introd.) . . . . *Anthriscus*

    5 Fruits linear-oblong to subglobose, usually prominently ribbed, nearly or quite beakless.

      6 Ribs of fruit broader than the intervals; bractlets of involucels conspicuous; leaves mostly ternate-bipinnate; leaflets consisting of broad, deeply incised segments; (introd.).

        7 Fruit broadly ovoid, nearly terete, about 3 mm long; bractlets linear; plant glabrous . . . . *Aethusa*

        7 Fruit linear-oblong, somewhat flattened laterally, to 7 mm long; bractlets ovate-lanceolate; plant minutely hirsute; (*C. temulum*) . . . . *Chaerophyllum*

      6 Ribs of fruit narrower than the intervals.

        8 Fruit reddish brown, subglobose, to 5 mm long, hard, the carpels scarcely separable; involucels consisting of 3 small linear-lanceolate bractlets; leaves 1–3-pinnate, the lowest ones simple or with obovate incised divisions, the upper ones 2–3-pinnately dissected into linear segments; plant glabrous; (introd.) . . . . *Coriandrum*

        8 Fruit elliptic to oblong, somewhat flattened laterally.

          9 Umbel 1–3-rayed; involucels consisting of several elliptic to narrowly obovate villous-margined bractlets; fruit elliptic or oblong, at least 4 mm long; leaves mostly ternate-bipinnate, glabrous or sparingly hispid on the margins and nerves; leaflets composed of broad deeply incised segments; stem glabrous or sparingly hispid, especially at the nodes; (*C. procumbens*; introd.) . . . . [*Chaerophyllum*]

          9 Umbel with at least 5 rays; involucels consisting of 1 or more linear bractlets; fruit oval-oblong; leaves and stem nearly or quite glabrous.

            10 Fruit evidently ribbed but not winged, 3 or 4 mm long, the oil-tubes solitary in the intervals, 2 on the commissure; leaves 2–3-pinnate into linear to linear-lanceolate or spatulate ultimate segments; root-crown not fibrous; (introd.) . . . . *Carum*

            10 Fruit about 5 or 6 mm long, the ribs narrowly winged, the oil-tubes up to 6 in the intervals and 8 on the commissure; leaves ternate, then pinnate-pinnatifid into relatively broad ultimate segments; root-crown coarsely fibrous . . . . *Ligusticum*

GROUP B (*see* p. 1154)

1 Leaves pinnate; fruit strongly flattened dorsally, the lateral ribs winged; (introd.).

2 Leaves 1-pinnate; leaflets finely serrate to near base and irregularly lobed or incised; involucre and involucels usually wanting .... *Pastinaca*
2 Leaves 2–3-pinnate; leaflets coarsely toothed near apex, entire at the cuneate base; involucre and involucels present .... *Levisticum*
1 Leaves (at least the cauline) ternate; leaflets finely to coarsely serrate.
  3 Fruit linear-oblong, over 1 cm long, its narrow ribs wingless; involucels usually absent; stem often villous at the nodes; (*O. occidentalis*; s B.C. and sw Alta.) .... *Osmorhiza*
  3 Fruit oblong or oblong-ovoid, at most 7 mm long.
    4 Fruit strongly compressed dorsally, the lateral wings about half as broad to nearly as broad as the body, the dorsal wings narrower; involucre consisting of several foliaceous toothed bracts sometimes as long as the rays; involucels consisting of well-developed, entire or irregularly few-toothed bractlets; (*A. dawsonii*; s B.C. and sw Alta.) .... *Angelica*
    4 Fruit nearly terete or only slightly compressed; involucre wanting; involucels consisting of a few short narrow bractlets.
      5 Central flower and fruit of each umbellet sessile; fruit somewhat compressed laterally, its filiform ribs wingless .... *Zizia*
      5 Central flower and fruit of each umbellet pedicelled; fruit nearly terete or slightly compressed dorsally, some or all of the ribs prominently winged; (s Ont.) .... *Thaspium*

GROUP C (*see* p. 1154)

1 Leaves 1-pinnate (*Angelica pinnata* may key out here).
  2 Fruits strongly flattened dorsally, the lateral ribs broadly winged, the dorsal ribs filiform; bracts of involucre deciduous or none.
    3 Fruit obovate, about 1 cm long, often somewhat pubescent, its dorsal ribs separated from below middle to apex by dark lines (oil-tubes); bractlets several, setaceous; leaflets 3–7, ovate to suborbicular, deeply lobed, scabrous; plant spreading-pubescent; (*H. sphondylium*; introd. from s Ont. to Nfld. and N.S.) .... *Heracleum*
    3 Fruit oblong, about 5 mm long, glabrous, its ribs not separated by dark lines; involucel a few linear bractlets or none; leaflets 5–9, linear and entire to elliptic and coarsely 3–several-toothed, they and the stem glabrous; (s Ont.) .... *Oxypolis*
  2 Fruits moderately flattened laterally, wingless, oval to subglobose, not more than 3 mm long.
    4 Umbels sessile or very short-peduncled, subtended by reduced foliage-leaves; bracts and bractlets none; leaflets of principal leaves 5–9, usually deeply 3-lobed into serrate or incised segments; plant glabrous; (introd. in N.S.) .... [*Apium*]
    4 Umbels long-peduncled, naked at base or subtended by an involucre of linear bracts.
      5 Involucre and involucels usually none; leaves strongly dimorphic, the few small upper ones linear or deeply pinnately dissected, the basal ones with 9–17 ovate to suborbicular, coarsely toothed or lobed leaflets; plant glabrous to puberulent or somewhat villous; (introd.) .... *Pimpinella*
      5 Involucre and involucels consisting of several conspicuous linear and lanceolate bracts and bractlets, respectively; plants glabrous.
        6 Fruit subglobose, barely 2 mm long, the corky-thickened wall obscuring the very slender ribs; bracts and bractlets foliaceous, often 3-lobed; leaflets 9–23, lanceolate to ovate, those of the lower leaves crenate or lobed (but submersed filiform-dissected leaves also sometimes present), those of the upper leaves commonly laciniate-incised or subpinnatifid; plant stoloniferous; (B.C.; s Ont. and sw ?Que.) .... *Berula*
        6 Fruit oval, 2 or 3 mm long, strongly corky-ribbed; leaflets 5–17, linear to ovate-lanceolate, sharply serrate (submersed ones, if present, bipinnately dissected into linear segments); plants nonstoloniferous; (transcontinental) .... *Sium*

1 Leaves 2–3-ternate or ternate-pinnate; bracts of involucre few or none.
  7 Fruit strongly flattened dorsally and with the lateral ribs broadly winged (except in *Angelica lucida*); sheaths of upper leaves often greatly dilated; stem coarse.
    8 Larger leaves 2-ternate, the usually 9 leaflets or segments oblong or obovate, sharply and irregularly serrate (often doubly so) and commonly 2–3-lobed; fruit 4 or 5 mm long; bractlets of involucre setaceous, deciduous; plant puberulent to glabrate; (introd.) .... *Imperatoria*
    8 Larger leaves commonly ternate-bipinnate, the many lanceolate to ovate leaflets or segments sharply serrate (often doubly so); fruit to about 1 cm long; bractlets several, linear; summit of stem and rays of inflorescence more or less densely puberulent, the plant otherwise glabrous .... *Angelica*
  7 Fruit only slightly flattened, wingless or very narrowly winged (*Angelica lucida* may key out here).
    9 Leaves 2–3-pinnate or ternate-bipinnate; fruit with broad rounded corky ribs; styles much longer than the depressed stylopodium.
      10 Primary lateral veins of leaflets directed to the marginal teeth; leaves 2–3-pinnate, the leaflets irregularly toothed; fruit oblong, truncate, to 3.5 mm long; stem fibrous-rooted, soft and weak, generally reclining, often rooting at the nodes; (SE Alaska and w B.C.) .... *Oenanthe*
      10 Primary lateral veins of leaflets tending to be directed to the sinuses (with a branch continuing to the teeth); fruit ovoid, to 4 mm long; base of stem tuberous-thickened and hollow, commonly with well-developed transverse partitions; roots fascicled, some usually tuberous-thickened .... *Cicuta*
    9 Leaves 1–2-ternate, their lateral veins ending in the teeth.
      11 Fruit 3 or 4 mm long, oblong-ovoid, slenderly ribbed; styles much longer than the conical stylopodium; bracts and bractlets usually none; petals with prolonged incurved tips; leaflets ovate to oblong, membranaceous, sharply serrate almost to base; plant loosely stoloniferous; (introd.) .... *Aegopodium*
      11 Fruit about 1 cm long, narrowly oblong, with prominent acute to narrowly winged ribs; styles about equalling the low-conical stylopodium; bracts 1–5, subulate; bractlets several, linear; leaflets obovate to rhombic, fleshy, coarsely serrate above the middle; plant from a large aromatic root .... *Ligusticum*

AEGOPODIUM L. [6034]

A. podagraria L.   Goutweed.   Herbe aux goutteux
Eurasian; introd. along roadsides and in waste places of N. America, as in sw B.C. (Vancouver), s Man. (Morden; DAO), Ont. (N to the Ottawa dist.), Que. (N to the Gaspé Pen. at Amqui, Matapédia Co.), Nfld., N.B., P.E.I., and N.S.
   Var. *variegatum* Bailey (leaves with broad white margins) is known as a garden-escape in E Que. (Amqui), N.B., and N.S.

AETHUSA L. [6048]

A. cynapium L.   Fool's-parsley
Eurasian; introd. along roadsides and in waste or cult. ground in N. America, as in sw B.C. (Vancouver Is.), Ont. (Cambridge (Galt), Waterloo Co.; OAC; reported from Hastings and Northumberland counties by John Macoun 1883), Que. (Boivin 1966b), St-Pierre and Miquelon (Rouleau 1956), N.B. (Macoun 1883), and N.S. (Shelburne and Halifax counties).

ANETHUM L. [6063]

A. graveolens L.   Dill
Asiatic; a garden-escape to roadsides and waste places in N. America, as in sw B.C. (Carter and Newcombe 1921; Vancouver Is.), Alta. (Boivin 1966b), s Man. (Stony Mountain; Morden), Ont. (N to the Ottawa dist.), and Que. (Montreal dist.).

ANGELICA L. [6082] Angelica. Angélique

(Ref.: M.E. Mathias and Lincoln Constance, N. Am. Flora 28B:192–202. 1945)

1 Fruit only slightly flattened, to about 9 mm long and 5 mm broad, the subequal ribs prominently corky-thickened and raised to a thin edge but scarcely winged; oil-tubes numerous, continuous about the seed, this loose within the pericarp at maturity; involucres deciduous; involucels of several conspicuous lanceolate to oblong bractlets; larger leaves commonly ternate-bipinnate, the ultimate segments ovate, irregularly incised-serrate and often 3-lobed; (Pacific and Atlantic coastal rocks and sands) . . . . . *A. lucida*

1 Fruit strongly flattened dorsally, the lateral ribs broadly winged; (plants mostly not essentially coastal).

  2 Involucre present at base of umbel, consisting of about 10 toothed to laciniate, oblanceolate, leafy bracts to about 2.5 cm long and nearly equalling the umbel-rays; involucels similar, to 8 mm long, surpassing the flowers (these distinctly yellowish when dried but reported to be pale greenish-yellow in life); ovaries glabrous; fruit to 7 mm long; oil-tubes few; umbel usually solitary; leaves deltoid in general outline, 1–3-ternate or 1–2-ternate-pinnate; leaflets to about 6 cm long and 3 cm broad, closely serrate or doubly-serrate; (SE B.C. and SW Alta.) . . . . *A. dawsonii*

  2 Involucre wanting (or occasionally a few sheaths or small deciduous bracts); flowers white or greenish white (rarely pinkish).

    3 Involucels subtending umbellets wanting (or occasionally a few small narrow bractlets); fruit to 6 or 7 mm long, the oil-tubes few (solitary in the intervals), the seed adhering to the pericarp.

      4 Leaves deltoid in general outline, the principal ones ternate, then 1–2-pinnate, their leaflets to about 1.5 dm long and 5 cm broad, coarsely serrate or doubly serrate and often irregularly few-cleft; ovaries glabrous; stout plant to 2 m tall; (S B.C. and S Alta.) . . . . *A. arguta*

      4 Leaves elliptic to oblong in general outline, pinnate to incompletely 2-pinnate, their leaflets to about 9 cm long and 3 cm broad, low-serrate; ovaries obscurely scabrous or minutely hispidulous; plant to 1 m tall; (?Alta.) . . . . [*A. pinnata*]

    3 Involucels consisting of conspicuous bractlets.

      5 Leaves 2-ternate or ternate-pinnate, the main divisions frequently reflexed (not directed forward as in other species), the rachis bent outwards at the point of insertion of the first pair of pinnae and commonly also at those of successive pairs; leaflets to 1 dm long and 6 cm broad, coarsely serrate to incised; ovaries minutely hispidulous; fruit to 4 mm long, the oil-tubes few; seed adhering to the pericarp; stout plant to about 12 dm tall; (S Alaska–B.C.–Alta.) . . . . *A. genuflexa*

      5 Leaves with the main divisions directed forward; rachis straight.

        6 Upper leaf-sheaths scarcely inflated, not prominently veined; central umbel not more than 1.5 dm broad; oil-tubes few; seed adhering to the pericarp; (introd.) . . . . *A. sylvestris*

        6 Upper sheaths strongly inflated and coarsely veined; central umbel to 3 dm broad; oil-tubes at least 25, continuous around the seed, this free in the pericarp at maturity.

          7 Lateral ribs of fruit barely overtopping the stylopodium (enlarged base of styles), the wire-like dorsal ribs much lower; (Ont. to Labrador, Nfld., and N.S.) . . . . *A. atropurpurea*

          7 Lateral and dorsal ribs of fruit subequal and distinctly overtopping the stylopodium; (E Que., Labrador, and Nfld.) . . . . *A. archangelica*

A. archangelica L.

/aST/E/GEwA/ (Hs) Rocky thickets and shores of E Que. (Ste-Anne-de-Beaupré, near Quebec City, and S Saguenay Co. of the Côte-Nord; CAN; GH; MT), Labrador (N to Makkovik, 55°10′N; CAN), and Nfld. (GH); W Greenland N to ca. 70°N, E Greenland N to ca. 66°30′N; Europe; W Asia. [*A. laurentiana* Fern.; *Archangelica officinalis* Hoffm.]. MAP: Hultén 1958: map 92, p. 101.

    Hultén also reports this species as introd. in the Montreal dist., Que., noting, "Distribution very uncertain on account of taxonomical difficulties and of cultivation." His map indicates no native area in N. America but the plant described as *A. laurentiana* Fern. (Rhodora 28(335):222. 1926) appears to justify the native range as outlined above.

A. arguta Nutt.
/T/W/ (Hs) Wet meadows, marshes, and bottomlands from s B.C. (N to Kamloops; CAN) and Alta. (Crowsnest Forest Reserve on the B.C.–Alta. boundary; Waterton Lakes; *A. lyallii* reported from near Lesser Slave L. by Raup 1934) to N Calif. [*A. lyallii* Wats.].

A. atropurpurea L.   Alexanders
/sT/EE/ (Hs) Moist thickets and wet ground from Ont. (N to the w James Bay watershed at ca. 53°N), James Bay (Charlton Is.), Que. (N to the E James Bay watershed at ca. 54°30′N, the Côte-Nord, Anticosti Is., and Gaspé Pen.), Labrador (N to Makkovik, 55°05′N; an early report from Nachvak, 59°07′N), Nfld. (GH), N.B. (Madawaska, Carleton, and Restigouche counties; ACAD; CAN; GH), P.E.I. (Tignish; West Prince), and N.S. (Shelburne, Inverness, and Victoria counties) to Ill., Ohio, and Md. [*Archangelica* Hoffm.].

A. dawsonii Wats.
/T/W/ (Hs) Moist or wet montane slopes from SE B.C. (Crowsnest Pass; Fernie; Flathead; 20 mi. N of Coleman) and sw Alta. (Waterton Lakes; Breitung 1957*b*) to Idaho and w Mont.

A. genuflexa Nutt.
/sT/W/eA/ (Hs) Moist places and swamps from the E Aleutian Is. and s Alaska (*see* Hultén 1947: map 899, p. 1199) through B.C. and w Alta. to N Calif.; E Asia. MAP: Hultén 1968*b*:705.

A. lucida L.
/ST/D/eA/ (Hs) Moist meadows, thickets, and coastal rocks and gravels: Aleutian Is., Alaska (N to Cape Lisburne at ca. 69°N), and the Yukon (N to on or near the Dist. Mackenzie boundary at ca. 54°N) through coastal B.C. to N Calif.; Ont. (sw James Bay watershed at 51°15′N); Charlton Is., James Bay; E Que. (St. Lawrence R. estuary from St-Jean-Port-Joli, L'Islet Co., to the Côte-Nord, Anticosti Is., and Gaspé Pen.) to Labrador (N to Hopedale, 55°27′N), Nfld., N.B. (St. John; CAN), P.E.I. (Wood Is., Queens Co.), N.S., and s N.Y.; E Asia; according to Joergensen, Soerensen, and Westergaard 1958, reports from s Greenland are based upon *A. archangelica*. [*Coelopleurum* Fern.; *Archangelica (C.; Pleurospermum) gmelinii* DC.; *C. longipes* C. & R.; *Ferula canadensis* L.; *Ligusticum (C.; Thaspium) actaeifolium* Michx.]. MAPS: Hultén 1968*b*:705; Porsild 1966: map 112 (*C. luc.*; indicating *C. gmel.* as a distinct western species), p. 80.

[*A. pinnata* Wats.]
[The inclusion of Alta. in the range of this species of the w U.S.A. (N to Mont. and Wyo.) by Rydberg (1922) requires confirmation.]

A. sylvestris L.
Eurasian; definitely known in N. America only from old fields and roadsides of NE N.S. (Louisbourg, Cape Breton Co., where taken by John Macoun in 1898 and again by J.S. Erskine in 1951; CAN; also reported as common around Sydney by Roland 1947). A collection in MT from Bonaventure, Gaspé Pen., E Que., has also been placed here but requires verification.

ANTHRISCUS Bernh.   [5938]   Beak-Chervil

1 Fruit covered with sharp warty tubercles, lance-ovoid, short-beaked, 3 or 4 mm long; umbels short-peduncled, their rays glabrous; bractlets of involucels several, lance-ovate, about 2 mm long; leaves minutely hispid beneath; stem glabrous . . . . *A. caucalis*
1 Fruit smooth; leaves glabrous or sparingly short-villous.
  2 Lateral umbels sessile or subsessile in the leaf-axils, their rays pubescent; fruit linear, the body 5 or 6 mm long, the beak about 3 mm long; bractlets of involucels narrowly lanceolate, 2 or 3 mm long; stem uniformly pubescent below . . . . *A. cerefolium*

2   Umbels all peduncled, their rays glabrous; fruit lanceolate, the body about 6 mm long, the beak about 1 mm long; bractlets of involucels lance-ovate, to 6 mm long; stem pubescent above the nodes .... *A. sylvestris*

A. caucalis Bieb.
Eurasian; introd. in waste places of N. America, as in sw B.C. (Eastham 1947; *A. vulgaris,* "Very abundant in Nanaimo and the surrounding district and apparently extending."), N.S. (Gleason 1958; not listed by Roland 1947), and se Va. (Fernald *in* Gray 1950; *A. scand.*). [*A. neglecta* Boiss. & Reut.; *A. scandicina* (Web.) Mansf.; *Scandix (A.; Chaerophyllum; Myrrhis) anthriscus* L.; *A. vulgaris* Pers., not Bernh.].

A. cerefolium (L.) Hoffm.
Eurasian; locally introd. along roadsides and in waste places in N. America, as in s Ont. (Niagara, Lincoln Co.; TRT) and Que. (near the mouth of the Matane R., Gaspé Pen., where taken by Forbes in 1904, and Cap-à-l'Aigle, Charlevoix Co., where taken by John Macoun in 1905; GH; reported from St. Helen's Is., Montreal, by Groh 1944*a*). [*Scandix (Cerefolium; Chaerophyllum; Selinum) cerefolium* L.; *Chaer. sativum* Lam.].

A. sylvestris (L.) Hoffm.   Cow-parsley
Eurasian; a garden-escape to fields and waste places in N. America, as in Ont. (Simcoe, Grey, Middlesex, Leeds, and Carleton counties), Que. (N to the Gaspé Pen. at Capucins, Matane Co.), Nfld., N.B. (Rothesay and St. John), and N.S. (near Sydney and Louisbourg, Cape Breton Co.; ACAD). [*Chaerophyllum* L.].

[APIUM L.]   [6004]

[A. graveolens L.]   Celery
[Eurasian; a casual garden-escape to waste places in N. America, as in N.S. (waste places on a farm at Black Duck L., Kings Co.; NSPM), where, however, scarcely establlshed.]

BERULA Hoffm.   [6038]

B. erecta (Huds.) Cov.
/T/X/E/   (Hs)   Wet ground and shallow water from s B.C. (collection in CAN from L. Okanagan, where taken by John Macoun in 1889; collection in V from Oliver, s of Penticton; reported from Kamloops by Macoun 1890, and from Popkum, near Yale, by Eastham 1947) and Wash. to Minn., s Ont. (near Port Colborne, Welland Co., where taken by Day in 1882 and apparently now extinct; Zenkert 1934), sw ?Que. (Huntingdon, near Montreal; R. Campbell, Can. Rec. Sci. 6(6):342–51. 1895), and N.Y., s to Baja Calif., N.Mex., Okla., and Fla.; Europe. [*Sium* Huds.; *S. (B.) angustifolium* L.].
   Our material appears wholly referable to var. *incisa* (Torr.) Cronq. (*Sium incisum* Torr.; *S. (B.) pusillum* Nutt.; leaves more or less dimorphic (rather than not markedly so), the lower ones with up to 21 elliptic to ovate, crenate to occasionally serrate or laciniate leaflets, the upper ones with smaller and relatively narrower, more sharply toothed or often irregularly incised to subpinnatifid leaflets; filiform-dissected submersed basal leaves also sometimes present).

BUPLEURUM L.   [5994]   Thoroughwax

1   Leaves linear to oblong-lanceolate, to about 1.5 dm long, acute, mostly clustered at the base of the stem and tapering to an obscure petiole, the few stem-leaves clasping by a rounded base; involucre consisting of up to 6 lanceolate to ovate, acute, foliaceous bracts to 1.5 cm long and 7 mm broad; bractlets of involucels to 5 mm long, acute, shorter than the yellow or purplish flowers; oil-tubes continuous about the seed and in each rib; perennial with a branching caudex surmounting a taproot; (Alaska–Yukon–Dist. Mackenzie–B.C.–sw Alta.) .... *B. ranunculoides*
1   Leaves ovate to oblong or obovate, the basal and lower cauline ones to 8 cm long and 5

cm broad, rounded at apex, subpetiolate or perfoliate at base, the numerous upper cauline leaves perfoliate; involucre wanting; bractlets of involucel to 12 mm long and 10 mm broad, acuminate, 2 or 3 times as long as the yellow flowers; oil-tubes of seed obscure or wanting; (introd.) . . . . *B. rotundifolium*

B. ranuculoides L.
/aST/W/EA/ (Hs (Ch)) Wet places or shallow water at low to fairly high elevations from the coasts of Alaska, the Yukon (Herschel Is.), and w Dist. Mackenzie to s Alaska–Yukon and from SE B.C. (South Kootenay Pass, on the B.C.–Alta. boundary) and sw Alta. (Waterton Lakes and Cardston; CAN) to Idaho and Wyo.; Eurasia. [Incl. var. *arcticum* Regel, *B. americanum* C. & R., *B. angulosum* C. & S. (not L.), and *B. triradiatum* Adams]. MAP: Hultén 1968b:698 (*B. tri.* ssp. *arct.*).

B. rotundifolium L.   Hare's-ear
Eurasian; introd. along roadsides and in waste places in N. America, as in sw B.C. (Victoria, Vancouver Is., where "introduced in bird seed"; V), Ont. (Ottawa; John Macoun 1886; not listed by Gillett 1958), Que. (Rouleau 1947), and N.B. (St. John, where taken on wharf-ballast by G.U. Hay in 1883; CAN).

CARUM L.   [6020]   Caraway.   Anis or Carvi

1   Umbels to 8 cm broad, flat-topped, with up to about 20 rays, the umbels and umbellets subtended by numerous linear to linear-lanceolate small bracts or bractlets; fruit with slender oil-tubes; ultimate leaf-segments to about 1.5 cm long; stem solid; perennial from a black globose tuber; (introd. in St-Pierre and Miquelon) . . . . [*C. bulbocastanum*]
1   Umbels mostly less than 4 cm broad, with usually less than 10 unequal rays, the umbels and umbellets naked at base or subtended by a solitary setaceous bract or bractlet; fruit with broad oil-tubes; ultimate leaf-segments much smaller and more crowded; biennial with a hollow stem from a fusiform taproot; (introd., transcontinental) . . . . *C. carvi*

[C. bulbocastanum Koch]   Earth-nut
[European; reported from St-Pierre and Miquelon by Rouleau (1956; ?established), the only apparent record of its occurrence in N. America. (*Bunium* L.).]

C. carvi L.   Caraway.   Anis canadien
Eurasian; introd. in neglected fields and waste places of N. America (particularly northwards), as in B.C. (N to Cariboo, ca. 53°N; CAN), Alta. (N to Lac la Biche, 54°46′N), Sask. (N to Prince Albert), Man. (N to York Factory, Hudson Bay, 57°N), Ont. (N to Fort Severn, Hudson Bay, ca. 56°N), Que. (N to the E James Bay watershed at ca. 51°30′N, L. St. John, the Côte-Nord, Anticosti Is., and Gaspé Pen.; also known from Akimiski Is., James Bay, ca. 53°N), Nfld., N.B., P.E.I., N.S., and sw Greenland.
    Forma *rhodanthum* Moore (petals pinkish rather than white) is known from E Que. (Rimouski, Rimouski Co.; type from St-Louis, Temiscouata Co.) and N.S. (Parrsboro, Cumberland Co.; A.E. Roland and W.G. Dore, Rhodora 44(525):337. 1942).

CAUCALIS L.   [5950]

C. microcarpa H. & A.   False Carrot
/t/W/ (T) Open moist slopes and streambanks from sw B.C. (Vancouver Is. and adjacent islands; CAN) to Baja Calif. and Mexico.

[CHAEROPHYLLUM L.]   [5935]   Chervil

1   Fruit oblong-ovoid, its slender ribs narrower than the intervals; umbel with not more than 3 primary rays; bractlets of involucels subtending the umbellets elliptic to narrowly obovate, villous on the margins; leaflets glabrous or sparingly hispid on the nerves and margins; stem glabrous or sparingly hispid (particularly at the nodes); (s Ont.) . . . . [*C. procumbens*]

1 Fruit linear-oblong, its rounded ribs broader than the intervals; umbels with at least 6 pubescent primary rays; bractlets of involucels narrowly ovate, acuminate; leaflets villous-hirsute on both sides; stem densely pubescent with short stiffish hairs . . . . . [*C. temulum*]

[C. procumbens (L.) Crantz]
[Moist woods and alluvial soil from Iowa to s Ont. (Seymour, Northumberland Co., and White Is. in the Detroit R. opposite Amherstburg, Essex Co., where taken by John Macoun in 1873 and 1882, respectively; CAN; apparently not taken since the latter date and extinct) and N.Y., s to Kans., Ark., Miss., and Ala. (*Scandix* L.).]

[C. temulum L.]
[Eurasian; the report from Stanbridge Station, Missisquoi Co., sw Que., by Frère Marie-Victorin (Ann. ACFAS 3:102. 1937; taken up by Raymond 1950*b*) is based upon *Anthriscus sylvestris,* relevant collections in CAN and MT.]

CICUTA L.   [6011]   Water-Hemlock.   Cicutaire

(Ref.: M.E. Mathias and Lincoln Constance 1942, and N. Am. Flora 28B:154–57. 1944)
1 Axils of upper leaves bearing clusters of bulblets; leaflets linear to linear-lanceolate, rarely over 4 mm broad, entire or sparingly salient-toothed; fruit about 2 mm long, the ribs broader than the narrow intervals; (transcontinental) . . . . *C. bulbifera*
1 Axils of leaves lacking bulblets.
  2 Fruits not noticeably narrowed to the commissure (the area of contact of their inner faces), to 4 mm long; lateral ribs of fruit much broader than the dorsal ribs; leaflets of the stem-leaves commonly at least 1 cm broad; (transcontinental) . . . . *C. maculata*
  2 Fruits distinctly narrowed to the commissure; lateral ribs of fruit about equalling the dorsal ribs, all of the ribs broader than the reddish- or purplish-brown intervals; pith of base of stem and of usually several of the tuberous-thickened roots separated by prominent cross-walls.
    3 Fruit oval to orbicular, at least as long as broad (to 4 mm long and 3 mm broad); umbels with up to 20 rays to 6 cm long; pedicels to 8 mm long; leaflets relatively broad, to 3.5 cm broad; (Alaska–B.C.–Alta.) . . . . *C. douglasii*
    3 Fruit elliptical, broader than long (to 2.2 mm long and 3 mm broad); umbels with at most about 14 rays to 8 cm long; pedicels to 12 mm long; leaflets of at least the stem-leaves linear to very narrowly lanceolate, commonly less than 3 mm broad; (Alaska–B.C. to James Bay) . . . . *C. mackenzieana*

C. bulbifera L.
/ST/X/ (Hs) Swamps and wet thickets from cent. Dist. Mackenzie (an isolated station at Norman Wells, 65°17′N; W.J. Cody, Can. Field-Nat. 74(2):92. 1960) and B.C.–Alta. to Sask. (N to L. Athabasca), Man. (N to Churchill), Ont. (N to the Winisk R. at ca. 55°N; *see* Hudson Bay–James Bay watershed map by Lepage 1966: map 15, p. 232), Que. (N to E James Bay at ca. 54°30′N, the Côte-Nord, Anticosti Is., and Gaspé Pen.), Labrador (N to the Hamilton R. basin), Nfld., N.B., P.E.I., and N.S., s to Oreg., Nebr., Ohio, and Va. MAP: Hultén 1968*b*:699.

C. douglasii (DC.) C. & R.
/ST/W/ (Hs) Marshes, streambanks, and ditches from Alaska (N to Fort Yukon, ca. 67°N; V.L. Harms, Can. Field-Nat. 83(3):254. 1969) and sw Dist. Mackenzie (J.W. Thieret, Can. Field-Nat. 75(3):118. 1961) through B.C. and w Alta. (reports from Sask. by Breitung 1957*a*, require confirmation) to s Calif., Mexico, N.Mex., and Colo. [*Sium* DC.; *C. californica* Gray; *C. occidentalis* and *C. vagans* Greene]. MAP: Hultén 1968*b*:699 (the extension into Sask. should presumably be deleted).

C. mackenzieana Raup
/ST/(X)/ (Hs) Marshes and swampy ground from the E Aleutian Is. and Alaska (N to ca. 67°N) to s-cent. Yukon, the Mackenzie R. Delta, Great Slave L., Sask. (N to the type locality at L.

Athabasca), Man. (N to Churchill), and James Bay (Ont. and Que., N to 54°22'N; *see* James Bay watershed map by Dutilly, Lepage, and Duman 1954: fig. 13, p. 99), S to B.C.–Alta.–Sask., S-cent. Man. (S to Wekusko L., about 90 mi NE of The Pas), and S James Bay. [*C. ?occidentalis sensu* Groentved 1937, not Greene; *C. virosa* of Canadian reports in part, not L.]. MAPS: Hultén 1968*b*:700; W.J. Cody, Nat. can. (Que.) 98(2): fig. 16, p. 149. 1971.

C. maculata L.   Spotted Cowbane, Musquash-root.   Carotte à Moreau
/sT/X/   (Hs)   Meadows, swampy ground, low thickets, and moist prairies from Alaska (an isolated station at Circle Hot Springs, ca. 66°N; CAN), the ?Yukon (Boivin 1966*b*), and Dist. Mackenzie (Fort Providence, W of Great Slave L.; CAN) to B.C.–Alta., Sask. (N to Prince Albert), Man. (N to the Nelson R. about 30 mi SW of York Factory, Hudson Bay), Ont. (N to Sandy L. at ca. 53°N, 93°W), Que. (N to the Côte-Nord, Anticosti Is., and Gaspé Pen.; concerning James Bay material, *see* Dutilly, Lepage, and Duman 1958:140), N.B., P.E.I., and N.S., S to Tex., Mo., Tenn., and N.C. [Incl. vars. *angustifolia* Hook. and *victorinii* Fern.; *C. virosa* of Canadian reports in part, not L.; *C. vir.* var. *mac.* (L.) C. & R.].

CNIDIUM Cusson   [6069]

1   Basal leaves 2(3)-pinnate, with linear to ovate-lanceolate ultimate segments; fruits to 3 mm long, the lateral ribs of the carpels slightly more winged than the dorsal ribs; (Alaska) . . . . *C. ajanense*
1   Basal leaves 2–3-pinnate, their ovate ultimate segments themselves deeply divided; fruits to 5 mm long, the carpels with 5 subequally-winged ribs; (Alaska–Yukon–NW Dist. Mackenzie) . . . . *C. cnidiifolium*

C. ajanense (Regel & Tiling) Drude
/ST/W/A/   (Hs)   Meadows in alpine or subalpine regions of N-cent. Alaska (a single station at Old Man Creek, ca. 66°N, 152°W; *see* Hultén 1947: map 896, p. 1199); Asia. [*Ligusticum* K.-Pol.; *Tilingia* Regel & Tiling]. MAP: Hultén 1968*b*:701.

C. cnidiifolium (Turcz.) Schischk.
/aS/W/eA/   (Hs)   Wet meadows, gravelly slopes, and streambanks of Alaska–Yukon (N to ca. 70°15'N; *see* Hultén 1947: map 898 (*Conio. cnid.*), p. 1199) and the coast of NW Dist. Mackenzie; E Asia. [*Selinum* Turcz.; *Conioselinum* Porsild; *S. (Conio.) dawsonii* C. & R.; *Laserpitium hirsutum sensu* Hooker 1832, perhaps not Lam.; *Conio. fischeri sensu* Hooker 1832, not Wimm. & Grab.]. MAP: Hultén 1968*b*:702.

CONIOSELINUM Hoffm.   [6081]

C. chinense (L.) BSP.   Hemlock-parsley
/ST/D/eA/   (Hs)   Thickets, open slopes, meadows, sandy shores, and wet woods: Aleutian Is. and coastal Alaska (N to Cape Lisburne, ca. 68°N; *see* Hultén 1947: map 897 (*C. benthamii*), p. 1199) through coastal B.C. to Calif.; Ont. (N to SW James Bay at 51°16'N) to Que. (St. Lawrence R. estuary from St-Jean-Port-Joli, L'Islet Co., to the Côte-Nord, Anticosti Is., and Gaspé Pen.; L. St. John), Labrador (N to Indian Harbour, 54°25'N), Nfld., N.B., P.E.I., and N.S., S to Mo., Ohio, and N.C.; E Asia. [*Athamanta* L.; *Cnidium* Spreng.; *Selinum (Conio.) benthamii* and *S. (Conio.) pacificum* Wats.; *S. hookeri* Wats.; *Conio. pumilum* Rose; *Ligusticum (Conio.) gmelinii* C. & S., not *Conio. gmel.* (Bray) Steud. nor *Coelopleurum gmel.* (DC.) Ledeb.; included in *C. vaginatum* (Spreng.) Thell. by Polunin 1959]. MAP: Hultén 1968*b*:704.

CONIUM L.   [5970]

C. maculatum L.   Poison Hemlock.   Cigue
Eurasian; introd. in waste places of N. America, as in S B.C. (Vancouver Is.; Lulu Is.; Langley; Salmon Arm), Sask. (McLean, near Regina; Breitung 1957*a*), Ont. (N to Ottawa), Que. (N to Grand Bay on the Saguenay R.), N.B. (Tracey Mills, Carleton Co.; ACAD), and N.S.

CORIANDRUM L.   [5953]

C. sativum L.   Coriander
European; introd. in waste places of N. America, as in Alaska (Boivin 1966*b*; not listed by Hultén 1947 and 1968*b*), Ont (N to Ottawa), SW Que. (N to the Montreal dist.), and N.S. (Lunenburg and Guysborough counties; ACAD).

CRYPTOTAENIA DC.   [6015]

C. canadensis (L.) DC.   Wild Chervil, Honewort
/T/EE/eA/   (Hs)   Rich woods and thickets from SE Man. (known only from Morden, about 55 mi S of Winnipeg, where apparently first taken by John Macoun in 1896 and as late as 1953 by the present writer; CAN; WIN) to Ont. (N to Ottawa), Que. (N to Kamouraska Co.), and N.B. (Carleton and Kings counties; NBM; not known from P.E.I. or N.S.), S to Tex., Ark., Ala., and Ga.; var. *japonica* (Hassk.) Makino in E Asia. [*Sison* L.; *Sium* Lam.; *Chaerophyllum* Crantz; *Conopodium* Koch; *Myrrhis* Gaertn.].

CYMOPTERIS Raf.   [6089]

(Ref.: M.E. Mathias and Lincoln Constance, N. Am. Flora 28B:170–85. 1945)
1   Flowers yellow; bractlets narrow, to 6 mm long; primary rays of umbel elongating
    unequally, the longer ones to over 7 cm long; involucre wanting; lateral wings of fruit
    about equalling or broader than the body; stems to 6 dm tall, subacaulescent or leafy near
    base; (S ?B.C.) .... [*C. terebinthinus*]
1   Flowers white or purplish; primary rays of umbel at most about 2 cm long; scapes to about
    3 dm tall.
    2   Bractlets subtending umbellets usually linear, entire, often membranaceous, equalling
        or surpassing the white florets; umbel compact, the primary rays rarely over 1 cm long,
        the pedicels of the umbellets about 1 mm long; involucre of primary umbel usually
        wanting, sometimes rudimentary; wings of fruit narrower than or about equalling the
        width of the body, narrowed toward base and sometimes acuminate at apex; (Alta. to
        SW Man.) .... *C. acaulis*
    2   Bractlets subtending umbellets ovate-oblong, white with a conspicuous green
        midnerve, often rather deeply cleft at apex, usually shorter than the white or purplish
        florets; umbel relatively open, the primary rays to about 2 cm long, the pedicels of the
        umbellets to about 1.5 cm long; involucre of primary umbel consisting of a low
        inconspicuous sheath or of conspicuous linear-oblong bracts, or wanting; wings of fruit
        about twice as broad as the body, conspicuously enlarged at base, narrowed toward
        the apex; (?Man.) .... [*C. montanus*]

C. acaulis (Pursh) Raf.
/T/WW/   (Grt)   Dry plains and valleys from Mont. to S Alta. (N to Lethbridge and Medicine Hat), Sask. (N to Carlton, about 40 mi SW of Prince Albert), SW Man. (N to Routledge, Brandon, and Carberry), and W Minn., S to Oreg., Colo., and Okla. [*Selinum* Pursh; incl. *C. glomeratus* Raf.; *Ferula ?palmella* Hook.]. MAP: M.E. Mathias, Ann. Mo. Bot. Gard. 17(3/4): fig. 4 (the occurrence in S Man. should be indicated), p. 235. 1930.

[C. montanus (Nutt.) T. & G.]
[The report of this species of the W U.S.A. (S S.Dak. to Colo. and Okla.) from S Man. by Lowe (1943; *Phell. mont.*) is probably based upon a 1909 Criddle collection from Aweme, SE of Brandon, referred to *Lomatium orientale* by Boivin (1968). The MAP by M.E. Mathias (Ann. Mo. Bot. Gard. 17(3/4): fig. 6, p. 239. 1930) indicates no Canadian stations. (*Phellopterus* Nutt.).]

[C. terebinthinus (Hook.) T. & G.]
[The report of this species of the W U.S.A. (N to Wash. and Mont.) from SE B.C. by John Macoun (1883; Kootenay Pass; this taken up by Henry 1915) requires clarification. The MAP by M.E. Mathias

(Ann. Mo. Bot. Gard. 17(3/4): fig. 8, p. 242. 1930) indicates no Canadian stations. (*Selinum* Hook.; *Laserpitium* Dougl.; *Pteryxia* C. & R.; *C. foeniculaceus* T. & G.).]

DAUCUS L. [6142] Carrot. Carotte

1   Involucral bracts scarious-margined below, the firm elongate segments filiform-subulate; primary rays of umbel to 7 cm long, the longer outer ones arching inwards to produce a concave mature umbel; umbellets usually with at least 20 flowers, the central flower often roseate or purple; fruit 3 or 4 mm long, broadest at the middle; relatively coarse biennial to about 12 dm tall, from a well-developed taproot, more or less spreading-hirsute throughout to subglabrous; (introd.) . . . . *D. carota*
1   Involucral bracts not scarious-margined, the segments linear or lanceolate, scarcely elongate; primary rays of umbel rarely over 4 cm long, ascending but straightish at maturity; umbellets with usually not more than 12 flowers, the flowers all white or sometimes purplish; fruit to 5 mm long, usually broadest below the middle; retrorsely pubescent annual to about 8 dm tall, from a slender taproot; (s B.C.) . . . . *D. pusillus*

D. carota L.   Wild Carrot, Queen Anne's-lace.   Carotte sauvage
Eurasian; introd. along roadsides and in dry fields and waste places of N. America, as in s B.C. (Vancouver Is., and adjacent islands and mainland), Sask. (Boivin 1966*b*; not listed by Breitung 1957*a*), Man. (known only from Brandon, where taken by John Macoun in 1896; CAN), Ont. (N to the Ottawa dist.), Que. (N to the Gaspé Pen. at York), Labrador (Goose Bay, 53°18′N; DAO), N.B., P.E.I., and N.S.

The typical form has a solitary roseate to purple flower near the centre of the umbel of otherwise white or whitish flowers, this resembling a small insect and presumably an adaptation to attract insects for pollination. Forma *epurpuratus* Farw., lacking this coloured flower, is reported from s Ont. by Landon (1960; Norfolk Co.). Forma *roseus* Millsp. (flowers all roseate or purplish) is reported from s Ont. by Gaiser and Moore (1966; Lambton Co.).

D. pusillus Michx.
/t/X/   (T)   Dry hills, prairies, and rocks from s B.C. (Vancouver Is. and adjacent islands and mainland) to Kans., Mo., and S.C., s to Baja Calif. and Fla. [*D. microphyllus* Presl].

ERIGENIA Nutt.   [5960]

E. bulbosa (Michx.) Nutt.   Harbinger-of-spring
/t/EE/   (Gst (Grt))   Rich deciduous woods from Wisc. to s Ont. (Lambton, Elgin, Middlesex, Oxford, Perth, Waterloo, Wellington, Peel, and York counties; *see* s Ont. map by Soper 1962: fig. 20, p. 33) and N.Y., s to Mo., Miss., and Ala.

ERYNGIUM L.   [5923]   Eryngo

1   Petals greenish; fruit less than 3 mm long; spinose involucral bracts to about 5 cm long, much surpassing the heads; lower leaves deltoid, deeply pinnately or ternate-pinnately divided to near the winged rachis, to about 2.5 dm long; upper leaves deeply ternately divided; (introd.) . . . . [*E. campestre*]
1   Petals blue or bluish; spinose involucral bracts rarely over 3 cm long, shorter than or not much surpassing the head.
    2   Blades of the basal leaves almost equalling or longer than their petioles, crenate to spinulose-serrate, oblong-oval, rounded or cordate at base, to about 1.5 dm long; upper leaves palmately lobed or divided; fruit 3 or 4 mm long; (introd.) . . . . *E. planum*
    2   Blades of the basal leaves much shorter than their septate-nodose petioles, spinulose-toothed or remotely spinulose-ciliate, more or less elliptic (or the blades often obsolete); upper leaves similar but sessile and reduced, often laciniate at base; fruit 2 or 3 mm long; (?B.C.) . . . . [*E. articulatum*]

[E. articulatum Hook.]
[The report of this species of the w U.S.A. (Wash. and Idaho to Calif.) from B.C. by Howell (noted by Henry 1915) requires confirmation.]

[E. campestre L.]
[European; reported as introd. in sw Que. by Rouleau (1947:61), where, however, scarcely established.]

E. planum L.
Eurasian; an occasional garden-escape in N. America, as in Ont. (collection in TRT from rocky ground at Lorne Park, Peel Co., where taken by Coventry in 1958; collection in CAN from Ottawa, where spreading on a hillside at the Central Experimental Farm and taken by Malte in 1915; reported from Cambridge (Galt), Waterloo Co., by Montgomery 1957). Reported from Que. and from B.C. to Sask. by Boivin (1968).

FOENICULUM Mill. [6062] Fennel

F. vulgare Mill. Sweet Fennel
European; a garden-escape to roadsides and dry fields in N. America, as in sw B.C. (Nanaimo, Vancouver Is.; CAN), ?Alta. (Colinton; Groh 1947), s Ont. (Lambton, Waterloo, and Welland counties), and sw Que. (garden at Chambly, near Montreal; MT; ?escaped).

GLEHNIA Schmidt [6085]

G. littoralis Schmidt
/sT/W/eA/ (Grt) Coastal dunes and sandy beaches from s Alaska (see Hultén 1947: map 901, p. 1199) and w B.C. (Queen Charlotte Is.; Vancouver Is.) to Calif.; E Asia. [Phellopterus Benth.]. MAP (aggregate species): Hultén 1968b:706.
    The N. American plant is referable to ssp. leiocarpa (Mathias) Hult. (G. lei. Mathias; fruits glabrous or with a few long hairs toward the tip rather than copiously hairy).

HERACLEUM L. [6122]

1 Principal leaves 1-ternate; plant more or less woolly; (transcontinental) . . . . H. lanatum
1 Principal leaves 1-pinnate (leaflets 3, 5, or 7); plant spreading-pubescent and somewhat
     scabrous; (introd.) . . . . H. sphondylium

H. lanatum Michx. Cow-parsnip. Berce
/ST/X/eA/ (Grt) Rich thickets, moist ground, and shores from the Aleutian Is. and cent. Alaska–Yukon to Great Slave L., Sask. (N to Prince Albert), Man. (N to about 10 mi s of Churchill), Ont. (N to Fort Severn, Hudson Bay, ca. 56°N), Que. (N to SE Hudson Bay at ca. 56°N, L. St. John, the Côte-Nord, Anticosti Is., and Gaspé Pen.), Labrador (N to Attikamagen L. at ca. 55°N, 67°W, and Sandwich Bay, ca. 53°30'N), Nfld., N.B., P.E.I., and N.S., s to Calif., N.Mex., Kans., Ohio, and Ga.; E Asia. [H. maximum Bartr.]. MAPS: Hultén 1968b:707; Raup 1947: pl. 31; Marcel Raymond, Ann. ACFAS 15:118. 1949.

H. sphondylium L. Hogweed
Eurasian; locally introd. along roadsides and in fields and waste places in N. America, as in s Ont. (Grey, Wentworth, York, and Perth counties), Que. (Quebec City; Amqui, sw Gaspé Pen.), Nfld. (near St. John's; GH), N.S. (Truro, Colchester Co.; ACAD; reported from Cape Breton Is. by Fernald in Gray 1950), and ?Labrador (Boivin 1966b).

HYDROCOTYLE L. [5893] Water-Pennywort

(Ref.: M.E. Mathias and Lincoln Constance, N. Am. Flora 28B:51–58. 1944)
1 Leaves centrally peltate (lacking a sinus extending to the junction with the petiole),
     suborbicular to reniform, crenate or shallowly lobed; inflorescence long-peduncled.

2 Inflorescence usually a simple open many-flowered subglobose umbel to 3 cm broad, commonly surpassing the leaves; pedicels filiform; fruits deeply notched at base, to 2 mm long and 3 mm broad, the dorsal ribs obtuse; (sw N.S.) .... *H. umbellata*

2 Inflorescence a simple or branching "spike" commonly shorter than the leaves, bearing usually 2 or more few-flowered whorls of sessile or subsessile flowers; fruits subtruncate to rounded or barely notched at base, to 2.5 mm long and 4 mm broad, the ribs acute; (sw B.C.) .... *H. verticillata*

1 Leaves not peltate (a deep sinus extending to the junction with the petiole, the petiole thus marginally attached); flowers whitish.

3 Umbels sessile or nearly so in the leaf-axils; fruits about 1.5 mm broad, on pedicels less than 1 mm long; leaves shallowly 6–10-lobed; (Ont. to Nfld. and N.S.) ... *H. americana*

3 Umbels on well-developed peduncles much shorter than the subtending petioles; fruits to 3 mm broad, on pedicels to 3 mm long; leaves 5–6-lobed nearly or quite to the middle; (introd. on Vancouver Is.) .... [*H. ranunculoides*]

H. americana L.
/T/EE/ (Hrr) Moist meadows and woods from Minn. to Ont. (N to Batchawana Bay at the SE end of L. Superior and the Ottawa dist.), Que. (N to Grosse-Ile, about 30 mi NE of Quebec City), sw Nfld. (reported as abundant at the mouth of Barachoix Brook, near Stephenville, by R.B. Kennedy, Rhodora 32(373):4. 1930; GH; MT), N.B., P.E.I., and N.S., s to Tenn. and N.C.

[*H. ranunculoides* L. f.]
[Native in marshes, ponds, and wet ground of the U.S.A. (N to Wash. and Pa.); known in Canada only through an 1887 collection in CAN by John Macoun on wharf-ballast at Nanaimo, Vancouver Is., sw B.C., where probably introd. but apparently not established. (*H. umbellata sensu* Macoun 1890, not L., the report based upon the above-noted Nanaimo plant).]

H. umbellata L.
/T/(X)/ (Hrr) Wet ground, ditches, and margins of ponds from Oreg. to Minn., Mich., Ohio, N.Y., and sw N.S. (St. John L., Yarmouth Co., where taken by Fernald et al. in 1921; GH; CAN), s to Calif., Mexico, Tex., and Fla.; tropical America.

H. verticillata Thunb.
/t/X/ (Hrr) Swamps, shores, and wet ground from sw B.C. (Coquitlam, near Vancouver; Henry 1915, this report accepted by Boivin 1966b) and ?Oreg. (Fernald *in* Gray 1950; not listed by Hitchcock et al. 1961) to Utah, Okla., Mo., and Mass., s to Mexico, Tex., and Fla.; tropical America. [*H. vulgaris sensu* Henry 1915, not L.].

IMPERATORIA L.   [6116]

I. ostruthium L.   Masterwort
European; locally introd. in waste places of N. America, as in Mich., s Ont. (Blair, Waterloo Co.; GH), St-Pierre and Miquelon (Rouleau 1956), and ?Nfld. (Hooker 1832; not listed by Rouleau 1956).

LEVISTICUM Hill   [6083]

L. officinale Koch   Lovage.   Livêche
European; a local garden-escape in N. America, as in Sask. (Langham, NW of Saskatoon), Ont. (L. Nipigon and Lambton, Elgin, Grey, Durham, Ontario, and Lanark counties), Que. (Montreal dist. and Montmagny and Charlevoix counties), and N.S. (L. Annis, Yarmouth Co.; GH). [*Ligusticum (Lev.) levisticum* L.].

LIGUSTICUM L.   [6071]   Lovage

(Ref.: M.E. Mathias and Lincoln Constance, N. Am. Flora 28B:143–48. 1944)
1 Leaves 2-ternate into broad, crenate to coarsely serrate or occasionally incised leaflets;

fruits to 8 mm long, with 1–3 oil-tubes in the intervals between the ribs; (B.C.; James Bay to Labrador, Nfld., and N.S.) . . . . *L. scothicum*
1 Leaves 1-ternate, then pinnate-pinnatifid; fruits to 6 mm long, with 3 or more oil-tubes in the intervals; (B.C.).
  2 Plants to about 7 dm tall, scapose or with 1 or 2 much reduced stem-leaves; terminal umbel with less than 15 rays; ribs of fruit narrowly winged.
    3 Plant glabrous throughout; ultimate segments of leaflets rather broadly oblong . . . .
    . . . . [*L. grayii*]
    3 Plant puberulent and minutely scabrous in the inflorescence; leaves more finely dissected, the ultimate segments generally rather narrowly lanceolate, their margins sometimes minutely scabrous; (Queen Charlotte Is., B.C.) . . . . *L. calderi*
  2 Plants to over 1 m tall, with 1 or more well-developed stem-leaves, these glabrous or minutely scabrous; terminal umbel with usually more than 15 rays.
    4 Ribs of fruit wingless; leaflets irregularly toothed to deeply incised into broadly lanceolate to ovate ultimate segments . . . . [*L. apiifolium*]
    4 Ribs of fruit narrowly winged; leaflets more regularly pinnatifid into linear to lanceolate ultimate segments; (s B.C.) . . . . *L. canbyi*

[L. apiifolium (Nutt.) Gray]
[Reports of this species of the w U.S.A. (Wash. to Calif.) from B.C. by John Macoun (1886; taken up by Henry 1915) are based upon *L. canbyi,* relevant collections in CAN. (*Cynapium* Nutt.).]

L. calderi Mathias & Constance
/T/W/ (Hs) "Queen Charlotte Islands, British Columbia, rocky cliffs, open slopes, and edge of coniferous forests, from near sea level to 3400 feet elevation." (M.E. Mathias and Lincoln Constance, Bull. Torrey Bot. Club 86(6):374. 1959; type from Tasu Sound, Moresby Is., Queen Charlotte Is.). [Perhaps finally to be merged with *L. canbyi*].

L. canbyi C. & R.
/T/W/ (Hs) Moist or wet meadows, streambanks, and boggy montane slopes from s B.C. (N to Griffin L., near Kamloops, and Glacier, in Rogers Pass; CAN) to Oreg., Idaho, and Mont. [*L. apiifolium* of B.C. reports (*see* above); *L. grayii sensu* John Macoun 1890, J.M. Macoun 1895, and Henry 1915, not C. & R., and *L. scopulorum sensu* J.M. Macoun 1895, not Gray, the relevant collections in CAN].

[L. grayii C. & R.]
[Concerning reports of this species of the w U.S.A. (N to Wash., Idaho, and Mont.) from B.C., see *L. canbyi*.]

L. scothicum L.   Scotch or Beach Lovage
/ST/D (coastal)/GEeA/ (Hs) Coastal rocks and salt marshes (ranges of Canadian taxa outlined below; known in the U.S.A. only along the Atlantic coast s to s N.Y. and Conn.); Greenland (*see* below); Iceland; coasts of N Europe and E Asia. MAPS and synonymy: *see* below.
1 Primary umbels mostly less than 6 cm broad, convex-topped; mericarps of fruit to 2.5 mm broad; leaflets of the lower leaves relatively small, their teeth blunt to broadly rounded; ultimate veins of the leaflets mostly confluent and forming a closed reticulum; [*L. hultenii* Fern., the type from Atka, Aleutian Is.; Aleutian Is.–Alaska (N to the Seward Pen.; *see* Hultén 1947: map 894, p. 1199 (*L. hult.*)) and coastal B.C. (s to Queen Charlotte Is. and Prince Rupert, ca. 54°20′N); coastal E Asia; MAPS: Hultén 1968b:702, and 1958: map 276, p. 295] . . . . ssp. *hultenii* (Fern.) Calder and Taylor
1 Primary umbels to 1 dm broad, flat-topped; mericarps of fruit to 4 mm broad; leaflets of the lower leaves relatively large, their teeth commonly acute to acuminate; ultimate veins of the leaflets free in large part; [cent. Ont. (w James Bay N to ca. 53°N), Que. (E James Bay N to SE Hudson Bay at ca. 56°10′N; St. Lawrence R. estuary from Berthier-en-Bas, Montmagny Co., to the Côte-Nord, Anticosti Is., Gaspé Pen., and Magdalen Is.), Labrador (N to Tikkoatokok Bay at ca. 57°N), Nfld., N.B., P.E.I., and N.S.; w Greenland N to ca.

65°45′N, E Greenland N to 60°35′N; Iceland; N Europe; MAPS: on the above-noted maps by Hultén] . . . . ssp. *scothicum*

LILAEOPSIS Greene [6047]

1   Peduncles to 6 cm long, equalling or surpassing the 3–6-jointed leaves, these to 6 cm long and 3 mm broad; umbel with at most 9 flowers; dorsal ribs of fruit acute; (w N.S.) . . . .
. . . . *L. chinensis*
1   Peduncles to 4.5 cm long, weak, shorter than the 5–11-jointed leaves, these to 1.5 dm long and 4 mm broad; umbel with up to 12 flowers; dorsal ribs of fruit obscure, the lateral ones broad; (sw B.C.) . . . . *L. occidentalis*

L. chinensis (L.) Ktze.
/T/EE/   (Hrr)   Brackish coastal marshes and tidal shores from sw N.S. (Tusket R., Yarmouth Co., and near Port Medway, Queens Co.; ACAD; CAN; GH) to Fla. and Miss. [*Hydrocotyle* L.; *H. (Crantzia; Lil.) lineata* Michx.]. MAP: Fassett 1928: fig. 1, pl. 12.

L. occidentalis C. & R.
/t/W/   (Hrr (Hel))   Coastal marshes and sandy or muddy beaches and shores from sw B.C. (Vancouver Is. and adjacent islands and mainland; CAN; reports from s Alaska require confirmation) to cent. Calif. [*L. lineata* var. *occ.* (C. & R.) Jeps.; *Crantzia lineata sensu* John Macoun 1890, not (Michx.) Nutt., relevant collections in CAN].

LOMATIUM Raf. [6117] Desert Parsley, Biscuit-root

(Ref.: Mathias 1938; M.E. Mathias and Lincoln Constance, N. Am. Flora 28B:222–58. 1945)
1   Leaves either all cauline or both cauline and basal (*L. brandegei* and perhaps certain other species may sometimes key out here); flowers yellow; (s B.C. and sw Alta.).
  2   Fruits broadly elliptic (each lateral wing nearly as broad as the seed-bearing body), usually glabrous at maturity; leaves glabrous or puberulent, dissected into linear segments; stems to over 6 dm tall.
    3   Leaves very finely ternate-pinnately dissected, the ultimate rather crowded segments usually less than 5 mm long and 1 mm broad; petioles to 1 dm long, all except those of some of the basal leaves sheathing the stem; bractlets of involucels well developed, subherbaceous to subscarious, obovate to elliptic or suborbicular, to 5 mm long, the tip often shallowly toothed or cleft; fruit to 11 mm long and 6 mm broad; plant glabrous or short-pubescent; (s B.C.) . . . . *L. utriculatum*
    3   Leaves somewhat more regularly 2-ternately dissected, the ultimate, relatively few and remote segments to over 1 dm long; petioles to about 2 dm long, sheathing only below the middle or near the base of the stem; bractlets of involucels filiform, about equalling the pedicels; plant usually hirtellous-puberulent; (ssp. *platycarpum*; s B.C.) . . . . *L. triternatum*
  2   Fruit relatively narrow, the wing seldom more than half the width of the body.
    4   Leaves small (the blade at most 7 cm long), granular-scaberulous to subglabrous, ternate-pinnately dissected into small narrow segments usually not over 4 mm long; fruit to 8 mm long, granular-scaberulous; stem usually less than 3 dm tall; (s B.C. and sw Alta.) . . . . *L. sandbergii*
    4   Leaves larger, the linear segments remote, to over 8 cm long; fruits usually glabrous; stem to 8 dm tall.
      5   Fruits at most 12 mm long; involucels wanting; plant glabrous; (s B.C.) . . . .
. . . . *L. ambiguum*
      5   Fruits to over 1.5 cm long; involucels consisting of several filiform bractlets about equalling the pedicels; leaves usually finely hirtellous-puberulent; (s B.C. and sw Alta.) . . . . *L. triternatum*
1   Leaves all basal or low-cauline.
  6   Flowers purple; lateral wings of fruit narrow and more or less corky-thickened; leaves

ternate-pinnately dissected into rather crowded linear segments commonly not over 1 cm long and 1.5 mm broad; stems glabrous; (s B.C. to sw Sask.) . . . . *L. dissectum*

6 Flowers yellow or white (sometimes purple-white in *L. macrocarpum*; rarely purple in *L. gormanii*); lateral wings of fruit not corky-thickened.

    7 Leaves firm, glabrous and strongly blue-glaucous, 1–2-ternate, then pinnate, the ultimate well-defined leaflets veiny, often stalked, lanceolate or oblong to ovate or subrotund, entire or often dentate toward tip, to 9 cm long and 6 cm broad; flowers yellow; peduncles often conspicuously swollen at apex; involucels wanting; fruit oblong or elliptic, to 1.5 cm long, the wings to about half as broad as the body; plant to 9 dm tall; (s B.C.) . . . . *L. nudicaule*

    7 Leaves more copiously dissected and rather "fern"-like, the ultimate segments seldom over 5 mm broad.

        8 Flowers usually white (or yellowish white in *L. martindalei*; sometimes purple-tinged in *L. macrocarpum*; rarely purple in *L. gormanii*).

            9 Leaves more or less densely soft-hairy (hence greyish rather than shining green), 1–2-ternate, then 2-pinnate-pinnatifid.

                10 Bractlets subtending umbellets linear-lanceolate to narrowly obovate, about equalling the flowers, glabrous or sometimes slightly ciliate toward base; fruit oblong-ovate, at most 1 cm long, the wings narrower than the body; ultimate leaf-segments linear, to 12 mm long; plant soft-puberulent; (s Sask. and s Man.) . . . . *L. orientale*

                10 Bractlets subtending umbellets linear-lanceolate, equalling to much surpassing the flowers, more or less tomentose or villous; fruit narrowly oblong, to 2 cm long, the wings often broader than the body; ultimate leaf-segments linear or narrowly oblong, rarely over 7 mm long; plant densely tomentose or villous, or glabrate; (s B.C. to s Man.) . . . .
    . . . . *L. macrocarpum*

           9 Leaves glabrous or only slightly puberulent; fruit glabrous or merely granular-roughened; (s B.C.).

                11 Involucels subtending the umbellets consisting of a few inconspicuous narrow bractlets or wanting.

                    12 Scapes to about 1.5 dm tall, from a short globose-thickened taproot to 2 cm thick; flowers white, with purple anthers (rarely wholly purple); pedicels seldom as much as 3 mm long at maturity; rays of umbel unequal, to 4 cm long; fruits to 7 mm long, their wings up to half the total width; leaves variously dissected, sometimes into crowded ultimate segments less than 5 mm long, sometimes into less-crowded segments often over 1 cm long; (?B.C.) . . . . . . [*L. gormanii*]

                    12 Scapes to about 2 dm tall, from an elongate, less-thickened taproot, or the taproot with a deep-seated thickening; flowers white or ochroleucous; pedicels to over 1 cm long; rays of umbel equal or unequal, to 6 cm long; fruits to about 1.5 cm long, their wings equalling or narrower than the body; (B.C.) . . . . *L. martindalei*

                11 Involucels present and more conspicuous; oil-tubes usually more numerous; leaf-blades to about 1 dm long, 1–2-ternate, then 1–2-pinnate-pinnatifid into linear segments to 3 mm broad.

                    13 Pedicels to about 2 cm long in fruit; fruit to 6.5 mm long; bractlets of involucels to 5 mm long, often connate at base, sometimes 2-cleft; ultimate linear segments of leaves to 1 dm long; (?B.C.) . . . .
    . . . . [*L. farinosum*]

                    13 Pedicels at most 4 or 5 mm long in fruit; fruit to about 12 mm long; bractlets of involucels mostly 2 or 3 mm long; ultimate linear segments of leaves usually less than 5 cm long; (s B.C.) . . . . *L. geyeri*

        8 Flowers yellow (occasionally white in *L. leptocarpum*).

           14 Lateral wings of the fruit thick and corky, about 1 mm broad, much narrower than the seed-bearing body, this to 18 mm long, the oil-tubes obscure;

some of the flowers always sterile; leaves ternate into 2–3-pinnate-pinnatifid primary divisions; plants to over 1.5 m tall; (s B.C. to sw Sask.) .... *L. dissectum*

14 Lateral wings of fruit not corky-thickened, usually broader; oil-tubes usually more evident; flowers usually all fertile; plants at most about 8 dm tall.

15 Leaves sparingly to more usually rather densely hirtellous-puberulent throughout; bractlets of involucels usually conspicuous.

16 Fruit to 1 cm long, distinctly hirtellous-puberulent, the wings to half the width of the seed-bearing body, several oil-tubes usually present in the intervals and on the commissure; leaves ternate, then 2–3-pinnate-pinnatifid into small crowded segments mostly not over 3 mm long; scapes to about 3 dm tall; (B.C. to s Man.) .... *L. foeniculaceum*

16 Fruit to over 1.5 cm long, nearly or quite glabrous, the oil-tubes solitary in the intervals, 2 on the commissure; leaves 1–2-ternate or 5-ternate, then 1–2-pinnate into remote linear segments to over 1 dm long; scapes to about 8 dm tall; (s B.C. and s Alta.) ..... *L. triternatum*

15 Leaves (and fruit) usually nearly or quite glabrous (sometimes minutely granular-roughened).

17 Fruits usually about 4 times as long as broad, very narrowly or scarcely winged (wings to about 0.5 mm broad), to 13 mm long and 2 or 3 mm broad, the oil-tubes solitary in the intervals, up to 4 on the commissure; pedicels mostly 1 or 2 mm long at maturity; bractlets of involucel linear-attenuate, to over 7 mm long; leaf-blades to about 1.5 dm long, primarily 1–2-ternate, then 2–3-pinnate-pinnatifid into filiform to linear, very unequal segments to 5 cm long and 2 mm broad; (?B.C.) .... [*L. leptocarpum*]

17 Fruits broader in outline, the wing narrower than to about equalling the seed-bearing body; ultimate leaf-segments relatively short and broad.

18 Fruit to about 16 mm long, the oil-tubes solitary in the intervals, 2 on the commissure, the pedicels over 1 cm long; bractlets of involucel few and inconspicuous or wanting; leaf-blades to about 5 cm long, 1–2-pinnate-pinnatifid (the primary separation some-times ternate); scapes to about 2 dm tall; root elongate and not much thickened or with a deep-seated thickening; (s B.C.) .... *L. martindalei*

18 Fruit to about 12 mm long, the inconspicuous oil-tubes up to 3 or 4 in the intervals and 5 or 6 on the commissure, the pedicels commonly not over 3 or 4 mm long; umbellets subtended by involucels; scapes to over 6 dm tall.

19 Scapes from a usually strongly tuberous-thickened root; rays of umbel to 1 dm long; fruits erect or spreading (or the lowermost ones somewhat deflexed), rather broadly elliptic to oblong, on pedicels mostly not over 3 mm long; leaves pinnately or ternate-pinnately dissected, highly variable as to the crowding and width of the ultimate segments; (sw Sask.) .... *L. cous*

19 Scapes from a long stout taproot crowned by a compactly branched woody caudex; rays of umbel to 5 cm long; fruits mostly deflexed, rather narrowly elliptic-oblong, on pedicels to 4(5) mm long; leaves ternate-pinnately dissected into ultimate segments to 5 cm long and 8 mm broad; (s B.C.) .... *L. brandegei*

L. ambiguum (Nutt.) C. & R.

/T/W/ (Grt) Rocky flats and slopes up to moderate elevations from s B.C. (chiefly valleys of the Dry Interior N to Kamloops, Chase, and Sicamous) to Oreg., Utah, and Wyo. [*Eulophus* and *Peucedanum* Nutt.; *Cogswellia* Jones].

L. brandegei (C. & R.) Macbr.
/T/W/ (Hs) Open or wooded slopes from the foothills to fairly high elevations from s B.C. (Manning Provincial Park, about 30 mi SE of Hope; Ashnola Range, SW of Penticton; CAN; V) to Wash. [*Peucedanum* C. & R.].

L. cous (Wats.) C. & R.
/T/WW/ (Grt) Prairies, sagebrush plains, and rocky slopes from Wash. to Mont. and sw Sask. (Cypress Hills; Breitung 1957a), s to Oreg., Wyo., and S.Dak. [*Peucedanum* Wats.; *Cogswellia* Jones; *L. (C.; P.) montanum* C. & R.].

L. dissectum (Nutt.) Math. & Const.
/T/W/ (Grt) Dry prairies and meadows, rocky slopes, and talus at low to moderate elevations from B.C. (N to Quesnel, ca. 53°N; V) to sw Alta. (N to Banff) and sw Sask. (Cypress Hills; Breitung 1957a), s to s Calif. and Colo. [*Leptotaenia* Nutt.; *Ferula dissoluta* Wats.].
    Some of our material is referable to var. *multifidum* (Nutt.) Math. & Const. *(Lept. (Fer.) mult.* Nutt.; *Lept. purpurea sensu* Rydberg 1922, at least in part, not (Wats.) C. & R., which is *Lom. columbianum* Math. & Const. of the w U.S.A.; fruits mostly on pedicels at least 4 mm long (to 2 cm) rather than sessile or short-pedicelled; ultimate leaf-segments averaging somewhat narrower than those of the typical form).

[*L. farinosum* (Hook.) C. & R.]
[This species of the w U.S.A. (Wash., Idaho, and Mont.) is noted by Henry (1915) as "Attributed to B.C. by Howell." No supporting voucher-specimens have been located. (*Peucedanum* Hook.; *Cogswellia* Jones).]

L. foeniculaceum (Nutt.) C. & R.
/T/WW/ (Hr) Dry plains and valleys at low to moderate (occasionally fairly high) elevations from E-cent. B.C. (Fort St. John, ca. 56°10'N), Alta. (N to the Peace River dist.; Herb. V), s Sask. (Cypress Hills; Fort Qu'Appelle; Moose Mountain; Maple Creek; Old Wives Creek), and s Man. (N to Miniota, about 50 mi NW of Brandon) to Nev., Ariz., Tex., and Mo. [*Ferula* and *Peucedanum* Nutt.; *Cogswellia* C. & R.; *Pastinaca* Spreng.; *L. (C.; Peuc.) villosum* Raf.]. MAP: W.L. Theobald, Brittonia 18(1): fig. 5, p. 12. 1966.
    Var. *macdougalii* (C. & R.) Cronquist (*L. mac.* C. & R.; *L. (C.) jonesii* C. & R.; plant usually fairly densely puberulent throughout rather than only moderately so; bractlets of the involucels less markedly connate than those of the typical form) is accredited to Alta. by Rydberg (1922).

L. geyeri (Wats.) C. & R.
/t/W/ (Grt) Plains, foothills, and valleys at low to moderate elevations from s B.C. (valleys of the Dry Interior N to Merritt and Vernon, E to Creston) to Wash. and Idaho. [*Peucedanum* Wats.; *Cogswellia* Jones; *P. bicolor sensu* J.M. Macoun 1894, not Wats., the relevant collection in CAN].

[*L. gormanii* (Howell) C. & R.]
[A collection of this species of the w U.S.A. (N to Wash. and Idaho) in the herbarium of Manning Provincial Park, about 30 mi SE of Hope, B.C., has been placed here but requires confirmation. (*Peucedanum* Howell).]

[*L. leptocarpum* (T. & G.) C. & R.]
[A species of the w U.S.A. (N to Wash. and Idaho) which, according to Henry (1915; *L. amb.* var. *lept.*), should be searched for in B.C. (*L. ambiguum* var. *lept.* (T. & G.) Jeps.).]

L. macrocarpum (Nutt.) C. & R.
/T/WW/ (Grt) Rocky hills and plains from B.C. (N to Cariboo, 52°51'N; CAN) to Alta. (N to Grande Prairie, 55°10'N), Sask. (N to Saskatoon; Breitung 1957a), and Man. (N to Steeprock, about 100 mi N of Portage la Prairie), s to Calif., Utah, Colo., and S.Dak. [*Peucedanum* Nutt.; *Cogswellia* Jones].

L. martindalei C. & R.
/T/W/ (Grt) Usually at fairly high elevations in dry meadows and rocky slopes from sw B.C. (Mt. Arrowsmith and Castlecrag Mt., Vancouver Is.; Mt. Cheam, near Agassiz; CAN; V) to Wash. and Oreg. [*Peucedanum* C. & R.; *Cogswellia* Jones].
   The typical form (flowers white or ochroleucous; fruits averaging less than twice as long as broad) is apparently restricted to the w U.S.A., most of our material being referable to var. *angustatum* C. & R. (flowers as in the typical form but fruits averaging 2 or 3 times as long as broad). Var. *flavum* (Jones) Cronquist (fruits as in var. *angustatum* but the flowers yellow) is reported from B.C. by Hitchcock et al. (1961).

L. nudicaule (Pursh) C. & R.
/t/W/ (Hr) Dry open or slightly wooded places at low to moderate elevations from s B.C. (Vancouver Is. and adjacent islands and mainland N to Spences Bridge, E to Manning Provincial Park, SE of Hope; the report from sw Alta. by Hitchcock et al. 1961, requires confirmation) to Calif. and Utah. [*Smyrnium* Pursh; *Cogswellia* Jones; *Peucedanum* Nutt.; *P. latifolium* Nutt.; *Seseli (P.) leiocarpum* Hook.].

L. orientale C. & R.
/T/WW/ (Grt) Dry prairies and open rocky slopes from Mont. to Sask. (Bienfait, about 120 mi SE of Regina; Breitung 1957*a*) and s Man. (Minto, about 20 mi s of Brandon; Aweme, about 20 mi SE of Brandon), s to Ariz., N.Mex., Kans., and Iowa. [*Cogswellia* Jones; *Peucedanum* Blank.].

L. sandbergii C. & R.
/T/W/ (Grt) Rocky slopes and ridges at moderate to high elevations from SE B.C. ("Passes of the Rockies. Kootenay"; Henry 1915) and sw Alta. (Waterton Lakes; Breitung 1957*b*) to Idaho and w Mont. [*Cogswellia* Jones; *Peucedanum* C. & R.].

L. triternatum (Pursh) C. & R.
/T/W/ (Hs) Dry to fairly moist open slopes and meadows at low to moderate elevations from s B.C. (N to Lillooet, about 70 mi w of Kamloops) and sw Alta. (N to Banff; CAN) to Calif., Utah, and Colo. [*Seseli* Pursh; *Cogswellia* Jones; *Eulophus* and *Peucedanum* Nutt.; *L. simplex* var. *leptophyllum* (Hook.) Math.].
   A collection in CAN from Penticton, Okanagan Valley, B.C., has been referred by Rydberg to ssp. *platycarpum* (Torr.) Cronquist (*L. plat.* Torr.; *Peucedanum (L.; Cogswellia) simplex* Nutt.; fruit relatively broad, each wing nearly equalling or broader than the seed-bearing body rather than seldom over half the width; secondary separation of the leaf commonly resulting in only 3(5) linear ultimate segments on each of the primary divisions rather than in a more or less pinnate primary division).

L. utriculatum (Nutt.) C. & R.
/t/W/ (Hs) Prairies and open rocky places from sw B.C. (Vancouver Is. and adjacent islands and mainland E to the lower Fraser Valley near Yale; CAN; V) to s Calif. [*Peucedanum* Nutt.; *Cogswellia* Jones].

MUSINEON Raf. [5972]

M. divaricatum (Pursh) Nutt.
/T/WW/ (Grt) Plains, valleys, and foothills from Mont. to s Alta. (N to Lethbridge; CAN), Sask. (N to Langham, about 20 mi NW of Saskatoon; reported from Carlton House, about 35 mi sw of Prince Albert, by John Macoun 1883), and s Man. (N to The Narrows of L. Manitoba, ca. 51°N; Macoun 1883), s to Nev. and Nebr. [*Seseli* Pursh; incl. the more or less scabrous var. *hookeri* T. & G. (*M. angustifolium* and *M. trachyspermum* Nutt.)]. MAP: M.E. Mathias and Lincoln Constance, Ann. Mo. Bot. Gard. 17(3/4): fig. 5 (the occurrence in s Man. should be indicated), p. 237. 1930.

[MYRRHIS Mill.] [5942]

[M. odorata (L.) Scop.] Myrrh
[European; a garden-escape in s Ont. (St. Thomas, Elgin Co.) and N.S. (Wolfville, Kings Co.; ACAD), where, however, scarcely established. (*Scandix* L.).]

OENANTHE L. [6046] Water Dropwort

O. sarmentosa Presl
/T/W/ (Hsr) Damp thickets and low wet places from SE Alaska (*see* Hultén 1947: map 893, p. 1198) through w B.C. (Vancouver Is. and adjacent islands and mainland E to the Chilliwack Valley; CAN; V; reported E to the lower Fraser Valley by Henry 1915; type from Nootka Sound, Vancouver Is.) to cent. Calif.

OSMORHIZA Raf. [5941] Sweet Cicely. Osmorhize

(Ref.: Constance and Shan 1948; M.E. Mathias and Lincoln Constance, N. Am. Flora 28B:105–09. 1944)
1 Fruit smooth, obtuse at base, to 2 cm long, the stylopodium (including the barely differentiated styles) to 1.2 mm long; flowers yellow; involucels usually wanting; stems clustered, often villous at the nodes; (B.C. and sw Alta.) . . . . *O. occidentalis*
1 Fruit bristly, attenuate at base; flowers greenish white (sometimes pink or purple); stems mostly solitary.
  2 Umbellets subtended by an involucel of narrow bractlets (these often deciduous at maturity); styles (including the barely differentiated stylopodium) to 4 mm long, nearly straight and parallel.
    3 Styles (with stylopodium) at most 1.5 mm long; stipules glabrous except for ciliate margins; (Ont. to Nfld. and N.S.) . . . . *O. claytonii*
    3 Styles (with stylopodium) at least 2 mm long; stipules densely long-hairy near the margins; (s Alta. to N.S.) . . . . *O. longistylis*
  2 Umbellets mostly lacking subtending bractlets; styles (including the abruptly differentiated stylopodium) minute, at most about 1 mm long, finally strongly outward-curved; (transcontinental).
    4 Fruit convexly narrowed to summit, the body at most 12 mm long; stylopodium broader than long, with the style not more than 0.5 mm long; pedicels and rays of umbel generally widely spreading . . . . *O. depauperata*
    4 Fruit concavely narrowed at summit; pedicels and rays of umbel mostly spreading-ascending . . . . *O. chilensis*

O. chilensis H. & A.
/sT/(X)/ (Hs) Woodlands at low to moderate elevations from the E Aleutian Is. and s Alaska to B.C., sw Alta. (Waterton Lakes; near Pincher Creek; Cypress Hills), sw Sask. (Cypress Hills; Breitung 1957a), s ?Man. (Hultén's below-noted map indicates an area near the s end of L. Winnipeg), Ont. (N shore of L. Superior; Manitoulin Is. and Bruce Pen., L. Huron), Que. (N to Anticosti Is. and the Gaspé Pen.), Nfld. (St. John Bay; GH), N.B. (Charlotte and Restigouche counties; CAN; not known from P.E.I.), and N.S. (Annapolis, Kings, Victoria, and Inverness counties), s to s Calif., Colo., S.Dak., Wisc., Mich., and Maine; S. America. [*Washingtonia* (*O.; Uraspermum*) *brevipes* C. & R.; *O.* (*Scandix; W.; U.*) *divaricata* Nutt.; *W.* (*O.*) *intermedia* Rydb.; *O.* (*S.; W.*) *nuda* Torr.]. MAPS: Hultén 1968b:697; Constance and Shan 1948: fig. 5, p. 136; Lincoln Constance, Quart. Rev. Biol. 38: fig. 1, p. 113. 1963; Fernald 1925: map 21 (*O. divar.*), p. 255.

Some of the Alaska–B.C.–Alta. material is referable to var. *purpurea* (C. & R.) Boivin (*O. (Washingtonia) purp.* C. & R., the type from Sitka, Alaska; flowers usually pink to purple rather than greenish white; fruits to 13 mm long rather than up to 2 cm, the depressed stylopodium generally broader than long rather than usually at least as long as broad). MAPS: (*O. purp.*): Hultén 1968b:697; on the above-noted map by Constance and Shan.

O. claytonii (Michx.) Clarke
/T/EE/ (Hs) Moist woods and clearings from E S.Dak. and Minn. to Ont. (N to L. Kapuskasing at 48°32'N; CAN), Que. (N to Anticosti Is. and the Gaspé Pen.), Nfld. (Boivin 1966*b*; not listed by Rouleau 1956), N.B., P.E.I., and N.S., S to Ark., Ala., and N.C. [*Myrrhis* Michx.; *Chaerophyllum* Pers.; *Washingtonia* Britt.; *O. brevistylis* DC.]. MAP: Constance and Shan 1948: fig. 4, p. 132.

The reports from Crane L., Sask., by Constance and Shan, and from Winnipeg, Man., by Lowe (1943) are based upon *O. longistylis* (relevant collections in CAN and WIN, respectively), as also, probably, the report from Riding Mt., Man., by Scoggan (1957).

O. depauperata Phil.
/sT/X/ (Hs) Moist woods and clearings from SE Alaska, SW Dist. Mackenzie (Mt. Coty, 60°18'N; W.J. Cody, Can. Field-Nat. 77(2):120. 1963), and B.C.–Alta. to Sask. (N to McKague, 52°37'N), S Man. (N to L. Winnipeg at ca. 52°N), Ont. (N to Longlac, N of L. Superior, 49°47'N), Que. (N to Rupert House, James Bay, ca. 51°30'N, the Côte-Nord, Anticosti Is., and Gaspé Pen.), S Labrador (Forteau, 51°28'N; Capitan Is.), Nfld., N.B., and N.S. (not known from P.E.I.), S to Calif., N.Mex., S.Dak., Minn., Mich., and Vt.; S. America. [*Washingtonia (O.) obtusa* C. & R.; *O. chilensis* var. *cupressimontana* Boivin]. MAPS: Hultén 1968*b*:698; Constance and Shan 1948: fig. 5 (*O. ob.*; the occurrence in S Labrador should be indicated), p. 136; Lincoln Constance, Quart. Rev. Biol. 38: fig. 1, p. 113. 1963.

O. longistylis (Torr.) DC. Anise-root
/T/(X)/ (Hs) Moist woods and clearings from S Alta. (Medicine Hat; GH) to Sask. (N to Pike L., near Saskatoon; Breitung 1957*a*), S Man. (N to Fort Ellice, about 75 mi NW of Brandon; CAN), Ont. (N to the Nipigon R. N of L. Superior; CAN), Que. (N to the Gaspé Pen. at Matapédia; GH), N.B., and N.S. (not known from P.E.I.), S to N.Mex., Tex., and Ga. [*Myrrhis* Torr.; *Washingtonia* Britt.; *O. aristata* var. *long.* (Torr.) Boivin]. MAP: Constance and Shan 1948: fig. 4 (the above Alta. and Ont. stations should be indicated), p. 132.

Some of the collections from S Ont., N.B., and N.S. are referable to var. *brachycoma* Blake (petioles, stem (except for the nodes), and at least the bases of the branches densely spreading-puberulent rather than glabrous).

O. occidentalis (Nutt.) Torr.
/T/W/ (Hs) Thickets and open slopes at low to moderate elevations from S B.C. (N to Kaslo, Kootenay L., about 35 mi NE of Nelson) and SW Alta. (Crowsnest Pass; Waterton Lakes; near Pincher Creek) to Calif. and Colo. [*Glycosma* Nutt.; *Myrrhis* B. & H.]. MAPS: Constance and Shan 1948: fig. 1, p. 118; Lincoln Constance, Quart. Rev. Biol. 38: fig. 1, p. 113. 1963.

[OXYPOLIS Raf.] [6107] Hog-fennel

[*O. rigidior* (L.) C. & R.] Cowbane, Water-Dropwort
[This species of the E U.S.A. (bogs, swamps, wet woods, and damp rocks from Minn. to N.Y., S to La. and Fla.) is known from Canada through several collections between 1886 and 1901 in S Ont. (Essex and Welland counties; DAO; TRT; *see* S Ont. map by Soper 1962: fig. 29, p. 45), where, however, apparently not taken since that time and almost certainly extinct. (*Sium* L.; *Archemora (Tiedemannia) rigida* DC.).]

PASTINACA L. [6120]

P. sativa L. Parsnip. Panais
European; introd. in fields, thickets, and waste places of N. America, as in Alaska (Manly Hot Springs), the Yukon (Boivin 1966*b*), B.C. (N to Spences Bridge), S Alta. (Belly R.; CAN), Sask. (N to McKague, 52°37'N), Man. (N to The Pas), Ont. (N to Big Trout L. at ca. 53°45'N, 90°W), Que. (N to the Gaspé Pen.), Nfld., N.B., P.E.I., and N.S. [*Peucedanum* Benth.]. MAP: Hultén 1968*b*:706.

PERIDERIDIA Reichenb. [6036]

1 Fruit suborbicular, 2.5–3.5 mm long and nearly or quite as broad; involucre wanting or of 1

or 2 (occasionally 6–8) setaceous bracts; basal leaves simply pinnate with 2–5 pairs of pinnae (or the lower ones again pinnate); plant to over 12 dm tall, the root solitary or occasionally fascicled; (s B.C. to sw Sask.) .... *P. gairdneri*
1   Fruit oblong-ovate, to 6 mm long and 2 mm broad; involucre usually of 6–10 linear-lanceolate bracts (or sometimes only 1 or 2 setaceous bracts, or wanting); basal leaves ternate-pinnately compound (bipinnate to tripinnate or ternate-pinnate); plant mostly less than 6 dm tall, the roots more often fascicled .... [*P. oregana*]

P. gairdneri (H. & A.) Mathias   Yampah, Squaw-root
/T/WW/   (Grt)   Dry or wet meadows and woods at low to moderate elevations from B.C. (N to the mouth of the Dean R., ca. 51°20′N; CAN), s Alta. (N to Macleod, w of Lethbridge), and sw Sask. (Cypress Hills, where first taken by John Macoun in 1880; CAN) to s Calif., ?N.Mex., Colo., and S.Dak. [*Atenia* H. & A.; *Carum* Gray]. MAP: Tsan-Ing Chuang and Lincoln Constance, Univ. Calif. Pub. Bot. 55: fig. 25, p. 60. 1969.
According to the above-noted map, the typical form is confined to the southern two-thirds of Calif., the northern plant being ssp. *borealis* Chuang & Constance (*Endosmia (Atenia) montana* Nutt.; differing from the typical form in its stricter habit, its 2 or 3 tubers (rather than 1), and its 5–7-veined (rather than single-veined) petals).

[*P. oregana* (Wats.) Mathias]   Eppaw
[Reports of this species of the w U.S.A. (s Wash. to Calif.) from sw B.C. and s Alta. by John Macoun (1890) are based upon *P. gairdneri*, relevant collections in CAN. The MAP by Chuang and Constance (loc. cit., fig. 18, p. 46) indicates no Canadian stations. (*Carum* Wats.).]

[PETROSELINUM Hoffm.]   [6006]

[*P. crispum* (Mill.) Mansf.]   Common Parsley
[European; an occasional garden-escape in N. America, as in sw B.C. (Vancouver Is.; Lulu Is.; New Westminster), Ont. (Kemptville, about 35 mi s of Ottawa; Groh 1946), and Nfld. (near St. John's; GH), but scarcely established in our area. (*Apium* Mill.; *A. petroselinum* L.; *P. hortense* Hoffm.).]

PIMPINELLA L.   [6033]

P. saxifraga L.   Burnet-Saxifrage
Eurasian; introd. in fields and along roadsides and shores in N. America, as in Ont. (taken by Frankton in 1952 at Ottawa from a large colony on a grassy railway embankment; DAO), Nfld. (Quiddy Viddy; GH), N.B. (York, Carleton, and Gloucester counties), and N.S. (Pubnico, Yarmouth Co.; NSAC). [Incl. var. *dissecta* Wallr.].

PODISTERA Wats.   [6059]

1   Fruit to 5 mm long and 3 mm broad, with 2 or 3 oil-tubes in the intervals between the ribs; flowers purplish; involucre consisting of numerous linear-lanceolate bracts to 1 cm long, the involucels similar to the involucre, surpassing the pedicels but shorter than the fruits; leaflets ovate to orbicular, to 1.5 cm long and broad, coarsely lobed or incised; (Alaska) .... *P. macounii*
1   Fruit to about 3 mm long and 1.5 mm broad, with a solitary oil-tube below each filiform rib; flowers unknown; involucre consisting of 1 or more small linear entire bracts, or perhaps sometimes wanting; involucels consisting of about 5 linear-acuminate entire purplish bractlets to 5 mm long, these connate at base, about equalling the fruit; leaflets narrowly lanceolate to orbicular, apiculate, entire; (the Yukon) .... *P. yukonensis*

P. macounii (C. & R.) Math. & Const.
/Ss/W/EA/   (Hr)   Dry tundra and stony slopes at low to moderate elevations: Alaska (N to ca. 66°N; *see* Hultén 1947: map 895 (*L. mut. alp.*), p. 1199); mts. of Europe; Asia. [*Ligusticum mac.* C. & R., the type from Cape Vancouver, Alaska; *Ligusticella* Math. & Const.; *Orumbella* C. & R.; *Ligusticum mutellinoides* ssp. *alpinum* (Ledeb.) Thell.]. MAP: Hultén 1968b:703 (*L. mut.*).

P. yukonensis Math. & Const.
/S/W/ (Hr) Stony montane slopes in w-cent. Yukon (type from the Little Klondike R. between 63° and 64°N; CAN; *see* M.E. Mathias and Lincoln Constance, Bull. Torrey Bot. Club 77(2):136–38. 1950). MAP: Hultén 1968*b*:703 (also indicating a possible occurrence on or near the Yukon boundary in E-cent. Alaska).

SANICULA L.   [5918]   Sanicle, Black Snakeroot.   Sanicle

(Ref.: Shan and Constance 1951; M.E. Mathias and Lincoln Constance, N. Am. Flora 28B:63–71. 1944)
1   Basal leaves pinnately or ternate-pinnately cleft or divided; umbels with 3–5 rays; plants to 5 or 6 dm tall; (B.C.).
  2   Principal leaves 1–2-pinnatifid, the rachis-wing toothed, the blade to about 13 cm long and 12 cm broad, its segments toothed or cleft; rays of umbel to about 1.5 dm long; umbellets about 20-flowered, the bractlets of their involucels to 2.5 mm long, slightly united at base; petals purplish red to dark purple; styles twice as long as the calyx, divergent; fruits up to 10 (the umbellets with about equal numbers of staminate and perfect flowers) . . . . *S. bipinnatifida*
  2   Principal leaves ternate-pinnately decompound, the blade to about 4 cm long and 3.5 cm broad, its primary divisions tending to be pinnatifid, the lowest pair separated from the terminal segment or segments by a narrow entire rachis; rays of umbel to 8 cm long; umbellets mostly 10–15-flowered, the bractlets of their involucels about 1 mm long, strongly united at base; petals yellow; styles about 3 times as long as the calyx, recurved; fruits rarely more than 5 (the umbellets with more staminate than perfect flowers) . . . . *S. graveolens*
1   Basal leaves ternately or palmately cleft or divided.
  3   Bractlets subtending umbellets bright yellow and mostly much longer than the umbellets; basal leaves palmately 3-parted (the divisions coarsely dentate to laciniate, with unequal mucronate teeth or lanceolate lobes), somewhat succulent, triangular-ovate to orbicular, to 6.5 cm long and 9 cm broad, spreading to form a prostrate rosette; plants bright yellowish-green, the very short stem many-branched from near the base into scapose branches to 3 dm long; (sw B.C.) . . . . *S. arctopoides*
  3   Bractlets subtending umbellets minute and inconspicuous.
    4   Leaves simple, the basal ones deltoid to rotund-cordate or -reniform in outline, to 13 cm long and 18 cm broad, deeply palmately 3–5-lobed, their segments singly or doubly serrate; umbels with usually 3 or 4 rays; petals yellow or greenish yellow; styles shorter than the bristles of the short-stalked fruit, this to 5 mm long; umbellets mostly 8–13-flowered, commonly with more perfect than staminate flowers; involucels consisting of 5 bractlets 1 or 2 mm long; stem solitary from a thickened taproot; (w B.C.) . . . . *S. crassicaulis*
    4   Basal leaves palmately compound, with 3–5 leaflets; umbels with rarely more than 3 rays; involucels minute; stems solitary or few together from a rootstock.
      5   Styles much surpassing the bristles of the fruit, recurved-spreading; staminate florets in separate umbellets or mixed with the perfect ones; branches of inflorescence strongly ascending; leaves usually 5-foliolate.
        6   Fruit short-stalked, less than 5 mm long; sepals of staminate florets soft, deltoid, at most 1 mm long; petals yellowish green; branches of inflorescence capillary; (Ont. to N.S.) . . . . *S. gregaria*
        6   Fruit sessile, at least 5 mm long at maturity; sepals of staminate florets rigid, lance-subulate, to 2 mm long; petals whitish green; branches of inflorescence relatively stout; (transcontinental) . . . . *S. marilandica*
      5   Styles shorter than the bristles of the fruit and hidden by them; staminate florets mixed with the pistillate; petals white; branches of the inflorescence divergent; leaves 3-foliolate (but the lateral leaflets often deeply cleft).
        7   Fruit short-stalked, its connivent sepals hidden by the bristles; pedicels of staminate florets only slightly longer than the calyx; (s Ont.) . . . . . . *S. canadensis*

7 Fruit sessile, its beak of connivent sepals prolonged beyond the bristles; pedicels of staminate florets 3 or 4 times as long as the calyx; (s Ont. to N.B.) .... *S. trifoliata*

S. arctopoides H. & A.
/t/W/ (Hs) Open coastal bluffs and headlands from sw B.C. (Vancouver Is. and adjacent islands; CAN; V; the Dawson collection from Queen Charlotte Is. noted by John Macoun 1883, probably relates to *S. crassicaulis,* the only species listed by Calder and Taylor 1968) to Calif. [*S. howellii* C. & R.]. MAP: Shan and Constance 1951: fig. 51, p. 66.

S. bipinnatifida Dougl.
/t/W/ (Hs) Open or sparsely wooded slopes and drier meadows from sw B.C. (between Victoria and Nanaimo, Vancouver Is.; CAN) to Baja Calif. MAP: Shan and Constance 1951: fig. 51, p. 66.

S. canadensis L.
/T/EE/ (Hs) Dry open woods from Nebr. to Minn., s Ont. (N to Waterloo, Middlesex, and Peterborough counties; CAN; OAC; TRT; reported N to Ottawa by John Macoun 1883, but not listed by Gillett 1958), N.H., and Mass., s to Tex. and Fla. [Incl. var. *grandis* Fern.; *S. marilandica* var. *can.* (L.) Torr.]. MAP: Shan and Constance 1951: fig. 45, p. 44.

S. crassicaulis Poepp.
/t/W/ (Hs) Moist or dry woods from w B.C. (Queen Charlotte Is.; Vancouver Is.) to Baja Calif.; Chile. [*S. menziesii* H. & A.]. MAPS: Shan and Constance 1951: fig. 50, p. 60; Lincoln Constance, Quart. Rev. Biol. 38: fig. 2, p. 114. 1963.

S. graveolens Poepp.
/T/W/ (Hs) Open or lightly wooded places at low to moderate elevations from B.C. (N to Ootsa Lake, about 150 mi w of Prince George at ca. 53°15′N; CAN) and Mont. to s Calif.; S. America. [*S. nevadensis* Wats.; *S. septentrionalis* Greene]. MAPS: Shan and Constance 1951: fig. 51, p. 66; Lincoln Constance, Quart. Rev. Biol. 38: fig. 2, p. 114. 1963.

S. gregaria Bickn.
/T/EE/ (Hs) Rich woods and thickets from Minn. to Ont. (N to the Ottawa dist.), Que. (N to l'Ange-Gardien, NE of Quebec City; *see* Que. map by Doyon and Lavoie 1966: fig. 5, p. 816), w N.B. (near Woodstock, Carleton Co.; CAN; not known from P.E.I.), and N.S. (Hants and Kings counties), s to Kans., Mo., Ala., and Fla. MAP: Shan and Constance 1951: fig. 45, p. 44.

S. marilandica L.
/T/X/ (Hs) Woods, thickets, meadows, and shores from B.C. (N to Hazelton, ca. 55°30′N; an isolated station at Liard Hot Springs, ca. 59°23′N) to Alta. (N to Peace River, 56°14′N), Sask. (N to Prince Albert), Man. (N to Cross L., NE of L. Winnipeg), Ont. (N to the w James Bay watershed at ca. 53°N), Que. (N to Anticosti Is. and the Gaspé Pen.), Nfld., N.B., P.E.I., and N.S., s to N.Mex., Colo., Mo., and Fla. MAP: Shan and Constance 1951: fig. 45 (somewhat incomplete northwards), p. 44.

S. trifoliata Bickn.
/T/EE/ (Hs) Rich woods from Minn. to Ont. (N to the Ottawa dist.; type from Amherstburg, Essex Co.), Que. (N to the Gaspé Pen. at Gaspé Basin; CAN), and w N.B. (Hampton, Kings Co.; NBM; not known from P.E.I. or N.S.), s to Tenn., Va., and New Eng. MAP: Shan and Constance 1951: fig. 45 (the occurrence in N.B. should be indicated), p. 44.

SCANDIX L. [5939]

S. pecten-veneris L. Shepherd's-needle, Lady's-comb
European; locally introd. along roadsides and in waste places of N. America, as in s B.C. (Victoria and Kootenay L.; Eastham 1947), Sask. (Golburn; Breitung 1957a), and s Ont. (an 1893 collection near Kingston by McMorine; Montgomery 1957).

SIUM L. [6038]

S. suave Walt. Water-parsnip. Berle
/ST/X/A/ (Hs (Hel)) Swamps, wet thickets, shores, and shallow water from Alaska (N to ca. 67°N; not known from the Yukon) to Great Bear L., Great Slave L., L. Athabasca (Alta. and Sask.), Man. (N to the Cochrane R. at ca. 59°N; CAN), Ont. (N to the Shamattawa R. at 54°24′N), Que. (N to E James Bay at 53°50′N, L. Mistassini, and the Gaspé Pen.), Nfld., N.B., P.E.I., and N.S., s to Calif., Kans., Ohio, and Fla.; Asia. [*S. cicutaefolium* Schrank; *S. (Apium) lineare* Michx.; *S. latifolium* and *Ammi ?majus sensu* Hooker 1832, not L.]. MAP: Hultén 1968*b*:700.

TAENIDIA Drude [6031]

T. integerrima (L.) Drude Yellow Pimpernel
/T/EE/ (Hp) Dry woods, thickets, and rocky hillsides from Minn. to s Ont. (N to Huron, Wellington, Peel, York, and Ontario counties) and sw Que. (N to Anticosti Is.; Marie-Victorin and Rolland-Germain 1969), s to Tex., La., Miss., Ala., and Ga. [*Smyrnium* L.; *Pimpinella* Gray; *Zizia* DC.].

THASPIUM Nutt. [6076] Meadow-parsnip

1 Stem pubescent around the upper nodes with minute stiffish hairs, to 12 dm tall; basal and principal stem-leaves 2-pinnate or ternate-pinnate, the lanceolate to ovate leaflets coarsely toothed to deeply incised (var. *angustifolium*) into linear or oblong segments; (s Ont.) .... *T. barbinode*
1 Stem glabrous or nearly so, to 8 dm tall; basal leaves mostly simple (occasionally ternate like the stem-leaves), finely crenate, broadly ovate, usually cordate at base; principal stem-leaves usually pinnate with 3 or 5 lanceolate to ovate, crenate leaflets (rarely the lateral leaflets themselves 2–3-lobed) .... [*T. trifoliatum*]

T. barbinode (Michx.) Nutt.
/t/EE/ (Hs) Rich woods, thickets, and rocky hillsides from Minn. to s Ont. (Essex, Kent, Lambton, Middlesex, Norfolk, Welland, Lincoln, and Wellington counties; reports from Man. are probably based upon *Zizia aurea*) and N.Y., s to Okla., Ark., and Ga. [*Ligusticum* Michx.; *Thapsia* Prov.].
   Var. *angustifolium* C. & R. (leaflets deeply incised into linear or oblanceolate segments rather than merely coarsely toothed or cleft) is known from s Ont. (Pelee Is., Essex Co., where taken by John Macoun in 1892; GH).

[*T. trifoliatum* (L.) Gray]
[Reports of this species of the E U.S.A. (N to Minn. and N.Y.) from Canada require confirmation, most or all of them apparently being based upon the habitally very similar *Zizia aptera. (Thapsia* L.; *Thaspium aureum* var. *atropurpureum* (Desv.) C. & R.).]

TORILIS Adans. [5945] Hedge-parsley

1 Umbels congested and head-like, on peduncles at most 1 cm long arising opposite the leaves, the 2 or 3 rays very short; involucre wanting; bractlets of the involucel surpassing the flowers, these about 1 mm broad, pinkish, their petals subequal; fruit 2 or 3 mm thick, the outer carpels with straight (rarely hooked) prickles, the inner carpels merely tuberculate; leaves 1–2-pinnate, the ultimate lobes linear-lanceolate; stems to 3.5 dm long, commonly prostrate; (introd. in N.B.) .... [*T. nodosa*]
1 Umbels long-peduncled, with usually more numerous rays; flowers commonly 2 or 3 mm broad, their petals unequal; stems erect, usually longer.
   2 Umbels to 4 cm thick, with up to 12 rays, subtended by 4 or more unequal bracts up to nearly as long as the rays; bractlets about equalling the pedicels of the flowers composing the umbellets; flowers 2 or 3 mm broad, pinkish- or purplish-white; fruit 3 or

4 mm thick, armed with hooked prickles; styles glabrous, recurved in fruit; leaves 1–3-pinnate, their lanceolate to ovate primary segments to 2 cm long, pinnatifid and serrate; (introd. in Ont.) . . . . *T. japonica*

2 Umbels to 2.5 cm thick, with usually less than 8 rays, bractless or subtended by a solitary bract; bractlets densely hispid, about equalling the umbellets; flowers about 2 mm broad, white or pinkish; fruit 4 or 5 mm thick, its prickles curved but not hooked, thickened at tip; styles more or less hairy, spreading in fruit; leaves 1–2-pinnate, their lanceolate primary segments to 3 cm long, pinnatifid or coarsely serrate; (introd. in sw B.C.) . . . . [*T. arvensis*]

[T. arvensis (Huds.) Link]
[Eurasian; locally introd. in the w U.S.A. but scarcely established (Hitchcock et al. 1961:522). Known in Canada only from sw B.C. (Agassiz, where taken by Glendenning in 1950; Herb. V). (*Caucalis* Huds.).]

T. japonica (Houtt.) DC.
Eurasian; introd. in open woods and waste places of N. America, as in Ont. (near Mt. Hope, Wentworth Co.; TRT; reported from near Niagara Falls, Welland Co., by Montgomery 1957, and from the Ottawa dist. by Gillett 1958). [*Caucalis* Houtt.; *Tordylium (C.; Torilis) anthriscus* L.].

[T. nodosa (L.) Gaertn.]
[Eurasian; locally introd. in waste places of N. America, as in sw N.B. (known only from an 1882 collection by Vroom at St. Stephen, Charlotte Co., where scarcely established; CAN; NBM). (*Tordylium* L.; *Caucalis* Scop.).]

ZIZIA Koch [6008]

1 Basal leaves mostly simple, ovate to rotund, deeply cordate at base, finely crenate; median and upper leaves with 3 or 5 lanceolate to ovate, finely serrate, coriaceous leaflets; (B.C. to Que.) . . . . *Z. aptera*

1 Leaves (except the uppermost) 2–3-ternate, membranaceous; leaflets lanceolate to ovate, sharply serrate; (Sask. to N.S.) . . . . *Z. aurea*

Z. aptera (Gray) Fern.
/sT/X/ (Hs) Moist low ground (often alkaline) from sw Yukon (Boivin 1966*b*) and B.C. to Alta. (N to Fort Vermilion, 58°24′N), Sask. (N to Prince Albert), Man. (N to Clearwater L., about 25 mi NW of The Pas), Ont. (probably native in Essex, Lambton, Huron, and Bruce counties; introd. along railways along the N shore of L. Superior and at Kapuskasing; also probably introd. in the Laurentide region NW of Montreal in Labelle and Montcalm counties, Que.), and R.I., S to Oreg., Nev., Colo., Mo., Ala., and Ga. [*Thaspium trifoliatum* var. *apt.* Gray; *Z. cordata* of Canadian reports, not *Smyrnium (Z.) cordatum* Walt., which is *Thaspium trifoliatum* (L.) Gray].

Z. aurea (L.) Koch Golden Alexanders
/T/EE/ (Hs) Damp thickets, meadows, and shores from Sask. (Strongfield, about 55 mi S of Saskatoon; Breitung 1957*a*) to Man. (N to Fairford, about 110 mi N of Portage la Prairie), Ont. (N to Russell Co.; TRT; also introd. along a railway near Thunder Bay), Que. (N to Montmagny, Quebec, and Temiscouata counties), N.B., and N.S. (not known from P.E.I.), S to Tex., Mo., Tenn., and Ga. [*Smyrnium* L.; *Thaspium* Nutt.].

Forma *obtusifolia* (Bissell) Fern. (leaflets obovate to subrotund and obtuse rather than lanceolate to ovate and acuminate) is reported from SE Man. by Löve and Bernard (1959; Otterburne, about 30 mi S of Winnipeg).

CORNACEAE (Dogwood Family)

CORNUS L. [6159] Dogwood, Cornel. Cornouiller

Herbs or more commonly shrubs or small trees with entire, simple, exstipulate, usually opposite leaves. Flowers small, regular, perfect, epigynous, in dense or open cymes (these sometimes subtended by a usually 4-bracted petaloid involucre). Calyx minutely 4-toothed. Petals and stamens each 4. Style 1. Ovary inferior. Fruit a 2-seeded drupe.

(Ref.: H.W. Rickett, N. Am. Flora 28B:299–311. 1945)
1 Flowers in dense head-like clusters subtended by 4 or more large, white or pink, petaloid bracts; mature drupes bright red.
  2 Shrub or tree to over 10 m tall; petaloid bracts to 6 cm long; flowers (and the hard ellipsoid drupes) sessile.
    3 Petaloid bracts often more than 4, short-acuminate at apex; flowers yellowish green or red, up to about 75 in a cluster to 2 cm broad, usually at least 20 drupes maturing in a cluster to 3 cm broad; (s B.C.) .... *C. nuttallii*
    3 Petaloid bracts 4, obcordate, the retuse summit indurated and blackish; flowers yellowish, at most about 30 in a cluster to 1.5 cm broad, usually not more than 6 drupes maturing; (s Ont.) .... *C. florida*
  2 Herbs at most about 2 dm tall, from slender cord-like rhizomes; flowers short-pedicelled; drupes soft; petaloid bracts acute or acuminate.
    4 Leaves sessile, elliptic, obtuse, to 5 cm long, in several distant pairs along the stem, with usually 2 pairs of lateral veins from near the base; petaloid bracts about 1 cm long; petals uniformly dark purple; ovary sparsely strigose; drupes ellipsoid; (Alaska–Dist. Mackenzie–B.C.; Que., Labrador, Nfld., and N.S.) .... *C. suecica*
    4 Leaves short-petioled, ovate-oblong to rhombic, acute or acutish, to 9 cm long, the principal ones pseudo-verticillate at the top of the stem, subpinnately veined, the lower leaves much reduced or scale-like; petaloid bracts to 2 cm long; petals never entirely dark purple; ovary copiously strigose; drupes globose; (transcontinental) .... *C. canadensis*
1 Flowers white or creamy, in open cymes; bracts minute or none; mature drupes blue or white.
  5 Leaves alternate (sometimes obscurely so through crowding at the ends of the branches), elliptic- to rhombic-ovate, pale beneath, on slender petioles to 6 cm long; cyme flattish-topped; drupes bluish black, with a bloom; plant with a characteristic "pagoda"-like appearance, the branches often in irregular horizontal layers; pith white; (s Man. to Nfld. and N.S.) .... *C. alternifolia*
  5 Leaves opposite, their petioles rarely over 2 cm long.
    6 Drupes blue or bluish; cymes flattish or somewhat round-topped.
      7 Pith of younger branches white; leaves with 7–9 pairs of veins, broadly ovate to rotund, woolly beneath; (SE Man. to N.S.) .... *C. rugosa*
      7 Pith of younger branches brown; leaves with at most 6 pairs of veins, minutely pubescent beneath; (Ont. to w N.B.) .... *C. amomum*
    6 Drupes white or lead-coloured.
      8 Leaves with 5–7 pairs of veins, broadly lanceolate to broadly ovate; cymes flattish-topped; younger branches deep red; pith white; (transcontinental) ....
        .... *C. stolonifera*
      8 Leaves with 3 or 4 pairs of veins; branches greyish.
        9 Leaves elliptic to ovate, spreading-pilose beneath, usually scabrous above; branches with brown pith; cymes round-topped, broader than high; (s Ont.) ......
        .... *C. drummondii*
        9 Leaves oblong-lanceolate to elliptic or narrowly ovate, appressed-pubescent beneath, glabrous above; branches with pale-brown pith (or pith of younger branches often white); cymes higher than broad, panicle-like, with bright-red pedicels; (Man. to sw Que.) .... *C. racemosa*

C. alternifolia L. f.   Pagoda-Dogwood, Green Osier
/T/EE/   (Mc)   Dry woods and rocky slopes from s Man. (N to Riding Mt.; DAO) to Ont. (N to the Kaministikwia R. w of Thunder Bay), Que. (N to Anticosti Is. and the Gaspé Pen.), Nfld., N.B., P.E.I., and N.S., s to Mo., Ala., and Ga. [*Svida* Small; *C. riparia* Raf.]. MAPS: Hough 1947:359 (the occurrence in Man. should be indicated); Hosie 1969:298.

C. amomum Mill.   Silky Dogwood
/T/EE/   (N (Mc))   Damp thickets and swamps from N.Dak. to Ont. (N to Renfrew and Carleton counties), Que. (N to the Montreal dist.), and w N.B. (St. John R. system in Carleton Co.; ACAD; not known from P.E.I. or N.S.), s to Okla., Ark., Ala., and Ga. [*Svida* Small, in part; *C. lanuginosa* Michx.; *C. sericea* L., in part].

Most of our material appears referable to var. *schuetzeana* (Mey.) Rickett (*C. obliqua* Raf.; *C. purpusii* Koehne; leaves relatively narrow, cuneate at base rather than rounded, glaucous and minutely white-papillate beneath rather than green and minutely rusty-pubescent beneath).

C. canadensis L.   Bunchberry, Dwarf Cornel.   Quatre-temps or Rougets
/aST/X/GeA/   (Hpr (Ch))   Woods and damp openings from the E Aleutian Is., Alaska (N to ca. 70°N), and s-cent. Yukon to Great Bear L., Great Slave L., L. Athabasca (Alta. and Sask.), northernmost Man.–Ont., Que. (N to s Ungava Bay and the Côte-Nord), Labrador (N to Okak, 57°40′N), Nfld., N.B., P.E.I., and N.S., s to Calif., N.Mex., S.Dak., Ohio, Pa., and N.J.; southernmost Greenland; E Asia. MAPS and synonymy: *see* below (citations of minor forms largely based upon those by Ernest Lepage (Nat. can (Que.) 73 (1946), 77 (1950), 78 (1951), 82 (1955), and 85 (1958)).

A hybrid with *C. suecica* (*C. canadensis* var. *intermedia* Farr. (*C. int.* (Farr) Calder & Taylor); *C. (Chamaepericlymenum) unalaschkensis* Ledeb.) occurs locally nearly throughout the N. American area, being known from Alaska–Yukon–B.C., ?Ont. (tentatively reported from the SE shore of L. Superior by Hosie 1938), Que. (SE Hudson Bay; Knob Lake, 54°48′N; Côte-Nord; Bic, Rimouski Co.; Cacouna, Temiscouata Co.; Shickshock Mts. of the Gaspé Pen.), Labrador (N to Cutthroat Harbour, ca. 57°30′N), Nfld., and N.S. (Yarmouth Co. and St. Paul Is.; ACAD). MAPS: Hultén 1968b:709; J.A. Calder and R.L. Taylor, Can. J. Bot. 43(11): fig. 1, p. 1396. 1965.

1   At least some of the involucral bracts green and foliaceous (rather than white or roseate).
   2   Involucral bracts all foliaceous and about twice as large as in the typical form; [type
      from Alaska; also known from E James Bay] . . . . f. *virescens* Lepage
   2   Involucre with normal petaloid bracts in addition to numerous foliaceous ones; [type
      from Nominingue, Labelle Co., Que.] . . . . f. *foliolosa* Lepage
1   Involucral bracts all petaloid.
   3   Involucral bracts roseate to rose-purple; [f. *rosea* Fern.; transcontinental; MAP: Raup
      1947: pl. 31] . . . . f. *purpurascens* (Miyabe & Tatewaki) Hara
   3   Involucral bracts white or greenish white.
      4   Stem and base of leaves pubescent with short crisped hairs; [type from Chimo,
         s Ungava Bay, Que.; also known from Alaska, E James Bay, and Labrador] . . . .
         . . . . f. *dutillyi* Lepage
      4   Stem (at least the nodes) and base of leaves appressed-pubescent.
         5   Stem bearing 2 or 3 inflorescences on distinct peduncles; [Que.: E James Bay
            and Pointe-au-Père, Rimouski Co.] . . . . f. *florulenta* Lakela
         5   Stem bearing a solitary inflorescence.
            6   Fertile stems lacking leaves; [type from Nominingue, Labelle Co., Que.] . . . .
               . . . . f. *aphylla* Lepage
            6   Fertile stems bearing leaves.
               7   Leaves not whorled.
                  8   Plant-axis conspicuously prolonged beyond the uppermost leaves;
                    [Ont. to s Labrador and N.S.] . . . . f. *elongata* Peck
                  8   Plant-axis not conspicuously prolonged.
                    9   Leaves opposite; [type from a branch oí the Roggan R.,
                      Ungava; also known from E James Bay] . . . . f. *bifoliata* Lepage
                    9   Leaves more or less 1-sided on the stem; [type from L. Manik,

Ungava; also known from w and e James Bay and Rimouski
Co., Que.] . . . . f. *secunda* Lepage
7 Leaves whorled (or apparently so).
10 Leaves of the whorl unequal, 2 or 4 of them smaller and narrower
than the remaining pair; [var. *alpestris* House; Que.] . . . .
. . . . f. *alpestris* (House) Lepage
10 Leaves of the whorl subequal.
11 Foliage-whorls 2, one consisting of larger leaves than the
other.
12 Small-leaved whorl the uppermost; [type from Goose Bay,
Labrador; also known from N.S.] . . . . f. *medeoloides* Lepage
12 Small-leaved whorl the lowermost; [type from Oka, Que.;
also known from Ont. and N.S.] . . . . f. *infraverticillata* Lepage
11 Foliage-whorl solitary.
13 Stem branched; [type from Goose Bay, Labrador; also
known from Ont., Que., and N.S.] . . . . f. *ramosa* Lepage
13 Stem simple.
14 Leaves white-tipped; [type from near Nominingue,
Labelle Co., Que.] . . . . f. *albomacula* Lepage
14 Leaves uniformly green.
15 Involucral bracts 6–30; [type from Nominingue,
Labelle Co., Que.] . . . . f. *ornata* Lepage
15 Involucral bracts 4.
16 Peduncle bearing (above the whorl of stem-
leaves) a pair of connate leaves forming a cup-
like structure; [type from near Nominingue,
Labelle Co., Que.; also known from N.B.] . . . .
. . . . f. *connatifolia* Lepage
16 Peduncle naked; [*Chamaepericlymenum*
Aschers. & Graebn.; *Cornella* Rydb.; *Cynoxylon*
Schaffn.; transcontinental; MAPS: Hultén
1968*b*:709; Raup 1947: pl. 31; Meusel 1943:
fig. 3b (incomplete)] . . . . f. *canadensis*

C. drummondii Meyer
/t/EE/ (Ms) Damp woods, thickets, and shores from Nebr. to Ill., Ohio, and s Ont. (Essex, Kent, and Norfolk counties; *see* s Ont. map by Soper and Heimburger 1961:28), s to e Tex. and Miss. [*C. asperifolia* of auth., not Michx.]. MAPS: H.W. Rickett, Am. Midl. Nat. 27(1): fig. 1, p. 260. 1942; the northern part of the range in the map for *C. asperifolia* by Hough 1947:361, evidently applies here.

C. florida L. Eastern Flowering Dogwood
/t/EE/ (Ms) Acidic woods from Kans. to Ohio, s Ont. (N to N Lambton and Halton counties; *see* s Ont. maps by Fox and Soper 1952: map 10, p. 83, and Soper and Heimburger 1961:29), and Maine, s to Mexico, Tex., and Fla. [*Cynoxylon* Raf.]. MAPS: *Atlas of Canada* 1957: map 12, sheet 38; Hosie 1969:294; Fowells 1965:162; Preston 1961:330; Hough 1947:357; Munns 1938: map 159, p. 163; Little 1971: map 124-N.

C. nuttallii Audubon Pacific or Western Flowering Dogwood
/t/W/ (Ms) Moist woods and streambanks from s B.C. (Vancouver Is. and mainland NE to Vernon; CAN; provincial floral emblem) to s Calif. and Idaho. [*Cynoxylon* Shafer]. MAPS: Hosie 1969:296; Preston 1961:330.

C. racemosa Lam.
/T/EE/ (N (Mc)) Thickets and streambanks from SE Man. (Dufferin; Roseisle; Otterburne; Emerson) to Ont. (N to the Ottawa dist.; *see* s Ont. map by Soper and Heimburger 1961:31), SW Que. (N to Pontiac and Nicolet counties and the Montreal dist.), and Maine, s to Okla., Mo., Ky., and Md. [*Thelycrania* Löve & Bernard; *C. paniculata* L'Hér.].

C. rugosa Lam.   Round-leaved Dogwood.   Bois de Calumet
/T/EE/   (N (Mc))   Dry woods and rocky slopes from SE Man. (N to Victoria Beach, about 55 mi NE of Winnipeg) to Ont. (N to near Thunder Bay and New Liskeard, 47°31′N), Que. (N to the Gaspé Pen.), N.B., and N.S. (reports from P.E.I. require confirmation), S to Iowa, Ohio, and Va. [*Svida* Rydb.; *Thelycrania* Pojark.; *C. circinata* L'Hér.; *C. tomentulosa* Michx.].
Forma *eucycla* Fern. (leaves scarcely pointed rather than abruptly apiculate) is known from the type locality, Iona, Victoria Co., N.S. Collections in MT and TRT from Carp, near Ottawa, Ont., have been referred to × *C. slavinii* Rehd. (*C. rugosa* × *C. stolonifera*).

C. stolonifera Michx.   Red Osier.   Hart rouge
/ST/X/   (Mc)   Moist woods, thickets, and shores, the aggregate species from N-cent. Alaska–Yukon–Dist. Mackenzie to Great Bear L., L. Athabasca (Alta. and Sask.), Man. (N to Gillam, about 165 mi S of Churchill), Ont. (N to the Fawn R. at ca. 55°30′N, 88°W), Que. (N to the E James Bay watershed at ca. 54°N and the Côte-Nord), Labrador (N to the Hamilton R. basin), Nfld., N.B., P.E.I., and N.S., S to Calif., Mexico, N.Mex., Nebr., Ohio, W.Va., and New Eng. MAPS and synonymy: *see* below.

1   Stone grooved lengthwise; petals often over 3 mm long; styles 2 or 3 mm long; pubescence various, but often conspicuously spreading or the hairs curled; [S B.C.] . . . . . .
. . . . var. *occidentalis* (T. & G.) Hitchc.

  2   Plant copiously pubescent, the inflorescence characteristically spreading-hairy; [*C. sericea* var. *occid.* T. & G.; *C. occid.* (T. & G.) Cov.; *C. (Svida) pubescens* Nutt.; MAPS (both as *C. occid.*): Preston 1947:254; H.W. Rickett, Brittonia 5(2): fig. 1, p. 150. 1944] . . . . f. *occidentalis*

  2   Plant subglabrous to sparingly appressed- or spreading-hairy, the inflorescence characteristically appressed-hairy; [*C. calif.* Meyer; *C. stol. (pub.)* var. *calif.* (Mey.) McMinn] . . . . f. *californica* (Mey.) Hitchc.

1   Stone smooth; petals usually 2 or 3 mm long; styles 1 or 2 mm long; pubescence usually entirely appressed-strigose . . . . var. *stolonifera*

  3   Drupes blue, drying reddish; [W and E James Bay, the type from Attawapiskat, W James Bay, Ont.] . . . . f. *azurea* Lepage

  3   Drupes white or lead-colour (rarely with a bluish flush).

    4   Drupe oblong, the nutlets about twice as long as broad; [type from along the Albany R., W James Bay, Ont.; S ?Man.: *see* Löve and Bernard 1959:416, under *Thelycrania ?alba*] . . . . f. *dolichocarpa* Lepage

    4   Drupe globose or subglobose, the nutlets as broad as or broader than long.

      5   Lower leaf-surfaces densely soft-pilose with spreading or woolly hairs; [*C. baileyi* Coult. & Evans, the type from Nipigon, Ont.; essentially transcontinental; MAP: Rickett, loc. cit., fig. 1 (very incomplete northwards), p. 150] . . . .
. . . . f. *baileyi* (C. & E.) Rickett

      5   Lower leaf-surfaces appressed-pubescent to glabrate.

        6   Pubescence of inflorescence and twigs consisting of a dense tomentum; [*Svida (C.) interior* Rydb.; Alaska–B.C. to W James Bay, Ont.; MAP: Rickett, loc. cit., fig. 1, p. 150] . . . . f. *interior* (Rydb.) Rickett

        6   Pubescence of inflorescence and twigs minute and mostly appressed.

          7   Leaves elliptic-lanceolate; [type from the Harricanaw R., Que., S of James Bay; also known from James Bay, Ont., and the Cypress Hills of SW Sask.] . . . . f. *angustior* Lepage

          7   Leaves ovate.

            8   Stems repent, forming large clones; leaves thin; inflorescence few-flowered; [type from Hemmingford, Huntingdon Co., Que.] . . . .
. . . . f. *repens* Vict.

            8   Stems erect; leaves relatively firm; flowers numerous; [var. *riparia* (Rydb.) Visher, not *C. riparia* Raf.; *Svida* Rydb.; *Thelycrania* Pojark.; *C. alba* of Canadian reports, not L.; *C. alba* ssp. *stol.* (Michx.) Wang.; (the report of *C. ?alba* f. *argento-marginata* (Rehd.) Pojark., the leaves with broad white margins, from S Man. by Löve

and Bernard 1959, may eventually prove referable to *C. stol.*); *C. sericea* L. in part; *C. instoloneus* Nels.; *C. stricta* of Canadian reports, not Lam.; *C. sanguinea* Marsh., not L.; transcontinental; MAPS: Rickett, loc. cit., fig. 1, p. 150; Raup 1947: pl. 31; Hultén 1968*b*:708] .... f. *stolonifera*

C. suecica L.

/aST/(X)/GEA/   (Hpr (Ch))   Woods, marshes, and bogs: Aleutian Is., Alaska (N to ca. 64°N; *see* Hultén 1947: map 905, p. 1200), and coastal B.C. (an early report from Vancouver Is.; a confirmatory collection in 1927 by Hardy along the Douglas Channel, ca. 53°N); cent. Dist. Mackenzie (Munn L., N of Great Slave L. at 63°35'N, 110°02'W; Porsild and Cody 1968; DAO); Que. (Hudson Bay–Ungava Bay watershed between ca. 57° and 60°N; St. Lawrence R. estuary from Kamouraska, Kamouraska Co., to the Côte-Nord, Anticosti Is., Gaspé Pen., and Magdalen Is.) to Labrador (N to ca. 59°N), Nfld., and N.S. (St. Paul Is.; Canso, Guysborough Co.; Scatari Is., Cape Breton Co.); W Greenland N to ca. 68°45'N, E Greenland N to 61°37'N; Iceland; N Europe; NE Asia. MAPS: Hultén 1968*b*:708, and 1958: map 238, p. 257; Meusel 1943: fig. 3b (incomplete for N. America); Tolmatchev 1932: fig. 3 (incomplete for N. America), p. 20.

Forma *semivirescens* Vict. (some of the flowers transformed into green leaves longer than the petaloid bracts) is known from the type locality, Cacouna, Temiscouata Co., E Que.

CLETHRACEAE (White Alder Family)

CLETHRA L. [6165]

Shrub to 3 m tall. Leaves obovate-oblong, to about 1 dm long, obtuse or subacute, finely and sharply serrate above the middle, gradually tapering to a petiole to about 12 mm long. Flowers perfect, 5-merous, polypetalous, hypogynous, in erect, densely short-pubescent racemes to about 2 dm long. Petals 5, white, obovate, about 8 mm long. Stamens 10, the anthers sagittate, becoming inverted, opening by a pair of pores at the apparent apex (morphological base). Fruit a pubescent capsule to 3 mm thick.

C. alnifolia L.   Sweet Pepperbush
/T/EE/   (N)   Swamps and moist woods, mostly near the coast, from sw N.S. (reported by Pierre Taschereau, Can. Field-Nat. 83(2):166. 1969, from the shore of Belliveau L., Belliveau Cove, Digby Co., N.S., where taken by him in 1968 to add a new family to the flora of Canada; ACAD; DAO; NSPM) to Fla. and E Tex.

Because of the very recent addition of this family to our flora, it has not been keyed out in the key to families except for placing it to key out along with Ericaceae. It differs from most genera of that family in its polypetalous flowers (petals distinct rather than united). From the polypetalous *Cladothamnus* and *Ledum,* it differs in its finely and sharply serrate leaves (rather than entire), its flowers being in racemes rather than solitary (*Cladothamnus*) or in umbel-like clusters (*Ledum*).

PYROLACEAE (Wintergreen Family)

(Ref.: Szczawinski 1962; P.A. Rydberg, N. Am. Flora 29:11–18; 21–32. 1914)
Low herbaceous or slightly shrubby plants with simple, entire or shallowly toothed, alternate, subverticillate, or basal, exstipulate leaves (or these reduced to scales or bracts). Flowers white, pink, or roseate, regular or nearly so, hypogynous. Sepals and petals usually 5. Stamens usually 10 (8 in *Hemitomes* and often in *Hypopitys*), the anthers mostly inverted or deflexed on their filaments and opening by a pair of pores at the apparent apex (the morphological base) or nearly their full length by longitudinal slits, awnless (except in *Pterospora*). Ovary superior. Fruit a usually 5-locular capsule. (Incl. Monotropaceae; often included in the Ericaceae).

1 Plant evergreen, with broad green leaves (except *Pyrola picta* f. *aphylla*); petals 5, distinct; anthers awnless, opening by a pair of pores at the apparent apex (morphological base).
  2 Flower solitary, waxy-white, 1 or 2 cm broad, on a 1–2-bracted scape usually less than 1 dm tall; anthers 2-horned; leaves all basal, subrotund, finely serrate, to 3 cm long; (transcontinental) . . . . *Moneses*
  2 Flowers several in terminal racemes or corymbs.
    3 Stem uniformly leafy; flowers corymbose, their styles nearly obsolete; anther-filaments dilated, hairy . . . . *Chimaphila*
    3 Stems leafy only near the base; flowers racemose, with definite styles; anther-filaments slender, glabrous . . . . *Pyrola*
1 Plants lacking green leaves (these reduced to often coloured scales or bracts), fleshy stemmed, saprophytic or root-parasitic.
  4 Corolla none; anthers awnless, inverted and opening by a pair of pores or a continuous slit at the apparent apex; inflorescence a terminal elongate spike-like raceme to 2 dm long, the bracts surpassing the flowers; stems striped longitudinally with white and pink, to 4 dm tall and 1 cm thick, bearing linear-lanceolate, pinkish to yellow-brown, scale-like leaves; (B.C.) . . . . *Allotropa*
  4 Corolla present; anthers opening nearly their full length by longitudinal slits; flowers usually surpassing their subtending bracts; stems not striped.
    5 Corolla gamopetalous, the petals united for more than half their length from base; flowers several to many in terminal racemes.
      6 Flowers on slender recurved pedicels in an elongate and open raceme; corolla globose-urn-shaped (strongly constricted below the 5 short spreading lobes), glabrous, to 8 mm long, pale yellow; sepals 5; stamens 10, the anthers with a pair of slender deflexed awns on the back; plant clammy-pubescent, to 1 m tall; (B.C. to P.E.I.) . . . . *Pterospora*
      6 Flowers short-pedicelled in a congested head- or spike-like raceme; corolla cup-shaped or open-campanulate, pink or flesh-colour, larger; anthers awnless; stems to about 2 dm tall, not clammy.
        7 Sepals 5; corolla 5-lobed, about 1 cm long; stamens 10; stems to about 12 cm tall . . . . *[Monotropsis]*
        7 Sepals 2–4; corolla usually 4-lobed, to 2 cm long; stamens usually 8; stems to about 2 dm tall; (sw B.C.) . . . . *Hemitomes*
    5 Corolla polypetalous, consisting of usually 4 or 5 (sometimes 3 or 6) separate petals.
      8 Flower solitary, waxy-white, curved to one side or even drooping; sepals often wanting; petals commonly 5; entire plant waxy-white (sometimes pinkish), aging or drying black, essentially glabrous, the clustered stems to about 2.5 dm tall; (transcontinental) . . . . *Monotropa*
      8 Flowers several in terminal bracted racemes, not waxy-white; sepals and petals commonly each 4 (or 5 in the terminal flower).
        9 Corolla not densely hairy within, it and the sepals fimbriate-pectinate; anthers linear, 2 or 3 mm long, their filaments glabrous or minutely puberulent; capsule subglobose, essentially glabrous, the placentation axile

(seeds borne along the central column); stems commonly not over about 1 dm tall . . . . *[Pleuricospora]*

9 Corolla usually densely hairy within, it and the sepals erose to lacerate-fimbriate; anthers oval, about 1 mm long, their filaments hairy; capsule oblong-ovoid, hairy, its placentation parietal (seeds borne on the inner wall); stems to 2.5 dm tall; (transcontinental) . . . . *Hypopitys*

ALLOTROPA T. & G.   [6168]

A. virgata T. & G.   Candystick, Sugarstick
/t/W/   (Gp (root-parasite))   Deep humus of coniferous forests from sw B.C. (Vancouver Is. and adjacent islands and mainland ᴇ to Chilliwack L.; CAN; V; *see* B.C. map by Szczawinski 1962:19; the report from Klondike, the Yukon, by John Macoun, Ottawa Naturalist 13(9):214. 1899, requires confirmation) to coastal Calif.

CHIMAPHILA Pursh   [6166]   Prince's Pine, Pipsissewa

1 Upper surface of leaves strongly variegated with white along the veins, the blades lanceolate to ovate-lanceolate, obtusish or rounded at base, sharply but remotely serrate; flowers not more than 5 (sometimes solitary), to 18 mm broad; (s Ont.) . . . . *C. maculata*
1 Upper surface of leaves not variegated; flowers somewhat smaller.
  2 Flowers rarely more than 3 (sometimes solitary); anther-filaments hairy over the entire swollen portion of the base; leaves narrowly to broadly elliptic (generally broadest near the middle), entire to sharply serrate, to 6 cm long; (s B.C.) . . . . *C. menziesii*
  2 Flowers usually more than 3; anther-filaments merely ciliate on the swollen base; leaves narrowly to broadly oblanceolate (broadest above the middle), finely and closely serrate; (transcontinental) . . . . *C. umbellata*

C. maculata (L.) Pursh   Spotted Wintergreen
/t/EE/   (Hpr (Ch; evergreen))   Dry woods from Ill. to Mich., s Ont. (Kent, Norfolk, Welland, Lincoln, Middlesex, Wentworth, and York counties and Baysville, ᴇ of L. Muskoka; CAN; TRT; John Macoun 1884 and 1890), and N.H., s to Tenn., Ala., and Ga. [*Pyrola* L.].

C. menziesii (R. Br.) Spreng.
/t/W/   (Hpr (Ch; evergreen))   Moist coniferous woods from s B.C. (ɴ to Kimsquit, about 45 mi ɴᴇ of Ocean Falls at ca. 52°30'N; CAN; ᴇ to Kootenay L.; *see* B.C. map by Szczawinski 1962:53) to s Calif., ?Idaho, and ?Mont. [*Pyrola* R. Br.].

C. umbellata (L.) Barton   Prince's Pine, Pipsissewa.   Herbe à clé
/sT/X/EA/   (Hpr (Ch; evergreen))   Coniferous woods, the aggregate species from sᴇ Alaska and B.C.–Alta. to L. Athabasca (Alta. and Sask.), Man. (ɴ to Flin Flon, 54°46'N), Ont. (ɴ to Sandy L. at ca. 53°N, 93°W), Que. (ɴ to Ville-Marie, 47°20'N, and the Gaspé Pen.), Nfld. (Boivin 1966*b*, confirming a report by Reeks 1873; not listed by Rouleau 1956), N.B., P.E.I., and N.S., s to s Calif., N.Mex., Colo., Minn., Mich., Ohio, and Ga.; Eurasia. MAP and synonymy: *see* below.
1 Leaves blunt-toothed, mostly less than 4 cm long; inflorescence subumbellate; calyx-lobes longer than broad; [*Pyrola umb.* L.; Eurasia only; MAP (aggregate species): Hultén 1968*b*:710] . . . . [var. *umbellata*]
1 Leaves sharp-toothed, to over 7 cm long; inflorescence racemose.
  2 Leaves very obscurely veined beneath; calyx-lobes longer than broad; capsules to 7.5 mm thick; peduncles often ascending; [*C. occ.* Rydb.; Alaska–B.C. to w James Bay, Ont.] . . . . var. *occidentalis* (Rydb.) Blake
  2 Leaves conspicuously veined beneath; calyx-lobes usually broader than long; capsules to 6 mm thick; peduncles commonly recurved; [*C. (Pyrola) corymbosa* Pursh; ᴇ Man. to Nfld. and N.S., the type from Bathurst, N.B.] . . . . var. *cisatlantica* Blake

HEMITOMES Gray [6173]

H. congestum Gray Gnome-plant
/t/W/ (Gp (root-parasite)) Deep humus of coastal coniferous forests from sw B.C. (Vancouver Is. and adjacent islands; Grouse Mt., Vancouver; New Denver, Kootenay Valley; *see* B.C. map by Szczawinski 1962:73) to Calif. [*Newberrya* Torr.].

HYPOPITYS Hill [6169]

H. monotropa Crantz Pinesap, False Beech-drops
/sT/X/EA/ (Gp (root-parasite)) Humus of chiefly coniferous forests from the Alaska Panhandle and B.C. to sw Alta. (Crowsnest Forest Reserve; Waterton), sw Sask. (Cypress Hills; Breitung 1957a; not known from Man.), Ont. (N to near Timmins, 48°28′N), Que. (N to the E James Bay watershed at 52°15′N, the Côte-Nord, Anticosti Is., and Gaspé Pen.), Nfld., N.B., P.E.I., and N.S., s to Calif., Mexico, and Fla.; Eurasia. [*H. brevis* Small; *Monotropa (H.) fimbriata* Gray; *M. (H.) hypopitys* L.; *M. (H.) lanuginosa* Michx.; *H. (M.) latisquama* Rydb.]. MAP: Hultén 1968b:715 (*Mon. hyp.*).

MONESES Salisb. [6167]

M. uniflora (L.) Gray One-flowered Pyrola
/ST/X/EA/ (Hr (evergreen)) Cool mossy woods from N Alaska–Yukon (N to ca. 65°N) and the Mackenzie R. Delta to Great Bear L., Great Slave L., L. Athabasca (Alta. and Sask.), Man. (N to 7 mi N of Churchill), northernmost Ont., Que. (N to E Hudson Bay at ca 56°30′N, the Côte-Nord, Anticosti Is., and Gaspé Pen.), Labrador (N to near Nain, 56°33′N), Nfld., N.B., P.E.I., and N.S., s to Calif., N.Mex., Minn., Mich., Pa., and New Eng.; Eurasia. [*Pyrola* L.; *M. grandiflora* Gray; *M. reticulata* Nutt.]. MAPS: Hultén 1968b:714; Raup 1947: pl. 31.

MONOTROPA L. [6169]

M. uniflora L. Indian-pipe
/sT/X/A/ (Gp (root-parasite)) In humus of deep shaded woods from the southernmost Alaska Panhandle and B.C. to L. Athabasca (Alta. and Sask.), Man. (N to Norway House, off the NE end of L. Winnipeg), Ont. (N to Sandy L. at ca. 53°N), Que. (N to the E James Bay watershed at ca. 52°15′N, the Côte-Nord, Anticosti Is., and Gaspé Pen.), Labrador (Goose Bay, 53°18′N), Nfld., N.B., P.E.I., and N.S., s to Calif., Mexico, and Fla.; Cent. America; Asia. MAP: Hultén 1968b:715.

[MONOTROPSIS Schwein.] [6172]

[M. odorata Ell.]
[The report of this species of the E U.S.A. (N to W.Va. and Md.) from the Don Valley, near Toronto, Ont., by C.W. Armstrong (Biol. Rev. Ont. 1(2):43. 1894) is probably referable to the habitally similar *Hypopitys monotropa*.]

[PLEURICOSPORA Gray] [6174]

[P. fimbriolata Gray] Fringed Pinesap
[According to Szczawinski (1962), the report of this species of the w U.S.A. (Wash. to NW Calif.) from sw B.C. by J.K. Henry (Ottawa Naturalist 31(5–6):55. 1917; Beaufort Range, near Horn L., Vancouver Is.; this taken up by Carter and Newcombe 1921, and Eastham 1947) is based upon *Hemitomes congestum*, the relevant collection having been examined by him.]

PTEROSPORA Nutt. [6170]

P. andromedea Nutt. Pine-drops
/T/X/ (Gp (root-parasite)) Deep humus in coniferous forests from B.C. (N to ca. 55°N; *see* B.C. map by Szczawinski 1962:115; concerning a report from s Alaska, *see* Hultén 1948:1215) to s Alta.

(Waterton Lakes; Pincher Creek; Cypress Hills), sw Sask. (Cypress Hills; Breitung 1957a; not known from Man.), Ont. (N to the Ottawa dist.; type from Niagara Falls, Welland Co.), Que. (N to Quebec City; DAO), and P.E.I. (?extinct; collection in CAN from Prospect Creek, where taken by John Macoun in 1888; not known from N.B. or N.S.), s to s Calif., Mexico, Mich., N.Y., and Vt.

PYROLA L. [6167] Pyrola, Wintergreen

1 Racemes 1-sided; corolla campanulate, usually longer than broad; style straight, to 9 mm long, surpassing the capsule and surrounded by the connivent stamens; leaves elliptic to ovate or suborbicular, longer than their petioles; (transcontinental) . . . . *P. secunda*
1 Racemes spiral; corolla broader than long.
  2 Style 1 or 2 mm long, straight or nearly so, lacking a collar or ring below the broad peltate 5-lobed stigma, surrounded by the connivent stamens; anthers lacking a pair of horn-like terminal tubes; corolla subglobose (the tips of the white, pink, or roseate petals incurving and nearly meeting), less than 1 cm broad; leaves dull, broadly elliptic to suborbicular, to about 4 cm long, shorter than to about equalling their petioles; (transcontinental) . . . . *P. minor*
  2 Style 3 mm long or more, deflexed at base and upwardly arched, with a distinct collar or ring below the stigma; stamens not connivent; anthers terminated by a pair of horn-like, oblique or bent tubes produced beyond the pores at the apex (morphological base, the anthers in *Pyrola* inverted on their filaments); petals spreading or loosely converging.
    3 Leaves deep green but greyish-mottled along the main veins above, ovate to elliptic-rotund, to 7 cm long (lacking in f. *aphylla*); petals greenish white or yellowish to purplish; (s B.C. and sw Alta.) . . . . *P. picta*
    3 Leaves not mottled.
      4 Leaves mostly spatulate or oblanceolate to rhombic-elliptic, tapering to an acute base, the blade to 6 cm long and not much more than half as broad as long (seldom over 2.5 cm broad), pale green or bluish green; petals cream to greenish white; (s B.C.) . . . . *P. dentata*
      4 Leaves broadly elliptic to orbicular, the blades mostly well over half as broad as long, often rounded at base.
        5 Calyx-lobes rounded or obtuse, not over 2 mm long; corolla greenish white, the petals converging; anthers terminated by tubes up to 0.8 mm long; leaf-blade usually shorter than the petiole; (transcontinental) . . . . *P. virens*
        5 Calyx-lobes lanceolate to ovate, acutish; petals loosely spreading; tubes terminating the anthers very short or obsolete; leaf-blade about equalling or longer than the petiole.
          6 Leaf-blades elliptic to obovate, thin, commonly longer than the petioles; petals white or creamy, more than 4 times as long as the calyx-lobes; bracts of raceme linear-subulate; (transcontinental) . . . . *P. elliptica*
          6 Leaf-blades firm and more or less coriaceous, often shorter than the petioles; petals not more than 3 times as long as the calyx-lobes; bracts of raceme lanceolate to ovate.
            7 Anthers bright lemon-yellow, at most 2.3 mm long, rounded at base, their locules barely constricted above, their filaments almost filiform at summit; petals to 11 mm long, creamy white or pinkish, thin, translucent, strongly veined, with a pale whitish margin on drying; (transcontinental) . . . . *P. grandiflora*
            7 Anthers to about 3.5 mm long, mostly with a short pointed tip at base, their locules definitely constricted above into a neck, their filaments flat; petals less strongly veined, lacking a pale whitish margin on drying.
              8 Anthers deep golden-yellow; petals milk-white or creamy, thick and leathery; calyx-lobes firm, 3–5-nerved nearly to tip; leaf-blades lustrous, not cordate at base; (Man. to Nfld. and N.S.) . . . .
              . . . . *P. rotundifolia*

8   Anthers deep purplish-red; petals deep pink or crimson, rather
     thin, about 7 mm long; calyx-lobes thin, essentially nerveless;
     (transcontinental) .... *P. asarifolia*

P. asarifolia Michx.   Pink Pyrola or Wintergreen
/ST/X/A/   (Hrr (evergreen))   Moist woods from the Aleutian Is., Alaska (N to ca. 68°N), and cent.
Yukon to the Mackenzie R. Delta, Great Bear L., Great Slave L., L. Athabasca (Alta. and Sask.),
Man. (N to York Factory, Hudson Bay, 57°N; reports N to Churchill require confirmation),
northernmost Ont., Que. (N to S Ungava Bay and the Côte-Nord), Nfld., N.B., P.E.I., and N.S., S to
Oreg., N.Mex., S.Dak., Minn., Ind., and New Eng.; Asia. MAPS and synonymy: *see* below.
1   Leaves obovate to orbicular, rounded to subcuneate at base, lustreless or dull;
     [*P. rotundifolia* vars. *incarnata* (Fisch.) DC. (*P. inc.* Fisch.), *purpurea* Bunge, and
     uliginosa (Torr.) Gray (*P. ulig.* Torr.); *P. bracteata* var. *hillii* Henry; transcontinental,
     largely replacing the typical form northwards; MAPS: Hultén 1958: map 124, p. 143, and
     1968b:711; Raup 1947: pl. 32 (var. *incarnata*; a dot should be added for Chimo, S
     Ungava Bay, N Que.)] .... var. *purpurea* (Bunge) Fern.
1   Leaves reniform, cordate-based, lustrous; [*P. rotundifolia* var. *asar.* (Michx.) Hook.; *P.
     bracteata* Hook. (*P. rot.* var. *br.* (Hook.) Gray); *P. elata* Nutt.; transcontinental but less
     northern than var. *purpurea*; MAP: on the above-noted 1958 map by Hultén; Hultén
     1968b:711] .... var. *asarifolia*

P. dentata Sm.
/t/W/   (Hrr (evergreen))   Coniferous forest from SW B.C. (type probably from Nootka, Vancouver
Is., according to Hitchcock et al. 1959; known on the adjacent mainland N to Lytton, ca. 50°15′N;
see B.C. map by Szczawinski 1962:125) to Calif. and Wyo. [*P. picta* ssp. *dent.* (Sm.) Piper; *P.
chimophiloides* Greene].

P. elliptica Nutt.   Shinleaf
/sT/X/eA/   (Hrr (Ch; evergreen))   Dry or moist woods from B.C. (N to ca. 54°15′N; *see* B.C map
by Szczawinski 1962:127) to L. Athabasca (Alta. and Sask.), Man. (N to about 25 mi N of The Pas),
Ont. (N to W James Bay at ca. 51°30′N), Que. (N to L. St. John, the Côte-Nord, and Gaspé Pen.; not
known from Anticosti Is.), Nfld., N.B., P.E.I., and N.S., S to ?Idaho, N.Mex., S.Dak., Ohio, Pa., and
New Eng.; Japan. [*Thelaia* Alef.].

P. grandiflora Radius   Arctic Pyrola or Wintergreen
/AST/X/GEA/   (Hrr (evergreen))   Tundra and peaty slopes, the aggregate species from the
coasts of Alaska–Yukon–Dist. Mackenzie–Dist. Keewatin to Banks Is., S Ellesmere Is., and
northernmost Ungava–Labrador (type from Labrador), S to N B.C. (S to ca. 56°N; *see* B.C. map by
Szczawinski 1962:130) and the mts. of SW Alta., Sask. (L. Athabasca), Man. (S to Gillam, about 165
mi S of Churchill), Ont. (S to NW James Bay at ca. 55°N), Que. (S to E James Bay at ca. 52°30′N,
Knob Lake, 54°48′N, and the Shickshock Mts. of the Gaspé Pen.), S Labrador, and Nfld.; W
Greenland N to ca. 80°N, E Greenland N to ca. 75°N; Iceland; N Eurasia. MAPS and synonymy: *see*
below.
1   Sepals lanceolate, acute, entire; [*P. gormanii* Rydb., the type from Dry Gulch, the Yukon] ......
     .... var. *gormanii* (Rydb.) Porsild
1   Sepals elliptic, usually obtuse or rounded at summit, crenulate.
     2   Leaf-blade shorter than the petiole; flowers numerous, rarely over 1.5 cm broad, the
          deltoid sepals entire; [*P. canadensis* Andres; Alaska–Yukon–Dist. Mackenzie (type
          from the Slave R.), L. Athabasca (Alta. and Sask.), and N Man. (Nueltin L.)] ....
          .... var. *canadensis* (Andres) Porsild
     2   Leaf-blade equalling the petiole; flowers few, at least 2 cm broad, the lanceolate
          sepals erose at apex; [*P. borealis* Rydb.; *P. chlorantha* var. ?*occidentalis* (R. Br.)
          Gray (*P. occ.* R. Br.); *P. groenlandica* and *P. pumila* Hornem.; *P. rotundifolia* var.
          pumila Hornem.; transcontinental; MAPS (aggregate species): Young 1971: fig. 14,
          p. 88; Eric Hultén 1968b:712, 1958: map 124, p. 142, and Sven. Bot. Tidskr. 43(2–3):
          fig. 2, p. 394. 1949; Porsild 1957: map 263, p. 193; Böcher 1954: fig. 43, p. 167]. A

hybrid with *P. minor* (× *P. media* Sw.) is known from two localities on s Disko Is., w Greenland .... var. *grandiflora*

P. minor L.
/aST/X/GEA/ (Hrr (evergreen)) Cool woods and thickets from the Aleutian Is., cent. Alaska–Yukon, and the Mackenzie R. Delta to Great Bear L., L. Athabasca (Alta. and Sask.), s Dist. Keewatin, northernmost Man.–Ont., Que. (N to Ungava Bay and the Côte-Nord), Labrador (N to Okak, 57°40′N), Nfld., N.B., P.E.I., and N.S., s to s Calif., N.Mex., Minn., and New Eng.; W and E Greenland N to ca. 70°N; Iceland; Eurasia. [*Braxilia* House; *Erxlebenia* Rydb.; incl. vars. *conferta* C. & S. (*P. con.* (C. & S.) Fisch.) and *parvifolia* Boivin]. MAPS: Hultén 1968*b*:713; Porsild 1957: map 264, p. 193, and 1951*b*: fig. 3, p. 142; Raup 1947: pl. 32.

P. picta Sm.
/T/W/ (Hrr (evergreen)) Coniferous forests from s B.C. (Vancouver Is. and adjacent islands and mainland N to ca. 52°N; *see* B.C. map by Szczawinski 1962:135) and sw Alta. (Waterton Lakes; Breitung 1957*b*) to s Calif. and Colo. [Incl. var. *integra* Gray].

Some of the s B.C. material (*see* B.C. map by Szczawinski 1962:119) is referable to f. *aphylla* (Sm.) Camp (*P. aphylla* Sm.; flowering stems lacking green leaves, the sterile branches occasionally with 1 or 2 leaves in addition to numerous scales).

P. rotundifolia L.   Muguet des Bois
/T/EE/EA/ (Hrr (evergreen)) Damp woods, thickets, bogs, and barrens (ranges of Canadian taxa outlined below), s to Minn., Ky., and N.C.; Eurasia. MAP and synonymy: *see* below.
1   Sepals to 3 mm long; petals less than 8 mm long; anthers less than 3 mm long; [var. *arenaria* Mert. & Koch; E Que. (Côte-Nord; Anticosti Is.; Gaspé Pen.), Nfld., and N.S.; E Greenland (CAN); Eurasia; MAP: Hultén 1958: map 123, p. 142] .... var. *rotundifolia*
1   Sepals at least 3 mm long; petals commonly about 8 mm long; anthers at least 3 mm long; [*P. americana* Sweet; s Man. (Sandilands Forest Reserve, SE of Winnipeg; DAO; reported from the Spruce Woods Forest Reserve, SE of Brandon, by R.D. Bird, Ecology 8(2):207–20. 1927; reports from farther north are probably based upon *P. grandiflora*), Ont. (N to the Algoma Dist. N of L. Huron), Que. (N to the Côte-Nord and Gaspé Pen.), Nfld. (Quarry Brook; DAO), N.B., P.E.I., and N.S.; MAP: on the above-noted map by Hultén] .... var. *americana* (Sweet) Fern.

P. secunda L.   One-sided Pyrola or Wintergreen
/aST/X/GEA/ (Hrr (Ch; evergreen)) Dry or moist woods, the aggregate species from N Alaska–Yukon and the coasts of Dist. Mackenzie–Dist. Keewatin to Man. (N to Nueltin L.), Ont. (N to s Hudson Bay at ca. 55°N), Que. (N to Ungava Bay and the Côte-Nord), Labrador (N to Hebron, 58°13′N), Nfld., N.B., P.E.I., and N.S., s to s Calif., Mexico, S.Dak., Ohio, and Va.; W Greenland; Iceland; Eurasia. MAPS and synonymy: *see* below.
1   Racemes with at most about 10 flowers; petals creamy white; style to 6 mm long; basal bracts ovate to oblong, only slightly involute; leaves ovate to orbicular, often rounded at summit, scarcely lustrous; [var. *minor* Gray; var. *pumila* Paine; essentially the range of the typical form but gradually replacing it northwards and known from W Greenland (between ca. 67° and 70°N) and Iceland; Asia (apparently absent from Europe); MAPS: Hultén 1968*b*:714; Porsild 1957: map 262, p. 193] .... var. *obtusata* Turcz.
1   Racemes with up to about 20 flowers; petals greenish yellow; style to 9 mm long; basal bracts of scape lanceolate, strongly involute; leaves elliptic to ovate, lustrous .... var. *secunda*
2   Leaves rounded at summit; [W and E James Bay and Nfld.] .... f. *eucycla* Fern.
2   Leaves mucronate at summit; [*Ramischia* Garcke; *Orthilia* House; somewhat more southern than var. *obtusata* and merging with it northwards, the aggregate species ranging throughout our area from B.C. to Nfld. and N.S., N to the coasts of Alaska–Yukon–Dist. Mackenzie–NW Victoria Is.–N Ungava–N Labrador; MAPS: Raup 1947: pl. 32 (aggregate species); Hultén 1968*b*:713] .... f. *secunda*

P. virens Schweigger

/ST/X/EA/ (Hrr (evergreen)) Chiefly coniferous forests, the aggregate species from Alaska-Yukon (N to ca. 64°N) and Dist. Mackenzie (N to Campbell L. at 68°14'N) to L. Athabasca (Alta. and Sask.), Man. (N to Reindeer L. at 57°37'N), Ont. (N to Big Trout L. at ca. 53°45'N, 90°W), Que. (N to SE Hudson Bay at ca. 56°10'N, the Côte-Nord, Anticosti Is., and Gaspé Pen.), Labrador (N to the Hamilton R. basin; reports from farther north probably refer to *P. grandiflora*), Nfld., N.B., P.E.I., and N.S., s to Calif., Ariz., S.Dak., and Pa.; Eurasia. MAPS and synonymy: *see* below.

1   Anthers 3 or 4 mm long; petals to 9 mm long; mature style to 1 cm long; [*P. convoluta*
    Bart.; s Ont.: Fernald *in* Gray 1950] . . . . var. *convoluta* (Bart.) Fern.
1   Anthers 2 or 3 mm long; petals and mature style at most 7 mm long . . . . var. *virens*
    2   Leaves mostly cuneate at base, rounded to truncate at summit, to 2.5 cm broad, few
        or even wanting; [Ont., Que., N.B., P.E.I., and N.S.] . . . . f. *paucifolia* Fern.
    2   Leaves rounded to subcordate at base, rounded at summit, to 3.5 cm broad; [*P.*
        chlorantha Sw.; transcontinental; MAPS (aggregate species): Hultén 1968*b*:712 (*P.*
        chlor.); Raup 1947: pl. 32] . . . . f. *virens*

ERICACEAE (Heath Family)

(Ref.: Szczawinski 1962; J.K. Small, N. Am. Flora 29:33–102. 1914)
Low to medium-sized shrubs with simple, entire or shallowly toothed leaves, these opposite or alternate (rarely whorled). Flowers regular or nearly so, perfect, 4–5-merous, hypogynous or sometimes epigynous, commonly gamopetalous (petals distinct in *Cladothamnus* and *Ledum*). Stamens the same number as the petals or corolla-lobes or twice as many, free from the corolla or borne at its base on a hypogynous disk, the anthers mostly upright and opening by terminal pores, often appendaged on the back with a pair of single or double, slender, recurved, horn-like awns. Style 1. Ovary superior or inferior. Fruit a capsule, berry, or drupe. (Including Vacciniaceae).

1 Fruit a fleshy, mealy, or juicy berry or drupe.
  2 Ovary superior, not adnate to the calyx-tube, the calyx remaining dry and subtending the base of the fruit; flowers pink or white.
    3 Tree or tall shrub commonly 5 or 6 m tall (to 30 m); bark smooth, the older portions dark brownish-red, exfoliating, the younger bark chartreuse, aging to deep red; leaves elliptic to oblong-ovate, coriaceous, glabrous and shining, entire or serrate, to about 1.5 dm long; flowers white, in large compound terminal racemes; corolla urn-shaped, to 7 mm long; anthers opening by a pair of slit-like terminal pores, with a pair of short recurved horn-like awns on the back; berries orange to red, about 1 cm thick; (s B.C.) . . . . *Arbutus*
    3 Low, commonly trailing shrubs (*Arctostaphylos columbiana* sometimes over 1 m tall).
      4 Fruit a white-pulpy berry with numerous seeds; corolla salverform, the slender tube to 1.5 cm long, the ovate lobes to 8 mm long; anthers opening by longitudinal slits, unappendaged; leaves oval to suborbicular, entire, to 1 dm long, rounded or cordate at base, on petioles about half as long as the blade; (s Man. to Nfld. and N.S.) . . . . *Epigaea*
      4 Fruit a red to purplish-black fleshy or mealy drupe with 1–5 nutlets; corolla urn-shaped or subglobose, at most 6 mm long, the lobes very short; anthers deflexed on their filaments, opening by a pair of terminal (morphologically basal) pores, with a pair of short recurved horn-like awns on the back; leaves oblanceolate to elliptic or ovate . . . . *Arctostaphylos*
  2 Ovary inferior, mostly adnate throughout to the calyx-tube (only basally adnate in *Gaultheria hispidula*), the fruit, with this exception, crowned by the calyx-teeth; anthers opening by a pair of terminal pores.
    5 Fruit a dark-blue to black drupe with 10 seed-like nutlets; corolla white or pink, tubular or conic, the short lobes erect or outwardly curved; anthers unappendaged; leaves oblanceolate to obovate, entire, to about 5 cm long, resin-dotted at least beneath; (Ont. to Nfld. and N.S.) . . . . *Gaylussacia*
    5 Fruit a berry with numerous small seeds; leaves not resin-dotted.
      6 Ovary technically superior, but the many-seeded capsule surrounded at least basally by the pulpy calyx and forming a fleshy, white, red, or purplish to bluish-black false berry (ovary partly truly inferior in *G. hispidula*); corolla urceolate to urceolate-campanulate, white or pinkish; anthers with or without dorsal appendages; leaves evergreen, leathery and shining, entire or sharply serrulate . . . . *Gaultheria*
      6 Ovary truly inferior, the fruit a true berry.
        7 Corolla 4-cleft nearly to base, pink or roseate, the linear lobes reflexed; flowers solitary on slender, recurving or nodding, terminal or axillary, commonly 2-bracted pedicels; anthers exserted, unappendaged, with very long terminal tubes; berries pink or red, acid; leaves ovate or elliptic-oblong, entire, at most about 17 mm long, leathery, evergreen; stems filiform and creeping . . . . *Oxycoccus*
        7 Corolla usually urceolate or globose, its 4 or 5 lobes mostly much shorter than the tube (about equalling it only in *V. stamineum* and *V. vitis-idaea*);

flowers solitary or in racemes; anthers included or only slightly exserted, with or without dorsal appendages; leaves often longer, entire or toothed, deciduous or evergreen; stems prostrate to erect .... *Vaccinium*

1 Fruit a dry capsule; leaves entire or nearly so.

  8 Leaves at most about 1 cm long, subulate or scale-like to linear, narrowly elliptic, or lance-oblong, entire, glabrous or puberulent to minutely tomentose beneath or merely ciliate, evergreen, usually crowded and overlapping.

    9 Sepals usually 4, brightly coloured and petaloid, scarious, much longer than the tubular, deeply 4-lobed corolla, closely subtended by 2 or 3 pairs of opposite bracts simulating a calyx; flowers roseate (sometimes white), sessile or subsessile in the axils of the upper opposite leaves, forming an elongate spike-like terminal inflorescence; stamens usually 8, each anther bearing a pair of deflexed horn-like appendages on the back, each locule opening by a long slit; capsule 4-valved, enclosed by the finally scarious, persistent perianth; leaves lance-oblong, opposite or often 4-ranked, 2–4 mm long, sessile, the larger ones auricled at base; tufted shrub to 1 m tall; (introd. in sandy coastal soils) .... *Calluna*

    9 Sepals 5, green, much shorter than the commonly 5-lobed corolla; inflorescence not spike-like; (arctic, subarctic, and alpine regions).

      10 Leaves at most 5 mm long, subulate or linear-oblong (then alternate) or scale-like (then opposite and often 4-ranked), mostly sessile (short-petioled in *C. stelleriana*); flowers white (sometimes pink), solitary and usually nodding on erect, axillary or terminal pedicels; corolla urceolate or campanulate; stamens 10, each anther opening by a pair of terminal pores and bearing a pair of slender recurved horn-like awns on the back; capsule 5-locular, loculicidal .... *Cassiope*

      10 Leaves mostly over 5 mm long, coriaceous and revolute-margined; flowers pedicelled in the axils of crowded upper leaves, forming few-flowered, terminal, umbel-like clusters; anthers unappendaged; capsules septicidal.

        11 Leaves all or mostly opposite, narrowly elliptic, entire, to 8 mm long, minutely tomentose beneath along the midvein and margins, their petioles 1 or 2 mm long; flowers short-pedicelled, erect; corolla white or pink, broadly campanulate, 3 or 4 mm long, the lobes about equalling the tube; stamens 5, the anthers opening lengthwise; capsules 2–3-locular; diffusely bushy-branched shrub rarely over 1 dm tall; (transcontinental) .... *Loiseleuria*

        11 Leaves alternate, linear, minutely serrulate, to about 1.5 cm long, their short petioles leaving a raised peg-like scar; flowers slender-pedicelled, spreading or nodding; corolla 5–8 mm long, shallowly lobed; stamens usually 10, the anthers opening by a pair of terminal pores; matted shrubs to 3 or 4 dm tall .... *Phyllodoce*

  8 Leaves mostly longer and less crowded and often with relatively broader blades.

    12 Petals distinct, (4)5(6), the more or less rotate corolla lacking a basal tube; anthers unappendaged on the back; capsules septicidal; leaves alternate, entire.

      13 Flowers salmon-pink to copper-colour, to about 3 cm broad, solitary at the ends of short leafy shoots; anthers opening by longitudinal slits; style recurved; leaves pale green, glabrous and somewhat glaucous, oblanceolate to elliptic-oblanceolate, mucronate, subsessile, to about 5 cm long; (s Alaska– w B.C.) .... *Cladothamnus*

      13 Flowers white, to about 1.5 cm broad, in terminal umbel-like clusters; anthers opening by terminal pores; style nearly straight; leaves felted or woolly beneath .... *Ledum*

    12 Petals united at least toward base; flowers in terminal (sometimes also lateral) racemes, umbels, or corymbs.

      14 Capsules loculicidal; flowers white or pinkish, to 7 or 8 mm long; anthers opening by a pair of terminal pores; leaves alternate, entire, coriaceous, to about 5 cm long.

        15 Leaves plane, densely scurfy (especially beneath) with minute brown

scales; flowers in elongate, 1-sided, leafy-bracted racemes, on pedicels
to 4 mm long; corolla white, cylindric-urceolate; anthers unappendaged
but each locule prolonged into a slender awn-like tube; (transcontinental) . . . .
. . . . *Chamaedaphne*

15 Leaves strongly revolute-margined, glaucous or white-puberulent beneath,
not scurfy; flowers in short racemes or umbel-like clusters; corolla white or
pink, ovoid-urceolate; anthers with a pair of recurved horn-like awns on the
back . . . . *Andromeda*

14 Capsules septicidal; anthers unappendaged.

16 Leaves minutely but closely serrulate with bristle-tipped teeth, thin,
elliptic-ovate to elliptic-obovate, alternate, to about 6 cm long; flowers in
umbel-like clusters on shoots of the previous year, appearing with the
leaves; corolla urceolate, 4-lobed, yellowish red, to 8 mm long; anthers
opening by a pair of terminal pores; capsule 4-locular; (Alaska–Yukon–
B.C.-sw Alta.) . . . . *Menziesia*

16 Leaves entire, often coriaceous; corolla 5-lobed, commonly pink to deep
rose-purple (sometimes white or ochroleucous); capsule 5-locular.

17 Corolla rotate, saucer-shaped, to about 2.5 cm broad, deeply parted,
each lobe with a pair of basal pouches into which the longitudinally
dehiscent anthers fit in the bud; leaves opposite or alternate . . . . *Kalmia*

17 Corolla broadly campanulate to tubular-campanulate, often larger and
very showy; anthers opening by a pair of terminal pores; leaves
alternate . . . . *Rhododendron*

ANDROMEDA L.   [6199]   Andromeda

1   Leaves to about 5 cm long, whitened beneath with a close minute puberulence; flowers at
most 6 mm long, they and the capsules in rather dense nodding clusters on curved
branchlets; pedicels mostly less than 1 cm long; (Man. to Nfld. and N.S.) . . . . . . *A. glaucophylla*

1   Leaves mostly less than 3 cm long, whitened beneath with a glaucous varnish-like
coating; flowers to 7 mm long, they and the capsules in erect lax clusters on erect or
ascending branchlets; pedicels to 2 cm long; (transcontinental) . . . . *A. polifolia*

A. glaucophylla Link   Bog-Rosemary
/aST/EE/G/   (N (Ch; evergreen))   Peaty or wet places from Man. (N to Gillam, about 165 mi s of
Churchill) to Ont. (N to the Fawn R. at ca. 54°35'N), Que. (N to s Ungava Bay, the Côte-Nord,
Anticosti Is., and Gaspé Pen.), Labrador (N to Windy Tickle, ca. 55°45'N), Nfld., N.B., P.E.I., and
N.S., s to Minn., Ind., and N.J.; sw Greenland. [*A. canescens* Small; *A. polifolia* vars. *gl.* (Link) DC.
and *latifolia* Ait.].

Var. *iodandra* Fern. (corolla broader than long, the anthers purple or purplish brown rather than
pale brown) is known from the type locality, Table Mt., w Nfld. A purported hybrid with *A. polifolia*
(× *A. jamesiana* Lepage; *A. pol.* var. *jam.* (Lepage) Boivin) is known from the type locality, Lake
R., w James Bay watershed, Ont., and from the E James Bay watershed, Que.

A. polifolia L.
/aST/X/GEA/   (Ch (N; evergreen))   Acid peat bogs and margins of pools from the coasts of
Alaska–Yukon–Dist. Mackenzie–Dist. Keewatin to Que. (N to Wolstenholme, w Hudson Strait) and
Labrador (N to Hebron, 58°13'N), s to ?Wash., s B.C. (the B.C. map by Szczawinski extending the
range southwards beyond that indicated in Raup's map), Alta. (s to Nestow, 54°14'N; CAN), Sask.
(s to McKague, 52°37'N), Man. (s to Cowan, NE of Duck Mt.; CAN), Ont. (s to Fort Severn, Hudson
Bay, ca. 56°N; the report from Wellington Co., s Ont., by Stroud 1941, the presumed basis of the
listing of the species by Soper 1949, is probably based upon *A. glaucophylla*), Que. (s to s James
Bay), and s Labrador (not known from the Atlantic Provinces); w Greenland N to ca. 68°N; Eurasia.
MAPS: Hultén 1968b:727; Raup 1947: pl. 32.

A form with the leaves green beneath, rather than glaucous, has been named var. *concolor*
Boivin (Can. Field-Nat. 65(1):16. 1951; type from Kodiak Is., Alaska).

ARBUTUS L.   [6211]

A. menziesii Pursh   Arbutus, Madrona
/t/W/   (Ms (evergreen))   West of the Cascades, chiefly in drier areas, from sw B.C. (Vancouver Is. and adjacent islands; Vancouver dist. and Skagit R., E of Chilliwack; *see* B.C. map by Szczawinski 1962:26) to Baja Calif. [*A. procera* Dougl.]. MAPS: Hosie 1969:300; Canada Department of Northern Affairs and Natural Resources 1956:270; Fowells 1965:90; Preston 1961:332; Little 1971: map 100-W.

According to M.L. Fernald (Rhodora 28(328):51. 1926), the report of the European *A. unedo* L., Strawberry-tree, from Nfld. by Cormack (1856; near Bonaventure, Trinity Bay) is possibly based upon *Viburnum cassinoides*.

ARCTOSTAPHYLOS Adans.   [6212]   Bearberry, Manzanita

1   Erect or spreading shrub commonly over 1 m tall; leaves lanceolate to elliptic or ovate, entire, evergreen, finely greyish-puberulent especially beneath, to about 5 cm long, rounded or abruptly tapering to petioles to about 5 mm long; twigs and petioles greyish-puberulent to tomentose and usually distinctly bristly with glandular or eglandular hairs; (sw B.C.) . . . . *A. columbiana*
1   Prostrate shrubs, often rooting along the branches; leaves mostly oblanceolate to cuneate-obovate, to about 3 cm long, gradually tapering to winged petioles; (transcontinental).
   2   Leaves entire, obscurely veined, leathery, evergreen; branchlets pubescent; fruit dry and mealy, dull red. . . . . *A. uva-ursi*
   2   Leaves crenate-serrate, plane to deeply veiny-rugose, less leathery, deciduous to highly marcescent; branchlets glabrous; fruit fleshy . . . . *A. alpina*

A. alpina (L.) Spreng.   Alpine Bearberry
/aST/X/GEA/   (Ch)   Rocky or gravelly tundra and barrens from the coasts of Alaska–Yukon–Dist. Mackenzie–Dist. Keewatin to N Banks Is., S Victoria Is., cent. Baffin Is., and northernmost Ungava–Labrador, S to B.C. (S to Yoho National Park at ca. 52°N; *see* B.C. map by Szczawinski 1962:30), Alta. (S to Columbia Icefield, Jasper National Park), Sask. (known only from L. Athabasca), Man. (S to near The Pas), Ont. (S to the w James Bay watershed at ca. 53°N), Que. (S to SE James Bay, L. Mistassini, the Côte-Nord, Anticosti Is., and Shickshock Mts. of the Gaspé Pen.; not known from the Maritime Provinces), Nfld., and the mts. of Maine and N.H.; w Greenland N to ca. 70°N, E Greenland N to ca. 75°N; Eurasia. [*Arbutus* L.; *Arctous* Nied.; *Mairania* Desv.]. MAPS: Hultén 1968*b*:730; Porsild 1957: map 271, p. 194; Meusel 1943: fig. 7d (incomplete for N. America).

Ssp. *rubra* (Rehd. & Wils.) Hult. (*A. rubra (ruber)* (Rehd. & Wils.) Fern.; *Arctous erythrocarpa* Small; leaves not ciliate rather than bristly-ciliate toward base and on the petioles; drupes scarlet rather than purple to purplish black) occurs essentially throughout the range of the typical form but is somewhat more southern, being known in the Canadian Arctic Archipelago only from sw Victoria Is. and w Baffin Is. near the Arctic Circle, and absent in Greenland and Europe. MAPS (*A. rubra*): Hultén 1968*b*:730; Porsild 1957: map 272, p. 194; Raup 1947: pl. 33. A purported hybrid between *A. alpina* and *A. rubra* has been named × *A. victorinii* Rolland-Germain (*see* Marie-Victorin and Rolland-Germain 1969).

A. columbiana Piper   Hairy Manzanita
/t/W/   (N (Mc))   Along the Pacific coast from sw B.C. (Vancouver Is. and adjacent mainland N to ca. 50°N; *see* B.C. map by Szczawinski 1962:32) to Calif. [*A. tomentosa* of auth., not (Pursh) Lindl.].

A hybrid with *A. uva-ursi* (× *A. media* Greene) is reported from sw B.C. by J.M. Macoun (1898; Nanaimo, Vancouver Is.).

A. uva-ursi (L.) Spreng.   Common Bearberry, Sandberry, Kinnikinnick.   Bousserole or Raisin d'ours
/aST/X/GEA/   (Ch (evergreen))   Exposed rocks and sands from the E Aleutian Is. and

Alaska–Yukon–Dist. Mackenzie (N to ca. 69°30′N) to L. Athabasca (Alta. and Sask.), SW Dist. Keewatin, northernmost Man.–Ont., Que. (N to E James Bay at 53°43′N, L. Mistassini, the Côte-Nord, Anticosti Is., and Gaspé Pen.), Labrador (N to the Hamilton R. basin), Nfld., N.B., P.E.I., and N.S., S to N Calif., N.Mex., S.Dak., Ill., Minn., Va., and New Eng.; W Greenland at ca. 68°N; Iceland; N Eurasia. [*Arbutus* L.; *Uva-ursi* Britt.; incl. var. *pacifica* Hult.]. MAPS (aggregate species): Hultén 1968*b*:729; Böcher 1954: fig. 58, p. 218.

The typical form has the young branchlets finely viscid-tomentose, becoming glabrate. Var. *adenotricha* Fern. & Macbr. (young branchlets finely viscid-tomentose but also bearing long viscid hairs intermixed with black stipitate glands; type from Golden, B.C.) occurs essentially throughout the N. American area, as shown in MAPS by Hultén (1968*b*:729) and Raup (1947: pl. 33). Var. *coactilis* Fern. & Macbr. (young branchlets minutely and permanently white-tomentose but not viscid) also occurs essentially throughout the N. American area. (*See* J.G. Packer, Can. J. Bot. 45(9):1768. 1967).

CALLUNA Salisb. [6236]

C. vulgaris (L.) Hull   Scotch Heather, Ling
European; a very local garden-escape to peaty or damp sandy places in N. America, as in SW B.C. (Lulu Is.; Eastham 1947), SW Que. (Ste-Marguerite; MT), Nfld. (CAN; GH), St-Pierre and Miquelon (Rouleau 1956), and N.S. (Yarmouth, Kings, Pictou, Halifax, and Inverness counties and Sable Is.). [*Erica* L.; *C. atlantica* Seem.]. MAP: Hultén 1958: map 117, p. 137.

CASSIOPE D. Don [6197] White or Mountain Heather

1 Corolla-lobes about equalling or longer than the tube; style ovoid or conic; flowers mostly solitary at the ends of the branches; leaves alternate.
  2 Sepals acute; corolla to 5 mm long, its lobes about equalling the tube; stamens to 1.5 mm long; pedicels capillary, to about 2 cm long, the flowers nodding; leaves subulate, sharp-pointed, appressed; (SW Dist. Mackenzie to S Baffin Is., Labrador, Nfld., and mts. of E Que.) . . . . *C. hypnoides*
  2 Sepals obtuse; corolla 5 or 6 mm long, its lobes commonly longer than the tube; stamens to 3 mm long; pedicels stout, at most about 1 cm long, the flowers erect; leaves linear-oblanceolate or narrowly oblong, obtuse or acutish, spreading, short-petioled; (Alaska–Yukon and mts. of B.C.) . . . . *C. stelleriana*
1 Corolla-lobes not over half the length of the tube; style slender, scarcely thickened at base; flowers usually several on each branch, solitary on usually subterminal capillary pedicels to over 2.5 cm long; leaves thick and scale-like, opposite, appressed, sessile.
  3 Leaves prominently grooved for nearly the entire length of the lower (outer and rounded) surface, lanceolate, markedly puberulent and ciliolate at least near the base; (transcontinental in arctic, subarctic, and alpine regions) . . . . *C. tetragona*
  3 Leaves grooved only at the extreme base of the lower surface, glabrous (or merely ciliolate, or minutely puberulent only at the extreme base); (western species).
    4 Leaves 2 or 3 mm long (stem, with leaves, about 2 mm thick), scarious-margined, neither ciliolate nor distinctly 4-ranked; (S Alaska–W B.C.) . . . . *C. lycopodioides*
    4 Leaves to 5 mm long (stem, with leaves, about 4 mm thick), not scarious-margined, distinctly 4-ranked, the young ones ciliolate near tip; (SE Alaska–B.C.–SW Alta.) . . . . *C. mertensiana*

C. hypnoides (L.) Don   Moss-Heather
/aST/EE/GEA/ (Ch (evergreen)) Rocky tundra and alpine or subalpine slopes from SE-cent. Dist. Mackenzie (Maguse L., E of Great Slave L.; CAN) and SE Dist. Keewatin (Thelon Game Sanctuary; CAN) to cent. Baffin Is. and northernmost Ungava–Labrador, S to Que. (S to E Hudson Bay at ca. 59°30′N and the Knob Lake dist. at ca. 55°N; isolated in the Shickshock Mts. of the Gaspé Pen.), Labrador (S to ca. 54°N), Nfld., and the mts. of ?N.Y., Maine, and N.H.; W and E Greenland N to ca. 74°N; Iceland; Spitsbergen; Scandinavia; NW Siberia. [*Andromeda* L.; *Harrimanella* Cov.]. MAPS: Hultén 1958: map 15, p. 35 (noting, also, a 1914 total-area map by Rikli);

Porsild 1957: map 267, p. 194; Löve and Löve 1956*b*: fig. 8, p. 146 (*Harrimanella*); Böcher 1954: fig. 28 (top), p. 111; Raymond 1950*b*: fig. 7, p. 16.

C. lycopodioides (Pall.) Don
/sT/W/eA/   (Ch (evergreen))   Montane slopes of the Aleutian Is., s Alaska (N to ca. 61°N; *see* Hultén 1948: map 927, p. 1335), and w B.C. (s to Smithers, ca. 54°30′N; *see* B.C. map by Szczawinski 1962:39); E Asia. [*Andromeda* Pall.; incl. ssp. *cristapilosa* Calder & Taylor].

C. mertensiana (Bong.) Don
/sT/W/   (Ch (evergreen))   Montane slopes, usually above timberline, from the Alaska Panhandle (N to ca. 60°N; *see* Hultén 1948: map 928, p. 1335; type from Sitka) through B.C. and sw Alta. (N to 53°54′N; CAN) to Calif., Nev., and Mont. [*Andromeda* Bong.; *A. cupressina* Hook.]. MAP: Hultén 1968*b*:725.

C. stelleriana (Pall.) DC.   Moss-Heather
/ST/W/eA/   (Ch (evergreen))   Alpine heaths, meadows, and bogs from the E Aleutian Is., Alaska (N to the Seward Pen. at ca. 65°N), and sw Yukon through B.C. to Mt. Rainier, Wash.; E Siberia and Japan. [*Andromeda* Pall.]. MAP: Hultén 1968*b*:726.

C. tetragona (L.) Don   Arctic White Heather
/AST/X/GEA/   (Ch (evergreen))   Dry heaths, rocky tundra, and montane slopes from the coasts of Alaska–Yukon–Dist. Mackenzie–Dist. Keewatin to northernmost Ellesmere Is. and Baffin Is., s in the West through B.C. and the mts. of sw Alta. (N to near Obed, about 70 mi NE of Jasper; CAN) to Wash. and Mont., farther eastwards s to s Dist. Mackenzie–Dist. Keewatin, N Que. (s to s Ungava Bay), and N Labrador (s to ca. 58°N; CAN; reported s to Hopedale, 55°27′N, by Delabarre 1902); w and E Greenland N of 65°N; Spitsbergen; N Scandinavia; N Asia. [*Andromeda* L.]. MAPS: Hultén 1968*b*:724; Porsild 1957: map 268, p. 194; Böcher 1954: fig. 43, p. 167; Raup 1947: pl. 32.
   Some of the western material is referable to var. *saximontana* (Small) Hitchc. (*C. sax.* Small, the type from Banff, Alta.; flowers usually not over 5 mm long, on pedicels seldom over twice the length of the subtending leaves, rather than to 7 mm long and on proportionately longer pedicels). MAP: Hultén 1968*b*:725.

CHAMAEDAPHNE Moench   [6200]

C. calyculata (L.) Moench   Leather-leaf, Cassandra.   Faux Bluets or Petit-Daphné
/ST/X/EA/   (N (Ch; evergreen))   Peat bogs and margins of acidic ponds from N Alaska–Yukon–Dist. Mackenzie (N to ca. 69°30′N) to Great Bear L., Great Slave L., L. Athabasca (Alta. and Sask.), s Dist. Keewatin, northernmost Man.–Ont., Que. (N to Ungava Bay and the Côte-Nord), Labrador (N to Hopedale, 55°27′N), Nfld., N.B., P.E.I., and N.S., s to NE B.C. (s to ca. 56°N; *see* B.C. map by Szczawinski 1962:50), Alta. (s to Smith, SE of Lesser Slave L.; CAN), Sask. (s to Crooked River, 52°51′N; Breitung 1957*a*), s Man., Iowa, Ohio, and Ga.; Eurasia. [*Andromeda* L.; *Cassandra* Don]. MAP: Hultén 1968*b*:727.
   According to Fernald *in* Gray (1950), the N. American plant differs from the typical form of Eurasia according to the following characters:
1   Calyx-lobes acuminate, up to half as long as the urceolate corolla; leaves averaging about
      3 times as long as broad; [Eurasia] . . . . [var. *calyculata*]
1   Calyx-lobes blunt to acute; [N. America; var. *ang.* also in E Asia].
    2   Calyx-lobes blunt to acute or acutish, about 1/3 as long as the subcylindric corolla;
          leaves to 4 times as long as broad; [*A. ang.* (Ait.) Pursh] . . . . . . var. *angustifolia* (Ait.) Rehd.
    2   Calyx-lobes blunt, up to half as long as the urceolate-cylindric corolla; leaves
          averaging about twice as long as broad; [*Andromeda cal.* var. *lat.* Ait., the type from
          Nfld.] . . . . var. *latifolia* (Ait.) Fern.

CLADOTHAMNUS Bong.   [6181]

C. pyrolaeflorus Bong.   Copper-flower
/sT/W/   (N (Mc))   Moist forests and along streams near the coast from s Alaska (N to ca. 61°N;

type from Sitka) through W B.C. (*see* B.C. map by Szczawinski 1962:59) to NW Oreg. [*Tolmiea occidentalis* Hook.]. MAP: Hultén 1968*b*:717.

EPIGAEA L.   [6205]   Trailing Arbutus, Ground Laurel

E. repens L.   Mayflower.   Fleur de Mai
/T/EE/   (Ch (evergreen))   Sandy or peaty woods and clearings from S Man. (N to near Dauphin, N of Riding Mt.; WIN; reports from Sask. require confirmation) to Ont. (N to Sioux Lookout, about 175 mi NW of Thunder Bay; OAC), Que. (N to the E James Bay watershed at ca. 54°N, L. Mistassini, the Côte-Nord, Anticosti Is., and Gaspé Pen.; reports from Labrador may refer to the Côte-Nord, Que.), Nfld., N.B., P.E.I., and N.S. (provincial floral emblem), S to Iowa, Ill., Miss., and Fla.

All of our material may evidently be referred to var. *glabrifolia* Fern. (leaves glabrous or soon so on both surfaces (except sometimes along the nerves beneath), rather than scabrous and persistently setose or pilose on both surfaces; type from Middleton, Annapolis Co., N.S.).

GAULTHERIA L.   [6206]   Wintergreen

1   Leaves mostly less than 1 cm long, elliptic to obovate, entire, bristly beneath, their
    margins revolute; flowers 4-merous, less than 3 mm long, solitary in the leaf-axils of
    the slender, trailing and creeping, matted stems; anthers awnless; fruit white;
    (transcontinental) .... *G. hispidula*
1   Leaves usually more than 1 cm long, their margins not revolute; flowers usually 5-merous
    and at least 3 mm long, borne on aerial branches; fruit red to purplish or bluish-black.
    2   Leaves clustered near the summit of the flowering-branches (these to 1.5 dm tall),
        elliptic to narrowly obovate, to 5 cm long, commonly minutely serrate with
        bristle-tipped teeth; flowers solitary in the upper axils or in short racemes; corolla
        urn-shaped or barrel-shaped (the lobes short), to 1 cm long; anthers with 2 double
        horns; fruit bright red; (SE Man. to Nfld. and N.S.) .... *G. procumbens*
    2   Leaves regularly distributed along the stems and branches; (B.C.; *G. humifusa* also in
        Alta.).
        3   Flowers rather numerous in elongate, terminal and subterminal, axillary, bracted
            racemes (rachises puberulent and glandular-hirsute); corolla campanulate-
            urceolate (the very short lobes spreading or reflexed), glandular-puberulent, to 1
            cm long; calyx glandular-pilose; anthers with 4 slender apical awns; fruit purplish,
            to 1 cm thick; leaves ovate to ovate-elliptic, to about 9 cm long, sharply serrulate;
            stems pilose or hirsute; (Alaska–B.C.) .... *G. shallon*
        3   Flowers solitary in the leaf-axils, short-pedicelled; corolla campanulate, prominently
            lobed, glabrous, at most 5 mm long; anthers awnless; fruit red or reddish, to 8 mm
            thick.
            4   Calyx glabrous, about equalling the corolla-tube; leaves generally oval,
                rounded or obtuse at apex, averaging less than 2 cm long, entire or
                inconspicuously serrulate; stems glabrous or finely puberulent (sometimes with
                longer reddish pilosity); (B.C. and SW Alta.) .... *G. humifusa*
            4   Calyx copiously brownish-reddish-pilose, about half the length of the corolla-
                tube; leaves more ovate in outline, acute, averaging about 3 cm long, usually
                conspicuously serrulate; stems puberulent and copiously brownish-pilose;
                (S B.C.) .... *G. ovatifolia*

G. hispidula (L.) Muhl.   Creeping Snowberry, Moxieplum.   Oeufs de perdrix
/sT/X/   (Ch (evergreen))   Moist coniferous forests from B.C. (N to ca. 56°N; *see* B.C. map by Szczawinski 1962:62) to Alta. (N to Lesser Slave L.; CAN), Sask. (N to Windrum L. at ca. 56°N; CAN), Man. (N to Flin Flon, 54°46′N), Ont. (N to Big Trout L. at ca. 53°45′N, 90°W; CAN), Que. (N to the Swampy R. at 56°34′N and the Côte-Nord), Labrador (N to Rigolet, 54°11′N), Nfld., N.B., P.E.I., and N.S., S to N Idaho, Minn., Mich., Pa., and N.C. [*Vaccinium* L.; *Chiogenes* T. & G.; *Oxycoccus* Pers.; *G. (C.) serpyllifolia* Pursh].

G. humifusa (Graham) Rydb.   Alpine or Matted Wintergreen
/T/W/   (Ch (evergreen))   Moist or wet subalpine to alpine meadows and slopes from s B.C. (N to ca. 52°N; Leena Hämet-Ahti, Ann. Bot. Fenn. 2(2):158. 1965) and sw Alta. (N to L. Agnes, 51°25′N; CAN) to N Calif. and Colo. [*Vaccinium* Graham; *G. myrsinites* Hook.].

G. ovatifolia Gray   Oregon Wintergreen
/T/W/   (Ch (evergreen))   Coniferous forests to alpine bogs from s B.C. (N to ca. 51°30′N; *see* B.C. map by Szczawinski 1962:67) to N Calif., Idaho, and ?Mont.

G. procumbens L.   Checkerberry, Teaberry.   Thé des bois
/T/EE/   (Hpr (evergreen))   Dry or moist woods (chiefly coniferous) from SE Man. (N to Victoria Beach, about 55 mi NE of Winnipeg; CAN) to Ont. (N to the N shore of L. Superior; MT), Que. (N to the Gaspé Pen. at Nouvelle, Bonaventure Co.), Nfld., N.B., P.E.I., and N.S., s to Minn., Wisc., Ala., and Ga.

G. shallon Pursh   Salal
/T/W/   (N (evergreen))   Coniferous forests (chiefly coastal) from SE Alaska (*see* Hultén 1948: map 934, p. 1335) and B.C. (N to Queen Charlotte Is. and Prince Rupert; isolated stations in the s Kootenay Valley; *see* B.C. map by Szczawinski 1962:70) to s Calif.; introd. in England. MAP: Hultén 1968*b*:728.
   The similar *G. miqueliana* Takeda of E Asia is known from the westernmost Aleutian Is. (Kiska Is.; *see* Hultén 1948: map 933, p. 1335):

1   Racemes many-flowered; flowers to 8 mm long; fruit purplish black; leaves to 9 cm long, acute; creeping to erect shrub to 12 dm tall . . . . *G. shallon*
1   Racemes with rarely over 6 flowers, these to about 5 mm long; fruit white; leaves to about 3.5 cm long, obtuse; decumbent dwarf shrub; [w Aleutian Is. and E Asia; MAP: Hultén 1968*b*:728] . . . . *G. miqueliana* Takeda

GAYLUSSACIA HBK.   [6215]   Huckleberry

1   Plant pubescent and copiously stipitate-glandular, the sepals also glandular-ciliate; leaves thick and rather firm, dark green and shining above, glandular beneath; bracts of raceme leaf-like and persistent; (Atlantic Provinces) . . . . *G. dumosa*
1   Plant essentially glabrous, more or less beset with sessile resin-dots; sepals not ciliate; leaves relatively thin and pale, dull; bracts small and deciduous.
   2   Leaves resin-dotted on both surfaces; flowers usually longer than their pedicels, in short conspicuously glandular racemes; (Ont. to Nfld. and N.S.) . . . . *G. baccata*
   2   Leaves resin-dotted only beneath; flowers usually shorter than their pedicels, in open sparingly glandular racemes . . . . [*G. frondosa*]

G. baccata (Wang.) Koch   Black Huckleberry.   Gueules noires
/T/EE/   (N (Ch))   Dry to moist woods, thickets, and clearings from Ont. (N to Manitoulin Is., N L. Huron, and Renfrew, Russell, and Carleton counties; *see* s Ont. map by Soper and Heimburger 1961:91; reports from Sask. and Man. require confirmation) to Que. (N to Rivière-du-Loup, Temiscouata Co., and Cap-Jaseux, Chicoutimi Co., 48°26′N), Nfld., St-Pierre and Miquelon, N.B., P.E.I., and N.S., s to La. and Ga. [*Andromeda* Wang.; *Vaccinium* (*G.*) *resinosum* Ait.; *G. frondosa* of Ont. reports, not (L.) T. & G., relevant collections in CAN; *V. corymbosum sensu* Fowler 1885, as to the Grand Lake, N.B., plant, not L., the relevant collection in NBM; *A.* (*Lyonia*) *?ligustricina sensu* Hooker 1834, not *V. lig.* L.; *A. ?paniculata sensu* Pursh 1814, not L.; *A.* (*Leucothoë*) *racemosa sensu* Pursh 1814, Hooker 1834, and John Macoun 1884, not L.].

G. dumosa (Andr.) T. & G.   Dwarf Huckleberry
/T/E/   (N)   Dry barrens, sphagnous bogs, and pinelands from Nfld., St-Pierre and Miquelon, N.B., P.E.I., and N.S. to Miss. and Fla. [*Vaccinium* Andr.].
   Our material is referable to var. *bigeloviana* Fern. *(V. (G.) hirtellum* of Canadian reports, not Ait.; leaves and bracts more generally and persistently glandular above (as well as beneath) than those of the typical form).

[G. frondosa (L.) T. & G.]   Dangleberry, Blue-tangle
[Reports of this species of the E U.S.A. (N to N.Y., N.H., and Mass.) from S Ont. by J.M. Macoun (1903), Zenkert (1934), and Soper (1949) are probably all based upon *G. baccata*, relevant collections in CAN. (*Vaccinium* L.; *V. glaucum* Young).]

KALMIA L.   [6192]   Laurel

1   Corymbs lateral; calyces and pedicels glandular-puberulent; corolla to 12 mm broad, deep rose-pink or crimson; capsules at most 5 mm thick; leaves flat, thin, short-petioled, elliptic-lanceolate to oblong, mostly opposite (or whorled in 3's), glabrate, to about 5 cm long; branchlets terete; (Ont. to Labrador, Nfld., and N.S.) . . . . *K. angustifolia*
1   Corymbs terminal; capsules to 8 mm thick; leaves coriaceous.
   2   Leaves mostly alternate, petioled, narrowly to broadly elliptic, to about 8 cm long, flat; branchlets terete; corolla to 3 cm broad, normally pink; calyces and pedicels viscid-pilose; capsules glandular; seeds oblong . . . . [*K. latifolia*]
   2   Leaves opposite (rarely in 3's), subsessile, linear to lanceolate, usually less than 3 cm long, white-puberulent beneath, their margins often revolute; branchlets 2-edged; corolla at most 2 cm broad, deep pink to crimson; calyces, pedicels, and capsules glabrous; seeds linear; (transcontinental) . . . . *K. polifolia*

K. angustifolia L.   Lambkill, Sheep-Laurel.   Crevard de moutons
/sT/EE/   (N (Ch; evergreen))   Sterile soil, old pastures, and barrens from Ont. (N to the SW James Bay watershed at ca. 51°30'N) to Que. (N to the E James Bay watershed at 53°47'N, L. Mistassini, the Côte-Nord, Anticosti Is., and Gaspé Pen.), Labrador (N to the Hamilton R. basin), Nfld., N.B., P.E.I., and N.S., S to Mich., Va., and Ga.

The report from E Man. by Lowe (1943) requires confirmation. Forma *candida* Fern. (flowers white rather than crimson to deep rose-pink; type, as first collection cited, from St. John's, Nfld.) is known from Nfld. and Mass.

[K. latifolia L.]   Mountain-Laurel, Calico-bush
[This species of the E U.S.A. (N to Ind. and New Eng.) is reported from Canada by A. Michaux (1803) and Pursh (1814), possibly a result of the somewhat flexible use of that geographical designation in their day. However, Gleason (1958) ascribes it to "N.B. to Ont." and an 1863 collection by Kennedy in GH from Napierville, SE of Montreal, Que., has been placed here. The supposed occurrence in Ont. and N.B. is indicated in MAPS by Preston (1961:334) and Munns (1938: map 161, p. 165) but no Canadian stations are indicated in the map by E.A. Kurmes (Am. Midl. Nat. 77(2): fig. 1, p. 525. 1967). If once a member of our flora, it is apparently now extinct. Reports from Saguenay Co., E Que., by Saint-Cyr (1887), from Anticosti Is., E Que., by B. Billings (Ann. Bot. Soc. Can. 1:59. 1861), and from S Labrador by Billings, Stearns (1884), and Waghorne (1898) probably refer to *K. angustifolia*.]

K. polifolia Wang.   Pale, Bog-, or Swamp-Laurel
/ST/X/   (N (Ch; evergreen))   Peat bogs and wet acidic meadows, the aggregate species from SE Alaska and the Yukon (N to ca. 64°N) to Great Bear L., Great Slave L., L. Athabasca (Alta. and Sask.), S Dist. Keewatin, northernmost Man.–Ont.), Que. (N to Ungava Bay and the Côte-Nord), Labrador (N to Nachvak, 59°07'N), Nfld. (the type material, according to Hitchcock et al. 1959, being "Garden specimens from plants introduced by Banks from Newf.", presumably grown at Berlin), N.B., P.E.I., and N.S., S to Oreg., Colo., Minn., Pa., and N.J. MAPS and synonymy: *see* below.
1   Leaves mostly not over 2 cm long, up to half as broad as long; flowers mostly not over about 1 cm long; plant usually not over 1 dm tall; [*K. glauca* var. *mic.* Hook.; *K. mic.* (Hook.) Heller; southernmost Yukon–SE Alaska (*see* Hultén 1948: map 921, p. 1334) and mts. of B.C. and SW Alta. (Waterton Lakes; Breitung 1957b); S Dist. Keewatin at Baralzon L., 60°N; perhaps throughout the range of the typical form; MAP: Hultén 1968b:721] . . . .
. . . . var. *microphylla* (Hook.) Rehd.
1   Leaves to 4 cm long, less than half as broad as long, strongly revolute; flowers to 18 mm broad; plant usually over 2 dm tall . . . . var. *polifolia*

2   Flowers white; [type from Trinity South, Nfld.] . . . . f. *leucantha* Schofield & Smith
2   Flowers deep pink to crimson; [*K. glauca* l'Hér. and its var. *rosmarinifolia* Pursh; *K. occidentalis* Small; transcontinental; MAP: Hultén 1968b:721] . . . . f. *polifolia*

LEDUM L.   [6183]   Labrador-tea.   Lédon

1   Leaves elliptic to oblong-ovate or ovate, to about 6 cm long, on petioles to over 1 cm long, the lower surface greenish or greyish, finely felted and copiously sprinkled with minute shining resin-granules; anther-filaments densely hairy on the lower half; (B.C.–Alta.) . . . . *L. glandulosum*
1   Leaves linear to oblong, sessile or short-petioled, normally densely rust-woolly beneath; (transcontinental).
   2   Leaves linear to linear-oblong, less than 1.5 cm long; stamens usually 10, their filaments pubescent below the middle; capsules 3 or 4 mm long and nearly as thick; (arctic and subarctic regions) . . . . *L. palustre*
   2   Leaves linear-oblong to oblong, to about 6 cm long; stamens usually 5, 6, or 7, their filaments mostly glabrous; capsules at least 5 mm long, usually about twice as long as thick . . . . *L. groenlandicum*

L. glandulosum Nutt.   Trapper's-tea
/T/W/   (N (evergreen))   Wet montane meadows and open woods (ranges of Canadian taxa outlined below), s to Calif. and NW Wyo.
According to D.B. Savile (Can. J. Bot. 47(7): B.C.–Alta. maps, fig. 1, p. 1093, and fig. 2, p. 1094. 1969), only the southernmost B.C.–Alta. material is referable to typical *L. glandulosum*, the more northern records representing a hybrid with *L. groenlandicum*.
1   Leaves plane or slightly revolute, usually averaging about twice as long as broad; capsules to 3(4.5) mm long, subglobose; [*L. groenlandicum* ssp. *gl.* (Nutt.) Löve & Löve; mts of B.C. (N to ca. 52°15′N; *see* B.C. map by Szczawinski 1962:86) and SW Alta. (N to Banff; CAN)] . . . . var. *glandulosum*
1   Leaves strongly revolute, averaging less than 1 cm broad, to over 4 times as long as broad; capsules ovoid, to 5.5 mm long; [*L. col.* Piper; Vancouver Is.: Szczawinski 1962] . . . . var. *columbianum* (Piper) Hitchc.

L. groenlandicum Oeder   Labrador-tea
/aST/X/G/   (N (Ch; evergreen))   Peat bogs and acid soils from Alaska–Yukon (N to ca. 68°N) and the coast of W Dist. Mackenzie to Great Bear L., Great Slave L., cent. Dist. Keewatin, northernmost Ont., Que. (N to Ungava Bay at ca. 59°30′N), Labrador (N to Hebron, 58°12′N), Nfld., N.B., P.E.I., and N.S., s to Oreg., Minn., Pa., and N.J.; s Greenland N to near the Arctic Circle (type locality). [*L. palustre* ssp. *gr.* (Oeder) Hult.; *L. pal.* vars. *dilatatum* Gray and *latifolium* (Ait.) Michx. (*L. lat.* Ait.); *L. canadense* Lodd.; *L. pacificum* Small]. MAPS: Hultén 1968b:718 (*L. pal.* ssp. *gr.*); Raup 1947: pl. 32.
Forma *denudatum* Vict. & Rousseau (*L. pal.* var. *lat.* f. *den.* (V. & R.) Boivin; leaves relatively soft, glabrous beneath except for long hairs along the veins rather than white- or rusty-woolly beneath) is known from the type locality, a cedar swamp N of Bic Mt., St-Fabien, Rimouski Co., E Que. The map by D.B. Savile (Can. J. Bot. 47(7): fig. 3, p. 1095. 1969) indicates the location of hybrids with *L. palustre* (ssp. *decumbens*) in Alaska and NW Canada.

L. palustre L.
/aST/X/GEA/   (N (Ch; evergreen))   Heaths, barrens, and dry rocky places at low to fairly high elevations from the coasts of Alaska–Yukon–Dist. Mackenzie–Dist. Keewatin to Victoria Is., cent. Baffin Is., and northernmost Ungava–Labrador, s to N B.C. (s to ca. 59°N; *see* B.C. map by Szczawinski 1962:83), Great Slave L., L. Athabasca (Alta. and Sask.), Man. (s to Gillam, about 165 mi s of Churchill), Ont. (s to NW James Bay at ca. 55°N), Que. (s to NE James Bay at 54°25′N and L. Marymac at ca. 57°N, 53°45′W), and Labrador (s to ca. 53°N); W Greenland N to ca. 74°N; Eurasia. [Incl. vars. *angustifolium* Hook. and *decumbens* Ait. (*L. dec.* (Ait.) Lodd.)]. MAPS: Hultén 1968b:717; Porsild 1957: map 265 (*L. dec.*), p. 194; Meusel 1943: fig. 20b (incomplete for N. America).

A possible hybrid with *Rhododendron lapponicum (R. vanhoeffenii* Abrom.; flowers pink, with 10 stamens rather than 5) is reported from w Greenland by Böcher, Holmen, and Jacobsen (1966).

LOISELEURIA Desv. [6189]

L. procumbens (L.) Desv. Alpine Azalea
/aST/X/GEA/ (Ch (evergreen)) Peaty or rocky tundra and slopes at low to high elevations from the Aleutian Is. and the coasts of Alaska–Yukon–Dist. Mackenzie–Dist. Keewatin to Baffin Is. (N to the Arctic Circle) and northernmost Ungava–Labrador, s to Wash., B.C. (Vancouver Is. and adjacent mainland; see B.C. map by Szczawinski 1962:720), ?Alta. (Fernald *in* Gray 1950; not listed by Moss 1959), northernmost Sask. (near Hasbala L. and Patterson L. at 59°55′N, where taken by Argus in 1963; not listed by Breitung 1957a), NE Man. (s to Churchill; CAN; reported s to Gillam, about 165 mi s of Churchill, by Lowe 1943; not known from Ont.), islands in James Bay s to ca. 52°N, Que. (s to E James Bay at ca. 54°N, the Côte-Nord, and Shickshock Mts. of the Gaspé Pen.; reported from Anticosti Is. by Verrill 1865, and from Cap-à-l'Aigle, Charlevoix Co., by R. Campbell, Can. Rec. Sci. 6(6):342–51. 1895), s Labrador, Nfld., N.S. (Kingsport, Kings Co.; GH; not known from N.B. or P.E.I.), and the mts. of Maine and N.H.; w Greenland N to ca. 74°N, E Greenland N to ca. 68°N; Iceland; Eurasia. [*Azalea* L.] MAPS: Hultén 1968b:720, and 1958: map 182, p. 201 (noting, also, a 1947 total-area map by Firbas); Porsild 1957: map 266, p. 194; Meusel 1943: fig. 2d (incomplete for N. America).

MENZIESIA Sm. [6185]

M. ferruginea Sm. Fool's-Huckleberry, False Azalea
/sT/W/ (N) Moist woods and streambanks at low to fairly high elevations from Alaska (N to ca. 64°N; *see* Hultén 1948: map 919, p. 134) and southernmost Yukon through B.C. (*see* B.C. map by Szczawinski 1962:94) and w Alta. (N to the Smoky R. at ca. 56°N) to N Calif. and Wyo. [Var. *glabella* (Gray) Peck (*M. glabella* Gray; *see* J.C. Hickman and M.P. Johnson, Madroño 20(1):1–11. 1969); *M. globularis* Hook.]. MAPS (aggregate species): Hultén 1968b:720; Hickman and Johnson, loc. cit., fig. 1, p. 2.

OXYCOCCUS Adans. [6216] Cranberry. Atoças

(Ref.: A.E. Porsild, Can. Field-Nat. 52(8):116–17. 1938)
1 Pedicels essentially glabrous; leaves ovate, broadest toward base.
  2 Pedicels red, bearing a pair of red scaly bracts below the middle; berry about 6 mm
    thick, pale pink; leaves at most about 6 mm long and 2 mm broad, strongly revolute,
    pointed; stem filiform, red or reddish brown; (transcontinental) . . . . *O. microcarpus*
  2 Pedicels 1–8, often lateral; berry to 12 mm thick, with a bloom; leaves to 8 mm long
    and 3 mm broad, flat or but slightly revolute; stem stouter, dark brown or black; (B.C.;
    Sask.; s Ont. to Nfld.) . . . . *O. ovalifolius*
1 Pedicels pubescent; leaves elliptic, broadest near the middle.
  3 Berry to 2 cm thick; pedicels 1–10, lateral, becoming evenly curved, bearing 1 or 2
    leaf-like green bracts; leaves to 17 mm long and 8 mm broad, flat or very slightly
    revolute, rounded at apex, evenly and symmetrically arranged; stems relatively stout,
    extensively trailing; (Ont. to Nfld. and N.S.) . . . . *O. macrocarpus*
  3 Berry at most 12 mm thick; pedicels terminal, abruptly curved toward tip, bearing a
    pair of scaly bracts; leaves smaller, pointed; stems slender, creeping, brown or black;
    (B.C. to N.S.) . . . . *O. quadripetalus*

O. macrocarpus (Ait.) Pers. Large or American Cranberry. Gros Atocas
/T/EE/ (Ch (evergreen)) Peat bogs and acid swamps from Minn. to Ont. (N to the N shore of L. Superior; reports from Sask. and Man. require confirmation), Que. (N to L. St. John, Anticosti Is., and the Gaspé Pen.; not known from the Côte-Nord), Nfld., N.B., P.E.I., and N.S., s to Ark., Ill., Ohio, and N.C. [*Vaccinium* Ait.].

O. microcarpus Turcz.
/ST/X/EA/ (Ch (evergreen)) Sphagnous bogs from the E Aleutian Is., N Alaska–Yukon (N to ca. 69°N), and the Mackenzie R. Delta to Great Bear L., Great Slave L., L. Athabasca (Alta. and Sask.), S-cent. Dist. Keewatin, northernmost Ont., Que. (N to S Ungava Bay and the Côte-Nord), and Labrador (N to ca. 55°N; not known from the Atlantic Provinces), S to B.C. (S to Queen Charlotte Is. and the mainland to ca. 54°N; *see* B.C. map by Szczawinski 1962:104), Alta. (S to near Coalspur, 53°11′N; CAN), Sask. (S to near Windrum L. at ca. 56°N; CAN), Man. (S to L. Winnipegosis at ca. 53°N), Ont. (S to the W James Bay watershed at 53°10′N), James Bay (Charlton Is., ca. 52°N), Que. (S to the Côte-Nord), and S Labrador; Iceland; Eurasia. [*Vaccinium* Hook. f.; *V. (O.) oxycoccus* of American auth. in part; *O. vulgaris* Pursh in part]. MAPS: Hultén 1968*b*:735; Raup 1947: pl 33.

O. ovalifolius (Michx.) Porsild
/sT/(X)/eA/ (Ch (evergreen)) Sphagnous bogs from ?Alaska (Rydberg 1922; not listed by Hultén 1948 and 1968*b*) and B.C. (Vancouver Is. and adjacent islands and mainland E to the Skagit R.; CAN; not known from Alta.) to Sask. (Breitung 1957*a*; not known from Man.), Ont. (Bruce Pen., L. Huron; Krotkov 1940), Que. (Côte-Nord and Gaspé Pen.; GH), Nfld. (GH), and N.S. (St. Paul Is.; GH; not known from N.B. or P.E.I.), S to Oreg., Minn., Mich., and N.C.; E Asia. [*Vaccinium oxycoccus* vars. *ovalifolius* Michx. and *intermedius* Gray (*O. int.* (Gray) Rydb.); *V. ox.* of Canadian reports in small part, not L.].

The above-noted range is subject to much revision as further studies are made of the distribution of *O. microcarpus, O. ovalifolius,* and *O. quadripetalus,* these included in most N. American reports in the *Vaccinium oxycoccus* complex.

O. quadripetalus Gilib.
/aST/X/GEA/ (Ch (evergreen)) Sphagnous bogs (ranges of Canadian taxa outlined below), S in the U.S.A. to an uncertain limit through general inclusion in the *Vaccinium oxycoccus* complex; W Greenland; Eurasia. MAP and synonymy: *see* below.
1 Leaves to 1 cm long and 5 mm broad, flat or slightly revolute; pedicels 1–4; [*Vaccinium (O.) oxycoccus* L. in part; B.C. (N to ca. 56°N; *see* B.C. map by Szczawinski 1962:106) and SW Dist. Mackenzie to N-cent. Sask. (N to Ile-à-la Crosse), cent. Man. (N to Oxford L. at ca. 55°N), Ont. (N to Big Trout L. at ca. 53°45′N, 90°W), Que. (N to E James Bay at ca. 54°N, L. Chibougamau at ca. 49°45′N, 74°W, and the Gaspé Pen.), N.B., and N.S.; reported from Copper and Bering Is., E Asia; MAP (aggregate species, as *O. palustris*): Hultén 1968*b*:736] .... var. *quadripetalus*
1 Leaves at most about 5 mm long and 3 mm broad, strongly revolute, somewhat 1-sided along the stem; pedicels 1 or 2; [*O. palustris (Vaccinium oxycoccus)* Pers. f. *microphylla* Lange, the type from Greenland; cent. Ont. (Swan L. at ca. 53°45′N, 91°30′W), South Twin Is., James Bay, ca. 53°N, Que. (E James Bay N to SE Hudson Bay at ca. 56°10′N; Ungava Bay; Arntfield, 48°12′N; Taschereau, 48°40′N; Montmorency Falls, near Quebec City; St-Fabien, Rimouski Co.; Côte-Nord; Anticosti Is.; Gaspé Pen.; Magdalen Is.), Labrador (N to Rigolet, 54°25′N), Nfld., N.B., P.E.I., and N.S.; W Greenland N to 64°10′N] ......
.... var. *microphyllus* (Lange) M.P. Porsild

PHYLLODOCE Salisb. [6193] Mountain-Heather

1 Corolla bright yellow, ovoid, glabrous or glandular externally; (Alaska–Yukon–Dist. Mackenzie–B.C.–Alta.) .... *P. aleutica*
1 Corolla roseate or purplish (or blue when dried), glabrous externally.
  2 Corolla companulate, roseate; sepals narrowly to rather broadly ovate, blunt, glabrous except for the minutely ciliate margins; (Alaska–Yukon–Dist. Mackenzie–B.C.– Alta.) .... *P. empetriformis*
  2 Corolla urceolate, purplish; sepals lanceolate, acutish, glandular-puberulent; (transcontinental) .... *P. caerulea*

P. aleutica (Spreng.) Heller Yellow Mountain-Heather
/ST/W/eA/ (Ch (evergreen)) Dry tundra and rocky places up to high elevations (ranges of Canadian taxa outlined below), S to Oreg. and Wyo.; E Asia. MAPS and synonymy: *see* below.

1 Corolla and anther-filaments glabrous; [*Menziesia al.* Spreng., the type from the Aleutian Is.; *Bryanthus* Gray; Aleutian Is. and s Alaska (N to ca. 63°N); E Asia; MAP: Hultén 1968b:723] . . . . ssp. *aleutica*

1 Corolla glandular externally; anther-filaments more or less pubescent toward base; [*Menziesia (Bryanthus; P.) gl.* Hook., the type from "mountains north of the Smoking River, lat. 56°", Alta.; Aleutian Is., s Alaska, sw Yukon, sw Dist. Mackenzie (N to Brintnell L. at ca. 62°N; CAN), and mts. of B.C. (*see* B.C. map by Szczawinski 1962:112) and sw Alta. (N to Jasper; CAN); MAPS: Porsild 1966: map 113 (*P. gl.*), p. 81; Raup 1947: pl. 32 (*P. gl.*); Hultén 1968b:724. A hybrid with *P. empetriformis* (× *P. intermedia* (Hook.) Camp; *Menziesia int.* Hook.) is known from SE B.C. (Skagit R., Field, and L. O'Hara; CAN)] . . . . ssp. *glanduliflora* (Hook.) Hult.

P. caerulea (L.) Bab.   Purple Mountain-Heather
/aST/X/GEA/   (Ch (evergreen))   Rocky tundra and slopes at low to high elevations from Alaska (N to the Seward Pen.; not known from the Yukon) and the coast of Dist. Mackenzie at Coronation Gulf to SE Dist. Keewatin, cent. Baffin Is., and northernmost Ungava–Labrador, s to s Alaska–Dist. Mackenzie–Dist. Keewatin, N Man. (MacLeod L. at ca. 59°11′N; Seal R., w of Churchill; CAN), Que. (s to SE Hudson Bay at ca. 55°10′N, Knob Lake, 54°48′N, the Côte-Nord, and Shickshock Mts. of the Gaspé Pen.; not known from Anticosti Is.), s Labrador, Nfld., N.S. (Victoria Co.; E.C. Smith and J.S. Erskine, Rhodora 56(671):249. 1954; not known from N.B. or P.E.I.), and mts. of Maine and N.H.; w and E Greenland N to ca. 75°N; Iceland; Eurasia. [*Andromeda* L.; *Menziesia* Sw.; *Bryanthus taxifolius* Gray]. MAPS: Hultén 1968b:723, and 1958: map 209, p. 229; Porsild 1957: map 269, p. 194; W.J. Cody, Can. Field-Nat. 67(3): fig. 1, p. 131. 1953; Cain 1944: fig. 19, p. 162.

P. empetriformis (Sw.) Don   Pink Mountain-Heather
/ST/W/   (Ch (evergreen))   Mountains at moderate to high elevations from SE Alaska, the Yukon (N to ca. 65°N), and sw Dist. Mackenzie (Macmillan Pass; Brintnell L.) through the mts. of B.C. (*see* B.C. map by Szczawinski 1962:109) and sw Alta. (Waterton Lakes; Breitung 1957b) to Calif., Idaho, and Mont. [*Menziesia* Sw.; *Bryanthus* Gray; *M. (P.) grahamii* Hook.]. MAP: Hultén 1968b:722.

RHODODENDRON L.   [6184]   Rhododendron, Rosebay.   Rhodora

1 Leaves thick and leathery, evergreen; corolla campanulate, nearly regular.
  2 Stems depressed and mat-forming; leaves elliptic, 1 or 2 cm long, strongly scurfy beneath; clusters few-flowered; corolla bright purple, about 1.5 cm broad; capsules less than 1 cm long, scurfy; (transcontinental in arctic and alpine regions) . . . .
  . . . . *R. lapponicum*
  2 Stems to over 5 m tall; leaves oblong-elliptic to -obovate, to about 2 dm long, glabrous or obscurely scurfy beneath; clusters many-flowered; corolla pale pink to deep rose-purplish, usually at least 2.5 cm broad; capsules 1 or 2 cm long.
    3 Pedicels glabrous; calyx minute, its lobes less than 1 mm long; corolla-lobes crisped-undulate; ovary hairy; leaves oblong-elliptic, broadest near the middle; (s B.C.) . . . . *R. macrophyllum*
    3 Pedicels (and ovary) stipitate-glandular; calyx-lobes to 4 mm long; corolla-lobes entire or somewhat wavy-undulate; leaves oblong-obovate, broadest above the middle . . . . [*R. maximum*]
1 Leaves rather thin, mostly deciduous, more or less pubescent (at least on the veins beneath or when young), elliptic to oblong or obovate.
  4 Calyx-lobes to over 1 cm long, oblong, foliaceous; corolla rotate-campanulate, nearly regular; stamens 10; capsules about 1 cm long.
    5 Corolla white or ochroleucous, glabrous externally, at most 2 cm long, not divided to base on the lower (outer) side; calyx-lobes to 12 mm long; flowers solitary or few together in axillary or lateral clusters; capsules thick-walled; leaves to 9 cm long, acutish, broadest at or only slightly above the middle; shrub to 2 m tall; (B.C.– sw Alta.) . . . . *R. albiflorum*
    5 Corolla rose-purple, pubescent externally, to about 2.5 cm long, divided to base on the lower side; calyx-lobes to about 2 cm long; flowers solitary or in pairs (rarely 3)

terminating leafy shoots; capsules thin-walled; leaves to about 5 cm long, broadest
well above the middle; shrub usually less than 2 dm tall; (Alaska) .... *R. camtschaticum*
4   Calyx-lobes very small or obsolete, at most 1 or 2 mm long; capsules to over 1.5 cm
long.
    6   Corolla pale to deep rose-purple, glabrous externally, to 3 cm long, with a very
       short tube, 2-lipped nearly to base (upper lip 3-lobed; lower lip of 2 oblong-linear,
       nearly distinct, recurving petals); stamens 10, about equalling the style and
       corolla; capsules to 1.5 cm long, glaucous-puberulent; shrub to 1 m tall; (Que.
       to Nfld. and N.S.) .... *R. canadense*
    6   Corolla whitish to bright pink (or drying pinkish blue), pubescent externally, the
       tube to over 2 cm long, the lobes shorter than or about equalling the tube; stamens
       5, much shorter than the style, this to 5 cm long; capsules to 2.5 cm long, setose or
       glandular-setose; shrub to about 3 m tall .... [*R. roseum*]

R. albiflorum Hook. White Rhododendron
/T/W/ (N) Chiefly montane in wet places and along streams from B.C. (N to Fairy L. at ca. 57°N;
see B.C. map by Szczawinski 1962:145) and sw Alta. (N to 53°54'N; CAN) to Oreg. and w Mont.
[*Azalea* Ktze.; *Azaliastrum* Rydb.; *Cladothamnus campanulatus* Greene; incl. f. *poikilon* Henry, the
three anterior petals dotted with orange or yellow rather than uniformly white, the type from B.C.].

R. camtschaticum Pallas Kamchatka Rhododendron
/Ss/W/eA/ (N) Alpine meadows (occasionally in subalpine woods) of the Aleutian Is. and
Alaska (N to the Seward Pen.); NE Asia. MAPS and synonymy: *see* below.
1   Petals ciliate, pubescent on the back; leaves of the sterile basal shoots not glandular-
ciliate (those of the flowering shoots often so); [*Therorhodion* Small; MAPS: Hultén
1968b:719, and 1948: map 917a, p. 1334] .... ssp. *camtschaticum*
1   Petals not ciliate, glabrous on the back; leaves of the sterile shoots glandular-ciliate;
[*Ther. gl.* Standl., the type from Imuruk Bay, Alaska; MAPS: Hultén 1968b:719, and 1948:
map 917b, p. 1334] .... ssp. *glandulosum* (Standl.) Hult.

R. canadense (L.) Torr. Rhodora
/T/E/ (N) Bogs, damp thickets, and acidic barrens and rocky slopes from Que. (N to the
Côte-Nord, Anticosti Is., and Gaspé Pen.), Nfld., N.B., P.E.I., and N.S. to NE Pa. and N N.J.
[*Rhodora* L.; *Rhododendron rhodora* Don; *Azalea (Rhododendron) ?viscosa sensu* Pursh 1814,
as to the Canadian plant, and McSwain and Bain 1891, not L.; *A. (Rhododendron) ?nudiflorum (A.
periclymenoides* Michx.) of Canadian reports, not L.].

    The report from Hebron, Labrador, 58°12'N, by A.P. Coleman (Geol. Surv. Can. Mem. 124:55.
1921) is undoubtedly erroneous. Forma *albiflora* (Rand & Redf.) Rehd. (flowers white rather than
pale to deep rose-purple) is known from Que. (Boivin 1966b) and N.S. (Lindsay 1878). Forma
viridifolium Fern. (leaves dark green and shining above, barely pilose beneath, rather than
grey-green or glaucous and more or less pilose on both surfaces) is known from the type locality,
Arcadia, Yarmouth Co., N.S.

R. lapponicum (L.) Wahl. Lapland Rosebay
/aST/X/GEA/ (Ch (evergreen)) Rocky barrens and subalpine woods from the coasts of
Alaska–Yukon–Dist. Mackenzie–Dist. Keewatin to Banks Is., E Devon Is., Baffin Is., and
northernmost Ungava–Labrador, S to N B.C. (S to ca. 58°45'N; *see* B.C. map by Szczawinski
1962:148), Great Slave L., N Man. (S to Churchill; the report from Gillam, about 165 mi S of
Churchill, by Lowe 1943, requires confirmation), Ont. (coasts of Hudson Bay–James Bay S to
54°22'N), Que. (S to S James Bay, L. Mistassini, and the serpentine plateau of Mt. Albert, Gaspé
Pen.), Nfld., and the mts. of N.Y., Maine, and N.H.; isolated in the mts. of sw Alta. (Opabin Creek,
52°15'N, near Banff; CAN); w Greenland N to ca. 79°N, E Greenland N to ca. 75°N; Scandinavia; N
Asia. [*Azalea* L.]. MAPS: Hultén 1968b:718, and 1958: map 181, p. 201; Porsild 1957: map 270,
p. 194; Gjaerevoll 1963: fig. 2, p. 263.

R. macrophyllum Don California Rhododendron
/t/W/ (Mc (evergreen)) Rocky places from sw B.C. (Vancouver Is.; Skagit R.; Manning

Provincial Park and vicinity, about 30 mi SE of Hope) to N Calif. [*R. californicum* Hook.]. MAP: the Canadian area is shown in the B.C. map by Szczawinski 1962:150.

[The similar *R. chrysanthum* Pall. (*R. aureum* Georgi) of E Asia has been reported from Sitka, Alaska (Hultén 1948), this requiring confirmation. However, there are collections in CAN from the Commander Is. and Bering Is., U.S.S.R., the western continuation of the Aleutian Is., lending credence to the Sitka report.]

[R. maximum L.]   Great Laurel, Rosebay
[This species of the E U.S.A. (swampy ground and damp woods from Ohio to N.Y., New Eng., and Va.) is accredited to S Ont. by Fernald *in* Gray (1950; perhaps on the basis of its tentative report from Norfolk Co. by John Macoun 1884), who also cites N.S. as a former locality. Neither Soper (1949) nor Landon (1960) list it for S Ont. However, it is reported from Sheet Harbour, Halifax Co., N.S., by Lindsay (1878) and George Lawson (Proc. & Trans. N.S. Inst. Sci. 4:172–78. 1878), where believed by Lawson to be native. There is a collection in DAO from Highlands National Park, South Ingonish, Cape Breton Is., where perhaps introd. According to H.W. Vogelmann and L.A. Charette (Rhodora 65(761):22. 1963), it has been found in N Vt. within 8 miles of the Que. boundary. MAPS (the S Ont. area indicated on the last three should apparently be deleted): H.H. Iltis, Castanea 21:118. 1956; Preston 1961:334; Hough 1947:369; Munns 1938: map 160, p. 164.]

[R. roseum (Loisel.) Rehd.]   Early Azalea, Election-pink
[The inclusion of SW Que. in the range of this species of the E U.S.A. (Maine to Mo., Tenn., and Va.) by Fernald *in* Gray (1950) requires confirmation. It is not listed by Marie-Victorin (1935). (*Azalea* Loisel.).]

VACCINIUM L.   [6216]   Blueberry, Bilberry, Huckleberry, Cranberry.   Airelle or Atocas

1   Anthers awnless.
  2   Leaves leathery and evergreen, glossy green above, paler beneath; corolla pink or reddish.
    3   Leaves elliptic to obovate-oblong, rounded at both ends, entire, subsessile, less than 2 cm long, sparsely dotted with erect black glands beneath; flowers few in small terminal clusters; calyx-lobes glandular-ciliate; corolla 4-lobed nearly to the middle; berries red; stems slender and creeping; (transcontinental) . . . . *V. vitis-idaea*
    3   Leaves narrowly to broadly ovate, acute, short-petioled, their thickened margins sharply serrulate; flowers in axillary clusters; calyx-lobes glabrous; corolla shallowly 5-lobed; berries deep purplish-black and glaucous, or blackish and shining; shrub to 4 m tall; (S B.C.) . . . . *V. ovatum*
  2   Leaves thinner and deciduous; flowers in dense terminal racemes; corolla with 5 very short lobes; berries blue or black; branching shrubs.
    4   Leaves entire (*V. angustifolium* var. *integrifolium* and typical *V. corymbosum* will key out here).
      5   Leaves at most about 4 cm long, elliptic-lanceolate to oval; stems less than 1 m tall; berries blue, glaucous.
        6   Branchlets and lower leaf-surfaces copiously velvety-pilose; leaves at most about 3 cm long and half as broad; corolla at most 6 mm long, white or pink-tinged; (transcontinental) . . . . *V. myrtilloides*
        6   Branchlets and leaves glabrous (pubescent in var. *crinitum*), the leaves glaucous beneath, to 5 cm long and 1/2–2/3 as broad; corolla to 8 mm long, greenish or purplish; (S Ont. and W N.S.) . . . . *V. vacillans*
      5   Leaves to over 7 cm long; stems usually over 1 m tall; berries blue to black.
        7   Leaves heavily downy or woolly beneath, unexpanded at anthesis, to 6 cm long and 3 cm broad; corolla yellowish or greenish-white tinged with purple, to 8 mm long; berries shining black, without bloom; (S Ont. and S Que.) . . . .
            . . . . *V. atrococcum*
        7   Leaves glabrous or sparingly pubescent beneath, half-grown at anthesis, to 8 cm long and 4 cm broad; corolla white or pinkish, to 12 mm long; berries blue to blue-black, with a bloom; (Ont. to N.S.) . . . . *V. corymbosum*

    4   Leaves normally serrulate, spinulose-serrulate, or ciliate-serrulate (entire only in *V. angustifolium* var. *integrifolium* and typical *V. corymbosum*).

        8   Leaves of fertile branches to 8 cm long and 4 cm broad, essentially glabrous; stems to 4 m tall; corolla to 12 mm long, white or pinkish; berries to 12 mm thick; (Ont to N.S.) . . . . *V. corymbosum*

        8   Leaves of fertile branches at most about 5 cm long; stems less than 1 m tall.

            9   Leaves to 5 cm long and 3.5 cm broad; stem to 9 dm tall; corolla greenish or purplish, to 8 mm long; berries dark blue, with bloom; (s Ont. and w N.S.) . . . . *V. vacillans*

            9   Leaves to about 3.5 cm long and 1.5 cm broad; stem to about 6 dm tall; corolla white or pink-tinged; (Man. to Labrador, Nfld., and N.S.) . . . .
. . . . *V. angustifolium*

1   Anthers bearing a pair of long horn-like awns on the back at the base of the terminal tubules; leaves deciduous.

  10 Flowers pendulous on filiform leafy-bracted jointed pedicels in loose racemes or small panicles on specialized branches (bracts much smaller than the foliage-leaves); corolla open-campanulate, greenish or purplish, 5-lobed nearly to middle; anthers and style long-exserted; branchlets and at least the lower surfaces of the entire-margined leaves more or less pubescent; berry commonly yellowish or greenish; (s Ont.) . . . .
. . . . *V. stamineum*

  10 Flowers solitary in the leaf-axils or in clusters of up to 4 from scaly buds; corolla urceolate to subglobose, the lobes very short; stamens and style included or the style slightly exserted.

    11 Flowers in clusters of up to 4 from scaly buds; calyx-lobes triangular-ovate, as long as or longer than broad, persistent on the mature berry; corolla pinkish, or white with pink lobes, to about 6 mm long; berries glaucous-blue; leaves oblanceolate to obovate, entire, typically firm and strongly reticulate-veiny beneath, to 3 cm long; twigs not angled; (transcontinental) . . . . *V. uliginosum*

    11 Flowers solitary in the leaf-axils, normally 5-merous; calyx-lobes usually rudimentary and deciduous (or nearly as long as broad in *V. parvifolium*), the calyx-tube forming a crowning ring on the berry; leaves relatively thin; twigs often angled.

      12 Leaves entire or only obscurely serrulate below the middle, glabrous or somewhat puberulent (especially when young).

        13 Berry bright red, to 9 mm thick; corolla pale, waxy, yellowish pink, about 4 mm long; leaves oblong-elliptic to oval, to 2.5 cm long, entire (sometimes serrulate on juvenile growth); twigs green, sharply angled; (SE Alaska–B.C.) . . . . *V. parvifolium*

        13 Berry bluish- or purplish-black, to 1 cm thick; corolla about 7 mm long; leaves ovate-elliptic to elliptic-obovate, to over 4 cm long; young twigs yellowish green, more or less angled.

          14 Fruiting pedicels often over 1 cm long, nearly straight, somewhat enlarged immediately below the ovary; corolla bronzy-pink, usually broader than long; style slightly exserted; leaves usually puberulent beneath and sparsely glandular-hirsute on the midvein beneath, the lateral veins not prominent; (Alaska–B.C.) . . . . *V. alaskaense*

          14 Fruiting pedicels mostly much less than 1 cm long, rather strongly curved but not enlarged immediately below the ovary; corolla pink, usually longer than broad; style usually included; leaves glabrous, not glandular along the midvein, the lateral veins prominent; (transcontinental) . . . . *V. ovalifolium*

      12 Leaves distinctly serrulate at least above the middle.

        15 Berry bright (but not deep) red, to 5 mm thick; corolla pinkish, about 4 mm long; pedicels less than 3 mm long; leaves usually less than 1.5 cm long, narrowly to broadly lanceolate; twigs bright green or yellow-green, sharply angled, very numerous; plant usually more or less matted, to about 2.5 dm tall; (B.C. and Alta.) . . . . *V. scoparium*

        15 Berry blue to deep bluish- or purplish-black (sometimes dark red in *V.*

myrtillus), to 8 or 9 mm thick; corolla commonly over 5 mm long; leaves commonly at least 2 cm long; plants often taller.

    16 Leaves oblanceolate to obovate (broadest above the middle; *V. membranaceum* var. *rigidum* may key out here), cuneate-based, more or less serrulate along the upper half (the margins of the lower half entire or indistinctly serrulate); corolla whitish or pink, to 6 or 7 mm long; berries glaucous-blue; twigs inconspicuously angled or terete.

        17 Corolla narrowly urn-shaped, twice as long as thick; filaments longer than the anthers (including the tubules); leaves oblanceolate, averaging not over about 2 cm long, strongly reticulate beneath, not glaucous; (transcontinental) . . . . *V. caespitosum*

        17 Corolla subglobose, less than twice as long as thick; filaments shorter than the anthers; leaves relatively broader, averaging about 3 cm long, glaucous beneath; (sw B.C.) . . . . *V. deliciosum*

    16 Leaves elliptic-lanceolate to elliptic or ovate (commonly broadest below the middle), their margins serrulate nearly throughout or at least along the lower half.

        18 Pedicels 2 or 3 mm long; corolla pinkish; berries dark red to bluish; leaves mostly less than 3 cm long; twigs greenish and strongly angled, usually puberulent; (B.C. and Alta.) . . . . *V. myrtillus*

        18 Pedicels to over 5 mm long; leaves often longer; twigs not strongly angled, glabrous or slightly puberulent.

            19 Branches terete; leaves elliptic, to about 3.5 cm long, slightly lustrous, strongly veiny beneath; berry blue-black; pedicels to 5 mm long; (E Que. and Nfld) . . . . *V. nubigenum*

            19 Branches 4-angled; leaves ovate, to 7 cm long, bright green; berry purplish to black; pedicels to 12 mm long; (s Dist. Mackenzie, B.C., and Alta.; Ont.) . . . . *V. membranaceum*

V. alaskaense Howell

/sT/W/ (N) Coastal woods from s Alaska (*see* Hultén 1948: map 938, p. 1336) through w B.C. (*see* B.C. map by Szczawinski 1962:156; chiefly coastal but isolated stations inland to ca. 126°W) to NW Oreg. [Alternative spellings: *alaskaensis* and *alaskensis*; *V. ovalifolium* Bong., not Sm.; *V. oblatum* Henry]. MAP: Hultén 1968b:734.

V. angustifolium Ait.   Low or Late Sweet Blueberry.   Bluets

/ST/X/ (N (Ch)) Peat bogs and open barrens, the aggregate species from Man. (N to Bear L. at ca. 55°N, 97°W; WIN) to Ont. (N to the N shore of L. Superior), Que. (N to the Larch R. at 57°35′N and the Côte-Nord), Labrador (N to Attikamagan L. at ca. 55°N; reported N to Nain, 56°32′N), Nfld., N.B., P.E.I., and N.S., s to Iowa, Ohio, and Va.

1 Leaves entire; [Que. (N to the Larch R. at ca. 57°N; type from Fort George, James Bay) and w Labrador] . . . . var. *integrifolium* Lepage

1 Leaves serrulate.

    2 Leaves spreading-pubescent beneath, especially along the nerves; berries blue, with a bloom; [Que. to Nfld. and N.S.] . . . . var. *hypolasium* Fern.

    2 Leaves glabrous or merely minutely pilose beneath at base of midrib.

        3 Leaves blue-green, glaucous beneath; berries usually black and without a bloom; [*V. pensilvanicum* var. *nigrum* Wood; *V. nigrum* (Wood) Britt.; *V. brittonii* Porter; SE Man. to Nfld. and N.S.] . . . . var. *nigrum* (Wood) Dole

        3 Leaves bright green on both sides; berries blue, with bloom.

            4 Leaves lanceolate to oblong, to 3 or 4 cm long; corolla to 1 cm long; [*V. (Cyanococcus) pen.* Lam.; *V. lamarckii* Camp; Man. (reports from Sask. require confirmation) to Labrador, Nfld., and N.S.] . . . . var. *laevifolium* House

            4 Leaves narrowly lanceolate, mostly less than 2 cm long; corolla 5 or 6 mm long; [*Cyanococcus* Rydb.; *V. pen.* var. *angust.* (Ait.) Gray; incl. *V. boreale* Hall & Aalders; Ont. to Labrador, Nfld., and N.S.; type from Nfld.–Labrador] . . . .
        . . . . var. *angustifolium*

V. atrococcum (Gray) Heller   Black Highbush-Blueberry
/T/EE/   (N)   Swamps, moist woods, and barrens from Ind. to Ont. (N to the Mer Bleue, near Ottawa), SW Que. (N to the Montreal dist.; MT), and New Eng., S to Ark. and N Fla. [*V. corymbosum* var. *atro.* Gray].

V. caespitosum Michx.   Dwarf Bilberry or Huckleberry
/ST/X/   (Ch)   Moist tundra, gravelly or rocky shores, woods, and clearings at low to high elevations from cent. Alaska–Yukon (N to ca. 64°30'N) and SW Dist. Mackenzie to B.C.–Alta., Sask. (N to Medstead, 53°19'N; Breitung 1957a), Man. (N to Duck Mt.; CAN), Ont. (N to the W James Bay watershed at 51°34'N), Que. (N to Ungava Bay, the Côte-Nord, Anticosti Is., and Gaspé Pen.), Labrador (N to Anatolik, 56°39'N), Nfld., N.B. (upper St. John R. system), and N.S. (Kings, Victoria, and Inverness counties; not known from P.E.I.), S to Calif., Colo., Minn., and New Eng. [Incl. vars. *angustifolium* Gray and *cuneifolium* Nutt. and *V. paludicola* Camp]. MAPS: Hultén 1968b:732; Porsild 1966: map 114, p. 81.

V. corymbosum L.   Highbush-Blueberry
/T/EE/   (N (Mc))   Swamps and moist woods from Wisc. to Ont. (N to the Ottawa dist.; *see* S Ont. map by Soper and Heimburger 1961:96), Que. (N to near Quebec City; Raymond 1950b), N.B. (Tower Hill, Charlotte Co.; NSPM; the report from Grand Lake, Queens Co., by Fowler 1885, is based upon *Gaylussacia baccata,* the relevant collection in NBM; not known from P.E.I.), and N.S., S to Ind., Ohio, Pa., and N.J.
1   Leaves entire, green on both sides; [range of the species] .... f. *corymbosum*
1   Leaves finely serrulate or ciliate-serrulate; [S Ont., SW Que., and N.S.].
    2   Leaves green on both sides; [var. *amoenum sensu* Gray, not *V. am.* Ait., basionym; *V. albiflorum* Hook.] .... f. *albiflorum* (Hook.) Camp
    2   Leaves glaucous beneath; [var. *pallidum sensu* Gray, not *V. pal.* Ait., basionym] ....
    .... f. *glabrum* (Gray) Camp

V. deliciosum Piper   Blue Huckleberry, Cascade Bilberry
/T/W/   (Ch)   Montane forests and slopes at moderate to high elevations from SW B.C. (Vancouver Is. and adjacent mainland E to Manning Provincial Park, about 30 mi SE of Hope; *see* B.C. map by Szczawinski 1962:161) to N Oreg.

V. membranaceum Dougl.   Mountain-Huckleberry
/sT/D/   (N)   Thickets and montane slopes (ranges of Canadian taxa outlined below), S to N Calif., Idaho, and Wyo. MAPS and synonymy: *see* below.
1   Leaves mostly ovate or ovate-oblong, gradually acute to acuminate at apex, to 5 cm long; corolla longer than broad; [*V. myrtilloides* var. *macrophyllum* Hook.; *V. mac.* (Hook.) Piper; S Dist. Mackenzie, B.C. (*see* B.C. map by Szczawinski 1962:166), and Alta. (N to ca. 55°N); Ont. (Bruce Pen., L. Huron; a collection in GH from Little Pigeon R. near Thunder Bay may also belong here); MAPS: Fernald 1935: map 7, p. 211; Hultén 1968b:732 (aggregate species)] .... var. *membranaceum*
1   Leaves oblong-obovate, abruptly acute to rounded at apex, usually somewhat shorter; corolla as broad as long; [*V. myrtilloides* var. *rig.* Hook.; *V. globulare* Rydb.; SE B.C. (N to ca. 52°N; *see* B.C. map for *V. globulare* by Szczawinski 1962:163) and Alta. (Waterton Lakes; Silver City; Lesser Slave L. at ca. 55°20'N); MAP: on the above-noted map by Fernald] .... var. *rigidum* (Hook.) Fern.

V. myrtilloides Michx.   Sour-top- or Velvet-leaf-Blueberry.   Bluets
/sT/X/   (N (Ch))   Peat bogs, moist woods, and clearings from S Dist. Mackenzie (near Fort Smith, ca. 60°N; W.J. Cody, Can. Field-Nat. 70(3):121. 1956; reported N to Great Bear L. by Hooker 1834) and B.C.–Alta. to Sask. (N to L. Athabasca), Man. (N to Nueltin L. at 59°43'N), Ont. (N to the Fawn R. at ca. 54°N, 88°W), Que. (N to the E James Bay watershed at 53°45'N, the Côte-Nord, Anticosti Is., and Gaspé Pen.; type probably from near L. Mistassini), Labrador (N to Rigolet, 54°11'N; CAN; GH), Nfld., N.B., P.E.I., and N.S., S to Mont., Iowa, and Va. [*V. angustifolium* var. *myrt.* (Michx.) House; *V. (Cyanococcus) canadense* Kalm].

Forma *chiococcum* (Deane) Fern. (berries whitish rather than blue) is tentatively reported from N.B. by Boivin (1966*b*; as *V. angustifolium* var. *myrt.* f. *chio.*).

V. myrtillus L.   Dwarf Bilberry
/aST/W/GEA/   (N (Mc))   Moist slopes at moderate to high elevations from SE B.C. (Kootenay and Columbia valleys N to ca. 51°30′N; see B.C. map by Szczawinski 1962:171) and SW Alta. (N to Banff; CAN) to Wash. and N.Mex.; SW Greenland; Eurasia. [*V. oreophilum* Rydb.].

V. nubigenum Fern.
/T/E/   (N)   Rocky or peaty slopes and barrens of E Que. (Shickshock Mts. of the Gaspé Pen.; type from Tabletop Mt.) and N Nfld. [Perhaps a hybrid between *V. membranaceum* and some other species]. MAP: Fernald 1933: map 23, p. 279.

V. ovalifolium Sm.   Tall Huckleberry
/sT/(X)/eA/   (N)   Peats, thickets, and open woods from the Aleutian Is., S Alaska (N to ca. 62°30′N), and southernmost Yukon (Bennett L.) through B.C. (W to near the SW Alta. boundary but not yet reported from Alta.) to Oreg. and Mont., with isolated areas on L. Superior (Ont. and Mich.) and from Que. (Laurentide Provincial Park in Charlevoix Co.; Shickshock Mts. of the Gaspé Pen.) to SE Labrador (N to Chateau, ca. 52°N), Nfld., and N.S. (Victoria Co., Cape Breton Is.; ACAD; CAN; MT); E Asia. [*V. chamissonis* Bong.]. MAPS: Hultén 1968*b*:733; Raymond 1950*b*: fig. 4, p. 12; Wynne-Edwards 1937: map 2, p. 24; Fernald 1925: map 10, p. 253.
    A hybrid with *V. parvifolium* is reported from SW B.C. by J.M. Macoun (1913; Strathcona Park, Vancouver Is.; CAN).

V. ovatum Pursh   Evergreen Huckleberry
/t/W/   (N (Mc; evergreen))   Woods and rocky slopes near the coast from SW B.C. (Prince Rupert dist. at ca. 54°N; Vancouver Is. and adjacent islands and mainland; see B.C. map by Szczawinski 1962:179) to S Calif. [*Metagonia* Nutt.].

V. parvifolium Sm.   Red Bilberry or Huckleberry
/T/W/   (N (Mc))   Woods and slopes from SE Alaska (see Hultén 1948: map 941, p. 1336) through B.C. (chiefly coastal but an isolated station in the Kootenay Valley; see B.C. map by Szczawinski 1962:181) to cent. Calif. MAP: Hultén 1968*b*:733.

V. scoparium Leiberg   Grouseberry, Whortleberry
/T/WW/   (N)   Mountains usually at rather high elevations from SE B.C. (Lillooet and Manning Provincial Park eastwards; N to ca. 52°30′N; see B.C. map by Szczawinski 1962:184) and SW Alta. (N to Jasper; CAN) to N Calif., Colo., and S.Dak. [*V. erythrocarpum* Rydb.].

V. stamineum L.   Deerberry, Squaw-Huckleberry
/t/EE/   (N)   Dry woods, thickets, and clearings from Mo. to Ohio, S Ont. (Welland, Lincoln, and Leeds counties; see S Ont. map by Soper and Heimburger 1961:92), N.Y., and Mass., S to La. and Fla. [*Polycodium* Greene; *V. album sensu* Pursh 1814, not L., which is *Symphoricarpos albus*].

V. uliginosum L.   Alpine Bilberry, Bog-Blueberry
/AST/X/GEA/   (Ch (N))   Peat bogs and rocky barrens and tundra at low to high elevations, the aggregate species from the Aleutian Is. and coasts of Alaska–Yukon–Dist. Mackenzie–Dist. Keewatin to Banks Is., Baffin Is., Ellesmere Is. (N to ca. 80°N), northernmost Ungava–Labrador, Nfld., N.B. (Bald Mt., Nipisiquit R. in ?Gloucester Co.; NBM), P.E.I. (Waterford, Prince Co.; MT), and N.S. (Cape Breton Is. and St. Paul Is.), S along the Pacific coast through B.C. to N Calif., farther eastwards S to L. Athabasca (Alta. and Sask.), N Man. (S to Gillam, about 165 mi S of Churchill), Ont. (coasts of Hudson Bay–James Bay; N shore of L. Superior), N Mich., N.Y., and Vt.; nearly circumgreenlandic; Iceland; Spitsbergen; N Eurasia. MAPS and synonymy: see below (the treatment based upon that by S.B. Young, Rhodora 72(792):439–59. 1970):
1   Leaves pubescent; shrub diffuse, spreading and stoloniferous; [*V. pubescens* Wormsk.; *V. (Myrtillus) ulig.* var. *pub.* (Wormsk.) Hornem.; *V. ulig.* vars. *alpinum* Bigel. and *lange-anum* Malte; essentially the range of the species; MAPS: Young, loc. cit., fig. 2, p. 449;

Raup 1947: pl. 33 (var. *alp.*); Hultén 1968*b*:734 (ssp. *alp.*); Porsild 1957: map 273 (var.
alp.), p. 195] . . . . ssp. *pubescens* (Wormsk.) Young
1 Leaves glabrous or only sparsely puberulent.
  2 Shrub dwarf, prostrate or matted; leaves mostly less than 1 cm long, the dead ones
    often persistent for several years; [*V. gaultherioides* Bigel., in part; *V. ulig.*
    microphyllum Lange; *V. microphyllum* (Lange) Hagerup; N part of the N. American-
    European range; MAPS: Young, loc. cit., fig. 3, p. 451; Hultén 1968*b*:735 (ssp. *micro.*);
    A. Löve 1950: fig. 15 (*V. micro.*), p. 48] . . . . ssp. *gaultherioides* (Bigel.) Young
  2 Shrubs usually taller and more robust.
    3 Stomata present on both leaf-surfaces; [w N. America] . . . . ssp. *occidentale*
      4 Plant of low, creeping habit; leaves mucronate or cuspidate at apex; [*V.*
        salicinum Cham.; Aleutian Is. and coast of s Alaska; MAP: Young, loc. cit.,
        fig. 6 (triangles), p. 454] . . . . var. *salicinum* (Cham.) Hultén
      4 Plant usually erect; leaves merely mucronulate at the obtuse or rounded apex;
        [*V. occidentale* Gray; SE Alaska through coastal B.C. to Calif., Ariz., and
        ?N.Mex.; MAP: Young, loc. cit., fig. 6 (dots), p. 454] . . . . var. *occidentale* (Gray) Hara
    3 Stomata present only on the lower leaf-surfaces.
      5 Fruit variable in shape (often subcylindrical), highly palatable; [Alaska, NW
        Dist. Mackenzie, and E Asia; MAP: Young, loc. cit., fig. 4, p. 452] . . . .
        . . . . ssp. *pedris* (Harshberger) Young
      5 Fruit spherical, relatively insipid and tasteless; [SW Greenland; Iceland; NW
        Europe; E ?N. America; MAP: Young, loc. cit., fig. 5, p. 453; the map by Raup
        1947: pl. 33, refers chiefly to the above taxa] . . . . ssp. *uliginosum*

V. vacillans Torr. Low Blueberry, Sugar-Huckleberry
/T/EE/ (N) Dry open woods, thickets, and clearings from Iowa to Ohio, s Ont. (Essex, Norfolk,
Lincoln, Welland, Wentworth, and York counties; *see* s Ont. map by Soper and Heimburger
1961:94; also reported from London, Middlesex Co., by John Macoun 1884, and a collection in TRT
from Wellington Co. has been placed here), N.Y., N.S. (Butler's L., Yarmouth Co.; Fernald 1921;
GH; CAN), and Maine, s to Mo. and Ga. [*V. torreyanum* Camp].

Collections in CAN and GH from near Leamington, Essex Co., s Ont., are referable to var.
crinitum Fern. (young branchlets and lower leaf-surfaces pubescent rather than the plant essentially
glabrous).

V. vitis-idaea L. Rock-Cranberry, Mountain-Cranberry. Berris, Graines rouges, or Pommes de
terre
/aST/X/GEA/ (Ch (evergreen)) Rocky tundra and peaty or rocky barrens from the Aleutian Is.
and coasts of Alaska–Yukon–Dist. Mackenzie–Dist. Keewatin to Victoria Is., Baffin Is. (N to near the
Arctic Circle), and northernmost Ungava–Labrador, s to B.C. (s to ca. 52°15′N; *see* B.C. map by
Szczawinski 1962:189), Alta. (s to the Elbow R. and Banff; CAN), Sask. (s to McKague, 52°37′N;
CAN), s Man., N Minn., Ont. (s to the N shore of L. Superior), Que. (s to near Quebec City), Nfld.,
N.B., P.E.I., N.S., and New Eng.; w Greenland N to ca. 78°N; Iceland; N Eurasia. [*Vitis-idaea* Britt.;
incl. var. *minus* Lodd. and its f. *pyricarpum* Lepage]. MAPS: Eric Hultén 1968*b*:731, and Sven. Bot.
Tidskr. 43(2–3): fig. 5, p. 397. 1949; Porsild 1957: map 274, p. 195; Raup 1947: pl. 33; Meusel
1943: fig. 20c.

DIAPENSIACEAE (Diapensia Family)

DIAPENSIA L.   [6273]

Dwarf matted evergreen shrub with crowded, cartilaginous, narrowly spatulate, entire, mostly opposite leaves less than 1.5 cm long. Flowers white, regular, perfect, gamopetalous, 5-merous, hypogynous, about 1 cm long, solitary on terminal peduncles. Stamens 5, alternating with the 5 corolla-lobes, their filaments adnate to the corolla-tube nearly as far as the sinuses. Style 1. Ovary superior. Fruit a subglobose capsule about 5 mm long.

D. lapponica L.
/AST/EE/GEA/   (Ch (evergreen))   Tundra, rocks, and gravels at low to high elevations (ranges of Canadian taxa outlined below), s in the East to the mts. of New Eng.; Greenland; Iceland; N Eurasia. MAPS and synonymy: *see* below.

1   Leaves narrowly spatulate, to 1.5 cm long, rather weakly reticulate, merging into obscure petioles; [coast of Dist. Mackenzie at Coronation Gulf to Baffin Is. at ca. 70°N (an isolated station on Ellesmere Is. at ca. 80°10′N) and northernmost Ungava–Labrador, s to SE Dist. Mackenzie, s Dist. Keewatin, N Man. (Baralzon L. at 60°N; not known from Sask. or Ont.), Que. (s to E James Bay at Cape Jones, 54°37′N, Knob Lake, 54°48′N, the Côte-Nord, and the Shickshock Mts. of the Gaspé Pen.), s Labrador, Nfld., and N.S. (Upper Salmon R., Victoria Co., Cape Breton Is.); w Greenland N to ca. 76°N, E Greenland N to ca. 71°N; Iceland; N Europe; NW Siberia; MAPS: Hultén 1968b:736, and 1958: map 204, p. 223; Porsild 1957: map 275, p. 195, and 1955: fig. 16, p. 51; *Atlas of Canada* 1957: map 5, sheet 38] . . . . ssp. *lapponica*

1   Leaves mostly obovate, generally shorter, more strongly reticulate, and with a deeper furrow along the median vein than in the typical form . . . . ssp. *obovata* (Schm.) Hult.

   2   Corolla white; [*D. obovata* (Schm.) Nakai; Alaska–Yukon (*see* Hultén 1948: map 944, p. 1336) and NW Dist. Mackenzie (Richardson Mts. w of the Mackenzie R. Delta; CAN); E Asia; MAPS: on the above-noted maps by Hultén] . . . . var. *obovata*

   2   Corolla roseate; [type from Hatcher Pass in the Talkeetna Mts. N of Anchorage, Alaska] . . . . var. *rosea* Hult.

PRIMULACEAE (Primrose Family)

Herbs with simple, entire or shallowly toothed leaves, these variously disposed on the stem or all basal. Flowers regular, perfect, hypogynous (somewhat epigynous in *Samolus*), gamopetalous. Calyx-lobes, corolla-lobes, and stamens commonly 5, sometimes 3, 4, or 6 (rarely 8; the corolla wanting in *Glaux*). Style and stigma each 1. Ovary superior. Fruit a longitudinally dehiscent or circumscissile capsule.

1 Caespitose, matted to cushion-forming perennials at most 1.5 dm tall; stems prostrate or ascending, dichotomously branched, ending in rosettes of small, entire, minutely scabrous-ciliolate or shallowly few-toothed leaves, the leaves persisting after withering; flowers rose-pink to bright red or magenta-purple, solitary or in umbels terminating leafless peduncles; (western species) . . . . *Douglasia*
1 Plants not caespitose or, if so, the corollas white or the flowering-stems leafy.
  2 Plant aquatic, the submersed leaves dissected into linear segments, scattered on the floating and rooting hollow stems and crowded at the base of the cluster of peduncles; peduncles bearing 3–10 small whitish flowers at the nodes, the internodes strongly inflated . . . . *[Hottonia]*
  2 Plants terrestrial; leaves entire or merely shallowly toothed.
    3 Plants scapose, the leaves in a basal rosette; flowers in terminal umbels.
      4 Corolla deep purple to rose or almost white, its lobes to 2 cm long, reflexed from the base; anthers forming an exserted cone, the very short filaments free or more or less united into a short tube; leaves linear-oblanceolate to obovate, entire, glabrous; (B.C. to S Man.) . . . . *Dodecatheon*
      4 Corolla with spreading or ascending lobes; stamens distinct, included; leaves entire or very shallowly toothed.
        5 Corolla white or creamy, constricted at the throat, its tube shorter than the calyx; style very short; leaves subsessile . . . . *Androsace*
        5 Corolla usually lilac or bluish purple, open at the throat, its tube equalling or surpassing the calyx; style filiform, elongate; leaves wing-petioled; perennials . . . . *Primula*
    3 Plants with leafy stems; flowers axillary or racemose.
      6 Stem with scale-like alternate leaves below and a single terminal whorl of large, lanceolate, entire or minutely crenulate ones; corolla white, mostly 7-parted; flowers slender-peduncled in the axils of the upper leaves . . . . . . *Trientalis*
      6 Stem leafy throughout, the leaves entire; corolla 4–6-parted (wanting in *Glaux*, with a petaloid 5-parted calyx).
        7 Leaves alternate, oblong or obovate.
          8 Flowers pink, mostly 4-merous, about 1 mm broad, nearly sessile in the axils of the subsessile leaves, these at most about 1 cm long; capsule free from the calyx-tube, subglobose, about 2 mm thick, circumscissile, often bearing the withered corolla on the deciduous top; low annual commonly less than 1 dm tall; (B.C. to Sask.; ?Man.; N.S.) . . . . *Centunculus*
          8 Flowers white, 5-merous, 2 or 3 mm broad, on slender pedicels 1 or 2 cm long in lax terminal racemes; leaves to over 5 cm long, the basal ones in a rosette; capsule fused below to the calyx-tube, globose, to about 5 mm long, dehiscent near the summit by 5 valves; perennial to about 6 dm tall (?B.C.; Ont. to N.S.) . . . . *Samolus*
        7 Leaves opposite or whorled.
          9 Plant fleshy; leaves opposite, to 1.5 cm long, bearing solitary, subsessile, white to crimson flowers about 4 mm long in their axils; petals none, the 5 calyx-lobes petaloid; capsule 5-valved; (transcontinental in saline and alkaline habitats) . . . . *Glaux*
          9 Plants scarcely fleshy, chiefly of fresh habitats; petals present.
            10 Flowers usually scarlet to brick-red, varying to white, 5-merous,

solitary on long slender pedicels in the leaf-axils; sepals 3 or 4 mm long, the petals about the same length or somewhat longer; capsule subglobose, circumscissile near the middle; leaves elliptic to ovate, sessile, opposite, 1 or 2 cm long; annual with 4-angled stems to about 3 dm tall; (introd.) . . . . *Anagallis*

10 Flowers yellow, larger; capsule globose or ovoid, dehiscent by longitudinal valves.

    11 Corolla-lobes entire (or merely minutely glandular-ciliate in *L. punctata*), neither cuspidate nor rolled about their stamens, often dark-dotted or -streaked; anthers oblong or oval, all fertile, their filaments commonly united at base; flowers on ascending or divergent peduncles or pedicels; leaves often punctate . . . . *Lysimachia*

    11 Corolla-lobes erose and cuspidate-tipped, each separately rolled around its stamen, neither dotted nor streaked; anthers linear, 5 fertile ones alternating with 5 staminodia, their filaments distinct to base; flowers nodding; leaves not punctate . . . .
. . . . *Steironema*

ANAGALLIS L. [6338] Pimpernel

1 Stem squarish, gland-dotted, not rooting at the nodes, to about 3 dm tall; leaves lanceolate to ovate, sessile, to about 3 cm long; corolla scarlet to brick-red (varying to white), rotate, less than twice as long as the calyx; pedicels about equalling the leaves; capsule about 5 mm thick; (introd., transcontinental) . . . . *A. arvensis*

1 Stem roundish, rooting at the nodes, to about 1.5 dm tall; leaves ovate to suborbicular, short-petioled, about 5 mm long; corolla pink, funnelform, 2 or 3 times longer than the calyx; pedicels much surpassing the leaves; capsule about 3 mm thick; (introd. in St-Pierre and Miquelon) . . . . [*A. tenella*]

A. arvensis L.   Scarlet Pimpernel
Eurasian; a garden-escape to waste sandy fields in N. America, as in sw B.C. (Vancouver Is. and adjacent islands; Vancouver), Alta. (Boivin 1966*b*), Ont. (N to the Ottawa dist.), Que. (N to near Quebec City), Nfld., St-Pierre and Miquelon, N.B., P.E.I., and N.S.; sw Greenland.

Forma *caerulea* (Schreb.) Baumg. (flowers blue rather than scarlet to brick-red) is reported from Ont. by Boivin (1966*b*).

[*A. tenella* L.]   Bog Pimpernel
[European; reported from St-Pierre and Miquelon by Rouleau (1956), where presumably introd. but not established. (*Lysimachia* L.). MAP: Hultén 1958: map 130, p. 149 (noting other earlier total-area maps).]

ANDROSACE L. [6321]

(Ref.: Robbins 1944)

1 Stoloniferous perennial, forming mats to 1 dm broad, the nodes and tips of the stolons producing globular rosettes of narrowly to broadly oblanceolate entire leaves to 1.5 cm long, these stiffly ciliate, otherwise glabrous or loosely villous; scapes single from each rosette, to 12 cm tall, villous with curly jointed hairs, each terminated by a compact umbel bearing at most 8 flowers, the pedicels usually shorter than the flowers; calyx about 3 mm long; corolla to 12 mm broad, white or creamy with a yellow eye (often aging pink); (Alaska and w Canada) . . . . *A. chamaejasme*

1 Annuals or biennials with a taproot; leaves linear-lanceolate to oblanceolate, oblong, or oblong-obovate, to over 3 cm long; flowers white, relatively numerous; corolla at most 4 mm broad, only slightly surpassing the calyx, this to 4 or 5 mm long.

    2 Bracts subtending umbel broad and foliaceous, oblanceolate to oblong, to 1 cm long,

less than 4 times as long as broad; calyx-lobes about equalling the tube; corolla about
2.5 mm broad, nearly included in the calyx-tube; scapes minutely stellate-pubescent,
rarely over 7 cm tall; (B.C. to w Ont.) . . . . *A. occidentalis*
2 Bracts subulate to narrowly lanceolate, commonly less than 5 mm long and at least 4
times as long as broad; calyx-lobes about half the length of the tube; corolla to 4 mm
broad, slightly surpassing the calyx; (transcontinental) . . . . *A. septentrionalis*

A. chamaejasme Host
/aST/W/EA/ (Ch (Hrr)) Tundra and rocky slopes at low to fairly high elevations from the
Aleutian Is. and coasts of Alaska–Yukon–Dist. Mackenzie (E to Coronation Gulf) through B.C. and
sw Alta. (N to Jasper; CAN) to Mont.; Eurasia. [Incl. var. *arctica* Knuth, ssp. *andersonii* Hult., ssp.
lehmanniana (Spreng.) Hult. (*A. lehm.* Spreng.), *A. (Drosace) albertina* Rydb., and *A. (D.) carinata*
Torr.]. MAPS: Porsild 1957: map 278 (var. *arct.*), p. 195; combine the maps by Hultén 1968*b*:745
(ssp. *and.*) and p. 744 (ssp. *lehm.*).

A. occidentalis Pursh
/T/WW/ (T) Dry sands, gravels, prairies, and rocky woods from s B.C. (Dry Interior N to Lytton,
Spences Bridge, and Kamloops; CAN; Eastham 1947) to s Alta. (Banff and Belly R.; CAN), s Sask.
(Mortlach, about 65 mi w of Regina, and Radville, about 70 mi s of Regina; Breitung 1957*a*), s Man.
(N to Roblin, s of Duck Mt.), and w Ont. (island in Lake of the Woods, where taken by John Macoun
in 1872; CAN, detd. St. John), s to N Calif., N.Mex., Tex., Ark., and Ind. [*A. simplex* Rydb.].

A. septentrionalis L.
/AST/X/GEA/ (T) Dry rocky, sandy, or gravelly places at low to fairly high elevations, the
aggregate species from the coasts of Alaska–Yukon–Dist. Mackenzie–Dist. Keewatin to N
Ellesmere Is. (an isolated station at ca. 79°30′N) and N Baffin Is., s through B.C.–Alta.–Sask.–Man.
to Calif., Ariz., N.Mex., and S.Dak., farther eastwards s to Ont. (coasts of Hudson Bay–James Bay),
Que. (coasts of Hudson Bay–James Bay; Chimo, s Ungava Bay; Bic, Rimouski Co.; Côte-Nord;
Gaspé Pen.; not known from Anticosti Is. or the Maritime Provinces), and w Nfld.; an isolated
station in w Greenland at ca. 78°N; Eurasia. MAPS and synonymy (together with a distinguishing key
to the scarcely separable *A. alaskana*): *see* below.
1 Umbels 1(2, 3)-flowered; [s Alaska, the type from Popof Is.; MAPS: Hultén 1948: map 957,
p. 1337, and 1968*b*:746] . . . . *A. alaskana* Cov. & Standl.
1 Umbels many-flowered . . . . *A. septentrionalis*
2 Scape usually solitary, strongly developed, strictly erect, to about 2.5 dm tall; pedicels
slender, numerous, the central ones straight and ascending, the lateral ones
arched-ascending.
3 Pedicels not glandular; [*A. gormanii* Greene; transcontinental; MAPS (aggregate
species): Hultén 1968*b*:745; Porsild 1957: map 277, p. 195; Fernald 1933: map 5,
p. 82; Wynne-Edwards 1937: map 2, p. 24] . . . . var. *septentrionalis*
3 Pedicels bearing dark stipitate glands; [w U.S.A.] . . . .
. . . . [var. *glandulosa* (Woot. & Standl.) St. John]
2 Scapes commonly several, many of them of nearly equal development; pedicels often
coarser and less numerous, many of them divergent.
4 Scapes at least 5 times longer than the pedicels; [essentially transcontinental;
type from the Mingan Is., E Que.] . . . . var. *robusta* St. John
4 Scapes shorter than the pedicels or at most 3 times their length.
5 Scapes over 1 dm tall, about twice the length of the slender, flexuous, widely
spreading, often very numerous pedicels; [*A. subulifera* (Gray) Rydb.; *A.
pinetorum* Greene; the Yukon–B.C. to Man.] . . . . var. *subulifera* Gray
5 Scapes less than 1 dm tall; plants dark green or reddish.
6 Scapes 2 or 3 cm tall; pedicels rarely more than 6, short and stout; [*A.
arguta* Greene; *A. subumbellata* (Nels.) Small; mts. of B.C. and Alta.] . . . .
. . . . var. *subumbellata* Nels.
6 Scapes mostly over 4 cm tall; pedicels numerous.
7 Calyx-lobes and base of calyx-tube essentially glabrous; [*A. diffusa*
Small; Alaska, B.C., Alta., and Sask.] . . . . var. *diffusa* (Small) Knuth

7 Calyx-lobes and base of calyx-tube densely short-stellate; [*A. puber-ulenta* Rydb.; B.C. to Man.] . . . . var. *puberulenta* (Rydb.) Knuth

CENTUNCULUS L.   [6339]

C. minimus L.   Chaffweed
/T/X/EA/   (T)   Fresh or brackish shores from s B.C. (N to Kamloops; Henry 1915) to SE Alta. (Empress), Sask. (Mortlach; Reed L.; Long L.; near Johnston L. at ca. 54°30′N; Cory, near Saskatoon; Bad Hills, ca. 51°30′N), ?Man. (Rossburn; Lowe 1943; not known from Ont. or Que.), and N.S. (Sable Is.; CAN; not known from N.B.; reports from P.E.I. refer to *Tillaea aquatica*, according to D.S. Erskine 1960), s to Baja Calif., Mexico, and Fla.; S. America; Europe; India; N Africa. MAP: Hultén 1958: map 255, p. 275.

DODECATHEON L.   [6341]   Shooting Star, American Cowslip

1 Corolla to 2 cm long, the lobes creamy white; sepals 3 or 4 mm long; capsules to 1 cm long, dehiscent to the tip by valves, protruding beyond the dried and persistent stamens and corolla; anther-filaments less than 1 mm long, deep reddish-purple; leaf-blades oblong-lanceolate to oblong or ovate, sinuate to sharply dentate or undulate-dentate, broadly rounded or truncate to cordate at base, glabrous, abruptly contracted to petioles of about the same length; scapes glabrous or sparsely glandular-hairy in the inflor-escence; (s B.C.) . . . . *D. dentatum*
1 Corolla-lobes commonly magenta to various shades of violet, red, or purple; corolla deciduous with the stamens as the enlarging capsule forces them off.
  2 Leaf-blades more or less spatulate to deltoid-elliptic or ovate, usually less than twice as long as broad, commonly rounded or truncate at base and abruptly narrowed to the petioles; filaments deep reddish-purple; capsules to about 12 mm long, operculate (the extreme style-bearing tip coming off like a lid); inflorescence glandular.
    3 Filament-tube 2–4 mm long; plant from a cluster of more or less tuberous-thickened roots, lacking rhizomes but with numerous "rice-grain" bulblets at anthesis; (s B.C.) . . . . *D. hendersonii*
    3 Filament-tube very short or almost obsolete; plants from a commonly (?always) horizontal rhizome (this erect, when present in other species); (Alaska–Yukon–Dist. Mackenzie–N B.C.) . . . . *D. frigidum*
  2 Leaf-blades lanceolate to oblong-lanceolate, oblanceolate, or spatulate (sometimes even obovate, but usually at least 3 times as long as broad), gradually narrowed to the winged petiole; bulblets wanting.
    4 Stigma conspicuously capitate (about twice the width of the middle part of the style); anther-filaments scarcely 1 mm long, free nearly or quite to base, deep reddish-purple; capsules to 11 mm long, opening from the top by valves (but the tip operculate); inflorescence (and leaves) often more or less densely glandular-hairy; (Alaska–B.C.) . . . . *D. jeffreyi*
    4 Stigma only slightly enlarged or only slightly thicker than the style; anther-filaments usually yellow.
      5 Capsule to 12 mm long, the tip operculate (the extreme style-bearing tip coming off like a lid); anther-filaments usually not over 1 mm long, free or united, their connectives transversely wrinkled; (B.C. to Sask.) . . . . *D. conjugens*
      5 Capsule to 1.5 cm long, dehiscent to the tip by valves; anther-filaments united into a tube to 3 mm long, their connectives smooth or only somewhat longitudinally wrinkled when dried; (Alaska–B.C. to Man.) . . . . *D. pauciflorum*

D. conjugens Greene
/T/W/   (Hr)   Seepage areas in sagebrush plains up to alpine meadows from SE B.C. (N to Natal, near the SW Alta. boundary), Alta. (N to ca. 53°N), and SW Sask. (Cypress Hills, Robsart, and Simmie; Breitung 1957a) to N Calif. and Wyo. [*D. cylindrocarpum* Rydb.; incl. var. *viscidum* (Piper) Mason (*D. visc.* Piper; *D. con.* var. *?beamishii* Boivin and its f. *lacteum* Boivin), the pubescent or

glandular-pubescent extreme to which most or all of our material appears referable]. MAPS: Katherine Beamish, Bull. Torrey Bot. Club. 82(5): map 1, p. 360. 1955; H.J. Thompson, Contrib. Dudley Herb. 4(5): fig. 9a, p. 101. 1953.

D. dentatum Hook.
/T/W/ (Hr) Streambanks and shaded moist slopes from S B.C. (Manning Provincial Park, about 30 mi SE of Hope; Princeton; Penticton) to Oreg. and Idaho. [*D. meadia* vars. *dentatum* (Hook.) Gray and *latilobum* Gray]. MAP: H.J. Thompson, Contrib. Dudley Herb. 4(5): fig. 13, p. 122. 1953.

D. frigidum C. & S.
/aSs/W/eA/ (Hr) Meadows and heaths at low to high elevations from the coasts of Alaska–Yukon (*see* Hultén 1948: map 961, p. 1338) and the Mackenzie R. Delta to northernmost B.C. (S to Dease L. at ca. 58°30′N; CAN) and SW Dist. Mackenzie. [*D. meadia* var. *fr.* (C. & S.) Gray]. MAPS: Hultén 1968*b*:749; H.J. Thompson, Contrib. Dudley Herb. 4(5): fig. 9b, p. 101. 1953; W.J. Cody, Nat. can. (Que.) 98(2): fig. 9, p. 149. 1971.

D. hendersonii Gray
/t/W/ (Hr) Moist prairies and open woods from SW B.C. (Vancouver Is. and Yale, lower Fraser Valley; CAN) to S Calif. [*D. meadia* var. *hend.* (Gray) Brandg.; *D. integrifolium* var. *latifolium* Hook.]. MAP: H.J. Thompson, Contrib. Dudley Herb. 4(5): fig. 3a, p. 85. 1953.

D. jeffreyi van Houtte
/sT/W/ (Hrr) Wet meadows and streambanks from S Alaska (N to ca. 60°30′N; *see* Hultén 1948: map 964 (*D. viv.*), p. 1338) through coastal B.C. (Queen Charlotte Is.; Prince Rupert; Vancouver Is.; Mittlenatch Is.) to Calif., Idaho, and Mont. [*D. meadia* var. *lancifolium* Gray; *D. viviparum* Greene]. MAPS: Hultén 1968*b*:749; H.J. Thompson, Contrib. Dudley Herb. 4(5): fig. 10, p. 108. 1953.

D. pauciflorum (Durand) Greene
/ST/WW/ (Hr) Open woods, meadows, prairies, moist slopes, and saline places along the coast from cent. Alaska–Yukon (N to ca. 65°N) and the Mackenzie R. Delta to Great Slave L., B.C.–Alta., Sask. (N to near Prince Albert), and S Man. (N to Eriksdale, about 60 mi N of Portage la Prairie), S to N Calif., Mexico, and Tex. MAPS and synonymy: see below.

1 Plants usually not over 5 cm tall, bearing only 1 or 2 flowers; [*D. watsonii* Tidestr.; *D. pulchellum (radicatum)* var. *wat.* (Tidestr.) Boivin; *D. uniflorum* Rydb.; SW B.C.: Mt. Arrowsmith, Vancouver Is.; MAP: H.J. Thompson, Contrib. Dudley Herb. 4(5): fig. 12a, p. 117. 1953] . . . . var. *watsonii* (Tidestr.) Hitchc.
1 Plants taller and usually with more flowers.
  2 Plant glandular-pubescent throughout; [*D. cusickii* Greene and its var. *album* Suksd.; *D. pulchellum* var. *album* (Suksd.) Boivin; *D. puberulentum* Heller; B.C. (N to Queen Charlotte Is. and Kamloops) and Alta. (Seebe); MAPS: Thompson, loc. cit., fig. 14, p. 123; Katherine Beamish, Bull. Torrey Bot. Club 82(5): map 1, p. 360. 1955] . . . .
   . . . . var. *cusickii* (Greene) Mason
  2 Plant glabrous to sparingly pubescent.
    3 Leaves ovate-lanceolate to ovate, rather abruptly narrowed to the petioles; [*D. macrocarpum (pulchellum)* var. *al.* Hult., the type from Port Hobron, Alaska; *D. meadia (radicatum)* var. *mac.* Gray (*D. mac.* (Gray) Knuth) in large part; incl. *D. superbum* Pennell & Stair; Alaska–B.C.; MAPS: Beamish, loc. cit., map 2 (*D. rad.* ssp. *mac.*), p. 360; combine the maps by Hultén 1968*b*:747 (*D. pul.* ssp. *sup.*) and p. 748 (*D. pul.* ssp. *al.*)] . . . . var. *alaskanum* (Hult.) Hitchc.
    3 Leaves mostly oblanceolate to spatulate, narrowed gradually to the petioles; [*D. meadia (pulchellum; salinum)* vars. *pauciflora* Dur. and *puberula* Nutt. (*D. puberulum* (Nutt.) Piper); *D. ?radicatum* Greene and its var. *sinuatum* Rydb.; *D. integrifolium* var. *vulgare* Hook.; Alaska–Yukon–B.C. to S Man.; MAPS: Hultén 1968*b*:748 (*D. pul.* ssp. *pauc.*); Porsild 1966: map 116, p. 81; Thompson, loc. cit., fig. 12b (*D. rad.*), p. 117; Beamish, loc. cit., map 2 (*D. rad.*), p. 360; N.C. Fassett, Am. Midl. Nat. 31(2): map 1, p. 460. 1944] . . . . var. *pauciflorum*

DOUGLASIA Lindl. [6318]

1 Flowers usually at least 3 in umbels terminating peduncles to 7 cm long, each umbel subtended by a whorl of at least 4 (up to 10) lanceolate to ovate bracts to 8 mm long; pedicels from very short to 1 or 2 cm long; leaves to 2 cm long and 6 mm broad; plants mat-forming; (B.C. and SW Alta.) . . . . *D. laevigata*
1 Flowers usually solitary, subtended by 1 or 2 bracts; leaves rarely over 1 cm long and 2 mm broad.
  2 Leaves linear-subulate to linear-oblong (broadest below the middle); plant to 1 dm tall; (SW Alta.) . . . . *D. montana*
  2 Leaves oblanceolate (broadest above the middle); plant to 5 cm tall; (Alaska–Yukon–W Dist. Mackenzie).
    3 Upper leaf-surface pubescent with simple hairs . . . . *D. ochotensis*
    3 Upper leaf-surface glabrous or pubescent with forked and branched hairs . . . . *D. arctica*

D. arctica Hook.
/aSs/W/ (Ch) Tundra and rocky places at low to fairly high elevations: Alaska (N to ca. 65°30′N), the Yukon (N to Herschel Is.), and NW Dist. Mackenzie (type from between the Mackenzie and Coppermine rivers). MAPS: Hultén 1968*b*:743; Porsild 1966: map 117B, p. 81.
Some of our material is referable to var. *gormanii* (Const.) Boivin (*D. gor.* Const., the type from near Fort Selkirk, the Yukon; leaves pubescent above with forked and branched hairs rather than glabrous above). MAPS (*D. gor.*): Hultén 1968*b*:743; Porsild 1966: map 117A, p. 81.

D. laevigata Gray
/T/W/ (Ch) Moist coastal bluffs to alpine talus and ridges from SW B.C. (Strathcona Park, Vancouver Is.; Herb. V, detd. Constance) and SW Alta. to Oreg. [Incl. var. *ciliolata* Const.; *D. nivalis* Lindl., the type from the Rocky Mts. of SW Alta., the basis of the above Alta. report].

D. montana Gray
/T/W/ (Ch) Foothills to alpine talus and ridges from SW Alta. (Waterton Lakes; Breitung 1957*b*) to N Wyo.

D. ochotensis (Willd.) Hult.
/aS/W/eA/ (Ch) Stony montane slopes of N Alaska and NW Dist. Mackenzie (between ca. 65° and 69°30′N); NE Siberia. [*Androsace* Willd.]. MAP: Hultén 1968*b*:742.

GLAUX L. [6337]

G. maritima L. Saltwort, Sea-Milkwort
/sT/(X)/EA/ (Hpr) Saline or brackish coastal marshes and sands and alkaline regions of the interior from S Alaska–Yukon (N to ca. 61°N; *see* Hultén 1948: map 967, p. 1338) and Great Slave L. to B.C.–Alta., Sask. (N to Prince Albert), Man. (N to Dawson Bay, L. Winnipegosis, ca. 53°N), Ont. (SW James Bay watershed N to ca. 53°N), James Bay (Charlton and Manawanan islands), Que. (SE James Bay watershed N to ca. 53°N; St. Lawrence R. estuary from St-Jean-Port-Joli, l'Islet Co., to the Côte-Nord, Anticosti Is., and Gaspé Pen.), Nfld., N.B., P.E.I., and N.S., S in the West to Calif. and N.Mex. and along the Atlantic coast to Va.; Eurasia. [Incl. vars. *angustifolia* and *macrophylla* Boivin]. MAP: Hultén 1968*b*:752.
Some of our material is referable to var. *obtusifolia* Fern. (leaves round-tipped and relatively broad; capsules to 4 mm thick rather than at most about 2.5 mm; branches usually strongly ascending rather than often prostrate). MAP: Potter 1932: map 4 (incomplete), p. 72.

[HOTTONIA L.] [6327]

[*H. inflata* Ell.] Featherfoil, Water-Violet
[The report of this species of the E U.S.A. (N to Mo., Ohio, and New Eng.) from P.E.I. by McSwain and Bain (1891) may relate to a casual introduction or to some other aquatic with finely dissected leaves.]

LYSIMACHIA L. [6330] Loosestrife

(Ref.: J.D. Ray 1956)
1 Stems creeping; leaves opposite, short-petioled, suborbicular, to about 3 cm long; flowers
   solitary in the leaf-axils, the petals dotted with dark red; (introd.) . . . . *L. nummularia*
1 Stems erect or ascending.
   2 Flowers white, in a slender terminal spike, the bracts longer than the pedicels;
      leaves alternate, ovate-lanceolate, tapering at both ends, their margins revolute;
      (introd.) . . . . [*L. clethroides*]
   2 Flowers yellow or orangish; leaves mostly opposite or whorled.
      3 Corolla uniformly yellow; (introd.).
         4 Calyx-segments at most about 5 mm long, dark-margined; flowers in
            leafy-bracted panicles; corolla-lobes entire; leaves opposite or whorled . . . .
            . . . . *L. vulgaris*
         4 Calyx-segments to 1 cm long, green throughout; flowers whorled in the upper
            leaf-axils or on short branches; corolla-lobes glandular-ciliolate; leaves chiefly
            in whorls of 3 or 4 . . . . *L. punctata*
      3 Corolla usually dark-dotted or -streaked.
         5 Flowers borne in racemes; leaves opposite (rarely some alternate).
            6 Racemes borne on short peduncles from the axils of the 2 or 3 pairs of
               longer leaves near midstem; flowers crowded, subsessile; corolla-lobes
               linear, much shorter than the stamens, these with free filaments; (B.C.
               to N.S.) . . . . *L. thyrsiflora*
            6 Racemes terminal, relatively loose, the flowers slender-pedicelled; corolla-
               lobes lance-oblong, at least as long as the stamens, these with filaments
               connate at base; (Man. to Labrador, Nfld., and N.S.) . . . . *L. terrestris*
         5 Flowers borne in the axils of normal foliage-leaves; leaves mostly in whorls of
            4 or 5; (Ont. to N.B.) . . . . *L. quadrifolia*

[*L. clethroides* Duby]
[Asiatic; introd. in Que. (Sillery, near Quebec City), where probably a casual garden-escape but not
established.]

L. nummularia L.   Moneywort, Creeping Jenny.   Monnayère
European; introd. in grasslands and along shores and moist roadsides. in N. America, as in sw B.C.
(Vancouver Is.; Miss M.C. Melburn, personal communication, Sept. 26, 1967), Ont. (N to the Ottawa
dist.), Que. (N to near Quebec City), Nfld. (cemetery at St. John's; GH), N.B. (St. Andrews,
Charlotte Co.; GH), P.E.I. (Charlottetown; MT), and N.S. MAP: J.D. Ray 1956: map 6, p. 116.

L. punctata L.   Garden-Loosestrife
Eurasian; introd. along damp roadsides and in swampy places in N. America, as in sw B.C.
(Nanaimo, Vancouver Is.; CAN), Ont. (N to the Ottawa dist.), Que. (N to the Gaspé Pen. at Matane),
Nfld. (Birchy Cove, Bay of Islands; GH), N.B., P.E.I., and N.S. [Incl. var. *verticillata* (Bieb.) Boiss.].
MAP: J.D. Ray 1956: map 7, p. 117.

L. quadrifolia L.   Whorled Loosestrife
/T/EE/   (Hpr)   Woods, thickets, and shores from Ont. (N to Chalk River, Renfrew Co.; OAC) to
sw Que. (N to Kazabazua, Gatineau Co.; OAC, detd. Calder) and s N.B. (Kings and St. John
counties and Grand Manan Is.; NBM; not known from P.E.I. or N.S.), s to Ill., Tenn., Ala., and Ga.
[*L. racemosa* Lam.; *L. stricta* Ait.]. MAP: J.D. Ray 1956: map 8 (the occurrence in s N.B. should be
indicated), p. 118.

    × *L. producta* (Gray) Fern. (*L. stricta* var. *pro.* Gray; *L. foliosa* Small), a widely distributed and
often abundant hybrid between *L. quadrifolia* and *L. terrestris* (leaves opposite, as in *L. terr.*, or in
whorls of 4 or 5, as in *L. quad.;* lower flowers subtended by normal foliage-leaves, as in *L. quad.*,
these gradually reduced to bracts above, as in *L. terr.*), is known from Que. (N to gravel flats at the
mouth of the Matapédia R., sw Gaspé Pen.; MT; J. Rousseau 1931) and is reported from s Ont. by
Soper (1949). MAP: J.D. Ray 1956: map 9 (incomplete northwards), p. 119.

L. terrestris (L.) BSP.   Yellow or Swamp-Loosestrife, Swamp-candles
/T/EE/   (Hpr)   Bogs, swamps, and shores from SE Man. (N to Sasaginnigak L., about 125 mi NE of Winnipeg; reports from Sask. require confirmation) to Ont. (N to the SW James Bay watershed at ca. 53°N), Que. (N to the E James Bay watershed at 53°50'N, the Côte-Nord, Anticosti Is., and Gaspé Pen.), Labrador (N to the Hamilton R. basin), Nfld., N.B., P.E.I., and N.S., S to Iowa, Ky., and Ga.; introd. in SW B.C. (Eastham 1947; Ucleulet, Vancouver Is., where brought in with cranberry plants from the East). [*Viscum* L.]. MAP: J.D. Ray 1956: map 10 (the occurrence in Man. and S Labrador should be indicated), p. 120.

The typical form has long sterile shoots that are often bulblet-bearing in their leaf-axils. The form lacking bulblets has been named f. *florifera* Boivin (type from the shores of the Kaministikwia R. w of Thunder Bay, Ont.; general throughout the range). × *L. commixta* Fern., a frequent hybrid combining in various degrees the characters of *L. terrestris* and *L. thyrsiflora,* is known from Ont. (N to the SW James Bay watershed at ca. 51°30'N), Que. (N to the Côte-Nord), N.B. (Shediac, Westmorland Co.; GH), P.E.I. (North L., Kings Co.; CAN; GH), and N.S. (Baddeck, Victoria Co.; MT). MAP: J.D. Ray 1956: map 9 (incomplete northwards), p. 119.

L. thyrsiflora L.   Tufted Loosestrife
/ST/X/EA/   (Hel)   Swamps, marshes, and bogs from the w Aleutian Is. (Attu Is.) and Alaska (N to ca. 66°30'N; not known from the Yukon) to the Mackenzie R. Delta (CAN), Great Slave L., L. Athabasca (Alta. and Sask.), Man. (N to Churchill; Schofield 1959), Ont. (N to the Fawn R. at ca. 54°N, 89°W), Que. (N to E James Bay at ca. 54°N and the Gaspé Pen.; reported from Anticosti Is. by Schmitt 1904), N.B., P.E.I., and N.S., S to Calif., Colo., Mo., and W.Va.; Eurasia. [*Naumburgia* Duby]. MAPS: Hultén 1968b:750; J.D. Ray 1956: map 11, p. 121.

L. vulgaris L.   Garden-Loosestrife
Eurasian; a garden-escape to roadsides and wet places in N. America, as in S Ont. (Kent, Northumberland, and York counties), SW Que. (N to near Quebec City; MT), N.B. (Kings and Restigouche counties; NBM), P.E.I. (Charlottetown; CAN), and N.S. MAP: J.D. Ray 1956: map 7 (the occurrence in N.B. should be indicated), p. 117.

PRIMULA L.   [6315]   Primrose, Cowslip.   Primevère

1   Corolla yellow (rarely pink); calyx to over 1.5 cm long; leaves obovate-spatulate to ovate-oblong, irregularly crenate or serrate, to over 1.5 dm long; (introd.).
 2   Scape to 3 dm tall, puberulent, bearing up to 30 long-pedicelled flowers; leaves abruptly narrowed or rounded to long winged petioles, more or less puberulent on both sides; calyx finely pubescent, its teeth to 3 mm long; corolla-limb to 1.5 cm broad .... *P. veris*
 2   Scape none or very small, the flowers solitary on long shaggy-hairy pedicels, to 3 cm broad; leaves gradually narrowed to a sessile or very short-petioled base, pubescent beneath, glabrous above except on the veins; calyx shaggy-hairy, its teeth to 6 mm long; corolla-limb to 3 cm broad .... [*P. vulgaris*]
1   Corolla pale pink or lilac (sometimes white or nearly so) to bluish purple.
 3   Lobes of corolla entire or only slightly emarginate; corolla violet with a lavender eye, its limb to 2 cm broad; leaves to over 1.5 dm long and up to 2.5 cm broad, commonly acutish, efarinose, entire or rather irregularly and obscurely toothed, rather fleshy, their lateral nerves inconspicuous; scapes stoutish and to about 2.5 dm tall, often farinose in the inflorescence; (Alaska–Yukon–w Dist. Mackenzie) ...... *P. tschuktschorum*
 3   Lobes of corolla distinctly emarginate or 2-lobed, obcordate.
  4   Leaves with up to over 9 large teeth above the middle of the blade, cuneate-obovate, gradually narrowed to petioles equalling or up to 3 times as long as their efarinose blades, these to about 3 cm long and 2.5 cm broad; calyx cleft up to 2/3 to base; corolla to 2.5 cm broad, its lobes deeply 2-lobed; capsule ovoid to globose, slightly shorter than the calyx; bracts subulate; plant to about 3 dm tall, glabrous or faintly puberulent but scarcely farinose; (Alaska–N B.C.) ...... *P. cuneifolia*
  4   Leaves entire or only shallowly toothed; calyx rarely cleft more than midway to base; corolla-lobes usually relatively less deeply lobed.

5 Bracts of involucre oblong or narrowly obovate, obtuse or abruptly contracted at tip, their bases often prolonged into narrow auricles to 1.5 mm long; calyx to 8 mm long; corolla-limb lilac, to 2 cm broad; capsules to 1.5 mm thick, to twice the length of the calyx; leaf-blades elliptic, ovate, or suborbicular, entire or with scattered small teeth, to 2 cm broad, shorter than to about equalling their slender petioles; plant efarinose; (Alaska–Yukon–N B.C.) .... *P. sibirica*

5 Bracts of involucre lance-subulate to lanceolate or lance-oblong, their bases at most merely gibbous-saccate.

   6 Corolla-limb to 2 cm broad (*P. mistassinica* may sometimes be sought here), whitish to bright violet; calyx to 6 mm long, the lobes about equalling the tube; capsule usually half again as long as the calyx; leaves to 4.5 cm long (including the margined petiole) and 1 cm broad, their blades linear-oblanceolate to cuneate-obovate, entire to coarsely serrate; scapes to 1.5 dm tall; plants glabrous (or the lower leaf-surfaces and the upper part of the scape sometimes slightly farinose); (Alaska to Dist. Mackenzie) ....
.... *P. borealis*

   6 Corolla-limb at most about 1.5 cm broad.

      7 Leaves entire or obscurely undulate-dentate, efarinose, the blades spatulate to oblong or obovate, to 1.5 cm broad, equalling or shorter than their slender petioles; calyx to 6 mm long, cleft to 1/3 its length; corolla-limb white to deep lilac or purplish, to 9 mm broad; capsules to about 2 mm thick, 2 or 3 times as long as the calyx; plant efarinose; (transcontinental) .... *P. egaliksensis*

      7 Leaves mostly distinctly toothed, the blades commonly longer than their margined petioles; mature capsules generally broader.

         8 Scape comparatively slender; involucral bracts at most 6 mm long, rarely saccate or gibbous at base; calyx-lobes about equalling the tube; stigma or tips of anthers exserted from the shrivelled corolla; capsule 2 or 3 mm thick; seeds nearly smooth; leaves very rarely farinose; (transcontinental) .... *P. mistassinica*

         8 Scape comparatively stout; involucral bracts to 14 mm long; anthers and stigma not exserted from the shrivelled lilac or violet corolla; capsule to 5 mm thick; seeds roughened or reticulate.

           9 Involucral bracts lanceolate to linear-oblong, flat, broadly gibbous at base; leaves strongly farinose beneath; calyx and summit of scape strongly farinose; calyx-lobes about half as long as the tube; (B.C. to Hudson Bay and James Bay) ....
.... *P. incana*

           9 Involucral bracts subulate or involute above the dilated, commonly saccate base.

              10 Leaves green beneath, rarely a little farinose, subentire or obscurely undulate-dentate, to about 4 cm long; calyx at most 6 mm long, green or lightly farinose, the lobes about half as long as the tube; corolla-limb at most 8 mm broad, the lobes to 3 mm broad, their terminal segments not over 1 mm long; (essentially transcontinental) .... *P. stricta*

              10 Leaves commonly strongly farinose beneath, mostly dentate, to about 13 cm long; calyx to 11 mm long, usually strongly farinose, the lobes about equalling the tube; corolla-limb at least 9 mm broad, the lobes over 3 mm broad, their terminal segments over 1.5 mm long; (Ont. to Labrador, Nfld., and N.S.) .... *P. laurentiana*

P. borealis Duby

/aSs/W/eA/ (Hr) Saline shores of Alaska–Yukon–NW Dist. Mackenzie, with isolated stations in SW Yukon (Porsild 1966), ?Banks Is. (*see* M.L. Fernald, Rhodora 30(353):95. 1928), and SE ?B.C. (Emerald L., near Field; *see* Hultén 1948:1269); NE Siberia. MAPS and synonymy: *see* below.

1 Leaves densely yellow-farinose beneath, to about 2 cm long, narrowed to short wing-margined petioles; [*P. ajanensis* Busch; islands, coasts, and riverbanks of W Alaska; MAP: Hultén 1948: map 945b, p. 1336] . . . . var. *ajanensis* (Busch) Hult.
1 Leaves efarinose or only sparingly farinose, to 4 cm long, the blade usually shorter than the petiole . . . . var. *borealis*
 2 Flowers white; [*P. parviflora* Duby; type from the delta of the Anderson R., NW Dist. Mackenzie] . . . . f. *albiflora* Cody
 2 Flowers lilac; [*P. chamissonis* Busch; *P. tenuis* Small; coasts and islands of W and N Alaska (type from the Seward Pen.) and Herschel Is., the Yukon (*see* Hultén 1948: map 945a, p. 1336), with additional stations in SW Yukon and along the coast of NW Dist. Mackenzie, and reported from Banks Is.; MAPS (aggregate species): Porsild 1966: map 118, p. 81; Hultén 1968b:742] . . . . f. *borealis*

P. cuneifolia Ledeb.
/Ss/W/eA/ (Hr) Wet meadows of the Aleutian Is., Alaska (N to ca. 66°30′N; *see* Hultén 1948: maps 946a and 946b, p. 1336), and N B.C. (White Pass, near the Alaska boundary; Herb. V; reported from Mt. Rapho, 56°13′N, by J.M. Macoun 1895); NE Asia. [Incl. the reduced extreme, ssp. *saxifragifolia* (Lehm.) Sm. & Forrest (*P. sax.* Lehm., the type from Unalaska Is., Aleutian Is.), to which most of our material has been referred]. MAP: combine the maps by Hultén 1968b:739 (ssp. *cun.* and ssp. *sax.*).

P. egaliksensis Wormsk. Greenland Primrose
/aST/X/G/ (Hr) Meadows and wet calcareous shores from the E Aleutian Is. and coasts of Alaska–Yukon–Dist. Mackenzie (E to Bernard Harbour, SW of Victoria Is.) to Great Bear L., SE Dist. Keewatin, N Man. (Churchill S to the Nelson R.), Ont. (coasts of James Bay–Hudson Bay), Que. (coasts of James Bay–Hudson Bay; S Ungava Bay; Côte-Nord), Labrador (N to Mugford, 57°48′N; not known from the Maritime Provinces), and N Nfld., S in the West to SE B.C. (Kicking Horse Pass; CAN) and the mts. of SW Alta. (Columbia Icefield, about 60 mi SE of Jasper; CAN); W Greenland (type locality) N to ca. 68°N; NE ?Siberia (Hultén 1948). MAP: Hultén 1968b:741.
 Forma *violacea* Fern. (*P. groenlandica* (Warming) Sm. & Forrest; corolla deep lilac to violet rather than white) occurs throughout the range.

P. incana Jones
/ST/(X)/ (Hr) Meadows, bogs, and damp places from NE-cent. Alaska (N to ca. 66°30′N), SW Yukon, and cent. Dist. Mackenzie (N to Norman Wells, 65°17′N; W.J. Cody, Can. Field-Nat. 74(2):94. 1960) to Great Slave L., Sask (N to Prince Albert), and Man. (N to York Factory, Hudson Bay, 57°N), S in the West through B.C.–Alta. to Idaho, Utah, and Colo., farther eastwards with isolated stations along the James Bay–Hudson Bay coasts of Ont. and on Charlton Is., James Bay (CAN). MAP: Hultén 1968b:741.

P. laurentiana Fern. Bird's-eye-Primrose
/ST/EE/ (Hr) Meadows, ledges, and cliffs (chiefly calcareous) from N Ont. (James Bay–Hudson Bay watershed N to ca. 55°N; *see* E Canada map by Lepage 1966: map 16, p. 236) to Que. (James Bay N to ca. 53°N; Ungava Bay watershed N to ca. 57°50′N; St. Lawrence R. estuary from Kamouraska Co. to the Côte-Nord, Anticosti Is., and Gaspé Pen.; type from Bic, Rimouski Co.), Labrador (N to the Hamilton R. basin), Nfld., N.S. (Yarmouth, Digby, Annapolis, and Kings counties; not known from N.B. or P.E.I.), and Maine. [*P. farinosa* of most or all E Canada reports, not L., and its var. *macropoda* Fern., not *P. mac.* Craib; *P. scotica* Hook., in part].
 Forma *chlorophylla* Fern. (leaves green rather than whitish- or yellowish-farinose beneath; type from Nfld.) occurs essentially throughout the range.

P. mistassinica Michx.
/ST/X/ (Hr) Calcareous or clayey shores, meadows, and ledges from W-cent. Yukon (N to ca. 65°30′N; near the Alaska boundary here but not yet known from Alaska) to Great Bear L., Great Slave L., L. Athabasca (Alta. and Sask.), Man. (N to the Churchill R. at ca. 57°20′N), Ont. (N to Fort Severn, Hudson Bay, ca. 56°N), Que. (N to the E James Bay watershed at 53°35′N, the type locality near L. Mistassini, and the St. Lawrence R. estuary from near Quebec City to the Côte-Nord,

Anticosti Is., and Gaspé Pen.; reported from Beauceville, Beauce Co., SW Que., by Raymond 1950b), Labrador (N to the Hamilton R. basin), Nfld., N.B. (St. John and Restigouche River systems), and N.S. (St. Paul Is. and Colchester, Inverness, and Victoria counties; not known from P.E.I.), S to S B.C.–Alta., Iowa, S Ont., and Maine. MAP and synonymy: *see* below.

1   Corolla-limb at most 1 cm broad, lacking a conspicuous yellow eye; [Cambridge (Galt), Waterloo Co., S Ont.; M.L. Fernald, Rhodora 30(353):91. 1928] . . . . . var. *noveboracensis* Fern.
1   Corolla-limb to 2 cm broad, with a conspicuous yellow eye . . . . var. *mistassinica*
    2   Corolla milk-white; [E Que., Nfld. (type locality), and N.S.] . . . . f. *leucantha* Fern.
    2   Corolla lilac or pale pink to bluish-purple; [*P. intercedens* Fern.; *P. hornemanniana* Hook.; *P. maccalliana* Wieg.; *P. pusilla* Goldie; transcontinental; MAP (aggregate species): Hultén 1968b:740] . . . . f. *mistassinica*

P. sibirica Jacq.
/ST/W/EA/   (Hr)   Wet meadows of Alaska (N to ca. 69°N), SW Yukon (N to ca. 64°N), and northernmost B.C. (Atlin, ca. 59°35′N; Herb. V; the report from Bernard Harbour, Dist. Mackenzie, by Macoun and Holm 1921, is based upon *P. egaliksensis,* the relevant collection in CAN); N Europe; Asia. MAPS: Hultén 1968b:738; Porsild 1966: map 119, p. 81.

P. stricta Hornem.
/aST/X/GEA/   (Hr)   Shores and moist places (often saline) from the coasts of Alaska–Yukon–Dist. Mackenzie–Dist. Keewatin to N Banks Is., Victoria Is., and northernmost Ungava–Labrador, S to N B.C. (Haines Road at ca. 59°30′N; Herb. V; isolated in the mts. of SW Alta. according to M.L. Fernald, Rhodora 30(352):67. 1928; reports from Sask. require confirmation), Great Slave L., NE Man. (Gillam to Churchill), S James Bay (Ont. and Que.), and S Labrador; W and E Greenland N to ca. 73°N; Iceland; Scandinavia; NW Asia. MAPS: Hultén 1968b:740, and 1958: map 171, p. 191; Porsild 1957: map 276, p. 195.

P. tschuktschorum Kjellm.
/Ss/W/eA/   (Hr)   Wet meadows and streambanks (ranges of N. American taxa outlined below); E Asia. MAPS and synonymy: *see* below.

1   Plants dwarf, the scapes not greatly elongating in fruit; leaves linear to lance-linear, entire; umbels usually 2–3-flowered . . . . ssp. *tschuktschorum*
    2   Corolla-lobes oblong, obtuse; [Little Diomede Is., Alaska; A.E. Porsild, Can. Field-Nat. 79(2):87. 1965; MAP (incl. var. *ber.*): Hultén 1968b:737] . . . . var. *tschuktschorum*
    2   Corolla-lobes linear-cuneate, distinctly notched (or rarely cleft to base); [type from St. Lawrence Is., Alaska] . . . . var. *beringensis* Porsild
1   Plants mostly over 2 dm tall, the scapes much elongated in fruit; leaves spatulate or oblanceolate to oblong, entire to distinctly toothed; [*P. arctica* Koidz.; MAP: Hultén 1968b:738].
    3   Umbels with up to over 10 flowers; calyx cleft halfway to base; leaves to 2 cm broad; scapes to 4 dm tall and 6 mm thick; [*P. eximia* and *P. macounii* Greene; *P. nivalis sensu* Hultén 1948, not Pall.; Alaska (Little Diomede Is.; type of *P. eximia* from St. Paul Is.) and the Yukon (at ca. 63°50′N)] . . . . ssp. *eximia* (Greene) Porsild
    3   Umbels with rarely more than 5 flowers; calyx cleft nearly to base; leaves to 1 cm broad; scapes to 2 dm tall and 4 mm thick; [cent. Alaska (type from the White Mts.) to W Dist. Mackenzie (Richardson Mts.)] . . . . ssp. *cairnesiana* Porsild

P. veris L.   English Cowslip
Eurasian; a garden-escape to meadows and waste places and locally established in N. America, as in SW B.C. (Victoria, Vancouver Is.; John Macoun 1884), S Ont. (Point Clark, Bruce Co.; OAC, "established"), Que. (Brandy Pot Is., near Cacouna; CAN; Ulverton, Richmond Co.), Nfld. and St-Pierre and Miquelon (Rouleau 1956; ?established), N.B. (Boivin 1966b), and N.S. (North Sydney, Cape Breton Is., where "Well established in meadows"; Macoun 1884; CAN). [Incl. *P. officinalis* Jacq.].

    A collection in ACAD from West Gore, Hants Co., N.S., has been referred to *P. polyantha* Mill., this, according to Bailey (1949a), being probably a hybrid of *P. elatior, P. veris,* and *P. vulgaris* parentage.

[P. vulgaris Huds.]   Primrose
[Eurasian; reported from SW B.C. by John Macoun (1884; "Well established in meadows in the vicinity of Victoria, Vancouver Island.") but there are apparently no other records of its occurrence in Canada since that time.]

SAMOLUS L.   [6328]

S. parviflorus Raf.   Water-Pimpernel, Brookweed
/T/X/   (Hs)   Wet soils and shallow water from SE ?Alaska (Hultén 1948) and ?B.C. (Fernald *in* Gray 1950; not known from Alta.–Sask.–Man.) to Ill., Mich., Ont. (N to the Ottawa dist.; TRT; Gillett 1958), SW Que. (N to the Montreal dist.; Raymond 1950b, as *S. valerandii*), N.B. (Kent, Westmorland, and Northumberland counties; GH; S.F. Blake, Rhodora 20(234):106. 1918), P.E.I. (Selkirk, Kings Co.; CAN; D.S. Erskine 1960), and N.S., S to Calif., Mexico, and Fla.; tropical America. [*S. floribundus* HBK.; *S. valerandii* of Canadian reports, not L.; *S. val.* var. *americanus* Gray]. MAP (NE area): Fassett 1928: fig. 1, pl. 11.

STEIRONEMA Raf.   [6631 and 6630 (*Lysimachia*)]

1   Stem-leaves firm, linear, sessile or nearly so, obscurely veined, their often somewhat revolute margins smooth or sparingly ciliate at the very base; lowest leaves oblong or spatulate; fruiting calyx to 7 mm long, the lobes to 5 mm long; corolla-lobes conspicuously pointed; stems 4-angled, from a thickish caudex; (SE Man. and S Ont.) ....
.... *S. quadriflorum*
1   Stem-leaves softer, linear to ovate, their pinnate veins evident; corolla-lobes erose and cuspidate.
  2   Stem-leaves broadly lanceolate to ovate, acuminate, usually over 3 cm broad, broadly rounded to subcordate at base, all on long ciliate-fringed petioles; calyx-lobes to 9 mm long; capsule shorter than to equalling the mature calyx; (B.C. to N.S.) .... *S. ciliatum*
  2   Stem-leaves (at least the middle and upper ones) linear to lanceolate or narrowly oblong, attenuate, bristly-ciliate at the sessile base, the relatively broader basal leaves often forming a rosette; calyx-lobes to 7 mm long; capsule shorter than the calyx; (S Alta. to Que.) .... *S. lanceolatum*

S. ciliatum (L.) Raf.   Fringed Loosestrife
/T/X/   (Hpr)   Moist soils, thickets, and shores from S B.C. (N to Kamloops; CAN) to Alta. (N to Fort Saskatchewan; CAN), Sask. (N to near Prince Albert; CAN), Man. (N to Hill L., N of L. Winnipeg; CAN), Ont. (N to Moose Factory, SW James Bay, 51°16'N), Que. (N to L. St. John and the SW Gaspé Pen. at Matapédia; reported from the Côte-Nord by Saint-Cyr 1887), N.B., P.E.I., and N.S., S to Oreg., Colo., Tex., and Fla. [*Lysimachia* L.; *S. pumilum* Greene]. MAP: J.D. Ray 1956: map 1 (*L. cil.*), p. 111.
    Forma *elongata* Löve and Bernard (leaves relatively narrow, truncate at base rather than rounded or cordate) is known from the type locality, near Otterburne, about 30 mi S of Winnipeg, Man.

S. lanceolatum (Walt.) Gray
/T/X/   (Hp(r))   Thickets, swamps, shores, and dry or moist open woods (ranges of Canadian taxa outlined below), S to Wash., Ariz., N.Mex., Tex., La., Miss., and Fla. MAPS and synonymy: *see* below.
1   Flowering stems usually less than 4 mm thick at base, often 4-angled above, arising from slender cord-like stolons and rhizomes; middle and upper leaves sessile or subsessile, paler beneath, bristly-ciliate at base; lower leaves usually persistent; sepals essentially nerveless; [*Lysimachia* Walt.; *L. (S.) heterophylla* Michx.; E U.S.A. only, the reports from Missisquoi Co., SW Que., by C.H. Knowlton, Rhodora 35(415):251. 1933, referring to the following taxon (relevant collection in GH) and from S Ont. by John Macoun 1884, referring to *S. quadriflorum* (relevant collection in Herb. US); MAP: J.D. Ray 1956: map 3 (*L. lanc.*), p. 113] .... [var. *lanceolatum*]

1 Flowering stems usually over 4 mm thick at base, not 4-angled above, lacking basal stolons; middle and upper leaves distinctly petioled, scarcely paler beneath, rarely ciliate; lower leaves not persistent; sepals distinctly nerved; [*Lysimachia (S.) hybrida* Michx.; *L. lanc.* ssp. *hyb.* (Michx.) Ray; SW Alta. (Camrose), Sask. (N to Saskatoon and Humboldt), S Man. (Portage la Prairie; Macgregor; 10 mi N of Brandon), Ont. (N to the Ottawa dist.), and Que. (N to Montreal and Duparquet, 48°30'N); MAP: J.D. Ray 1956: map 4 (*L. lanc.* ssp. *hybr.*), p. 114] . . . . var. *hybridum* (Michx.) Gray

S. quadriflorum (Sims) Hitchc.
/T/EE/ (Hpr) Calcareous swamps, wet meadows, and shores from SE Man. (reported from the Winnipeg Valley by J.D. Ray 1956, on the basis of an 1859 Bourgeau collection in GH, and from Otterburne, about 30 mi S of Winnipeg, by Löve and Bernard 1959; the report from Foxwarren, Man., by Lowe 1943, is based upon *Gratiola neglecta,* the relevant collection in WIN) to Ont. (N to Bruce, Wellington, and York counties), N.Y., and Mass., S to Mo., Ill., Ky., and Va. [*Lysimachia* Sims; *L. (Steironema) longifolia* Pursh; *S. (L.) revoluta* Raf.]. MAP: J.D. Ray 1956: map 5 (*L. quad.*), p. 115.

TRIENTALIS L.   [6333]   Star-flower

1 Leaves of the terminal cluster to about 1 dm long, the others all reduced to minute scaly bracts; rhizome tuberous at apex, the tuber 1 or 2 cm long and up to 6 mm thick at apex, usually erect; flowers white, pink, or roseate; (transcontinental) . . . . *T. borealis*
1 Leaves of the terminal cluster usually less than 5 cm long, the lower ones greatly reduced but more or less foliaceous; rhizome horizontal, slightly thickened toward apex but not tuberous; flowers white; (Alaska–Yukon–Dist. Mackenzie–B.C.–Alta.) . . . . *T. europaea*

T. borealis Raf.   American Star-flower
/ST/X/ (Gst) Woods, prairies, and peaty slopes, the aggregate species from S-cent. Yukon (Keno, near Mayo, ca. 64°N; Porsild 1951a) and B.C. to L. Athabasca (Alta. and Sask.), Man. (N to Reindeer L. at 57°37'N; CAN), Ont. (N to the Fawn R. at ca. 54°N, 89°W), Que. (N to Ungava Bay at ca. 58°N and the Côte-Nord), Labrador (N to Hebron, 58°12'N), Nfld., N.B., P.E.I., and N.S., S to Calif., Idaho, Minn., N Ga., and Va. MAPS and synonymy: *see* below.
1 Leaves broadly elliptic-ovate, acute or abruptly acuminate; flowers pinkish or roseate; [*T. lat.* Hook.; *T. europaea* var. *lat.* (Hook.) Torr.; the Yukon–W B.C.; MAP: Hultén 1968b:750] . . . . . .
. . . . ssp. *latifolia* (Hook.) Hult.
1 Leaves lanceolate, acuminate; flowers white . . . . ssp. *borealis* var. *borealis*
  2 Stem with up to 5 leafy branches, these each with up to 3 whorls of reduced leaves; [Que., the type from St-Paulin, Maskinongé Co.; also known from the E James Bay watershed at 52°37'N and the Gaspé Pen.), S Labrador (Goose Bay; RIM), Nfld. (Rouleau 1956), and P.E.I. (Charlottetown; D.S. Erskine 1960)] . . . . f. *ramosa* Vict.
  2 Stem unbranched.
    3 Stem with 2 or 3 whorls of somewhat reduced leaves in addition to the normal terminal whorl; [Man. (Herb Lake, about 80 mi NE of The Pas; CAN), S Labrador (Goose Bay; CAN), and Que. (type from Lac Desmarais, about 220 mi NW of Montreal; also known from the E James Bay watershed)] . . . .
. . . . f. *pluriverticillata* Vict. & Rolland
    3 Stem bearing a single whorl of foliage-leaves at summit.
      4 Leaves narrowly lanceolate; [var. *ten.* House; Que.: S Ungava Bay and the E James Bay watershed; RIM] . . . . f. *tenuifolia* (House) Lepage
      4 Leaves broadly lanceolate; [*T. europaea* var. *americana* Pers.; *T. amer.* (Pers.) Pursh; transcontinental; MAPS: Meusel 1943: fig. 20d (incomplete); Hultén 1968b:750] . . . . f. *borealis*

T. europaea L.   European Star-Flower
/ST/W/EA/ (Hpr) Woods and subalpine meadows from the Aleutian Is., Alaska (N to ca. 68°N), the Yukon (N to ca. 65°N), and SW Dist. Mackenzie (N to Great Slave L.) through B.C.–Alta. to S Oreg. and N Idaho; Eurasia. MAP: Hultén 1968b:751.

Most of our material is referable to ssp. *arctica* (Fisch.) Hult. (*T. arctica* Fisch., the type material from Alaska and Kamchatka; *T. aleutica* Tatew.). MAPS: Hultén 1968*b*:751; Porsild 1966: map 120, p. 81; Raup 1930: map 23 (*T. arct.*), p. 203.

PLUMBAGINACEAE (Leadwort or Plumbago Family)

Scapose herbs with entire basal leaves. Flowers regular, perfect, hypogynous. Calyx-lobes, petals (distinct or united only at base), and stamens each 5, the stamens inserted on the corolla-tube or at its base and opposite the corolla-lobes. Styles 5. Ovary superior. Fruit dry, 1-seeded, indehiscent or circumscissile.

1   Flowers pink to purple, in dense terminal hemispheric clusters to 2.5 cm broad; leaves linear, greyish green, to about 8 cm long; scapes to about 3 dm tall; (essentially trans-continental) . . . . *Armeria*

1   Flowers lavender, solitary or in pairs along one side of a spike, the spikes in an open panicle; leaves spatulate to lance-obovate; scapes to about 6 dm tall; (E Que. to Nfld. and N.S.) . . . . *Limonium*

ARMERIA Willd. [6350] Thrift

A. maritima (Mill.) Willd.

/AST/X/GEA/ (Ch (Hr)) Cliffs, tundra, gravelly barrens, and shores (ranges of Canadian taxa outlined below), s in the West to s Calif. (an isolated station in Colo.), farther eastwards s to L. Athabasca, s Dist. Keewatin, James Bay, E Que., and Nfld.; circumgreenlandic; Iceland; Eurasia. MAPS and synonymy: *see* below.

1   Outer involucral bracts usually not more than half as long as the obtuse inner ones; leaves flat, rarely over 6 cm long; [*A. (Statice) sib.* Turcz.; *A. (S.) vulgaris* var. *sib.* (Turcz.) Rosenv.; *S. maritima* var. *sib.* (Turcz.) Simmons; N Alaska; s Ellesmere Is. to Chesterfield Inlet and s Baffin Is.; MAP: G.H. Lawrence, Am. Midl. Nat. 37: fig. 1, p. 758. 1947] . . . . var. *sibirica* (Turcz.) Lawrence

1   Outer involucral bracts more than half as long as the inner ones; leaves to over 1.5 dm long.

    2   Calyx-tube glabrous; leaves relatively broad, at most about 5 cm long but usually at least 2 mm broad (to 3.5 mm); [*Statice interior* Raup; type from L. Athabasca, Sask.] . . . . var. *interior* (Raup) Lawrence

    2   Calyx-tube pubescent at least on the vertical ribs; leaves usually over 5 cm long (if shorter, then less than 2 mm broad).

        3   Inner involucral bracts obtuse, the sheath formed by the two outer, reflexed, connate bracts shorter than the diameter of the head; calyx with pubescent cross-ribs, the spaces between the vertical ribs also usually pubescent; scapes rarely over 2 dm tall; [*A. (Statice) lab.* Wallr.; *A. scabra* ssp. *lab.* (Wallr.) Inversen; *St. (A.) lab.* vars. *genuina* and *submutica* Blake and f. *glabriscapa* and *pubiscapa* Blake of the latter taxon; Banks Is. and Great Bear L. to Ellesmere Is., Labrador (presumed type locality), Nfld., and the serpentine plateau of Mt. Albert, Gaspé Pen., E Que.; MAPS: Lawrence, loc. cit., fig. 1, p. 758; Porsild 1957: map 279, p. 195, and 1955: fig. 22, p. 173; Hultén 1958: map 88, p. 107; H.G. Baker, Evolution 7:126. 1953] . . . . var. *labradorica* (Wallr.) Lawrence

        3   Inner involucral bracts acute, mucronate, or obtuse (but if obtuse, the spaces between the ribs glabrous); scapes often over 2 dm tall, the terminal sheath usually longer than the diameter of the head.

            4   Leaves mostly not over 1 mm broad; outer (and often the inner) involucral bracts usually mucronate; spaces between the calyx-ribs glabrous or pubescent; [*Statice* Mill.; *A. vulgaris* var. *mar.* (Mill.) Rosenv.; *A. elongata* var. *mar.* (Mill.) Skottsb.; s Greenland; MAPS: Lawrence, loc. cit., fig. 1, p. 758; Porsild 1955: fig. 22, p. 173; Hultén 1958: map 88, p. 107] . . . . var. *maritima*

            4   Leaves mostly at least 1.5 mm broad; spaces between the calyx-ribs glabrous.

                5   Outer involucral bracts lance-attenuate, acute or cuspidate, surpassing the inner ones; leaves glabrous; [*A. andina* var. *cal.* Boiss.; *A. macloviana* ssp. *cal.* (Boiss.) Iversen; *Statice arctica* var. *cal.* (Boiss.) Blake;

Vancouver Is., B.C.; MAPS: Lawrence, loc. cit., fig. 1, p. 758; Porsild 1955: fig. 22, p. 173; Hultén 1958: map 88, p. 107] .... var. *californica* (Boiss.) Lawrence

5 Outer involucral bracts ovate to obovate, obtuse, usually shorter than the inner ones; leaves often minutely ciliate below the middle; [*A. vulgaris* Willd. and its var. *purp.* Mert. & Koch f. *arctica* Cham. in part, and ssp. *arct.* (Cham.) Hult.; *A. arctica* (Cham.) Wallr. in part; *A. scabra* ssp. *arct.* (Cham.) Iversen; *A. campestris* var. *chamissoi* Wallr.; *Statice arct.* var. *genuina* Blake; Alaska and NW Dist. Mackenzie; ?Vancouver Is.; MAPS: Lawrence, loc. cit., fig. 1, p. 758; Hultén 1968*b*:752, and 1958: map 88 (ssp. *arct.*), p. 107; Porsild 1955: fig. 22 (ssp. *arct.*; the Vancouver Is. plant being referred entirely to ssp. *californica*), p. 173] .... var. *purpurea* (Mert. & Koch) Lawrence

LIMONIUM Mill. [6351]

L. carolinianum (Walt.) Britt.   Sea-Lavender, Marsh-Rosemary

/T/EE/ (Hs) Coastal salt marshes from E Que. (St. Lawrence R. estuary from St-Jean-Port-Joli, l'Islet Co., to Charlevoix and Rimouski counties and the Gaspé Pen.) to Nfld., N.B., P.E.I., and N.S., S to Fla. and Tex. [*Statice* Walt.; *S. limonium* var. *car.* (Walt.) Gray; incl. *L. nashii* Small and *L. trichogonum* Blake]. MAP: Hultén 1958: map 272, p. 291.

Reports of the European *S. limonium* L. (*L. vulgare* Mill.) from Canada are mostly referable here, it being known only from cent. Sask. (Big Muddy) and S Ont. (in a weedy section of a cemetery at Lansing, York Co., where "growing without cultivation"; TRT).

OLEACEAE (Olive Family)

Shrubs with simple entire leaves or trees with pinnately compound leaves (the 5–11 leaflets subentire or shallowly toothed). Flowers perfect or unisexual, regular, hypogynous. Calyx-lobes small or wanting. Corolla wanting or well developed and 4-lobed. Stamens 2. Style 1, the stigma 2-cleft. Ovary superior. Fruit a berry, capsule, or samara.

1 Leaves pinnately compound, the leaflets entire or shallowly serrate; flowers very small, usually unisexual, in racemes or panicles in the axils of the leaf-scars; corolla none; fruit a dry winged narrowly elliptic or oblanceolate samara; trees . . . . *Fraxinus*
1 Leaves simple; flowers conspicuous, with well-developed 4-lobed corollas, perfect; shrubs; (introd.).
  2 Fruit a blackish drupe-like hard berry with 1 or 2 seeds; corolla white, narrowly campanulate; leaves ovate- to oblong-lanceolate, firm, obtusish, entire, tapering to a short petiole . . . . *Ligustrum*
  2 Fruit a 2-locular capsule with winged seeds.
    3 Flowers 1–3 in the leaf-axils, to 2.5 cm long, bright yellow with a slight greenish tinge, the revolute corolla-lobes much longer than the tube; seeds numerous; leaves generally lanceolate to elliptic-oblong, cuneate at base, usually serrate only above the middle (sometimes entire), their petioles to 12 mm long . . . . [*Forsythia*]
    3 Flowers numerous in panicles, less than 1.5 cm long, usually lilac (sometimes white), the horizontally spreading corolla-lobes no longer than the tube; seeds 4; leaves ovate, truncate or subcordate at base, entire, long-petioled . . . . *Syringa*

[FORSYTHIA Vahl] [6421]

[F. viridissima Lindl.] Golden-bells
[Asiatic; spreading slightly or persisting about old places in N. America but scarcely established, as in s Ont. (near London, Middlesex Co., where growing along the borders of a woods.)]

FRAXINUS L. [6420] Ash. Frêne

1 Calyx a mere ring or none; fruits linear-oblong to oblong, winged nearly to the bluntish base; leaflets 7–11(13), serrate.
  2 Twigs 4-angled or narrowly winged; lateral leaflets short-stalked, green on both sides; calyx minute, deciduous; (s Ont.) . . . . *F. quadrangulata*
  2 Twigs terete; leaflets sessile; calyx none.
    3 Base of leaflets rusty-tomentose along the midrib; flowers dioecious; (SE Man. to Nfld. and N.S.) . . . . *F. nigra*
    3 Base of leaflets not tomentose; flowers polygamous; (introd.) . . . . *F. excelsior*
1 Calyx present, persisting at the base of the fruit; fruits tapering below, only the upper half winged; leaflets 5–9.
  4 Wing extending only along the upper third of the fruit, the free part of the wing (above the tip of the fruit) longer than the seed-containing body; leaflets distinctly whitened beneath, undulate or serrate, the stalks of all but the terminal one nearly wingless; (Ont. to N.S.) . . . . *F. americana*
  4 Wing extending to the middle of the fruit, the free part above the tip about equalling the seed-containing body; lateral leaflets sessile or on short winged stalks.
    5 Leaflets more or less whitened beneath, entire or crenate-serrate, short-acute to acuminate; samaras oblanceolate, to 5 cm long and 9 mm broad; (sw B.C.; ?extinct) . . . . [*F. latifolia*]
    5 Leaflets scarcely whitened beneath, taper-pointed; samaras linear-oblanceolate, to over 6 cm long but rarely over 6 mm broad; (Sask. to N.S.) . . . . *F. pennsylvanica*

F. americana L. White Ash. Frêne blanc or Franc-frêne
/T/EE/ (Mg) Rich woods from Minn. to Ont. (N to the SE shore of L. Superior and the Ottawa dist.), Que. (N to the Gaspé Pen. at Mont-St-Pierre; MT), N.B., P.E.I., and N.S., s to Tex. and Fla.

[*F. canadensis* Gaertn.; *F. epiptera* Michx.]. MAPS: Hosie 1969:304; Fowells 1965:191; Gleason and Cronquist 1964: fig. 14.7, p. 161; Preston 1961:346; Canada Department of Northern Affairs and Natural Resources 1956:274; Hough 1947:391; Munns 1938: map 165, p. 169; Little 1971: map 126-N.

Forma *iodocarpa* Fern. (fruits purple rather than brownish yellow) is known from Ont. (Boivin 1966*b*) and SW Que. (N to the Montreal dist.; GH; MT). Var. *microcarpa* Gray (the small-fruited extreme with fruits at most about 2.5 cm long at maturity) is reported from S Canada by Fernald *in* Gray (1950).

F. excelsior L.   European Ash
Eurasian; occasionally spreading to roadsides, railway-embankments, and waste ground and becoming established in N. America, as in Nfld. (sandy and gravelly banks of the Waterford R., near St. John's; CAN; GH) and N.S. (Wolfville, Kings Co.; ACAD; reported from Pictou, Dartmouth, and Bridgewater by M.L. Fernald, Rhodora 50(596):214. 1948).

[F. latifolia Benth.]
[This species of the W U.S.A. (Wash. to Calif.) is known from Canada only through early collections in SW B.C. (Cloverdale, near Vancouver; Victoria, Vancouver Is.), where taken by John Macoun in 1887 and 1893 and apparently now extinct. (*F. oregona* Nutt.). MAP: Preston 1961:352 (indicating the occurrence on the SW B.C. mainland, where perhaps once native).]

F. nigra Marsh.   Black Ash.   Frêne noir or Frêne gras
/T/EE/   (Ms)   Swamps and shores from SE Man. (N to Sasaginnigak L., about 125 mi NE of Winnipeg) to Ont. (N to the W James Bay watershed at ca. 53°N; Dutilly, Lepage, and Duman 1954), Que. (N to Anticosti Is. and the Gaspé Pen.), Nfld., N.B., P.E.I., and N.S., S to N.Dak., Iowa, Ohio, and Del. [*F. sambucifolia* Lam.]. MAPS: Little 1971: map 129-N; Hosie 1969:310; Fowells 1965:182; Preston 1961:350; Canada Department of Northern Affairs and Natural Resources 1956:276; Hough 1947:385; Munns 1938: map 164, p. 168.

F. pennsylvanica Marsh.   Red Ash.   Frêne rouge
/T/(X)/   (Ms)   Low grounds and shores, the aggregate species from Mont. and SE ?Alta. to S Sask. (N to near Moose Jaw), Man. (N to The Pas), Ont. (N to White River, N of L. Superior, 48°35′N), Que. (N to L. St. John and Cabano, Temiscouata Co.), N.B., P.E.I., and N.S., S to E Tex. and N Fla. MAPS and synonymy: *see* below.
1   Plant essentially glabrous; leaflets serrate; fruits to about 4.5 cm long . . . .
. . . . var. *subintegerrima* (Vahl) Fern.
  2   Fruits green or yellowish green; [var. *lanceolata* (Borkh.) Sarg.; *F. lanceolata* Borkh.;
      F. juglandifolia var. *subint.* Vahl; *F. viridis* Michx. f.; Sask. to Que.; planted in P.E.I.
      and N.S, where possibly spreading; MAPS (all as var. *lanceolata* or *F. lanceolata* and
      apparently erroneously indicating an occurrence in SE Alta.): Canada Department of
      Northern Affairs and Natural Resources 1956:282; Munns 1938: map 168, p. 172;
      Preston 1961:348; Hough 1947:394] . . . . f. *subintegerrima*
  2   Fruits purplish; [type from Ottawa, Ont.; TRT; DAO; Boivin 1966*b*] . . . . f. *scotica* Boivin
1   Petioles, panicles, twigs, and lower leaf-surfaces conspicuously pubescent.
  3   Leaflets serrate; fruit usually less than 4 cm long . . . . var. *austinii* Fern.
    4   Fruits green or yellowish green; [Incl. f. *megaphylla* Vict. & Rousseau; Man. to
        N.B. and N.S.] . . . . f. *austinii*
    4   Fruits purplish; [type from Grand L., Queens Co., N.B.; DAO; Boivin 1966*b*] . . . .
        . . . . f. *colorata* Boivin
  3   Leaflets entire or merely undulate; fruit to over 7 cm long . . . . var. *pennsylvanica*
    5   Fruits green or yellowish green; [*F. campestris* Britt.; *F. pubescens* Lam.; S Ont.
        and SW Que., reports from elsewhere in Canada referring to the above varieties;
        MAPS (some for the aggregate species, the others excluding var. *subint.*): Fowells
        1965:185; Hosie 1969:306; Canada Department of Northern Affairs and Natural
        Resources 1956:280; Preston 1961:348; Munns 1938: map 167, p. 171; Hough
        1947:396; Little 1971: map 130-N] . . . . f. *pennsylvanica*

5   Fruits purplish; [type from Longueuil, near Montreal, Que,; MT] ....
.... f. *erythrocarpa* Vict. & Rousseau

F. quadrangulata Michx.   Blue Ash
/t/EE/   (Ms)   Dry or moist rich woods from Wisc. and Mich. to s Ont. (Pelee Point and islands of the Erie Archipelago, Essex Co.; reported from Middlesex Co. by Fox and Soper 1953 (*see* their s Ont. map, fig. 24, p. 27), and from Lambton Co. by Gaiser and Moore 1966), s to Okla., Ark., and Ala. MAPS: Preston 1961:352; Hough 1947:389; Munns 1938: map 163, p. 167; Little 1971: map 128-E; Hosie 1969:308.

LIGUSTRUM L.   [6436]

L. vulgare L.   Privet, Prim
European; spreading to thickets and open woods in N. America, as in sw B.C. (Elk L., Vancouver Is., where growing along a roadside ditch; Herb. V), s Ont. (Lambton, Welland, Lincoln, and Wellington counties), Nfld. (Rouleau 1956; ?established), and N.S. (border of a woods near Curry's Corner, Hants Co.; ACAD).

SYRINGA L.   [6423]

S. vulgaris L.   Lilac.   Lilas
European; often persisting or well established as an escape to roadsides and waste places in N. America, as in Sask. (Boivin 1966*b*), Ont. (N to the Ottawa dist.), Que. (Marie-Victorin 1935), Nfld. (Rouleau 1956), ?N.B. (Boivin 1966*b*), P.E.I. (Herbert Groh, Sci. Agric. 7(10):392. 1927), and N.S. (Yarmouth; CAN).

GENTIANACEAE (Gentian Family)

(Ref.: Gillett 1963*b*)

Smooth herbs with leaves commonly simple, sessile, and entire (but leaves 3-foliolate in *Menyanthes,* shallowly crenate in *Nymphoides,* and long-petioled in *Menyanthes, Fauria,* and *Nymphoides*; species of other genera sometimes with petioled basal leaves). Flowers regular, perfect, gamopetalous, commonly 4–5-merous (5–12-merous in *Sabatia*). Stamens as many as the corolla-lobes and alternating with them, inserted on the corolla-tube. Style 1, bearing 2 stigmas. Ovary superior or partially inferior. Fruit a 1-locular, usually 2-valved, many-seeded capsule. (Including Menyanthaceae).

1 Leaves reduced to minute subulate opposite or alternate scales; flowers white, yellowish, or purplish, 4-merous, at most 7 mm long, in slender terminal panicles; (eastern species) . . . . *Bartonia*
1 Leaves with normal blades.
  2 Leaves either compound or floating, alternate along thick creeping rootstocks; flowers 5-merous; aquatic or marshland perennials.
    3 Leaves 3-foliolate, long-petioled, the elliptic to obovate leaflets sessile or nearly so; flowers racemose, to over 1.5 cm broad; corolla white to roseate, its lobes conspicuously fringed on the upper surface; ovary partially inferior; (trans-continental) . . . . *Menyanthes*
    3 Leaves simple, floating, cordate-ovate, shallowly crenate, on long petiole-like stems bearing near the summit and beneath the solitary leaf umbel-like clusters of white or creamy flowers, these at most about 1 cm broad; corolla-lobes naked or merely ciliate; ovary superior; (Ont. to Nfld. and N.S.) . . . . *Nymphoides*
  2 Leaves simple, entire, not floating.
    4 Corolla rotate, the lobes longer than the tube.
      5 Leaves all basal, the blades cordate-ovate to broadly reniform, to 12 cm broad (usually considerably broader than long), finely to coarsely crenate, on petioles to 3 dm long; scapes to 5 dm tall, naked, the terminal cyme loosely flowered; corolla white, the tube to 4 mm long, the lobes to 6 mm long (their midnerve and usually their margins with erect toothed flanges running lengthwise); ovary partially inferior; perennial with thick fleshy rhizomes; (s Alaska–w B.C.) . . . . *Fauria*
      5 Leaves opposite on the stem (whorled in *Frasera caroliniensis*; basal tufts also often present), longer than broad, sessile; ovary superior.
        6 Corolla-lobes bearing 2 fringed crown-scales near base within, typically blue or bluish purple; stigma none or short and thick; flowers long-pedicelled; leaves opposite; plants glabrous.
          7 Leaves narrowly lanceolate to oblanceolate or spatulate, to 3 cm long, the 4 or 5 (rarely 6 or 7) acute lobes to 1.5 cm long; style none, the stigmas decurrent down the upper half of the ovary; flowers solitary in the leaf-axils and at the tip of the stem; slender annuals or biennials to 2.5 dm tall; (transcontinental) . . . . *Lomatogonium*
          7 Leaves mostly basal, oblong-elliptic to obovate, to over 1 dm long, the basal ones narrowed to winged petioles about as long as the blade; corolla bluish purple, spotted with green or white, the 5 broadly rounded, often erose lobes about 1 cm long; style less than 1 mm long; flowers solitary in the leaf-axils or the inflorescence sometimes thyrsoid; rather stout perennials from short thick rhizomes; (Alaska–B.C.) . . . . *Swertia*
        6 Corolla-lobes with only 1 or no crown-scales at base; style slender (sometimes deciduous).
          8 Corolla 4-lobed, dark blue to purplish or greenish yellow dotted with purple; inflorescence thyrsoid or paniculate; leaves opposite or whorled; (s Ont.) . . . . *Frasera*

8   Corolla 5–12-lobed, typically roseate with a yellowish eye; inflorescence cymose, the flowers mostly solitary on alternate or opposite upper branches; leaves opposite; (N.S.) . . . . *Sabatia*

4   Corolla salverform to tubular or campanulate, the tube equalling or longer than the lobes; flowers solitary or cymose; leaves opposite.

    9   Corolla usually 4-spurred at base, 4-cleft, purplish green, to 1.5 cm long; lower leaves spatulate to oblanceolate, the upper ones lanceolate to ovate; (transcontinental) . . . . *Halenia*

    9   Corolla spurless, whitish or yellowish to blue, violet-blue, or reddish purple; flowers 4–5-merous.

        10   Corolla salverform, the tube at most about 2 mm thick and about 12 mm long, the abruptly horizontally divergent lobes to 6 mm long; style filiform; anthers spirally twisted after flowering; flowers yellowish to deep pink-red . . . . *Centaurium*

        10   Corolla tubular, funnelform, or nearly campanulate, the tube usually over 3 mm thick, its lobes erect to horizontally spreading; style short or none; anthers remaining straight; flowers solitary or cymose.

            11   Corolla-lobes firm, mostly erect to spreading-ascending (or finally horizontally spreading in *G. nivalis*), entire, alternating with thin, commonly toothed, cleft, or fringed plaits in the sinuses (the fringed sinus-plaits much larger than the corolla-lobes in *G. andrewsii* and not to be confused with them); nectar-glands in a whorl at the base of the ovary, not inserted on the base of the corolla-tube; capsules stipitate; seeds often winged; flowers usually 5-merous (those of *G. aquatica, G. nivalis,* and *G. prostrata* often 4-merous), sessile or very short-pedicelled at the summit of the stem or in the upper axils, each subtended by 1 or 2 basal bracts; chiefly perennials, often with clustered stems . . . . *Gentiana*

            11   Corolla-lobes more or less spreading, shallowly to deeply fringed at least on the lower lateral margins, lacking plaits in the sinuses; nectar-glands located in a whorl on the base of the corolla-tube, alternating with the stamens; capsules sessile or stipitate; seeds wingless; flowers 4–5-merous, usually distinctly pedicelled, lacking basal bracts; annuals or biennials . . . . *Gentianella*

BARTONIA Muhl. [6501]

1   Scale-leaves chiefly alternate below the inflorescence, the lower stem-nodes only slightly reduced; corolla white, to 7 mm long, its lanceolate lobes entire, acute, non-apiculate; capsule dehiscing by terminal separation of the short style; (Nfld., St-Pierre and Miquelon, N.B., and N.S.) . . . . *B. paniculata*

1   Scale-leaves chiefly opposite below the inflorescence, progressively crowded toward the purple base; corolla greenish yellow, usually less than 4 mm long, its oblong lobes apiculate at the erose or entire apex; capsule dehiscing below the usually elongate style; (s Ont. to St-Pierre and Miquelon and N.S.) . . . . *B. virginica*

B. paniculata (Michx.) Muhl.   Screw-stem
/T/EE/   (Hp)   Wet peat and sand from St-Pierre and Miquelon and Nfld. to s N.B. (Grand Manan Is., Charlotte Co.; GH; not known from P.E.I.) and N.S., s to E Tex. and Fla. [*Centaurella* Michx.; *B. lanceolata* Small; *B. virginica* var. *pan.* (Michx.) Boivin; incl. vars. *intermedia* and *sabulonensis* Fern.]. MAPS: J.M. Gillett, Rhodora 61(722): map 2, p. 51. 1959; *Atlas of Canada* 1957: map 14, sheet 38; Braun 1937: fig. 12, p. 197; Fernald 1933: map 24, p. 281.

All our material is referred by Gillett to the northern phase, ssp. *iodandra* (Robins.) Gillett (*B. iodandra* Rob., the type from Nfld.; plant relatively stout, the stem purple throughout rather than green or essentially so; anthers usually purple rather than yellow (if yellow, the filaments often purple), sometimes apiculate rather than always rounded at apex, to 1 mm long rather than at most about 0.5 mm).

B. virginica (L.) BSP.
/T/EE/ (Hp) Sphagnous bogs, peaty and sandy shores, and dry thickets from Minn. to Ohio, S Ont. (Norfolk Co.; OAC; reported from Lambton Co. by Dodge 1915), SW Que. (Montreal dist.), St-Pierre and Miquelon, and SW N.S. (not known from N.B. or P.E.I.), S to La. and Fla. [*Sagina* L.; *B. tenella* Muhl.]. MAPS: J.M. Gillett 1963*b*: fig. 34, p. 81, and Rhodora 61(722): map 1, p. 51. 1959.

CENTAURIUM Hill [6496] Centaury

(Ref.: Gillett 1963*b*)
1 Basal leaves several, often forming a tuft or rosette, rather strongly veined from the base; flowers nearly sessile, usually numerous in compact terminal cymes of the cymose inflorescence; calyx membranous in the sinuses; corolla yellowish to pinkish-red; anthers to 2 mm long; plants to 5 dm tall; (introd.) . . . . *C. erythraea*
1 Basal leaves usually well spaced and not forming rosettes, only the main vein conspicuous; flowers often pedicelled, usually few, white to deep pink or pink-red; anthers usually less than 1.5 mm long; plants rarely over 3 dm tall.
  2 Calyx membranous in the sinuses; (introd.) . . . . *C. pulchellum*
  2 Calyx not membranous in the sinuses; (SW B.C.).
    3 Pedicels usually over 2 cm long and longer than the central flowers . . . . . . *C. exaltatum*
    3 Pedicels all less than 2 cm long and usually much shorter than the flowers . . . .
    . . . . *C. muhlenbergii*

C. erythraea Raf.
Eurasian; introd. in meadows, fields, ditches, and waste places in N. America, as in SW B.C. (Vancouver Is. and adjacent islands; V; Carter and Newcombe 1921), S Ont. (Norfolk and Welland counties), SW Que. (Owl's Head Mt., Brome Co.; CAN), and N.S. (Sable Is.; GH). [*Gentiana (Erythraea) centaurium* of auth., not L.; *C. minus* of auth., not Moench; *C. umbellatum* of auth., not Gilib.]. MAP: Gillett 1963*b*: fig. 33, p. 78.

C. exaltatum (Griseb.) Wight
/t/WW/ (T) Moist places (often around hot springs and alkaline lakes) from southernmost B.C. (margins of a saline pond at Osoyoos, near the U.S.A. boundary about 30 mi S of Penticton; V) to Nebr., S to Calif. and Colo. [*Cicendia* Griseb.].

C. muhlenbergii (Griseb.) Wight
/t/W/ (T) Moist soil from SW B.C. (Stanley Park, Vancouver; Henry 1915; Gillett 1963*b*) to Calif. and Nev. [*Erythraea* Griseb.].

C. pulchellum (Sw.) Druce
Eurasian; introd. in fields and waste places of N. America, as in SE N.B. (Cape Tourmentine, where taken by Soeur Ste-Marie in 1947 and noted as common in a field; QFA). [*Gentiana* Sw.].

FAURIA Franchet [6542]

F. crista-galli (Menzies) Makino Deer-Cabbage
/sT/W/eA/ (Grh (Hel)) Sphagnous bogs, swamps, and wet ground from S Alaska (*see* Hultén 1948: map 984, p. 1340) through coastal B.C. to N Wash.; Japan. [*Menyanthes cr.* Menzies, the type from Prince William Sound, S Alaska; *Nephrophyllidium* Gilg; *Villarsia* Griseb.]. MAPS: Hultén 1968*b*:762; Gillett 1963*b*: fig. 37, p. 87.

FRASERA Walt. [6512]

1 Stem-leaves opposite, linear, prominently 3-nerved from the base, commonly with thickened narrow white margins, the upper ones reduced (principal leaves in a basal tuft, to 3 dm long and 2 cm broad, linear-oblanceolate to narrowly spatulate); inflorescence a narrow, open or congested, mostly interrupted thyrse; calyx-lobes dark blue or purplish;

corolla pale to fairly dark bluish or purplish, the lobes about 1 cm long; perennial to about 7 dm tall, from a branching caudex; (s ?B.C.) . . . . *[F. albicaulis]*

1   Stem-leaves mostly in whorls of 4, lance-oblong (the lowest ones spatulate), to 4 dm long and about 1 dm broad, not white-margined, the midrib prominent, the rest of the venation reticulate; inflorescence a loose pyramidal panicle, the lower branches to 1.5 dm long; calyx green; corolla greenish yellow dotted with purple, the lobes to about 1.5 cm long; glabrous biennial or triennial to about 3 m tall; (s Ont.) . . . . *F. caroliniensis*

[F. albicaulis Dougl.]
[The inclusion of s B.C. in the range of this species of w N. America (Wash. and Mont. to Calif. and Nev.) by Hitchcock et al. (1959) requires confirmation. (*F. nitida* var. *alb.* (Dougl.) Card; *Leucocraspedum* Rydb.; *Swertia* Ktze.). MAP: H.H. Card, Ann. Mo. Bot. Gard. 18(2): fig. 2, p. 252. 1931 (indicating no Canadian stations).]

F. caroliniensis Walt.   Columbo
/t/EE/   (Hs)   Rich woods and dryish meadows (often calcareous) from Wisc. to Mich., s Ont. (Lambton, Brant, Lincoln, Oxford, Waterloo, and Wentworth counties; *see* s Ont. map by Soper 1962: fig. 21, p. 34), and N.Y., s to La. and Ga. [*Swertia* Ktze.]. MAP: Gillett 1963*b*: fig. 16, p. 43.

GENTIANA L.   [6509]   Gentian.   Gentiane

(Ref.: Gillett 1963*b*; Pringle 1967).
1   Annuals, the stem usually branched from the base or throughout (rarely simple), at most about 2.5 dm tall, often with basal rosettes of leaves; stem-leaves rarely over 1.5 cm long; flowers 4–5-merous (evidently only 5-merous in *G. douglasiana*), less than 2 cm long, the tube greenish below, becoming blue or purple above and on the lobes; seeds wingless.
    2   Flowers terminal and solitary, 4–5-merous; stem branched from the base.
        3   Leaves green-margined, recurving; ovary-stalk extremely elongate (the capsule often completely exserted from the marcescent corolla); calyx to 12 mm long; corolla to 18 mm long; stems to 2.5 dm tall; (Alaska–Yukon–Dist. Mackenzie–B.C.–Alta.) . . . . *G. prostrata*
        3   Leaves white-margined, scarcely recurving; ovary-stalk very short (the subsessile capsule included in the corolla); calyx to 8 mm long; corolla to 10 mm long; stems less than 1 dm tall; (sw Alta. and s Sask.) . . . . *G. aquatica*
    2   Flowers mostly in terminal simple cymes, the stem usually branched throughout (rarely simple); ovary-stalk short or none (the sessile or subsessile capsule included in the corolla).
        4   Corolla about 1 cm long, its lobes 1/3 as long as the tube; calyx campanulate, its lobes 1/2 as long as the tube; seeds light tan, smooth; (Alaska–B.C.) . . . .
            . . . . *G. douglasiana*
        4   Corolla to 2.5 cm long, its lobes about 1/5 as long as the tube; calyx tubular, its lobes 1/4 as long as the tube; seeds dark brown, reticulate; (Labrador and Greenland) . . . . *G. nivalis*
1   Perennials, the 1–several usually simple stems arising from a stout rootstock or from horizontal rhizomes, often taller; stem-leaves mostly longer (but the lower ones usually becoming smaller and somewhat scale-like, basal rosettes present only in *G. algida* and *G. glauca*); flowers 5-merous, usually at least 3 cm long (at most 2 cm long in *G. glauca*); seeds often winged.
    5   Corolla yellowish green (rarely white; *G. glauca* f. *chlorantha* will key out here), to 5 cm long, the tube flecked with purple; seeds with irregular hyaline wings; leaves linear to linear-oblanceolate, the basal ones forming a loose rosette; stem to 2 dm tall, from a vertical rootstock; (Alaska–Yukon) . . . . *G. algida*
    5   Corolla blue, purple, or white (sometimes greenish-mottled).
        6   Calyx consisting of two roseate spathes about half as long as the corolla-tube, its lobes being mere teeth 1 or 2 mm long (2 teeth on one spathe, 3 on the other); corolla to about 3.5 cm long, its lobes reniform; seeds wingless; leaves elliptic to

ovate, the lower ones progressively reduced; stem to 3 dm tall, from a horizontal rhizome; (Alaska–B.C.) .... G. platypetala

6   Calyx tubular, its lobes usually well developed.

7   Stem-leaves at most about 1 cm long, not reduced down the stem, the basal leaves to 2 cm long and forming rosettes (separate winter rosettes also arising from the slender horizontal rhizome); corolla to 2 cm long; seeds irregularly wavy-winged; stem to 1.5 dm tall; (Alaska–Yukon–Dist. Mackenzie–B.C.–w Alta.) .... G. glauca

7   Stem-leaves longer, the lower ones reduced and more or less scale-like, not forming rosettes; corolla usually at least 3 cm long; seeds wingless or with a flat wing; stems commonly taller.

8   Flowers mostly solitary and terminal or 3 in a terminal cyme, to 4 cm long; sinus-plaits between the corolla-lobes deeply 2-toothed to somewhat lacerate; capsule sessile; seeds wingless; leaves ovate to obovate or ovate-rotund, mostly semicordate and sheathing at base, commonly not over 2.5 cm long, rarely as much as twice as long as broad; (B.C. and Alta.) .... G. calycosa

8   Flowers commonly more numerous, or also axillary; capsule stipitate; leaves mostly narrowly to broadly lanceolate and at least twice as long as broad.

9   Sinus-plaits between the corolla-lobes nearly or quite entire; corolla to 4.5 cm long; seeds fusiform, wingless; leaves mostly oblong-lanceolate, to 6 cm long and about 2 cm broad; stems to over 1 m tall; (w B.C.) .... G. sceptrum

9   Sinus-plaits variously toothed or fringed; seeds flattened and wing-margined.

10   Anthers free or promptly separating; corolla-lobes often more or less spreading at maturity; seeds mostly restricted to the region of the sutures of the capsule; leaves firm, to about 5 cm long; stems to 5 dm tall.

11   Leaves not more than 13 pairs below the inflorescence; calyx-tube to 7 mm long, the lobes unequal (the shortest ones often minute) and not over half the length of the tube; corolla to 3 cm long, its lobes not much surpassing the laciniate appendages; (B.C. to Man.) .... G. affinis

11   Leaves up to 19 pairs below the inflorescence; calyx-tube about 1 cm long, its subulate to linear-lanceolate lobes subequal and about equalling the tube; corolla to 4.5 cm long, its lobes at least twice as long as the 2-cleft appendages; (s Man. and s Ont.) .... G. puberulenta

10   Anthers at anthesis cohering in a tube or cone around the style; corolla-lobes erect or ascending; seeds covering the inner walls of the capsule; leaves mostly softer and longer.

12   Calyx-lobes smooth or at most scabrous-margined; corolla to over 4 cm long, its lobes 1 or 2 mm longer than the 1–3-toothed sinus-plaits; (s Man. to s Labrador and N.B.) .... G. linearis

12   Calyx-lobes distinctly ciliate; sinus-plaits of the corolla 2–3-cleft or fringed.

13   Calyx-lobes oblanceolate; corolla slightly open, to 5 cm long, its lobes to over 7 mm long; leaves obtuse to acute but usually not acuminate; (?Ont) .... [G. saponaria]

13   Calyx-lobes lanceolate to obovate or orbicular; corolla completely closed, rarely over 4.5 cm long; involucral and upper leaves acuminate.

14   Corolla-lobes reduced to a minute projection much surpassed by the fringed sinus-plaits; (Sask to Que.) .... G. andrewsii

14 Corolla-lobes larger, rounded and up to 2 or 3 mm long.
  15 Calyx-lobes lanceolate; sinus-plaits fringed; (var.
     dakotica; Sask. and Man.) .... *G. andrewsii*
  15 Calyx-lobes obovate to orbicular; sinus-plaits 2–3-
     cleft; (SW Que.) .... *G. clausa*

G. affinis Griseb.

/T/WW/ (Hp) Meadows and damp places at low to fairly high elevations from SE B.C. (Cranbrook; South Kootenay Pass and Crowsnest Pass; there is an 1856 McTavish collection in CAN, purportedly from Fort Good Hope, S Dist. Mackenzie, but more likely wrongly labelled) to S Alta. (N to Red Deer; CAN), S Sask. (N to near Battleford; CAN), and S Man. (N to near Grandview, S of Duck Mt.; a collection in WIN from Churchill perhaps reflects a casual introduction there), S to Calif., Ariz., Colo., and S.Dak. [*Dasystephana* Rydb.; *Pneumonanthe* Greene; *G. (D.) interrupta* Greene; *G. (D.) oregana* Engelm.; incl. *G. (D.; P.) forwoodii* Gray, with calyx-lobes nearly or quite lacking]. MAP: Gillett 1963*b*: fig. 9, p. 28.

G. algida Pallas

/ST/W/A/ (Hs) Meadows and stony slopes at low to fairly high elevations in the Aleutian Is., Alaska (N to the Seward Pen.), and w Yukon (N to ca. 65°N; *see* Hultén 1948: map 971, p. 1339), with a disjunct southern area from Mont. to Utah and Colo.; Asia. [*G. (Dasystephana) romanzovii* Ledeb.; *G. frigida* of auth., not Haenke]. MAPS: Hultén 1968*b*:754: Gillett 1963*b*: fig. 1, p. 11.

G. andrewsii Griseb. Closed Gentian

/T/EE/ (Hp) Moist places and shaded woods, the aggregate species from SE Sask. (Zeneta, about 35 mi SE of Yorkton; Breitung 1957*a*) to Man. (N to Dauphin), Ont. (N to Sioux Lookout, about 170 mi NW of Thunder Bay), and Que. (N to Hull and Montreal; *see* Que. map by Raymond 1950*b*: fig. 26, p. 67; reported N to Quebec City by John Macoun 1884; reports from the Maritime Provinces probably refer chiefly to *G. linearis,* from Nfld. possibly to a species of *Gentianella*), S to S.Dak., Iowa, Ill., and Va. MAPS and synonymy (together with a distinguishing key to *G. alba,* reported from Canada): *see* below.

1 Calyx-lobes with bracket-shaped keels decurrent on the tube, their margins minutely
  denticulate but not ciliate; corolla slightly open, to 5.5 cm long, basically white (the veins
  and some of the veinlets green, the outer edges of the lobes sometimes tinged with
  red-violet, the inside base of the tube streaked with grey-green to purple); [*G.
  (Dasystephana; Pneumonanthe) flavida* Gray; reported from Northumberland Co., S Ont.,
  by Pringle 1967 (perhaps on the basis of the report from Heely Falls by John Macoun
  1884) but Macoun's other Ont. citations are chiefly or wholly based upon *G. linearis* and
  its ssp. *rubricaulis* (relevant collections in CAN) and the Northumberland plant (MTMG;
  taken by Macoun in 1891) may be the white-flowered *G. andrewsii* f. *albiflora;* MAP:
  Pringle 1967: fig. 15, p. 20] .... [*G. alba* Muhl.]
1 Calyx-lobes not keeled, their margins ciliate; corolla completely closed, to 4.5 cm long,
  typically blue (becoming blue-violet) above a whitish base (the tube blue-striped within) ....
  .... *G. andrewsii*
  2 Corolla-lobes to 3 mm long, triangular to rounded, often mucronate; [Sask. and Man.;
    Pringle 1967] .... var. *dakotica* Nels.
  2 Corolla-lobes reduced to minute projections less than 1 mm long .... var. *andrewsii*
    3 Corolla white or roseate.
      4 Corolla white; [S Man. (Winnipeg) and S Ont. (Norfolk, Waterloo, and Carleton
        counties)] .... f. *albiflora* Britt.
      4 Corolla roseate; [type from Ile-Bizard, near Montreal, Que.] ....
        .... f. *rhodantha* Rouleau & Kucyniak
    3 Corolla predominantly blue or blue-violet; [*Dasystephana* Small; range of the
      species; MAPS: Pringle 1967: fig. 14, p. 16; Gillett 1963*b*: fig. 5, p. 20]. Pringle
      notes that *G. billingtonii* Farw. (type from Lambton Co., S Ont.) may be, as
      suggested by Farwell, a hybrid between *G. andrewsii* and *G. puberulenta* ....
      .... f. *andrewsii*

G. aquatica L.

/T/W/A/ (T) Bogs, sandy flats, and meadows at low to fairly high elevations from sw Alta. (N to the Banff dist.; reports from Alaska–Dist. Mackenzie–B.C. by Boivin 1966*b*, result from his inclusion of *G. prostrata* in this taxon) and s Sask. (*G. fremontii* reported from Mortlach, NW of Moose Jaw, by Breitung 1957*a*) to Colo.; Asia. [*G. fremontii* Torr.; *G. humilis* Stev., not Salisb.]. MAP: Gillett 1963*b*: fig. 11, p. 32.

G. calycosa Griseb.

/T/W/ (Hp) Alpine and subalpine slopes and meadows from SE B.C. (King Edward Peak, Kootenay dist.; DAO) and sw Alta. (Waterton Lakes; DAO; Breitung 1957*b*) to Calif. [*Dasystephana* Rydb.].

Our material is referable to var. *obtusiloba* (Rydb.) Hitchc. (*Dasystephana ob.* Rydb.; calyx-lobes similar to the tube in texture rather than thick and fleshy; flowers commonly confined to the terminal cluster rather than additional flowers often present on peduncles to 3 cm long from the upper 1 or 2 nodes).

G. clausa Raf.   Closed Gentian

/T/EE/ (Hp) Meadows, thickets, and borders of rich woods from Minn. to sw Que. (Shefford, about 50 mi E of Montreal; MT; Pringle 1967) and New Eng., s to Mo., Tenn., and N.C. [*Dasystephana* Heller; *Pneumonanthe* Greene]. MAP: Pringle 1967: fig. 15, p. 20.

G. douglasiana Bong.

/sT/W/ (T) Bogs and wet meadows from s Alaska (N to ca. 62°N; type from Sitka) through coastal B.C. to Wash. [Incl. f. *maculata* Boivin]. MAPS: Hultén 1968*b*:757; Gillett 1963*b*: fig. 12, p. 35.

G. glauca Pallas

/aST/W/eA/ (Grh) Tundra and alpine meadows from northernmost Alaska, N Yukon, and NW Dist. Mackenzie through B.C. and the mts. of sw Alta. (N to Jasper) to Mont.; E Asia. [*Dasystephana* Rydb.]. MAPS: Hultén 1968*b*:757; Gillett 1963*b*: fig. 2, p. 13; Raup 1947: pl. 33.

Forma *chlorantha* Jordal (flowers greenish yellow rather than blue or blue-green) is known from the type locality in the Brooks Range, Alaska.

G. linearis Froel.   Closed Gentian

/T/EE/ (Hp) Damp or wet places from Man. (N to Riding Mt.) to Ont. (N to the N shore of L. Superior near Thunder Bay and the w James Bay watershed at ca. 53°N), Que. (N to the E James Bay watershed at ca. 53°50'N, the Côte-Nord, and Gaspé Pen.), Labrador (N to the Hamilton R. basin), and N.B. (York, Kings, Kent, and Charlotte counties; not known from P.E.I. or N.S.), s to Nebr., Minn., Pa., and W.Va. [*Dasystephana (Pneumonanthe)* Britt.; *G. saponaria* var. *lin.* (Froel.) Griseb.; *G. ?ochroleuca sensu* Hooker 1838, not Froel.; *G. ?pneumonanthe sensu* A. Michaux 1803, and Pursh 1814, not L.]. MAPS: Gillett 1963*b*: fig. 6, p. 23; Pringle 1967: fig. 15, p. 20.

Some of our material is referable to var. *rubricaulis* (Schw.) Gillett (*G. rub.* Schw.; incl. *G. lin.* vars. *lanceolata* and *latifolia* Gray; leaves pale green rather than dark green, the involucral ones relatively broad and enveloping the calyces rather than spreading; calyx-lobes hyaline except at the tip rather than not hyaline).

G. nivalis L.   Snow Gentian

/aST/E/GEwA/ (Hs) Dry to moist turf, gravels, and ledges: Labrador (between ca. 55°N and Komaktorvik Bay at 59°22'N; w and E Greenland N to ca. 73°N; Iceland; Europe; Asia Minor. [*G. propinqua sensu* Delabarre 1902, as to the Saglek Bay plant, the relevant collection in GH]. MAPS: Hultén 1958: map 97, p. 117; Porsild 1951*b*: fig. 8, p. 143; Gillett 1963*b*: fig. 13, p. 36.

Forma *albiflora* (Lange) Gillett (var. *alb.* Lange, the type from Greenland; flowers white rather than predominantly blue) is known from Greenland.

G. platypetala Griseb.

/sT/W/ (Hp) Grassy slopes at low to moderate elevations from s Alaska (N to ca. 61°30'N; type from Sitka) to w-cent. B.C. (s to Queen Charlotte Is. and the adjacent mainland). [*Pneumonanthe*

Greene; *G. covillei* Nels. & Macbr.; *G. gormanii* Howell]. MAPS: Hultén 1968*b*:754; Gillett 1963*b*: fig. 3, p. 15, and fig. 4, p. 17.

G. prostrata Haenke   Moss Gentian
/ST/W/EA/   (T)   Damp soils and ledges from the E Aleutian Is., N Alaska–Yukon (N to ca. 69°30′N), and W Dist. Mackenzie (Nahanni Butte, ca. 63°N, 127°W; W.J. Cody, Can. Field-Nat. 77(4):227. 1963) through B.C. and the mts. of SW Alta. (N to Jasper) to N Calif., Utah, and Colo.; Eurasia. [*Chondrophylla* And.; *C. americana* (Engelm.) Nels.]. MAPS: Hultén 1968*b*:758; Gillett 1963*b*: fig. 10, p. 31.

G. puberulenta Pringle
/T/EE/   (Hp)   Damp soils from S Man. (N to McCreary, E of Riding Mt.; reports of *G. (Dasystephana) puberula* from Sask. by Rydberg 1922 and 1932, require confirmation with respect to the present species) and S Ont. (Toronto; CAN; *G. puberula* reported from Squirrel Is., Lambton Co., by Dodge 1915) to N.Y., S to Nebr., La., Ky., and Md. [*G. (Dasystephana) puberula* of Canadian reports, not Michx.]. MAPS: Pringle 1967: fig. 14, p. 16; Gillett 1963*b*: fig. 8 (as *D. puberula*), p. 26.

[*G. saponaria* L.]   Soapwort-Gentian
[Reports of this species of the E U.S.A. (N to Minn. and N.Y.) from S Ont. by John Macoun (1884), Dodge (1915), and Soper (1949) are probably based upon *G. andrewsii* or *G. puberulenta,* neither Gillett (1963*b*) nor Pringle (1967; MAP, fig. 14, p. 16) recording it for Canada. (*Dasystephana* Small; *G. puberula* Michx., not *sensu* most Canadian auth.).]

G. sceptrum Griseb.
/T/W/   (Hp)   Bogs and wet places (chiefly near the coast) from W B.C. (N to Prince Rupert, ca. 54°N; Herb. V, detd. Gillett) to NW Calif. [*Pneumonanthe* Greene]. MAPS: Gillett 1963*b*: fig. 3, p. 15, and fig. 4, p. 17.

GENTIANELLA Moench   [6509]   Gentian.   Gentiane

(Ref.: Gillett 1957 and 1963*b*)
1   Flowers 4-merous, commonly over 2.5 cm long; corolla normally blue or blue-violet, its lobes more or less deeply ciliate-fringed marginally at least near the base; calyx-lobes with thin hyaline margins, the sinuses with a small inner membrane at base; anthers distinctly longer than broad, attached in the upper third to the filaments; seeds papillate, borne over the entire inner surface of the capsule.
  2   Base of calyx and the glossy purple calyx-heels smooth or wrinkled but not papillate; corolla-lobes shallowly fringed in the lower half, the apex merely erose or dentate; upper leaves usually obtuse; (essentially transcontinental) .... *G. detonsa*
  2   Base of calyx near top of pedicel and usually at least one pair of green or purple calyx-heels minutely papillate, the lobe-margins also occasionally papillate; upper leaves acute; (B.C. to Que.) .... *G. crinita*
1   Flowers 4–5-merous, rarely over 2.5 cm long; corolla-lobes not marginally fringed; calyx-lobes with green margins, the sinuses lacking an inner membrane; anthers slightly longer than broad, attached at about the middle to the filaments; seeds smooth, chiefly in 2 rows along the margin of each suture.
  3   Pedicels surpassing the adjacent internodes; corolla blue to white, 4–5-merous, about 1 cm long, the lobes each bearing a pair of minute fringed scales within at base; calyx to 1 cm long, the tube nearly obsolete; median and upper leaves elliptic-ovate to ovate, obtuse; (essentially transcontinental) .... *G. tenella*
  3   Pedicels shorter than the adjacent internodes; corolla to 2 cm long.
    4   Lobes of corolla each subtended within at base by a rather densely ciliate-fringed hyaline scale; corolla 4–5-lobed, the tips of the lobes obtuse to acute (but not mucronate or bristle-tipped).
      5   Calyx-tube shorter than or at most equalling the acute to obtuse lobes; scales at base of corolla-lobes free at base; corolla lilac to pale blue (atypically

cream-colour); terminal flowers about equalling the lateral ones; (transcontinental) .... *G. amarella*

5 Calyx-tube longer than the rounded, often auricled lobes; scales at base of
corolla-lobes united at base; corolla blue; terminal flowers larger than the
lateral ones; (Aleutian Is.) .... *G. auriculata*

4 Lobes of corolla naked within at base (lacking a ciliate-fringed scale), mucronate
or bristle-tipped; corolla lilac to blue or blue-violet (atypically greenish white).

6 Stems to 8 dm tall, with up to 15 pairs of rather crowded ovate-lanceolate
leaves to over 5 cm long and 3 cm broad, these rounded to cordate-clasping
at base; flowers 4–5-merous, to 2 cm long, those of the branches scarcely
reduced; calyx-lobes subequal; pedicels usually shorter than the flowers;
(s Ont.) .... *G. quinquefolia*

6 Stems mostly not over 3 dm tall, simple or branched from the base, rarely
bearing more than 5 pairs of rather remote, elliptic to spatulate, non-clasping
leaves to about 3.5 cm long and 13 mm broad; flowers of the ascending basal
branches often reduced; calyx-lobes usually distinctly unequal in length or
shape.

7 Corolla to 2 cm long, it and the calyx 4-lobed; outer calyx-lobes relatively
broad and foliaceous; flowers solitary and axillary or in loose simple cymes
(the cymes occasionally aggregated), rarely subtended by a pair of
bract-like leaves; (essentially transcontinental) .... *G. propinqua*

7 Corolla to 11 mm long, it and the calyx usually 5-lobed; calyx-lobes
irregular but scarcely foliaceous; flowers in compact terminal aggregate
cymes or in a compact head, subtended by the upper pair or pairs of
bract-like leaves; (w Greenland) .... *G. aurea*

G. amarella (L.) Börner   Felwort
/ST/X/EA/   (T)   Meadows, beaches, and moist places from the Aleutian Is. and cent.
Alaska–Yukon–Dist. Mackenzie (N to ca. 65°30′N) to Great Bear L., Great Slave L., L. Athabasca
(Alta. and Sask.), Man. (N to the Churchill R. at ca. 57°15′N), northernmost Ont., Que. (N to s
Ungava Bay and the Côte-Nord), Labrador (N to ca. 54°N), Nfld., N.B., and N.S. (not known from
P.E.I.), s to Calif., Mexico, N.Mex., S.Dak., and Vt.; southernmost ?Greenland; Iceland; N Eurasia.
[*Gentiana* L.]. MAPS (aggregate species): Hultén 1968b:759; Gillett 1963b: fig. 29, p. 70, and 1957:
fig. 7, p. 252.

Our plant has been separated, chiefly on the basis of its somewhat smaller flowers, as ssp. *acuta*
(Michx.) Gillett (*Gentiana (Amarella) acuta* Michx., the type from near Tadoussac, Saguenay Co., E
Que., and its f. *albescens* Lepage and vars. *nana* Engelm., *stricta* Griseb., and *strictiflora* Rydb.
(*Gentiana strictiflora* (Rydb.) Nels.); *Gentiana anisosepala* Greene, *G. plebeja* Cham., and *G.
tenuis* Griseb.; *Amarella anisosepala*, *A. conferta*, and *A. macounii* Greene). M.L. Fernald
(Rhodora 19(224):150. 1917) notes that Michaux's type material from Tadoussac consists of plants
with ochroleucous (rather than lilac or violet) flowers, distinguishing this colour-phase as *Gentiana
amarella* f. *michauxiana* Fern. In conformity with the present treatment, it may be known as
Gentianella amarella f. *michauxiana* (Fern.) Scoggan.

G. aurea (L.) Sm.   Golden Gentian
/aSs/-/GE/   (T)   Coasts of w Greenland N to ca. 70°30′N, of E Greenland N to ca. 65°N; Iceland;
N Scandinavia. [*Gentiana* L.; *Gentiana involucrata* Rottb.]. MAP: Hultén 1958: map 67, p. 87.

G. auriculata (Pallas) Gillett
/s/W/eA/   (T)   Known in N. America only from subalpine meadows of the westernmost Aleutian
Is. (Attu Is.; *see* Hultén 1948: map 973, p. 1339); E Asia. [*Gentiana* Pallas; *Amarella* Greene]. MAP:
Hultén 1968b:758.

G. crinita (Froel.) Don   Fringed Gentian
/sT/X/   (T)   Moist places, the aggregate species from s Yukon (Mackintosh; CAN) and sw Dist.
Mackenzie (Fort Providence, sw of Great Slave L.; J.W. Thieret, Can. Field-Nat. 75(3):118. 1961) to
B.C.–Alta., Sask. (N to Prince Albert), Man. (N to Grand Rapids, near the NW end of L. Winnipeg),

Ont. (N to the W James Bay watershed at ca. 53°N), Que. (N to E James Bay at 52°N and the Gaspé Pen.), N.Y., and Maine, S to Mont., N.Dak., Iowa, Ohio, and N.C. MAPS and synonymy: *see* below.

1   Corolla to 6 cm long, the upper half of its lobes with a fringe to 5 mm long, the summit also often short-ciliate; calyx-lobes prominently keeled.

    2   Median leaves mostly lanceolate-ovate to ovate, rounded to subcordate at base; [*Gentiana crinita* Froel.; Man. to SW Que.; MAPS: Gillett 1957: fig. 1A, p. 214, and 1963*b*: fig. 21, p. 55, fig. 22, p. 57, fig. 23, p. 59, and fig. 24, p. 61] . . . . ssp. *crinita*

    2   Median leaves linear to linear-lanceolate; [*Gentiana (Anthopogon) procera* Holm, the type from near Sarnia, Lambton Co., S Ont.; *Gentiana barbata (crinita)* var. *browniana* Hook.; S Man. and S Ont.; MAPS: on the above-noted maps by Gillett] . . . .
. . . . ssp. *procera* (Holm) Gillett

1   Corolla at most about 4 cm long, the lobes short-fringed, the summit rarely ciliate; calyx-lobes lacking prominent keels.

    3   Corolla-lobes orbicular; stigmas flabelliform, on styles to 1.5 mm long; [*Gentiana vict.* Fern., the type from near Quebec City, Que., from where it extends in the intercotidal zone of the St. Lawrence R. estuary to Portneuf, Deschambault, and l'Islet counties; MAPS: on the above-noted maps by Gillett] . . . . ssp. *victorinii* (Fern.) Gillett

    3   Corolla-lobes usually oblong; stigmas reniform, sessile or nearly so; [*Gentiana (Anthopogon) mac.* Holm, the type from Cardston, SW Alta.; *Gentiana gaspensis* Vict. and *G. tonsa* (Lunell) Vict. (*Gentiana detonsa* var. *tonsa* Lunell); *A. tonsum* (Lunell) Rydb.; the Yukon–S Dist. Mackenzie–B.C. to E Que.; MAPS: on the above-noted maps by Gillett] . . . . ssp. *macounii* (Holm) Gillett

G. detonsa (Rottb.) Don   Fringed Gentian

/aST/X/GEA/   (T)   Meadows, bogs, and moist ground, the aggregate species from Alaska (N to near the Arctic Circle), the Yukon (N to ca. 63°), and the coast of Dist. Mackenzie (E to Coronation Gulf) to N Alta. (Wood Buffalo National Park; not known from B.C., Sask., or Man.), N Ont. (coasts of James Bay–Hudson Bay N to Winisk, 55°16′N), Que. (coasts of James Bay–Hudson Bay N to ca. 56°N; Côte-Nord; Anticosti Is.), and W Nfld., S in the West to Calif., Ariz., and Mexico; W Greenland N to ca. 66°N, E Greenland at ca. 72°N; Iceland; Eurasia. MAPS and synonymy: *see* below.

1   Seeds with rounded ends, the elongate, inflated or collapsed and scale-like whitish papillae restricted to the ends (occasionally absent); calyx-tube attenuate to the pedicel; corolla less than 1 cm broad at the constricted or scarcely expanded orifice.

    2   Stems to over 1.5 dm tall, simple or branched from the base; basal rosette poorly developed or reduced to a single pair of leaves; [*Gentiana (Anthopogon) detonsa* Rottb. and its vars. *barbata* (Froel.) Griseb. and *groenlandicum* Vict.; *Gentiana barbata* Froel., *G. richardsonii* Porsild, and *G. serrata* Gunn.; NW Alaska and NW Dist. Mackenzie; Greenland; MAPS: Hultén 1968*b*:755 (*Gentiana det.*); Gillett 1957: fig. 1A, p. 214, and 1963*b*: fig. 18, p. 49, fig. 19, p. 51, and fig. 20, p. 52; Porsild 1966: map 121, p. 82] . . . . ssp. *detonsa*

    2   Stems less than 1 dm tall, simple or the branches arising from the axils of stem-leaves (rarely from the base); basal rosette well developed; [Alaska–Yukon, the type from S Alaska; MAPS: on the above-noted maps by Gillett and Porsild] . . . . . ssp. *yukonensis* Gillett

1   Seeds irregularly angled, the rounded, inflated, light-brown papillae distributed over most of the surface; calyx-tube abruptly constricted to the pedicel.

    3   Flowers to 7 cm long and about 1.5 cm broad at the orifice, the keels of the calyx prominent, especially at the base of the tube; [*Gentiana elegans* Nels.; *Gentiana (Anthopogon) thermalis* Ktze.; reported from Field, B.C., by Eastham 1947, who, however, also includes Dist. Mackenzie in the general range; the MAP given by Gillett 1957: fig. 1A, p. 214, indicates an occurrence in the W U.S.A. only; the above B.C. citation is probably based upon *Gentianella crinita* ssp. *macounii*] . . . .
. . . . [ssp. *elegans* (Nels.) Gillett]

    3   Flowers rarely over 4 cm long and 12 mm broad at the orifice, the keels of the calyx not prominent; leaves elliptic to spatulate.

        4   Stems usually less than 2 dm tall, profusely branched from the base; corolla-lobes at most about half the length of the tube; [*Gentiana nesophila* Holm, the type from Anticosti Is., E Que.; Ont. (coasts of James Bay–Hudson Bay N to Fort Severn, ca.

56°N), Que., and w Nfld.; MAPS: on the above-noted maps by Gillett] . . . .
. . . . ssp. *nesophila* (Holm) Gillett

4 Stems mostly 3–6 dm tall and branched above, the branches commonly purplish; corolla-lobes about equalling the tube; [*Gentiana raupii* Porsild; s Alaska–Yukon–Dist. Mackenzie (type from the Mackenzie R.) to Wood Buffalo National Park, N Alta.; s Hudson Bay–James Bay, Ont.; MAPS: on the above-noted maps by Gillett and Porsild; Hultén 1968*b*:756 (*Gentiana raupii*); W.J. Cody, Nat. can. (Que.) 98(2): fig. 27 (*Gentiana raupii*), p. 155. 1971 . . . . ssp. *raupii* (Porsild) Gillett

G. propinqua (Richards.) Gillett
/aST/X/eA/ (T) Moist turfy, gravelly, or rocky places, the aggregate species from the coasts of Alaska–Yukon–Dist. Mackenzie (E to Coronation Gulf) to Banks Is. and Victoria Is., s in the West through B.C.–Alta. to Mont., farther eastwards known from the coasts of James Bay–Hudson Bay (N to Churchill, Man.), SE Labrador (GH; Fernald *in* Gray 1950), E Que. (Gaspé Pen. at Anse-Pleureuse, near Mont-Louis; GH; MT), and NW Nfld.; NE Siberia (Chukch Pen.). MAPS and synonymy: *see* below.

Concerning ssp. *aleutica,* Hultén (1968*a*) believes that the merging of *G. aleutica* with *G. propinqua* is untenable, it being more closely related to *G. aurea.*

1 Terminal flowers about 1 cm long, not much larger than the lateral ones; corolla-lobes obtuse and somewhat denticulate, white to pale violet; [*Gentiana aleutica* C. & S., the type from Unalaska, Aleutian Is.; Aleutian Is.–s Alaska; MAPS: Hultén 1968*b*:760, and 1948: map 970 (*G. aleut.*), p. 1339; Gillett 1957: fig. 4, p. 241, and 1963*b*: fig. 25, p. 63] . . . .
. . . . ssp. *aleutica* (C. & S.) Gillett

1 Terminal flowers to 2 cm long, conspicuously larger than the lateral ones; corolla-lobes sharp-pointed . . . . ssp. *propinqua* var. *propinqua*

2 Plant green; flowers white (drying ochroleucous); [B.C.: type from about 9 mi NE of Skeena Crossing, ca. 55°10'N; also known from Mi 250 of the Richardson Highway] . . . .
. . . . f. *acyanea* Gillett

2 Plant markedly purple-tinged; flowers blue; [*Gentiana* Rich.; *Amarella* Greene; *Gentiana arctophila* Griseb.; transcontinental; MAPS: on the above-noted maps by Gillett; Hultén 1968*b*:760; the maps for *G. arctophila* by Hultén 1968*b*:761, Porsild 1957: map 280, p. 195, and Raup 1947: pl. 33, are also applicable here] . . . . . . f. *propinqua*

G. quinquefolia (L.) Small Stiff Gentian, Gall-of-the-earth
/t/EE/ (T) Moist soils from s Ont. (N to Toronto, York Co.; Gillett 1957; reported N to Frontenac Co. by John Macoun 1884) to N.Y. and Maine, s to Tenn. and N Fla. [*Gentiana* L.; *Gentiana quinqueflora* Lam.; incl. *G. occidentalis* (Gray) Small]. MAPS: Gillett 1963*b*: fig. 27, p. 67, and fig. 28, p. 68; 1957: fig. 5, p. 244.

G. tenella (Rottb.) Börner
/aST/X/GEA/ (T) Moist tundra and meadows at low to fairly high elevations from Alaska (N to ca. 69°N), SW Yukon, and the coasts of Dist. Mackenzie–Dist. Keewatin to Southampton Is. and N Que. (Hudson Bay coast at ca. 60°N; w Ungava Bay), with a disjunct area in the w U.S.A. from Calif. and Wyo. to N Mexico; w Greenland N to 68°21'N, E Greenland N to ca. 77°N; Iceland; Spitsbergen; Eurasia. [*Amarella* Cock.; *Gentiana* Rottb. and its var. *occidentalis* Rousseau & Raymond; *Lomatogonium* Löve & Löve; incl. ssp. *pribilofii* Gillett]. MAPS: Hultén 1968*b*:756, and 1958: map 224, p. 243; Gillett 1963*b*: fig. 30, p. 73, and 1957: fig. 9, p. 263; Porsild 1966: map 122, p. 82, and 1957 (1964 revision): map 341, p. 203.

HALENIA Borkh. [6513]

H. deflexa (Sm.) Griseb. Spurred Gentian
/sT/X/ (T) Damp woods, thickets, and open places, the aggregate species from B.C. (N to Cariboo, 52°51'N; CAN) to Alta. (N to Valleyview, 55°04'N; CAN), Sask. (N to Cumberland House, ca. 54°N), Man. (N to Cross Lake, NE of L. Winnipeg), Ont. (N to the Severn R. at ca. 55°40'N), Que. (N to E James Bay at 52°37'N and the Côte-Nord), SE Labrador (N to Battle Harbour, 52°16'N),

Nfld., N.B., and N.S. (not known from P.E.I.), s to Mont., S.Dak., Mich., N.Y., and New Eng. and the mts of Mexico. MAPS and synonymy: *see* below.

1   Flowers rarely over 1 cm long; upper leaves ovate or cordate-ovate; stems usually less
    than 2 dm tall, the principal internodes rarely to 5 cm long; [*H. brentoniana* Griseb.; E
    Que. (Côte-Nord and Magdalen Is.), SE Labrador (N to Cartwright, 53°42′N), Nfld., and
    N.S. (Cape Breton Is.); MAP: Caroline Allen, Ann. Mo. Bot. Gard. 20(1): fig. 2, facing
    p. 132. 1933] . . . . var. *brentoniana* (Griseb.) Gray
1   Flowers to 1.5 cm long; upper leaves oblong-lanceolate to ovate, acuminate; stems to
    about 9 dm tall, the principal internodes to 1 dm long . . . . var. *deflexa*
    2   Corollas spurless (or the earliest ones sometimes spurred); [*H. heterantha* Griseb.;
        Caribou Is., SE Labrador, at 53°13′N] . . . . f. *heterantha* (Griseb.) Fern.
    2   Corollas all spurred; [*Tetragonanthus* Ktze.; *Swertia* Sm.; *S. corniculata sensu* A.
        Michaux 1803, and Pursh 1814, not L.; transcontinental; MAPS: on the above-noted
        map by Allen (somewhat incomplete northwards); Gillett 1963*b*: fig. 17, p. 45] . . . . f. *deflexa*

LOMATOGONIUM A. Br.   [6117]

L. rotatum (L.) Fries   Marsh Felwort
/aST/X/GEA/   (T)   Wet meadows and shores from the coasts of Alaska–Yukon–Dist. Mackenzie to Victoria Is., SE Dist. Keewatin, NE Man. (Churchill and York Factory; an isolated station "along drainage ditch from sewage treatment lagoon" near Gilbert Plains, N of Riding Mt., where taken by James Parker in 1970 and probably introd. by waterfowl), James Bay (Ont. and Que. coasts N to ca. 55°N), s Ungava Bay, E Que. (St. Lawrence R. estuary from St-Roch-des-Aulnets, l'Islet Co., to the Côte-Nord, Anticosti Is., Gaspé Pen., and Magdalen Is.), Labrador (N to 58°05′N), Nfld., ?N.B. (Boivin 1966*b*; not known from P.E.I. or N.S.), and SE Maine, s in the West through B.C.–Alta. to Idaho, Wyo., and N.Mex.; isolated on alkaline flats in Sask. (Sutherland, Mortlach, Chaplin, Little Arm Creek, and Vonda L.; Breitung 1957*a*); W Greenland N to ca. 70°N; Iceland; NE Europe; Asia. MAPS and synonymy: *see* below.

1   Flowers white; [essentially the range of the species; type from Greenland] . . . .
    . . . . f. *albiflorum* Polunin
1   Flowers porcelain-blue.
    2   Middle and upper leaves linear; flowers relatively small and numerous, with linear
        calyx-lobes; [*Pleurogyne rotata* var. *ten.* Griseb.; *P. fontana* Nels.; N-cent. Alaska–
        cent. Yukon–SW Dist. Mackenzie–N B.C.–Alta.–S Sask.; E Que. (Anticosti Is.);
        MAP: Porsild 1966: map 123, p. 82] . . . . f. *tenuifolium* (Griseb.) Fern.
    2   Middle and upper leaves relatively broad; flowers generally larger and fewer, with
        lanceolate calyx-lobes.
        3   Leaves mostly obtuse.
            4   Leaves oblong-lanceolate; [E Que.: Côte-Nord and Anticosti Is.] . . . .
                . . . . f. *americanum* (Griseb.) Fern.
            4   Leaves oval; [type from Amherst Is., Magdalen Is., E Que.; M.L. Fernald,
                Rhodora 21(251):197. 1919] . . . . f. *ovalifolium* Fern.
        3   Leaves lance-acuminate; [*Swertia* L.; *Pleurogyne* Griseb.; *S. pusilla* Pursh; *P.
            carinthiaca* var. *pus.* (Pursh) Gray; transcontinental; MAPS: Hultén 1968*b*:761;
            Porsild 1957: map 281, p. 196, and 1966: map 123, p. 82; Fernald 1925: map 67,
            p. 339, and 1929: map 19, p. 1497 (both incomplete northwards)] . . . . f. *rotatum*

MENYANTHES L.   [6543]

M. trifoliata L.   Buckbean, Bogbean.   Herbe à canards
/aST/X/GEA/   (Grh (Hel))   Bogs and shallow ponds from the Aleutian Is. and N Alaska–Yukon–W Dist. Mackenzie (N to ca. 70°N) to Great Bear L., Great Slave L., L. Athabasca (Alta. and Sask.), cent. Dist. Keewatin, Man. (N to Churchill), northernmost Ont., Que. (N to Ungava Bay and the Côte-Nord), Labrador (N to Cape Harrigan, 55°51′N), Nfld., N.B., P.E.I., and N.S., s to Calif., Colo., Mo., and Del.; W Greenland N to 68°44′N; Iceland; Eurasia. MAP: Hultén 1968*b*:763.

Most of our material is referable to var. *minor* Raf. (*M. verna* Raf.; corolla uniformly white over the inner surface or roseate at the tips of the lobes, the relatively short beard most abundant on the lower half of the lobes, rather than corolla often pink or roseate over most of the inner surface, this nearly covered by the long beard).

NYMPHOIDES Hill   [6545]

N. cordata (Ell.) Fern.   Floating-heart.   Faux-Nymphéa
/T/EE/   (HH)   Ponds and slow streams from Ont. (N to Whitewater L., near Sudbury, and Chalk River, Renfrew Co.) to Que. (N to Lac Desmarais, Labelle Co., about 220 mi NW of Montreal), Nfld., N.B., and N.S. (not known from P.E.I.), S to La. and Fla. [*Villarsia* Ell.; *N. (Limnanthemum) lacunosum* of Canadian reports, not *V. lac.* Vent., basionym]. MAP: Gillett 1963*b*: fig. 38, p. 89.

SABATIA Adans.   [6494]   Sabatia

1   Calyx- and corolla-lobes commonly 5, the latter 1 or 2 cm long; middle and upper leaves narrowly to broadly cordate-ovate, clasping; lower leaves broadly oblong to rotund; biennial with a strongly 4-angled stem and mostly opposite flowering-branches ....
.... [*S. angularis*]
1   Calyx- and corolla-lobes 8 or more, the latter to 3 cm long; stem-leaves lance-acuminate, mucronate, at most subclasping, obscurely 1–3-nerved; rosette-leaves oblanceolate, acuminate; perennial with stem less strongly angled and flowering-branches mostly alternate; (N.S.) .... *S. kennedyana*

[S. angularis (L.) Pursh]   Rose-pink, Bitter-bloom
["Open woods, clearings, prairies and fields, Fla. to La. and Okla., n. to se. N.Y., s. Ont., Mich., Wisc. and Mo.; casually adv. in Mass." (Fernald *in* Gray 1950). The inclusion of S Ont. may be based upon an early report by Gray noted by John Macoun (1884; "This species as a Canadian plant is unknown to me."). The MAP by R.L. Wilbur (Rhodora 57(673): map 3, p. 21. 1955) indicates a northward extension to the S shore of L. Erie in Ohio and N.Y. It may once have occurred on the Ont. side of the lake but, if so, is now probably extinct.]

S. kennedyana Fern.   Plymouth Gentian
/T/E/   (Hsr)   Sandy or peaty margins of ponds: SW N.S. (several localities in Yarmouth Co.; GH; CAN; ACAD); Mass.; R.I. MAP: R.L. Wilbur, Rhodora 57(673): map 14, p. 98. 1955.
The typical form has roseate, narrowly cuneate-obovate corolla-lobes, their margins not overlapping. Forma *eucycla* Fern. (corolla-lobes roseate but broadly obovate, their margins overlapping) is known from the shores of Tusket L., Gavelton, Yarmouth Co., N.S., the type locality. Forma *candida* Fern. (corolla-lobes white) is known from the same locality.

SWERTIA L.   [6512]

S. perennis L.
/sT/W/EA/   (Hs)   Subalpine or alpine meadows and moist places from the E Aleutian Is. and Alaska (N to ca. 63°N) through B.C. to Calif. and N.Mex.; Eurasia. [*S. obtusa* Ledeb.]. MAPS: Hultén 1968*b*:762; Gillett 1963*b*: fig. 15, p. 41.

APOCYNACEAE (Dogbane Family)

Herbs with milky juice and simple opposite entire exstipulate leaves. Flowers regular, perfect, hypogynous, gamopetalous, solitary or in cymes. Calyx-lobes, corolla-lobes, and stamens each 5, the stamens inserted on the corolla-tube and alternating with its lobes. Style or stigma 1. Ovaries 2, superior. Fruit a pair of slender many-seeded follicles.

1   Flowers numerous in terminal and axillary cymes, the white to pink, relatively short-lobed corollas not over 1 cm broad; anthers slightly adherent to the sessile stigma and prolonged into a cone beyond it; seeds with a coma or tuft of silky hairs at apex; stems erect or depressed; (transcontinental species) .... *Apocynum*
1   Flowers solitary on axillary peduncles, white to mauve or blue-purple, mostly 2 cm broad or more; anthers separate; style slender; seeds naked; leaves evergreen; stems trailing or sprawling and mat-forming; (introd.) .... *Vinca*

APOCYNUM L.   [6684]   Dogbane, Indian Hemp.   Apocyn

(Ref.: R.E. Woodson 1930, and N. Am. Flora 29:188–92. 1938; Boivin 1966*a*)
1   Corolla greenish white or white, tubular or ovoid, less than 5 mm long and usually less than twice as long as the calyx, its lobes erect or but slightly spreading; seeds commonly 4 or 5 mm long; leaves ascending .... *A. cannabinum*
1   Corolla pink or white with pink stripes, campanulate, usually over 5 mm long, often more than twice as long as the calyx, the lobes spreading or reflexed; seeds 3 or 4 mm long, their comas 1 or 2 cm long.
  2   Calyx usually less than half as long as the corolla, its lobes lanceolate to deltoid or ovate, often obtuse; corolla-lobes soon recurving; cymes both terminal and lateral, flowering simultaneously, the flowers to 9 mm long, soon nodding; leaves mostly spreading or drooping .... *A. androsaemifolium*
  2   Calyx usually at least half as long as the corolla, its narrowly lanceolate lobes acute to acuminate; corolla-lobes not recurving; cymes terminal, the central one flowering first, the flowers at most 7 mm long, not nodding; leaves often ascending .... *A. medium*

A. androsaemifolium L.   Spreading Dogbane.   Herbe à la puce
/ST/X/   (Grh (Hp))   Dry thickets and fields, the aggregate species from cent. Alaska–Yukon (N to ca. 65°30′N) to Great Bear L., Great Slave L., L. Athabasca (Alta. and Sask.), Man. (N to Gillam, about 165 mi S of Churchill; an 1881 report by Bell N to York Factory), Ont. (N to Sandy L. at ca. 53°N, 93°W), Que. (N to the Bell R. at 49°36′N, Anticosti Is., and the Gaspé Pen.), Nfld., N.B., P.E.I., and N.S., S through B.C.–Alta. to Calif., N Mexico, Tex., and Ga. MAPS and synonymy: *see* below.
1   Pods drooping .... ssp. *androsaemifolium*
  2   Leaves glabrous beneath; [var. *glabrum* Macoun; S B.C. (*see* B.C. map by Boivin 1966*a*: pl. 1, p. 121); Que., N.B., and N.S.; MAPS (aggregate species, the areas apart from those here indicated for var. *androsaemifolium* referring chiefly to var. *incanum*): Porsild 1966: map 124, p. 82; Woodson 1930:175 (incomplete northwards); Hultén 1968*b*:763] .... var. *androsaemifolium*
  2   Leaves more or less pubescent beneath.
    3   Branches and upper leaf-surfaces glabrous or nearly so; [*A. rhomboideum* and *scopulorum* Greene; transcontinental] .... var. *incanum* A. DC.
    3   Branches and upper leaf-surfaces (and also the calyces) densely puberulent; [*A. griseum* Greene, the type from Trail, B.C.; also reported by Boivin from Wood Mt., the lower Fraser valley, and Arrow Lakes, S B.C.; MAP: on the above-noted B.C. map by Boivin] .... var. *griseum* (Greene) Bég. & Bel.
1   Pods ascending to erect .... ssp. *pumilum* (Gray) Boivin
  4   Plant glabrous; [*A. pumilum* (Gray) Greene; S B.C. (*see* the above-noted B.C. map by Boivin) and the Cypress Hills of SE Alta.] .... var. *pumilum*
  4   Plant pubescent at least on the lower leaf-surfaces.
    5   Plant pubescent only on the lower leaf-surfaces; [S B.C. and SW Alta.; MAP: on the

above-noted B.C. map by Boivin] .... var. *woodsonii* Boivin
5   Plant pubescent or puberulent on the branches and both surfaces of the leaves;
    [*A. tomentellum* Greene; Vancouver Is.; MAP: on the above-noted B.C. map by
    Boivin] .... var. *tomentellum* (Greene) Boivin

A. cannabinum L.   Indian Hemp
/sT/X/   (Grt (Hp))   Thickets, borders of woods, and rocky or gravelly shores, the aggregate
species from SW Dist. Mackenzie (near Fort Simpson, ca. 62°N; Raup 1947) and B.C.–Alta. to Sask.
(N to Prince Albert), Man. (N to Victoria Beach, about 60 mi NE of Winnipeg), Ont. (N to the Fawn R.
at ca. 54°40′N, 89°W), Que. (N to the Nottaway R. at 50°52′N, Anticosti Is., and the Gaspé Pen.),
Nfld., N.B., and N.S. (not known from P.E.I.), S to Calif., Tex., and Fla. MAPS and synonymy: *see
below.*
1   Plant variously pubescent (either on the lower or both leaf-surfaces and/or in the
    inflorescence); [var. *pubescens* (Mitchell) DC.; S Ont. (N to the N shore of L. Ontario);
    MAP: Boivin 1966*a*: pl. 1, p. 121; the map by Woodson 1930:175 (incomplete northwards)
    for the aggregate species refers chiefly to the following taxa] .... var. *cannabinum*
1   Plant nearly or quite glabrous.
    2   Leaves petioled, cuneate to broadly rounded at base; [*A. album* Greene; S Ont., S and
        E Que., and N.S.; MAP: on the above-noted map by Boivin] .... var. *glaberrimum* DC.
    2   Median stem-leaves sessile or subsessile, more or less cordate and often clasping at
        base .... var. *hypericifolium* (Ait.) Gray
        3   Stems ascending or erect; [*A. hypericifolium* Ait.; *A. sibiricum* Jacq. and its vars.
            cordigerum and *salinum* (Greene) Fern.; *A. cordigerum, A. salinum,* and *A.
            suksdorfii* Greene; transcontinental; MAP: on the above-noted map by Woodson
            (somewhat incomplete northwards)] .... f. *hypericifolium*
        3   Stems depressed and matted; [*A. hyper.* f. *arenarium* Gates; *A. sib.* f. *aren.*
            (Gates) Fern.; more or less throughout the range of f. *hyper.* but more common
            eastwards] .... f. *arenarium* (Gates) Boivin

A. medium Greene
/T/X/   (Grh (Hp))   Thickets and shores from S B.C. (Enderby, near Salmon Arm; Okanagan) to
Alta. (N to Fort Saskatchewan; not known from Sask. or Man.), Ont. (N to Mattice, 49°37′N), Que. (N
to Anticosti Is. and the Gaspé Pen.), Nfld., N.B., and N.S. (not known from P.E.I.), S to N.Mex.,
Tex., Mo., Tenn., and Va. MAP: Woodson 1930:175.
This taxon comprises a polymorphous series combining in various degrees the characters of *A.
androsaemifolium* and *A. cannabinum* and is evidently of hybrid origin through this parentage.

VINCA L.   [6598]   Periwinkle.   Pervenche

1   Flowers to 5 cm broad, up to 4 on a flowering stem; calyx-lobes subulate, ciliate; follicles
    to 5 cm long, each with 1 or 2 seeds; leaves ovate or deltoid-ovate, mostly broadly
    rounded to subcordate at base, marginally ciliate, otherwise glabrous (or pubescent
    along the nerves), to 7 cm long, on petioles to about 1 cm long; trailing stems rooting
    only at the tips .... *V. major*
1   Flowers to 3 cm broad, nearly always solitary near the top of the flowering stems;
    calyx-lobes lanceolate, glabrous; follicles to 2.5 cm long, each with up to 4 seeds; leaves
    lance-elliptic, coriaceous, glabrous except along the midvein, to 5 cm long, very
    short-petioled; trailing stems rooting at the nodes .... *V. minor*

V. major L.   Large Periwinkle
European; a garden-escape to roadsides and borders of woods in N. America, as in SW B.C.
(Vancouver Is. and adjacent mainland; CAN).

V. minor L.   Common Periwinkle
Eurasian; a garden-escape to roadsides and waste places in N. America, as in SW B.C. (Vancouver
Is. and adjacent mainland), Ont. (N to the Ottawa dist.), Que. (N to the Quebec City dist.), N.B., and
N.S.

ASCLEPIADACEAE (Milkweed Family)

Perennial herbs with usually milky juice. Leaves simple and entire, alternate, opposite, or whorled. Flowers regular, perfect, hypogynous, gamopetalous (the petals at least united toward base), with a central crown (corona) of 5 fleshy hooded bodies or a 5-angled ring on the tube of stamens. Anthers 5, adherent to the stigma, united by their filaments. Styles and ovaries 2, the latter superior. Fruit a pair of many-seeded follicles, the seeds tipped by a coma or tuft of silky hairs. Inflorescence usually a simple umbel.

1   Stem not twining; leaves alternate, opposite, or whorled; corolla-crown consisting of 5
    fleshy hooded bodies . . . . *Asclepias*
1   Stem more or less twining; leaves lance-ovate, opposite; corolla-crown consisting of a
    5-lobed ring or disk; (introd.) . . . . *Cynanchum*

ASCLEPIAS L.   [6791]   Milkweed.   Asclépiade

1   Horns of corolla-hood lacking or rudimentary; flowers greenish; pods lance-attenuate.
    2   Corolla-crown borne on a short column, shorter than the anthers; umbels mostly
        lateral, long-peduncled; leaves mostly alternate or spirally arranged, crowded and
        very numerous, linear to narrowly lanceolate.
        3   Pedicels and pods densely short-spreading-hairy; flowers to 8.5 mm long, the
            hoods reaching scarcely beyond the base of the anther-wings; (s Ont.) . . . . . . *A. hirtella*
        3   Pedicels and pods minutely incurved-puberulent; flowers about 7 mm long, the
            hoods reaching to the projecting centre of the anther-wings . . . . [*A. longifolia*]
    2   Corolla-crown sessile, about equalling the anthers; leaves mostly opposite.
        4   Plant spreading-hirsute; umbel solitary, terminal, peduncled; expanded flowers to
            8 mm long; leaves at most 8 pairs, lanceolate to lance-ovate; stems to 3 dm long;
            (s Man. and s Ont.) . . . . *A. lanuginosa*
        4   Plant minutely pilose or glabrate; umbels mostly lateral, subsessile; expanded
            flowers to 13 mm long; leaves up to 12 pairs, linear to oval or oblong; stems to 8
            dm long; (Alta. to Ont.) . . . . *A. viridiflora*
1   Incurved horns present, one arising from the base of each of the 5 hooded bodies of the
    corolla-crown; umbels peduncled.
    5   Plant hirsute with rather stiff spreading hairs; juice not milky; corolla greenish orange
        to orange-red; pods lance-fusiform, to 12 cm long, on deflexed pedicels; leaves
        sessile or nearly so, mostly irregularly alternate, linear-lanceolate to lanceolate; (Ont.
        and sw Que.) . . . . *A. tuberosa*
    5   Plant glabrous or soft-pubescent; juice milky; leaves opposite or whorled.
        6   Leaves chiefly in whorls of 3 or more; pods on erect pedicels.
            7   Leaves sessile, narrowly linear, 3–6 in a whorl; flowers greenish white; pods
                slenderly fusiform; (se Sask. to s Ont.) . . . . *A. verticillata*
            7   Leaves petioled, ovate-lanceolate to ovate, in 1 or 2 whorls of 4 below 1 or 2
                pairs of opposite ones; flowers pale pink; pods lance-attenuate; (s Ont.) . . . .
                . . . . *A. quadrifolia*
        6   Leaves opposite, more or less petioled.
            8   Flowers dark purple; horns not surpassing the hoods; pods lance-ovoid, to 1.5
                dm long, on downy-pubescent deflexed pedicels; leaves elliptic or ovate-
                oblong, minutely velvety-downy beneath; (s Ont.) . . . . *A. purpurascens*
            8   Flowers not dark purple.
                9   Fruiting pedicels erect; flowers pink to rose-purple; horns surpassing the
                    hoods; pods lance-fusiform, long-attenuate at apex, to 9 cm long; (Man. to
                    N.S.) . . . . *A. incarnata*
                9   Fruiting pedicels deflexed.
                    10 Leaves thin, ovate, tapering at both ends, acuminate; flowers whitish to
                        pale dull-purple; hoods more or less tubular, their lateral margins
                        almost parallel; horns conspicuously exserted; pods lance-fusiform, to
                        1.5 dm long; (s Ont.) . . . . *A. exaltata*

10 Leaves thicker, firm, rounded or acute at apex; hoods scoop-shaped, their lateral margins divergent, not surpassed by the horns; pods thick-lanceolate or slenderly ovoid.

  11 Leaves oblong or oblong-ovate, cordate and subsessile at base, obtuse or rounded at the minutely apiculate summit, their lateral veins almost at right angles to the midrib; flowers pale dull-purple; plant glabrous; (S Ont.) . . . . *A. sullivantii*

  11 Leaves tapering or rounded to subcordate at base, distinctly petioled, their lateral veins more or less ascending; flowers greenish-white to -purple; plant pubescent.

    12 Hoods at least 11 mm long, abruptly narrowed above the base into a narrowly oblong tongue; pods white-woolly, beset with soft spine-like protuberances; plant densely white-woolly; (B.C. to Man.) . . . . *A. speciosa*

    12 Hoods at most 8 mm long, the summit ovate or oblong; plant less pubescent.

      13 Stem coarse, to 2 m tall; leaves to about 2.5 dm long, their veins not strongly ascending; pods to about 3 cm thick, beset with soft spine-like protuberances; (S Man. to N.S.) . . . .
. . . . *A. syriaca*

      13 Stem slender, to 6 dm tall; leaves less than 1 dm long, their lateral veins ascending at an angle of about 45°; pods about 2 cm thick, merely thinly pubescent; (Alta. to w Ont.) . . . . *A. ovalifolia*

A. exaltata L.

/T/EE/ (Grh) Rich woods and clearings from Minn. to S Ont. (N to Wellington, Simcoe, and Leeds counties) and S Maine, S to Iowa, Ill., Ky., and Ga. [*A. phytolaccoides* Pursh; *A. nivea* Sims, not L.]. MAP: Woodson 1947: map 8, p. 17.

A. hirtella (Pennell) Woodson

/t/EE/ (Hp) Prairies, fields, and waste places from Wisc. to S Ont. (Essex, Lambton, and Huron counties; CAN; TRT) and W.Va., S to E Kans., Ark., and Ala. [*Acerates* Pennell].

A. incarnata L. Swamp-Milkweed

/T/(X)/ (Grh (Hp)) Swamps, wet thickets, and shores, the aggregate species from Wyo. to S Man. (N to Camper, about 75 mi N of Portage la Prairie, and Victoria Beach, about 50 mi NE of Winnipeg; CAN; WIN), Ont. (N to Lake of the Woods and Renfrew and Carleton counties), Que. (N to Bellerive, about 80 mi NW of Montreal, and Grosse-Ile, near Quebec City in Montmagny Co.), N.B., P.E.I. (Avondale, Queens Co.; PEI), and N.S., S to N.Mex., Tex., La., Tenn., and S.C. MAP and synonymy: *see* below.

1 Stem and lower leaf-surfaces essentially glabrous.

  2 Leaves oblong-lanceolate, to 2 dm long, tapering to tip . . . . var. *incarnata*

    3 Corolla pink to rose-purple; [s Man. to N.B. and N.S.; MAP (aggregate species): Woodson 1947: map 3, p. 16] . . . . f. *incarnata*

    3 Corolla white, the calyx roseate; [type from Harrietsville, Middlesex Co., S Ont.] . . . .
. . . . f. *rosea* Boivin

  2 Leaves ovate-oblong, less than 1 dm long, obtuse or subacute; [N.S.: Yarmouth, Lunenburg, Halifax (type from Shubenacadie), Inverness, and Victoria counties] . . . .
. . . . var. *neoscotica* Fern.

1 Stem and lower leaf-surfaces copiously short-pilose; leaves broadly lanceolate to elliptic, to 1.8 dm long; [*A. pulchra* Ehrh.; Ont., N.B., and N.S.] . . . . var. *pulchra* (Ehrh.) Pers.

A. lanuginosa Nutt.

/T/(X)/ (Grh (Hp)) Dry prairies and sandy fields from Mont. to S Man. (Sidney, E of Brandon, and Aweme, SE of Brandon) and S Ont. (Pelee Point, Essex Co., and Grand Bend, Huron Co.; QUK; TRT), S to Wyo., Kans., Iowa, and Ill.

[A. longifolia Michx.]
[The report of this species of the E U.S.A. (N to Del. and Va.) from S Ont. by John Macoun (1886; *Acerates long.*) is based upon *Asclepias viridiflora* var. *lanceolata,* the relevant collections in CAN. (*Acerates* Ell.).]

A. ovalifolia Dcne.
/T/WW/ (Grh) Prairies and aspen-oak parklands from Alta. (N to Peace River, 56°14'N), Sask. (N to Hudson Bay Junction, 52°52'N), S Man. (N to Fort Ellice, about 70 mi NW of Brandon), and W Ont. (Boivin 1966b) to Nebr., Iowa, and Ill. [*A. variegata* var. *minor* Hook.].
    Forma *verticillata* Löve & Bernard (upper leaves whorled rather than opposite) is known from Fort Saskatchewan, Alta., and the type locality near Otterburne, Man., about 30 mi S of Winnipeg.

A. purpurascens L.   Purple Milkweed
/t/EE/ (Grh (Hp)) Dry to moist thickets, woods, and openings from S.Dak. to Minn., S Ont. (Lambton Co. along the Detroit R. and on Walpole Is.; Sandwich, Essex Co.; Rondeau Provincial Park, Kent Co.; CAN; Dodge 1914; John Macoun 1884), and N.H., S to Okla., Ark., Miss., Tenn., and N.C. MAP: Woodson 1947: map 6, p. 16.

A. quadrifolia Jacq.
/t/EE/ (Grh (Hp)) Dry woods from Minn. to S Ont. (Niagara Falls, Welland Co.; CAN; reported from Prince Edward and Lennox-Addington counties by John Macoun 1884) and N.H., S to Kans., Ark., Ala., and N.C. MAP: Woodson 1947: map 4, p. 16.

A. speciosa Torr.   Showy Milkweed
/T/WW/ (Grh (Hp)) Prairies and aspen-oak parklands from B.C. (N to Kamloops) to S Alta. (N to Lethbridge and Medicine Hat), S Sask. (N to Dundurn, about 30 mi S of Saskatoon), and S Man. (N to Ochre River, about 80 mi N of Brandon), S to Calif., Tex., and Mo. [*A. douglasii* Hook.].
    The type is from the Canadian Rocky Mountains. A hybrid with *A. syriaca* is reported from SE Man. by Löve and Bernard (1959; Otterburne).

A. sullivantii Engelm.
/t/EE/ (Grh) Rich low grounds and prairies from Minn. to S Ont. (near Windsor, Essex Co.; Squirrel Is. and Walpole Is., Lambton Co.; OAC; MT; CAN), S to Kans., Mo., Ill., Ind., and Ohio.

A. syriaca L.   Common Milkweed.   Herbe à coton or Petits cochons
/T/EE/ (Grh (Gr)) Thickets, fields, and roadsides from S Man. (N to Victoria Beach, about 55 mi NE of Winnipeg) to Ont. (N to Renfrew and Carleton counties), Que. (N to the Gaspé Pen. at Matapédia), N.B., P.E.I., and N.S., S to Kans. and Ga. [*A. cornuti* Dcne.]. MAPS: Dominique Doyon, Soc. Qué. Prot. Plant. Rapp. 40: fig. 2, p. 96. 1958; Herbert Groh, Sci. Agric. 23: fig. 1, p. 626. 1943.
    Forma *leucantha* Dore (flowers whitish rather than dull purple; type from Renfrew, Ont.) is known from several Ont. counties and from Papineau Co., SW Que. Forma *polyphylla* Boivin (leaves all in whorls of 3 rather than opposite) is reported from E Ont. and SW ?Que. by Boivin (1966b).

A. tuberosa L.   Butterfly-weed, Pleurisy-root
/T/(X)/ (Hp) Dry open soil from Minn. to S Ont. (N to Constance Bay, about 20 mi W of Ottawa), SW Que. (Boivin 1966b), and New Eng., S to Ariz., Colo., Tex., and Fla. [Incl. var. *interior* (Woodson) Shinners]. MAP: Woodson 1947: map 2, p. 16.

A. verticillata L.
/T/EE/ (Hp) Dry woods and open sterile soil from SE Sask. (Estevan, about 125 mi SE of Regina) to S Man. (N to Brandon and Winnipeg) and S Ont. (Essex, Norfolk, and Welland counties), S to Tex., Mexico, and Fla.

A. viridiflora Raf.   Green Milkweed
/T/X/ (Grh (Hp)) Dry sands, prairies, woods, and openings (ranges of Canadian taxa outlined below), S to N.Mex., Tex., and Fla.

1   Leaves oval or oblong, obtuse or merely short-pointed; [*Acerates* Eat.; var. *?obovata* (Ell.) Torr.; Alta. (Milk River) to Sask. (N to Saskatoon), Man. (N to St-Lazare, about 75 mi NW of Brandon), and S Ont. (near Sandwich, Essex Co.)] .... var. *viridiflora*
1   Leaves narrower, acute or acuminate.
   2   Leaves lanceolate to narrowly ovate, usually at least 1.5 cm broad; [*Acerates vir.* var. *lanc.* (Ives) Gray; S Sask. (Touchwood Hills; Qu'Appelle Valley), S Man. (N to St-Lazare), and S Ont. (N to Great Cloche Is., N L. Huron); sterile specimens also taken by the writer between Lytton and Spences Bridge, S B.C., where probably introd. on roadside gravel banks] .... var. *lanceolata* (Ives) Torr.
   2   Leaves linear to linear-lanceolate, at most about 1 cm broad; [*Acerates vir.* var. *linearis* Gray; Alta. (Moss 1959), Sask. (Saskatoon and Wadena; Breitung 1957a), S Man. (N to near St-Lazare), and S Ont. (Lambton, Kent, and Norfolk counties)] ....
   .... var. *linearis* (Gray) Fern.

CYNANCHUM L.   [6834]   Swallow-wort

1   Corolla greenish white, glabrous within; stems suberect .... *C. vincetoxicum*
1   Corolla purplish.
   2   Corolla-lobes maroon or pinkish, twice as long as broad, glabrous within; peduncles at least 2 cm long .... *C. medium*
   2   Corolla-lobes dark purple, as broad as long, pubescent within; peduncles mostly about 1 cm long .... *C. nigrum*

C. medium R. Br.
European; apparently known in N. America only from SW B.C. (Victoria, Vancouver Is.; "In cultivated grounds", where taken by Fletcher in 1885; CAN) and Ont. (Welland, Wentworth, York, Ontario, Durham, Northumberland, Frontenac, and Carleton counties; CAN; DAO; OAC; R.J. Moore, Can. Field-Nat. 73(3):146. 1959). [*Vincetoxicum* Dcne.].

C. nigrum (L.) Pers.
European; a garden-escape to fields, roadsides, and waste places in N. America, as in SW Que. (Montreal; DAO; according to Moore, loc. cit., reports from B.C. and Ont. are based upon *C. medium*). [*Asclepias* L.; *Vincetoxicum* Moench].

C. vincetoxicum (L.) Pers.
European; a garden-escape to fields and waste places in N. America, as in S Ont. (Niagara Falls, Welland Co.; TRT). [*Asclepias* L.; *Vincetoxicum officinale* (L.) Moench].

CONVOLVULACEAE (Morning-glory Family)

Commonly stem-twining herbs (often with some milky juice) with simple petioled alternate leaves (margins usually entire or merely shallowly undulate; leaves often deeply 2-lobed or pectinately divided in *Ipomoea*, reduced to minute scales in *Cuscuta*). Flowers regular, perfect, gamopetalous, hypogynous. Sepals or calyx-lobes, corolla-lobes, and stamens each 5, the stamens alternating with the corolla-lobes. Ovary superior. Fruit a globular or plump 2–6-seeded capsule. (Including Cuscutaceae).

1   Plants yellow to orange-brown, twining and parasitic on other plants; leaves reduced to
      minute scales; corolla very small, subglobose or globose-ovoid; styles 2; stamens
      inserted at the sinuses of the corolla . . . . *Cuscuta*
1   Plants with normal green leaves and large salverform to campanulate flowers; style
      solitary; stamens inserted near base of corolla.
      2   Stigma 1, capitate or of 2 or 3 globular lobes; leaves cordate-ovate or deeply 3-lobed
            (*I. hederacea*) or pectinately divided (*I. quamoclit*); (introd. in S Ont.) . . . . *Ipomoea*
      2   Stigmas 2, linear-filiform to slenderly ovoid; leaves mostly cordate-ovate and more or
            less hastate (reniform in *C. soldanella*) . . . . *Convolvulus*

CONVOLVULUS L.   [6993] Bindweed. Liseron

1   Calyx bractless at base (but the peduncles with a pair of small opposite linear bracts at
      the point of origin of the slender pedicel or pair of pedicels); stigmas filiform; corolla white
      or reddish-tinged, to 2 cm long; leaves very variable, to 6 cm long, from almost linear to
      oblong or ovate, obtuse to acute, their bases sagittate, hastate, or cordate; plants trailing
      or slightly climbing by twining; (introd.) . . . . *C. arvensis*
1   Calyx enclosed by a pair of large, often cordate, basal bracts; stigmas oval or oblong;
      peduncles usually 1-flowered.
      2   Flowers commonly "double", with narrow wavy bright-pink petals 2 or 3 cm long;
            leaves hastate, to 8 cm long, the basal lobes acute; stems twining; plant pubescent;
            (introd.) . . . . *C. japonicus*
      2   Flowers normal, commonly at least 3 cm long.
            3   Leaf-blades reniform, rounded or conspicuously retuse at apex, tipped with a short
                  mucro, broader than long, on petioles up to 3 times as long as the blade; corolla
                  pinkish purple, rarely over 5 cm long; plant glabrous, fleshy, from deep rootstocks,
                  the stems creeping but not twining; (S B.C.) . . . . *C. soldanella*
            3   Leaf-blades rather narrowly to broadly cordate-ovate in outline and more or less
                  hastate-based.
                  4   Stems erect or creeping or merely twining at the tip, the basal leaves much
                        smaller than the upper ones; flowers 1–4 from the lower and median axils,
                        white, creamy white, or pink-tinged; petioles at most barely half as long as the
                        leaf-blade; (Ont. to N.S.) . . . . *C. spithamaeus*
                  4   Stems twining or trailing, the lower leaves only slightly reduced; flowers
                        numerous, from many axils, white to roseate; petioles relatively long; (trans-
                        continental) . . . . *C. sepium*

C. arvensis L.   Field-Bindweed, Cornbine
Eurasian; introd. in old fields and waste places in N. America, as in S B.C. (N to Kamloops), Alta. (N to Fort Saskatchewan), Sask. (Breitung 1957a), Man. (N to the Winnipeg dist.), Ont. (N to the N shore of L. Superior; Baldwin 1958), Que. (N to the Gaspé Pen.), N.B., P.E.I. (D.S. Erskine 1960), and N.S. [Incl. var. *obtusifolius* Choisy; *C. ambigens* House].

C. japonicus Thunb.   Rose Glorybind
Asiatic; an occasional garden-escape in N. America, as in Ont. (N to near Westmeath, Renfrew Co.; Montgomery 1957) and sw Que. (Montreal; MT). [*Calystegia pubescens* Lindl.].

C. sepium L.   Hedge-Bindweed, Wild Morning-glory.   Gloire du matin
/T/X/   (Hp (Grh))   Thickets, shores, and waste places, the aggregate species from S B.C. (Queen Charlotte Is.; Vancouver Is. and the adjacent mainland N to Vancouver and Nelson) to Alta. (N to Fort Saskatchewan; CAN), Sask. (N to Nipawin, 53°22'N), Man. (N to Porcupine Mt.), Ont. (N to Renison, SW of James Bay at ca. 51°N; Hustich 1955), Que. (N to Anticosti Is. and the Gaspé Pen.), Nfld., N.B., P.E.I., and N.S., S to Calif., N.Mex., Tex., and Fla. (evidently at least partly introd. in the W U.S.A.).

1   Blades of principal leaves (above the sinus) mostly more than twice as long as broad, commonly not over 3 cm broad; flowers white; peduncles shorter than the subtending leaves; [*C. repens* L.; *Calystegia sepium* var. *pubescens* Gray; reported from Alta., Sask., and Man. by John Macoun 1884, and there is a collection in GH from Baie-du-Vin Is., Northumberland Co., E N.B.] . . . . var. *repens* (L.) Gray

1   Blades of principal leaves (above the sinus) mostly 2/3 as broad as long or broader, to 1 dm broad.
  2   Leaf-sinus with nearly parallel margins; flowers white, on densely pubescent peduncles about equalling the petioles; [*C. frat.* Mack. & Bush; *C. interior* House; *C. macounii* Greene; Sask. (N to Regina; Breitung 1957a)] . . . .
  . . . . var. *fraternifolius* Mackenz. & Bush
  2   Leaf-sinus with divergent margins.
    3   Leaves round-oval to subrotund, with rounded summit and lobes; bracts obtuse, at most about 2 cm long; flowers white, less than 5 cm long; [introd. near Yarmouth, N.S.] . . . . var. *dumetorum* Pospichal
    3   Leaves broadly ovate, pointed, longer than broad; bracts enclosing the calyx acutish, to 3.5 cm long; flowers white to roseate, to 8 cm long, on peduncles often surpassing the leaves . . . . var. *sepium*
      4   Corolla roseate; [Sask., Ont., Que., and P.E.I.] . . . . f. *coloratus* Lange
      4   Corolla white.
        5   Leaves, petioles, and stem soft-pilose; [S Man. (near Winnipeg) and along the coasts of E Que. (Rimouski Co.; Gaspé Pen; Anticosti Is.; Magdalen Is.), N.B. (Wolf Is., Charlotte Co.), and N.S. (Victoria Co., Cape Breton Is.)] . . . . .
        . . . . f. *malacophyllus* Fern.
        5   Plant glabrous or soon glabrate; [*Volvulus* Junger; *Calystegia* R. Br.; incl. vars. *americanus* Sims (*Con. amer.* (Sims) Greene) and *communis* Tryon; transcontinental] . . . . f. *sepium*

C. soldanella L.   Beach Morning-glory
/t/W/EeA/   (Hp (Grh))   Coastal beaches and sand dunes from SW B.C. (Queen Charlotte Is.; Vancouver Is. and adjacent islands; CAN; V) to Calif.; Europe; E Asia; New Zealand. [*Calystegia* R. Br.; *Con. reniformis* R. Br.].

C. spithamaeus L.   Low Bindweed
/T/EE/   (Hp (Grh))   Sandy or rocky open places and thin woods from Minn. to Ont. (N to Thunder Bay; John Macoun 1884), Que. (N to St-Maurice, near Trois-Rivières), and N.S. (Janvrin Is., Richmond Co.; DAO, detd. Breitung; not known from N.B. or P.E.I.), S to Iowa, Tenn., Ala., and Ga. [*Calystegia* Pursh; *Volvulus* Ktze.; *Con. stans* Michx.; *Cal. tomentosa* Pursh].

      CUSCUTA L.   [6968]   Dodder, Love-tangle, Coral-vine.   Cuscute

(Ref.: Yuncker 1932)
1   Stigmas linear or narrowly ovoid; flowers mostly 5-merous; capsules globose or depressed-globose, irregularly circumscissile near base; scales at base of stamens fringed; (introd.).
  2   Styles much longer than the ovary, they and the stamens finally exserted from the short-cylindric corolla; flowers whitish or pinkish, the corolla-tube often surpassing the calyx, its lobes spreading; calyx-lobes rather narrowly ovate . . . . C. *epithymum*
  2   Styles about equalling the ovary, they and the stamens included in the urn-shaped corolla; flowers yellowish, the corolla-tube about equalling the calyx.

3   Corolla-lobes erect; calyx-lobes membranous . . . . *C. epilinum*
3   Corolla-lobes spreading at maturity; calyx-lobes rather fleshy, the tips turgid and slightly recurved (lower part of calyx commonly shining in herbarium specimens) . . . .
. . . . [*C. approximata*]
1   Stigmas capitate; capsule indehiscent or irregularly rupturing.
   4   Sepals distinct, they and the corolla-lobes obtuse or rounded at summit; flowers sessile or subsessile in dense rope-like masses, each subtended by up to 5 suborbicular overlapping bracts; (?Ont.–Que.) . . . . [*C. compacta*]
   4   Sepals united at least at base; flowers not bracted.
      5   Flowers all or mostly 4-merous, in globose clusters; corolla-lobes erect.
         6   Corolla-lobes obtuse or rounded, erect, much shorter than the slender tube, the withered corolla capping the capsule; calyx-lobes free nearly to base, much shorter than corolla-tube, very unequal, the outer greatly overlapping the inner; styles equalling or longer than the capsule; scales regularly short-toothed along the margins and summit; (B.C. to N.S.) . . . . *C. cephalanthi*
         6   Corolla-lobes acute; calyx-lobes about equalling the corolla-tube; scales 2-cleft or reduced to a few teeth.
            7   Calyx-lobes acute; corolla slenderly campanulate, its lobes with incurved tips, the withered corolla capping the capsule and soon cast off; scales consisting of 2-cleft few-toothed narrow halves; (s Sask. and s Man.) . . . .
. . . . *C. coryli*
            7   Calyx-lobes very obtuse; corolla broader, its lobes with erect tips, the withered corolla basal; scales 2-cleft or reduced to a few teeth; (s Ont. and sw Que.) . . . . *C. polygonorum*
      5   Flowers all or mostly 5-merous; corolla-lobes more or less spreading.
         8   Capsule depressed-globose, without a thickened summit; flowers short-pedicelled in loose clusters; corolla-lobes acute, wide-spreading, the tips often inflexed; withered corolla basal; (B.C. to N.S.) . . . . *C. pentagona*
         8   Capsule globose-ovoid, commonly longer than broad, thickened at the base of the styles; tips of corolla-lobes not inflexed; withered corolla capping or surrounding the upper part of the capsule.
            9   Corolla-lobes and calyx-lobes each ovate-lanceolate and acute to acuminate; flowers in umbellate-cymose clusters; (sw B.C.) . . . . *C. salina*
            9   Corolla-lobes and calyx-lobes each ovate and obtuse; flowers in rather open panicled cymes.
               10   Corolla-lobes widely spreading in anthesis; scales oblong, fringed nearly to base, nearly as long as the corolla-tube; styles 2/3 as long as the capsule; (s Man. to N.S.) . . . . *C. gronovii*
               10   Corolla-lobes only slightly spreading in anthesis; scales 2-cleft, long-fringed at summit, about half as long as the corolla-tube; styles 1/4 as long as the capsule; (Alta. to Man.) . . . . *C. umbrosa*

[C. approximata Bab.]
[Eurasian; not yet known from Canada but, according to Hitchcock et al. (1959), doing considerable damage to leguminous crops, especially alfalfa, in the w U.S.A.]

C. cephalanthi Engelm.
/T/X/   (T)   Parasitic on various shrubs and coarse herbs from s B.C. (Fernald *in* Gray 1950) to Alta. (Medicine Hat; CAN; not known from Sask.), s Man. (Shellmouth, about 100 mi NW of Brandon; DAO, detd. Yuncker; not known from Ont., Que., N.B., or P.E.I.), and N.S. (Halifax and Lunenburg counties; CAN; GH; NSPM), s to Oreg., N.Mex., Tex., and Va. [*C. tenuiflora* Engelm.].

[C. compacta Juss.]
[Parasitic on a great variety of shrubs and coarse herbs in the E U.S.A. Reports from s Ont. by John Macoun (1884; Malden, Essex Co.) and Que. by Rouleau (1947) require confirmation.]

C. coryli Engelm.

/T/(X)/ (T) Parasitic on various shrubs and herbs from Mont. to s Sask. (Mortlach, about 60 mi w of Regina; Breitung 1957a), s Man. (Winnipeg; DAO, detd. Anderson and Yuncker), and New Eng., s to Ariz., Tex., and N.C.

C. epilinum Weihe Flax-Dodder. Cuscute du lin

Eurasian; very injurious to cult. flax in N. America, as in ?Man. ("While no specimens are at hand it is understood that the dodder of flax in Manitoba is this species, and that it occurs little, if at all, on other hosts"; Groh 1944a), Ont. (Cambridge (Galt), Waterloo Co.; CAN), and sw Que. (collections in MT from Soulanges, near Montreal; in GH from St-Alphonse, Shefford Co.; and in TRT from "Lower Canada", where taken by Pringle in 1880, verified by Anderson and Gaertner).

C. epithymum L. Clover-Dodder. Cuscute du thym

European; very injurious in clover-fields in N. America, as in s B.C. (N to Armstrong, about 15 mi N of Vernon; Herb. V), Ont. (N to the Ottawa dist.; Gillett 1958), and N.B. (Ingleside, Kings Co.; G.U. Hay, Nat. Hist. Soc. N.B. Bull. 12:70. 1894). [*C. europaea* var. *ep.* L.].

C. gronovii Willd.

/T/(X)/ (T) Parasitic on many hosts from ?B.C. (Boivin 1966b; this and his Alta.-Sask. reports may refer to other species treated here but evidently included by him in *C. gronovii*) and s Man. (N to Victoria Beach, about 50 mi NE of L. Winnipeg; Herb. Man. Prov. Mus., Winnipeg) to Ont. (N to the Ottawa dist.), Que. (N to near Quebec City), N.B., P.E.I., and N.S., s to Ariz., Tex., and Fla. [Incl. var. *latiflora* Engelm.].

C. pentagona Engelm.

/T/X/ (T) On many hosts, especially legumes, from sw B.C. (Alberni and Elgin, Vancouver Is.; CAN; not known from Alta.) to Sask. (Oliver, about 65 mi sw of Saskatoon; Breitung 1957a, as *C. campestris*), Man. (N to Winnipeg; CAN), Ont. (N to Stormont and Carleton counties), Que. (N to the Montreal dist.), Nfld. (Topsail; DAO), N.B. (Grand Lake, Victoria Co.; CAN), P.E.I. (Charlottetown; NSPM), and N.S. (Kings and Annapolis counties), s to s Calif., Mexico, Tex., and Fla.; tropical America. [Incl. var. *calycina* Engelm. (*C. arvensis* var. *cal.* Engelm.; *C. campestris* Yuncker) and *C. arvensis* Beyrich and its var. *pent.* Engelm.].

C. polygonorum Engelm.

/T/EE/ (T) Parasitic on *Polygonum, Penthorum, Lycopus,* and other herbs from Minn. to s Ont. (near Amherstburg, Essex Co.; CAN and TRT, detd. Anderson) and sw Que. (Lachine, near Montreal; GH, detd. Yuncker), s to Kans., Mo., Tenn., and Del. [*C. chlorocarpa* Engelm.].

C. salina Engelm.

/t/W/ (T) Mostly on chenopodiaceous species from sw B.C. (Vancouver Is. and adjacent islands and mainland; CAN; V) to Calif. and Mexico. [Incl. var. *squamigera* (Engelm.) Yuncker (*C. californica* var. *sq.* Engelm.; *C. sq.* (Engelm.) Piper)].

C. umbrosa Beyrich

/T/WW/ (T) Parasitic on various shrubs and herbs from Alta. (Moss 1959; Edmonton) to Sask. (N to Nipawin, 53°22′N; Breitung 1957a) and Man. (N to Dauphin, N of Riding Mt.), s to Utah and N.Mex. [*C. curta* (Engelm.) Rydb.; *C. megalocarpa* Rydb.; *C. planiflora* of Sask. reports, not Engelm.].

IPOMOEA L. [7003] Morning-glory

1 Leaves pectinately divided to near midrib into parallel linear-filiform segments; corolla scarlet-red, the very narrow tube to 4 cm long, the limb rotate; stamens and style exserted; peduncles with 1 or 2 flowers . . . . [*I. quamoclit*]
1 Leaves entire or at most deeply 3-lobed; corolla funnelform to nearly campanulate; stamens and style not exserted.

2 Stigma unlobed or with at most 2 lobes; capsule 2-locular; sepals essentially
   glabrous, firm, oblong-ovate, obtuse; corolla to 8 cm long, white (with purple in the
   tube); peduncles with up to 5 flowers, mostly longer than the petioles; leaves
   cordate-ovate or panduriform (fiddle-shaped), 2-lobed at base .... *I. pandurata*
2 Stigma 3-lobed; capsule 3-locular; sepals herbaceous, hirsute.
   3 Leaves often deeply 3-lobed; sepals with long linear tips; corolla less than 5 cm
     long, sky-blue when fresh, changing to pink or purple, the tube white; flowers
     sessile or short-pedicelled .... *I. hederacea*
   3 Leaves merely rather shallowly 2-lobed at base; sepals with short acutish tips;
     corolla to 7 cm long, white, blue, red, or purple; flowers mostly long-pedicelled ....
     .... *I. purpurea*

I. hederacea (L.) Jacq.
Tropical America; introd. into cult. fields, roadsides, and waste places in N. America, as in s Ont.
(Essex, Lambton, Welland, York, and Perth counties). [*Convolvulus* L.].

I. pandurata (L.) Mey.   Wild Potato-vine, Man-of-the-earth
/t/EE/   (Grt)   Dry open or partly shaded soil from Kans. to Mo., Ill., Mich., Ohio, s Ont. (Pelee
Point and Pelee Island, Essex Co.; Queenston, Lincoln Co.; CAN; TRT), N.Y., and Conn., s to Tex.
and Fla. [*Convolvulus* L.].

I. purpurea (L.) Roth   Common Morning-glory
Tropical America; a garden-escape to fields and roadsides in N. America, as in Ont. (N to Ottawa),
sw Que. (N to Hull and Montreal), and N.S. (Lindsay 1878; John Macoun 1884). [*Convolvulus* L.].

[I. quamoclit L.]   Starglory
[Tropical America; a local garden-escape in N. America, as in s Ont. (J.M. Macoun 1897; near
Niagara Falls, Welland Co.), where perhaps once established but now extinct. (*Quamoclit* Britt.; *Q.
pennata* (Desr.) Voigt; *Q. vulgaris* Choisy).]

POLEMONIACEAE (Polemonium Family)

Herbs with simple or pinnately compound, alternate or opposite leaves. Flowers regular, perfect, hypogynous, gamopetalous. Calyx-lobes, corolla-lobes, and stamens each 5, the stamens alternating with the corolla-lobes. Style 1. Ovary superior. Fruit a 3-locular capsule.

1 Leaves chiefly opposite (uppermost ones sometimes alternate), sessile or very short-petioled; calyx-tube with usually conspicuous hyaline membranes separating the firm green portions terminating in the teeth (membranes inconspicuous in *Linanthus bicolor*).

  2 Leaves palmately cleft nearly to base into up to 7 (sometimes 9) linear spinose-tipped segments 1 or 2 mm long; corolla commonly funnelform, the anther-filaments about equally inserted in its throat; seeds often several in each locule, often becoming mucilaginous when moistened or emitting spiralling threads; annuals rarely over 3 dm tall; (B.C. to sw Sask.) .... *Linanthus*

  2 Leaves entire or nearly so; corolla salverform (with a slender tube and a rotate limb), the anther-filaments very unequally inserted near the top of the tube; seeds usually 1 in each locule.

    3 Annual at most about 3 dm tall, subsimple to much branched, puberulent or glandular-puberulent at least above; leaves linear to elliptic (or the lower ones obovate), to about 5 cm long and 8 mm broad, the upper leaves regularly alternate; flowers mostly in pairs at the ends of the stem and branches, one subsessile, the other short-pedicelled; corolla salverform, to 1.5 cm long, the tube white or yellowish, the short spreading lobes lavender to pink; seeds becoming mucilaginous when moistened or emitting spiralling threads; (s B.C.) ...... *Microsteris*

    3 Perennials (except *P. drummondii*); only the uppermost leaves sometimes alternate; seeds remaining unchanged when moistened .... *Phlox*

1 Leaves all or chiefly alternate (lowermost ones sometimes opposite).

  4 Calyx-tube of essentially uniform texture throughout, not ruptured by the developing capsule.

    5 Leaves simple, mostly entire (to deeply pinnatifid in *C. heterophylla* but not separated into definite leaflets); calyx-tube chartaceous at anthesis; filaments mostly unequally inserted in the corolla-throat (equally in *C. tenella*); seeds mucilaginous when moistened or emitting spiralling threads; annuals .... *Collomia*

    5 Leaves pinnately compound, the leaflets chiefly entire (palmately divided in *P. viscosum*); calyx-tube herbaceous at anthesis; filaments equally inserted in the corolla-throat; seeds sometimes mucilaginous when moistened; chiefly perennials (*P. micranthum* annual) .... *Polemonium*

  4 Calyx-tube with conspicuous hyaline membranes separating the firm green portions terminating in the teeth, usually ruptured by the developing capsule; filaments about equally inserted in the corolla-throat; leaves mostly deeply divided or cleft.

    6 Leaves commonly about 1 cm long, sessile, palmately divided to near base into up to 7 rigid linear-subulate spinulose-tipped segments (the central segment the largest) and bearing axillary fascicles; dead leaves usually persisting through one or more seasons; flowers solitary in the axils (upper axils often crowded, simulating a spike), to 2.5 cm long, commonly white (but lavender-tinted outside and varying to yellowish or almost salmon-colour); seeds not mucilaginous when moistened; shrub to 6 dm tall, more or less densely branched, sweetly aromatic, glabrous to puberulent or glandular; (s B.C.) .... *Leptodactylon*

    6 Leaves longer, pinnately lobed or divided (entire or often trifid in *Gilia minutiflora*); flowers borne in basically cymose inflorescences; seeds commonly mucilaginous when moistened.

      7 Calyx-lobes equal or nearly so, simple, they and the leaf-segments merely sharp-pointed or spinulose-tipped; annuals or perennials .... *Gilia*

      7 Calyx-lobes mostly unequal (or subequal in *N. squarrosa*), the larger ones often 3-forked, the lobes and the leaf-segments with whitish spine-tips about 1

mm long; flowers in dense terminal leafy-bracted capitate clusters; corolla-lobes at most 3 mm long; annuals; (s B.C. to s Sask.) .... *Navarretia*

COLLOMIA Nutt. [7015]

1   At least the lower leaves pinnatifid, the lobes themselves often toothed or subpinnatifid (the blade to 3.5 cm long and 2 cm broad; middle and upper leaves progressively less cleft); corolla white to pink or lavender, much surpassing the calyx, the lobes 3 or 4 mm long; locules 2–3-seeded; stems often several from the base and often bushy-branched, viscid-villous at least above; (sw B.C.) .... *C. heterophylla*
1   All of the leaves entire; locules 1-seeded.
   2   Corolla at least 2 cm long, much longer than the calyx, generally yellowish or salmon-colour, the lobes to 1 cm long; calyx-lobes to 4 mm long at anthesis; leaves narrowly to broadly lanceolate; stems commonly solitary and simple, to 1 m tall; (s B.C.) .... *C. grandiflora*
   2   Corolla at most 1.5 cm long, white to pale lavender, pink, or bluish, the lobes at most 3 mm long; calyx-lobes at most 3 mm long at anthesis.
      3   Flowers numerous in leafy-bracted capitate clusters terminating the stem or its branches, much longer than the calyces; anther-filaments unequally inserted in the corolla-throat; leaves linear to lanceolate, to 7 cm long and 13 mm broad; stem to 6 dm tall, simple or branched, glandular or viscid-pubescent above, finely puberulent or subglabrous below; (B.C. to w Ont.; introd. eastwards) .... *C. linearis*
      3   Flowers solitary in the axils (the plant floriferous nearly to base), commonly less than twice as long as the calyces; anther-filaments equally inserted in the corolla-throat; leaves linear or nearly so, to 5 cm long and 5 mm broad; stem rarely over 1.5 dm tall, stipitate-glandular at least above, freely branched; (?B.C.) .... [*C. tenella*]

C. grandiflora Dougl.
/T/W/ (T) Open or lightly wooded dry places at low to moderate elevations from s B.C. (Vancouver Is. and adjacent islands and mainland n to Lytton, Kelowna, and Nelson) to Calif., Ariz., and Wyo. [*Gilia* Gray].

C. heterophylla Hook.
/t/W/ (T) Streambanks and woods at lower elevations from sw B.C. (Vancouver Is. and adjacent islands and mainland e to Yale, lower Fraser Valley) to Calif. and n Idaho. [*Gilia* Dougl.; *Navarretia* Benth.].

C. linearis Nutt.
/T/WW/ (T) Dry to moist, open or lightly shaded places at low to moderate elevations, the natural range very difficult to determine because of the weedy nature of the species but apparently from s B.C.–Alta.–Sask.–Man. to w Ont. (Ingolf, near the Man. boundary), s to Calif., N.Mex., Nebr., and Wisc.; apparently introd. elsewhere, as in Alaska–Yukon (n to the Alaska–Yukon boundary at ca. 67°N), s Dist. Mackenzie, n Alta. (L. Mamawi, se of L. Athabasca), Man. (Churchill to Gillam, about 165 mi s of Churchill), Ont. (L. Superior region eastwards), Que. (n to the s Gaspé Pen.), N.B., P.E.I., and N.S. (*Gilia* Gray). MAPS: Hultén 1968*b*:765; E.T. Wherry, Bartonia 18: fig. 1 (Man. to e Que. and N.B.), p. 54. 1936.

[C. tenella Gray]
[The report of this species of the w U.S.A. (Wash. to Nev. and Utah) from B.C. by Rydberg (1922) requires confirmation. (*Gilia ten.* (Gray) Nels. & Macbr., not Benth., which is *Linanthus bicolor* (Nutt.) Greene; *G. leptotes* Gray).]

GILIA R. & P. [7016]

1   Corolla normally bright red or scarlet (sometimes pink or white), to 3 or 4 cm long, the

tube much longer than the calyx, the limb to over 1 cm broad; flowers in a terminal raceme or thyrsoid panicle; stems to over 1 m tall.

  2   Plant pubescent (at least the calyces, pedicels, and upper part of stem stipitate-glandular and usually copiously villous with long spreading crisped white hairs); leaf-segments narrowly linear; (s B.C.) . . . . *G. aggregata*

  2   Plant glabrous; leaf-segments filiform; (introd.) . . . . *G. rubra*

1 Corolla white to blue, at most about 1 cm long, the tube not much surpassing the calyx, the limb commonly less than 5 mm broad.

  3   Flowers numerous in dense cymose clusters terminating the stem and branches; anther-filaments attached at the corolla-sinuses; leaves deeply 1–2-pinnatifid into narrow ultimate segments.

      4   Perennial, branched and woody at base, with several ascending or erect stems commonly not over 2 dm tall, more or less arachnoid-tomentose or in large part glabrous; corolla white, to 6 mm long, the lobes 2 mm long; (?B.C.) . . . . . . [*G. congesta*]

      4   Annual, the stem erect or sparingly branched, glabrous to stipitate-glandular or obscurely floccose, to 1 m tall; corolla light bluish, the slender lobes about equalling the tube; (introd. in Alaska–Yukon–B.C.) . . . . *G. capitata*

  3   Flowers scattered, not in dense clusters; annuals; (s B.C.).

      5   Leaves borne chiefly at or near the base of the stem, the basal ones oblanceolate, to 7 cm long and 2 cm broad, usually deeply pinnatifid, the lobes entire or few-toothed; flowers blue or blue-lavender, to 11 mm long, borne in an open sparingly bracted terminal cymose inflorescence; filaments inserted at the sinuses of the corolla; plant to 4 dm tall, stipitate-glandular at least above; (?B.C.) . . . . [*G. sinuata*]

      5   Leaves well distributed along the stem and branches, linear and entire (but some of the lower leaves often trifid with narrow segments), to 4 cm long but scarcely 1 mm broad; flowers white or pale bluish, to 7 mm long, solitary opposite the leaves or in their axils or terminating axillary branches; filaments inserted well below the sinuses of the corolla; plant to 6 dm tall, subglabrous to densely and finely stipitate-glandular or glandular-scaberulous; (s B.C.) . . . . *G. minutiflora*

G. aggregata (Pursh) Spreng.   Scarlet Gilia, Sky-rocket
/t/W/ (Hs) Dryish meadows and open or lightly wooded, often rocky slopes from s B.C. (valleys of the Dry Interior N to Armstrong, about 15 mi N of Vernon) to Calif. and Mexico. [*Cantua* Pursh; *Ipomopsis* Grant].

G. capitata Sims
Native in the w U.S.A. from Wash. to Calif.; introd. in sw Alaska, the Yukon (Dawson), and B.C. (Vancouver Is., Sicamous, Nelson, and along the Sikanni R. at ca. 57°14'N; CAN). MAP: Hultén 1968*b*:766.

    Var. *achilleaefolia* (Benth.) Mason (*G. ach.* Benth.; corolla to about 12 mm long rather than about 8 mm, the flowers in loose cymes to loosely capitate rather than in dense spherical heads) is known as an occasionally introd. weed in Alaska–B.C.

[*G. congesta* Hook.]
[The tentative inclusion of B.C. in the range of this species of the w U.S.A. (N to Oreg. and N.Dak.) by Rydberg (1922) requires clarification.]

G. minutiflora Benth.
/t/W/ (T) Dry, sandy sagebrush plains and valleys from s B.C. (Dry Interior N to Cache Creek and Kamloops; CAN) to Oreg. and Idaho. [*Ipomopsis* Grant].

G. rubra (L.) Heller   Standing-cypress
Native in the U.S.A. from Okla. and N.C. to Tex. and Fla.; introd. elsewhere, as in s Ont. (near Sarnia, Lambton Co.; Port Dover, Norfolk Co.). [*Ipomopsis* Wherry]. MAP: E.T. Wherry, Bartonia 18: fig. 2, p. 57. 1936.

[G. sinuata Dougl.]
[The report of this species of the w U.S.A. (N to Wash. and Mont.) from B.C. by Boivin (1966b) may be based upon the report of *G. inconspicua* from that province by John Macoun (1884; taken up by Henry 1915). However, Hitchcock et al. (1959) note that "The identity of the original material of *G. inconspicua* (Smith) Sweet is still doubtfull, but it is probably not the same as *G. sinuata*." (*G. ?inconspicua* (Sm.) Sweet).]

LEPTODACTYLON H. & A. [7016]

L. pungens (Torr.) Nutt.
/t/WW/ (N (Ch)) Dry sandy or rocky plains and slopes at low to moderate elevations from s B.C. (Dry Interior in the Okanagan Valley from Keremeos and Osoyoos, near the U.S.A. boundary, N to Summerland, about 10 mi N of Penticton; CAN) to Mont. and w Nebr., s to Baja Calif. and N.Mex. [*Cantua* Torr.; *Gilia (Navarretia) pungens* (Torr.) Benth., not Dougl., which is *N. squarrosa; Phlox (G.; L.) hookeri* Dougl.].

LINANTHUS BENTH. [7016]

1 Flowers subsessile in a dense terminal leafy cluster; corolla salverform, to about 1.5 cm long, the slender tube much longer than the calyx, the abruptly spreading, deep-pink to purplish lobes 2 or 3 mm long; calyx firm, to 1 cm long, the membranes between the narrow acerose segments small and inconspicuous, not extending to the base; seeds several in each locule; leaves rather firm, the harshly ciliate segments to nearly 2 cm long; stem puberulent, to 1.5 dm tall, sometimes branched at base; (sw B.C.) . . . . *L. bicolor*
1 Flowers long-pedicelled in the leaf-axils and terminating the ultimate branches; corolla white to lavender or light blue, to about 1 cm long, the more or less funnelform tube not much surpassing the calyx, the lobes about equalling to distinctly longer than the tube; membranes between the relatively herbaceous segments of the calyx prominent and extending nearly or quite to the base; plants glabrous to scabrous-puberulent, to 3 dm tall.
  2 Seeds solitary in each locule of the capsule and filling it; corolla at most 2.5 mm long, less than 1.5 times as long as the calyx, glabrous, its lobes about equalling the tube; calyx to 3 mm long at maturity; anther-filaments glabrous; leaf-segments to 1.5 cm long; (s ?B.C.) . . . . [*L. harknessii*]
  2 Seeds 2 or more per locule; corolla-lobes usually distinctly longer than the tube; leaf-segments to 2 cm long.
    3 Corolla commonly not over 5 mm long, mostly not more than twice as long as the calyx, the tube usually hairy within near the insertion of the glabrous or subglabrous filaments; (s B.C. to sw Sask.) . . . . *L. septentrionalis*
    3 Corolla to 1 cm long, to 3.5 times as long as the calyx, the tube generally glabrous or nearly so; filaments generally hairy at base . . . . [*L. pharnaceoides*]

L. bicolor (Nutt.) Greene
/t/W/ (T) Open or thinly wooded slopes at low elevations from sw B.C. (s Vancouver Is. and Saltspring Is.; CAN; V) to s Calif. [*Leptosiphon* Nutt.; *Gilia* Piper; *G. tenella* Benth., not *Collomia tenella* Gray].
Our material is referable to the coastal ecotype, var. *minimus* (Mason) Cronq. (corolla at most about 1.5 cm long rather than to 3 cm, scarcely bicoloured rather than usually prominently so).

[*L. harknessii* (Curran) Greene]
[The report of this species of the w U.S.A. (Wash. and Idaho to Calif. and Nev.) from B.C. by Eastham (1947; Nicola and Toad Mt., Nelson) requires confirmation. (*Gilia* Curran).]

[*L. pharnaceoides* (Benth.) Greene]
[The report of this species of the w U.S.A. (Wash. and Idaho to Calif.) from s Alta. by J.M. Macoun (1897) is based upon *L. septentrionalis* (the relevant collections in CAN), to which the report from s B.C. by John Macoun (1884) also probably refers. (*Gilia* Benth.; *G. liniflora* var. *ph.* (Benth.) Gray).]

L. septentrionalis Mason
/T/W/ (T) Dry meadows, sagebrush plains, and slopes at low to moderate elevations from s B.C. (valleys of the Dry Interior N to Kamloops; CAN), s Alta. (Milk River, Medicine Hat, and Cypress Hills; CAN), and sw Sask. (Cypress Hills and Nashlyn, s of Cypress L.; Breitung 1957a) to Calif. and Colo.

MICROSTERIS Greene [7015]

M. gracilis (Hook.) Greene
/t/W/ (T) Dry to moderately moist open places from s B.C. (probably native N to Lytton and possibly N to Revelstoke and Cariboo, but perhaps introd. at Prince George, ca. 54°N, as in s Alaska–Yukon; in B.C., extending E to Flathead, near the Alta. boundary, but the report from Alta. by Moss 1959, requires confirmation) and Mont. to Baja Calif. and N.Mex.; S. America. [*Gilia* Hook.; *Phlox* Greene; *Collomia* Dougl.; *C. micrantha* Kell.; *M. stricta* Greene]. MAP: Hultén 1968b:766.

Some of our material is referable to var. *humilior* (Hook.) Cronq. (*Collomia (Phlox)* gr. var. *hum.* Hook.; *M. humilis* and *M. glabella* Greene; plant bushy-branched from near the base rather than usually only above the middle; corolla usually less than 1 cm long rather than to 1.5 cm, its lobes generally 1 or 2 mm long rather than up to 4 mm).

NAVARRETIA R. & P. [7016]

1 Stem subglabrous to puberulent (often becoming more or less villous in the inflorescence); leaves at most 3.5 cm long, the terminal segment more or less elongate; petal-traces simple up to about the level of the sinuses or beyond, so that each corolla-lobe (usually less than 2 mm long) receives only 1 principal vein; calyx commonly less than 7 mm long, the teeth more or less unequal; stigmas normally 2; capsule 2-locular, indehiscent or tardily opening by disintegration of the lower lateral walls; seeds rarely more than 6 in a locule; (s B.C. to s Sask.) .... *N. minima*
1 Stem copiously glandular-hairy, to 4 dm tall, simple or moderately branched; leaves firm, to 6 cm long, the terminal segment not elongate; corolla pale to deep blue, to 12 mm long, the petal-traces forked well down into the tube so that each lobe (2 or 3 mm long) receives 3 veins; calyx at least 9 mm long, the teeth subequal; stigmas 3; capsule 3-locular, dehiscent by 3 valves from the top downwards; seeds commonly 8 or 9 in each locule; (sw B.C.) .... *N. squarrosa*

N. minima Nutt.
/T/W/ (T) Moist meadows and lowlands (ranges of Canadian taxa outlined below), s to Calif., Ariz., and Colo.
1 Corolla white, to 6 mm long, its lobes about 1 mm long and half as broad, each with a single unbranched midvein; seeds 1 (sometimes 2, rarely 3) in each locule; leaves and bracts relatively soft; stem to 1 dm tall; [*Gilia* Gray; s B.C. (Boivin 1966b), s Alta. (Empress; Milk River), and s Sask. (8 localities cited by Breitung 1957a)] .... var. *minima*
1 Corolla white to pale blue, its lobes only slightly or scarcely longer than broad, with branched midvein; seeds up to 6 in each locule; [*Aegochloa (N.; Gilia) int.* Benth.; *N. propinqua* Suksd.; sw B.C.: in a park at Victoria, Vancouver Is., where perhaps introd.; Herb. V] .... var. *intertexta* (Benth.) Boivin

N. squarrosa (Esch.) H. & A. Skunkweed
/t/W/ (T) Open valleys and hillsides from sw B.C. (Vancouver Is. and Saltspring Is.; CAN; V) to Calif. [*Hoitzia* Esch.; *Gilia* H. & A.; *G. pungens* Dougl., not (Torr.) Benth.].

PHLOX L. [7014] Phlox

(Ref.: Wherry 1955)
1 Low loosely tufted mat-forming perennials from a much-branched subligneous base, the stems decumbent or ascending; leaves crowded, more or less rigid, at most about 2.5 cm

long, persistent, they and the calyx-lobes cuspidate or spinulose-tipped; flowers white, pink, lavender, or light blue; styles united below for well over half their length.

2 Calyces and leaves (at least along the margins at base) more or less arachnoid or woolly-villous with somewhat entangled hairs; corolla-tube commonly surpassing the calyx (or often shorter than the calyx in an eastern phase of *P. hoodii*); corolla-lobes entire or only shallowly notched or crenate at summit.

   3 Leaves to about 2 cm long and 2 mm broad, rather soft, glabrous except for the often arachnoid-ciliate basal margins; (s B.C. and sw Alta.) .... *P. diffusa*

   3 Leaves mostly not over 1 cm long, rigid and pungent.

      4 Leaves at most about 0.5 mm broad, more or less arachnoid over the surfaces at least when young, the whole plant often with a silvery cast; (Alta. to Man.). .... *P. hoodii*

      4 Leaves commonly about 1 mm broad, arachnoid-ciliate along the basal margins, otherwise glabrous, the whole plant usually greener; (w arctic and subarctic regions) .... *P. richardsonii*

2 Calyces (and whole plant) more or less pubescent but not at all arachnoid or woolly-villous (villous in *P. sibirica* but the hairs then straightish and not entangled).

   5 Leaves mostly lanceolate to elliptic and over 2 mm broad (up to 5 mm), to about 2.5 cm long, usually ciliate below middle, otherwise glabrous, their margins rather markedly whitened with a narrow cartilaginous thickening; (s Alta. and s Sask.) .... *P. alyssifolia*

   5 Leaves mostly narrower, subulate or linear, their margins less markedly or scarcely whitened-cartilaginous.

      6 Plant copiously long-villous, rarely glandular; leaves relatively soft, commonly 2 or 3 mm broad, villous at least when young; corolla-tube about equalling the calyx, the corolla-lobes often notched or emarginate and somewhat crenate at summit; (Alaska to Dist. Mackenzie) .... *P. sibirica*

      6 Plants usually more or less pubescent in the inflorescence with gland-tipped hairs, not long-villous; leaves firm and pungent, narrowly linear, mostly less than 1 mm broad, often short-ciliate toward base, otherwise glabrous; corolla-tube markedly surpassing the calyx.

         7 Leaves mostly strongly ascending and often somewhat incurving; corolla-lobes entire or only shallowly emarginate; (?B.C.) .... [*P. caespitosa*]

         7 Leaves mostly more or less squarrose-recurving; corolla-lobes with an emargination 1 or 2 mm deep; (Ont.; introd. eastwards) .... *P. subulata*

1 Herbs with erect or decumbent flowering stems and relatively large deciduous leaves with flat blades; flowers rather short-pedicelled in terminal and subterminal cymes.

   8 Leaves alternate (only the lowermost opposite), lanceolate to broadly ovate or oblong, acute, narrowed to base or clasping, to about 8 cm long; flowers rose-red to red-purple, to about 2.5 cm broad; styles united rarely over half their length from base, they and the stamens included; glandular-viscid annual; (introd.) .... *P. drummondii*

   8 Leaves all opposite (or the uppermost sometimes alternate); perennials.

      9 Styles united well over half their length from base, they and the stamens about equalling the corolla-tube and visible at its throat; styles equalling or longer than the fruiting calyx.

         10 Leaves elliptic-lanceolate to -ovate, their lateral veins conspicuous beneath, their margins minutely bristly-ciliate; corolla normally pinkish purple, the tube usually sparingly pubescent outside; calyx to 1 cm long, with awn-tipped lobes; (introd.) .... *P. paniculata*

         10 Leaves narrowly linear to broadly lanceolate, their lateral veins obscure; corolla-tube glabrous; calyx-lobes not awn-tipped.

            11 Leaves linear, to about 8 cm long and 2.5(4) mm broad; flowers rather few, loosely cymose; corolla white to pink; calyx commonly about 1 cm long, the lobes usually longer than the tube; plants from an eventual taproot, but often branched and creeping below the ground level, often woody at base, glabrous to strongly glandular (especially in the inflorescence) or hairy; (s B.C.) .... *P. longifolia*

11 Leaves mostly narrowly to broadly lanceolate (upper ones often broader), long-acuminate, to 12 cm long; flowers numerous in a cylindric short-branched inflorescence usually at least twice as long as thick; corolla red-violet or purple; calyx to 8 mm long, the lobes at anthesis shorter than the tube; stem to 9 dm tall, strongly purple-spotted; inflorescence densely puberulent with nonglandular hairs, the plant otherwise essentially glabrous; (introd.) .... *P. maculata*

9 Styles rarely united more than half their length from base, they and the stamens rarely surpassing the middle of the corolla-tube; styles much shorter than the villous fruiting calyx, the teeth of the latter subulate-tipped.

12 Leaves lanceolate to narrowly ovate, obtusish; corolla-lobes blue to purple; calyx densely glandular; stem with elongate sterile rooting basal shoots; (Ont. and sw Que.) .... *P. divaricata*

12 Leaves linear to narrowly lanceolate, attenuate; corolla-lobes white to pink or violet; stem lacking decumbent sterile basal shoots.

13 Plants shrubby toward base (the woody branches often prostrate), from a woody taproot, generally glandular or glandular-hairy above (at least on the pedicels and calyces), often glabrous below, to 4 dm tall; corolla white to pink, the tube not much surpassing the calyx, the lobes nearly always deeply notched at apex; (s B.C.) .... *P. speciosa*

13 Plants not shrubby, to 8 dm tall; corolla white to pink or violet, the tube about twice as long as the calyx, the lobes entire; (s Ont.; introd. westwards) .... *P. pilosa*

P. alyssifolia Greene
/T/WW/ (Ch) Dry open places at low to moderate elevations from Mont. to sw Alta. (Fort MacLeod, Pincher Creek, and Whisky Gap; CAN) and s Sask. (Bengough; Rockglen; type from near Wood Mountain), s to Colo. and S.Dak. MAP: Wherry 1955:136.

[*P. caespitosa* Nutt.]
[Reports of this species of the w U.S.A. (Wash. and Mont. to Oreg. and Idaho) from B.C. by Henry (1915) and Eastham (1947) require confirmation, probably being based upon *P. diffusa*. (*P. douglasii* Hook.; *P. rigida* Benth.). MAP: Wherry 1955:141.]

P. diffusa Benth.
/T/W/ (Ch) Open or wooded rocky slopes at low to moderate elevations from s B.C. (N to the Marble Range NW of Clinton at ca. 51°N) and s Alta. (N to the Hand Hills at ca. 51°30′N) to Calif. [*P. douglasii* var. *dif*. (Benth.) Gray].
Our material is referable to var. *longistylis* (Wherry) Peck (styles to 1 cm long rather than at most 4.5 mm). MAP: Wherry 1955:160.

P. divaricata L.   Blue Phlox
/T/EE/ (Ch) Damp to dry open woods and rocky slopes from Ill. to Mich., Ont. (N to the Ottawa dist.), sw Que. (N to the Montreal dist.; reported N to near Quebec City by John Macoun 1886), and NW Vt., s to Ala. and S.C.; cult. and locally natzd. elsewhere (as perhaps in the N part of the Canadian area). [*P. canadensis* Sweet]. MAPS: E.T. Wherry 1955:42, and Bartonia 12: fig. 1, p. 27. 1930.
Forma *albiflora* Farw. (corolla white rather than blue or purple) is known from s Ont. (Lambton, Elgin, Norfolk, and Bruce counties; CAN; OAC; TRT).

P. drummondii Hook.   Annual Phlox
A native of Texas; a garden-escape elsewhere, as in s Ont. (Point Edward, Lambton Co., where reported by Dodge 1915, as inclined to persist in sand as a garden-escape; MICH) and s N.B. (in a woods at St. John, where taken by Warner in 1892; NBM).

P. hoodii Richards.   Moss-pink
/T/WW/ (Ch) Dry prairies, sagebrush plains, and foothills from Wash. to Alta. (N to ca. 53°N

according to Porsild 1966, who refers reports from Alaska–Yukon by Hultén 1948 and 1968b, to *P. richardsonii*), Sask. (N to the type locality near Carlton House, about 35 mi SW of Prince Albert; CAN, photograph), and S Man. (N to Bield, S of Duck Mt.), S to Calif., Utah, and Wyo. [*P. canescens* T. & G.; *P. muscoides* Nutt.]. MAPS: Wherry 1955:165; Hultén 1968b:764 (the Alaska–Yukon area should evidently be deleted).

P. longifolia Nutt.
/t/W/ (Ch (Hp)) Dry rocky places at low to high elevations from S B.C. (Dry Interior in the Okanagan Valley from Keremeos and Osoyoos, near the U.S.A. boundary, N to Summerland, about 10 mi N of Penticton) and Mont. to S Calif. and N.Mex. [*P. humilis* Dougl.; incl. var. *linearifolia* (Hook.) Brand (*P. speciosa* var. *lin.* Hook.; *P. lin.* (Hook.) Gray) and ssp. *calva* Wherry]. MAP: Wherry 1955:89.

P. maculata L.   Wild Sweet William
Native in the E U.S.A. (N to Minn. and Conn.); introd. or a garden-escape elsewhere, as in S Ont. (Muskoka Dist., Georgian Bay, L. Huron, where "persisting after cult. and perhaps spreading"; TRT), SW Que. (N to near Quebec City in Portneuf Co.; MT), and N.S. (Parrsboro, Cumberland Co.; Groh and Frankton 1949b). [*P. odorata* Sweet; incl. *P. suaveolens* Ait.]. MAPS: E.T. Wherry 1955:115, and Bartonia 14: fig. 2, p. 22. 1932.
    Forma *immaculata* Fern. (flowers white or pale and unspotted rather than purple-spotted) is reported from SW Que. by Boivin (1966b).
    P. stolonifera Sims, Creeping Phlox, is apparently known from Canada only through a 1962 collection by Bahr in SW Que. (along a road through mixed woods near Bolton Centre, Brome Co.; MTMG; CAN). It is a native of the E U.S.A. (Ohio to Pa., S to N Ga.) and occasionally spread from cult. elsewhere, as probably in the case of the Que. plant. It may be distinguished from other long-styled phloxes by its elongate stolons.

P. paniculata L.   Fall-pink, Perennial Pink
Native in the E U.S.A. (N to Iowa and N.Y.); a garden-escape elsewhere, as in S Ont. (Kent, Lambton, Lincoln, Waterloo, Wellington, and York counties), Que. (Boivin 1966b), N.B. (Norton, near St. John, where taken by Hay in 1877; NBM), and N.S. (a dump at Port Williams, Kings Co.; ACAD). MAPS: E.T. Wherry 1955:121, and Bartonia 15: fig. 2, p. 21. 1933.

P. pilosa L.
/t/EE/ (Hp) Dry open soils and sandy woods from E Kans. to Mich., S Ont. (Essex, Lambton, and Welland counties; CAN; OAC), N.Y., and N.J., S to Tex. and Fla. [Incl. var. *virens* (Michx.) Wherry; *P. aristata* Michx.]. MAPS: E.T. Wherry 1955:52, and Bartonia 12: fig. 2, p. 39. 1930.
    Var. *fulgida* Wherry (pedicels and calyces whitish-villous with glandless hairs rather than copiously stipitate-glandular) is reported from Sask. by Fernald *in* Gray (1950; this taken up by D.A. Levin, Brittonia 18:145. 1966; not listed by Breitung 1957a) and there is a collection in CAN from S Man. (St. Vital Park, Winnipeg). It is probably a casual garden-escape in our area. MAPS: on the above-noted maps by Wherry.

P. richardsonii Hook.
/aS/W/ (Ch) Rocky slopes from the coasts of Alaska and Dist. Mackenzie (E to near Coronation Gulf; type from the "Arctic sea-coast") to N Banks Is. and Victoria Is., S to S-cent. Alaska–Yukon. [*P. sibirica* ssp. *rich.* (Hook.) Hult.]. MAPS: Porsild 1957: map 283, p. 196; Hultén 1968b:765 (*P. sib.* ssp. *rich.*); Wherry 1955:127.

P. sibirica L.
/aSs/W/EA/ (Ch) Calcareous slopes at low to fairly high elevations in Alaska–Yukon (N to ca. 69°30′N) and NW Dist. Mackenzie (Richardson Mts., W of the Mackenzie R. Delta; CAN); NE Europe; Asia. [*P. borealis* Wherry; *P. alaskensis* Jordal]. MAPS: Wherry 1955:127 (*P. bor.*); Hultén 1968b:764.

P. speciosa Pursh
/t/W/ (Hp (Ch)) Sagebrush plains and wooded slopes from S B.C. (Dry Interior in the Okanagan

Valley N to Summerland, about 10 mi N of Penticton; Wherry 1955; Eastham 1947) and Mont. to Calif. [*P. occidentalis* Dur.]. MAP: Wherry 1955:13 (*P. speciosa occ.* (Dur.) Wherry).

P. subulata L.   Moss- or Mountain-Phlox, Moss-pink
/T/EE/   (Ch)   Dry sands, gravels, and rocks from Mich. to S Ont. (apparently native in Lambton, Norfolk, and Haldimand counties; apparently a garden-escape in Waterloo, Wellington, and Dundas counties, as also near Montreal, Que., and in Digby and Kings counties, N.S.), N.Y., and N.J., s to Tenn. and N.C. MAPS: E.T. Wherry 1955:70, and Bartonia 11: fig. 2, p. 20. 1929.

According to Fernald *in* Gray (1950), the plant of Norfolk Co., s Ont., is referable to var. *brittonii* (Small) Wherry (*P. brittonii* Small; plant relatively small-dimensioned, the hairs of the inflorescence usually gland-tipped rather than glandless). Forma *albiflora* Britt. (flowers white rather than roseate or purplish) is reported from S Ont. by Boivin (1966*b*).

POLEMONIUM L.   [7017]   Greek Valerian, Jacob's-ladder

(Ref.: Davidson 1950)
1   Leaflets numerous, mostly or all cleft nearly or quite to base into up to 5 segments (to 6 mm long and 3 mm broad) and thus appearing verticillate; corolla violet-blue, its limb definitely shorter than the tube; stamens and style included; inflorescence densely cymose-capitate, not much elongating in fruit; plant commonly not over 2 dm tall, densely stipitate-glandular or viscid-woolly, from a much-branched caudex above a stout taproot; (S B.C.–Alta.) . . . . *P. viscosum*
1   Leaflets undivided.
   2   Corolla ordinarily white, to 5 mm long, equalling or more commonly a little shorter than the calyx; stamens and style included; flowers terminal and solitary (but soon appearing leaf-opposed through elongation of the stem); stem rather uniformly leafy, the leaves with up to 7 pairs of leaflets (to 9 mm long and 4 mm broad); slender taprooted annual to 3 dm tall, often freely branched and loosely ascending (sometimes nearly prostrate), generally glandular-puberulent throughout; (S B.C.) . . . .
   . . . . *P. micranthum*
   2   Corolla blue or blue-violet to pale purple or purplish blue, much surpassing the calyx; perennials.
      3   Stamens and style strongly exserted from the corolla; inflorescence a compact thyrsiform panicle, its short branches sparingly short-pubescent but scarcely viscid; longer (lower) leaves with up to 10 pairs of narrowly lanceolate to narrowly ovate sharply acute leaflets; stem leafy, to about 1 m tall, simple, erect, glabrous except in the inflorescence, from a horizontal rhizome; (introd.) . . . . *P. van-bruntiae*
      3   Stamens and style included in or only slightly exserted from the corolla; branches of inflorescence usually more or less glandular or glandular-villous.
         4   Inflorescence cymose-paniculate (the ultimate flower-clusters cymose), loose and open, its long branches and the slender pedicels (at anthesis as long as or longer than the calyces) minutely glandular-puberulent; capsules stipitate; corolla rarely over 1.5 cm long; leaves predominantly cauline, the longer (lower) ones with at most 8 pairs of elliptic to oval acutish leaflets; stems essentially glabrous up to the inflorescence, to 6 dm tall; (introd.) . . . . *P. reptans*
         4   Inflorescence compact (head-like or thyrsiform), the pedicels at anthesis mostly shorter than the calyces, the ultimate cymes crowded and dense; capsules sessile; leaves predominantly basal, their leaflets commonly more numerous; stem and branches of inflorescence copiously pubescent and more or less viscid.
            5   Habit spreading, the ascending stems usually several from a much-branched caudex, commonly less than 3 dm tall; leaflets commonly less than 1 cm long and 4 mm broad; (B.C.–Alta.).
               6   Corolla funnelform or tubular-funnelform, often only half as broad as long, the tube commonly a little surpassing the calyx (the latter to 8.5 mm long at anthesis); plant to 1.5 dm tall, densely stipitate-glandular or viscid-pubescent . . . . *P. elegans*

6 Corolla more broadly campanulate, commonly as broad as or broader than long, the tube not surpassing the calyx; plants more or less strongly glandular or viscid in the inflorescence, elsewhere usually less so or subglabrate . . . . *P. pulcherrimum*

5 Habit erect, the stems solitary from the upturned end of a short simple horizontal rhizome; corolla broadly campanulate.

7 Stem usually not over 2 dm tall, more or less scapose, the leaves nearly all basal or 1 or 2 (sometimes 3) reduced ones above; leaflets narrowly to broadly elliptic, obtusish, commonly less than 1 cm long, less than twice as long as broad; corolla to 2.5 cm long; calyx densely glandular-pilose; cymes commonly solitary or 1 or 2 additional ones from the upper axils; (w arctic and subarctic regions) . . . . *P. boreale*

7 Stem to about 1 m tall, commonly bearing several reduced leaves; leaflets narrowly lanceolate to elliptic, acute, to about 4 cm long, more than twice as long as broad; corolla rarely over 1.5 cm long; calyx sparingly pubescent; cymes lax, elongate, usually more numerous; (B.C. and Alta.; introd. in E Que. and Nfld.) . . . . *P. caeruleum*

P. boreale Adams

/aSs/WW/GEA/ (Hsr (Ch)) Dryish meadows, sandy tundra, and calcareous slopes at low to fairly high elevations from the coasts of Alaska–Yukon–NW Dist. Mackenzie to N Banks Is., Melville Is., and ?Cornwallis Is., S to S Alaska–Yukon–W Dist. Mackenzie; E Greenland between ca. 74° and 76°N.; Spitsbergen; Scandinavia; N Asia. [*P. richardsonii* Graham; *P. ?humile* Willd., not Salisb. nor Turcz.; *P. villosum* Sweet, not Rud.; *P. lanatum* of auth., perhaps not Pallas; incl. the coarse extreme, ssp. *macranthum* (Cham.) Hult. (*P. humile (pulchellum)* var. *mac.* Cham.)]. MAPS: Porsild 1957: map 282, p. 196; Davidson 1950: map 6 (somewhat incomplete for N. America), p. 278; combine the maps by Hultén 1968b:768 (ssp. *bor.* and ssp. *mac.*).

Var. *villosissimum* Hult. (the inflorescence strongly white-villous rather than only moderately so) is known from McKinley Park, Alaska, the type locality.

P. caeruleum L. Jacob's-ladder

/aST/WW/EA/ (Hs (Ch)) Swampy ground and wet meadows (ranges of Canadian taxa outlined below), S to Calif. and Colo.; isolated in cedar swamps of N Minn.; Eurasia. MAPS and synonymy: *see* below.

1 Stamens included in the corolla and markedly surpassed by the style; [*P. occidentale* Greene and its var. *intermedium* Brand (*P. intermedium* (Brand) Rydb.); *P. acutiflorum* ssp. *occ.* (Greene) Hult.; S Alaska and S-cent. Yukon through the mts. of B.C. and W Alta.; reports from Sask. by Rydberg 1922 and 1932, are discounted by Breitung 1957a; MAP: Davidson 1950: map 1, p. 273] . . . . ssp. *occidentale* (Greene) Davidson

1 Stamens slightly exserted beyond the corolla and not markedly surpassed by the style.

2 Corolla-lobes commonly ciliate or erose along the margins, acute; [*P. vill.* Rud.; *P. acutiflorum* Willd. and its f. *lacteum* Lepage; Aleutian Is. and northernmost Alaska–Yukon–Dist. Mackenzie to B.C. (S to Chilcotin, ca. 52°N) and S Alta. (S to the Macleod R. at ca. 49°N); MAPS: on the above-noted map by Davidson; Hultén 1968b:767 (*P. acut.*)] . . . . ssp. *villosum* (Rud.) Brand

2 Corolla-lobes glabrous, obtuse; [a garden-escape in N. America; known in Canada only from E Que. (Gaspé Pen.) and a persistent planting in an abandoned cemetery near the Straits of Belle Isle, NW Nfld.; MAP: on the above-noted map by Davidson (Eurasian area of ssp. *vulgare*)] . . . . ssp. *caeruleum*

P. elegans Greene

/T/W/ (Hs (Ch)) Open rocky places at high elevations from S B.C. (Skagit Valley near Chilliwack, where taken by J.M. Macoun in 1905; CAN, verified by both Wherry and Davidson; reported from Vancouver Is. and Queen Charlotte Is. by Carter and Newcombe 1921, but not listed for Queen Charlotte Is. by Calder and Taylor 1968; a collection in V from between Hope and Princeton has also been placed here; the reports from South Kootenay Pass, SE B.C., and from Sheep Mt., Waterton Lakes, SW Alta., by J.M. Macoun 1906, are based upon *P. viscosum*, the relevant collections in CAN) to Wash. and N Calif. MAP: Davidson 1950: map 9 (somewhat incomplete northwards), p. 281.

P. micranthum Benth.
/t/W/ (T) Dry or moist open places in the plains (often with sagebrush) and foothills from S B.C. (Dry Interior N to Boston Bar, Lytton, Merritt, Nicola, Kamloops, and Vernon; CAN; V; the report from Vancouver Is. by John Macoun 1884, requires confirmation) and Mont. to Calif. and Utah; S. America. [*Polemoniella* Heller]. MAP: Davidson 1950: map 5, p. 277.

P. pulcherrimum Hook.
/ST/W/ (Hp (Ch)) Dry or moist rocky places at low to high elevations from Alaska (N to ca. 67°30'N), the Yukon (N to ca. 64°N), and NW Dist. Mackenzie to Great Bear L., S through B.C. and SW Alta. (N to Banff; CAN) to Calif. and Colo. The type is an early collection by Drummond in the "Highest Rocky Mountains", presumably in B.C.–Alta. [*P. caeruleum* var. *pul.* Hook.; *P. lindleyi* Wherry; *P. fasciculatum* and *P. rotatum* Eastw.; *P. humile* var. *pulchellum* Gray]. MAPS: Hultén 1968*b*:769; Davidson 1950: map 8, p. 280.

P. reptans L.
Native in the E U.S.A. (N to Minn. and N.Y.); a garden-escape elsewhere, as in S Ont. (collection in MT from London, Middlesex Co.; reported from Wellington Co. by Stroud 1941) and SW Que. (Montreal). [*P. humile* Salisb., not Willd. nor Turcz.]. MAP: Davidson 1950: map 3 (indicating no Que. stations), p. 281; the maps by Cain 1944: fig. 63, p. 462, and E.T. Wherry, Bartonia 17: fig. 2, p. 9. 1935, show no Canadian stations).

P. van-bruntiae Britt.   Jacob's-ladder
Native in the E U.S.A. (N.Y. and Vt. to W.Va.) and perhaps once a member of the native flora of N.B., it being reported by G.U. Hay (Nat. Hist. Soc. N.B., Bull. 5:43. 1886) from along a cool stream near Trout L., Charlotte Co. It is elsewhere known in our area apparently only from rubbish heaps, gardens, and waste places of Que. (Arthabasca, Kamouraska, and Rimouski counties and the Gaspé Pen. at Gaspé Basin). [*P. caeruleum* ssp. *van-br.* (Britt.) Davidson]. MAPS: Davidson 1950: map 1, p. 273; Cain 1944: fig. 63, p. 462; E.T. Wherry, Bartonia 17: fig. 1, p. 6. 1935.

P. viscosum Nutt.   Skunkweed
/T/W/ (Hs (Ch)) Open rocky places at high elevations from S B.C. (Mt. Garibaldi, NE of Vancouver; Lillooet; Manning Provincial Park, SE of Hope; South Kootenay Pass) and SW Alta. (Waterton Lakes; CAN) to Oreg., Ariz., and N.Mex. [*P. confertum* Gray]. MAP: Davidson 1950: map 9, p. 281.

Forma *leucanthum* Williams (flowers white rather than blue) is reported from SW Alta. by Breitung (1957*b*: Waterton Lakes).

HYDROPHYLLACEAE (Waterleaf Family)

Herbs with alternate, commonly simple (sometimes compound), moderately lobed to deeply pinnatifid leaves. Flowers regular, perfect, hypogynous, gamopetalous. Sepals or calyx-lobes, corolla-lobes, and stamens each 5, the latter inserted on the corolla-tube and alternating with its lobes. Style 1. Ovary superior. Fruit a 1-locular 2-valved capsule.

1 Flowers white or lavender, solitary in or opposite the leaf-axils (or some in a loose
few-flowered raceme-like terminal cyme); stamens included; leaves coarsely pinnatifid;
taprooted annuals.
  2 Calyx with lanceolate, spreading to reflexed auricles to 3 mm long at the sinuses;
  seeds with a partial or complete, persistent or deciduous outer covering over the
  seed-coat; (S B.C. and SW Alta.) . . . . *Nemophila*
  2 Calyx not auricled in the sinuses, about equalling the corolla; seeds lacking a covering
  over the seed-coat; leaves deeply pinnatifid into up to 13 oblong, commonly coarsely
  cut-toothed segments, at least the lower ones opposite; (Alta. to Man.) . . . . *Ellisia*
1 Flowers commonly numerous in spicate to racemose, often helicoid cymes.
  3 Leaves chiefly basal, reniform-orbicular, palmately veined, rather shallowly lobed or
  coarsely toothed; flowers white; stigma entire or nearly so; stamens included; more or
  less fibrous-rooted perennials; (Alaska–B.C.–SW Alta.) . . . . *Romanzoffia*
  3 Leaves chiefly cauline, wholly or mostly alternate, commonly deeply lobed or parted
  (sometimes entire in *Phacelia*); stigma more or less lobed.
    4 Inflorescence consisting of subdichotomously branched cymes without a central
    axis; stamens exserted; fibrous-rooted perennials of moist habitats . . . . . . *Hydrophyllum*
    4 Inflorescence consisting of one or more helicoid cymes with a central axis (the
    cymes often aggregated into a compound, often thyrsoid inflorescence); stamens
    included or exserted; taprooted annuals or perennials of often dry habitats . . . . *Phacelia*

ELLISIA L. [7023]

E. nyctelea L.
/T/X/ (T) Moist shaded places from Mont. to S Alta. (Medicine Hat; CAN), Sask. (N to Langham, about 25 mi NW of Saskatoon; Breitung 1957a), and S Man. (N to Dauphin, N of Riding Mt.), S to N.Mex., Tex., Okla., and Mich., and along the Atlantic slope from E Pa. and N.J. to N.C.; a casual adventive E to New Eng. [*Macrocalyx* Ktze.; *Nyctelea* Britt.]. MAP: Lincoln Constance, Rhodora 42(494): map 1, p. 37. 1940.

HYDROPHYLLUM L. [7021] Waterleaf

(Ref.: Constance 1942)
1 Leaves subrotund in outline, 5–9-lobed, cordate at base.
  2 Leaves palmately veined; calyx-lobes narrowly linear, sparingly ciliate, otherwise
  glabrous, sometimes with minute teeth in the sinuses; corolla white to purplish;
  stamens long-exserted; peduncles mostly shorter than the petioles; plant nearly
  glabrous; fibrous-rooted perennial from a long scaly-toothed rhizome; (S Ont. and
  SW Que.) . . . . *H. canadense*
  2 Leaves pinnately veined; calyx-lobes lanceolate, bristly-hairy, a small reflexed lobe
  in each sinus; corolla lavender to violet; stamens exserted only 1–3 mm; peduncles
  much surpassing the petioles; plant abundantly hirsute; taprooted biennial; (S Ont.) . . . .
  . . . . *H. appendiculatum*
1 Leaves oblong or ovate in outline, deeply pinnately divided into up to 11 or more
segments, the basal pair or pairs of segments commonly distinct and themselves deeply
lobed; rhizomatous perennials with usually more or less pubescent stems.
  3 Segments of leaves entire (or toothed or divided only near apex); corolla commonly
  lavender or purplish blue (sometimes white); anthers at most 1.3 mm long; peduncles
  seldom over 5 cm long and well surpassed by the leaves, often reflexed in fruit;

rhizome very short, bearing a cluster of finger-like roots; plant to about 4 dm tall; (s B.C. and sw Alta.) . . . . *H. capitatum*

3   Segments of leaves rather copiously cut or coarsely toothed; anthers to over 2 mm long; peduncles commonly surpassing their subtending leaves; rhizome usually conspicuous, often scaly; plants usually taller.

4   Leaf-blades evidently longer than broad, with up to 15 primary segments, the lower segments remote, only the upper ones confluent; cymes subcapitate in flower, the pedicels rarely over 7 mm long; corolla white to lavender or purplish; calyx-lobes abundantly. soft-hairy on the back, weakly ciliate; (sw B.C.) . . . . . *H. fendleri*

4   Leaf-blades scarcely longer than broad, the principal divisions usually 5, all but the lowermost pair commonly confluent; pedicels to 12 mm long.

5   Stems retrorse-hispid above, harsh to the touch; corolla to 7 mm long, green or greenish white (sometimes cream-colour); calyx-lobes strigillose or glabrate on the back, strongly hispid-ciliate; (s Vancouver Is.) . . . . *H. tenuipes*

5   Stems glabrate to strigose above; corolla to 1 cm long, white to lavender or purple; calyx-lobes strigillose on the back, weakly ciliate; (se Man. to Que.) . . . .
. . . . *H. virginianum*

H. appendiculatum Michx.
/t/EE/   (Hs)   Rich woods from Minn. to s Ont. (Essex, Middlesex, Oxford, and Norfolk counties), s to e Kans., Mo., Tenn., and Pa. MAP: Constance 1942:728.

H. canadense L.
/T/E/   (Grh)   Rich moist woods from s Ont. (N to Grey, Simcoe, and Durham counties) to sw Que. (Boivin 1966b), Vt., and Mass., s to e Mo. and n Ala. MAP: Constance 1942:728.

H. capitatum Dougl.
/T/W/   (Hs)   Thickets, woods, and moist open slopes at low to fairly high elevations from s B.C. (N to Spences Bridge, about 45 mi sw of Kamloops) and sw Alta. (Waterton Lakes; Castle River Forest Station, sw of Lethbridge) to Calif. and Colo. [Incl. var. *pumilum* Hook. and its subvar. *densum* Brand]. MAP: Constance 1942:727.

H. fendleri (Gray) Heller
/T/W/   (Hs)   Thickets and moist open places at low to fairly high elevations from s B.C. (valleys of the Skagit and Chilliwack rivers; Mt. Cheam, near Agassiz; Manning Provincial Park, se of Hope; near Lytton; Marble Range, nw of Clinton), Idaho, and Wyo. to n Calif. and N.Mex. [*H. occidentale* var. *fend.* Gray]. MAP: Constance 1942:721.

Our material is evidently all referable to var. *albifrons* (Heller) Macbr. (*H. alb.* Heller and its var. *eualbifrons* subvar. *pendulum* Brand; pubescence relatively soft and abundant, the leaves merely strigose above, softly white-hairy beneath, rather than scabrous above and not white-hairy beneath, the sepals softly long-ciliate marginally and villous-puberulent on the back rather than bristly-ciliate marginally and essentially glabrous above). MAP: on the above-noted map by Constance.

H. tenuipes Heller
/t/W/   (Hs)   Moist woods at lower elevations from sw B.C. (Goldstream, s Vancouver Is., where taken by John Macoun in 1887 and 1908; CAN, verified by Constance) to n Calif. MAP: Constance 1942:727.

H. virginianum L.   John's-cabbage
/T/EE/   (Hsr)   Rich moist woods and clearings from se Man. (Selkirk, about 15 mi ne of Winnipeg; WIN) to Ont. (N to the Ottawa dist.), Que. (N to Ste-Anne-de-la-Pocatière, Kamouraska Co.; QSA; this a slight extension northwards beyond the St-Jean-Port-Joli, l'Islet Co., station indicated in the Que. map by Doyon and Lavoie 1966: fig. 4, p. 815; not known from the Maritime Provinces; a puzzling report from Nfld. by Waghorne 1898, probably referable to some other species), and New Eng., s to Kans., Ark., Tenn., and Va. [*H. virginicum* L.]. MAP: Constance 1942:716.

NEMOPHILA Nutt. [7022]

1 Leaves all alternate, the blade commonly with a narrow rachis and 2 pairs of lateral lobes to 2 cm long and 7 mm broad; corolla shorter than the calyx, this with stiffly ciliate lobes and auricles 1 or 2 mm long; seeds mostly solitary; (s B.C. and sw Alta.) . . . . *N. breviflora*
1 Leaves all opposite (or some of the upper ones alternate); corolla equalling or surpassing the calyx; seeds mostly 2 or more.
    2 Corolla mostly over 1.5 cm broad, white or whitish and conspicuously flecked with blackish purple; auricles of calyx 1 or 2 mm long; style to over 2.5 mm long; leaf-blades with mostly 3 or 4 pairs of lobes to about 8 mm long, the upper lobes larger than the lower and sometimes with 1 or 2 teeth; (introd.) . . . . [*N. menziesii*]
    2 Corolla at most 6 mm broad; style to 1.5 mm long; leaves commonly with 2 or 3 pairs of lateral lobes, the lower pair as large as or larger than the upper ones; (sw B.C.).
        3 Appendages of the calyx well developed, to 3 mm long, at least half as long as the lobes at maturity (these to 4 mm long); leaves rather deeply pinnatifid, the rachis mostly 1–4 mm broad . . . . *N. pedunculata*
        3 Appendages of the calyx 1 or 2 mm long, seldom half as long as the lobes (these 2 or 3 mm long); leaves often less deeply cleft . . . . *N. parviflora*

N. breviflora Gray
/T/W/ (T) Thickets, woods, and open slopes at low to moderate elevations from s B.C. (Nicola; Sophie Mt., near Rossland; Toad Mt., near Nelson; Mt. Brest, near Penticton; Marble Range, NW of Clinton; Botanie L., sw of Spences Bridge) and sw Alta. (Waterton Lakes; Breitung 1957b) to Calif. and Colo. [*Viticella* Macbr.]. MAP: Lincoln Constance, Univ. Calif. Publ. Bot. 19(10): map 2, p. 356. 1941.

[N. menziesii H. & A.]
[Native in the w U.S.A. (Oreg. to N Baja Calif.) and occasionally introd. elsewhere, as in SE ?Alaska (a report from Sitka noted by Hultén 1948) and sw ?B.C. (reported from Victoria, Vancouver Is., by Henry 1915). MAPS: Hultén 1968b:769; Constance, loc. cit., map 4 (w U.S.A. only), p. 366.]

N. parviflora Dougl.
/t/W/ (T) Usually on wooded or open but shaded slopes at low to moderate elevations from sw B.C. (Vancouver Is. and adjacent islands and mainland; CAN; V) and Idaho to Calif. and Utah. [*Viticella* Macbr.; *N. micrantha* and *N. pustulata* Eastw.]. MAP: Constance, loc. cit., map 2, p. 356.

N. pedunculata Dougl.
/t/W/ (T) Moist meadows and lowlands (occasionally at moderate elevations) from sw B.C. (several localities on Vancouver Is.; CAN; GH; V) and Idaho to Baja Calif. [*Viticella* Macbr.; *N. sepulta* Parish]. MAP: Constance, loc. cit., map 3, p. 362.

PHACELIA Juss. [7025] Scorpion-weed

1 Leaves entire or some of the larger ones with a large entire terminal segment and 1 or 2 (4) pairs of smaller entire lobes or distinct leaflets at base.
    2 Stamens about equalling the corolla; corolla blue-lavender, very broadly campanulate, the limb to over 1.5 cm broad; seeds usually at least 6; leaves all cauline (lower ones reduced), they or their segments linear to linear-oblanceolate; nonglandular annual to about 5 dm tall, the simple or freely branched erect stem densely puberulent; (B.C. and sw Alta.) . . . . *P. linearis*
    2 Stamens conspicuously exserted; corolla dull whitish, the limb less than 8 mm broad; seeds commonly only 1 or 2; leaves (or their segments) lanceolate or elliptic, the largest ones at the base of the stem or in a basal tuft.
        3 Leaves all entire (or sometimes some of them with a small pair of lobes at base), more or less silvery with a short dense pubescence (longer bristles, if present, relatively sparse); inflorescence commonly nonglandular; stems usually several from a branched caudex surmounting a taproot; (B.C. and sw Alta.) . . . . *P. hastata*

3 Leaves (at least some of the middle and lower ones) usually with 1 or 2(4) pairs of lateral lobes or leaflets at base, often somewhat greyish but scarcely silvery, often markedly spreading-bristly; inflorescence commonly glandular-puberulent; stems typically solitary and erect from a taproot (or surrounded by several shorter ascending stems); (?B.C.) . . . . [*P. heterophylla*]

1 Leaves coarsely toothed to 1–2-pinnatifid or pinnately lobed (pinnately compound in *P. campanularia*), lacking a large entire terminal segment.

4 Stamens not exserted beyond the lavender or whitish corolla, the latter only slightly longer than the calyx, to 4 mm long and 3 mm broad (but the calyx becoming firm and veiny, strongly accrescent in fruit and finally up to 8 mm long); seeds 4 or fewer, prominently pitted-reticulate; inflorescence dense even at maturity; leaves to 9 cm long (including the short petiole) and 2.5 cm broad, pinnatifid or pinnately lobed, the lobes toothed; glandular-hairy annual to 2(3) dm tall; (introd.) . . . . *P. thermalis*

4 Stamens more or less exserted beyond the corolla.

5 Middle and upper stem-leaves markedly reduced, the principal leaves persistent at the base of the stem or in basal tufts; inflorescence a capitate or thyrsoid aggregation of short helicoid cymes, not evidently glandular; corolla dark blue to blue-purple or purple, hairy on both surfaces; stamens long-exserted; perennial with a taproot surmounted by a branching caudex.

6 Leaves pinnately lobed or merely coarsely toothed; corolla deciduous; plants long-hairy in the short compact inflorescence, otherwise green, sparsely strigose or hirsute and conspicuously to obscurely glandular; (B.C. and sw Alta.) . . . . *P. lyallii*

6 Leaves more or less pinnatifid or bipinnatifid; corolla persistent and surrounding the fruit; inflorescence virgate; plants commonly densely silky or loosely woolly throughout, not evidently glandular.

7 Pubescence velvety, consisting of spreading hairs; (Alaska–Yukon) . . . *P. mollis*

7 Pubescence silky, consisting of appressed hairs; (s B.C. and sw Alta.) . . . . *P. sericea*

5 Stem rather uniformly leafy throughout; helicoid cymes more or less corymbosely disposed.

8 Leaves pinnately compound with sessile or subsessile coarsely toothed (or cleft and again toothed) leaflets, the lower leaflets remote, the upper ones confluent; corolla white to lavender or dull cream; anther-filaments glabrous; strongly glandular-hairy and odoriferous perennial with numerous coarse but weak and brittle, simple or branched, prostrate or weakly ascending stems from a branched caudex; (Vancouver Is.) . . . . [*P. ramosissima*]

8 Leaves simple; corolla usually deeper blue; annuals or biennials.

9 Plant strongly viscid-glandular; corolla deep blue, with glabrous append-ages at the base of the anther-filaments; leaves cordate-ovate, coarsely dentate; (introd. in Alta.) . . . . [*P. campanularia*]

9 Plants nonglandular or only moderately glandular.

10 Corolla lacking appendages at the base of the anther-filaments (these densely villous on the lower half), its lobes conspicuously fringed; calyx-lobes strongly divergent at maturity, stiffly long-ciliate, otherwise essentially glabrous; racemes soon loosening; seeds at most 4; leaves with up to 5 pairs of lanceolate to oblong, often falcate, sharply acute, entire lobes; plant sparingly strigose, not at all glandular; (s Ont.) . . . . *P. purshii*

10 Corolla with ridge-like appendages adnate to the base of the glabrous anther-filaments, its lobes entire; calyx-lobes rather densely bristly over the back, ascending at maturity; racemes scarcely loosening; leaves more divided, 2-pinnatifid or pinnate-pinnatifid; plants often markedly glandular.

11 Anthers much exserted from the corolla (to a distance about equalling the corolla-length); seeds at most 4; plant moderately

spreading-hispid, rough to the touch; annual without basal tufts of
leaves; (introd. from B.C. to Ont.) . . . . *P. tanacetifolia*
11 Anthers only moderately exserted; seeds numerous; plant rather
soft-hairy; annual or biennial with basal tufts of leaves; (Alaska–
B.C. to w Ont.) . . . . *P. franklinii*

[P. campanularia Gray]
[This Californian species is reported as introd. in Alta. by Boivin (1966*b*; Fort Saskatchewan), where probably a casual garden-escape but not established.]

P. franklinii (R. Br.) Gray
/ST/(X)/ (Hs) Meadows, gravels, and disturbed areas at low to moderate elevations from w-cent. Alaska (N to near Fairbanks, ca. 65°N), sw Yukon, and Great Bear L. to Great Slave L., L. Athabasca (Alta. and Sask.), Man. (N to Churchill), and w-cent. Ont. (Nipigon R. and the N shore of L. Superior E to near Schreiber), s to B.C., s Idaho, Wyo., N Minn., and N Mich. [*Eutoca fr.* R. Br., the type from along the Churchill R. in N Sask.]. MAPS: Hultén 1968*b*:770; Porsild 1966: map 125, p. 82; J.M. Gillett, Rhodora 62(740): fig. 4, p. 215. 1960.

Forma *variegata* Boivin (corolla white, streaked with purple, rather than uniformly purple) is known from the type locality, Round L., near Thunder Bay, Ont.

P. hastata Dougl.
/T/WW/ (Hs) Dry sandy and rocky places at low to high elevations, the aggregate species from s B.C. (Mt. Arrowsmith, Vancouver Is.; mainland, chiefly in the Dry Interior, N to Lillooet, Clinton, Vernon, and Fairmont Hot Springs) and sw Alta. (Crowsnest Pass and South Kootenay Pass, both on the B.C.–Alta. boundary; Waterton Lakes; CAN) to Calif., Colo., and w Nebr.
1 Stems ascending to suberect, mostly over 1.5 dm tall; leaves generally distinctly silvery
with a fine, loose, short pubescence; plants of relatively low elevations.
2 Some of the leaves generally with a pair of small lateral lobes; calyx often relatively
stiffly-long-hispid; [w U.S.A. only, reports from B.C. or Alta. referring to one or other
of the following taxa; MAP: L.R. Heckard, Univ. Calif. Publ. Bot. 32(1): fig. 14, p. 76.
1960] . . . . [var. *hastata*]
2 Leaves usually all entire (occasionally some with a pair of small lateral lobes); calyx
usually somewhat less hispid; [*P. leucophylla* Torr.; *P. magellanica* f. *angustifolia*
Brand; *P. circinata sensu* John Macoun 1886, not Jacq.; s B.C. and sw Alta.] . . . .
. . . . var. *leucophylla* (Torr.) Cronq.
1 Stems either more of less prostrate, or less than 1.5 dm tall, or the leaves distinctly
greener.
3 Flowers light lavender to dull purplish; leaves usually all entire; stem prostrate or
merely ascending at tip; plant of high elevations; [*P. alpina* Rydb.; B.C.: Rydberg
1922] . . . . var. *alpina* (Rydb.) Cronq.
3 Flowers mostly dull whitish; stems often ascending to suberect.
4 Dwarf alpine and subalpine plants generally not over 1.5(2) dm tall, often
prostrate; leaves mostly entire; [w U.S.A. only] . . . . [var. *compacta* (Brand) Cronq.]
4 Plants to over 5 dm tall, usually of moderate elevations; larger leaves often with a
pair of lateral lobes; [*P. leptosepala* Rydb., the type from Vermilion Lake, B.C.;
also known from Mt. Arrowsmith, Vancouver Is., from Manning Provincial Park,
about 30 mi SE of Hope, and from between Hope and Princeton, B.C.] . . . .
. . . . var. *leptosepala* (Rydb.) Cronq.

[P. heterophylla Pursh]
[Reports of this species of the w U.S.A. (N to Wash. and Mont.) from B.C. require confirmation. A collection in CAN from near Ymir, about 15 mi s of Nelson, has been placed here by Constance, one of the specimens on the sheet being named a hybrid with *P. hastata* by Calder. However, the MAP by L.R. Heckard (Univ. Calif. Publ. Bot. 32(1): fig. 13, p. 66. 1960) indicates no Canadian stations and several collections in CAN have been referred to *P. hastata* by Calder. (Incl. var. *pseudohispida* (Brand) Cronq. and *P. mutabilis* Greene).]

P. linearis (Pursh) Holz.
/T/W/ (T) Dry open places in the sagebrush plains and foothills from s B.C. (Vancouver Is. and adjacent islands; mainland, chiefly in the Dry Interior, N to Lac la Hache, ca. 54°10′N; CAN) and sw Alta. (Crowsnest Pass, on the B.C.–Alta. boundary; Milk River; Sweet Grass Hills; CAN) to N Calif., Utah, and Wyo. [*Hydrophyllum* Pursh; *Eutoca (P.) menziesii* R. Br.].

P. lyallii (Gray) Rydb.
/T/W/ (Hs) Talus slopes and rock crevices at high elevations from SE B.C. (North and South Kootenay Passes, on the B.C.–Alta. boundary; John Macoun 1884) and sw Alta. (Sheep Mt., Waterton Lakes, where taken by Macoun in 1895; CAN, verified by both Constance and Gillett) to w Mont. [*P. sericea* var. *ly.* Gray, the type from "Rocky Mts. at lat. 49 degrees"]. MAP: J.M. Gillett, Rhodora 62(740): fig. 4, p. 215. 1960.

P. mollis Macbr.
/Ss/W/ (Hs) Dry slopes and roadsides of E Alaska (N to ca. 65°N) and sw Yukon (N to ca. 63°N; type from Coffee Creek). [*P. sericea sensu* Porsild 1951a, not (Grah.) Gray, the relevant collections in CAN]. MAPS: Hultén 1968b:770; J.M. Gillett, Rhodora 62(740): fig. 4, p. 215. 1960.

P. purshii Buckl. Miami-mist
/t/EE/ (T) Rich woods, clearings, and fields (sometimes becoming a weed) from Wisc. to Ohio, s Ont. (reported by Dodge 1914, as apparently native on Chicken Is. of the Erie Archipelago, s Essex Co.; an 1898 collection by Wm. Macoun in CAN from the Central Experimental Farm, Ottawa, where introd. in clover), and Pa., s to Okla. and Ala. MAPS (the Essex Co., Ont., station should be indicated): Lincoln Constance, Contrib. Gray Herb. Harv. Univ. 168: fig. 5, p. 30. 1949; Am. Acad. Arts & Sci. Proc. 78: fig. 6, p. 146. 1950.

[*P. ramosissima* Dougl.]
[This species of the w U.S.A. (N to Wash. and Idaho) is known from Canada only through an 1893 collection in CAN by John Macoun on ballast heaps at Nanaimo, Vancouver Is., sw B.C., where apparently not taken since that time and not established. (*P. hispida sensu* J.M. Macoun 1895, not Gray).]

P. sericea (Graham) Gray
/T/W/ (Hs) Open or wooded rocky slopes and ledges at moderate to high elevations from s B.C. (Mt. Arrowsmith, Vancouver Is.; mainland N to the Marble Range, NW of Clinton, and Rogers Pass; CAN) and sw Alta. (N to Banff; CAN) to N Calif. and Colo. [*Eutoca* Graham; incl. var. *caespitosa* Brand and *P. ciliosa* Rydb.; *P. idahoensis sensu* Eastham 1947 (excluding the Creston, B.C., report, this referable to *P. tanacetifolia*), not Henderson]. MAP: J.M. Gillett, Rhodora 62(740): fig. 5, p. 219. 1960.

P. tanacetifolia Benth. Fiddle-neck
This Californian species is reported by Groh (1944a) as introd. in B.C. (Baldonnel and Montney in the Peace River dist.) and Alta. (Kevisville; Grande Prairie). It is also known from Creston, B.C. (Herb. V), s Man. (Experimental Farm, Brandon, where taken by John Macoun in 1896; CAN), and s Ont. (Toronto; TRT). [*P. idahoensis sensu* Eastham 1947, as to the Creston, B.C., plant, not Henderson].

P. thermalis Greene
This species of the w U.S.A. (SE Oreg. and Idaho to Calif.) is reported from s Sask. by Breitung (1957a; Val Marie, about 70 mi S of Swift Current, where introd. and presumably established in ditches). [*P. glandulifera sensu* Fraser and Russell 1944, not Piper, the relevant collection in SCS].

ROMANZOFFIA Cham. [7031]

1 Plants at most about 1 dm tall, bearing well-developed brown-woolly tubers at the base, more or less evidently glandular-villous; inflorescence compact, scarcely surpassing the leaves, the mature pedicels less than 1 cm long; (sw B.C.) .... *R. tracyi*

1 Plants to about 3 dm tall, lacking basal tubers but the petioles strongly dilated and overlapping to form a scaly crown.
  2 Inflorescence open, the slender pedicels longer than the calyces; corolla about 3 times as long as the calyx; leaves glabrous on both surfaces except for a viscid marginal ciliation; (s Alaska–B.C.–Alta.) .... *R. sitchensis*
  2 Inflorescence compact, the pedicels about as long as the calyces; corolla not much surpassing the calyx; (Aleutian Is.–Alaska) .... *R. unalaschcensis*

R. sitchensis Bong.
/sT/W/ (Hr (Hs)) Moist places at low to fairly high elevations from s Alaska (N to ca. 61°N; type from Sitka) through B.C. and sw Alta. (Waterton Lakes; Banff; Jasper) to N Calif. and Idaho. [*R. glauca, R. macounii,* and *R. rubella* Greene; *R. minima* Brand]. MAP: Hultén 1968*b*:771.

According to Hultén (1968*a*), a report of the very similar *R. suksdorfii* Greene (differing in its red-woolly tubers) from Alaska by Peck is most probably a mistake.

R. tracyi Jeps.
/t/W/ (Gb) Coastal rocks and bluffs from sw B.C. (Vancouver Is. and Jansen's Is., where taken by John Macoun in 1909; CAN) to N Calif. [*R. unalaschcensis sensu* J.M. Macoun 1913, not Cham.].

R. unalaschcensis Cham.
/s/W/ (Hr) Moist places in the Aleutian Is. (type from Unalaska) and sw Alaska (N to ca. 59°N). [*Saxifraga nutans* Don]. MAPS: Hultén 1968*b*:771, and 1948: map 995, p. 1341.

Some of our material is referable to var. *glabriuscula* Hult. (leaves essentially glabrous rather than viscid-pubescent beneath; type from Unimak Is., Aleutian Is.).

BORAGINACEAE (Borage Family)

Herbs (commonly rough-hairy) with simple entire alternate leaves. Flowers usually regular (corolla somewhat 2-lipped in *Echium* and *Lycopsis*), perfect, hypogynous, gamopetalous, 5-merous, in commonly scorpioid (more or less tip-coiled) cymes. Stamens 5, inserted on the corolla-tube and alternating with its lobes. Style 1, entire or 2-cleft. Ovary superior. Fruit commonly consisting of 4 (sometimes 2, rarely only 1) 1-seeded nutlets surrounding the base of the style. (Including Heliotropaceae).

1 Nutlets armed with hooked or barbed prickles; corolla regular, salverform or broadly funnelform; stigma solitary and simple; leaves alternate.
  2 Nutlets covered by the prickles, depressed and spreading, attached obliquely to the receptacle by the apical third of the inner face . . . . *Cynoglossum*
  2 Nutlets prickly chiefly on their margins (or the intermarginal prickles much reduced); corolla blue or white.
    3 Nutlets spreading radially in pairs from the broad low gynobase at maturity, the prominent margin upturned, often inflexed near the middle; corolla white, the limb at most 1 mm broad; leaves linear, to 2 cm long and 1 mm broad; prostrate or weakly ascending strigose annual, often rosette-branched at base; (s B.C.) . . . .
. . . . *Pectocarya*
    3 Nutlets erect or nearly so on the elevated receptacle; corolla blue or white.
      4 Pedicels short, the fruiting calyces erect or ascending; receptacle subulate, the nutlets attached to it along nearly their entire inner angle; corolla 3 or 4 mm broad; racemes evidently bracted . . . . *Lappula*
      4 Pedicels slender, the fruiting calyces reflexed; receptacle pyramidal, the nutlets attached to it along the middle third of the inner face; racemes naked or nearly so . . . . *Hackelia*
1 Nutlets unarmed (at most tuberculate or muricate).
  5 Flowers solitary or at most 4 in the leaf-axils, blue, about 3 mm long and broad, on short recurved pedicels; calyx surpassing the corolla, in fruit greatly enlarged, strongly veiny, with 5 broad flat irregularly toothed lobes, about 1 cm long and 1.5 cm broad; stigma solitary and simple; leaves oblanceolate, subopposite or more or less whorled, to about 6 cm long; (introd.) . . . . *Asperugo*
  5 Flowers in racemiform or spike-like terminal cymes, these mostly simple and curved or coiled before anthesis, occasionally freely branched; leaves chiefly or all alternate (or the lower ones subopposite in *Plagiobothrys*).
    6 Attachment of nutlets surrounded by a tumid annular ring, the rather strongly convex-based nutlets leaving a pit upon the flat or low-convex receptacle; (introd.).
      7 Corolla rotate, with a very short tube, clear blue, to 2.5 cm broad; anthers dark, exserted, forming a column around the simple style, their filaments dilated and prolonged into spur-like appendages appressed to and about half as long as the anthers; principal leaves oblong to obovate, the lower ones petioled; plant harshly villous-hirsute . . . . *Borago*
    7 Corolla tubular, salverform, or funnelform, with a well-developed tube surpassing the anthers; filaments not appendaged.
      8 Corolla blackish red, its throat lacking appendages but with villous spots; annular ring at base of nutlets coarsely papillose-dentate; style 2-cleft or stigma 2-lobed . . . . *[Nonea]*
      8 Corolla-throat with appendages formed by intrusions of tissue; stigma solitary and simple; plants rough-hairy.
        9 Corolla to 18 mm long, dull blue or yellow, thick-tubular, regular, the 5 lobes erect or recurving at tip, the appendages in its campanulate throat lanceolate, acute, denticulate; cymes bractless . . . . *Symphytum*
        9 Corolla commonly smaller, funnelform, blue, the appendages in its ill-defined throat deltoid or oblong; cymes leafy-bracted.

10 Corolla-limb to 11 mm broad, regular (the lobes equal); corolla-tube straight, distinctly surpassing the calyx, the scales in its throat merely pubescent; calyx-lobes about equalling or a little longer than the tube; biennial or short-lived perennial .... *Anchusa*

10 Corolla-limb about 5 mm broad, oblique or slightly irregular (the lobes slightly unequal); corolla-tube abruptly bent near the middle, barely surpassing the calyx, the scales in its throat densely bristly-hirsute; calyx-lobes much longer than the tube; annual .... *Lycopsis*

6 Attachment of nutlets lacking an annular ring, the nutlets leaving no evident pit on falling.

11 Corolla irregular (its lobes very unequal), blue, to 2 cm long, its throat unappendaged; stamens unequal, strongly exserted; stigma 2-lobed; plant very rough-hirsute; (introd.) .... *Echium*

11 Corolla regular or nearly so.

12 Nutlets 2, each commonly 2-celled; anthers sagittate, their auricles usually appendaged; leaves broad, deeply cordate, sessile; (introd.) .... [*Cerinthe*]

12 Nutlets 4 (sometimes only 1 or 2 ripening for each flower); anthers usually neither sagittate nor appendaged (sagittate only in *Onosmodium*); leaves not cordate.

13 Corolla normally blue or rose-pink.

14 Low densely caespitose silky-pubescent plants with small firm linear-oblanceolate leaves chiefly in basal tufts at the ends of the caudex-branches; nutlets bordered by elongate, often jagged teeth; (Alaska–Yukon–Dist. Mackenzie) .... *Eritrichium*

14 Plants otherwise.

15 Leaves all petioled, ovate, with acuminate callous-tipped apices, rounded or (the lower ones) cordate at base; cymes short, very lax, bractless; flowers bright blue, to 1.5 cm broad; calyx appressed-hairy; nutlets attached near apex; short-hairy long-stoloniferous perennial to 3 dm tall; (*O. verna*; introd.) .... [*Omphalodes*]

15 At least the upper leaves sessile; nutlets attached at or below the middle.

16 Calyx glabrous or with appressed straight hairs, or merely with minutely ciliate-margined lobes; nutlets attached laterally to the receptacle at or below their middle; inflorescence often leafy-bracted .... *Mertensia*

16 Calyx pubescent at least at base with minute, straight or hooked, gland-tipped or glandless hairs; nutlets smooth and shining, attached by a small basilateral scar to the broad low receptacle; inflorescence usually completely bractless.

17 Nutlets nearly perfectly tetrahedronal, all 4 sides subequal and flat, each nearly the shape of an equilateral triangle; corolla-lobes imbricate (overlapping) in the bud; (introd.) .... [*Trigonotis*]

17 Nutlets asymmetrical, the sides rounded and not having the shape of an equilateral triangle; corolla-lobes convolute (rolled together lengthwise) in the bud ... *Myosotis*

13 Corolla white, greenish white, yellow, or orange-yellow (rarely bluish-tinged).

18 Leaves chiefly opposite (alternate in *P. tenellus*), linear or linear-spatulate, strigose or hirsute, to about 4 cm long; corolla white, with a yellow eye, about 1.5 mm long, the limb to 5 mm broad; (B.C. to sw Man.) .... *Plagiobothrys*

18 Leaves all or chiefly alternate.

19 Plants glabrous and usually more or less glaucous; corolla white or tinged with blue.

  20 Lowermost leaves reduced and scale-like, the others fleshy, oblanceolate to narrowly obovate, sessile or short-petioled; flowers to 9 mm broad, sessile in 1-sided scorpioid paired spikes terminating the common peduncle; ovary merely shallowly lobed, the style wanting, the sessile stigma expanded into a disk as broad as the ovary around the low central cone; nutlets only tardily separating or cohering in pairs; taprooted annual or short-lived perennial, the prostrate or ascending stems to 6 dm long; (s Alta. to sw Man.) . . . . *Heliotropium*

  20 Lowermost leaves spatulate or oblong-spatulate, slender-petioled; flowers to over 12 mm broad, long-pedicelled in elongate racemes; ovary 4-lobed, the style arising between the lobes; nutlets attached near apex; erect annual to about 3 dm tall; (*O. linifolia*; introd.) . . . . [*Omphalodes*]

19 Plants more or less pubescent, usually rather bristly-hirsute; ovary 4-parted from above, forming 4 nutlets around the base of the definite style.

  21 Corolla tubular, greenish white, its acute or acuminate lobes erect; style long-exserted from the corolla-tube; anthers sagittate; nutlets smooth or pitted, broadly attached at base to the broad low receptacle; (s Alta. to s Ont.) . . . . *Onosmodium*

  21 Corolla salverform to broadly funnelform, its lobes more or less spreading; style included or short-exserted; anthers not sagittate.

    22 Flowers yellow or orange-yellow.

      23 Nutlets tuberculate or muricate, with a narrow scar of attachment on one side below middle, the receptacle short-pyramidal; inflorescence bractless; (western species) . . . . *Amsinckia*

      23 Nutlets smooth or minutely pitted, with a broad scar of attachment at the truncate base, the receptacle flat or depressed; inflorescence leafy-bracted . . . . *Lithospermum*

    22 Flowers white (sometimes greenish white in *Lithospermum*).

      24 Stigmas 2 or style 2-cleft; nutlets with a broad scar of attachment at the truncate base, the receptacle flat or depressed; inflorescence leafy-bracted . . . . *Lithospermum*

      24 Stigma solitary and uncleft; scar of attachment of nutlets small.

        25 Racemes (spike-like) bracted at least below; corolla white, salverform, the lobes imbricate in the bud; nutlets lacking a dorsimarginal ridge, attached ventrally for most of their length to the elongate central axis (the scar commonly appearing as an elongate, closed or narrowly open groove); leaves at most about 5 cm long; plant copiously pustulate-hispid; (western species) . . . . *Cryptantha*

        25 Racemes usually completely bractless; corolla rotate or short-salverform, the lobes convolute

in the bud; nutlets with a dorsimarginal ridge complete around the base, smooth and shining, attached by a small basilateral scar to the low broad receptacle, smooth; leaves commonly larger .... *Myosotis*

AMSINCKIA Lehm.   [7082]   Fiddle-neck, Tarweed

1   Two of the 5 sepals of most or all of the flowers united laterally to above middle (often nearly to tip); stamens inserted above the middle of the corolla-tube.

    2   Plants maritime; leaves often erose-denticulate, to about 5 cm long and 12 mm broad; sepals to 8 mm long; corolla-tube about 10-nerved below the insertion of the stamens; nutlets to 2.5 mm long, rather weakly tuberculate or rugose; (B.C.) .... *A. spectabilis*

    2   Plants not maritime; leaves entire, to about 1 dm long and 1.5 cm broad; sepals to 14 mm long; corolla-tube usually about 20-nerved below the insertion of the stamens; nutlets to 3.5 mm long, strongly tuberculate-checkered; (?Man.) .... [*A. tesselata*]

1   Sepals all distinct, to 10 or 12 mm long; corolla-tube about 10-nerved below the insertion of the stamens; nutlets strongly muricate or tuberculate; leaves entire; plants not maritime; (B.C.).

    3   Corolla-throat obstructed by hairy scale-like appendages; stamens inserted below the middle of the corolla-tube; sepals to 1 cm long; nutlets to 3 mm long; leaves to 1 dm long and 1.5 cm broad .... *A. lycopsoides*

    3   Corolla-throat open, glabrous, the scales scarcely developed; stamens inserted above the middle of the corolla-tube.

        4   Stem copiously pubescent throughout with softer, more or less retrorse, short hairs beneath the longer setae; leaves to about 12 cm long but rarely over 1 cm broad, their setae mostly ascending; corolla to 8 mm long; (?B.C.) .... [*A. retrorsa*]

        4   Stem nearly or quite lacking softer, shorter hairs below the setae except often in or near the inflorescence; leaves to 15 cm long and 2 cm broad, their setae often widely spreading .... *A. menziesii*

A. lycopsoides Lehm.
/t/W/   (T)   Dry open flats and slopes (often in disturbed soil) from s B.C. (Vancouver Is. and adjacent islands; Douglas, near Vancouver; CAN; introd. in Alaska–Yukon N to ca. 65°N) and W Mont. to Calif. and ?Nev. [*A. barbata* Greene]. MAP: Hultén 1968b:778.

A. menziesii (Lehm.) Nels. & Macbr.
/t/W/   (T)   Dry to moist fields and open ground from s B.C. (Vancouver Is. and adjacent islands, Manning Provincial Park, SE of Hope, Osoyoos, Grand Forks, and Flathead; ?introd. farther N, as at Revelstoke, Quesnel, and near Atlin; introd. in Alaska–Yukon N to ca. 65°30'N and in Man.) to Calif. and Nev. [*Echium* Lehm.; *A. borealis, A canadensis, A. foliosa,* and *A. micrantha* Suksd.; *A. idahoensis* Jones; *A. microcalyx* Brand; incl. *A. intermedia* F. & M. (*A. macounii* Brand)]. MAP: Hultén 1968b:778.

[A. retrorsa Suksd.]
[The inclusion of s B.C. in the range of this species of the W U.S.A. (Idaho to s Calif. and Utah) by Hitchcock et al. (1959) requires confirmation.]

A. spectabilis F. & M.
/t/W/   (T)   Sandy and gravelly beaches from W B.C. (Queen Charlotte Is. and Vancouver Is.; reports from SE Alaska require confirmation) to N Baja Calif. [*A. lycopsoides (spectabilis)* var. *bracteosa* Gray; *Lithospermum lycop.* Lehm., not *A. lycop.* Lehm.].

[A. tesselata Gray]
[The report of this species of the W U.S.A. (Wash. and Idaho to N Baja Calif. and Ariz.) from Neepawa, s Man., by Scoggan (1957) requires confirmation, perhaps being based upon *A. menziesii* or, if correctly identified, probably introd. but scarcely established.]

ANCHUSA L.   [7093]

A. officinalis L.   Common Alkanet
European; introd. along roadsides and in waste places in N. America, as in S B.C. (Brentwood, Vancouver Is.; Herb. V, detd. Jones; reported from Winfield, near Osoyoos, by Boivin 1966*b*).

ASPERUGO L.   [7084]

A. procumbens L.   Madwort
Eurasian; introd. in waste places of N. America but mostly casual and not long-persistent, as in S Alaska (N to ca. 61°30′N), the Yukon (Boivin 1966*b*), B.C. (N to Armstrong, N of Vernon; Eastham 1947), Alta. (Banff; CAN), Man. (a large infestation near Manitou, about 50 mi S of Portage la Prairie), and Ont. (N to Ottawa); W Greenland at ca. 69°N. MAP: Hultén 1968*b*:779.

BORAGO L.   [7091]

B. officinalis L.   Borage
European; an occasional garden-escape to waste places in N. America, as in S B.C. (Vancouver Is.; McMurdo, Kootenay Valley), Alta. (Moss 1959), Sask. (Delisle, about 25 mi SW of Saskatoon; Breitung 1957*a*), S Man. (Brandon; Carberry; Portage la Prairie; Ninette, about 30 mi SE of Brandon), Ont. (Lambton, Waterloo, Perth, York, Victoria, and Northumberland counties), Que. (N to Lévis), St-Pierre and Miquelon (GH), N.B. (Fowler 1885), and N.S. (Kings, Halifax, and Inverness counties).

[CERINTHE (Tourn.) L.]   [7116]

[C. major L.]   Honeywort
[European; occasionally cult. in N. America and found escaping from a garden at Loretteville, near Quebec City, Que., by Pease in 1904 (GH; I.M. Johnston, Contrib. Gray Herb. Harv. Univ. 70:44. 1924), where, however, scarcely established.]

CRYPTANTHA Lehm.   [7081]

(Ref.: Payson 1927)
1   Relatively coarse perennials with a conspicuous tuft of basal leaves; corolla-limb
    commonly over 4 mm broad; calyx persistent; spikes aggregated into a terminal
    irregularly bracteate thyrse, often elongate and distinct at maturity; (*Oreocarya*).
    2   Nutlets smooth; corolla-tube distinctly surpassing the calyx; (?B.C.) . . . . [*C. leucophaea*]
    2   Nutlets more or less roughened, at least dorsally; corolla-tube about equalling the
        calyx; (S B.C. to S Sask.) . . . . *C. nubigena*
1   Relatively slender annuals lacking a conspicuous tuft of basal leaves; corolla-limb (except
    in *C. intermedia*) rarely over 2.5 mm broad.
    3   Calyx circumscissile a little below the middle, the scarious cup-shaped basal portion
        persistent, the more herbaceous upper portion deciduous; flowers solitary in the
        closely crowded upper axils, not forming elongate spikes; plants forming cushions at
        most about 6 cm high; (?B.C.) . . . . [*C. circumscissa*]
    3   Calyx divided nearly to base, not circumscissile, of uniform texture and usually finally
        deciduous; flowers borne in naked, finally elongate, unilateral helicoid false spikes,
        these not closely aggregated; plants with a more or less evident central axis, not
        cushion-forming; (*Krynitzkia*).
        4   Corolla-limb commonly not over 4 mm broad; nutlets 1 or 2, smooth and shining;
            (Vancouver Is.) . . . . [*C. intermedia*]
        4   Corolla-limb at most 2.5 mm broad.
            5   Nutlets (or some of them) with evident tubercles or spiculate papillae on the
                dorsal surface; (S B.C.) . . . . *C. ambigua*
            5   Nutlets all smooth or inconspicuously granular, not at all tuberculate.

6 Nutlets solitary; hairs of the calyx tending to be curved or more or less
hooked at tip; corolla-limb to 2.5 mm broad; (s ?B.C.) . . . . [*C. flaccida*]
6 Nutlets normally 4 (sometimes 2 or 3); hairs of the calyx straight or nearly
so.
   7 Nutlets obliquely compressed, the scar lying near one margin;
     corolla-limb 1 or 2 mm broad; (s B.C.) . . . . *C. affinis*
   7 Nutlets symmetrical, the scar median on the ventral face; corolla-limb
     about 1 mm broad.
      8 Nutlets lanceolate, at most 0.7 mm broad; (B.C. to s Sask.) . . . .
       . . . . *C. fendleri*
      8 Nutlets ovate, to 1.2 mm broad; (s B.C.) . . . . *C. torreyana*

C. affinis (Gray) Greene
/t/W/ (T) Dry to moist, open or thinly wooded places at low to moderate elevations from s B.C. (Chase L., near Kamloops, where taken by John Macoun in 1889; CAN; the report from Anarchist Mt., near Osoyoos, by Eastham 1947, is based upon *C. ambigua,* the relevant collection in CAN; other collections also require confirmation) and Mont. to s Calif., Nev., and Wyo. [*Krynitzkia* Gray; *C. confusa* Rydb.].

C. ambigua (Gray) Greene
/t/W/ (T) Dry open places at low to moderate elevations from s B.C. (Dry Interior at Keremeos and Anarchist Mt., Osoyoos, both near the U.S.A. boundary s of Penticton; the report from Nanaimo, Vancouver Is., by J.M. Macoun (1895; taken up by Carter and Newcombe 1921, and Eastham 1947) is based upon *C. intermedia,* the relevant collection in CAN) and Mont. to Calif. and Colo. [*Krynitzkia* Gray; *Eritrichium muriculatum* var. *amb.* Gray; *C. affinis sensu* Carter and Newcombe 1921, and Eastham 1947, not (Gray) Greene].

[C. circumscissa (H. & A.) Johnston]
[The report of this species of the w U.S.A. (Wash. to Baja Calif. and Ariz.; S. America) from s B.C. by Henry (1915; Spences Bridge, as *Pito. circ.*) requires clarification, possibly being based upon an 1889 collection in CAN by John Macoun of *C. torreyana,* distributed as the Californian *Eritrichium (C.; Plagiobothrys) torreyi* Gray. (*Lithospermum* H. & A.; *Eritrichium* and *Krynitzkia* Gray; *Greeneocharis* Rydb.; *Pitocalyx* Torr.).]

C. fendleri (Gray) Greene
/T/WW/ (T) Sand dunes and very sandy soil from B.C. (Boivin 1966*b*: Vancouver Is.; also tentatively reporting it from se Alaska, where possibly introd.) to s Alta. (Dunmore, Medicine Hat, Milk River, and along the Red Deer River; CAN) and s Sask. (Crane Lake, Moose Jaw, Mortlach, Swift Current, Neinan, and Webb; Breitung 1957*a*), s to Nev., N.Mex., and Nebr. [*Krynitzkia* Gray; *C. minima* Rydb. in part; *C. kelseyana* Greene in part; *C. crassisepala* of Sask. reports according to Breitung 1957*a*, not (T. & G.) Greene (the report of which from Stony Mountain, se Man., by Lowe 1943, also requires clarification)].

[C. flaccida (Dougl.) Greene]
[The report of this species of the w U.S.A. (Wash. to s Calif.) from s B.C. by Henry (1915; Dry Interior) may possibly refer to *C. fendleri,* Boivin (1967*a*) noting that a so-named 1852 collection from Vancouver Is. by Jeffrey is so referable. Henry's citation appears to be based upon that of *Eritrichium leiocarpum* (F. & M.) Wats. from "Dry ground of interior British Columbia" by John Macoun (1884). (*Myosotis* Dougl.).]

[C. intermedia (Gray) Greene]
[Native in the w U.S.A. (Wash. and Idaho to Calif.); known from Canada through an 1893 collection by John Macoun in sw B.C. (ballast-heaps at Nanaimo, Vancouver Is.; CAN, distributed as *Krynitzkia ambigua,* rev. Payson), where probably not established. A collection in Herb. V from Trail, B.C., requires further study. (*Eritrichium* and *Krynitzkia* Gray; incl. *C. grandiflora* Rydb.).]

[C. leucophaea (Dougl.) Payson]
[A species of the w U.S.A. (Wash. and Oreg.), the tentative report of which from s B.C. by John Macoun (1884) was later (1890) amended by him to "Not yet found by Canadian collectors." (*Myosotis* Dougl.; *Oreocarya* Greene).]

C. nubigena (Greene) Payson
/T/W/ (Hs (Ch)) Dry open slopes and rocky places at moderate to high elevations (ranges of Canadian taxa outlined below), s to Calif. and Idaho.
1   Corolla-limb mostly 8–12 mm broad; basal leaves spatulate or rather broadly lanceolate, seldom conspicuously spreading-bristly; nutlets lance-ovate to ovate, more or less roughened on both sides; [*Oreocarya (C.) cel.* Eastw.; *O. affinis* Greene, not *C. affinis* (Gray) Greene; *Cynoglossum (Myosotis; Oreocarya; Eritrichium; Krynitzkia) glomeratum* Pursh, not *Cryptantha glomerata* Lehm.; *C. bradburiana* Pays.; *K. ?sericea sensu* John Macoun 1890, perhaps not Gray; s B.C. (Kootenay L.) and sw Alta. (Fort Macleod, Pincher Creek, and Lethbridge; CAN); concerning reports from Man., *see* Scoggan 1957] . . . . var. *celosioides* (Eastw.) Boivin
1   Corolla-limb rarely as much as 1 cm broad; basal leaves often narrower and markedly spreading-bristly.
    2   Nutlets roughened ventrally as well as dorsally; [*Oreocarya (C.) mac.* Eastw.; *O. (C.) interrupta* Greene; *O. (C.) spiculifera* Piper; s B.C. (Similkameen R.), s Alta. (Dunmore; Fort Macleod; Pincher Creek; Milk River; Cardston; Medicine Hat), and s Sask.; concerning reports from Man., *see* Scoggan 1957; MAP: Hultén 1968b:777 (*C. spic.*)] . . . . var. *macounii* (Eastw.) Boivin
    2   Nutlets smooth or nearly so ventrally, slightly or moderately roughened dorsally; [*Oreocarya nub.* Greene; *C. sobolifera* Pays., this reported from Waterton Lakes, sw Alta., by Breitung 1957b] . . . . var. *nubigena*

C. torreyana (Gray) Greene
/t/W/ (T) Dry places at low to moderate elevations from s B.C. (Fraser–Thompson Valley between Yale and Spences Bridge, Okanagan Range between Princeton and Penticton, and between Grand Forks and Cascade, w of Rossland; CAN; introd. at Skagway, SE Alaska) and w Mont. to Calif., Utah, and Wyo. [*Krynitzkia* Gray; *Plagiobothrys torreyi sensu* John Macoun 1890, not Gray, the relevant collections in CAN]. MAP: Hultén 1968b:777.

CYNOGLOSSUM L.   [7064]   Hound's-tongue, Beggar's-lice.   Langue de chien

1   Dorsal surface of nutlets flattened or sunken, surrounded by a raised margin, the nutlets (to 7 mm long) overtopped by the firm beak-like style; plant soft-pubescent, leafy to summit, the inflorescence of numerous false racemes in the leaf-axils or terminating short axillary branches; corolla dull reddish-purple (rarely white); upper leaves sessile by a rounded or slightly cordate base; (introd.) . . . . *C. officinale*
1   Dorsal surface of nutlets rounded and rimless, the nutlets hiding the delicate style; plant villous-hispid or -hirsute, leafy chiefly below the middle, the inflorescence terminal, with a common naked peduncle.
    2   Stem glabrous or nearly so; all of the leaves (except the scaly sub-basal bracts) long-petioled, the blades broadly rounded to truncate or subcordate at base; corolla blue or violet, to 1.5 cm broad; nutlets to about 1 cm long; (s ?B.C.) . . . . [*C. grande*]
    2   Stem rough-hairy; upper leaves clasping by a cordate base.
        3   Corolla blue, lilac, or white, to 12 mm broad, its lobes somewhat overlapping near the middle; flowering calyx at least 3 mm long; nutlets to 9 mm long; larger petioled leaves to over 1 dm broad . . . . [*C. virginianum*]
        3   Corolla blue, at most 8 mm broad, its lobes not overlapping; flowering calyx less than 3 mm long; nutlets to 5 mm long; larger petioled leaves to about 8 cm broad; (transcontinental) . . . . *C. boreale*

C. boreale Fern.   Northern Wild Comfrey
/T/X/ (Hs) Rich woods and thickets from B.C. (N to Prince George and Vanderhoof; an isolated

station at Liard Hot Springs, 59°25'N) to Alta. (N to Edmonton; CAN), Sask. (Prince Albert; Bernard Boivin, Nat. can. (Que.) 87(2):32. 1960, the first record), Man. (N to Grand Rapids, near the NW end of L. Winnipeg), Ont. (N to the Nipigon R. N of L. Superior and southernmost James Bay), Que. (N to the SE James Bay watershed, Anticosti Is., and the Gaspé Pen.; type from along the Petite-Cascapédia R., Gaspé Pen.), Nfld., N.B., and N.S. (Kings and Hants counties; not known from P.E.I.), S to Iowa, Wisc., Ind., N.Y., and Conn. [*C. occidentale sensu* John Macoun 1890, not Gray, the relevant collection in CAN].

[C. grande Dougl.]
[The report of this species of the W U.S.A. (Wash. to Calif.) from Kootenay, S B.C., by Henry (1915; taken up by I.M. Johnston, Contrib. Gray Herb. Harv. Univ. 70:33. 1924) requires confirmation.]

C. officinale L.   Common Hound's-tongue.   Langue de chien
Eurasian; introd. in pastures and waste places of N. America, as in S B.C. (N to Armstrong, about 15 mi N of Vernon), Alta. (Waterton Lakes; Breitung 1957b), Sask. (Alameda, about 130 mi SE of Regina, and Wolseley, about 55 mi E of Regina; Breitung 1957a), Man. (N to Rossburn, about 70 mi NW of Brandon), Ont. (N to the N shore of L. Superior), Que. (N to the Gaspé Pen. at Mont-Louis), N.B., and N.S. (Boivin 1966b; not known from P.E.I.).

[C. virginianum L.]   Wild Comfrey
[Reports of this species of the E U.S.A. (N to Mo., Ohio, and N.J.) from Ont., Que., and N.B. by John Macoun (1884; taken up for Ont. by Soper 1949) are chiefly or wholly based upon *C. boreale*, relevant collections in CAN, etc.]

ECHIUM L.   [7118]   Viper's Bugloss.   Vipérine

1   Leaves relatively soft, with prominent lateral veins, the upper leaves cordate at base; flowers red, becoming purplish blue; stamens 2; (introd. in S Man., S Ont., and Nfld.) . . . . *E. lycopsis*
1   Leaves very harsh, lacking evident lateral veins, the upper ones rounded at base; flowers blue; stamens 4; (transcontinental weed) . . . . *E. vulgare*

E. lycopsis L.   Purple Viper's Bugloss
European; known in Canada (no U.S.A. reports have been seen) from S Man. (wasteland at Brandon; CAN; DAO), S Ont. (wasteland at Vineland, Lincoln Co.; OAC), and Nfld. (Murray's Pond, near St. John's, where a bad weed in a garden; GH). [*E. plantagineum* L.].

E. vulgare L.   Blueweed, Blue Devil.   Vipérine
Eurasian; introd. along roadsides and in dry fields and waste places of N. America, as in B.C. (N to Cariboo, about 40 mi SE of Williams Lake), Alta. (Moss 1959), Sask. (near Regina; Breitung 1957a), Man. (Boivin 1966b), Ont. (N to the N shore of L. Superior), Que. (N to the Gaspé Pen.), Nfld., N.B., P.E.I., and N.S.
   Forma *albiflorum* Hoffm. (flowers white rather than brilliant blue) is reported from Ont. by Boivin (1966b). Forma *roseum* Zimm. (flowers pink) is known from Ont. (Guelph, Wellington Co.; Silver L., near Perth, Perth Co.; Ottawa dist.). Var. *pustulatum* (Sibth. & Sm.) Rouy (pubescence of leaves coarse and pustular-based rather than slender and nonpustular; inflorescence relatively open, to about 3 dm thick, rather than elongate and less than 1 dm thick) is ascribed to Que. and N.S. by Gleason (1958).

ERITRICHIUM Schrad.   [7074]

1   Plant loosely grey-villous, the leaf-pubescence generally concentrated toward the apex and tending to project as an apical tuft or fringe; corolla to 7 mm broad; (Alaska–Yukon) . . . . *E. nanum*
1   Plant closely strigose, greener, the leaf-pubescence not projecting beyond the apex as a conspicuous tuft; corolla to over 7 mm broad; (Alaska–Yukon–Dist. Mackenzie) . . . . *E. rupestre*

E. nanum (Vill.) Schrad.

/aSs/W/A/ (Ch (Hs)) Sandy tundra and montane slopes up to fairly high elevations in Alaska–Yukon (N to the arctic coast); arctic Asia. [*Myosotis* Vill.; *Omphalodes* Gray; incl. the relatively pubescent phase, var. *elongatum* (Rydb.) Cr.]. MAP: combine the maps by Hultén 1968*b*:773 (*E. aret.*) and 774 (*E. cham.*).

Our material is referable to var. *aretioides* (Cham.) Herder (*Myosotis (E.) aret.* Cham.; *E. villosum* var. *aret.* (Cham.) Gray; *Omphalodes nana* var. *aret.* (Cham.) Gray; incl. *E. chamissonis* A. DC.), differing from the typical form in the much more elongate teeth of the nutlets.

E. rupestre (Pall.) Bunge

/S/W/A/ (Ch (Hs)) Rock crevices up to fairly high elevations in Alaska (N to ca. 70°30′N), SW Yukon (at 62°05′N; CAN), and NW Dist. Mackenzie (Porsild and Cody 1968); Asia. [*Myosotis* Pallas; incl. *E. splendens* Kearney]. MAPS (*E. spl.*): Hultén 1968*b*:775; Olav Gjaerevoll, K. Nor. Vidensk. Selsk. Skr. 10: fig. 6, p. 28. 1967.

HACKELIA Opiz [7073] Stickseed, Beggar's-lice

1 Corolla-limb at most about 3 mm broad; dorsal surface of nutlets 2 or 3 mm long; prickles of nutlets nearly subulate; flowers of individual racemes mostly subtended by small bracts; stem-leaves lanceolate to ovate-oblong, spreading, relatively soft; annuals, biennials, or short-lived perennials with mostly solitary stems.

  2 Fruit (cluster of 4 nutlets) globose, the nutlets nearly uniformly prickly over the whole back; (Ont. and Que.) .... *H. virginiana*

  2 Fruit broadly pyramidal; prickles of nutlets confined chiefly or wholly to the margins (2 or 3 poorly developed intramarginal ones rarely present); (Dist. Mackenzie–B.C. to N.B.) .... *H. deflexa*

1 Corolla-limb mostly at least 4 mm broad; dorsal surface of nutlets mostly 3–5 mm long; prickles of nutlets lanceolate, very flat and often confluent at base; flowers of individual racemes mostly bractless; leaves firm (except in *H. jessicae*).

  3 Intramarginal prickles of nutlets wanting or rarely 1 or 2, the marginal prickles free nearly or quite to base; corolla-limb blue; scales in the corolla-throat only minutely papillate; larger (lower) stem-leaves to about 2 dm long and 3 cm broad; basal leaves usually barely equalling the stem-leaves and soon withering; robust biennial (sometimes short-lived perennial) with few or often solitary stems; (B.C. to Ont.) .... *H. floribunda*

  3 Intramarginal prickles several or many (but considerably smaller than the marginal ones and sometimes poorly developed); basal leaves well developed, mostly larger than the stem-leaves and usually persistent (sometimes relatively small and deciduous in *H. jessicae*); perennials with usually several stems.

    4 Marginal prickles ordinarily united for at least 1/3 of their length, forming a distinct cup-like border to the nutlet; (?B.C.).

      5 Corolla light blue (often withering pink), the limb commonly at least 8 mm broad, the scales in the throat short-papillate-hairy or merely papillate; stem-leaves nearly linear; basal leaves linear-oblanceolate, to about 1.5 dm long and 12 mm broad; plant copiously grey-strigose throughout .... [*H. ciliata*]

      5 Corolla ochroleucous or greenish-tinged, the limb mostly 4 or 5 mm broad, the scales in the throat essentially glabrous; stem-leaves linear to lanceolate; basal leaves oblanceolate or narrowly elliptic, to about 2 dm long and 1.5 cm broad; plant greener, strigose chiefly in the inflorescence .... [*H. hispida*]

    4 Marginal prickles free nearly or quite to base; corolla-limb usually at least 7 mm broad.

      6 Corolla blue with a yellow or whitish eye, the throat-scales only minutely papillate; basal leaves to about 3.5 dm long and 4 cm broad (sometimes smaller than the stem-leaves and soon deciduous); stem to about 8 mm thick toward base; (S B.C. and SW Alta.) .... *H. jessicae*

      6 Corolla predominantly white with a yellow eye (blue in *H. diffusa* var. *coerulescens*); stem at most about 5 mm thick toward base.

        7 Corolla marked with pale blue (commonly with 10 bluish marks toward the

centre and often lightly blue-veined in part), the scales in the throat villous-puberulent; pubescence of the stem largely or wholly appressed (antrorse above, retrorse below); (?B.C.) .... [*H. patens*]

7   Corolla not marked with blue (occasionally more evenly and very lightly washed with blue), the scales in the throat papillate-puberulent.

    8   Larger leaves as much as 1 cm broad; pubescence of the stem wholly appressed to partly spreading; (S B.C.) .... *H. arida*

    8   Larger leaves mostly over 1 cm broad; pubescence of the middle and lower part of the stem largely spreading; (S B.C. and SW Alta.) ... *H. diffusa*

H. arida (Piper) Johnston
/t/W/   (Hs)   Dry sagebrush plains and open or lightly wooded dry slopes from southernmost B.C. (Dry Interior at Rock Creek, near the U.S.A. boundary about 50 mi SE of Penticton; Boivin 1966*b*) to Wash. [*Lappula* Piper].

[H. ciliata (Dougl.) Johnston]
[Native in Wash. and to be looked for in S B.C.–Alta. (*see* John Macoun 1884). (*Cynoglossum* Dougl.; *Lappula* Greene).]

H. deflexa (Wahl.) Opiz
/sT/X/EA/   (Hs (bien. or T))   Thickets, open woods, and rocky places (often calcareous) from SW Dist. Mackenzie (N to Fort Simpson, ca. 62°N; W.J. Cody, Can. Field-Nat. 75(2):65. 1961) and B.C.–Alta. to Sask. (N to Saskatoon), Man. (N to The Pas), Ont. (N to the N shore of L. Superior), Que. (N to Anticosti Is.), and N N.B. (Restigouche Co.; not known from P.E.I. or N.S.), S to Wash., Idaho, Colo., Kans., Iowa, Wisc., and Vt.; Eurasia. [*Myosotis* Wahl.; *Echinospermum* Lehm.; *Cynoglossum* Roth; *Lappula* Garcke]. MAP: Hultén 1937*b*: fig. 14, p. 129.

Our plant is referable to var. *americana* (Gray) Fern. & Johnston (*Echinospermum (Lappula) defl.* var. *amer.* Gray, the type from Sask; *see* M.L. Fernald, Rhodora 40(477):342. 1938); *H. (L.) amer.* (Gray) Fern.; *L. (H.) leptophylla* Rydb.; *L. besseyi* Rydb.; corolla to 3 mm broad rather than 5 or 6 mm; nutlets rather sharply pebbled, their marginal prickles mostly distinct and linear-subulate rather than bluntly pebbled and with an elevated wing-margin formed by the lanceolate flat bases of the prickles).

H. diffusa (Lehm.) Johnston
/T/W/   (Hs)   Cliffs and talus slopes from S B.C. (N to Kamloops; CAN) and SW Alta. (N to Banff; CAN) to Oreg. [*Echinospermum* Lehm.; *Lappula* Greene].

H. floribunda (Lehm.) Johnston
/T/(X)/   (Hs)   Thickets, meadows, and moist places at low to moderate elevations from B.C. (N to Telegraph Trail, ca. 54°N; CAN) to Alta. (N to Banff and Calgary; CAN), Sask. (N to Saskatoon; Breitung 1957*a*), S Man. (N to Victoria Beach, about 55 mi NE of Winnipeg; Lowe 1943), and S Ont. (type locality given as "Lake Pentanguishene to the Rocky Mts.", Pentanguishene being located in Simcoe Co., SE of Georgian Bay, L. Huron), S to Calif., N.Mex., and Minn. [*Echinospermum* Lehm.; *Lappula* Greene].

[H. hispida (Gray) Johnston]
[The report of this species of the w U.S.A. (Wash., Oreg., and Idaho) from Spences Bridge, B.C., by Henry (1915) requires confirmation. (*Echinospermum diffusum* var. *hispidum* Gray; *Echinospermum* and *Lappula* Greene).]

H. jessicae (McGregor) Brand
/T/W/   (Hs)   Meadows, thickets, and open woods at low to fairly high elevations from S B.C. (N to the Marble Range, NW of Clinton; CAN; collection in Herb. V from Azouetta L., near Pine Pass at ca. 55°20′N, where perhaps introd., as in S Alaska–Yukon) and the mts. of SW Alta. (Waterton Lakes; Mt. Assiniboine, on the B.C.–Alta. boundary at 50°52′N; CAN) to Calif., Utah, and Wyo. [*Lappula* McGregor; *H. leptophylla sensu* Hultén 1949, not (Rydb.) Johnst.]. MAP: Hultén 1968*b*:773.

[H. patens (Nutt.) Johnston]
[Reports of this species of the w U.S.A. (N to Idaho and sw Mont.) from B.C. by Eastham (1947) and from Alta. by Rydberg (1922) require confirmation. (*Rochelia* Nutt.; *Lappula coerulescens* Rydb.; *H. diffusa* var. *coer.* (Rydb.) Johnst.).]

H. virginiana (L.) Johnston
/T/EE/ (Hs) Rich woods and thickets from S.Dak. to Ont. (N to the Ottawa dist.; Gillett 1958), sw Que. (N to the Montreal dist.; MT; reported N to near Quebec City by John Macoun 1884, this perhaps based upon *H. deflexa*), and Maine, s to Okla., La., Ala., and Ga. [*Echinospermum* Lehm.; *Lappula* Greene].

HELIOTROPIUM L.   [7052]   Heliotrope

H. curassavicum L.   Seaside-Heliotrope
/T/X/ (T) Saline or alkaline flats and shores from Wash. to s Alta. (Fort Macleod and along the Red Deer R.; CAN), s Sask. (Crane Lake, Cypress Hills, and Gander L., sw of Swift Current; CAN; DAO), and sw Man. (Whitewater, Boissevain, Medora, Lyleton, and Deleau; CAN; DAO; WIN), s to Calif., Mexico, and Fla.; tropical America; introd. in the tropics of the Old World.

Our material is referable to var. *obovatum* A. DC. (*H. spathulatum* Rydb.; leaves obovate, obtuse, mostly not over 3 times as long as broad, rather than linear-lanceolate, acutish, and mostly over 3 times as long as broad).

LAPPULA Moench   [7073]   Stickseed.   Bardanette

1   Margins of nutlets with a double row of slender prickles, these not confluent at base; stem erect; (introd.) .... *L. mysotis*
1   Margins of nutlets with a single row of stout flattened prickles, these often confluent at base; stem at length diffusely branched, the branches often prostrate; (Alaska–B.C. to Man.) .... *L. redowskii*

L. myosotis Moench
Eurasian; introd. along roadsides and in fields and waste places in N. America, as in Alaska (N to ca. 67°N), the Yukon (N to ca. 64°N), B.C.–Alta., Sask. (N to Prince Albert), Man. (N to Churchill), Ont. (N to Moosonee, sw of James Bay at 51°16′N), Que. (N to L. St. John and the Côte-Nord), Labrador (N to Makkovik, 55°05′N), Nfld., N.B., P.E.I., and N.S. [*L. erecta* Nels.; *L. fremontii* Greene; *L. echinata* Gilib.; *Myosotis (Echinospermum; L.) lappula* L.]. MAP: Hultén 1968b:772.

L. redowskii (Hornem.) Greene
/ST/WW/A/ (T) Dry to moderately moist open soil (often in disturbed areas) from Alaska (N to ca. 67°N), the Yukon (N to ca. 64°N), and NW Dist. Mackenzie to B.C.–Alta., Sask. (N to Humboldt and Saskatoon), and s Man. (N to St. Lazare, about 75 mi NW of Brandon), s to Calif., Mexico, Tex., and Mo.; introd. eastwards to New Eng.; S. America; Asia. [*Myosotis* Hornem.; *Echinospermum* Lehm.; *E. (L.) red.* var. *occidentale* Wats.; *L. (E.) occ.* (Wats.) Greene; *L. heterosperma* and *L. montana* Greene; *E. patulum* Lehm.]. MAP (*L. occ.*): Hultén 1968b:772.

Forma *cupulata* (Gray) Scoggan (*Echinospermum red.* var. *cup.* Gray; *L. (E.) cup.* (Gray) Rydb.; *L. desertorum* Greene; *E. (L.) texanum* Scheele; at least some of the marginal prickles of the nutlets fused at base into a prominent cup-like border rather than distinct nearly to base) is known from B.C. (N to Telegraph Creek, ca. 57°50′N), s Alta. (Medicine Hat; CAN), and s Sask. (Cypress Hills and Wood Mountain; CAN). Forma *brachystyla* (Gray) Scoggan (*Echinospermum brachycentrum* var. *brachystylum* Gray, the type from Spences Bridge, B.C.; *L. brachystyla* (Gray) Macbr.; *L. anaplocarpa* Greene; prickles of nutlets nearly obsolete, the style shorter than or barely equalling the nutlets rather than surpassing them) is known only from the type locality.

LITHOSPERMUM L.   [7109]   Stoneseed, Gromwell, Puccoon.   Grémil

1   Corolla white, scarcely surpassing the calyx-lobes, lacking folds or crests in the throat; nutlets dull grey or pale brown, prominently tuberculate-roughened; leaves

linear to narrowly lanceolate, 1-nerved; annual or biennial from a slender taproot; (introd.) .... *L. arvense*
1 Corolla greenish white to yellow or orange-yellow, with 5 folds or crests in the throat (except in *L. ruderale*); nutlets ivory-white to pale brown, smooth or sparingly pitted, lustrous; perennials with thick taproots.
  2 Leaves lanceolate to ovate-lanceolate, the principal lateral veins evident beneath; corolla greenish white to pale yellow, about equalling the calyx; nutlets smooth or remotely pitted; stem commonly branched above.
    3 Leaves lanceolate to oblanceolate or oblong, acute-tipped, firm, mostly not over 1.5 cm broad; corolla greenish white or nearly white, 4 or 5 mm long; nutlets less than 3 mm long; (introd.) .... *L. officinale*
    3 Leaves lanceolate to ovate-lanceolate, distinctly acuminate, relatively thin, to about 4 cm broad; corolla yellowish white, commonly 6 or 7 mm long; nutlets to 3.5 mm long; (S Ont. and Que.) .... *L. latifolium*
  2 Leaves mostly narrower, linear to narrowly lanceolate, firm, lacking evident lateral veins; corolla yellow to orange, the long tube distinctly surpassing the calyx-lobes, the limb to over 2 cm broad; stem finally much branched from the base, the individual stems simple or finally with relatively short branches above; (roots often staining herbarium sheets purple).
    4 Leaves commonly 3 or 4 mm broad (but up to 8 mm) and narrowly linear; corolla orange-yellow, the tube to about 2 cm long, the limb to about 2 cm broad, its lobes erose-denticulate; later flowers cleistogamous; fruiting pedicels mostly arching; nutlets pitted, with a distinct collar-like flange below the basal constriction; plant canescent with a short appressed stiff pubescence; (B.C. to S Ont.) .... *L. incisum*
    4 Leaves commonly broader; corolla-lobes entire; cleistogamous flowers wanting; fruiting pedicels mostly erect; nutlets smooth, lacking a basal collar.
      5 Plant appressed-hirsute with short stiff hairs; corolla bright orange-yellow, its tube to nearly 1.5 cm long, bearded within at base, its limb to over 2 cm broad; calyx-lobes to 1.5 cm long; nutlets ivory-white, about 4 mm long; (Ont.) ....
       .... *L. caroliniense*
      5 Plants with stems copiously spreading-pubescent; corolla smaller, the tube not bearded within; calyx-lobes less than 1 cm long.
        6 Leaves narrowly oblong, rounded at summit, commonly not over 3 or 4 cm long and not obscuring the flowers; corolla bright yellow, with a prominent crest in the throat, the limb to 1.5 cm broad; nutlets 2 or 3 mm long; pubescence consisting of thin soft hairs; (Sask. to Ont.) .... *L. canescens*
        6 Leaves linear-lanceolate, tapering from below middle to the acuminate apex, to about 8 cm long, the upper ones very crowded and obscuring the flowers; corolla greenish or pale yellow, the crest in the throat obsolete, the limb less than 1 cm broad; nutlets to over 5 mm long; pubescence coarser and stiffer; (B.C. to SW Sask.) .... *L. ruderale*

L. arvense L.   Corn-Gromwell
Eurasian; introd. along sandy roadsides and in fields and waste places of N. America, as in S B.C. (Agassiz; Chilliwack; Armstrong; L. Okanagan), S Man. (Winnipeg; CAN), Ont. (N to the Ottawa dist.), Que. (N to the Montreal dist.), St-Pierre and Miquelon (Rouleau 1956), and N.S. (Cambridge Station, Kings Co.; ACAD; CAN; not listed by Roland 1947).

L. canescens (Michx.) Lehm.   Puccoon, Indian-paint
/T/EE/ (Hp) Dry or sandy prairies and open woods from Sask. (N to McKague, 52°37'N) to Man. (N to Grand Rapids, near the NW end of L. Winnipeg) and Ont. (N to the Kaministikwia R. W of Thunder Bay), S to Tex. and Ga. [*Batschia* Michx.; *B. conspicua* Rich.].

L. caroliniense (Walt.) MacM.   Puccoon
/T/X/ (Hp) Dry sandy woods and dunes from Mont. to Ont. (N to Constance Bay, about 20 mi W of Ottawa; A.J. Breitung, Nat. can. (Que.) 84(3/4):87. 1957) and SE Va., S to Mexico, Tex., and Fla. [*L. croceum* Fern.; *L. hirtum* (Muhl.) Lehm.; *L. gmelinii* (Michx.) Hitchc.].

L. incisum Lehm.
/T/(X)/ (Hp) Dry plains, prairies, and foothills from B.C. (N to Telegraph Trail, ca. 54°N; Henry 1915) to Alta. (N to near Peace River at ca. 56°15′N), Sask. (Moose Mountain Creek, Souris Plain, and River-that-Turns; CAN), s Man. (N to St. Lazare, about 75 mi NW of Brandon), and s Ont. (Essex, Lambton, Wentworth, and Halton counties), s to N Mexico, Tex., and Ind. [*Anonymos* Walt.; *L. angustifolium* Michx., not Forsk.; *L. linearifolium* Goldie; *L. (Batschia; Pentalophus) longiflorum* Spreng., not Salisb.; *L. (P.) mandanense* Spreng.].

L. latifolium Michx.
/T/EE/ (Hp) Rich woods and thickets from Minn. to s Ont. (N to Huron, Halton, and Durham counties) and sw Que. (N to Loretteville, near Quebec City; CAN), s to E Kans., Ark., and Tenn.

L. officinale L. Common Gromwell. Herbe aux perles or Graines de lutin
European; introd. in pastures and open places in N. America, as in s Man. (known only from High Bluff, near Portage la Prairie, where taken by Hargrave in 1883; MTMG; the report from Morden by Groh 1947, is based upon *Onosmodium molle* var. *hispidissimum,* the relevant collection in DAO; the report from St. Lazare in the undated supplement to Lowe's 1943 checklist is based upon *L. canescens,* the relevant collection in SASK), Ont. (N to the Ottawa dist.), Que. (N to the Gaspé Pen. at Nouvelle), and N.B. (Campbellton, Restigouche Co., and Lockart Mills, Carlton Co.; ACAD; NBM).

L. ruderale Dougl. Puccoon
/T/W/ (Hp) Dry places and foothills at low to moderate elevations from B.C. (N to Williams Lake; CAN) to s Alta. (N to Calgary) and sw Sask. (Cypress Hills; Breitung 1957a), s to N Calif. and Colo. [*L. pilosum* Nutt.].

LYCOPSIS L. [7094]

L. arvensis L. Bugloss. Chaudronnette
Eurasian; introd. in dry or sandy fields and waste places of N. America, as in Alta. (Fort Saskatchewan; CAN), Sask. (Indian Head; Killdeer, about 120 mi sw of Regina), s Man. (a heavy infestation in grainfields near Carberry; DAO; the report from Portage la Prairie by Groh 1944a, is based upon *Borago officinalis,* the relevant collection in DAO), Ont. (N to Ottawa), Que. (N to the Gaspé Pen. and Saguenay Co. of the Côte-Nord), Nfld., N.B., P.E.I., and N.S. [*Anchusa* Bieb.].

MERTENSIA Roth [7102] Lungwort, Bluebells

(Ref.: Williams 1937)
1 Stem prostrate; leaves thick and more or less fleshy, spatulate to ovate or obovate,
    sparingly to densely papillate above, they and the stem strongly purplish-pruinose; corolla
    rose-pink (becoming pale blue), about 8 mm long; nutlets smooth and shining,
    acute-angled, utricle-like (the outer coat becoming inflated and papery); (transcontinental
    on coastal sands) . . . . *M. maritima*
1 Stem erect or somewhat decumbent at base; corolla blue or purplish (commonly pink
    when young); nutlets dull (wrinkled when dry), obtuse-angled, not utricle-like; leaves often
    strigose above; plants neither strongly purplish-pruinose nor coastal.
    2 Stem relatively tall and robust, to over 1 m tall, its leaves with evident lateral veins.
        3 Leaves typically glabrous on both surfaces or merely papillate above, mostly
            subobtuse or acute; limb of corolla subequal to or slightly shorter than the tube.
            4 Corolla-tube to about 2 cm long, with a dense ring of hairs inside at the base;
                calyx to 1 cm long, the lobes eciliate; leaves eciliate; (s Ont. and s ?Que.) . . . .
                . . . . *M. virginica*
            4 Corolla-tube at most 8 mm long, glabrous or with crisped hairs within; calyx
                to about 4 mm long, the lobes ciliate or papillate on the margins; leaves
                ciliate . . . . [*M. ciliata*]
        3 Leaves pubescent on one or both surfaces, acuminate; limb of corolla distinctly

longer than the tube, this glabrous or sparsely pubescent within; calyx mostly 4 or
5 mm long, the lobes ciliate; (Alaska–B.C. to E James Bay) . . . . *M. paniculata*
2 Stem usually less than 4 dm tall, its leaves with a strong midrib but usually lacking
evident lateral veins.
  5 Roots shallow-seated, tuberous-thickened, easily detached; basal leaves rarely
developed in flowering plants; stem-leaves rather few, usually obtuse, oblong-
lanceolate to broadly ovate, mostly less than 3 times as long as broad, the
lowermost 1 or several strongly reduced (sometimes to a mere petiolar sheath);
corolla-limb usually noticeably shorter than the tube; (S B.C. and SW Alta.) . . . .
. . . . *M. longiflora*
  5 Roots deeper-seated and scarcely tuberous, firmly attached; leaves acute or
obtuse, basal ones mostly well developed.
    6 Corolla-limb longer than or subequal to the tube; basal leaves usually
lanceolate-ovate, to 14 cm long and 4 cm broad; stem-leaves linear to broadly
lanceolate or oblong-elliptic, to 1 dm long and 3 cm broad; (S Sask.) . . . .
. . . . *M. lanceolata*
    6 Corolla-limb slightly shorter than the tube.
      7 Stem to about 1.5 dm tall, its leaves to about 3.5 cm long and 11 mm
broad; (W arctic regions) . . . . *M. drummondii*
      7 Stem to 4 dm tall, its leaves to about 8 cm long and 1.5 cm broad;
(?B.C.) . . . . [*M. oblongifolia*]

[M. ciliata (Torr.) Don]
[The report of this species of the W U.S.A. (Oreg., Idaho, and Mont. to Calif. and Colo.) from Alaska
by Henry (1915) is probably based upon *M. paniculata,* as also his Rocky Mt. report (this
presumably based upon the citation from "South fork of Salmon River, near the Idaho boundary" by
J.M. Macoun 1906, the relevant collection in CAN). Williams (1937) cites no Canadian stations.
(*Pulmonaria* Torr.).]

M. drummondii (Lehm.) Don
/a/W/ (Hs) Sandy slopes of NW Alaska (N of 70°N), S Victoria Is., and the coast of Dist.
Mackenzie (E to near Coronation Gulf; probable type locality, it being given as "Arctic Sea-shore").
[*Lithospermum* Lehm.; *M. sibirica* var. *dr.* (Lehm.) Gray]. MAPS: Hultén 1968b:784; Porsild 1957:
map 285, p. 196; *Atlas of Canada* 1957: map 15, sheet 38; W.J. Cody, Nat. can. (Que.) 98(2):
fig. 25, p. 155. 1971.

M. lanceolata (Pursh) A. DC.
/T/WW/ (Hs) Prairies and slopes at low to moderate elevations from Mont. (possibly B.C.–Alta.;
Boivin 1966b) to S Sask. (Manoka, Hitchcock, Chaplin, and Dundura; Breitung 1957a; also reported
from Wood Mountain and Estevan by Williams 1937; the report from Riding Mt., Man., by Lowe
1943, is based upon *M. paniculata,* the relevant collection in WIN; his report from Tilston requires
clarification), S to N.Mex., Colo., and S.Dak. [*Pulmonaria* Pursh; *M. linearis* Greene].

M. longiflora Greene
/T/W/ (Grt) Sagebrush plains and foothills at low to moderate elevations from S B.C. (N to near
Cariboo, ca. 51°10′N; Williams 1937) and SW Alta. (Waterton Lakes; CAN) to N Calif. and Idaho. [*M.
horneri* Piper; *M. oblongifolia sensu* John Macoun 1884, Henry 1915, and Rydberg 1922, in part,
not (Nutt.) Don].
  Forma *alba* Boivin (flowers white rather than blue) is known from the type locality, Chase, B.C.,
about 30 mi NE of Kamloops.

M. maritima (L.) S.F. Gray Sea-Lungwort, Oysterleaf
/aST/X/GEA/ (Hp) Coastal sands and gravels from the Aleutian Is. and coasts of Alaska–
Yukon–Dist. Mackenzie–Dist. Keewatin to Banks Is., Devon Is., N Baffin Is., and northernmost
Ungava–Labrador, S along the Pacific coast to Vancouver Is., along the Hudson Bay–James Bay
coasts of Man.–Ont.–Que., and along the Atlantic seaboard from E Que. (St. Lawrence R. estuary
from Ile-aux-Coudres, about 50 mi NE of Quebec City, and St-Jean-Port-Joli, l'Islet Co., to the

Côte-Nord, Anticosti Is., and Gaspé Pen.) to s Labrador, Nfld., N.B., N.S. (not known from P.E.I.), and Mass.; w Greenland N to ca. 80°N, E Greenland N to 65°37'N; Iceland; Spitsbergen; N Europe; E Asia. [*Pulmonaria* L.; *Lithospermum* Lehm.; *P. parviflora* Michx.; incl. var. *tenella* Fries and the coarse extreme, *M. asiatica* (Takeda) Macbr.]. MAPS: Hultén 1968*b*:781, and 1958: map 275, p. 295; Porsild 1957: map 284, p. 196.

Forma *albiflora* Fern. (corolla whitish rather than rose-pink, fading blue) is known from Alaska (Boivin 1966*b*), E Que. (Mingan Is. of the Côte-Nord; Marie-Victorin and Rolland-Germain 1969), and N.S. (type from Sand Beach, Yarmouth Co.).

[M. oblongifolia (Nutt.) Don]
[The report of this species of the w U.S.A. (Wash. and Mont. to Calif.) from SE B.C. by Henry (1915: Columbia Valley; previously reported from "Mountains of Montana to the borders of British Columbia" by John Macoun 1884) is probably based upon collections in CAN from Trail, where taken by J.M. Macoun in 1902, these, however, being referable to *M. longiflora*. (*Pulmonaria* Nutt.; *Lithospermum marginatum* Lehm.).]

M. paniculata (Ait.) Don
/aST/X/ (Hs) Damp meadows, thickets, and woods, the aggregate species from Alaska (N to ca. 67°N) and the coasts of the Yukon–NW Dist. Mackenzie to L. Athabasca (Alta. and Sask.), Man. (N to Churchill), northernmost Ont., and w-cent. Que. (James Bay watershed N to ca. 52°N), s through B.C. to Oreg., Mont., N Iowa, N Mich., and w-cent. Ont. (N shore of L. Superior). MAPS and synonymy: see below.
1 Pedicels with appressed-strigose pubescence; calyx-lobes glabrous on the back; leaves pubescent on both sides, the hairs on their lower surfaces appressed; nutlets sometimes with long subspinose processes; [*M. east.* Macbr.; Alaska (type from Nome; see Hultén 1949: map 1008, p. 1468) and s-cent. Dist Mackenzie (reported from ca. 63°N, 127°W, by W.J. Cody, Can. Field-Nat. 77(4):227. 1963); MAP: Hultén 1968*b*:783] . . . .
. . . . var. *eastwoodiae* (Macbr.) Hult.
1 Pedicels and calyx-lobes typically spreading-pubescent; nutlets lacking subspinose processes.
   2 Anthers to 5 mm long; leaves rather consistently broadly ovate, scabrous above, coarsely and rather sparsely strigose or hirsute-strigose to glabrous beneath; stems mostly solitary; [*M. platyphylla* Heller; *M. subcordata* Greene; N ?B.C.: collections in V from White Pass, Atlin, and Nulki L. have been placed here] . . . .
. . . . var. *platyphylla* (Heller) Jones
   2 Anthers at most 3.5 mm long; leaves lanceolate to broadly ovate; stems usually numerous.
      3 Clayx-lobes in anthesis to 6 mm long; corolla-limb about 10 mm broad; [reported by J.F. Macbride, Contrib. Gray Herb. Harv. Univ. 48:6. 1916, from Alaska (type locality), B.C., Alta., and the Moose R. basin s of James Bay, Ont.; not recognized by Hultén 1949] . . . . var. *longisepala* Macbr.
      3 Calyx-lobes in anthesis 2 or 3 mm long; corolla-limb at most 8 mm broad.
         4 Leaves glabrous on both sides or minutely strigose above, or glabrous above and pubescent beneath; [var. *alaskana* (Britt.) Williams (*M. alaskana* Britt.); *M. pratensis* var. *borealis* Macbr.; Alaska–Yukon (see Hultén 1949: map 1010b, p. 1468) and NW Dist. Mackenzie (A.E. Porsild, Rhodora 41(487):283. 1939) to s B.C. (Salmo, near Trail; Williams 1937); MAP (var. *alaskana*): Hultén 1968*b*: 782] . . . . var. *borealis* (Macbr.) Williams
         4 Leaves appressed-strigose above, spreading-short-hirsute beneath; [*Pulmonaria paniculata* Ait., the type from Hudson Bay; *Lithospermum* Lehm.; *L. corymbosum* Lehm.; *M. laevigata* Piper; *M. pilosa* (Cham.) DC.; *M. sibirica* of Alaskan reports, not (Pursh) Don; Alaska–B.C. to James Bay, Que.; MAPS (aggregate species): Hultén 1968*b*:782; Raup 1947: pl. 34] . . . . var. *paniculata*

M. virginica (L.) Pers. Virginian Cowslip, Bluebells
/T/EE/ (Hs) Rich woods, clearings, and bottomlands from Minn. to s Ont. (Elgin, Lambton,

Middlesex, Wellington, Welland, and Lincoln counties) and sw ?Que. (Boivin 1966*b*), s to E Kans., Ark., Ala., and S.C. [*Pulmonaria* L.].

MYOSOTIS L.   [7100]   Forget-me-not

1   Hairs of calyx straight and closely appressed, nonglandular; fruiting pedicels divergent, longer than the calyces; perennials of moist or wet habitats.
   2   Corolla blue, the limb to 1 cm broad, the tube surpassing the calyx; calyx-lobes about half as long as the calyx-tube; style about equalling the calyx-tube; racemes usually bractless; stem angled, stoloniferous; (introd.) . . . . *M. scorpioides*
   2   Corolla pale blue, the limb at most 6 mm broad, the tube about as long as the calyx; calyx-lobes about as long as the calyx-tube; style much shorter than the calyx-tube; principal racemes usually with 1 or 2 bracts near base; stem not angled, nonstoloniferous; (B.C. and Alta.; Ont. to Nfld. and N.S.) . . . . *M. laxa*
1   Hairs of calyx (or some of them) divergent and hooked or glandular at tip; annuals or biennials, chiefly of dryish habitats.
   3   Fruiting pedicels equalling or longer than the calyces; calyx-lobes subequal; flowers blue or white; racemes bractless.
      4   Corolla 2 or 3 mm broad, its lobes spreading-ascending; calyx-lobes only slightly longer than the tube; (introd.) . . . . *M. arvensis*
      4   Corolla at least 4 mm broad, its lobes spreading horizontally; calyx oblique, its lobes distinctly longer than the tube; (B.C. and Alta.; introd. elsewhere) . . . . *M. sylvatica*
   3   Fruiting pedicels shorter than the calyces.
      5   Calyx-lobes unequal (2 long, 3 short); flowers white, 1 or 2 mm broad, on appressed-ascending pedicels; racemes usually bracted at base or as far up as the middle; (s B.C.; s Ont.) . . . . *M. verna*
      5   Calyx-lobes subequal, about as long as the tube; flowers finally blue or white; pedicels divergent; (introd.).
         6   Corolla blue, about 1.5 mm broad, its tube about equalling the calyx; style distinctly shorter than the nutlets; inflorescence commonly over 3/4 of the total height of the plant, the lower flowers subtended by foliaceous bracts . . . . *M. micrantha*
         6   Corolla yellow, changing to blue, about 2 mm broad, its tube surpassing the calyx; style commonly equalling or longer than the nutlets; inflorescence rarely over half the total height of the plant, only the lowest flower bracted . . . . *M. discolor*

M. arvensis (L.) Hill
Eurasian; a garden-escape to roadsides, fields, and waste places in N. America, as in ?Alaska (*see* Hultén 1968*a*), B.C. (Queen Charlotte Is.; Vancouver Is. and adjacent islands and mainland), Sask. (Bjorkdale, 52°43′N; DAO), Man. (near Brandon and Winnipeg), Ont. (N to Sault Ste. Marie and Ottawa), Que. (N to the Gaspé Pen. at St-Tharsicius, Matapédia Co.; reported from Anticosti Is. by Schmitt 1904), Nfld., N.B., P.E.I. (Charlottetown; CAN; GH), N.S., and sw Greenland (Polunin 1959). [*M. scorpioides* var. *arv.* L.; *M. caespitosa sensu* Lindsay 1878, in part, not Schultz, a relevant collection in NSPM].

M. discolor Pers.
European; locally introd. into fields and waste places in N. America, as in sw B.C. (Vancouver Is. and adjacent islands and mainland N to Cache Creek, about 60 mi w of Kamloops), Ont. (N to the Ottawa dist.), and N.S. (near Wolfville, Kings Co.; ACAD). [*M. versicolor* Sm.].

M. laxa Lehm.
/T/X/EA/   (Hp)   Wet ground and shallow water from B.C. (N to Queen Charlotte Is. and Prince Rupert, ca. 54°N; CAN) and Alta. (N to about 50 mi w of Edmonton; CAN) to N Calif. and Chile, then from Ont. (N to Moose Factory, near the sw tip of James Bay) to Que. (N to L. St. John, Anticosti Is., and the Gaspé Pen.), Nfld., N.B., P.E.I., and N.S., s to Tenn. and Ga.; Eurasia. [*M. caespitosa sensu* Lindsay 1878, in part, a relevant collection in NSPM].

M. micrantha Pallas
European; introd. along roadsides and in fields and waste places of N. America, as in s B.C. (Vancouver Is.; between Creston and Cranbrook; Herb. V), Alta. (Waterton Lakes), Ont. (N to the Ottawa dist.; Gillett 1958; TRT), and Que. (N to Grosse-Ile, NE of Quebec City; MT). [*M. stricta* Link; *M. collina* (*M. ramosissima* Rochel) *sensu* J.M. Macoun 1897, not Bab. nor Hoffm., the relevant collection in CAN].

M. scorpioides L.
Eurasian; introd. into wet grounds and quiet waters of N. America, as in s Alaska-Yukon (Hultén 1949; *M. pal.*), B.C. (Queen Charlotte Is.; Vancouver Is.; lower Fraser R.; Creston; Natal), Man. (Boivin 1966*b*), Ont. (N to the Ottawa dist.; Gillett 1958), Que. (N to Rivière-du-Loup, Temiscouata Co.; MT), St-Pierre and Miquelon, Nfld., P.E.I., and N.S. [Var. *palustris* L.; *M. pal.* (L.) Hill (or ?Lam.; *see* Hitchcock et al. 1959:231)]. MAP (*M. pal.*): Hultén 1968*b*:780.

M. sylvatica Hoffm.
/aST/WW/EA/ (Hs) Meadows and moist open slopes at low to high elevations (a garden escape to fields and waste places in the East), var. *alpestris* from N Alaska and the coasts of the Yukon and NW Dist. Mackenzie through B.C.–Alta. (the report from Ont. by Boivin 1966*b*, requires confirmation) to Idaho, Wyo., and the Black Hills of S.Dak.; Eurasia.
1   Fruiting pedicels up to twice the length of the calyces; nutlets dark brown; (introd.) . . . .
. . . . var. *sylvatica*
   2   Flowers blue; [*M. arvensis* var. *syl.* (Hoffm.) Pers.; introd. in Ont. (N to Hearst,
      49°42′N), Que. (N to the Gaspé Pen.), Labrador (Boivin 1966*b*), Nfld. (Rouleau 1956),
      N.B. (Fredericton; CAN), and N.S. (Halifax; NSPM)] . . . . f. *sylvatica*
   2   Flowers white or pinkish; [introd. in Que. and Nfld.; Boivin, 1966*b*] . . . . f. *lactaea* Boenn.
1   Fruiting pedicels about equalling the calyces (the lower ones slightly longer); nutlets
  black; [native] . . . . var. *alpestris* (Schm.) Koch
   3   Flowers blue; [*M. alpestris* Schm. and its ssp. *asiatica* Vestergr. (*M. asiatica*
      (Vestergr.) Schischk.); range as outlined above; MAP: Hultén 1968*b*:779] . . . . f. *alpestris*
   3   Flowers white or pinkish; [s Alaska (type from Port Vita) and sw Alta. (Waterton
      Lakes; Boivin 1966*b*)] . . . . f. *eyerdamii* Boivin

M. verna Nutt.
/T/X/ (Hs (bien. or T)) Dry or moist open woods, clearings, and prairies from s B.C. (Vancouver Is. and mainland N to Bella Coola, ca. 52°20′N; CAN; V) to Minn., s Ont. (Essex, Wellington, Welland, Lincoln, and Hastings counties), and New Eng., s to Oreg., Idaho, Tex., Okla., Tenn., and Fla. [Incl. *M. macrosperma* Engelm.; *M. virginica* of auth., not *Lycopsis virg.* L., basionym].

[NONEA (Nonnea) Medic.]   [7096]

[N. vesicaria (L.) Reichenb.]
[European; introd. or a garden-escape in s Alta. (Swalwell, 51°34′N; Boivin 1966*b*), where scarcely established (*see* Moss 1959:390).]

[OMPHALODES Mill.]   [7062]   Navelwort

1   Flowers white; nutlets with toothed inflexed margins; stem-leaves linear-lanceolate; erect,
  glabrous and slightly glaucous annual; (introd. in Ont.) . . . . [*O. linifolia*]
1   Flowers bright blue; nutlets with ciliate margins; stem-leaves ovate-lanceolate to ovate;
  long-stoloniferous, soft-pubescent perennial with ascending stems; (introd. in Que.) . . . .
. . . . [*O. verna*]

[O. linifolia (L.) Moench]
[European; an occasional garden-escape in N. America, as in Ont. (Boivin 1966*b*; Ottawa), where, however, scarcely established. The report from Labrador by Hooker (1829–40) requires clarification. (*Cynoglossum* L.).]

[O. verna Moench]   Creeping Navelwort
[European; an occasional garden-escape in N. America but scarcely established, as in Que. (along a street in Quebec City, where taken by Yves Desmarais in 1953; CAN).]

ONOSMODIUM Michx.   [7113]

O. molle Michx.   Marble-seed, False Gromwell
/T/(X)/   (Hs)   Sandy, gravelly, or rocky prairies, thickets, and open woods, the aggregate species from Mont. and s Alta. (Bow R.; Belly R.; Fort Macleod) to Sask. (Carnduff, about 150 mi SE of Regina), Man. (N to Brandon and Portage la Prairie), s Ont. (N to Huron, Waterloo, and Hastings counties), and N.Y., s to N.Mex., Tex., La., Tenn., and N.C.
1   Nutlets constricted just above the base to form an evident collar, mostly smoothish but dull, to 3.5 mm long; leaves to about 1.5 dm long and 4 cm broad; plant to over 1 m tall, relatively coarsely and conspicuously hairy; [*O. hispidissimum* Mack.; *O. hispidum* Michx. in part; *O. carolinianum* and *O. virginianum* of most Canadian reports, not *Lithospermum car.* Lam. nor *L. virginianum* L., respectively; s Man. and s Ont.] . . . .
. . . . var. *hispidissimum* (Mack.) Cronq.
1   Nutlets lacking an evident collar; plants commonly not over 7 dm tall, the leaves to about 8 cm long and 2 cm broad.
  2   Nutlets smooth and shining, 3.5–5 mm long; pubescence relatively coarse and loose; [*O. occidentale* Mack.; s Alta. to s Man.] . . . . var. *occidentale* (Mack.) Johnston
  2   Nutlets pitted and dull, at most 3 mm long; [E U.S.A.] . . . . [var. *molle*]

PECTOCARYA DC.   [7061]

P. linearis (R. & P.) DC.
/t/W/   (T)   Dry soil and sagebrush plains from s B.C. (Osoyoos, Lytton, and Spences Bridge; CAN; John Macoun 1884), Idaho, and Wyo. to Baja Calif. and Mexico; S. America. [*Cynoglossum* R. & P.].
  Our material is referable to var. *penicillata* (H. & A.) Jones (*Cynoglossum (P.) pen.* H. & A.; margins of nutlets bristly chiefly near the tip rather than nutlets bristly along the sides as well as near the tip, the central bristles commonly the stoutest).

PLAGIOBOTHRYS F. & M.   [7079]

1   All of the stem-leaves normally alternate (sometimes paired), few and scattered, they and the stem spreading-hirsute (the stem often also retrorsely strigose); basal leaves rosette-tufted, to 3 cm long and 8 mm broad; corolla-limb to 4 mm broad; scar located just below the middle of the nutlet, this thick-cruciform, tuberculate in rows; stems mostly several, erect or ascending from a definite taproot; (s B.C.) . . . . *P. tenellus*
1   At least some of the lower stem-leaves opposite, the leaves essentially all cauline, they and the commonly fibrous-rooted stems sparsely to moderately strigose; scar located near the base of the nutlet, this ovoid, more or less rugose or rugose-tuberculate (in *P. scouleri* often also with minute distally-branched bristles).
  2   Corolla-limb to 1 cm broad; spikes naked, tending to be paired; (Vancouver Is.) . . . .
. . . . *P. figuratus*
  2   Corolla-limb at most 4 mm broad; spikes seldom paired, often irregularly bracted below; (s B.C. to sw Man.) . . . . *P. scouleri*

P. figuratus (Piper) Johnston
/t/W/   (T)   Meadows and low ground from sw B.C. (Vancouver Is.; CAN; introd. in SE Alaska near Juneau) to sw Oreg. [*Allocarya* Piper; *P. hirtus* var. *fig.* (Piper) Johnst.]. MAP: Hultén 1968b:776 (*P. hirt.* var. *fig.*).

P. scouleri (H. & A.) Johnston
/T/WW/   (T)   Moist low ground at low to moderate elevations from s B.C. (Vancouver Is. and adjacent islands; Rossland; Nelson; Creston; Cranbrook) to sw Alta. (N to Jasper), s Sask. (N to

Saskatoon; Breitung 1957a), and sw Man. (Melita, about 60 mi sw of Brandon), s to Calif. and N.Mex. [*Myosotis* H. & A.; *Allocarya* Greene; *Eritrichium* A. DC.; *Krynitzkia* Gray; *M. (A.; E.; K.) ?californica* F. & M.; *A. (P.) media* Piper; *E. fulvum* (Hook.) DC.]. MAP: combine the maps by Hultén 1968b:775 (*P. orient.*) and 776 (*P. cog.*).

Material from sw B.C. (Vancouver Is. and adjacent islands) appears to belong to the typical form. Our other material, including plants introd. in the E Aleutian Is. and Alaska–Yukon, is referable to var. *penicillatus* (Greene) Cronq. (*Allocarya pen.* Greene; *A. (P.) cognata, cusickii, hispidula,* and *scopulorum* Greene; *A. divaricata* Piper; *Eritrichium californicum* var. *subglochidiatum* Gray; *P. ?leptocladus sensu* T.M.C. Taylor 1966b, perhaps not (Greene) Johnst.; incl. *A. plebeja* (C. & S.) Greene and *P. orientalis* (L.) Johnst.; *E. (A.; K.) chorisianum* of auth., perhaps not DC.; corolla-limb at most 2 mm broad rather than to 4 mm broad; stem prostrate to ascending rather than ascending to erect).

P. tenellus (Nutt.) Gray
/t/W/ (T) Dry open soil from sw B.C. (Vancouver Is.; CAN, several localities) and Idaho to Baja Calif., Nev., and ?Utah. [*Myosotis* Nutt.; *Eritrichium* Gray; *P. echinatus* Greene].

SYMPHYTUM L. [7090] Comfrey. Consoude

1 Leaves barely or not at all decurrent on the stem; corolla roseate (becoming bluish), its lobes erect; nutlets rough, constricted above the base; stem and pedicels bearing broad-based prickle-like hairs .... *S. asperum*
1 Upper leaves broadly decurrent by wings extending down the stem; corolla whitish, yellowish, or purple, its lobes recurving at tip; nutlets nearly smooth, not constricted above base; inflorescence and upper part of stem densely pilose-hispid, broad-based prickle-like hairs typically wanting .... *S. officinale*

S. asperum Lepechin Prickly Comfrey
Asiatic; introd. along roadsides and in waste places of N. America, as in s B.C. (N to Spences Bridge; CAN), s Man. (Brandon), Ont. (N to Ottawa), Que. (N to E of Quebec City in Montmorency Co.; MT), Nfld., N.B., P.E.I. (Charlottetown; GH), and N.S. [*S. asperrimum* Donn].

S. officinale L. Common Comfrey. Lange de vache or Herbe à la coupure
Eurasian; introd. along damp roadsides and in waste places in N. America, as in B.C. (Boivin 1966b), Alta. (Moss 1959), Ont. (N to Renfrew and Carleton counties; concerning reports from Man., *see* Scoggan 1957), Que. (N to Ste-Anne-de-la-Pocatière, Kamouraska Co.; RIM), Nfld. (near St. John's; GH), N.B., and N.S. (according to D.S. Erskine 1960, reports from P.E.I. may refer to *S. asperum*).

[TRIGONOTIS Stev.] [7103]

[*T. peduncularis* (Trev.) Benth.]
[Eurasian; known in N. America only through an 1893 collection by John Macoun from sw B.C. (Nanaimo, Vancouver Is.; CAN, distributed as *Myosotis arvensis*, referred here by I.M. Johnston, Contrib. Gray Herb. Harv. Univ. 70:46. 1924), Johnston stating that, "It is a roadside weed in Asia and probably introduced in America on ballast." (*Myosotis* Trev.).]

VERBENACEAE (Vervain Family)

Herbs with opposite, simple, toothed to incised, exstipulate leaves. Flowers more or less-2-lipped, perfect, gamopetalous, hypogynous, in single or often panicled spikes. Calyx- and corolla-lobes each 4 or 5. Stamens 4, didynamous (in 2 pairs of unequal length), inserted on the corolla-tube. Style 1. Ovary superior. Fruit separating at maturity into 2 or 4 nutlets.

1 Corolla 4-lobed, rather distinctly 2-lipped, white, pink, or bluish; calyx 2-keeled, its 2 or 4 teeth longer than the tube; fruit splitting into 2 nutlets; spikes at first subglobose, later elongating to about 3.5 cm, solitary on filiform peduncles 2 or 3 times the length of the subtending leaves, the peduncles 1 or 2 at each node; leaves narrowly elliptic to oblong-lanceolate, sharply serrate except at the cuneate base; stems weak, often rooting at the nodes; (s Ont.) . . . . *Lippia*
1 Corolla 5-lobed, only obscurely 2-lipped, white or purple; calyx unequally 5-ribbed and 5-angled, unequally 5-toothed, the teeth shorter than the tube; fruit splitting into 4 nutlets; stems mostly erect . . . . *Verbena*

LIPPIA L. [7145]

L. lanceolata Michx.　Fog-fruit
/t/X/　(Hpr)　Wet brackish to wet sands, ditches, and low ground from Nebr. to Minn., s Ont. (Leamington, Essex Co., where taken by John Macoun in 1892 as new to Canada; CAN; collection in TRT from Walpole Is., Lambton Co.), Pa., and N.J., s to s Calif., La., and Fla. [*Phyla* Greene; incl. var. *recognita* Fern. & Grisc., a coarser but doubtless completely intergrading phase].

VERBENA L. [7138] Vervain. Verveine

1 Leaves 3-lobed or pinnately incised and also serrate, hirsute; spikes lax, sessile, their divergent bracts much surpassing the calyces; corolla light blue to purple, 2 or 3 mm broad; stem low and diffusely branched, usually prostrate or decumbent; (B.C. to s Ont.) . . . . *V. bracteata*
1 Leaves merely toothed or shallowly incised (or deeply lobed only at base); floral bracts rarely surpassing the calyces; stem erect, simple or sparingly branched.
  2 Spikes numerous, in panicles; calyx 2 or 3 mm long; leaves distinctly petioled.
    3 Corolla white, about 2 mm broad; spikes slender, interrupted; leaves coarsely (often doubly) crenate-serrate; nutlets corrugated on the back; (SE Sask.; Ont. to N.B.) . . . . *V. urticifolia*
    3 Corolla violet-blue, at least 3 mm broad; spikes stiff and compact; leaves coarsely incised-serrate, sometimes hastately lobed at base; nutlets smooth or faintly striate; (B.C.; Sask. to N.B. and N.S.) . . . . *V. hastata*
  2 Spikes solitary or few, stiff; calyx 4 or 5 mm long; nutlets reticulate above, strongly striate below; leaves subsessile; (Ont. and s Que.).
    4 Leaves oblong to broadly obovate, pilose and prominently veined beneath; corolla to 9 mm broad; stem copiously pubescent . . . . *V. stricta*
    4 Leaves narrowly lanceolate or oblanceolate, often scabrous; corolla 5 or 6 mm broad; stem glabrous or sparingly strigose . . . . *V. simplex*

V. bracteata Lag. & Rodr.
/T/X/　(Hp (T))　Sandy prairies, fields, and roadsides and other disturbed habitats from s B.C. (N to Kamloops; CAN) to s Alta. (Crowsnest Pass, Waterton Lakes, Red Deer R., and Lethbridge; CAN), s Sask. (N to Moose Jaw; Breitung 1957a), s Man. (N to Wawanesa, about 20 mi SE of Brandon; CAN), s Ont. (probably native in sand at Point Edward, near Sarnia, Lambton Co.; John Macoun 1890; probably introd. along railway ballast and in freight yards at Strathroy, Middlesex Co., and Westmeath, Renfrew Co.), and Va. (adv. to New Eng.), s to Calif., Mexico, and Fla. [*V. bracteosa* Michx.].
　A 1903 collection by Dodge from Point Edward, Lambton Co., s Ont., is considered by J.M. Macoun (1906) as evidently a hybrid between *V. bracteata* and *V. stricta* (× *V. deamii* Moldenke).

However, Boivin (1966*b*) reports apparently the same collection as × *V. dodgei* Boivin (*V. bracteata* × *V. hastata*). A hybrid with *V. urticifolia* (× *V. perriana* Moldenke) is tentatively reported from Ont. by Boivin.

V. hastata L.   Blue Vervain, Simpler's-joy
/T/X/   (Hp)   Damp thickets, wet meadows, and shores from s B.C. (N to Vernon; CAN; not known from Alta.) to s Sask. (Wadena, about 120 mi E of Saskatoon; Breitung 1957a), Man. (N to Washow Bay, s of The Narrows, L. Winnipeg; CAN), Ont. (N to Sault Ste. Marie and Kapuskasing, 49°24′N), Que. (N to Grosse-Ile, NE of Quebec City), N.B., and N.S. (not known from P.E.I.), s to Calif., Tex., and N Fla. [*V. paniculata* Lam.].
      Forma *rosea* Cheney (corolla-limb pink rather than violet-blue) is reported from s Ont. by Landon (1960; Norfolk Co.). Collections in OAC from s Ont. have been identified by H.N. Moldenke as × *V. engelmannii* Moldenke (*V. hastata* × *V. urticifolia*; Walpole Is., Lambton Co.) and × *V. paniculatistricta* Engelm. (*V. hastata* × *V. stricta* × *V. rydbergii* Moldenke; Point Edward, Lambton Co.). Reports by Moldenke of the Californian *V. lasiostachys* Link and *V. robusta* Greene from SW B.C. are considered by Boivin (1967a) as probably based upon an error in locality.

V. simplex Lehm.
/T/EE/   (Hp)   Dry or sandy soil from Nebr. to Minn., Ont. (N to the Ottawa dist.; Gillett 1958), SW Que. (Montreal dist.; *see* map of the NE limits by Rouleau 1945: fig. 6, p. 161), and N.H., s to Kans., La., Miss., Ala., and Fla. [*V. angustifolia* Michx., not Mill.].

V. stricta Vent.   Hoary Vervain
/T/X/   (Hp)   Prairies, fields, and roadsides from Wash., Idaho, and Mont. to Ont. (N to the Ottawa dist.; Gillett 1958) and SW Que. (N to near Oka and Montreal), s to N.Mex., Tex., Okla., Ark., and Tenn.; adv. eastwards to New Eng. and Del. (our material may consist of both native and introd. material).
      Forma *albiflora* Wadmond (corolla white rather than purple) is known from SW Que. (between Oka and St-Eustache; MT, detd. H.N. Moldenke).

V. urticifolia L.   White Vervain
/T/EE/   (Hp)   Rich thickets and borders of woods from extreme SE Sask. (Gainsborough, about 140 mi SE of Regina; Breitung 1957a; not known from Man.) to Ont. (N to the Ottawa dist.; Gillett 1957), Que. (N to Cap-Tourmente, about 30 mi NE of Quebec City), and N.B. (York and Carleton counties; not known from P.E.I. or N.S.), s to Tex., La., Ala., and N Fla.

LABIATAE (Mint Family)

Herbs with square stems and simple, opposite, aromatic (mostly gland-dotted) leaves. Flowers perfect, hypogynous, gamopetalous, 5-merous, chiefly in axillary whorls, the corolla commonly distinctly 2-lipped (sometimes apparently 4-merous by fusion of 2 petals). Stamens 2 or 4 (2 long, 2 short), inserted on the corolla-tube. Style 1. Ovary superior. Fruit consisting of 4 seed-like nutlets surrounding the base of the style. (Lamiaceae).

1   Calyx-tube with a distinct cap-like protuberance on the upper side; flowers normally blue; stamens 4 .... *Scutellaria*
1   Calyx-tube lacking a cap-like protuberance.
   2   Corolla apparently 1-lipped, only the lower lip well developed; ovary not cleft to base, the nutlets laterally attached; style terminal; flowers to over 1 cm long; stamens 4; leaves lanceolate to ovate or obovate.
      3   Corolla normally bluish to purple, its lower lip 3-lobed (or 4-lobed by splitting of the middle lobe), its upper lip nearly truncate except for 2 short lobes; stem less than 3 dm tall; (introd.) .... *Ajuga*
      3   Corolla normally pink-purple or yellow, its lower lip apparently 5-lobed by fusion at its base of the 2 small lobes of the split upper lip; stem to about 1 m tall ..... *Teucrium*
   2   Corolla nearly regular to strongly 2-lipped (if the latter, the well-developed upper lip entire, notched, or 2-lobed, the lower lip 3-lobed, its middle lobe usually broadest and often notched); ovary cleft to base, the nutlets basally attached and the style basal (except in *Isanthus,* with nutlets laterally attached and style terminal).
      4   Calyx-teeth 10, subulate, strongly hooked at the apex, strongly recurving at maturity, somewhat shorter than the tube, this 4 or 5 mm long; stamens 4, they and the style included in the tube of the 2-lipped corolla; flowers about 6 mm long, white, in dense axillary clusters; leaves rugose, broadly elliptic to round-ovate or more or less fan-shaped, petioled, crenate; plant whitish-woolly, the stem prostrate to erect; (introd.) .... *Marrubium*
      4   Calyx-teeth not strongly hooked; stamens and style exserted beyond the corolla-tube (but not necessarily surpassing the lip).
         5   Fertile stamens 2.
            6   Calyx nearly regular, its lobes essentially alike at least in shape.
               7   Flowers slender-pedicelled in usually compound panicles from the 1 or 2 upper leaf-axils; corolla yellow, about 1.5 cm long, lemon-scented, strongly 2-lipped, the middle lobe of the lower lip lacerate-fringed; leaves ovate, to about 2 dm long, the lower ones on petioles to about 1 dm long, stem to over 1 m tall; (s Ont.) .... *Collinsonia*
               7   Flowers sessile or subsessile in dense terminal clusters or axillary whorls.
                  8   Corolla nearly regular, apparently 4-lobed, white, not over 5 mm long; flowers in axillary whorls; plants usually stoloniferous .... *Lycopus*
                  8   Corolla strongly 2-lipped, to 5 cm long; flowers in terminal clusters (also in axillary clusters in one species); leaves lanceolate to ovate, shallowly serrate, mostly petioled .... *Monarda*
            6   Calyx distinctly 2-lipped (*Collinsonia* may key out here).
               9   Flowers bluish, at most 5 mm long, in few-flowered whorls in the axils of normal foliage-leaves; calyx villous in the throat, gibbous on the lower side near base, strongly 13-nerved; annuals to about 4 dm tall .... *Hedeoma*
               9   Flowers to 2.5 cm long, in terminal racemes, spikes, or heads; calyx naked in the throat, not gibbous; perennials, often taller.
                  10   Calyx 13-nerved, its upper lip with 3 awned teeth, its lower lip with 2 nearly awnless teeth; flowers pale bluish-purple, about 12 mm long, numerous in dense whorls; (s Ont. and s Que.) .... *Blephilia*
                  10   Calyx 10–15-nerved, its upper lip entire or 3-lobed, its lower lip

deeply 2-lobed; flowers blue; whorls usually less densely flowered; (introd.) .... *Salvia*
- 5 Fertile stamens 4.
  - 11 Floral whorls in the axils of normal foliage-leaves, only the uppermost leaves and internodes of the inflorescence somewhat reduced.
    - 12 Stem prostrate and creeping, rooting at the nodes; leaves roundish or reniform, coarsely crenate, to 3 or 4 cm broad, often purple-tinged; flowers purplish blue, 1 or 2 cm long, usually 3 in a whorl; (introd.) .... *Glechoma*
    - 12 Stem erect or ascending.
      - 13 Calyx distinctly 2-lipped (nearly regular in linear-leaved species of *Satureja*); stamens covered by the upper corolla-lip.
        - 14 Calyx about 8 mm long (slightly larger in fruit), the 3 lobes of its flattened upper lip broader than long, short-pointed; corolla to 1.5 cm long, pale blue to white, its upper lip 2-lobed; leaves ovate, coarsely blunt-serrate, lemon-scented, the larger ones obtuse to truncate at base, to about 7 cm long; stem to about 1 m tall, finely canescent (at least above) and generally with some long spreading hairs and long-stipitate glands (at least toward summit); (introd.) .... *Melissa*
        - 14 Calyx to 1 cm long, the lobes of its upper lip narrowly triangular, long-acuminate; upper lip of the purple-red to nearly white corolla entire or merely notched .... *Satureja*
      - 13 Calyx regular or nearly so.
        - 15 Corolla nearly regular, its 4 or 5 lobes subequal; flowers small.
          - 16 Corolla pale blue, little surpassing the calyx, the 4 upper lobes ascending, the lower lobe deflexed; stamens not exserted from the corolla; flowers solitary in the leaf-axils or 2 or 3 together, forming a large leafy panicle, the filiform pedicels to about 1 cm long; nutlets strongly reticulate; leaves entire or nearly so; clammy-pubescent annual to 3 or 4 dm tall; (Ont and sw Que.) .... *Isanthus*
          - 16 Corolla whitish or pale purple, apparently 4-lobed (the upper lip entire or merely notched); stamens exserted from the corolla; flowers short-pedicelled, numerous in each whorl; nutlets smooth or roughened; leaves serrate, often sharply so; perennials with rhizomes or stolons .... *Mentha*
        - 15 Corolla distinctly 2-lipped, the concave, helmet-like, entire or emarginate upper lip covering the stamens; (introd.).
          - 17 Flowers purplish, about 1.5 mm long, on pedicels to 5 mm long; calyx strongly 10-nerved; anthers glabrous; leaves broadly ovate, coarsely crenate .... *Ballota*
          - 17 Flowers sessile; anthers pubescent (except in *Leonurus*).
            - 18 Calyx 5-nerved, its teeth slenderly acuminate but not spiny; corolla creamy white, pink, or purplish, over 1 cm long .... *Lamium*
            - 18 Calyx-teeth stiff and spine-like.
              - 19 Lower lip of the small purplish corolla lacking yellow protuberances; calyx-tube 5–10-nerved; leaves deeply lobed or parted .... *Leonurus*
              - 19 Lower lip of corolla (more than 1 cm long) with 2 yellow protuberances at base; calyx-tube strongly 10-nerved and usually with 10 intermediate nerves; leaves entire or coarsely toothed .... *Galeopsis*
  - 11 Floral whorls all or chiefly subtended by bracts conspicuously differing from the foliage-leaves, the internodes usually markedly reduced, the whole inflorescence head-like, spike-like, racemose, or paniculate.

20 Corolla strongly 2-lipped, the upper lip more or less concave and helmet-like, covering the stamens.
  21 Calyx regular or nearly so.
    22 Upper (inner) pair of stamens longer than the lower pair; calyx 15-nerved; leaves ovate-lanceolate to rhombic-ovate, rounded to cordate at base, coarsely serrate, to 1.5 dm long.
      23 Anthers paired, their locules strongly divergent, their filaments not exserted; corolla dull white, purple-dotted, at least 1 cm long; (introd.) .... *Nepeta*
      23 Anthers all separated, their locules nearly parallel, their filaments surpassing the corolla (this less than 1 cm long) .... *Agastache*
    22 Upper pair of stamens shorter than the lower pair; calyx 5–10-nerved; corolla pink to red or various shades of purple.
      24 Whorls 2-flowered; calyx membranaceous, obscurely 10-nerved; leaves narrowly lanceolate to narrowly oblong, sharply serrate, sessile or subsessile; plants glabrous or nearly so .... *Dracocephalum*
      24 Whorls mostly with 4 or more flowers; calyx firm, 5–10-nerved; plants more or less pubescent.
        25 Leaves mostly in a basal rosette, lanceolate to narrowly ovate, obtuse, coarsely crenate, those of the stem few and remote, reduced; corolla bright reddish-purple, to 1.5 cm long, with scattered hairs, the upper lip nearly flat; outer stamens not diverging from the corolla after anthesis; (introd.) .... *Betonica*
        25 Leaves cauline (the basal ones greatly reduced); corolla with an oblique ring of hairs within toward base, the upper lip rounded; outer stamens diverging laterally from the corolla after anthesis .... *Stachys*
  21 Calyx distinctly 2-lipped; corolla light blue to violet or purple; leaves petioled.
    26 Middle lobe of upper calyx-lip longer and much broader than the other 4 lobes; upper (inner) pair of stamens longer than the lower pair .... *Moldavica*
    26 Upper calyx-lip almost truncate, barely 3-lobed, the lower lip with 2 lance-acuminate lobes; upper pair of stamens shorter than the lower pair; leaves narrowly lanceolate to ovate-oblong, entire or obscurely crenate; (transcontinental) .... *Prunella*
20 Corolla nearly regular or distinctly 2-lipped (but the upper lip nearly flat, not helmet-like); at least 1 pair of stamens surpassing the corolla.
  27 Corolla nearly regular, very shallowly lobed; calyx regular or weakly 2-lipped; flowers small.
    28 Corolla (at least the tube) included in the calyx, lavender to blue or pale purple, usually glabrous; leaves often sharply serrate; perennials with rhizomes or stolons .... *Mentha*
    28 Corolla much surpassing the hirsute calyx, pubescent externally; leaves lanceolate to ovate.
      29 Leaves entire, subsessile or short-petioled, the blade to 3.5 cm long; corolla whitish to pink-purple, 1 or 2 cm long, 5-lobed, its slender lobes about half as long as the tube or longer; flowers in a solitary terminal head to 4 cm broad subtended by a distinct involucre of several rather dry, veiny, more or less purplish-tinged, ciliate bracts up to 1.5 cm long .... [*Monardella*]
      29 Leaves crenate-serrate, the blades to about 7 cm long, tapering to long petioles; corolla pale blue or purplish,

about 3 mm long, 4-lobed, the lobes broad and very short; flowers in terminal and upper-axillary spikes to 5 cm long, the spikes leafy-bracted and subtended by normal foliage-leaves; (introd.) .... *Elsholtzia*

27 Corolla distinctly 2-lipped; calyx pubescent; perennials.

  30 Calyx bearded in the throat; leaves entire or nearly so, short-petioled; corolla purplish; (introd.).

    31 Calyx about 2.5 mm long, regular; corolla about 7 mm long; leaves to about 3 cm long, mostly ovate or deltoid-ovate, broadly rounded to definite petioles, loosely hairy below, purplish; stems erect, from horizontal rhizomes ...... *Origanum*

    31 Calyx 3 or 4 mm long, 2-lipped, the upper lip 3-toothed, the lower with 2 subulate divisions; corolla rose-purple, to 6 mm long; leaves at most 12 mm long and 6 mm broad, tapering to rather short petioles, glabrous beneath, some-times sparingly pubescent above; stems partly creeping and rooting at the nodes .... *Thymus*

  30 Calyx naked in the throat, nearly regular; leaves subsessile or short-petioled (except in *Pycnanthemum incanum,* with 2-lipped calyx and petioles up to 1.5 cm long).

    32 Flowers blue-purple, about 1 cm long, in crowded clusters forming a spike; leaves lanceolate to oblanceolate, at most about 3 cm long, entire, often with smaller ones in their axils; stems finely puberulent, from a stout woody rhizome, to about 6 dm tall; (introd.) .... *Hyssopus*

    32 Flowers whitish, lilac, or purplish, small, in mostly dense heads disposed in open corymbs; leaves to 1 dm long, entire or shallowly toothed; (chiefly s Ont.) ..... *Pycnanthemum*

AGASTACHE Clayt. [7241] Giant Hyssop

(Ref.: Lint and Epling 1945)

1 Calyx puberulent at least at anthesis.

  2 Corolla blue, about 1 cm long; calyx to 7 mm long, its blue, triangular-lanceolate, acute teeth at most 2 mm long; leaves glaucous-whitened beneath with a fine dense puberulence, their petioles mostly less than 1.5 cm long; stem glabrous or nearly so; (B.C. to s Ont.; introd. eastwards) .... *A. foeniculum*

  2 Corolla whitish or light violet-purple, to 14 mm long; calyx about 1 cm long, its whitish to pinkish or lavender-purple, narrowly triangular-acuminate lobes to 5 mm long; leaves green and either glabrous or often very finely hispid-puberulent beneath, their petioles to 5 cm long; stem very finely retrorse-puberulent (sometimes glabrate); (s B.C.) .... *A. urticifolia*

1 Calyx glabrous, green or pink-tinged; leaves green and either glabrous or finely puberulent beneath, their petioles to about 6 cm long; stem (or at least the branches) more or less puberulent.

  3 Calyx about 6 mm long, its obtuse or subacute lance-ovate teeth at most 1.5 mm long; corolla pale greenish-yellow; stem sharply 4-angled; (Ont. and s Que.) .... *A. nepetoides*

  3 Calyx to 9 mm long, its acute triangular-lanceolate teeth at least 2 mm long; corolla purplish; stem obtusely 4-angled; (s Ont.) .... *A. scrophulariaefolia*

A. foeniculum (Pursh) Ktze. Blue Giant Hyssop

/T/WW/ (Hs) Dry thickets, plains, and prairies from s Dist. ?Mackenzie (an early Anderson collection in CAN from the Mackenzie R., possibly a confusion of locality) and B.C. (N to Prince George) to Alta. (N to McMurray, 56°44′N), Sask. (N to Prince Albert), Man. (N to Porcupine Mt.), and Ont. (probably native w to near Thunder Bay, L. Superior; probably introd. along roadsides and railways in the Algoma, Cochrane, and Timiskaming districts and in Renfrew and Carleton counties;

possibly native in Bruce and Wentworth counties, s Ont.), s to ?Wash., Colo., S.Dak., and Ill.; introd. eastwards in fields and waste places, as in E Ont. (*see* above), Que. (Timiskaming, Labelle, Argenteuil, Deux-Montagnes, Terrebonne, Laval, and Temiscouata counties), N.B. (Gloucester, Kings, and Charlotte counties), New Eng., and Del. [*Stachys* Pursh; *Hyssopus (A.) anethiodora* Nutt.; *Lophanthus anisatus* Benth.]. MAP: Lint and Epling 1945: map 1 (incomplete northwards), p. 213.

The typical form has deep-blue calyx-lobes, corolla-lobes, and styles. In the following forms, the corolla is uniformly white or pink-tinged. Forma *bernardii* Boivin (Nat. can. (Que.) 87(2):25. 1960; calyx-lobes roseate; corolla and anthers white or pinkish; style roseate or pale blue) is known from Sask. (Bjorkdale; Katepwa) and Que. (type from Nominingue, Labelle Co.). Forma *candidum* Boivin (loc. cit., p. 26; calyx-lobes, corolla, anthers, and style all white) is known from Sask. (type from McKague) and Man. (Riding Mt.).

A. nepetoides (L.) Ktze. Yellow Giant Hyssop
/T/EE/ (Hs) Rich thickets and borders of woods from S.Dak. to Ont. (N to Huron, Waterloo, and York counties; reported N to Casselman, Russell Co., by John Macoun 1884, but not listed by Gillett 1958) and SW Que. (N to the Montreal dist.; MT), S to Kans., Mo., Ky., and Ga. [*Hyssopus* L.; *Lophanthus* Benth.]. MAP: Lint and Epling 1945: map 1 (incomplete for Canada), p. 213.

A. scrophulariaefolia (Willd.) Ktze. Purple Giant Hyssop
/t/EE/ (Hs) Rich woods and thickets from S Ont. (Essex, Kent, and Lincoln counties; CAN; TRT) to Vt. and Mass., S to Kans., Mo., Ky., and N.C. [*Hyssopus* Willd.; *Lophanthus* Benth.]. MAP: Lint and Epling 1945: map 1, p. 213.

A. urticifolia (Benth.) Ktze. Western Giant Hyssop
/t/W/ (Hs) Thickets and open slopes at low to fairly high elevations from S B.C. (Dry Interior in the Okanagan and Columbia valleys near the U.S.A. boundary SE of Penticton between Greenwood and Rossland; CAN) to Calif. and Colo. [*Lophanthus* Benth.]. MAP: Lint and Epling 1945: map 1, p. 213.

AJUGA L. [7211] Bugleweed

1 Stems copiously soft-pubescent, tufted, nonstoloniferous; calyx to 8 mm long, villous, its teeth linear-lanceolate .... *A. genevensis*
1 Stems essentially glabrous, stoloniferous and mat-forming; calyx at most 6 mm long, sparingly villous, its teeth triangular-ovate .... *A. reptans*

A. genevensis L.
Eurasian; locally introd. as a garden-escape into fields and waste places in N. America, as in Ont. (Bracebridge, E of Georgian Bay, L. Huron; OAC; reported from Wellington Co. by Stroud 1941, and from l'Orignal, Prescott Co., about 50 mi E of Ottawa, by Groh and Frankton 1949*b*).

A. reptans L. Carpet Bugleweed
Eurasian; locally introd. along roadsides and in fields in N. America, as in Ont. (Ottawa; CAN), Que. (Montreal; MT), SE Nfld. (Salmonier, where "established in field"; GH), and P.E.I. (Hurst 1952).

BALLOTA L. [7279] Fetid Horehound

B. nigra L. Black Horehound
European; locally introd. into waste places in N. America, as in S Ont. (near London, Middlesex Co., where taken by John Dearness in 1888 and 1889; CAN).

BETONICA L. [7281]

B. officinalis L. Betony
European; a local garden-escape in N. America, as in P.E.I. (Brackley Beach, Queens Co., where taken in a fencerow by I.J. Bassett in 1950; DAO). [*Stachys* Trev.; *S. betonica* Benth.].

BLEPHILIA Raf. [7297]

1 Leaves of flowering stems narrowed to a subsessile base; lobes of lower calyx-lip extending beyond the sinuses of the upper lip; floral whorls often all approximate; plant somewhat downy; (S Ont.) . . . . *B. ciliata*
1 Leaves distinctly petioled, obtuse to subcordate at base; lobes of lower calyx-lip not reaching to the sinuses of the upper lip; whorls less crowded; plant more or less hirsute; (S Ont. and SW Que.) . . . . *B. hirsuta*

B. ciliata (L.) Benth.
/t/EE/ (Hpr) Dry woods, thickets, and clearings from Iowa to Wisc., S Ont. (Walpole Is., Lambton Co.; MICH; reported from Pelee Is., Essex Co., by Dodge 1914), and Vt., S to E Tex., Mo., Miss., and Ga. [*Monarda* L.].

B. hirsuta (Pursh) Benth.
/T/EE/ (Hpr) Moist shady places from Minn. to S Ont. (Lambton Co.; Dodge 1915), SW Que. (near Brome, Brome Co.; John Macoun 1884), and Vt., S to E Tex., Mo., Tenn., and Ga. [*Monarda* Pursh].

COLLINSONIA L. [7331] Horse-Balm

C. canadensis L. Richweed, Stoneroot
/t/EE/ (Gst (Grh)) Rich moist woods from Wisc. to S Ont. (N to Lambton, Wellington, and York counties; *see* S Ont. map by Soper 1962: fig. 29, p. 45), Vt., and Mass., S to Ark. and Fla.; (a puzzling report from Nfld. by Waghorne 1898).

DRACOCEPHALUM L. [7250 and 7257 (*Physostegia*)] False Dragonhead

1 Uppermost leaves commonly shorter than their subtended internodes, these often much longer than the median ones; calyx to 1 cm long, its teeth viscid-glandular and usually minutely puberulent; corolla to 3 cm long; (Man. to N.B. and ?N.S.) . . . . *D. virginianum*
1 Uppermost leaves mostly longer than their subtended internodes, these often shorter than the median ones; calyx rarely over 7 mm long.
  2 Uppermost leaves broadest near the middle and tapering at both ends; corolla to 2 cm long; (Alta. to W Ont.) . . . . *D. formosius*
  2 Uppermost leaves broadest near the broadly rounded base; corolla usually less than 1.5 cm long; (B.C. to W Ont.) . . . . *D. nuttallii*

D. formosius (Lunell) Rydb.
/sT/WW/ (Hp) Damp thickets and shores from N Alta. (N to the Slave R. at 59°31'N; not known from Sask.) to Man. (N to Gillam, about 165 mi S of Churchill) and W Ont. (Minnitaki L. and Abram L., near Kenora, and Lac Seul, about 85 mi NE of Kenora; CAN), S to ?(Nebr., Mo., and Ohio; range uncertain through confusion with *D. nuttallii*). [*Physostegia* Lunell; *P. ledinghamii* Boivin; *D. speciosum* of Canadian reports in part, not Sweet].

D. nuttallii Britt.
/T/WW/ (Hp) Damp thickets, meadows, and shores from B.C. (N to Spences Bridge; CAN) to Alta. (N to the Red Deer R.; CAN), Sask. (N to Tisdale, 52°51'N; Breitung 1957a), Man. (Virden and the Winnipeg dist.), and W Ont. (N shore of L. Superior near Thunder Bay; CAN), S to Oreg., Nebr., and Minn. [*Physostegia* Fassett; *P. parviflora* Nutt.].

D. virginianum L.
/T/EE/ (Hp (Grh)) Damp thickets, meadows, and shores from Man. (Doghead Point, N end of The Narrows, L. Winnipeg, where taken by John Macoun in 1872; CAN) to S Ont. (N to Wellington and York counties; CAN; TRT), Que. (N to NE of Quebec City in Montmagny Co.; MT), and N.B. (Grand Manan, Charlotte Co.; CAN; GH; reported from Pictou, N.S., by Lindsay 1878; not known

from P.E.I.), s to N.Dak., La., Tenn., and N.C. [*Physostegia* Benth. and its var. *elongata* Boivin; incl. the reduced extreme, *P. granulosa* Fassett, to which a collection from Mill Village, N.S., has been referred by Boivin].

ELSHOLTZIA Willd. [7334]

E. ciliata (Thunb.) Hylander
Asiatic; locally introd. along roadsides and in old fields in N. America, as in s Man. (Birds Hill, near Winnipeg; DAO; Groh and Frankton 1949*b*), Ont. (Aultsville, Stormont Co.; DAO; Dore and Gillett 1955), Que. (Montreal; Notre-Dame-du-Lac, Temiscouata Co.; John Macoun 1890), and nw N.B. (Boivin 1966*b*). [*E. cristata* Willd.; *E. patrinii* (Lepechin) Garcke].

GALEOPSIS L. [7270] Hemp-Nettle. Gratte or Galéopside

1 Stem softly appressed-pubescent or becoming glabrate, not swollen at the nodes;
corolla-tube often scarcely surpassing the calyx . . . . *G. ladanum*
1 Stem more or less hispid with retrorsely spreading hairs, swollen below the nodes.
2 Corolla to 2 cm long, pink, purple, or white (rarely pale yellow with a violet spot), the
tube rarely slightly surpassing the calyx . . . . *G. tetrahit*
2 Corolla to over 3 cm long, pale yellow and usually with a violet spot on the lower,
entire or slightly emarginate lip, the tube much surpassing the calyx . . . . *G. speciosa*

G. ladanum L. Red Hemp-Nettle
Eurasian; locally introd. into waste ground in N. America, as in Que. (Grosse-Ile, NE of Quebec City in Montmagny Co.; MT; CAN), St-Pierre and Miquelon, Nfld. (St. John's; Waghorne 1898), N.B. (wharf-ballast at St. John, where taken by G.U. Hay in 1881; NBM), and N.S. (Boivin 1966*b*).
Var. *latifolia* (Hoffm.) Wallr. (*G. lat.* Hoffm.; leaves ovate-oblong and distinctly serrate rather than linear to lanceolate and entire or nearly so; plant glandular at least in the inflorescence rather than nonglandular) is reported from St-Pierre and Miquelon by Fernald *in* Gray (1950).

G. speciosa Mill.
Eurasian; apparently known in N. America only from Alta. (near Edmonton; E. H. Moss, Blue Jay 19:33. 1961) and Que. (St-Zacharie, Dorchester Co.; Chambourd, L. St. John, Roberval Co.; St-Eloi, Temiscouata Co.; *see* Que. map by C. Rousseau 1968: map 138, p. 115).

G. tetrahit L. Common Hemp-Nettle. Ortie royale or Gratte
Eurasian; introd. along roadsides and in waste places and fields in N. America, as in Alaska (N to ca. 65°30′N), Dist. Mackenzie (Great Slave L.; W.J. Cody, Can. Field-Nat. 70(3):123. 1956), B.C.–Alta., Sask. (N to Waskesiu Lake, 53°55′N), Man. (N to Gillam, about 165 mi s of Churchill), Ont. (N to the w James Bay watershed at 52°11′N), Que. (N to the E James Bay watershed at 53°50′N, the Côte-Nord, Anticosti Is., and Gaspé Pen.), Labrador (N to Rigolet, 54°11′N), Nfld., N.B., P.E.I., and N.S. MAP and synonymy: *see* below.
1 Leaves blunt-tipped; [seashore litter at Ste-Anne-des-Monts, Gaspé Pen., E Que.; GH] . . . .
. . . . var. *arvensis* Schlecht.
1 Leaves acuminate.
2 Middle lobe of the lower corolla-lip entire, its network of dark markings restricted to the
base and never reaching the margin; [introd., transcontinental] . . . . var. *tetrahit*
2 Middle lobe of the lower corolla-lip deeply emarginate, its network of markings
reaching the margin, or the whole lip dark . . . . var. *bifida* (Boenn.) Lej. & Court.
3 Corolla white; [introd. in Man., Que., N.B., and N.S.] . . . . f. *albiflora* House
3 Corolla pink or purplish; [*G. bifida* Boenn.; introd., transcontinental; MAP (*G. bif.*):
Hultén 1968*b*:788] . . . . f. *bifida*

GLECHOMA L. [7249] Ground Ivy

G. hederacea L. Gill-over-the-ground, Run-away-Robin. Lierre terrestre
Eurasian; introd. along roadsides and in yards and grassy meadows in N. America, as in SE Alaska

(Wrangell), B.C. (Vancouver Is.; Saltspring Is.; Hope), Alta. (N to Fort Saskatchewan; CAN), Sask. (N to McKague, 52°37′N), Man. (N to The Pas), Ont. (N to the E shore of L. Superior at Michipicoten), Que. (N to the Gaspé Pen.), S Labrador (Battle Harbour, 52°16′N; CAN), Nfld., N.B., P.E.I., and N.S. [*Nepeta* Trev.; *N. glechoma* Benth.; *G. hirsuta* W. & K.; incl. var. *micrantha* Moricand (var. *parviflora* (Benth.) Druce)]. MAP: Hultén 1968*b*:786.

HEDEOMA Pers.   [7302]   Mock Pennyroyal

1   Leaves linear or nearly so, entire, sessile, nearly glabrous except along the hispid-ciliate
      margins, to about 2 cm long; calyx-teeth all subulate (upper lip cleft to or below middle)
      and hispid-ciliate; sterile anther-filaments minute or obsolete; stem simple or branched, to
      about 2 dm tall; (S Alta. to SW Que.) . . . . *H. hispida*
1   Leaves lanceolate to ovate or obovate, entire or serrulate, the principal ones
      short-petioled, to about 3 cm long; upper calyx-lip not cleft to below middle, its teeth
      narrowly triangular; lower calyx-teeth setaceous-subulate, ciliate; sterile anther-filaments
      manifest; stem usually branched, to about 4 dm tall; (Ont. to N.S.) . . . . *H. pulegioides*

H. hispida Pursh
/T/X/   (T)   Dry open places from Mont. to S Alta. (Fort Macleod and Red Deer Lakes), S Sask. (N to Elbow, about 50 mi S of Saskatoon; Breitung 1957*a*), ?Man. (Boivin 1966*b*), Ont. (N to the Ottawa dist.), Que. (N to Shawville, Pontiac Co., and Ste-Agathe, Terrebonne Co.), and N.Y., S to Colo., Tex., La., and Miss.; adv. eastwards to W New Eng. [Incl. f. *simplex* Lalonde, the stem simple rather than branched].

H. pulegioides (L.) Pers.   Pudding-grass
/T/EE/   (T)   Dry soil from S.Dak. to Minn., Ont. (N to Manitoulin Is., N L. Huron, and the Ottawa dist.), Que. (N to Kingsmere, Gatineau Co., and the Montreal dist.), N.B., and N.S. (not known from P.E.I.), S to E Kans., Ark., Tenn., and Fla. [*Cunila* L.; *Melissa* L.].

HYSSOPUS L.   [7313]

H. officinalis L.   Hyssop
Eurasian; introd. and locally abundant in dry pastures and along roadsides in N. America, as in Sask. (Carmel, about 50 mi E of Saskatoon; Breitung 1957*a*), Ont. (N to the Ottawa dist.), Que. (Pontiac, Gatineau, and Missisquoi counties), and N.S. (Walbrook, Kings Co.; CAN).
    Forma *albus* Alefeld (flowers white rather than blue-purple) is reported from Ont. and SW Que. by Boivin (1966*b*).

ISANTHUS Michx.   [7217]

I. brachiatus (L.) BSP.   Fluxweed
/T/EE/   (T)   Dry soil, often calcareous, from Minn. to Ont. (N to Arnprior, Renfrew Co., where evidently introd.; Groh and Frankton 1949*b*; presumably native in S Ont.), SW Que. (N to the Montreal dist.), and Vt., S to Tex., Ark., Tenn., and Ga. [*Trichostema* L.; *I. caeruleus* Michx.]. MAP: Harlan Lewis, Brittonia 5: fig. 6 (*Tric. br.*), p. 290. 1945.
    Var. *linearis* Fassett (Rhodora 35(420):388. 1933; leaves linear and 1-nerved rather than lanceolate to elliptic and 3-nerved from the base) is known from S Ont. (Bruce Pen., L. Huron; type from Cloche Pen., N L. Huron).

LAMIUM L.   [7271]   Dead-Nettle

1   Creeping-based perennials; corolla more than 1.5 cm long, its tube with a conspicuous
      ring of hairs within toward base; leaves all petioled; (introd.).
      2   Corolla creamy white, the tube about equalling the calyx; leaves ovate-lanceolate to
            ovate, coarsely serrate, the upper ones long-acuminate, to about 1 dm long   . . . . . *L. album*
      2   Corolla reddish purple, roseate, or whitish, the tube surpassing the calyx; leaves

ovate, crenate-dentate (often doubly so), often with an irregular white stripe along the midvein, the lower ones cordate at base, the upper ones broadly ovate, blunt, to about 6 cm long . . . . *L. maculatum*

1 Annuals or biennials without creeping stems; corolla pink or purplish, not over 1.5 cm long; (introd.).

   3 Calyx 5 or 6 mm long, densely clothed with white, more or less spreading hairs, the teeth usually somewhat shorter than the tube and connivent in fruit; corolla pinkish purple, either small and cleistogamous or much exserted and to about 1.5 cm long, the tube glabrous within; at least the upper bracteal leaves sessile and somewhat clasping, often broader than long (the lowest bracts rarely short-petioled) . . . .

    . . . . *L. amplexicaule*

   3 Calyx usually over 6 mm long, its teeth spreading; bracteal leaves not clasping.

      4 At least the upper bracteal leaves sessile (but not clasping); calyx to 12 mm long in flower, appressed-pubescent, the teeth longer than the tube; corolla-tube scarcely surpassing the calyx, with a faint ring of hairs within; (introd. in SW Greenland) . . . . *L. moluccellifolium*

      4 All of the bracteal leaves usually distinctly petioled.

         5 Leaves deep green or purplish, all cordate-ovate, shallowly crenate-serrate, the upper ones at most 2.5 cm long; corolla-tube with a ring of hairs within toward base . . . . *L. purpureum*

         5 Leaves pale green, irregularly incised-dentate, the upper ones ovate-deltoid to subrhombic, irregularly incised, to over 4 cm long, the small cordate-ovate lower ones crenate-dentate; corolla-tube naked within or with a faint ring of hairs toward base . . . . *L. hybridum*

L. album L.   Snowflake, White Dead-Nettle
Eurasian; introd. along roadsides and in old lawns and waste places in N. America, as in SE Alaska (Juneau; Hultén 1950), Sask. (Speers, about 50 mi NW of Saskatoon; Breitung 1957a), S Man. (Brandon), Ont. (near Cambridge (Galt), Waterloo Co.; Hamilton, Wentworth Co.), Que. (N to Bic, Rimouski Co.; GH), and N.B. (Chatham and Newcastle, Northumberland Co.). MAP: Hultén 1968b:788.

L. amplexicaule L.   Henbit
Eurasian; introd. along roadsides and in fallow fields and waste places in N. America, as in S Dist. Mackenzie (Fort Providence, ca. 61°20′N; W.J. Cody, Can. Field-Nat. 75(2):66. 1961), B.C. (N to Quesnel; Eastham 1947), Alta. (N to Fort Saskatchewan; CAN), Sask. (N to Marshall and Medstead, both W of Prince Albert; DAO), Man. (N to Minitonas, N of Duck Mt.; WIN), Ont. (N to Prescott and Ottawa; John Macoun 1884), Que. (N to Rupert House, James Bay, 51°29′N), SE Labrador (Forteau, 51°28′N; GH), Nfld., N.B., and N.S. (Truro; DAO); SW Greenland.
    Forma *clandestinum* (Reichenb.) Beck (the very small unopening corollas tubular and round-tipped) occurs throughout the range.

L. hybridum Vill.   Henbit
Eurasian; introd. into waste places and cult. soil in N. America, as in Que. (Rouleau 1947), Nfld. (GH; CAN), and P.E.I. (Charlottetown; GH; CAN). [*L. incisum* Willd.; *L. dissectum* With.].

L. maculatum L.   Spotted Dead-Nettle
European; introd. along roadsides and in waste ground in N. America, as in SW B.C. (Vancouver; Eastham 1947), S Ont. (Lambton, Huron, Middlesex, Waterloo, Wellington, and Bruce counties), SW Que. (Chambly, near Montreal; MT), and ?Nfld. (Waghorne 1898; taken up by Rouleau 1956).
    Forma *lacteum* (Wallr.) Beck (flowers white rather than rose-purple) is known from B.C. (Boivin 1966b) and S Ont. (Stratford, Perth Co.; John Macoun 1890).

L. moluccellifolium Fries
European; introd. into SW Greenland where, according to Polunin (1959), evidently fairly well established.

L. purpureum L.   Purple Dead-Nettle
Eurasian; introd. along roadsides and in waste places of N. America, as in sw B.C. (Vancouver Is. and adjacent islands and mainland; V), s Ont. (Essex, Lambton, Elgin, Middlesex, Waterloo, Norfolk, and Welland counties), Que. (Bonaventure Is., Gaspé Pen.; Groh 1944*b*), Nfld., P.E.I., N.S., and w Greenland (CAN).

LEONURUS L.   [7273]   Motherwort.   Agripaume

1   Upper lip of the pale-pink corolla densely white-villous; calyx strongly 5-angled, its two
    lower lobes strongly deflexed; lower leaves palmately lobed, the upper ones subentire or
    3-lobed .... *L. cardiaca*
1   Upper lip of the purplish corolla finely pubescent; calyx barely angled, the lower lobes
    scarcely deflexed; leaves all deeply 3-parted, the divisions themselves incised or cleft ....
    .... *L. sibiricus*

L. cardiaca L.   Common Motherwort
Eurasian; introd. into fields and waste places in N. America, as in B.C. (Boivin 1966*b*), se Sask. (Gainsborough, about 160 mi se of Regina; Breitung 1957*a*), Man. (N to Dauphin, N of Riding Mt.; DAO), Ont. (N to Carleton and Russell counties), Que. (N to Baie-St-Paul, Charlevoix Co.), N.B., P.E.I., and N.S. [Incl. var. *villosa* (Desf.) Benth.].

L. sibiricus L.
Eurasian; locally introd. along roadsides and in waste places in N. America, as in se Man. (Dufrost, s of Winnipeg; DAO; Groh 1947), ?Ont. (Boivin 1966*b*), and sw Que. (Montreal; CAN; J.M. Macoun 1907).

LYCOPUS L.   [7326]   Water-Horehound, Bugleweed.   Lycope

1   Calyx-teeth ovate to broadly triangular, blunt or subacute, less than 1 mm long, not
    surpassing the nutlets; corolla about twice as long as the calyx; leaves shallowly serrate.
    2   Stem glabrate or finely pubescent with ascending incurved hairs, usually tuberous at
        base and the stolons often tuberiferous; leaves light green, gradually narrowed at both
        ends, with a few low teeth and a subsessile base; corolla-lobes flaring; stamens
        exserted; nutlets topped by a merely undulate crest; (transcontinental) .... *L. uniflorus*
    2   Stem usually minutely pubescent, not tuberous at base, the stolons mostly
        nontuberiferous; leaves dark green or purplish, coarsely serrate, rather abruptly
        narrowed to a short-petioled base; corolla-lobes erect; stamens usually included; crest
        of nutlets deeply tuberculate .... [*L. virginicus*]
1   Calyx-teeth narrowly triangular, sharply acuminate, to 3 mm long and overtopping the
    nutlets.
    3   Leaves sessile at the obtuse or broadly rounded base, dark green, more or less
        ascending, scabrous, with up to 12 sharp teeth on each margin; stem from elongate
        tubers, its pubescence consisting of long multicellular hairs; corolla barely surpassing
        the calyx; (B.C. to Que.) .... *L. asper*
    3   Principal leaves rather abruptly narrowed to a petiolar base; tubers rarely developed.
        4   Leaves strigose above; corolla barely surpassing the calyx; (introd.) ..... *L. europaeus*
        4   Leaves essentially glabrous above.
            5   Principal leaves shallowly serrate; calyx-teeth relatively soft; corolla about
                twice as long as the calyx; summit of nutlets tuberculate; (s ?Ont.) .... [*L. rubellus*]
            5   Principal leaves coarsely toothed to incised or pinnatifid; calyx-teeth with rigid
                tips; corolla barely surpassing the calyx; summit of nutlets entire or only
                slightly undulate; (B.C. to Nfld. and N.S.) .... *L. americanus*

L. americanus Muhl.
/T/X/   (Hpr)   Marshes and moist low ground (ranges of Canadian taxa outlined below), s to Calif., Tex., and Fla. MAPS and synonymy: *see* below.
1   Stem-angles narrowly winged; lowermost leaves unlobed (but with their lower teeth

occasionally longer than the middle and upper ones); bracts and calyx-teeth not glandular-ciliate; nutlets with a thin ridge across the rounded upper surface and down the sides; [*L. laur.* Rolland-Germain, the type from Cap-Rouge, near Quebec City, Que.; fresh or brackish tidal flats of the St. Lawrence R. estuary, Que., in Lotbinière, Lévis, Québec, and Montmorency counties; MAP: Frère Rolland-Germain, Nat. can. (Que.) 72(7/8): fig. 3, p. 181. 1945] .... var. *laurentianus* (Rolland-Germain) Boivin

1 Stem-angles with a prominent ridge; lowermost leaves usually with one or more pairs of linear lobes; bracts and calyx-teeth glandular-ciliate; nutlets with a corky crest along the flattened outer edge of the upper surface and down the sides.

2 Leaves scabrous above; [S Ont.: Essex, Kent, and Norfolk counties] ....
.... var. *scabrifolius* Fern.

2 Leaves glabrous above; [*L. sinuatus* Ell.; *L. rubellus sensu* Dodge 1914, and Soper 1949, not Moench, a relevant collection in CAN; *L. ?obtusifolius* Benth., not Vahl; B.C. (N to Terrace, ca. 54°30′N), Alta. (N to Fort Saskatchewan), Sask. (N to McKague, 52°37′N), Man. (N to Gillam, about 165 mi S of Churchill), Ont. (N to SW James Bay at 51°38′N), Que. (N to the Bell R. at 49°43′N, L. St. John, and the Gaspé Pen.), Nfld., P.E.I., and N.S.; MAP (aggregate species): N.C. Henderson, Am. Midl. Nat. 68(1): fig. 8 (somewhat incomplete northwards), p. 114. 1962] .... var. *americanus*

L. asper Greene
/T/(X)/ (Gst) Marshes and shores of lakes and streams from B.C. (N to Cache Creek, about 50 mi NW of Kamloops; an isolated station at Circle Hot Springs, E-cent. Alaska, at ca. 65°N; CAN; *see* Hultén 1949: map 1016 (*L. luc.*), p. 1468) to Alta. (N to Fort Saskatchewan; CAN), Sask. (N to McKague, 52°37′N), Man. (N to Birch River, E of Porcupine Mt.; CAN), and the James Bay watershed of Ont.–Que. (N to 51°33′N), S to Calif., Colo., Mo., and Mich.; introd. elsewhere, as in Ont. (along a railway in the Kenora dist.; Essex and Kent counties (?native)). [*L. lucidus* of N. American reports in greater part, not Turcz.; *L. luc.* var. *americanus* Gray, not *L. americanus* Muhl.]. MAPS: Hultén 1968b:790 (*L. luc.* ssp. *amer.*); N.C. Henderson, Am. Midl. Nat. 68(1): fig. 17 (James Bay stations should be indicated), p. 127. 1962).

If merged with the Asiatic *L. lucidus,* the N. American plant may be distinguished as follows:

1 Nutlets with a smooth uniform crest; leaves acute at base; [Asia] .... [*L. lucidus*]

1 Nutlets with an undulate crest of up to 5 low rounded protuberances; leaves obtuse at base; [*L. asper* Greene; range as outlined above] .... ssp. *americanus* (Gray) Hult.

L. europaeus L.
Eurasian; introd. along roadsides, near dwellings, and in waste places in N. America, as in S Ont. (collections in DAO and TRT from Lincoln, Wellington, Wentworth, York, Northumberland, Prince Edward, Frontenac, Hastings, and Leeds counties), SW Que. (Beauharnois, l'Assomption, and Berthier counties; *see* S Que. map by C. Rousseau 1968: map 142, p. 118), and N.S. (wharf-ballast at Point Pleasant, Halifax Co.; ACAD; the reports from Windsor and Truro by Lindsay 1878, require confirmation, perhaps referring to *L. americanus,* his report from Halifax being apparently validated by the above collection). MAP: R.L. Stuckey and W.L. Phillips, Rhodora 72(791): fig. 1, p. 352. 1970.

Var. *mollis* (Kern.) Briq. (leaves copiously villous or pilose beneath rather than merely sparsely pubescent, relatively shallowly lobed or toothed) is known from S Ont. (Toronto; GH).

[*L. rubellus* Moench]
[Reports of this species of the E U.S.A. (Mich. to Ohio and New Eng., S to Tex. and Fla.) from S Ont. by Dodge (1914; Point Pelee, Essex Co.) and Soper (1949) require confirmation. A so-named 1901 collection by John Macoun in CAN from Sarnia, Lambton Co., has been referred to *L. americanus* by Boivin. However, it may eventually prove to be a Canadian species, the MAP by N.C. Henderson (Am. Midl. Nat. 68(1): fig. 11, p. 119. 1962) indicating a station in E Mich. very close to Sarnia.]

L. uniflorus Michx.
/T/X/eA/ (Hpr (Gst)) Marshes and banks of lakes and streams, the aggregate species from B.C. (N to Queen Charlotte Is. and Prince George, ca. 54°N; isolated stations around hot springs in cent. Alaska; *see* Hultén 1949: map 1017, p. 1468) to ?Alta. (the listing by Moss 1959, requires

confirmation), Sask. (N to L. Athabasca), Man. (N to Reindeer L. at 57°23'N), Ont. (N to the W James Bay watershed at ca. 53°N), Que. (N to the E James Bay watershed at ca. 54°15'N, the Côte-Nord, Anticosti Is., and Gaspé Pen.; type from between L. St. John and L. Mistassini), Labrador (Goose Bay, 53°19'N), Nfld., N.B., P.E.I., and N.S., S to N Calif., Idaho, Mont., Okla., Ark., and N.C.; E Asia. MAPS and synonymy: *see* below.

1 Leaves narrowly to broadly ovate; [N.S. (Yarmouth Co.; type from Sable Is.) and Nfld.
  (Rouleau 1956)] .... var. *ovatus* Fern. & St. John
1 Leaves lanceolate to lance-oblong .... var. *uniflorus*
  2 Stolons arising from the middle and upper leaf-axils as well as basally, finally
    tuberiferous; [Que. (Eastmain R., E James Bay, ca. 52°15'N; Dutilly, Lepage, and
    Duman 1958) and N.S. (type from North Sydney, Cape Breton Is.)] ..... f. *flagellaris* Fern.
  2 Stolons arising from the tuberous base of the stem; [*L. communis* and *L.*
    membranaceus Bickn.; *L. macrophyllus* Benth.; *L. pumilus* Vahl; *L. virginicus* var.
    pauciflorus Benth.; *L. virg.* of Canadian reports, not L.; range of the species; MAPS
    (aggregate species): N.C. Henderson, Am. Midl. Nat. 68(1): fig. 3, p. 108. 1962;
    Hultén 1968b:791] .... f. uniflorus

[*L. virginicus* L.]
[Reports of this species of the E U.S.A. (N to Nebr., Minn., and Maine) from N.S. by Cochran (1829) and Lindsay (1878) are based upon *L. uniflorus* (relevant collections in NSPM and MTMG), to which various other reports from Canada also probably refer. The MAP by N.C. Henderson (Am. Midl. Nat. 68(1): fig. 5, p. 110. 1962) indicates no Canadian stations.]

MARRUBIUM L. [7238] Horehound

M. vulgare L. Common Horehound
Eurasian; introd. in waste places of N. America, as in SE Alaska (Juneau; Hultén 1949), S B.C. (Vancouver Is. and adjacent islands; Keremeos; Osoyoos; Penticton; Grand Forks), Sask. (Boivin 1966b), S Ont. (N to Bruce, Grey, Middlesex, Wellington, York, and Prince Edward counties), Que. (Boivin 1966b), and N.S. (Bridgewater, Lunenburg Co.; Groh and Frankton 1949b; an early report by Cochran 1829). MAP: Hultén 1968b:785.

MELISSA L. [7304] Balm

M. officinalis L. Common Balm
Eurasian; a garden-escape to roadsides and waste places in N. America, as in SW B.C. (Vancouver Is. and adjacent mainland at Moodyville), S Man. (Brandon), and S Ont. (Essex, Huron, Grey, Middlesex, Welland, and Carleton counties).

MENTHA L. [7328] Mint. Menthe

1 Flowers in subglobose clusters in the axils of foliage-leaves much surpassing the flowers
  and separated by internodes of only gradually decreasing length; leaves lanceolate to
  ovate, more or less petioled, to about 7 cm long.
  2 Bracteal leaves much reduced in comparison with the foliage-leaves, these sharply
    serrate; calyx-tube essentially glabrous, its teeth pilose; stem more or less pubescent;
    (introd.) .... M. *cardiaca*
  2 Bracteal leaves not markedly reduced.
    3 Stem essentially glabrous; leaves coarsely and sharply serrate; calyx-tube
      glabrous throughout or sparingly pubescent toward summit, its teeth pilose;
      (introd.) .... M. *gentilis*
    3 Stem retrorse-pubescent on the angles, pubescent or glabrous on the sides;
      leaves serrate with low to salient teeth; calyx thinly pubescent to long-villous
      throughout; (transcontinental) .... M. *arvensis*
1 Flowers in terminal spikes with much reduced internodes or in head-like clusters, similar
  peduncled axillary inflorescences also often present; bracteal leaves (if present) narrow
  and little surpassing the flowers; (introd.).

4   Inflorescence an ovoid or subglobose head of rarely more than 3 whorls (often with axillary whorls below); leaves oval to ovate, petioled.
    5   Plant glabrous or nearly so, with characteristic lemon odour; stamens included; leaves oval or ovate .... *M. citrata*
    5   Plant pubescent with recurved hairs; stamens exserted; leaves round-ovate ....
    .... *M. aquatica*
4   Inflorescence a spike of several to many whorls.
    6   Leaves with petioles to about 1.5 cm long, glabrous or nearly so; stamens included; calyx 3 or 4 mm long, its tube glabrous, its teeth more or less pilose or hirsute; spikes about 1 cm thick (excluding the glabrous corollas) .... *M. piperita*
    6   Leaves sessile or subsessile (petioles less than 3 mm long); stamens usually exserted; calyx about 2 mm long.
        7   Calyx-tube glabrous, its teeth more or less pilose or hirsute; spikes about 6 mm thick (excluding the glabrous corollas); leaves essentially glabrous ....
        .... *M. spicata*
        7   Calyx minutely pubescent throughout; corollas hairy outside; leaves more or less pubescent; stem canescent or downy.
            8   Leaves oblong-lanceolate, acute, sharply serrate, to about 8 cm long; spike interrupted (especially toward base); corolla glabrous within ....
            .... *M. longifolia*
            8   Leaves broadly elliptic-oblong to broadly ovate or suborbicular, rounded to cordate at the often clasping base, crenate-dentate, at most about 5 cm long; spike compact; corolla with a ring of hairs within; (introd. in s Ont.) ....
            .... *M. rotundifolia*

M. aquatica L.   Water-Mint
Eurasian; locally introd. into wet places in N. America, as in sw Que. (Montreal and the shore of the St. Lawrence R. at Laprairie, opposite Montreal; MT) and ?N.S. (Truro, Colchester Co.; Lindsay 1878).
    Collections in ACAD and CAN from Cape Breton Co., N.S., have been referred by J.K. Morton to × *M. smithiana* Graham, a purported hybrid between *M. aquatica, M. arvensis,* and *M. spicata.*

M. arvensis L.   Common Mint
/ST/X/EA/   (Hpr (Grh))   Moist places at low to moderate elevations, the aggregate species (partly introd.) from Alaska–Yukon–w Dist. Mackenzie (N to ca. 66°30′N) to Great Bear L., L. Athabasca (Alta. and Sask.), Man. (N to Churchill), Ont. (N to the Fawn R. at 54°40′N), Que. (N to E James Bay at 52°12′N, the Côte-Nord, Anticosti Is., and Gaspé Pen.), s Labrador (N to the Hamilton R. basin), Nfld., N.B., P.E.I., and N.S., s to Calif., N.Mex., Mo., W.Va., and Del.; Eurasia. MAP and synonymy: *see* below.
1   Leaves elliptic to ovate, broadest chiefly below the middle, rounded to short petioles ....
.... var. *arvensis*
    2   Stem near the lowest flowering axils pubescent on the sides and angles; leaves pubescent.
        3   Angles of stem much more pubescent than the sides; [transcontinental, appar-ently both native and introd.; MAP (aggregate species): Hultén 1968*b*:792] ....
        .... f. *arvensis*
        3   Angles and sides of stem about equally pubescent.
            4   Hairs mostly entangled or spreading, with elongate cells, to 3.5 mm long; [var. *lanata* Piper; *M. lanata* (Piper) Rydb.; B.C.; Ont. to Nfld. and N.S.] ....
            .... f. *lanata* (Piper) Stewart
            4   Hairs appressed, with very short cells, at most 1.5 mm long; [Ont.; Fernald *in* Gray 1950] .... f. *puberula* Stewart
    2   Stem near the lowest flowering axils glabrous on the sides, only minutely pubescent on the angles; leaves becoming glabrate; [var. *glabra* Benth.; cent. Ont. (w James Bay at 54°12′N), Que. (E James Bay N to 53°32′N; Anticosti Is.; Kamouraska and Temiscouata counties; Gaspé Pen.; Magdalen Is.), Labrador (Goose Bay), Nfld., and N.S.] .... f. *glabra* (Benth.) Stewart

1   Leaves lanceolate to lance-oblong, broadest near or slightly above the middle, usually
    cuneate or attenuate to the petiole . . . . var. *villosa* (Benth.) Stewart
    5   Corolla white; sepals pale yellow; [var. *glabrata* f. *albiflora* Rouleau, the type from St.
        Helen's Is., Montreal, Que.] . . . . f. *albiflora* (Rouleau) Scoggan
    5   Corolla pink to light purple.
        6   Flowers in umbel-like clusters at the ends of axillary peduncles, each umbel
            subtended by a pair of bracts; [type from near Otterburne, SE Man.] . . . .
            . . . . f. *pedunculata* Löve & Bernard
        6   Flowers in nearly sessile clusters in the leaf-axils.
            7   Stem near the lowest flowering axils pubescent on the sides and angles;
                leaves pubescent.
                8   Angles of stem much more pubescent than the sides; [var. *canadensis* (L.)
                    Briq. (*M. canadensis* L.); *M. borealis* Michx.; *M. occidentalis* and *M.*
                    rubella Rydb.; transcontinental] . . . . f. *villosa*
                8   Angles and sides of stem about equally pubescent.
                    9   Hairs mostly entangled or spreading, with elongate cells, to 3.5 mm
                        long; [Alaska–B.C.] . . . . f. *lanigera* Stewart
                    9   Hairs appressed, with very short cells, at most 1.5 mm long; [Alaska,
                        B.C. (Vancouver Is.), Alta. (N to Wood Buffalo National Park at
                        59°31′N), and Sask. (L. Manitou)] . . . . f. *brevipilosa* Stewart
            7   Stem near the lowest flowering axils glabrous on the sides, only minutely
                pubescent on the angles; leaves becoming glabrate; [var. *glabrata* (Benth.)
                Fern.; *M. glabrior* (Hook.) Rydb.; *M. penardii* (Briq.) Rydb.; transcontinental] . . . .
                . . . . f. *glabrata* (Benth.) Stewart

M. cardiaca Baker
European; introd. along shores and in wet meadows and waste places in N. America, as in S B.C.
(Chilliwack and Kootenay; Herb. V), S Ont. (Point Edward, Lambton Co.; OAC), Que. (Boivin
1966*b*), Nfld. (Bay of Islands; GH), N.B. (McAdam, York Co.; Groh and Frankton 1949*b*), P.E.I.
(Queens and Prince counties; CAN; GH), and N.S. (Digby, Kings, and Lunenburg counties and
Boularderie Is., Cape Breton Is.; CAN; GH).
    This species is included in *M. gentilis* by Clapham, Tutin, and Warburg (1962), which they
consider to be a hybrid between *M. arvensis* and *M. spicata*. Boivin (1966*b*) distinguishes it as the
phenotype (outward morphological expression), nm. *cardiaca* (Baker) Boivin, of that species.

M. citrata Ehrh.   Lemon-scented Mint
European; introd. along shores and in wet meadows and waste places in N. America, as in B.C.
(Queen Charlotte Is.; Vancouver; Belrose), S Ont. (Lambton, Middlesex, and Grey counties), Que.
(Boivin 1966*b*), and N.S. (North Sydney, Cape Breton Is.; GH). [× *M. piperita* var. *cit.* (Ehrh.) Briq.;
× *M. pip.* nm. *cit.* (Ehrh.) Boivin].

M. gentilis L.   Red Mint
European; introd. into rich soil in N. America, as in Ont. (Toronto, where taken by Scott in 1897;
CAN; reported from Ottawa by Groh and Frankton 1949*b*), Que. (N to Argenteuil Co. and the
Montreal dist.; MT; CAN), P.E.I. (Charlottetown and Royalty Junction, Queens Co.; CAN; GH), and
N.S. (Shelburne and Digby counties). [*M. rubra sensu* J.R. Churchill, Rhodora 4(38):34. 1902, and
John Adams, Can. Field-Nat. 51(7):107. 1937, not Sm. (which is × *M. smithiana* Graham), the
relevant P.E.I. collection in GH; *M. ?sativa sensu* Lindsay 1878, John Macoun 1884, and Fowler
1885, perhaps not L.].
    This species is considered by Clapham, Tutin, and Warburg (1962) to be a hybrid between *M.*
arvensis and *M. spicata*. They also merge *M. sativa* L. with × *M. verticillata* L., which they
consider to be a hybrid between *M. aquatica* and *M. arvensis*. The N.S. plant (Digby, in CAN;
Harper L., Shelburne Co., in GH and CAN) is referable to f. *variegata* Moldenke (leaves strongly
variegated, blotched with white bands along the veins).

M. longifolia (L.) Huds.   Horse-Mint
Eurasian; introd. in thickets and along roadsides and damp shores in N. America, as in sw B.C. (reported by Eastham 1947, as forming large beds in fields and along railway tracks at Pemberton, about 70 mi NE of Vancouver), Ont.–Que. (Boivin 1966b), and N.S. (Tannery Pond, Wolfville, Kings Co.; ACAD and NSPM, distributed as *M. aquatica,* rev. D. Erskine). [*M. spicata* var. *long.* L.; *M. sylvestris* L.; incl. var. *mollissima* (Borkh.) Rouy].

M. piperita L.   Peppermint.   Menthe poivrée
European; introd. into wet meadows and along streams in N. America, as in SE Alaska (Hultén 1949), sw B.C. (Vancouver Is. and adjacent mainland at Sardis), Ont. (N to near Kenora; TRT), Que. (N to Ste-Anne-de-la-Pocatière, Kamouraska Co.; QSA), St-Pierre and Miquelon (Boivin 1966b), ?N.B. (John Macoun 1884; Fowler 1885), P.E.I. (Bothwell, Kings Co.; GH), and N.S.
This species is considered by Clapham, Tutin and Warburg (1962) to be a hybrid between *M. aquatica* and *M. spicata.*

M. rotundifolia (L.) Huds.   Apple-scented Mint
European; introd. along roadsides and in fields and waste places in N. America, as in s Ont. (near London, Middlesex Co., where taken by J. Dearness in 1901; CAN; reported from Norfolk Co. by Landon 1960, and from Walkerton, Bruce Co., by Groh and Frankton 1949b). [*M. spicata* var. *rot.* L.].

M. spicata L.   Spearmint.   Baume
Eurasian; introd. along roadsides and in meadows and waste places in N. America, as in cent. Alaska (Tanana Hot Springs; Hultén 1949), B.C. (Vancouver Is.; Chilliwack; Keremeos; Revelstoke), Sask. (Bjorkdale, 52°43′N), s Man. (Winnipeg), Ont. (N to the Ottawa dist.), Que. (N to the Gaspé Pen.; the report of *M. viridis* from Nfld. by Reeks 1873, may refer here or to *M. cardiaca*), N.B. (Restigouche; NBM), P.E.I. (Charlottetown; GH), and N.S. [*M. viridis* of auth., not L.]. MAP: Hultén 1968b:791.

MOLDAVICA Adans.   [7250]   Dragonhead

1   Whorls of flowers crowded in a dense terminal head, the bracts spinose-tipped; calyx about 1 cm long, the tube and spinose-tipped lobes subequal; (transcontinental) . . . . *M. parviflora*
1   Whorls of flowers in loose, elongate, interrupted racemes, the bracts entire or merely ciliate; calyx to about 8 mm long, the lobes distinctly shorter than the tube; (introd.) . . . . *M. thymiflora*

M. parviflora (Nutt.) Britt.
/ST/X/   (Hs)   Rocky or gravelly places (often in disturbed areas and other waste places) from N-cent. Alaska (N to near the Arctic Circle), the Yukon (N to ca. 64°30′N), and w Dist. Mackenzie (N to Norman Wells, ca. 65°N) to B.C.–Alta., Sask. (N to Windrum L. at ca. 56°N; CAN), Man. (N to Churchill), Ont. (N to the Fawn R. at ca. 54°N, 90°W; CAN), and Que. (Timiskaming; Montebello, Papineau Co.; St. Siméon and Ile-aux-Coudres, Charlevoix Co.; Grosse-Ile, Montmagny Co.; Bic, Rimouski Co.; Roberval, L. St. John), s to Oreg., Ariz., N.Mex., Nebr., Mo., and w N.Y.; introd. elsewhere (and probably in some of the above stations), as in N.S. (gravelly roadside 10 mi E of Amherst, Cumberland Co.; CAN). [*Dracocephalum* Nutt., not *Physostegia parv.* Nutt., which is *D. nuttallii* Britt.]. MAP: Hultén 1968b:787 (*D. parv.*).

M. thymiflora (L.) Rydb.
Eurasian; introd. into open woods, fields, and waste places in N. America, as in the Yukon (Boivin 1966b), s B.C. (Tulameen Valley, E of Hope; CAN), Alta. (N to Fort Vermilion, 58°24′N; Groh and Frankton 1948), Sask. (Lebret and Katepwa, both NE of Regina), s Man. (Brandon; Forrest; Portage la Prairie), Ont. (N to the Ottawa dist.; Gillett 1958), and Que. (Lionel Cinq-Mars et al., Nat. can. (Que.) 98(2):196. 1971; field at St-Jérôme, Lac St-Jean Co.). [*Dracocephalum* L.].

MONARDA L.   [7296]   Monarda, Horsemint

(Ref.: McClintock and Epling 1942)
1   Flower-clusters normally borne in the upper 2–4 axils as well as at the tip of the stem;
    stamens not surpassing the strongly arched upper corolla-lip; expanded part (throat) of
    the corolla-tube shorter than the cylindrical part; principal leaves lanceolate; stems closely
    retrorse-pubescent.
    2   Calyx densely villous in the throat, its narrowly triangular teeth about 1.5 mm long;
        corolla cream-colour to yellowish with purple spots; leaves closely greyish-pilose to
        -tomentulose beneath, those subtending the flower-clusters pointed but not awn-
        tipped; perennial with tough crowns; (introd.) . . . . *M. punctata*
    2   Calyx densely hirsute in the throat, its setaceous awn-like teeth to 7 or 8 mm long;
        corolla white or pink, dotted with purple; principal leaves thinly and evenly puberulent
        on both surfaces, glabrate, those subtending the flower-clusters oblong, densely
        pubescent, ending in long slender awns; annual without a strong crown-base; (?B.C.) . . . . . .
        . . . . [*M. citriodora*]
1   Flower-clusters normally solitary and terminal (or sometimes also in the upper 1 or 2
    axils); stamens surpassing the straightish upper corolla-lip; expanded part of corolla-tube
    as long as or longer than the cylindrical part; leaves ovate-lanceolate to ovate.
    3   Corolla dull white, yellowish white, or flesh-colour, dark-spotted, at most 3 cm long,
        the upper lip glabrous or nearly so; calyx usually heavily long-bearded in the throat;
        bracteal leaves green or whitish-tinged; stems glabrous or sparingly pubescent;
        (introd.) . . . . [*M. clinopodia*]
    3   Corolla lilac, roseate, purplish, or crimson; bracteal leaves pinkish to red or purple.
        4   Corolla to 4.5 cm long, vermilion to nearly scarlet, glabrous or nearly so, the upper
            lip about half as long as the tube; calyx glabrous or nearly so; bracteal leaves red
            (at least at base), eciliate or short-ciliate; stem glabrous or sparingly pilose
            especially at the nodes; (introd.) . . . . *M. didyma*
        4   Corolla 2 or 3 cm long, incurved-pubescent, the upper lip much shorter than the
            tube.
            5   Corolla roseate to reddish purple, the upper lip thinly pubescent at apex;
                calyx glabrate to villous in the throat; bracteal leaves purplish red to dark
                purple, stiffly pectinate-ciliate; stem glabrous or sparingly pubescent;
                (s ?Ont.) . . . . [*M. media*]
            5   Corolla lilac or pink, the upper lip densely villous at apex; calyx densely
                hirsute-bearded in the throat; bracteal leaves green or pink-tinged; stem
                usually more or less pubescent above; (B.C. to sw Que., probably partly
                introd.) . . . . *M. fistulosa*

[M. citriodora Cerv.]   Lemon-Mint
[The report of this species of the cent. U.S.A. (N to Kans. and Mo.; adv. eastwards to Mich., Tenn.,
and Ga.; *see* MAP by McClintock and Epling 1942: fig. 16, p. 188) from SE B.C. by Ulke (1935;
Wilmer, near Windermere) requires confirmation. If correctly identified, it was probably introd. at
that locality. (*M. dispersa* Small).]

[M. clinopodia L.]
[Native in the E U.S.A. (N to Ill. and N.Y.; *see* the MAP by McClintock and Epling 1942: fig. 16,
p. 188) and a garden-escape in New Eng. and s Ont. (Montgomery 1945; Huron Co., where
probably not established).]

M. didyma L.   Oswego-tea, Bee-balm
Native in the E U.S.A. (N to Mich. and N.Y.; *see* the MAP by McClintock and Epling 1942: fig. 12, p.
175) and a garden-escape elsewhere, as in s Ont. (N to Grey and York counties) and sw Que. (N to
near Shawinigan, St. Maurice Co.; MT).

M. fistulosa L.   Wild Bergamot
/sT/X/   (Hpr)   Dry thickets, clearings, and borders of woods (ranges of Canadian taxa outlined

below; the report from Dist. Mackenzie by Fernald *in* Gray 1950, requires confirmation), s to N Mexico, N.Mex., Ariz., Tex., La., Ala., and Ga. MAPS and synonymy: *see* below.

1 Leaves with petioles to 1.5 cm long; stem often branched above.
    2 Stem-pubescence consisting of spreading hairs; leaves spreading-villous on the nerves beneath .... var. *fistulosa*
        3 Corolla lilac or pink; [*M. rugosa* Ait.; reported from sw Que. by Fernald *in* Gray 1950; MAP: McClintock and Epling 1942: fig. 8 (triangles: "spreading pili only"), p. 166] .... f. *fistulosa*
        3 Corolla white; [s Ont.: Elgin, Lambton, Norfolk, and Stormont counties] ....
        .... f. *albescens* Farw.
    2 Stem-pubescence consisting of decurved hairs; leaves minutely pubescent, short-pilose, or glabrate beneath; [incl. vars. *longipetiolata* and *maheuxii* Boivin; *M. mollis* L.; *Origanum fistulosum* ?*canadense* Cornuti; Sask. (Fernald *in* Gray 1950), s Man. (N to Brandon and Winnipeg), s Ont. (N to Wellington Co.), and sw Que. (Boivin 1966*b*); MAP: on the above-noted map with *M. fistulosa* (but solid dots, "short curled pubescence"), *M. mollis* being cited in the synonymy of *M. fistulosa*] ....
    .... var. *mollis* (L.) Benth.
1 Leaves subsessile or with petioles rarely over 5 mm long; stem decurved-pubescent, usually simple .... var. *menthaefolia* (Graham) Fern.
    4 Corolla lilac or pink; [*M. menthaefolia* Graham, the type from Norway House, off the NE end of L. Winnipeg, Man.; B.C. (N to Bear Flats, ca. 56°N), Alta. (N to Wood Buffalo National Park at 59°07′N), Sask. (N to McKague, 52°37′N), and Man.; introd. along a railway embankment near Peninsula, N shore of L. Superior, Ont.; MAP: McClintock and Epling 1942: fig. 9 (*M. menth.*; incomplete northwards), p. 170] .... f. *menthaefolia*
    4 Corolla white; [Alta. (Waterton; Lacombe) and sw Man. (type from Waldheim)] ....
    .... f. *russellii* Boivin

[M. media Willd.]
[The reports of this species of the E U.S.A. (Ind. to N.Y., s to Tenn. and N.C.; a garden-escape eastwards to New Eng.; MAP: McClintock and Epling 1942: fig. 11, p. 174) from s Ont. by J.M. Macoun (1897; Wingham, Huron Co., as *M. fist.* var. *rubra*) and Dodge (1915; Kettle Point, Lambton Co.) require confirmation, as does the identity of a collection in MT from near Kingston, Frontenac Co. (*M. fistulosa* var. *rubra* Gray).]

M. punctata L.   Dotted Monarda, Horsemint
Native in the U.S.A. (N to Kans., Minn., and Vt.; introd. elsewhere, as in s Ont. (collections in OAC from along a railway at Point Edward, Lambton Co.; in TRT from a dry field near Cornwall, Stormont Co.), these referable to var. *villicaulis* Pennell (stem densely villous with relatively long spreading hairs rather than with short recurved hairs; leaves densely hairy beneath, the hairs concealing the glands, rather than minutely pubescent or glabrous beneath, the glands not concealed). MAP: McClintock and Epling 1942: fig. 14, p. 181.

[MONARDELLA Benth.]   [7316]

[M. odoratissima Benth.]
[A species of open, wet or dry, often rocky places at low to moderate elevations in the w U.S.A. (Wash. and Idaho to s Calif. and N.Mex.), apparently not yet known from Canada but to be searched for in s B.C.–Alta. (*Madronella* Greene).]

NEPETA L.   [7247]   Catmint

1 Stem and lower leaf-surfaces rather copiously white-downy; principal leaves ovate to ovate-oblong, coarsely dentate; calyx very pubescent; (introd., transcontinental) ..... *N. cataria*
1 Plant essentially glabrous; principal leaves oblong, finely crenate; (introd. in s Que.) ....
    .... *N. grandiflora*

N. cataria L.　Catnip.　Herbe à chats or Chataire
Eurasian; introd. along roadsides and in dooryards and waste places in N. America, as in SE Alaska (Sitka; Hultén 1949), S B.C. (Vancouver Is.; Kootenay L.), Alta. (N to the Peace River dist.; Raup 1934), S Sask. (Carnduff and Gainsborough; Breitung 1957a), S Man. (Coulter; Morden; Winnipeg), Ont. (N to the N shore of L. Superior and L. Timiskaming at ca. 47°30′N), Que. (N to the Gaspé Pen.), Nfld., N.B., P.E.I., and N.S.

N. grandiflora Bieb.
Eurasian; locally introd. along roadsides and in waste places in N. America, as in SW Que. (summit of Mt. Royal, Montreal, where taken by Bissell in 1914; GH).

ORIGANUM L.　[7315]

O. vulgare L.　Marjoram.　Origan
Eurasian; a garden-escape to roadsides, old fields, and open woods in N. America, as in SW B.C. (Henry 1915; Elgin, E of Vancouver), Ont. (N to the Ottawa dist.), SW Que. (N to the Montreal dist.), P.E.I. (New Glasgow and Rusticoville, Queens Co.; D.S. Erskine 1960), and N.S. (Hants and Pictou counties; ACAD; CAN).

PRUNELLA L.　[7254]　Selfheal

P. vulgaris L.　Heal-all, Carpenter-weed.　Herbe au charpentier
/ST/X/EA/　(Hsr)　Grasslands, fields, roadsides, and waste places (partly introd.), the aggregate species from the E Aleutian Is. and Alaska (N to ca. 65°N) to B.C.–Alta., Man. (N to Grand Rapids, near the NW end of L. Winnipeg; not known from Sask.), Ont. (N to the Fawn R. at ca. 54°40′N), Que. (N to the E James Bay watershed at 53°19′N, Anticosti Is., and the Gaspé Pen.), S Labrador (N to the Hamilton R. basin), Nfld., N.B., P.E.I., and N.S., S to Calif., N.Mex., Kans., and N.C.; Eurasia. MAP and synonymy: *see* below.
1　Principal stem-leaves ovate or ovate-oblong (averaging 1/2 as broad as long), broadly
　　cuneate or rounded at base; [Eurasian].
　　2　Plant sparingly hairy . . . . var. *vulgaris*
　　　　3　Corolla lavender, violet, or bluish; [*Brunella* L.; introd., transcontinental] . . . . . f. *vulgaris*
　　　　3　Corolla white; [introd. at Brookline, Mass.; not yet known from Canada but to be
　　　　　　searched for] . . . . [f. *albiflora* (Bogenh.) Britt.]
　　2　Plant densely white-villous-hispid on the stem, petioles, and often the lower leaf-
　　　　surfaces; [introd. in the E U.S.A.; not yet known from Canada] . . . . [var. *hispida* Benth.]
1　Principal stem-leaves lanceolate to ovate-oblong (averaging 1/3 as broad as long), more
　　narrowly cuneate to an acutish base; [apparently native].
　　4　Floral bracts with glabrous or sparingly short-ciliate margins.
　　　　5　Leaves and stem nearly or quite glabrous; bracts green or at most purple-tinged;
　　　　　　corolla violet; [B.C., the type from Vancouver Is.; also known from Howser Station
　　　　　　in the Selkirk Mts.] . . . . var. *calvescens* Fern.
　　　　5　Leaves pilose beneath; stem pilose; bracts mostly deep purple; corolla dark purple
　　　　　　or blackish purple; [known only from Calif.] . . . . [var. *atropurpurea* Fern.]
　　4　Floral bracts with margins copiously bristly-ciliate with long white hairs.
　　　　6　Bracts densely tomentose or lanate on the back, they and the calyces dark purple;
　　　　　　[Aleutian Is., the type from Unalaska] . . . . var. *aleutica* Fern.
　　　　6　Bracts glabrous or sparingly pilose on the back.
　　　　　　7　Ribs and margins of the calyx-lobes glabrous or very short-ciliate; [known only
　　　　　　　　from fresh tidal shores of the St. Lawrence R. estuary in Lotbinière, Lévis, and
　　　　　　　　Québec counties, Que., the type from near Lauzon, Lévis Co.] . . . .
　　　　　　　　. . . . . var. *rouleauiana* Vict.
　　　　　　7　Ribs and margins of calyx-lobes long-bristly-ciliate . . . . . var. *lanceolata* (Bart.) Fern.
　　　　　　　　8　Calyx purple.
　　　　　　　　　　9　Corolla pink; [S Ont. and SW Que.] . . . . f. *rhodantha* Fern.
　　　　　　　　　　9　Corolla lavender, violet, or bluish; [transcontinental; type from Port-à-
　　　　　　　　　　　　Port, Nfld.] . . . . f. *iodocalyx* Fern.

8 Calyx green or at most with purple-tinged margins.
  10 Corolla white; [B.C. (Vancouver Is.), E Que. (Gaspé Pen.), N.B.
    (Bathurst), and N.S.] .... f. *candida* Fern.
  10 Corolla lavender, violet, or bluish; [transcontinental; MAP: Hultén
    1968*b*:787] .... f. *lanceolata*

PYCNANTHEMUM Michx. [7317] Mountain-mint, Basil

(Ref.: Grant and Epling 1943)
1 Calyx 2-lipped, the 2 teeth of the lower lip to 1.5 mm long, distinctly longer than the 3
teeth of the upper lip, the teeth often with long flexuous bristle-like hairs near the apex;
heads loosely flowered, to 3.5 cm broad; leaves ovate-lanceolate to -oblong, whitened
beneath with a fine pubescence, remotely serrate, to 5.5 cm broad, on petioles to 1.5 cm
long; stem minutely incurved-pubescent, the upper internodes also with long straight
hairs; (S Ont.) .... *P. incanum*
1 Calyx regular, its 5 teeth subequal and lacking bristle-like hairs; heads densely flowered,
not over 2 cm broad; petioles at most 3 mm long.
  2 Leaves ovate-lanceolate to ovate, to 4 cm broad, subsessile or short-petioled,
    shallowly serrate, nearly glabrous; stem pubescent at least on the upper internodes;
    calyx minutely but densely pubescent, its teeth less than 1 mm long; heads to 1.5 cm
    broad; bracts whitened with minute pubescence .... [*P. muticum*]
  2 Leaves linear to lanceolate, not over 2 cm broad.
    3 Sides of stem glabrous; leaves entire, sessile; heads less than 1 cm broad;
      outermost bracts essentially glabrous above; (Ont. and SW Que.).
      4 Calyx-teeth 1 or 2 mm long, with firm sharp tips; leaves mostly less than 5 mm
        broad, with only 1 or 2 pairs of lateral veins, essentially glabrous .... *P. tenuifolium*
      4 Calyx-teeth less than 1 mm long, scarcely sharp-tipped; leaves to about 1 cm
        broad, with 3 or 4 pairs of lateral veins, glabrous or minutely pubescent
        beneath .... *P. virginianum*
    3 Sides of upper half of stem pubescent; leaves lanceolate, entire or obscurely
      toothed, short-petioled, to 2 cm broad and with up to 7 pairs of lateral veins;
      calyx-teeth barely 1 mm long; heads to 2 cm broad; outermost bracts densely
      velvety above.
      5 Upper internodes of stem minutely pilose on the sides and angles; leaves
        acute, glabrous or short-pilose on the larger veins beneath; inner bracts thinly
        pubescent, densely short-ciliate, the midvein conspicuous; (S Ont. and SW
        Que.) .... *P. verticillatum*
      5 Upper internodes densely hoary-villous on the sides and angles; leaves
        obtuse or subacute, finely and densely pubescent beneath; inner bracts
        densely canescent, the midvein concealed; (S Ont.) .... *P. pilosum*

P. incanum (L.) Michx.
/t/EE/ (Hpr) Dry woods and thickets from Ill. to S Ont. (near Hamilton, Wentworth Co., where
taken by Dickson in 1895 and 1897 and by Scott in 1900; CAN; TRT) and N.Y., S to Ala. and Fla.
[*Clinopodium* L.; *Origanum* Walt.; *P. muticum sensu* J.M. Macoun 1896, not (Michx.) Pers.]. MAPS:
Grant and Epling 1943: map 1, p. 199, and map 4, p. 207.

[*P. muticum* (Michx.) Pers.]
[The report of this species of the E U.S.A. (N to Ill., Mich., and Maine) from S Ont. by J.M. Macoun
(1896; near Hamilton) is based upon *P. incanum*, the above-noted Dickson collections in CAN.
[*Brachystemum* Michx.]. MAP: Grant and Epling 1943: map 12, p. 228.]

P. pilosum Nutt.
/t/EE/ (Hpr) Dry to moist woods, thickets, and clearings from Iowa to Mich. and S Ont. (Essex,
Lambton, and Kent counties; CAN; TRT), S to Okla., Ark., and Tenn.; introd. elsewhere, as from
Mass. to Pa. and in E Que. (Percé, Gaspé Co., where taken by F.E. Boys in 1955). MAPS: Grant and
Epling 1943: map 2, p. 201, and map 7, p. 217.

P. tenuifolium Schrad.

/t/EE/ (Hpr) Dry to moist or boggy ground and woods from Minn. to Mich., Ohio, s Ont. (apparently native in Essex, Lambton, Grey, and Welland counties; apparently introd. in waste ground of a churchyard near Prescott, Grenville Co.; CAN; TRT), sw ?Que. (the report by Boivin 1966b, may be based upon introd. plants), and New Eng., s to Tex. and Ga. [*P. flexuosum* (*P. linifolium* Pursh) of Canadian reports, not *Origanum flex.* Walt., basionym]. MAPS: Grant and Epling 1943: map 2, p. 201, and map 11, p. 225.

P. verticillatum (Michx.) Pers.

/T/EE/ (Hpr) Dry to moist meadows, thickets, and clearings from Mich. to s Ont. (Elgin and Middlesex counties; OAC; TRT) and sw Que. (Oka, Deux-Montagnes Co., and Rouville, Rougemont Co.; TRT), s to N.C. [*Brachystemum* Michx.; *P. virginianum* var. *vert.* (Michx.) Boivin]. The MAP by Grant and Epling (1943: map 9, p. 220) indicates no Canadian stations.

P. virginianum (L.) Durand & Jackson

/T/EE/ (Hpr) Dry to wet thickets, gravelly shores, and meadows from N.Dak. to Ont. (N to the Ottawa dist.; Gillett 1958) and sw Que. (N to the shores of the Ottawa R. in Pontiac Co. and of the St. Lawrence R. around Montreal), s to Kans., Mo., and N.C. [*Satureja* L.; *P. lanceolatum* Pursh; *Thymus virginicus* Pursh]. MAP: Grant and Epling 1943: map 10, p. 222.

SALVIA L. [7290] Sage

1 Calyx 3-lobed, the upper lip entire; corolla at most about 1 cm long, only slightly surpassing the calyx; flowers 1–3 at each node; leaves linear-lanceolate to lanceolate, entire or with a few low teeth, to about 5 cm long; bracteal leaves linear-lanceolate; annual with minutely recurved-pubescent stem; (introd.) .... *S. reflexa*

1 Calyx 5-lobed, the upper lip 3-toothed (at least minutely so); flowers usually more numerous; leaves usually larger; bracteal leaves ovate; perennials; (introd.).
  2 Flowers 12 or more at each node, to 1.5 cm long; calyx about 6 mm long; leaves broadly ovate to deltoid, coarsely and irregularly toothed, often lyrate at base, minutely hirsute on both sides; stem hispid .... *S. verticillata*
  2 Flowers usually less than 8 at each node.
    3 Corolla at most 12 mm long, only slightly surpassing the calyx; middle tooth of upper calyx-lip minute; flowers up to 4 at each node; leaves ovate-lanceolate, crenate, rugose, canescent beneath, finally glabrous above; stem softly villous ....
    .... *S. sylvestris*
    3 Corolla to over 2 cm long, 2 or 3 times as long as the calyx; flowers up to 8 at each node.
      4 Leaves chiefly basal; upper calyx-lip shorter than the lower one.
        5 Teeth of the upper calyx-lip minute and close together; corolla blue, to 2 cm long, the upper lip arched into a half-circle and laterally flattened, longer than the tube; lower anther-locule wanting or rudimentary; leaves irregularly serrate or crenate, acutish, blistery-rugose above, pubescent along the nerves beneath, to 12 cm long; (introd. in s Ont.) ...... [*S. pratensis*]
        5 Teeth of the upper calyx-lip conspicuous and widely separated; corolla violet, to 3 cm long, its straight upper lip much shorter than the broad lower one, this shorter than the tube; both anther-locules polleniferous; basal leaves commonly deeply lyrate-pinnatifid into rounded segments, rounded at summit, glabrous or sparingly strigose-hirsute, to 2 dm long .... [*S. lyrata*]
      4 Leaves chiefly cauline; upper calyx-lip about equalling the lower one, the calyx-teeth all short-awned.
        6 Bracteal leaves ovate-lanceolate, not much surpassing the calyces; racemes interrupted; lower corolla-lip longer than the upper one; leaves lanceolate to narrowly elliptic, finely crenate, tapering to base, canescent on both sides; (introd. in s Ont. and sw Que) .... [*S. officinalis*]
        6 Bracteal leaves round-ovate, often coloured; racemes more compact,

usually branching to form a panicle; lower corolla-lip shorter than the upper one; leaves ovate, coarsely and irregularly toothed, truncate to subcordate at base, pubescent on both sides; (introd. in S Ont.) . . . . *[S. sclarea]*

[S. lyrata L.]
[This species of the U.S.A. (Colo. to Mo., Ill., Pa., and Conn., S to Tex. and Fla.) is not yet known from Canada but should be searched for.]

[S. officinalis L.]   Common Sage
[Eurasian; an occasional garden-escape to dumps and waste places in N. America but scarcely established, as in S Ont. (Cartwright, Ontario Co., where taken by Scott in 1890; TRT) and SW Que. (Boivin 1966*b*).]

[S. pratensis L.]
[European; a garden-escape to fields, pastures, and waste places in N. America, as in S Ont. (fields near Conestoga, Waterloo Co., where taken by Stone in 1936 but probably not established; OAC; F.H. Montgomery, Can. Field-Nat. 62(2):88. 1948).]

S. reflexa Hornem.
Native in dry open soils of the U.S.A. (N to Mont. and Wisc.); introd. elsewhere, as in S Sask. (Moosomin, about 130 mi E of Regina; Breitung 1957*a*), S Man. (along a railway at Winnipeg; DAO), S Ont. (Peterborough, Prince Edward, Hastings, Wellington, and Grenville counties; CAN; MT; TRT), and SW Que. (Montebello, Papineau Co.).

[S. sclarea L.]
[European; a local garden-escape in N. America, as in S Ont. (Grey Co.; OAC; reported from Wellington Co. by Stroud 1941), where scarcely established.]

S. sylvestris L.
Eurasian; introd. along roadsides and in waste places in N. America, as in SW B.C. (collection in CAN from Nanaimo, Vancouver Is., where taken on ballast heaps by John Macoun in 1893 and distributed as *Hyptis verticillata* Jacq.; collection in V from Oliver, about 25 mi S of Penticton), Alta. (Moss 1959), S Man. (Ninette, about 30 mi SE of Brandon; CAN), and S Ont. (N to Frontenac Co. near the Renfrew Co. boundary). [*S. nemorosa* L.].

S. verticillata L.   Lilac Sage
Eurasian; locally introd. along roadsides and in fields and waste places in N. America, as in S Ont. (near Flesherton, Grey Co., where established for some time and spreading, according to Montgomery 1957; Harttington, Frontenac Co.; OAC).

SATUREJA L.   [7305]   Savory, Calamint.   Sariette

1   Leaves of flowering-stems linear to linear-lanceolate or -oblanceolate, entire or nearly
so, 1 or 2 cm long; calyx nearly regular; stems to about 4 dm tall.
  2   Stem essentially glabrous (except near the nodes); calyx bearded in the throat,
otherwise glabrous, its teeth about half as long as the tube; corolla at least 8 mm long;
stoloniferous perennial; (Ont.) . . . . *S. glabella*
  2   Stem finely pubescent; calyx naked in the throat, its ciliate teeth about equalling the
tube; corolla at most 7 mm long; annual; (introd.) . . . . *S. hortensis*
1   Leaves ovate-lanceolate to subrotund; stems more or less pubescent.
  3   Calyx about 5 mm long, with short subequal broadly deltoid teeth; corolla white or
purple-tinged, to 1 cm long, minutely pubescent outside; flowers paired at the nodes
(solitary in the axils) on pedicels to 1.5 cm long; leaves ovate to subrotund, to 3.5 cm
long, generally with a few blunt teeth; stems from a woody rhizome, prostrate and
freely rooting, often with short ascending branches; (B.C.) . . . . *S. douglasii*
  3   Calyx-teeth unequal, the calyx thus more or less 2-lipped; leaves ovate-lanceolate to
deltoid-ovate; stems erect or decumbent-based.

4 Flowers in dense terminal clusters (often, also, in the uppermost axils), subtended by elongate linear-subulate long-ciliate bracts; calyx hirsute throughout, scarcely bearded in the throat, about 1 cm long, its subequal teeth all subulate, its lower lip cleft to base; stems to about 6 dm tall; (B.C.; Ont. to Nfld. and N.S.) . . . . *S. vulgaris*

4 Flowers few to several in the axils of many or most of the foliage-leaves, subtended by minute bracts; calyx bearded in the throat, the 2 teeth of its lower lip subulate and much longer than the 3 triangular teeth of the upper lip; (introd.).

    5 Cymes sessile and overtopped by the subtending bracteal leaves, rarely with more than 3 flowers; calyx 5 or 6 mm long, bristly-ciliate on the nerves, gibbous at base; corolla at most 1 cm long; leaves elliptic to oblong, to about 12 mm long, entire or obscurely and remotely serrulate, scabrous; stem decumbent at base, nonstoloniferous, with reflexed soft pubescence, to about 4 dm tall . . . . *S. acinos*

    5 Cymes peduncled and surpassing the subtending bracteal leaves, commonly with 5 or more flowers; calyx scarcely gibbous at base; stem stoloniferous, to about 1 m tall . . . . [*S. calamintha*]

S. acinos (L.) Scheele Mother-of-Thyme
Eurasian; introd. along roadsides and in old fields and waste places in N. America, as in B.C. (Boivin 1966*b*), Ont. (N to Manitoulin Is., N L. Huron, and the Ottawa dist.), Que. (N to Cap-à-l'Aigle, Charlevoix Co.; Groh and Frankton 1949*b*), and P.E.I. (Prince Co.; ACAD). [*Thymus* L.; *Calamintha* Clairv.; *Acinos arvensis* (Lam.) Dandy; *A. thymoides* Moench].

[**S. calamintha** (L.) Scheele]
[Eurasian; introd. along roadsides and in waste places in N. America and reported from SW Que. by R. Campbell (Can. Rec. Sci. 6(6):342–51. 1895; Mt. Royal, Montreal), where scarcely established, if correctly identified. (Incl. vars. *nepeta* (L.) Briq. (*Calamintha nep.* (L.) Scheele) and *sylvatica* Briq.).]

S. douglasii (Benth.) Briq.
/t/W/ (Ch) In coniferous woods from S B.C. (Vancouver Is. and adjacent islands and mainland N to Salmon Arm, about 45 mi E of Kamloops; V; Hultén 1968*b*, notes a report from Juneau, SE Alaska, where probably introd. if correctly identified) and Idaho to S Calif. [*Thymus* Benth.; *Micromeria* Benth.; *T. (M.) chamissonis* Benth.]. MAP: Hultén 1968*b*:790.

S. glabella (Michx.) Briq.
/T/EE/ (Hpr) Damp calcareous cliffs, gravels, and silts from Minn. to Ont. (N to Cobden, Renfrew Co.; OAC) and N.Y., S to Tex., Ark., Ill., and Ohio. [*Cunila* Michx.; *Calamintha* Benth.].

Our material is referable to var. *angustifolia* (Torr.) Svenson (*Calamintha (S.) arkansana* Nutt.; *C. nuttallii* Benth.; *S. (Clinopodium) glabra* of auth., not *Hedeoma glabra* Nutt., basionym; differing from the typical form in having creeping leafy stolons, beardless or nearly beardless stem-nodes, relatively narrow leaves, and smaller flowers). Its f. *albiflora* Boivin (flowers white rather than blue-purple) is known from S Ont., the type locality.

S. hortensis L. Summer-Savory
European; a garden-escape in N. America, as in S Ont. (N to Wellington, York, Peel, and Hastings counties), Que. (N to Ste-Anne-de-la-Pocatière, Kamouraska Co.; DAO), N.B. (Kouchibouguac, Kent Co., and near St. Andrews, Charlotte Co.; CAN; NBM), and St-Pierre and Miquelon–Nfld. (Rouleau 1956).

S. vulgaris (L.) Fritsch Basil, Dogmint
/T/(X)/EA/ (Hpr) Woods, thickets, shores, and waste places (ranges of Canadian taxa outlined below, var. *neogaea* perhaps native in N. America), S to Colo., ?N.Mex., Minn., Ind., Tenn., and N.C.; Eurasia. MAP and synonymy: *see* below.

1 Leaves essentially glabrous except for long hairs on the nerves beneath; corolla whitish to pinkish; [S B.C. (New Westminster and Aldershot, where perhaps introd.); ?Man. (the

report from Norway House by John Macoun 1884, requires confirmation); perhaps both native and introd. from Ont. (N to near Thunder Bay) to Que. (type from Percé, Gaspé Pen.), Nfld., and N.S.; MAP: Hultén 1958: map 56, p. 75] . . . . var. *neogaea* Fern.
1   Leaves densely villous or almost velvety beneath, copiously strigose-villous above; corolla purple-red, to 1.5 cm long.
  2   Cymes longer than or only slightly surpassed by their subtending bracts; [*Clinopodium vulgare* var. *dim.* Simon; reported by Montgomery 1957, as introd. in a dense stand near Harriston, Wellington Co., s Ont.] . . . . var. *diminuta* (Simon) Fern. & Wieg.
  2   Cymes (except sometimes the upper ones) commonly much surpassed by their subtending bracts; [*Clinopodium* L.; *Calamintha clinopodium* Benth.; Eurasia only; MAP: on the above-noted map by Hultén] . . . . [var. *vulgaris*]

SCUTELLARIA L.   [7234]   Skullcap

1   Flowers in axillary racemes.
  2   Corolla to 2 cm long, curved upward from the calyx; leaf-petioles at most 1 cm long; lower bracteal leaves not much reduced; internodes of racemes to 2.5 cm long; (Ont. to N.B. and N.S.) . . . . *S. churchilliana*
  2   Corolla less than 1 cm long, nearly straight; leaf-petioles to 3 cm long; bracteal leaves much smaller than the foliage-leaves; internodes of racemes mostly less than 1 cm long; (transcontinental) . . . . *S. lateriflora*
1   Flowers solitary in the axils of ordinary foliage-leaves; leaf-petioles at most 4 mm long.
  3   Corolla at most 1 cm long; calyx to 3.5 mm long; leaves mostly ovate, rarely over twice as long as broad, subsessile or very short-petioled.
    4   Calyx copiously stipitate-glandular or glandular-viscid; leaves usually not over 1.5 cm long, entire or sparingly shallow-crenate, distinctly hirsute on the whole upper surface; stem to 3 dm tall, the sides more or less stipitate-glandular or becoming glabrate, the angles also with minute retrorse eglandular hairs; rhizomes producing a chain of elongate whitish tubers to 2 cm long; (Ont. and Que.) . . . . *S. parvula*
    4   Calyx minutely hirsute on the veins, otherwise glabrous; leaves to 5 cm long, very thin, remotely crenate-dentate, essentially glabrous above; stem to 5 dm tall, glabrous except at base; rhizomes filiform, not tuberiferous; (s Ont.) . . . . *S. nervosa*
  3   Corolla usually over 1 cm long; rhizomes or stolons filiform or slightly thickened but not tuberiferous.
    5   Stems to 1 m tall, minutely pilose especially along the angles above with descending hairs; leaves to 8 cm long and 3 cm broad, sessile or on thick petioles at most 4 mm long, incurved-puberulent beneath; corolla to 2.5 cm long, the palate papillate, not hairy; (transcontinental) . . . . *S. epilobiifolia*
    5   Stems minutely pilose or strigose especially along the angles above with incurved-ascending hairs.
      6   Corolla commonly less than 1.5 cm long; leaves to 8 cm long and 3.5 cm broad, narrowly to broadly ovate, acuminate, rounded or subcordate at base, coarsely toothed, minutely and sparsely pilose on the veins beneath, pinnately veined, on slender petioles to 9 mm long; (Ont. to N.B. and N.S.) . . . . *S. churchilliana*
      6   Corolla commonly 2 or 3 cm long, the palate with at least a few long flattened white hairs; leaves commonly about 2 or 3 cm long and 1 cm broad, lance-elliptic to oblong or nearly ovate, obtuse or rounded at apex, tapering or somewhat rounded at base, entire or the lower ones often obscurely toothed, minutely but densely puberulent on both surfaces, often 3–5-nerved from near the base, subsessile or short-petioled; (s B.C.) . . . . *S. angustifolia*

S. angustifolia Pursh
/t/W/   (Gst (Hpr))   Moist or dry meadows and rocky places from s B.C. (near Victoria, Vancouver

Is., where taken by James Fletcher in 1885; Pend-d'Oreille R., near the U.S.A. boundary s of Trail, where taken by J.M. Macoun in 1902; CAN) to s Calif. [*S. veronicifolia* Rydb.]. MAP: Epling 1942: map 10, p. 37.

S. churchilliana Fern.
/T/EE/ (Hpr) Sandy, gravelly, or alluvial shores and thickets from s Ont. (Lambton Co.; Gaiser and Moore 1966) to Que. (Rouville, Laprairie, Kamouraska, and Rimouski counties), w N.B. (St. John R. system; not known from P.E.I.), and N.S. (Lunenburg Co.; E.C. Smith and J.S. Erskine, Rhodora 56(671):250. 1954), s to s Maine. MAP: Epling 1942: map 2, p. 5.

S. epilobiifolia Hamilton   Common Skullcap
/ST/X/ (Hpr) Meadows, thickets, and shores from Alaska (N to ca. 67°N), the Yukon (N to ca. 64°N), and Great Bear L. to Great Slave L., L. Athabasca (Alta. and Sask.), Man. (N to about 10 mi s of Churchill), Ont. (N to the Fawn R. at ca. 54°40'N), Que. (N to the E James Bay watershed at 52°37'N and the Côte-Nord), s Labrador (N to the Hamilton R. basin), Nfld., N.B., P.E.I., and N.S., s to Calif., Ariz., N.Mex., Mo., and Del. [*S. galericulata* vars. *epil.* (Hamilt.) Jord. and *pubescens* Benth.]. MAPS: Epling 1942: map 2, p. 5; the N. American area in the map by Hultén 1968*b*, for *S. gal.* var. *pub.* applies here. (The very closely related *S. galericulata* L. of Eurasia differs chiefly in the consistently deeper and sharper pebbling of the nutlets).
Forma *albiflora* (Millsp.) Fern. (corolla white rather than blue, marked with white) is known from sw Que. (Iberville Co.; DAO) and Nfld. (Rouleau 1956). Forma *rosea* (Rand. & Redf.) Fern. (corolla pink) is reported from s Ont. by Boivin (1966*b*).

S. lateriflora L.   Mad-dog Skullcap
/sT/X/ (Hpr) Alluvial thickets and swampy ground from ?Alaska (*see* Hultén 1968*a*) and B.C. (N to Quesnel; V) to ?Alta. (Moss 1959), Sask. (near Trossachs, about 60 mi s of Regina; Breitung 1957*a*), Man. (N to Cross Lake, NE of L. Winnipeg), Ont. (N to the sw James Bay watershed at 52°11'N), Que. (N to the E James Bay watershed at 52°16'N and the Gaspé Pen.; reported from Anticosti Is. by John Macoun 1884), Nfld., N.B., P.E.I., and N.S., s to Calif. and Ga. [Incl. the small-flowered extreme, var. *grohii* Boivin; *S. ?canescens* (*S. incana* Biehler of the E U.S.A.) *sensu* Macoun 1884, not Nutt.]. MAP: Epling 1942: map 2, p. 5.
Forma *albiflora* (Farw.) Fern. (corolla white rather than blue) is reported from sw Que. by Boivin (1966*b*). Forma *rhodantha* Fern. (corolla pink) is known from the type locality near the mouth of the Dartmouth R., Gaspé Pen., E Que.

S. nervosa Pursh
/t/EE/ (Hpr) Moist woods and thickets from Ill. to s Ont. (near Kingsville, Essex Co., where taken by John Macoun in 1901; CAN; TRT) and N.J., s to La., Tenn., and N.C. MAP: Epling 1942: map 3, p. 14.
Our plant is referable to var. *calvifolia* Fern. (leaves glabrous above rather than copiously strigose).

S. parvula Michx.
/T/EE/ (Gst (Hpr)) Dry or moist sands and gravels (chiefly calcareous) from Iowa to Wisc., Ont. (N to Georgian Bay, L. Huron, and the Ottawa dist.; reports from Sask. and N.S. by John Macoun 1884, probably refer to *S. lateriflora*), Que. (N to Portneuf, Montmagny, and L'Islet counties; type from near Montreal, according to Ernest Rouleau, Rhodora 47(561):272. 1945), and Maine, s to Tex., Ala., and Ga. [Incl. the glabrescent extreme, var. *leonardii* (Epling) Fern. (*S. leonardii* Epling)]. MAP: Epling 1942: map 3, p. 14.

STACHYS L.   [7281]   Hedge-Nettle.   Épiaire

1   Plants densely white-woolly or tomentose; upper leaves sessile; (introd.).
2   Leaves conspicuously toothed, silky-pilose above, tomentose beneath, the larger ones rounded or cordate at base; calyx-teeth visible; stem relatively slender, pilose-tomentose; biennial .... *S. germanica*

    2   Leaves heavily woolly, narrowed at base, their teeth hidden; calyx-teeth hidden in
        wool; stem coarse, densely felted; matted perennial .... *S. olympica*
1   Plants not white-woolly.
    3   Rather small annual with fibrous roots and no stolons, the long-hirsute stem diffuse or
        decumbent, often branching from the base; leaves broadly ovate to suborbicular,
        cordate at base, obtuse or rounded at summit, strongly crenate, to about 4 cm long,
        the larger ones with petioles up to half as long as the blade; corolla pink, spotted with
        purple, only slightly surpassing the long-hirsute calyx; (introd.) .... *S. arvensis*
    3   Rather tall perennials with creeping rhizomes and tuberiferous stolons; corolla much
        surpassing the calyx.
        4   Leaves all petioled, the middle cauline ones with petioles mostly at least 1.5 cm
            long (to 4.5 cm); (B.C.).
            5   Corolla deep red-purple, the tube to 2.5 cm long; calyx to over 11 mm long,
                glandular-villous; plant to 1.5 m tall, inland as well as coastal .... *S. cooleyae*
            5   Corolla pink or pink-purple, the tube less than 1.5 cm long; calyx to 9 mm long,
                long-spreading-pubescent and sometimes also glandular; plant to 8 dm tall,
                chiefly near the coast .... [*S. mexicana*]
        4   Middle and upper stem-leaves sessile or on petioles rarely over 1 cm long.
            6   Middle and upper internodes of stem nearly or quite glabrous on the sides;
                surface of calyx glabrous, only the nerves commonly bristly-hairy, the teeth
                soon outwardly curving; (s Man. to Que.) .... *S. tenuifolia*
            6   Middle and upper internodes of stem pubescent on both the sides and the
                angles (rarely almost glabrous); surface of calyx pubescent; leaves normally
                sessile or short-petioled, mostly pubescent beneath; (transcontinental) ....
                .... *S. palustris*

S. arvensis L.
European; introd. in grasslands and waste places in N. America, as in B.C. (Boivin 1966*b*), N.S.
(Kings, Hants, Lunenburg, and Halifax counties), and ?P.E.I. (Hurst 1952).

S. cooleyae Heller
/T/W/ (Gst) Swamps and moist low ground from w B.C. (Queen Charlotte Is.; Vancouver Is.
and adjacent mainland N to Prince Rupert; CAN; type from Nanaimo, Vancouver Is.; the report of *S.
emersonii* (*S. mexicana* Benth.) from SE Alaska by Hultén (1949; 1968*b*) is considered referable to
S. cooleyae by Calder and Taylor 1968) to s Oreg. [*S. ciliata* ssp. *macrantha* Piper].

S. germanica L.
European; introd. along roadsides and in fields and waste places in N. America, as in Ont. (Guelph,
Wellington Co.; Snelgrove, Peel Co.; Pembroke, Renfrew Co.). [Incl. *S. italica* Mill.].

[S. mexicana Benth.]
[According to Calder and Taylor (1968), reports of this species of the w U.S.A. (Wash. to Calif.)
from B.C. all probably refer to *S. cooleyae,* most such reports being based upon the recognition of
S. ciliata var. *pubens* Gray (*S. pub.* (Gray) Heller) as identical with *S. mexicana.* They note
Cronquist's opinion, however, that the variety (type from the Fraser R., B.C.) appears to be a hybrid
between either *S. cooleyae* or *S. mexicana* and *S. palustris,* concluding that, at least in the case of
the B.C. plant, *S. cooleyae* is the parent with *S. palustris.*]

S. olympica Poir.   Woolly Hedge-Nettle, Lamb's-ears
Native in the Caucasus; introd. along roadsides and in pastures in N. America, as in s Ont. (York,
Bruce, and Grey counties) and sw Que. (Berthier, Berthier Co.; Groh and Frankton 1949*b*). [*S.
lanata* Jacq.].

S. palustris L.   Woundwort.   Crapaudine
/ST/X/EA/ (Gst) Meadows, streambanks, and other moist places, the aggregate species from
cent. Alaska–Yukon–Dist. Mackenzie to B.C.–Alta., Sask. (N to near Prince Albert), Man. (N to the
Churchill R. at ca. 57°N), Ont. (N to the Severn R. at ca. 55°30′N), Que. (N to the E James Bay

watershed at ca. 52°N, the Côte-Nord, and Gaspé Pen.), Nfld., N.B., P.E.I., and N.S., s to Ariz.,
N.Mex., Nebr., Ohio, and New Eng.; Eurasia. MAPS and synonymy: *see* below.
1 Calyx closely viscid-pilose; (introd.).
  2 Leaves lanceolate, acuminate.
    3 Leaves sessile or subsessile.
      4 Angles of stem retrorse-hirsute with hairs much longer than the appressed
        hairs of the sides; [s Man. (Löve and Bernard 1959) to Nfld. and N.S.] . . . .
        . . . . var. *palustris*
      4 Angles and sides of stem subequally densely long-retrorse-hirsute; [*S.
        segetum* Mutel; introd. in N.B. (Bathurst; Grand Manan Is.) and N.S. (Pictou,
        Pictou Co.)] . . . . var. *segetum* (Mutel) Grogn.
    3 Principal leaves slender-petioled; pubescence of stem as in the typical form;
      [introd. in s ?Man. (Löve and Bernard 1959), s Ont. (Fernald *in* Gray 1950), Que.,
      Nfld., N.B., and P.E.I.] . . . . var. *petiolata* Clos
  2 Leaves oblong or oblong-ovate, obtuse or subacute; [introd. in s ?Man. (Löve and
    Bernard 1959), s Ont. (St. Thomas, Elgin Co.), E Que. (York, Gaspé Co.), and P.E.I.
    (Charlottetown)] . . . . var. *elliptica* Clos
1 Calyx scarcely viscid, short-pilose below a long-hirsute pubescence of hairs to 3 mm long;
  leaves sessile or very short-petioled; (native).
  5 Principal leaves oblong to oblong-ovate, blunt or subacute; [*S. pilosa* Nutt.; *S.
    borealis* and *S. leibergii* Rydb.; *S. scopulorum* Greene; transcontinental; MAP:
    Hultén 1968*b*:789] . . . . var. *pilosa* (Nutt.) Fern.
  5 Principal leaves narrowly lanceolate to narrowly oblong, acuminate.
    6 Angles of stem with few or no long hairs, the sides minutely and retrorsely
      appressed-pubescent; [type from Sioux Lookout, Ont., about 170 mi NW of
      Thunder Bay] . . . . var. *macrocalyx* Jennings
    6 Angles of stem abundantly long-retrorse-hirsute.
      7 Sides of stem with dense long pubescence similar to that of the angles; [*S.
        homotricha* (Fern.) Rydb.; *S. velutina* of Canadian reports, not Willd.; *S.
        ambigua* of Canadian reports, not Sm.; essentially transcontinental; MAP (E
        area): M.L. Fernald, Rhodora 45(539): map 5, p. 467. 1943] . . . .
        . . . . var. *homotricha* Fern.
      7 Sides of stem short-pilose; [transcontinental; type from L. Nipigon, Ont.] . . . .
        . . . . var. *nipigonensis* Jennings

S. tenuifolia Willd.
/T/EE/ (Gst (Hpr)) Moist meadows, thickets, and shores, the aggregate species from s Man. (N
to Inwood, near the s end of L. Winnipeg) to Ont. (N to the Ottawa dist.), Que. (N to Montmorency
and Montmagny counties), N.Y., and N.H., s to E Tex., La., and S.C. MAPS and synonymy: *see*
below.
1 Calyx glabrous throughout; angles of upper internodes of stem glabrous, scabrous, or
  shortly retrorse-hispid; leaves glabrous, their petioles to 3 cm long; bracts subtending
  floral-whorls usually not ciliate; [reported from Otterburne, s Man., by Löve and Bernard
  1959; the MAP by M.L. Fernald, Rhodora 45(539): map 1, p. 467. 1943, indicates no
  Canadian stations] . . . . var. *tenuifolia*
1 Calyx often bristly on the angles; angles of upper internodes conspicuously retrorse-
  hispid with bristles to 2 mm long; leaves often strigose above and hispid on the veins
  beneath, sessile or the lower ones short-petioled; bracts bristly-ciliate.
  2 Principal leaves narrowly ovate to broadly oblong; [var. *aspera* of auth., not *S. aspera*
    Michx.; s Man. (Otterburne, about 30 mi s of Winnipeg; Löve and Bernard 1959), Ont.
    (N to Ottawa; Fernald, loc. cit.), and Que. (N to Montmorency and Montmagny
    counties); MAP: M.L. Fernald, Rhodora 45(539): map 3 (the s Man. station should be
    indicated), p. 467. 1943] . . . . var. *platyphylla* Fern.
  2 Principal leaves narrowly lanceolate to narrowly lance-oblong; [*S. hispida* Pursh (*S.
    palustris* var. *hisp*. (Pursh) Boivin and its f. *cleoniquei* Boivin); s Man. to sw Que.; the
    MAP by M.L. Fernald, Rhodora 45(539): map 4, p. 467. 1943, indicates no Canadian
    stations] . . . . var. *hispida* (Pursh) Fern.

TEUCRIUM L.   [7212]   Germander, Wood-Sage.   Germandrée

1   Leaves to 2 cm long, deeply divided into linear or oblong segments, on petioles to 1 cm long; calyx strongly saccate at base, its subequal short teeth deltoid; flowers in axillary whorls, reddish purple, on pedicels to 8 mm long; much branched, copiously short-glandular-villous annual; (introd.) . . . . [*T. botrys*]

1   Leaves merely shallowly toothed; calyx not saccate; flowers in terminal racemes; stoloniferous perennials.

    2   Flowers pale yellow, 1 or 2 at each node of the slender loose 1-sided raceme; calyx glabrous or glabrate, 5 or 6 mm long, the upper tooth broadly depressed-ovate and recurved, the other 4 teeth triangular and erect, subulate-tipped; leaves deltoid-lanceolate to -ovate, crenate, rugose, cordate or truncate at base, to 5 or 6 cm long, on petioles to about 1 cm long; plant short-pubescent; (introd.) . . . . *T. scorodonia*

    2   Flowers pink-purple, several at each node of the dense raceme; calyx to 9 mm long; petioles to 1.5 cm long.

        3   Stem villous; leaves white-villous beneath with spreading hairs; calyx viscid-villous with spreading hairs, its teeth all acute and subequal; (B.C.; Sask. to sw Que.) . . . .
. . . . *T. occidentale*

        3   Stem and lower leaf-surfaces appressed-pubescent with eglandular hairs; calyx hoary with short incurved hairs, the 3 teeth of its upper lip obtuse (or the middle one acutish) and shorter than the other 2 teeth; (Ont. to N.B. and N.S.) . . . .
. . . . *T. canadense*

[*T. botrys* L.]
[European; locally introd. and abundant in dry calcareous pastures in N. America, as in s Ont. (near London, Middlesex Co., where taken by John Dearness in 1888 and 1889 but evidently not found since that time; CAN).]

T. canadense L.   Wood-Sage, American Germander
/T/EE/   (Hpr)   Shores, thickets, and woods from Ont. (N to the Ottawa dist.) to Que. (N to near Quebec City; MT; John Macoun 1884), N.B., and N.S. (the report from P.E.I. by McSwain and Bain 1891, requires confirmation), s to Tex. and Fla. [*T. littorale* Bickn.].
    Some of our material is referable to var. *virginicum* (L.) Eat. (*T. virg.* L.; leaves relatively thin and broad, neither rugose nor markedly papillate above as in the typical form).

T. occidentale Gray
/T/X/   (Hpr)   Alluvial soils and wet places from s B.C. (N to Kamloops; CAN; not known from Alta.) to Sask. (Cypress Hills; Eagle Creek; Yorkton; Lumsden), Man. (N to Eriksdale, about 70 mi NW of Winnipeg; CAN), Ont. (N to the Ottawa dist.), sw Que. (N to Berthier, Berthier Co.), and Maine, s to Calif., Mexico, N.Mex., Kans., Ohio, and Pa. [*T. canadense* var. *occ.* (Gray) McCl. & Epling].
    Some of our material is referable to var. *boreale* (Bickn.) Fern. (*T. bor.* Bickn.; calyces and floral bracts with few or no short-stipitate glands, the typical form with these in addition to the longer villosity).

T. scorodonia L.   Wood-Sage, Germander-Sage
European; locally established in Ont. (Fernald *in* Gray 1950), sw Que. (Ste-Cécile-de-Whitton, Frontenac Co., where taken by James Fletcher in 1903; CAN), and Ohio.

THYMUS L.   [7319]   Thyme.   Thym

1   Stem below the inflorescence sharply 4-angled, only the angles long-hairy (2 opposite sides narrow and short-pubescent, the other 2 sides broader and glabrous); leaves ciliate at base, otherwise glabrous, their lateral veins slender and not prominent beneath when dried, their margins more or less upturned; inflorescence usually elongated and interrupted at base; plant tufted, the branches ascending; (garden-escape) . . . . . *T. pulegioides*

1 Stem obscurely angled; plants with long creeping branches, forming a mat; lateral
leaf-veins prominent beneath when dried; inflorescence usually capitate; leaves flat,
ciliate and also often somewhat pubescent above.
  2 Flowering stems equally short-pubescent on all sides; (introd.) .... *T. serpyllum*
  2 Flowering stems densely hairy on 2 opposite sides with hairs of varying length, the
  other 2 sides less hairy or glabrous; (Greenland) .... *T. arcticus*

T. arcticus (Durand) Ronniger
/aST/-/GE/   (Ch)   Dry open places at low to fairly high elevations in w and e Greenland, n to ca.
69°N; Iceland; Europe. [*T. praecox* ssp. *arct.* (Dur.) Jalas.; incl. *T. drucei* Ronn.]. MAPS (*T. drucei*):
Hultén 1958: map 76, p. 95; Löve & Löve 1956*b*: fig. 25, p. 230.

T. pulegioides L.
European; a garden-escape in sw B.C. (Stanley Park, Vancouver; J.M. Macoun 1895, as *T. cham.*),
sw Que. (Hemmingford, Huntingdon Co.; CAN; MT), and P.E.I. (Prince Co.; D.S. Erskine 1960). [*T.
chamaedrys* Fries; *T. serpyllum* ssp. *cham.* (Fr.) Vollman].

T. serpyllum L.   Creeping Thyme.   Serpolet
Eurasian; a garden-escape in N. America, as in B.C. (n to Terrace, ca. 54°30'N), Ont. (n to
Carleton Co.; OAC; not listed by Gillett 1958), Que. (n to the Gaspé Pen. at the mouth of the
Dartmouth R., where abundant when taken by the writer in 1940; CAN), and N.S. (Cumberland and
Pictou counties; CAN; ACAD).
    Var. *albus* Hort., a cult. form with white (rather than purplish) flowers, is reported from P.E.I. by
D.S. Erskine (1960; "white-flowered form").

NOTE

In addition to the above genera of Labiatae, species of two other genera (apparently restricted to
the e U.S.A. and not keyed out above) are reported from Canada by Hooker (1838), this stated to
be on the authority of Pursh (1814). These are common dittany, *Cunila mariana* L. (*C. origanoides*
(L.) Britt.), and bastard pennyroyal, *Trichostema dichotomum* L. The area actually given by Pursh
for the former is "New York to Carolina" and for the latter "Pennsylvania to Carolina". The only
explanation for Hooker's report appears to be a hasty reading of "Canada" for "Carolina".

SOLANACEAE (Nightshade Family)

Chiefly herbs (*Lycium* and *Solanum dulcamara* woody twiners) with alternate, commonly simple, entire to deeply pinnatifid leaves. Flowers perfect, regular or nearly so, hypogynous, gamopetalous, 5-merous. Stamens 5, inserted on the corolla-tube and alternating with its lobes. Style 1. Ovary superior. Fruit a commonly 2-locular capsule or berry.

1   Stems woody, sometimes climbing; leaves entire; flowers in axillary clusters of up to 8; calyx not enlarged in fruit; berry orange-red or scarlet; (introd.) .... *Lycium*
1   Stems herbaceous (or somewhat woody only at the base).
    2   Fruit a capsule; corolla campanulate, tubular, or funnelform; (introd.).
        3   Flowers and circumscissile capsules (enclosed in calyces) sessile and solitary in the leaf-axils, forming 1-sided leafy spikes; corolla dull yellow, strongly reticulate with purple veins, about 3 cm long and broad, purple in the throat; leaves oblong-ovate, to 2 dm long, very coarsely toothed or shallowly pinnate-lobed, the upper sessile, the lower short-petioled .... *Hyoscyamus*
        3   Flowers and capsules pedicelled; leaves entire to coarsely sinuate-toothed or undulate, more generally petioled.
            4   Capsule thorny or prickly, subtended by a flaring collar consisting of the base of the mature circumscissile tubular calyx; flowers solitary in the upper axils; corolla-lobes abruptly acuminate-toothed; leaves ovate, to 2 dm long ...... *Datura*
            4   Capsule unarmed.
                5   Capsule circumscissile (the top coming off like a lid); flowers solitary in the leaf-axils, glossy brown outside, dull olive-green within, to about 2.5 cm long; leaves narrowly to broadly ovate, to over 1.5 dm long and about 8 cm broad; (introd.) .... [*Scopolia*]
                5   Capsule dehiscing longitudinally.
                    6   Flowers in terminal racemes or panicles; corolla whitish to greenish-yellow (often turning purplish), regular, cylindrical; stamens equal .... *Nicotiana*
                    6   Flowers solitary in the leaf-axils; corolla white or pink, slightly oblique and 2-lipped, funnelform; one of the stamens much smaller than the others .... [*Petunia*]
    2   Fruit a berry.
        7   Mature calyx not enlarged and not enclosing the berry; corolla rotate or nearly so; anthers forming a tube around the style.
            8   Leaves mostly simple (compound only in *S. dulcamara* and *S. tuberosum*); anthers opening by terminal pores or short terminal clefts; fruit pulpy, at most 2.5 cm thick .... *Solanum*
            8   Leaves pinnate, with up to 9 ovate to oblong irregularly toothed leaflets; anthers tapering to a long sterile tip and opening by longitudinal slits; fruit juicy, thicker; (garden-escape) .... [*Lycopersicum*]
        7   Mature calyx inflated and wholly or partly enclosing the berry.
            9   Anthers forming a tube around the style, opening by terminal pores or clefts; corolla rotate; flowers cymose or umbellate, rarely solitary; (introd.) ...... *Solanum*
            9   Anthers separate, longitudinally dehiscent; flowers mostly solitary in or above the axils near the tip of the stem.
                10 Calyx consisting of nearly distinct sagittate-auricled sepals; corolla open-campanulate, pale blue, to 2.5 cm long and broad; leaves ovate-lanceolate to ovate, coarsely and unevenly toothed or lobed; plant glabrous; (introd.) .... *Nicandra*
                10 Calyx lobed only at summit, otherwise united, not auricled at base; plants glabrous or pubescent and often viscid-glandular.
                    11 Corolla rotate, white or blue-tinged, with a yellow eye, to 4 cm broad; flowers commonly 2 or more from the upper nodes; calyx scarcely angled, ribless, in fruit closely investing the berry; leaves ovate-

lanceolate to ovate, entire or nearly so; plant thinly villous and more or
less viscid; (Sask. to Que., partly or wholly introd.) .... *Chamaesaracha*

11 Corolla rotate-campanulate, greenish, yellowish white, yellow, or
scarlet, usually with a darker centre; flowers solitary in or above the
axils; calyx 5-angled and often 10-ribbed, in fruit inflated and loosely
covering the berry .... *Physalis*

CHAMAESARACHA Gray [7397]

C. grandiflora (Hook.) Fern. Large White-flowered Ground-cherry
/T/EE/ (T) Rocky or sandy fields, open woods, recent clearings, and shores from cent. Sask.
(Saskatchewan R. and Churchill R. systems between ca. 53°N and Ile-à-la-Crosse, 55°27'N;
Breitung 1957a) to Man. (N to Minatonas, N of Duck Mt.; DAO), Ont. (N to Sioux Lookout, about 170
mi NW of Thunder Bay), and Que. (N to Cap-à-l'Aigle and La Malbaie, Charlevoix Co., and
Cap-à-l'Orignal, near Bic, Rimouski Co.), S to Minn., Wisc., S Ont.–Que., and Vt.; closely related
species in S. America and China. [*Physalis gr.* Hook., the type locality, as first area cited by Hooker
1838, being "Sandy banks of the Saskatchawan"; *Leucophysalis* Rydb.].

Because of the very weedy nature of this species, its actual native area is very difficult to delimit.
Its apparent restriction in Sask. to the old fur-trade routes points to its introduction there in the days
of canoe-freighting, as may be the case with the N Ont. stations. The Que. stations also have a
weedy aspect and our material may be entirely introd. from the U.S.A., with the possible exception
of the S Ont.–Que. plant.

DATURA L. [7415] Jimsonweed, Thorn-apple. Pomme épineuse or Herbe aux sorciers

1 Capsule erect, dehiscent by 4 valves, to about 6 cm long; calyx-tube strongly prismatic
and narrowly 5-winged; corolla to 1 dm long, its margins 5-toothed; leaves coarsely
sinuate-toothed or angled; glabrous annual .... *D. stramonium*
1 Capsule inclined or nodding, opening irregularly; calyx-tube scarcely angled; leaves entire
to slightly angled or sinuate-toothed; soft-puberulent and somewhat glaucous-whitened
perennials.
  2 Corolla white, to 1.5 dm long, its limb with 5 attenuate-tipped lobes alternating with 5
  shorter lobes; capsule covered with very short spines or tubercles .... [*D. metel*]
  2 Corolla white, suffused with lavender or violet, to 2 dm long, its limb 10-toothed;
  capsule covered with slender spines .... *D. meteloides*

[D. metel L.]
[Tropical America; introd. along roadsides and in waste places of S Ont. (Kingston, Frontenac Co.,
where taken by Fowler and McMorine in 1900 but probably not established; Montgomery 1957) and
SW Que. (Boivin 1966b). (*D. innoxia* of auth., not Mill.)]

D. meteloides Dunal
Tropical America; introd. along roadsides and in waste places of S Ont. (Norfolk, Wellington, and
Ontario counties; OAC; TRT).

D. stramonium L. Stramonium
Asiatic; introd. into waste places in N. America, as in S B.C. (Vancouver Is., Saltspring Is., and the
Dry Interior; Herb. V), Alta. (Moss 1959), Sask. (Melfort, Senlac, and Shellbrook, about 25 mi W of
Prince Albert; Breitung 1957a), Ont. (N to the Ottawa dist.; John Macoun 1884), Que. (N to
Kamouraska Co.), N.B., P.E.I., and N.S.

Some of our material is referable to var. *tatula* (L.) Torr. (*D. tatula* L.; corolla pale violet rather
than white or nearly so; stem purplish rather than green).

HYOSCYAMUS L. [7396] Henbane

H. niger L. Black Henbane. Jusquiame
European; introd. along roadsides and in waste places of N. America, as in S B.C. (Agassiz, SW of

Chilliwack; Herb. V; reported from Vernon and Salmon Arm by Eastham 1947), Alta. (Fort Saskatchewan, Fort Macleod, and Banff; CAN), Sask. (N to Cudworth, 52°30′N; Breitung 1957a), Man. (N to Dauphin, N of Riding Mt.), Ont. (N to the Ottawa dist.), Que. (N to L. St. John and the Gaspé Pen.), N.B., P.E.I. (Summerside, Prince Co.; GH), and N.S.

LYCIUM L. [7379]

L. halimifolium Mill. Matrimony-vine
Eurasian; spreading from cult. to roadsides, fields, and waste places in N. America, as in S B.C. (Keremeos, SW of Penticton; CAN; reported from Nanaimo, Vancouver Is., and Kamloops by Groh and Frankton 1948), Alta. (Edmonton; CAN), S Sask. (Indian Head; CAN), Ont. (N to Durham and Frontenac counties), and N.S. [*L. vulgare* (Ait.) Dunal; incl. *L. chinense* Mill.].

[LYCOPERSICUM Mill.] [7407]

[L. esculentum Mill.] Tomato
[Tropical America; occasionally spontaneous on rubbish-heaps in N. America but not becoming established, as in SW B.C. (Ocean Park, near New Westminster; V), Ont. (Ottawa and the shores of the Nation R. near Ottawa; CAN; reported from Lambton Co. by Gaiser and Moore 1966), SW Que. (St. Helen's Is., Montreal; Rouleau 1945), and N.S. (Lindsay 1878). (*Solanum (Lycop.) lycopersicon* L.).]

NICANDRA Adans. [7377]

N. physalodes (L.) Pers. Apple-of-Peru
Peruvian; introd. near dwellings and in old gardens and waste places in N. America, as in Ont. (N to the Ottawa dist.), Que. (N to wharf-ballast at Rimouski, Rimouski Co.; MT; RIM), P.E.I. (Charlottetown; DAO), and N.S (Hants and Kings Counties; ACAD). [*Atropa* L.; *Physalodes* Britt.].

NICOTIANA L. [7434] Tobacco

1 Corolla funnelform, greenish, woolly outside, to 8 cm long, its throat somewhat swollen,
  its reniform acute spreading lobes becoming reddish; (the cultivated tobacco) .... [*N. tabacum*]
1 Corolla salverform (with a slender tube abruptly expanded into a flat limb).
  2 Corolla to about 1 dm long, white or greenish white (becoming purplish), the limb 2 or
    3 cm broad, its lobes ovate-lanceolate to ovate; capsule to 1.5 cm long; leaf-blades to
    2.5 dm long, the lower ones spatulate to elliptic, the upper ones linear to lanceolate;
    (introd. in S Ont. and S Que.) .... [*N. longiflora*]
  2 Corolla shorter, the limb to about 1.5 cm broad, its broadly rounded lobes apiculate;
    capsule to about 1 cm long.
    3 Corolla to 2 cm long, greenish-white or -yellow; principal leaf-blades ovate, to 2
      dm long; (introd. in S Ont.) .... [*N. rustica*]
    3 Corolla to 3.5 cm long, dingy white; principal leaf-blades elliptic to lance-ovate, to
      about 12 cm long; (S B.C.) .... *N. attenuata*

N. attenuata Torr. Coyote-Tobacco
/t/WW/ (T) Dry sandy open places from S B.C. (Dry Interior between Lytton and Spences Bridge; CAN; John Macoun 1884) and Idaho to Baja Calif. and Tex. MAP: T.H. Goodspeed, Chronica Botanica 16: fig. 12, p. 40. 1954.

[N. longiflora Cav.]
[Tropical America; an occasional garden-escape in N. America but scarcely established, as in S Ont. (Hamilton, Wentworth Co.; TRT) and SW Que. (Montreal; CAN; J.M. Macoun 1907).]

[N. rustica L.] Wild Tobacco
[Peruvian; formerly cult. by the Indians of N. America (and still apparently so, there being an interesting 1949 collection by Marius Barbeau in CAN, "Southern Ontario. In cultivation by

Indians"), and occasionally found in waste places, as in s Ont. (Belleville, Hastings Co., "Spontaneous in gardens", where taken by John Macoun in 1878; CAN; reported from Fort Erie, Welland Co., by Macoun 1884, and from Wellington Co. by F.H. Montgomery, Can. Field-Nat. 62(2):93. 1948), where scarcely established.]

[N. tabacum L.]   Tobacco
[Tropical America; occasionally escaped from cult. to waste places in N. America, as in s Ont. (Belleville, Hastings Co.; Boivin 1966*b*).]

[N. alata Link & Otto.]
[Var. *grandiflora* Comes of this South American species (not keyed out above) is reported from Que. as an ephemeral by C. Rousseau (Nat. can. (Que. 98(4):720. 1971; Ste-Foy, near Quebec City), where probably a garden-escape. It has very large, fragrant, villous flowers, the yellowish-green corolla-tube much dilated, the limb yellowish outside.
    A hybrid between it and *N. forgetiana* Sander (× *N. sanderae* Sander; corolla-tube greenish yellow tinged with rose, the limb with carmine-rose lobes) is also reported from the same locality by Rousseau.]

[PETUNIA Juss.]   [7436]   Petunia

1   Corolla less than 1 cm long and broad, blue or purple (the tube yellow), not much
    surpassing the foliaceous sepals; leaves fleshy, narrowly spatulate or oblanceolate, 1 or 2
    cm long and mostly less than 5 mm broad ....[*P. parviflora*]
1   Corolla larger; principal leaves broadly elliptic to oblong or ovate, to over 8 cm long.
    2   Corolla dull white, to 6 or 7 cm long and about as broad, its cylindric tube 3 or 4 times
        as long as the calyx ....[*P. axillaris*]
    2   Corolla white or variously coloured, the funnel-shaped tube relatively short and broad.
        3   Stems stout; flowers white to deep red-purple, variously striped and barred, to
            about 9 cm long and nearly as broad ....[*P. hybrida*]
        3   Stems very slender; flowers rose-red or violet, to about 4 cm long and broad ....
            ....[*P. violacea*]

[P. axillaris (Lam.) BSP.]   Large White Petunia
[Argentinian; an occasional garden-escape in N. America but scarcely established, as in s Ont. (Toronto; Boivin 1966*b*).]

[P. hybrida Vilm.]   Common Garden Petunia
[Apparently a series of hybrids between *P. axillaris* and *P. violacea*; reported from sw Que. by Rouleau (1945; St. Helen's Is., Montreal; MT) where, however, scarcely established.]

[P. parviflora Juss.]   Seaside-Petunia
[Tropical America; introd. along sea-beaches and in waste and cult. ground in N. America; known in Canada only through an 1883 collection in CAN by John Macoun from wharf-ballast at Pictou, N.S.]

[P. violacea Lindl.]
[See *P. hybrida*.]

PHYSALIS L.   [7401]   Ground-cherry.   Coqueret

(Ref.: Waterfall 1958)
1   Corolla whitish, about 2.5 cm broad, shallowly but distinctly lobed; fruiting calyx firm, to 5
    cm long, bright red to scarlet, drooping on peduncles 2 or 3 cm long; leaves
    ovate-rhombic; essentially glabrous perennial; (introd.) ....*P. alkekengi*
1   Corolla yellow or greenish yellow, with merely angulate margins; fruiting calyx green or
    brown.
    2   Plant densely pubescent with fine branching hairs; leaves elliptic, oval, or oblong,
        blunt, entire or somewhat undulate; corolla about 2 cm broad; anthers yellow, about

3 mm long; fruiting calyx 2 or 3 cm long; berry yellow or orange; perennial; (introd.) . . . .
. . . . [*P. viscosa*]

2   Plant glabrous or with simple or sparingly branched hairs.
   3   Annuals, lacking rhizomes; anthers blue, on slender filaments; (introd.).
      4   Plant glabrous or glabrate; corolla to 2.5 cm broad; fruiting calyx purple-
         veined, obscurely angled, rounded at base, with blunt triangular teeth, nearly
         filled by the viscid purplish berry; anthers about 3 mm long; leaves ovate or
         rhombic, cuneate or tapering at base, subentire to sinuate-toothed or
         prominently dentate . . . . *P. ixocarpa*
      4   Plant copiously pubescent; corolla at most 1 cm broad; fruiting calyx strongly
         angled, rounded or subcordate at base, with long narrow teeth, not filled by
         the yellow berry; anthers commonly less than 2 mm long; leaves broadly
         ovate, broadly rounded or cordate at base, entire or unevenly sinuate-dentate . . . . . .
         . . . . *P. pubescens*
   3   Perennials with stout rhizomes; calyx-teeth narrowly triangular, often acuminate;
      berry usually yellow.
      5   Anthers deep blue-purple, about 3 mm long, on slender filaments; fruiting
         pedicels not over 1 cm long; leaves rounded or cordate at base,
         long-acuminate; plant densely and finely villous . . . . [*P. peruviana*]
      5   Anthers yellow or light blue, their filaments at least 1/3 as broad as the
         anthers; fruiting pedicels to 3 cm long; leaves not long-acuminate.
         6   Anthers averaging about 2.5 mm long, their filaments not clavate; (s Man.
            to sw Que.) . . . . *P. virginiana*
         6   Anthers averaging about 4 mm long, their filaments often clavate; (Man. to
            N.S.) . . . . *P. heterophylla*

P. alkekengi L.   Chinese Lantern, Winter-cherry
Asiatic; a local garden-escape in N. America, as in Ont. (Ottawa, Smiths Falls, and Aultsville, Stormont Co.; DAO; TRT; Montgomery 1957).

P. heterophylla Nees   Cerise de terre sauvage
/T/EE/   (Grh)   Fields and open woods from Man. (Boivin 1966b; reports from Sask. require confirmation) to Ont. (N to Timiskaming, ca. 47°30′N), Que. (N to Ste-Anne-de-la-Pocatière, Kamouraska Co.; QSA), ?N.B. (a Deam report noted by Groh and Frankton 1949b; early P.E.I. reports require confirmation), and N.S. (Kings Co.; ACAD), s to E Tex., Okla., Ky., and Ga. [Incl. vars. *ambigua* (Gray) Rydb. and *nyctaginea* (Dunal) Rydb. and *P. lanceolata* Michx.].

P. ixocarpa Brotero   Tomatillo
A native of Mexico and the sw U.S.A.; a garden-escape elsewhere, as in Man. (Boivin 1966b), Ont. (Perth, Dundas, Renfrew, and Carleton counties), and sw Que. (Ile-aux-Alumettes, Pontiac Co.; GH, as *P. peruviana,* revised by Fernald).

[P. peruviana L.]   Cape-Gooseberry
[A native of S. America; the report from Ottawa, Ont., by John Macoun (1884) is based upon *P. ixocarpa,* the relevant collection in CAN.]

P. pubescens L.
Native in tropical America and the s U.S.A.; introd. elsewhere, as in ?B.C. (Cache Creek, about 45 mi w of Kamloops; John Macoun 1884), Man. (Winnipeg; DAO), Ont. (N to Ottawa), Que. (Wakefield, Gatineau Co.; Chambly; St. Helen's Is., Montreal), and N.B. (Bass River, Kent Co.; NBM).
   Some or all of our material is referable to var. *grisea* Waterfall (*P. pruinosa sensu* J.M. Macoun 1906, and later Canadian auth., not L., the relevant collection in CAN; leaves greyish rather than green, often with "mealy" or sessile granular glands).

P. virginiana Mill.
/T/(X)/   (Grh)   Dry sandy or rocky woods, clearings, meadows, and waste places from s Man. (N

to Makinak, about 80 mi N of Brandon, and Victoria Beach, about 55 mi NE of Winnipeg) to Ont. (N to the Ottawa dist.), SW Que. (Montreal dist.), and N.J., S to N Mexico, N.Mex., Tex., Ala., and Ga.

Some of our material is referable to var. *subglabrata* (Mack. & Bush) Waterfall (*P. sub.* M. & B.; *P. longifolia* var. *sub.* (M. & B.) Cronq.; *P. ?philadelphica* Lam.; plant nearly glabrous or with a few short ascending hairs rather than more or less villous with long multicellular hairs (if short-pubescent, the hairs directed downward)).

[P. viscosa L.]
[Tropical America; a collection in TRT from Toronto, Ont., requires confirmation, as do reports from Ont. by John Bell (Geol. Surv. Can., appendix to the reports for 1866–1869. 1870; Manitoulin Is., N L. Huron) and Gillett (1958; Ottawa dist.).]

[SCOPOLIA Schreb.]   [7393]

[S. carniolica Jacq.]
[European; apparently known from N. America only through a 1935 collection from SW Que. (Mt. Royal, Montreal; MT), where taken by Marie-Victorin and Rolland-Germain in 1935, the plant noted as naturalized in woods but apparently not found since that time. (*Hyoscyamus scopolia* L.).]

SOLANUM L.   [7407]   Nightshade.   Morelle

1   Stem, branches, pedicels, and principal leaf-veins prickly or spiny, they and the leaves more or less stellate-pubescent with branched hairs.
  2   Calyx-tube stellate-pubescent, unarmed, not wholly covering the yellow berry; corolla normally pale violet; leaves sinuate-lobed; perennial from a deep creeping rhizome; (S Ont.; introd. northwards) . . . . *S. carolinense*
  2   Calyx-tube spiny or prickly and covering the berry; leaves deeply pinnate-lobed, the segments themselves lobed or coarsely toothed; annuals; (introd.).
    3   Corolla yellow; one anther much larger and longer than the others, with an incurved beak; fruiting pedicels erect; calyx closely investing the berry; seeds coarsely rugose; pubescence consisting entirely of branched hairs . . . . *S. rostratum*
    3   Corolla violet or blue; anthers equal; fruiting pedicels spreading; calyx loosely covering the red berry; seeds minutely reticulate; pubescence consisting partly of simple gland-tipped hairs . . . . [*S. sisymbriifolium*]
1   Stem unarmed; berries naked, not wholly covered by the calyx (except at maturity in *S. sarrachoides*); plants glabrous, or pubescent with simple hairs; (chiefly introd.; *S. nigrum* and *S. triflorum* perhaps partly native).
  4   Stem woody below, tending to climb or scramble, from a rhizome; corolla violet or purple; berries red; leaves ovate, entire or with 1 or 2 basal divergent lobes . . . .
    . . . . *S. dulcamara*
  4   Stem herbaceous, not climbing; corolla white or purple-tinged.
    5   Leaves pinnately compound, the larger ovate stalked leaflets irregularly alternating with much smaller sessile ones; berries green or yellowish, rarely produced; plant tuberiferous; (garden-escape) . . . . [*S. tuberosum*]
    5   Leaves simple; annuals, without tubers.
      6   Leaves oblong, deeply pinnate-lobed, with rounded sinuses; berry green . . . .
      . . . . *S. triflorum*
      6   Leaves ovate, entire or merely sinuate (rarely bluntly lobed at base).
        7   Stem and leaves essentially glabrous; berries black.
          8   Berries dull black, to 13 mm thick; stone-cell concretions lacking or very small and not more than 1 or 2 per berry; inflorescence always subracemose; anthers mostly over 2 mm long; leaves thickish, opaque . . . . .
          . . . . *S. nigrum*
          8   Berries lustrous black, to about 9 mm thick, with usually at least 4 stone-cell concretions; inflorescence mostly umbelliform; anthers at most 2 mm long; leaves thin and translucent . . . . *S. americanum*

7   Stem and leaves copiously pubescent.
   9   Stem and leaves ashy with a dense appressed villosity or pilosity;
        flowers at most about 5 in an umbel-like inflorescence; calyx scarcely
        enlarged in fruit, the yellow to red berry standing above it . . . . . [*S. villosum*]
   9   Stem and leaves hirsute with spreading hairs; flowers up to 9 in a
        racemose-corymbose inflorescence; calyx greatly enlarged in fruit and
        covering much of the green berry . . . . *S. sarrachoides*

S. americanum Mill.
Native in the E U.S.A. (N to N.Dak. and Maine); introd. elsewhere, as in S Ont. (N to Georgian Bay,
L. Huron, and Grenville Co.; TRT) and SW Que. (Mont-St-Hilaire, about 20 mi NE of Montreal;
Maycock 1961). [Scarcely separable from *S. nigrum* and perhaps better treated as its var. *amer.*
(Mill.) Schulz].

S. carolinense L.   Horse-nettle, Ball-nettle
/t/EE/   (Grh (Gr))   Sandy openings, fields, and waste places from Nebr. to Ohio, S Ont. (probably
native in counties along L. Erie and in Grey Co., Bruce Pen., L. Huron; probably introd. farther
northwards, as at Ottawa), N.Y., and New Eng., S to Tex. and Fla.

S. dulcamara L.   Bittersweet, Nightshade.   Morelle douce-amère
Eurasian; introd. into thickets and clearings near dwellings in N. America, as in B.C. (N to Cariboo,
ca. 51°10′N), Alta. (Moss 1959), Man. (Winnipeg), Ont. (N to Ottawa), Que. (N to near Portneuf,
Portneuf Co.; see Que. map by Robert Joyal, Nat. can. (Que.) 97(5): map G, fig. 2, p. 564. 1970),
Nfld. (near St. John's; GH), N.B., P.E.I., and N.S.
   Forma *albiflorum* House (flowers white rather than violet to purple) is reported from S Ont. by
Boivin (1966b; Sparta, Elgin Co.). Var. *villosissimum* Desv. (plant copiously pubescent rather than
essentially glabrous) occurs throughout the area.

S. nigrum L.   Black Nightshade.   Tue-chien
Eurasian (but probably native in the E U.S.A. if *S. americanum* is merged with it); introd. along
roadsides and in thickets and waste places in N. America, as in S B.C. (Vancouver Is. and adjacent
islands; Agassiz, near Chilliwack), Alta. (N to Fort Vermilion, 58°24′N; Groh 1949), Sask. (Breitung
1957a), Man. (N to The Pas), Ont. (N to the Ottawa dist.), Que. (N to Buckingham and the Montreal
dist.), Nfld. (St. John's and Humber Valley; GH; MT), N.B. (Kent Co.; NBM), P.E.I. (Charlottetown;
Herbert Groh, Sci. Agric. 7(10):394. 1927), and N.S. [Incl. *S. interius* Rydb. and *S. nodiflorum*
Jacq.]. MAP: Hultén 1968b:792 (exclude the Alaskan stations; see *S. sarrachoides*).

S. rostratum Dunal   Buffalo-bur, Kansas-thistle
A native of the Great Plains of the U.S.A.; introd. elsewhere, as in S B.C. (Vancouver Is.; Lulu Is.;
Armstrong; Salmon Arm; Newgate, S of Fernie), SW Alta. (Fort Macleod; Groh and Frankton 1948),
Sask. (Swift Current; Breitung 1957a), Man. (N to Cormorant, about 45 mi NE of The Pas; DAO),
Ont. (N to Ottawa, where taken by James Fletcher in 1884), and ?P.E.I. (Charlottetown; Hurst
1952). [*Androcera* Rydb.].

S. sarrachoides Sendtner
A native of S. America; introd. elsewhere, as in cent. Alaska (Circle Hot Springs, the collection in
CAN distributed as *S. nigrum* var. *guineense* L., rev. I.J. Bassett; the Sitka, SE Alaska, collection
reported as *S. alatum* Moench by Hultén 1949, also probably belongs here), S B.C. (Vancouver Is.
and the mainland N to Kamloops, where taken by John Macoun in 1899), Alta. (Taber and Fort
Saskatchewan, both as *S. nigrum*, rev. Bassett), S Man. (Helston, near Brandon; WIN), and E Que.
(Rimouski, Rimouski Co.; CAN; RIM).

[S. sisymbriifolium Lam.]
[Tropical America; introd. elsewhere, as in S Ont. (an orchard in Lincoln Co.; OAC), where scarcely
established).]

S. triflorum Nutt.
Native in the w U.S.A. from the Cascades to the Great Plains (Hitchcock et al. 1959); introd. elsewhere, as in s B.C. (N to Savona, w of Kamloops, and Salmon Arm; Herb. V), Alta. (N to Fort Saskatchewan; CAN), Sask. (N to near Prince Albert, where taken by John Macoun in 1886; CAN), Man. (N to Dropmore, about 100 mi NW of Brandon), Ont. (N to Thunder Bay and Hawkesbury, Prescott Co.), and Que. (N to Taschereau, Abitibi Co., 48°40′N, E to L. St. Peter in Yamaska Co.).

[S. tuberosum L.] Potato. Patate
[A native of S. America; an occasional garden-escape to dumps and waste places elsewhere but never established, as in N B.C. (Liard Hot Springs, 59°25′N; CAN), s Man. (Morden, SW of Winnipeg; CAN), s Ont. (Wellington Co.; F.H. Montgomery, Can. Field-Nat. 62(2):94. 1948), Que. (N to the Gaspé Pen. at Ste-Flavie), Nfld. (Rouleau 1956), N.B. (Sussex, Kings Co.; CAN), P.E.I. (near Grand Tracadie, Queens Co.; ACAD), and N.S. (Sable Is.; CAN). See D.S. Correll, *The Potato and its Wild Relatives,* Texas Research Foundation, 1962, 606 pp.]

[S. villosum (L.) Mill.]
[Eurasian; reports from our area appear largely or wholly referable to *S. sarrachoides.* (*S. nigrum* var. *vill.* L.).]

SCROPHULARIACEAE (Figwort Family)

(Ref.: Pennell 1935)

Herbs with simple (some of the lower leaves 3-foliolate in *Tonella*), opposite or alternate (rarely whorled), exstipulate leaves. Flowers perfect, gamopetalous, hypogynous, typically 5-merous and more or less irregular or 2-lipped (the lower lip usually 3-lobed, the upper lip either 2-lobed or the lobes nearly or quite united to form a usually laterally compressed "helmet" or galea). Stamens sometimes 2 (5 in *Verbascum*) but more commonly 4 and didynamous (2 long, 2 short; a fifth sterile filament sometimes present), inserted on the corolla-tube and alternating with its lobes. Style 1. Ovary superior. Fruit a 2-locular, usually many-seeded capsule.

1 Stem-leaves mostly alternate (the lower ones sometimes opposite or whorled; leaves usually all basal in *Limosella*).
  2 Corolla spurred at base, strongly 2-lipped, the throat largely or entirely closed by a prominent palate on the lower lip; calyx 5-lobed, regular; stamens 4; stigmas wholly united; (chiefly introd.).
    3 Leaves palmately veined and 5–9-lobed, reniform-orbicular, long-petioled; corolla about 8 mm long, pale violet with a yellow palate; capsule opening by 2 pores; stem filiform, trailing or twining, the flowers solitary in its axils . . . . *Cymbalaria*
    3 Leaves pinnately veined, entire or with a few obscure teeth.
      4 Leaves broadly ovate to suborbicular, to 3 cm long; corolla about 1 cm long, yellow, the upper lip purple within; capsule opening by 2 pores; stem prostrate at base, the flowers solitary in its axils . . . . *Kickxia*
      4 Leaves linear to narrowly lanceolate; stem erect.
        5 Flowers axillary, on pedicels to about 1.5 cm long; corolla 5 or 6 mm long, blue-purple, with yellow on the depressed palate; capsule asymmetrical, opening by 2 pores; leaves linear, to about 2 cm long; plant glandular-pubescent . . . . *Chaenorrhinum*
        5 Flowers subsessile or short-pedicelled in terminal bracted racemes; capsule symmetrical, dehiscent by valves; plants glabrous . . . . *Linaria*
  2 Corolla not spurred.
    6 Leaves all basal or those of the stem much reduced, at least the middle and upper ones sessile and bract-like, the principal ones long-petioled in basal clusters.
      7 Leaves entire, usually all basal and forming dense clusters at the rooting nodes of the slender runners, their fleshy elliptic blades usually less than 2 cm long and 7 mm broad, much shorter than the slender petioles; flowers solitary on naked scape-like pedicels to about 3 cm tall (or sometimes short-pedicelled in a cluster subtended by a whorl of leaves); calyx regular, 2 or 3 mm long, its 5 triangular lobes less than half as long as the tube; corolla nearly regular, its tube about equalling the calyx, its lobes spreading; stamens 4; glabrous annuals of fresh or brackish shores and wet sands . . . . *Limosella*
      7 Leaves mostly distinctly toothed; flowers racemose; calyx 2–4-lobed; stamens 2.
        8 Plants glabrous (except for the more or less ciliate calyx-lobes), to about 4 dm tall, from fleshy, often creeping rhizomes; leaves subentire to shallowly toothed; flowers bluish, to about 8 mm long; calyx 2-lobed, to 8 mm long; capsules 2-seeded; (Alaska to Dist. Mackenzie) . . . . *Lagotis*
        8 Plants more or less pubescent (or *Besseya rubra* often nearly or quite glabrate in age); capsules several- or many-seeded.
          9 Stems to about 1.5 dm tall, copiously woolly with brownish hairs; principal leaf-blades mostly as broad as or broader than long, truncate to cordate at base, rather coarsely doubly-serrate or incised-serrate, woolly especially along the margins; calyx 4-lobed, the sepals distinct nearly to base; corolla pink to blue (sometimes wanting); (Alaska–Yukon–w Dist. Mackenzie) . . . . *Synthyris*
          9 Stems to over 4 dm tall, they and the leaves usually more or less whitish-pubescent (but not brownish-woolly); principal leaves tapering

to broadly rounded at base (rarely subcordate); corolla none; (S Alta. and sw Sask.) . . . . *Besseya*

6 Leaves more gradually reduced upwardly in comparison with the basal leaves (or the middle and upper ones sometimes the largest).

    10 Stamens 5; corolla normally yellow, nearly regular, rotate, the tube much shorter than the lobes; stigmas wholly united; calyx regular, deeply 5-lobed; flowers in elongate simple or panicled spikes or racemes; leaves lanceolate to lance-ovate; stem to about 2 m tall; (introd.) . . . . *Verbascum*

    10 Stamens 2 or 4.

        11 Stamens 2; stigmas wholly united; corolla-limb rotate, much longer than the very short tube, 4-lobed and weakly 2-lipped (the upper lip of 2 fused lobes); flowers whitish to purplish-blue; calyx 4-lobed, the lobes equal or the upper pair the shortest . . . . *Veronica*

        11 Stamens 4; corolla tubular to campanulate.

            12 Corolla weakly 2-lipped (the 5 lobes all directed forward), white, yellow, or purple; calyx deeply 5-lobed, its lobes distinctly unequal; stigmas 2-lobed; flowers pedicelled in long 1-sided racemes; (introd.) . . . . . . *Digitalis*

            12 Corolla distinctly 2-lipped; stigmas wholly united.

                13 Corolla with throat closed by a prominent palate, distinctly saccate at base, the 2 large lobes of the upper lip erect or reflexed; calyx deeply 5-lobed; capsule asymmetrical; leaves entire; (introd.) . . . . *[Antirrhinum]*

                13 Corolla with open throat, galeate (its mostly subentire upper lip laterally compressed or keeled, more or less arching and helmet-shaped).

                    14 Calyx mostly 5-lobed (sometimes 2-lobed or merely split in front, or subentire at the oblique summit); pollen-sacs similar in size and position; leaves toothed to dissected, often basal as well as cauline; chiefly perennials . . . . *Pedicularis*

                    14 Calyx 4-cleft or lateral pairs of its 4 lobes often partly or wholly connate; pollen-sacs unequally set, one fixed by the middle and appearing terminal on the filament, the other (sometimes reduced or obsolete) attached by its apex and pendulous or reflexed along the filament; leaves wholly cauline, entire to deeply few-cleft but rarely distinctly pinnatifid or marginally toothed.

                        15 Galea (hood or beak of the 2-lipped corolla) distinctly surpassing the lower lip; perennials, the erect or ascending stems commonly clustered from a woody caudex . . . . . *Castilleja*

                        15 Galea only slightly or not at all surpassing the lower lip; annuals, the slender stems simple or branched above . . . . *Orthocarpus*

1 Leaves chiefly opposite or whorled (or the upper bracteal ones often alternate); corolla spurless.

    16 Leaves (at least some of them) whorled, lanceolate to lance-ovate, acuminate, sharply serrate, petioled; corolla nearly regular, the limb 4-lobed; flowers in slender terminal spike-like racemes; stigmas wholly united; stamens 2; stem to about 2 m tall.

        17 Corolla tubular, white, pink, or purplish, about 8 mm long, the lobes much shorter than the tube; calyx usually deeply 5-parted, the 2 lobes of the lower lip the longest; capsule narrowly ovoid, 4 or 5 mm long, much longer than broad, the acute apex not emarginate; leaves in whorls of up to 7; (S Man. to w Ont.) . . . . *Veronicastrum*

        17 Corolla rotate, blue-violet, the lobes at most about 5 mm long, much longer than the tube; calyx deeply 4-parted; capsule broadly oblong, only slightly longer than broad, rounded or slightly emarginate at summit; leaves mostly opposite, but some in whorls of 3 or rarely 4 . . . . *Veronica*

    16 Leaves mostly opposite.

18 Calyx 4-lobed (or *Veronica latifolia* with a fifth very small sepal; or calyx
essentially entire to variously lobed in *Pedicularis*); stigmas wholly united.
  19 Stamens 2; plants not root-parasitic.
    20 Corolla pale yellow, strongly 2-lipped, the small ascending upper lip more
      or less saccate, the obovate lower lip inflated and slipper-like; flowers
      numerous, about 12 mm broad; leaves pinnately parted or divided into 2 or
      3 pairs of broadly lanceolate to ovate, toothed or pinnatifid segments;
      pilose annual; (introd.) .... [*Calceolaria*]
    20 Corolla with a spreading, 4-lobed and weakly 2-lipped limb; flowers whitish
      to purplish blue .... *Veronica*
  19 Stamens 4; corolla strongly 2-lipped, with a well-developed tube, galeate (the
    upper lip usually subentire and laterally compressed, arching and more or less
    helmet-shaped); plants frequently root-parasitic, often darkening in drying.
    21 Leaves palmately veined, coarsely serrate, ovate, seldom more than twice
      as long as broad; flowers sessile or subsessile; corolla whitish, bluish, or
      purplish, with darker bluish or purplish veins, not over 1 cm long; calyx
      nearly regular; annuals with slender stems .... *Euphrasia*
    21 Leaves pinnately veined or pinnatifid, relatively longer.
      22 Capsule oblique at summit, opening along the longer margin.
        23 Leaves entire or the upper ones coarsely toothed to lacerate near
          base; flowers at most about 12 mm long, solitary in the remote
          axils; corolla whitish with a yellow palate; calyx-lobes subulate, the
          upper pair slightly the longer; seeds not more than 4; (transconti-
          nental) .... *Melampyrum*
        23 Leaves shallowly to deeply pinnatifid; flowers larger, yellow, pink,
          red, or purple, in terminal spikes or racemes; calyx subentire and
          oblique or variously lobed; seeds numerous .... *Pedicularis*
      22 Capsule symmetrical or nearly so, opening along 2 margins; flowers in
        spike-like racemes; leaves sessile or subsessile.
        24 Corolla yellow or bronze-tinged, to about 2 cm long; fibrous-rooted
          annuals.
          25 Calyx conspicuously inflated in fruit, strongly flattened; seeds
            flat, broadly winged; leaves triangular-lanceolate to oblong,
            serrate or crenate-serrate, to about 6 cm long and 1.5 cm
            broad; plant not glandular-hairy .... *Rhinanthus*
          25 Calyx scarcely inflated in fruit, not markedly flattened; seeds
            turgid, wingless; leaves lanceolate to ovate, crenate-dentate, to
            4 cm long and 2 cm broad; plant glandular-hairy; (introd. in SW
            B.C.) .... *Parentucellia*
        24 Corolla purplish; seeds not winged.
          26 Annual with fibrous roots; corolla rose-red to red-purple,
            pubescent, about 1 cm long; capsule elliptic; leaves oblong-
            lanceolate, to 3 cm long, coarsely few-toothed; stem
            scabrous-pubescent, to about 4 dm tall; (introd.) .... *Odontites*
          26 Perennial from a somewhat woody rhizome; corolla rich dark
            purple, densely glandular-pubescent, to over 1.5 cm long; calyx
            purple; capsule ovoid; leaves ovate or oval, crenate-serrate,
            less than 3 cm long; stem viscid-villous, to about 2 dm tall; (NE
            Man. to Baffin Is. and Nfld.) .... *Bartsia*
18 Calyx 5-lobed.
  27 Leaves deeply pinnatifid or cleft to base; stamens 4.
    28 Some of the long-petioled lower leaves 3-foliolately compound, others
      deeply 3-parted (all of the leaves small, the blade not over 2 cm long;
      middle and upper leaves with progressively narrower segments; upper-
      most leaves sessile and almost entire; lowermost long-petioled leaves
      often merely toothed); corolla blue and white, subrotate, with a short tube
      and spreading, somewhat 2-lipped limb to 4 mm broad; stamens equal;

stems to 2.5 dm tall, slender and weak, often branched, glabrous throughout, or the pedicels sparingly stipitate-glandular below the flowers . . . . [*Tonella*]

28 Leaves more or less deeply pinnatifid but neither 3-foliolate nor pinnately dissected to base into distinct leaflets; stamens didynamous (2 long, 2 short).

    29 Capsule oblique at summit; corolla yellow or purple-tinged, strongly 2-lipped and galeate . . . . *Pedicularis*

    29 Capsule symmetrical; (s Ont.).

        30 Corolla pale lavender or greenish white, weakly 2-lipped, usually less than 5 mm long, not much surpassing the linear-lobed calyx; stigma 2-lobed; flowers axillary, on slender pedicels to about 1 cm long; leaves triangular-ovate in outline, to about 3 cm long; stem often decumbent at base, diffusely branched, to about 2 dm tall . . . . *Conobea*

        30 Corolla yellow or purple-tinged, somewhat irregular, at least 3 cm long, much surpassing the calyx; stigma unlobed; flowers in terminal racemes; leaves lanceolate to lance-ovate in outline; stem erect, to about 2 m tall . . . . *Aureolaria*

27 Leaves entire or toothed, not pinnatifid (but lower leaves of *Aureolaria virginica* often with 1 or 2 pairs of large obtuse lobes near base).

    31 Fertile stamens 2 (a second pair of rudimentary or well-developed sterile filaments also often present); stigmas 2-lobed; flowers pedicelled in the leaf-axils; corolla more or less 2-lipped; leaves entire or only obscurely toothed, sessile or subsessile; stem to about 4 dm tall.

        32 Sepals connate below into a 2-lipped, somewhat 5-grooved calyx 5 or 6 mm long at anthesis, the upper 3-toothed lobe the longest; corolla yellow, often with some maroon dots, to 9 mm long; fruit rather finely short-stipitate-glandular; leaves broadly linear to elliptic or oblong (the lower ones mostly oblanceolate), to 5 cm long and 1.5 cm broad, they and the calyx glandular-pubescent . . . . [*Mimetanthe*]

        32 Sepals free nearly or quite to base.

            33 Calyx subtended at its immediate base by a pair of sepal-like bracts; sepals lanceolate; corolla white to golden yellow; stem glabrous or glandular-puberulent above . . . . *Gratiola*

            33 Calyx naked at base; sepals linear; corolla white to pale lavender; glabrous annuals . . . . *Lindernia*

    31 Fertile stamens 4, a fifth sterile filament sometimes present.

        34 Leaves linear to linear-oblanceolate, entire, sessile or subsessile, 1-nerved; stigmas wholly united; annuals.

            35 Corolla weakly 2-lipped, pink, roseate, or purple, to over 3 cm long; sterile stamen none; flowers solitary in the axils of the upper leaves; calyx regular; leaves usually scabrous above . . . . *Gerardia*

            35 Corolla distinctly 2-lipped, about 5 mm long, cleft nearly to base between the lips, conspicuously gibbous on the upper side near base, the lips blue-violet (upper lip sometimes white); a sterile gland-like rudimentary stamen present; flowers axillary and in a terminal umbel-like group . . . . *Collinsia*

        34 Leaves broader, mostly pinnately several-veined.

            36 Corolla nearly regular or weakly 2-lipped.

                37 Flowers yellow or blue, solitary on long pedicels in the axils of ordinary finely to coarsely toothed foliage-leaves; sterile filaments none; stigma 2-lobed . . . . *Mimulus*

                37 Flowers sessile or pedicelled in the axils of bracts, these much smaller than the foliage-leaves, the whole inflorescence spicate, racemose, or paniculate; stigma entire.

38 Corolla yellow, to over 3 cm long, glabrous outside; calyx finely pubescent; sterile filaments none; capsule to 1.5 cm long, densely rusty-pubescent; flowers sessile or on stout pedicels not over 3 mm long, the inflorescence spike-like; leaves petioled, lance-ovate, the larger ones often with 1 or 2 pairs of large obtuse lobes near base; (*A. virginica*; S Ont.) .... *Aureolaria*

38 Corolla white to bluish-violet or purple; leaves not lobed.

  39 Flowers sessile, the inflorescence a terminal spike; corolla dark purple, pubescent, about 2 cm long; calyx pubescent, about 7 mm long; sterile filament none; capsule barely surpassing the calyx; leaves sessile, lance-ovate, coarsely sinuate-dentate, very scabrous; stem short-hirsute; (S Ont.) .... *Buchnera*

  39 Flowers subsessile to slender-pedicelled, the inflorescence racemose or paniculate; a long slender filament present in addition to the perfect stamens; capsule distinctly surpassing the calyx; leaves entire or shallowly toothed, those of the basal rosette petioled, those of the stem sessile and sometimes clasping .... *Penstemon*

36 Corolla distinctly 2-lipped.

  40 Flowers yellow or blue, solitary on long pedicels in the axils of ordinary finely to coarsely toothed foliage-leaves; sterile filament none; stigma 2-lobed; calyx regular or irregular, its lobes usually shorter than the tube .... *Mimulus*

  40 Flowers in a spicate or paniculate terminal inflorescence (if axillary in *Collinsia verna*, chiefly in 1–3 subterminal whorls of about 5 flowers each, forming an umbel-like cluster); a sterile filament present.

    41 Flowers sessile in a terminal spike; corolla over 2 cm long, white or partly greenish-yellow, often tinged with pink or purple; calyx regular, of separate obtuse or rounded, broadly elliptic and overlapping sepals; leaves shallowly serrate, subsessile or on short winged petioles; stem to about 2 m tall; (SE Man. to Nfld. and N.S.) .... *Chelone*

    41 Flowers smaller, slender-pedicelled.

      42 Flowers chiefly in 1–3 subterminal whorls of about 5 flowers each, forming an umbel-like cluster; corolla commonly at least 1 cm long, conspicuously gibbous on the upper side at base, the lower lip bright blue, the upper lip white varying to pale blue; calyx about half as long as the corolla, somewhat irregular, its narrowly triangular lobes longer than the tube; leaves to about 5 cm long, oblong-ovate or triangular-ovate, entire or obscurely serrate, the lower ones petioled, the others sessile; annuals to about 4 dm tall, finely glandular-puberulent above .... *Collinsia*

      42 Flowers irregularly paniculate; corolla rarely over 1 cm long, reddish brown or the lower lobe yellowish green, the median lobe of the lower lip deflexed; calyx regular, about 1/3 as long as the corolla, its broadly ovate or broadly triangular lobes about as long as the tube; leaves narrowly to broadly ovate, petioled, serrate or incised, to about 2 dm long; stem to over 3 m tall, square in cross-section; rhizomatous perennials ....
.... *Scrophularia*

[ANTIRRHINUM L.]   [7482]   Snapdragon

1   Calyx-lobes ovate, to 5 mm long; corolla 3 or 4 cm long, of various colours (often red-purple, rarely yellowish white); capsule about 1.5 cm long; seeds not cup-shaped; perennial .... [*A. majus*]
1   Calyx-lobes linear, to 2 cm long; corolla less than 1.5 cm long, pink-purple; capsule about 1 cm long; seeds appearing cup-shaped because of the broad incurved wing; annual .... [*A. orontium*]

[A. majus L.]   Common Snapdragon
[European; an occasional garden-escape to roadsides and waste places in N. America, but scarcely persistent, as in s Ont. (Wellington Co.; F.H. Montgomery, Can. Field-Nat. 62(2):92. 1948) and ?N.S. (Roland 1947).]

[A. orontium L.]   Lesser Snapdragon
[European; an occasional garden-escape to roadsides and waste places in N. America but scarcely established, as in s Alaska (Hultén 1968*b*), sw B.C. (Victoria, Vancouver Is.; John Macoun 1884), s Ont. (grainfield in Perth Co.; OAC; reported from Niagara Falls, Welland Co., by J.M. Macoun 1897), and P.E.I. (Pennell 1935). MAP: Hultén 1968*b*:793.]

AUREOLARIA Raf.   [7604]   False Foxglove

1   Calyces, pedicels, and capsules strongly stipitate-glandular; corolla yellow and commonly purple-tinged, glandular-pubescent outside; seeds wingless; leaves sessile or subsessile, rather finely 2-pinnatifid, minutely pubescent; annual; (s Ont.) .... *A. pedicularia*
1   Plant nonglandular; corolla uniformly yellow, glabrous outside; seeds winged; leaves petioled, at least the lower ones coarsely lobed; perennials; (s Ont.).
    2   Plant finely pubescent throughout; pedicels at most about 6 mm long; lower leaves coarsely sinuate or pinnatifid, the upper ones less deeply lobed or entire ...... *A. virginica*
    2   Plant essentially glabrous or leaves puberulent; pedicels to about 1.5 cm long; lower leaves deeply pinnate-lobed, the upper ones dentate or entire .... *A. flava*

A. flava (L.) Farw.
/t/EE/   (Hp)   Deciduous woods from Minn. to s Ont. (Essex, Lambton, Middlesex, Norfolk, Waterloo, York, Welland, and Lincoln counties) and sw Maine, s to Ala. and Ga. [*Gerardia* L.; *G. quercifolia* Pursh; *Agalinis* Boivin]. MAP: Pennell 1935: map 100, p. 393.

A. pedicularia (L.) Raf.
/t/EE/   (T)   Dry deciduous woods and clearings from Minn. to s Ont. (Essex, Lambton, Huron, Norfolk, Waterloo, Wellington, Brant, Lincoln, Welland, Wentworth, and York counties) and sw Maine, s to Ill., Ohio, and N.C. [*Gerardia* L.; *Agalinis* Blake; *Dasystoma* Benth.]. MAP: Pennell 1935: map 101, p. 398.
   Some of our material is referable to ssp. *ambigens* (Fern.) Farw. (incl. ssp. *intercedens* Pennell; stem more or less copiously stipitate-glandular above rather than nonglandular).

A. virginica (L.) Pennell   Downy Foxglove
/t/EE/   (Hp)   Dry deciduous woods from Minn. to s Ont. (Norfolk, Waterloo, Welland, and Lincoln counties; CAN; TRT; Zenkert 1934) and N.H., s to La., Ala., and Fla. [*Rhinanthus* L.; *Agalinis* Blake; *Gerardia* BSP.; *Dasystoma* Britt.; *D. pubescens* Benth.].

BARTSIA L.   [7645]

B. alpina L.   Velvet-bells, Alpine Bartsia
/aST/EE/GEA/   (Hp)   Moist turfy slopes (chiefly calcareous) from Man. (known only from Churchill) to Ont. (coasts of Hudson Bay–James Bay s to 53°25′N), Que. (coasts of Hudson Strait and Hudson Bay–James Bay s to 53°50′N; Knob Lake dist. at ca. 54°45′N), s Baffin Is. (N to ca.

56°N), the coast of Labrador, and NW Nfld.; W Greenland N to ca. 76°N, E Greenland N to 68°44'N; Iceland; Europe; Asia Minor. MAPS: Hultén 1958: map 21, p. 41; Porsild 1957: map 290, p. 197.

BESSEYA Rydb. [7583] Kitten-tails

1 Calyx 4(3)-lobed, the base cup-shaped and surrounding the ovary and stamens; inflorescence much elongate at maturity, the lower fruits becoming remote, the bracts distinctly narrowed toward base; basal leaves to 12 cm long, often subcordate at base; stem to 6 dm tall; (S ?B.C.) .... [*B. rubra*]
1 Calyx 2(3)-lobed; inflorescence remaining rather compact, its bracts seldom much narrowed below; basal leaves to about 7 cm long, seldom at all cordate; stem to 4 dm tall; (S Alta. and SW Sask.) .... *B. wyomingensis*

[B. rubra (Dougl.) Rydb.] Red Kitten-tails
[The report of this species of the W U.S.A. (Wash., Mont., and Oreg.) from SE B.C. by Henry (1915; Flathead R., SE of Fernie) requires confirmation, perhaps being based upon *B. wyomingensis*. The report from SW Sask. by John Macoun (1884; Fort Walsh, Cypress Hills) is based upon this latter species, the relevant collection in CAN. (*Gymnandra* Dougl.; *Synthyris* Benth.).]

B. wyomingensis (Nels.) Rydb.
/T/WW/ (Hs) Open slopes and dry meadows in the lowlands and foothills from S ?B.C. (see *B. rubra*), S Alta. (Crowsnest Pass, on the B.C.–Alta. boundary; Waterton Lakes; Milk River Ridge), and SW Sask. (DAO; Breitung 1957a) to Utah, Colo., and Nebr. [*Wulfenia* Nels.; *Synthyris* Heller; *W. (B.; S.) gymnocarpa* Nels.; *B. cinerea sensu* Breitung 1957a, not *Veronica cin.* Raf., basionym].

BUCHNERA L. [7622]

B. americana L. Blue-hearts
/t/EE/ (Gp) Moist sandy soil, prairies, and open woods (doubtless parasitic on the roots of various plants) from E Kans. to Mo., Mich., S Ont. (Squirrel Is., Ipperwash Beach, and Port Franks, Lambton Co.; CAN; TRT; *see* S Ont. map by Soper 1962: map 24, fig. 22, p. 36), N.Y., and N.J., S to Tex. and Fla. MAP: Pennell 1935: map 141 (the dot indicating a station at the S end of the Bruce Pen., L. Huron, should be deleted; *see* Soper 1962), p. 478.

[CALCEOLARIA L.] [7474]

[C. scabiosifolia R. & S.] Slipperwort
[A native of S. America; reported from Ont. by Boivin (1966b; Ottawa), where probably a garden-escape but scarcely established.]

CASTILLEJA Mutis [7631] Indian Paint-brush, Painted-cup

1 Root annual; stems usually solitary, erect, to 8 dm tall; leaves and bracts entire, linear-lanceolate, the bracts much longer than the flowers, the uppermost ones tipped with red or scarlet; corolla yellowish, to 2.5 cm long; calyx to 2 cm long; plant glandular-villous, of alkaline marshes and meadows; (B.C.) .... *C. exilis*
1 Root perennial, usually woody; stems clustered, often decumbent and sometimes rooting at the base; bracts usually 3–9-lobed or divided (if entire, relatively broad).
  2 Leaves mostly entire (or the upper ones below the floral bracts sometimes with 1(2) pairs of lateral lobes (other species, particularly *C. applegatei* and *C. hispida,* may often key out here); perennials with usually clustered stems on a woody caudex, the leaves all cauline.
    3 Inflorescence predominantly various shades of rose, red, or scarlet (atypically whitish or yellow).
      4 Galea short, at most about 6 mm long and 1/3 the length of the corolla-tube; lower corolla-lip to 3 mm long; corolla to 18 mm long; calyx to 16 mm long;

bracts mostly oval (the lowest ones lanceolate and with a pair of short linear lateral lobes); stems to about 5 dm tall; (B.C. to James Bay) . . . . *C. raupii*

4   Galea longer, commonly at least half the length of the corolla-tube and usually at least 4 times as long as the dark-green, thickened, lower corolla-lip.

    5   Calyx much more deeply cleft ventrally (the side away from the floral axis) than dorsally, usually more showy than the mostly 3-lobed bracts (inflorescence bright red or scarlet, occasionally yellow); corolla to over 4 cm long; leaves linear, usually glabrous; stems to over 7 dm tall, usually more or less hispid below but otherwise glabrous; (S ?B.C.) . . . . [*C. linariaefolia*]

    5   Calyx about equally cleft on both sides (or more deeply so dorsally); bracts more showy than the calyx.

        6   Bracts mostly more or less toothed or cleft into acute terminal lobes (rarely entire); inflorescence scarlet (rarely crimson); corolla to 4 cm long; leaves glabrous to puberulent or finely villous; stem to 8 dm tall, often branched, glabrous or short-pubescent; (Alaska–B.C. to w Ont.) . . . . *C. miniata*

        6   Bracts or their segments mostly obtuse or rounded at summit; inflorescence crimson (rarely purple; drying scarlet).

            7   Bracts mostly with 1 or 2 pairs of short lateral lobes; corolla to 3.5 cm long; plant to 3(4) dm tall, commonly glabrate or only obscurely viscid-villous in the inflorescence; (B.C. and Alta.) . . . . *C. rhexifolia*

            7   Bracts mostly entire (or the upper ones sometimes lobed); corolla to 3 cm long.

                8   Stems to 3 dm tall, viscid-villous; leaves viscid-puberulent or -villous; (?B.C.) . . . . [*C. elmeri*]

                8   Stems to 6 dm tall, glabrous or only slightly pilose below the inflorescence; leaves usually glabrous or glabrate (sometimes finely pubescent); (Alaska–B.C.) . . . . *C. hyetophila*

3   Inflorescence predominantly greenish white to yellow (sometimes varying to orange, red, or purple).

    9   Lower corolla-lip at most 1/5 the length of the galea (*C. lutescens* may sometimes key out here); leaves distinctly 3-ribbed; stems to about 6 dm tall; (B.C.).

        10 Bracts acute to attenuate; corolla to 2 cm long; leaves finely pubescent; stems finely pilose (villous-hirsute through the inflorescence); (N B.C.) . . . . *C. fulva*

        10 Bracts mostly obtuse or rounded at summit.

            11 Inflorescence elongate, predominantly yellow but varying to pink, orange, or even red; leaves villous-puberulent to glabrate; (B.C.–Alta.) . . . . *C. gracillima*

            11 Inflorescence relatively compact and more consistently yellowish; leaves roughish-appressed-pubescent; (Alaska–Yukon–B.C.) . . . . *C. unalaschcensis*

9   Lower corolla-lip commonly at least 1/4 the length of the galea; bracts mostly obtuse or rounded at summit; leaves often less distinctly ribbed.

    12 Plant strongly viscid-villous nearly throughout, rarely over 1 or 2 dm tall; corolla to 2.5 cm long (varying from yellow to red or purple); leaves mostly linear-lanceolate; (B.C. and sw Alta.) . . . . *C. occidentalis*

    12 Plants more or less pubescent but not strongly viscid-villous throughout, commonly to 5 or 6 dm tall.

        13 Bracts commonly 3–7-lobed; leaves linear or linear-lanceolate, scabrous-puberulent; stems puberulent to obscurely villous or hispid; (B.C. and sw Alta.) . . . . *C. lutescens*

        13 Bracts commonly entire (or some of them with 1 or 2 pairs of lateral lobes); (essentially transcontinental) . . . . *C. pallida*

2   Leaves (at least a relatively large number of the upper ones below the inflorescence) with 1 or more pairs of lateral lobes.

14 Inflorescence predominantly various shades of rose, red, crimson, or scarlet (occasionally whitish or yellow).
   15 Plants annual or biennial, with clustered, mostly entire, oblong or obovate rosette-leaves; stem-leaves soft-pubescent, commonly with 1 or 2 pairs of linear or narrowly oblong lobes; bracts commonly deeply 3-lobed, bright scarlet toward summit (yellow in f. *lutescens*); calyx deeply divided into 2 lateral halves, each half gradually widened to a broadly rounded, truncate, or barely emarginate (otherwise entire) summit; galea much less than half the length of the corolla-tube; stems to about 6 dm tall, more or less villous; (sw Sask. to Ont.) . . . . *C. coccinea*
   15 Plants perennial, the stems clustered on a woody caudex, the leaves all cauline; lateral halves of the calyx themselves deeply cleft; galea at least half the length of the corolla-tube (mostly about equalling it or even slightly longer).
      16 Pubescence consisting partly of long hispid multicellular hairs; (s B.C. and sw Alta.).
         17 Mid-blade (undivided portion) of leaves linear to linear-lanceolate, the pair of divergent elongate lobes not much narrower and arising from near or below its middle; corolla to 2.5 cm long; calyx to 2.5 cm long, its lobes rounded; inflorescence pale or deep rose to crimson; stems to 4 dm tall . . . . *C. angustifolia*
         17 Mid-blade of upper leaves lanceolate to ovate-lanceolate, most of the leaves entire but at least the uppermost ones with a subapical pair of short ascending lobes much narrower than the mid-blade; corolla to 4 cm long; calyx to 3 cm long; inflorescence bright red or scarlet; stems to 6 dm tall . . . . *C. hispida*
      16 Pubescence (when present) more villous than hispid, the sparse hairs slender and entangled (*C. hispida* may key out here).
         18 Stems and leaves more or less strongly glandular-viscid; corolla to 3.5 cm long; bracts 3–5-parted, mostly equalling or surpassing the subtended flowers, bright red, scarlet, or occasionally yellow; stems to 5 dm tall; (s ?B.C.) . . . . [*C. applegatei*]
         18 Stems and leaves scarcely or not at all glandular-viscid.
           19 Leaves finely villous, most of them with 1 or 2(3) pairs of relatively elongate lobes not much narrower than the mid-blade; bracts mostly 5-parted, much shorter than the subtended flowers at anthesis; inflorescence bright scarlet or crimson; corolla to over 3.5 cm long; (s B.C.) . . . . *C. rupicola*
           19 Leaves somewhat villous or glabrate, most of them with 1(2) pairs of rather short lateral lobes usually much narrower than the mid-blade; bracts 3–5-parted, about as long as the subtended flowers; corolla to 2.5 cm long; plants commonly blackening on drying; (Alaska–B.C. and sw Alta.) . . . . *C. parviflora*
14 Inflorescence usually predominantly yellow or yellowish (bracts green in *C. sessiliflora;* often purplish in *C. pallescens*; occasionally pinkish to reddish, crimson, or purplish in *C. angustifolia, C. cusickii, C. flava,* and *C. rustica;* bracts red-tipped in *C. suksdorfii*); perennials with stems usually clustered on a woody caudex (except *C. suksdorfii*), the leaves all cauline.
   20 Corolla-tube greatly elongated, 3 or 4 cm long, strongly curved, the galea about 1 cm long, the whole corolla greenish yellow, pinkish, or purplish; calyx to 4 cm long, yellowish; bracts green, leaf-like; leaves densely puberulent, the lower ones linear and entire, the upper ones broader but with linear lobes; stems to 4 dm tall, villous-tomentose; (s Sask. and s Man.) . . . . *C. sessiliflora*
   20 Corolla-tube usually less than 2 cm long and not strongly curved; bracts more or less petaloid.
      21 Plants of N B.C. near the Yukon boundary; corolla to 22 mm long, the galea to 8 mm long, about twice as long as the lower lip; leaves lance-linear, attenuate, the upper ones with 1 or 2 pairs of slender lateral

lobes mostly at least 1/3 the length of the leaf; stems to about 12 cm tall, appressed-pubescent or glabrate below, often spreading pubescent above and yellowish-hairy in the inflorescence; (Alaska–Yukon–Dist. Mackenzie–N B.C.) .... *C. hyperborea*

21 Plants chiefly of S B.C. (*C. cusickii* and *C. pallescens* also in S Alta.).

  22 Lower corolla-lip prominent, commonly at least 2/3 as long as the galea.

    23 Calyx-lobes broad and usually rounded at apex; bracts mostly obtuse and entire or with 1 or 2 pairs of short lateral lobes, viscid-villous or puberulent; stems to 6 dm tall, they and the leaves viscid-villous (rarely hispidulous); (B.C. and SW Alta.) .... *C. cusickii*

    23 Calyx-lobes usually acute; bracts mostly with 1 or 2 pairs of narrow lateral lobes, puberulent and ciliate.

      24 Lower corolla-lip distinctly pouched and pubescent; leaves densely puberulent; stems commonly not over 2 dm tall, densely puberulent with retrorse hairs or somewhat hispidulous; (B.C. and SW Alta.) .... *C. pallescens*

      24 Lower corolla-lip scarcely pouched, usually not puberulent or only obscurely so; leaves hispid or villous and sometimes glandular; stems to 4 dm tall, hispid or villous; (?B.C.) .... [*C. thompsonii*]

  22 Lower corolla-lip not prominent, at most about 1/2 as long as the galea; stems to 5 or 6 dm tall; (B.C.).

    25 Corolla to 5 cm long, its galea about equalling the tube; inflorescence at first compact, later elongating; bracts red-tipped above a yellow band; leaves densely puberulent; stems usually solitary, from a slender creeping base, obscurely villous or sometimes hispid or glabrate; (?B.C.) .... [*C. suksdorfii*]

    25 Corolla at most 2.5 cm long, the galea much shorter than the tube; flowers rather remote; stems clustered on a woody caudex.

      26 Primary lobes of calyx cleft into 2 linear obtuse lobes; leaves hispidulous to viscid-villous, the lower entire ones linear-lanceolate, the upper ones oblong-ovate or -obovate, with mostly 1–3 pairs of very short tooth-like lobes near apex; bracts golden yellow, similar to the upper leaves, oblong, obtuse, puberulent and more or less viscid-villous, nearly hiding the flowers; stems softly viscid-villous; (SW B.C.) .... *C. levisecta*

      26 Primary lobes of calyx cleft into 2 linear acute lobes; leaves puberulent, the lower ones linear, the upper ones commonly with a single pair of long linear lobes near or below the middle; bracts broader than the leaves but their blades commonly more deeply dissected (1 or 2 pairs of slender acute or acutish lobes) and not hiding the flowers.

        27 Calyx subequally cleft dorsally and ventrally; bracts puberulent or villous; leaves densely puberulent; stems appressed-puberulent to hispidulous or finely villous; (?B.C.) .... [*C. rustica*]

        27 Calyx less deeply cleft dorsally than ventrally; leaves crisp-puberulent; (S B.C.).

          28 Plant green, the crisp puberulence minute; bracts yellowish, crisp-puberulent; stem often nearly or quite glabrous .... *C. cervina*

          28 Plant greyish, the dense crisp puberulence usually longer; bracts yellowish or occasionally reddish, villous or hispid; stem finely pubescent with soft, often retrorse, crisped hairs .... *C. flava*

C. *angustifolia* (Nutt.) Don
/T/W/ (Hp) Dry hills and sagebrush plains from s B.C. (Vancouver Is. and mainland N to Lillooet and Salmon Arm; CAN) and sw Alta. (Waterton Lakes; CAN) to Oreg., Idaho, and Wyo. [*Euchroma* Nutt.; *E. bradburyi* Nutt.; *C. bennittii* Nels. & Macbr.].

[C. *applegatei* Fern.]
[Dry slopes at moderate to high elevations from Oreg., Idaho, and Wyo. to Calif. and Nev. A collection in Herb. V from Manning Provincial Park, about 30 mi SE of Hope, has been placed here but requires confirmation.]

C. *cervina* Greenm.
/T/W/ (Hp) Grasslands and open coniferous woods from s B.C. (N to Cache Creek, about 45 mi w of Kamloops, and Canal Flats, about 45 mi N of Cranbrook; CAN; type from Lower Arrow L., N of Trail) to N Wash. and N Idaho.

C. *coccinea* (L.) Spreng. Scarlet Painted-cup
/T/EE/ (T (Hs, bien.)) Peaty meadows, prairies, thickets, and damp sands and gravels from SE Sask. (Buchanan, about 120 mi NE of Regina; Breitung 1957a) to s Man. (N to Steeprock, about 100 mi N of Portage la Prairie), Ont. (N to Finland, about 70 mi s of Kenora, the Sault Ste. Marie dist., and the Bruce Pen., L. Huron), and s N.H., s to Okla., La., and Fla. [*Bartsia* L.]. MAP: Pennell 1935: map 155, p. 536.
Forma *alba* Farw. (floral bracts white rather than bright scarlet) is known from s Man. (Vivian, E of Winnipeg; DAO). Forma *lutescens* Farw. (floral bracts yellow) is known from Man. (N to Moosehorn, about 110 mi NW of Winnipeg) and s Ont. (Ferndale, Bruce Co.; TRT).

C. *cusickii* Greenm.
/T/W/ (Hp) Meadows and slopes at low to high elevations from s B.C. (N to Kamloops; V) and sw Alta. (between Waterton Lakes and Pincher Creek; Sweet Grass Hills, NW of Calgary; CAN; V) to Oreg., Idaho, and Mont. [*C. pallida* var. *camporum* Greenm. (*C. camp.* (Greenm.) Howell)].

[C. *elmeri* Fern.]
[According to Eastham (1947), a collection from Mt. Brent, near Penticton, has been placed here by Pennell. The genus is so critical, however, that further collections are desirable before accepting this species as a member of our flora.]

C. *exilis* Nels.
/t/W/ (T) Alkaline marshes and meadows from s-cent. B.C. (Boivin 1966b) and Mont. to Calif. and N.Mex.

C. *flava* Wats.
/t/W/ (Hp) Dry soils and sagebrush plains from s B.C. (Osoyoos L., where taken by Dawson in 1877, and Kamloops, where taken by John Macoun in 1889; CAN), Idaho, and Mont. to Nev. and Colo. [*C. brachyantha* Rydb.; *C. breviflora* Gray, not Benth.; *Orthocarpus tenuifolius sensu* John Macoun 1884, as to the Osoyoos L. plant, not (Pursh) Benth., the above relevant Dawson collection in CAN].

C. *fulva* Pennell
/sT/W/ (Hp) Known only from NE B.C. (between Sifton Pass, ca. 57°45′N, and Mt. Selwyn and the type locality, Hudson Hope, both ca. 56°N; CAN; Pennell 1934).

C. *gracillima* Rydb.
T/W/ (Hp) Wet meadows of Mont., Idaho, and Wyo. Collections in Herb. V from s B.C. (Kinbasket, about 60 mi N of Revelstoke, and Fairmont Hot Springs) have been referred to *C. ardifera* and Alta. is included in the range by Rydberg (1922). (*C. ardifera* Macbr. & Pays.).

C. *hispida* Benth.
/T/W/ (Hp) Grassy slopes and forest openings at low to moderate elevations from s B.C.

(Vancouver Is., Vancouver, Tulameen Valley near Princeton, Tranquille L. near Kamloops, Mt. Brent w of Penticton, Rossland, Trail, and Columbia Valley; CAN; V) and sw Alta. (Crowsnest Pass; Waterton Lakes; Three Hills, NE of Calgary; Okotoks, S of Calgary; CAN) to Oreg. and Idaho. [*C. angustifolia* var. *hisp.* (Benth.) Fern.; *C. remota* Greene; incl. ssp. *acuta* Pennell].

C. hyetophila Pennell
/sT/W/ (Hp) Moist places along the coasts of Alaska (N to ca. 61°30′N; type from Windham Bay) and B.C. (S to S Vancouver Is.). MAPS: Hultén 1968b:808; Pennell 1934: map 6 (incomplete), p. 538.
The scarcely separable *C. chrymactis* Pennell of SE Alaska (type from Glacier Bay) differs in its generally longer corollas, 3-lobed rather than entire floral bracts, and more compact inflorescence. MAP: Hultén 1968b:808.

C. hyperborea Pennell
/Ss/W/eA/ (Hp) Dry places at low to fairly high elevations from Alaska (N to ca. 69°30′N; type from the Sheenjek Valley), the Yukon (N to ca. 67°N), and the Mackenzie R. Delta to northernmost B.C. (Haines Road at ca. 59°30′N; CAN); NE Siberia. [*C. ?kuschei* Eastw.]. MAPS: Hultén 1968b:812; Pennell 1934: map 3 (somewhat incomplete), p. 531.
Three Yukon endemics (?microspecies) described by Pennell (1934) may be rather arbitrarily separated from *C. hyperborea* as follows:
1 Corolla to 13 mm long, the lower lip about 3/4 the length of the stout galea, this only
   slightly exserted beyond the calyx; [SW Yukon].
   2 Leaves linear; stems finely pubescent below, more coarsely pubescent in the
      inflorescence; [type from near Kluane L.; MAPS: Hultén 1949: map 1040, p. 1470;
      Pennell 1934: map 3, p. 531] .... [*C. muelleri* Pennell]
   2 Leaves lanceolate; stems densely villous; [type from Bear Creek; *C. ?annua* Pennell;
      MAPS: Hultén 1968b:811, and 1949: map 1045, p. 1471; Pennell 1934: map 3, p. 531] ......
      .... [*C. villosissima* Pennell]
1 Corolla to 2 cm long, the lower lip about 2/3 the length of the galea, this well exserted
   beyond the calyx.
   3 Leaves linear, entire or the upper ones with short lateral lobes; stems to 3.5 dm tall,
      they and the leaves canescent-pubescent; [the Yukon, the type from the Lewes R.,
      probably near Fort Selkirk; MAPS: Hultén 1968b:810, and 1949: map 1046, p. 1471;
      Pennell 1934: map 3, p. 531] .... [*C. yukonis* Pennell]
   3 Leaves lance-linear, at least the upper ones with long spreading lobes; stems shorter,
      finely pubescent to glabrate; [Alaska to w Dist. Mackenzie and N B.C.] ..... *C. hyperborea*

C. levisecta Greenm.
/t/W/ (Hp) Meadows and prairies from SW B.C. (collections in CAN and V from Vancouver Is. and Trial Is.; collection in CAN from Queen Charlotte Is., detd. Pennell, but not listed by Calder and Taylor 1968) to Oreg.

[*C. linariaefolia* Benth.]
[The inclusion of B.C. in the range of this species of the w U.S.A. (N to Oreg., Idaho, and Mont.) by Rydberg (1922) is probably based upon collections in CAN from Cascade and the Kettle R., both E of Grand Forks, where taken by J.M. Macoun in 1902, these referred to *C. lutescens* by Porsild.]

C. lutescens (Greenm.) Rydb.
/t/W/ (Hp) Grasslands and open coniferous woods from S B.C. (Vancouver Is.; Osoyoos; near Kamloops; Elko; Cascade, E of Grand Forks; Flathead) and SW Alta. (Waterton Lakes; Breitung 1957b) to Oreg. and Mont. [*C. pallida* var. *lut.* Greenm.; *C. linariaefolia sensu* Rydberg 1922, as to the B.C. part of the range, not Benth., relevant collections in CAN].

C. miniata Dougl.
/T/WW/ (Hp) Meadows and slopes at low to moderate elevations from the S Alaska Panhandle and B.C. to Alta. (N to Spirit River, 55°47′N), Sask. (N to ca. 54°N), Man. (N to 18 mi N of The Pas), and w Ont. (Boivin 1966b), S to Calif. and N.Mex. [*C. confusa* Greene; *C. dixonii* Fern.; incl. the

glabrous or glabrate *C. lanceifolia* and *C. magna* Rydb. and the pubescent extremes, *C. crispula* Piper and *C. tweedyi* Rydb.; *C. rhexifolia* of most or all reports from Sask. and Man., not Rydb.]. MAPS: Hultén 1968*b*:807; W.J. Cody, Can. Field-Nat. 70(3): fig. 1, p. 124. 1956.

C. occidentalis Torr.
/T/W/ (Hp) Slopes and meadows at high elevations from S B.C. (collection in CAN from Rogers Pass through the Selkirks between Revelstoke and Golden; reported from Yoho and Windermere by Ulke 1935) and SW Alta. (N to Jasper; CAN) to Utah and Colo.

C. pallescens (Gray) Greenm.
/t/W/ (Hp) Dry hills and sagebrush plains from SE B.C. (L. Osoyoos, near the U.S.A. boundary S of Penticton; Sidley, E of L. Osoyoos; Lower Arrow L., N of Trail; CAN) and SW Alta. (Waterton Lakes; Three Hills, NE of Calgary; Cardston; CAN) to Idaho, Mont., and Wyo. [*Orthocarpus* Gray].

C. pallida (L.) Spreng.
/aST/X/eA/ (Hp) Rocky, gravelly, or peaty places at low to fairly high elevations, the aggregate species from the coasts of Alaska–Yukon–Dist. Mackenzie–Dist. Keewatin to N Banks Is., Victoria Is., S Baffin Is., and northernmost Ungava–Labrador, S in the West through SE B.C. (Kicking Horse L.; CAN) and SW Alta. (Banff; CAN) to Utah, Colo., and the Black Hills of S.Dak., farther eastwards S to cent. Man. (Hayes R. from 100 mi SW of York Factory to York Factory; not known from Sask.), N Minn., cent. Ont. (N shore of L. Superior; coasts of Hudson Bay–James Bay), Que. (S to SE James Bay, Anticosti Is., and the Gaspé Pen.), Nfld., N.B. (Victoria, Madawaska, and Restigouche counties; not known from P.E.I. or N.S.), and New Eng.; E Asia. MAPS and synonymy: *see* below.

1 Floral bracts violet or reddish (sometimes with yellowish tips), moderately to copiously white-villous; corollas with purple margins; stems usually several in a tuft, rarely over 3 dm tall, pubescent throughout or glabrate near base; [*C. elegans* Malte, the type from the mouth of the Tree R., Coronation Gulf, coast of Dist. Mackenzie; incl. *C. mexiae* Eastw.; *C. unalaschcensis sensu* M.O. Malte, Rhodora 36(425):187. 1934, as to the Herschel Is., the Yukon, plant, not *C. pall. unal.* C. & S., according to Pennell 1934; NW part of the area; MAPS: Hultén 1968*b*:811 (*C. el.*); Porsild 1957: map 288, p. 196; Pennell 1934: map 1, p. 523] .... ssp. *elegans* (Malte) Pennell
1 Floral bracts yellow or ochroleucous; stems often taller.
  2 Floral bracts copiously villous with yellowish hairs; stems several, to 2 dm tall; [Alaska: type from the valley of the Chandalar R., N Alaska; a collection from Circle, E-cent. Alaska, may also belong here] .... ssp. *auricoma* Pennell
  2 Floral bracts glabrous to hirsute or white-villous; stems single or few, taller.
    3 Leaves glabrous; floral bracts glabrous or sparingly ciliate on the nerves and margins; stems essentially glabrous except in the inflorescence; [*C. sept.* Lindl., the type a single specimen developed from turf brought to England from Labrador; incl. *C. sept.* var. *micmacorum* Rousseau; *C. acuminata* of E N. America reports, not Spreng.; *C. ?sulphurea* Rydb.; Great Bear L. eastwards; MAP: Porsild 1957: map 287 (*C. sept.*), p. 196] .... ssp. *septentrionalis* (Lindl.) Scoggan
    3 Leaves and floral bracts usually distinctly pubescent; stems usually pubescent above the base.
      4 Pubescence relatively short and sparse, the stem usually appressed-pubescent, the leaves often glabrate; [*C. caud.* (Pennell) Rebr.; *C. pallida sensu* Hultén 1949, not (L.) Spreng.; *Bartsia ?acuminata* Pursh; Alaska (type from Port Clarence), the Yukon, and W Dist. Mackenzie; NE Asia; MAPS: Hultén 1968*b*:809 (*C. caud.*); Penell 1934: map 1, p. 523] .... ssp. *caudata* Pennell
      4 Pubescence denser and consisting of longer and more spreading hairs; [*Bartsia pallida* L.; Asia only] .... [ssp. *pallida*]

C. parviflora Bong.
/sT/W/ (Hp) Gravels, turfs, talus slopes, and subalpine and alpine meadows from S Alaska (N to ca. 61°N; type from Sitka) through B.C. and SW Alta. (N to Jasper) to Oreg. [Incl. *C. henryae* Pennell and *C. oreopola* Greenm.]. MAPS: Pennell 1934: map 4 (incl. *C. henryae;* incomplete), p. 535; combine the maps by Hultén 1968*b*:812 and 813 (*C. hen.*).

Var. *albida* (Pennell) Ownbey (floral bracts whitish or pinkish rather than deep rose to crimson) is known from s B.C. (Tranquille L., sw of Kamloops; CAN).

C. raupii Pennell
/aSs/(X)/ (Hp) Meadows and tundra from Alaska–Yukon (N to ca. 64°30′N) and the coast of w Dist. Mackenzie to Great Bear L., Great Slave L., L. Athabasca (Alta. and Sask.; type from the Caribou Mts. of N Alta. at 58°55′N), and cent. Dist. Keewatin (Aberdeen L., ca. 64°27′N, 99°W; CAN), s to northernmost B.C. (Liard Hot Springs, ca. 59°N), N Alta. (s to the Athabasca R. at 57°25′N; CAN), NE Man. (between York Factory and Churchill), N Ont. (coasts of Hudson Bay–James Bay s to 51°15′N), islands in James Bay, and w-cent. Que. (E James Bay coast between ca. 52° and 54°12′N). [Incl. ssp. *ursina* Pennell]. MAPS: Hultén 1968*b*:809; Pennell 1934: map 2, p. 530.

C. rhexifolia Rydb.
/T/W/ (Hp) Subalpine and alpine meadows and slopes from s B.C. (N to Lillooet and Kamloops; Herb. V; probably extending farther northwards but the area uncertain through confusion with other species, particularly *C. miniata*) and sw Alta. (N to Jasper) to Oreg., Utah, and Colo. [*C. lauta* Nels.; *C. purpurascens* Greenm.; *C. purpurascens* and *C. subpurpurascens* Rydb.].

C. rupicola Piper
/T/W/ (Hp) Cliffs and rocky slopes at moderate to high elevations from s B.C. (along the Skagit and Chilliwack rivers, where taken by J.M. Macoun between 1901 and 1906; CAN) to Oreg.

[C. rustica Piper]
[The report of this species of the w U.S.A. (Oreg., Idaho, and Mont.) from SE B.C. by Eastham (1947; Windermere, Columbia Valley, as *C. subcinerea*) requires confirmation, perhaps being referable to the closely related *C. cervina*, known from the same locality. (*C. subcinerea* Rydb.).]

C. sessiliflora Pursh Downy Painted-cup
/T/WW/ (Hp) Dry prairies and plains from Mont. to s Sask. (N to Indian Head, about 40 mi E of Regina) and s Man. (N to Birtle, about 60 mi NW of Brandon), s to Ariz., Tex., Mo., Ill., and Wisc. MAP: Pennell 1935: map 150, p. 524.
Forma *purpurina* Pennell (corolla purplish rather than yellowish white) is reported from sw Man. by Boivin (1966*b*).

[C. suksdorfii Gray]
[The report of this species of Wash. and Oreg. from s B.C. by J.M. Macoun (1906; Chilliwack Valley; this taken up by Henry 1915, and the probable basis of the inclusion of B.C. in the range by Rydberg 1922) is based upon *C. miniata*, relevant collections in CAN.]

[C. thompsonii Pennell]
[The inclusion of s B.C. in the range of this species of Wash. by Hitchcock et al. (1959) requires clarification.]

C. unalaschcensis (C. & S.) Malte
/Ss/W/ (Hp) Grassy places near the coast and subalpine meadows in the Aleutian Is., s Alaska–Yukon (N to ca. 62°N; the report from Herschel Is., the Yukon, by M.O. Malte, Rhodora 36(425):187. 1934, is referable to *C. pallida* ssp. *elegans* according to Pennell 1934), and coastal B.C. (s to Queen Charlotte Is.; CAN; DAO; V). [*C. pallida* unal. C. & S., the type from Unalaska, Aleutian Is.; *C. ?eximia* Eastw.; incl. ssp. *transnivalis* Pennell]. MAPS: Hultén 1968*b*:807; Pennell 1934: map 5, p. 536.

CHAENORRHINUM Reichenb. [7484]

C. minus (L.) Lange Dwarf Snapdragon
Eurasian; introd. along roadsides, railways (particularly common in cindery ballast), and waste places in N. America, as in s B.C. (Vancouver Is.; Agassiz; Elko), s Alta. (N to Edmonton), Sask. (N

to Prud'homme, s of Prince Albert), Man. (N to Gillam, about 165 mi s of Churchill), Ont. (N to Moosonee, sw James Bay, 51°16′N), Que. (N to the Gaspé Pen. at Matapédia), N.B., P.E.I., and N.S. [*Antirrhinum* L.; *Linaria* Desf.].

CHELONE L.   [7507]

C. glabra L.   Turtlehead, Balmony.   Tête de Tortue or Galane
/T/EE/   (Hp)   Wet thickets, streambanks, and marshy places, the aggregate species from SE Man. (Sandilands Forest Reserve and Shoal L.; WIN) to Ont. (N to Moose Factory, sw James Bay, 51°16′N), Que. (N to SE James Bay at ca. 51°30′N, L. St. John, and the Gaspé Pen.; reported from the Côte-Nord by Saint-Cyr 1887; not known from Anticosti Is.), Nfld., N.B., P.E.I., and N.S., s to Mo., Ala., and Ga. MAPS and synonymy: see below.

1   Corolla deep roseate or purple at summit and in throat; leaves broadly lanceolate to narrowly oval, relatively thin and long-petioled, to 6 cm broad; [*C. glabra* f. *rosea* Fern.; *C. montana* (Raf.) Pennell & Wherry; reported as frequent in Norfolk Co., s Ont., by Landon 1960; MAP: Pennell 1935: map 39 (the occurrence in s Ont. should be indicated), p. 185] . . . . var. *elatior* Raf.
1   Corolla whitish except for the pinkish summit or greenish-yellow lobes.
  2   Upper leaves scarcely reduced, rounded or subcordate at base; corolla whitish outside, the lips purplish within; [Ont. (N to James Bay) to Nfld. (type locality) and N.S.; MAP: on the above-noted map by Pennell] . . . . var. *dilatata* Fern. & Wieg.
  2   Upper leaves distinctly reduced, tapering to narrow bases.
    ·3   Corolla greenish yellow at summit, whitish within; leaves linear-lanceolate . . . .
    . . . . var. *linifolia* Coleman
      4   Lower leaf-surfaces minutely pubescent; [range of f. *linifolia*] . . . .
      . . . . f. *velutina* Pennell & Wherry
      4   Lower leaf-surfaces glabrous; [*C. linifolia* (Coleman) Pennell; E Man. (Sandilands Forest Reserve; Shoal L.) and Ont. (N to Thunder Bay); MAP: on the above-noted map by Pennell] . . . . f. *linifolia*
    3   Corolla creamy white to pinkish at summit or within the lobes; leaves lanceolate to ovate . . . . var. *glabra*
      5   Lower leaf-surfaces minutely pubescent; [*Chlonanthes tom.* Raf.; apparently throughout the range] . . . . f. *tomentosa* (Raf.) Pennell
      5   Lower leaf-surfaces glabrous; [incl. var. *elongata* Pennell & Wherry; Ont. to Nfld. and N.S.; MAP: on the above-noted map by Pennell] . . . . f. *glabra*

COLLINSIA Nutt.   [7503]   Blue-eyed Mary

1   Leaves relatively broad, the principal ones triangular- or oblong-ovate, widest immediately above the truncate or cordate-clasping base, entire or serrate with a few teeth; corolla to 12 mm long, the very gibbous throat much shorter than the lips, the upper lip white (varying to pale blue), the lower lip bright blue; upper pair of anther-filaments bearded at base; (s Ont.) . . . . *C. verna*
1   Leaves narrower, the principal ones nearly linear to narrowly elliptic or oblong, widest near or just below the middle; corolla-lips more uniformly blue; filaments all glabrous or nearly so.
  2   Corolla commonly less than 7 mm long, the tube bent at an oblique angle to the calyx and strongly gibbous on the upper side at the bend, longer than the lips; flowers on slender pedicels to 1.5 cm long, the lowest ones solitary; leaves commonly entire, the upper ones often whorled, the lowermost ones small, spatulate to rotund, commonly deciduous; (B.C. to Ont.) . . . . *C. parviflora*
  2   Corolla to over 1.5 cm long, the tube bent at about a right angle to the calyx and shortly spur-pouched at the bend, about equalling the lips; flowers rather short-pedicelled, the inflorescence often somewhat interrupted-thyrsoid; leaves often more distinctly toothed, the lowermost ones often better developed (with blades to 1.5 cm long); (s B.C.) . . . . *C. grandiflora*

C. grandiflora Lindl.   Blue-eyed Mary
/t/W/   (T)   Moist or dryish open flats and slopes at low to moderate elevations from sw B.C. (Vancouver Is. and adjacent mainland; CAN) to Calif.

C. parviflora Lindl.   Blue-lips
/sT/X/   (T)   Moist places at low to fairly high elevations from the N Alaska Panhandle and s Yukon through B.C. and sw Alta. (N to Banff) to s Calif., Colo., and S.Dak., farther eastwards known from s Sask. (Cypress Hills, Carlyle, and Little Birch L.; Breitung 1957a), s Man. (High L., about 80 mi E of Winnipeg; WIN), N Mich., Ont. (near Thunder Bay; Kenora; Elgin and Hastings counties), and w Vt. [*C. grandiflora* var. *pusilla* Gray; *C. "pauciflora" sensu* Hooker 1838, orthographic error; *C. tenella* (Pursh) Piper, not Benth.]. MAP: Hultén 1968b:794.

C. verna Nutt.   Blue-eyed Mary
/t/EE/   (T)   Rich woods and thickets from E Iowa to Wisc., s Ont. (Middlesex, Oxford, and Welland counties; CAN; TRT), and N.Y., s to E Kans., Ark., Ky., and Va. MAP: Pennell 1935: map 81, p. 295.

CONOBEA Aubl.   [7545]

C. multifida (Michx.) Benth.
/t/EE/   (T)   Wet sandy, gravelly, or loamy shores from Kans. to Iowa and southernmost Ont. (Pelee Is., Essex Co., where taken by John Macoun in 1892; CAN; ?extinct), s to Tex., La., Ala., and Ga. [*Capraria* Michx.; *Leucospora* Nutt.]. MAP (*Leuc. mult.*): Pennell 1935: map 20, p. 105.

CYMBALARIA Hill   [7478]

C. muralis Baumg.   Coliseum-Ivy, Ivy-leaved Toadflax
European; a garden-escape to roadsides and waste places in N. America, as in sw B.C. (Nanaimo, Vancouver Is.), Ont. (N to the Ottawa dist.), sw Que. (Marie-Victorin 1935), N.B. (wharf-ballast at St. John; NBM), and N.S. (Yarmouth; ACAD). [*Antirrhinum (Linaria) cymbalaria* L.].

DIGITALIS L.   [7593]   Foxglove

1   Corolla normally purple (sometimes white), with deeper purple spots on a white background within the lower part of the tube, to 5 cm long, essentially glabrous outside except for the ciliate lobes; calyx-lobes broadly ovate; leaf-blades ovate, crenate-dentate, pubescent beneath, all but the uppermost ones long-petioled; (introd.) . . . . *D. purpurea*
1   Corolla predominantly creamy-white or yellow, lined or spotted with brown, violet, or purple, more or less glandular-pubescent on both faces; calyx-lobes linear to narrowly lanceolate; leaves narrowly lanceolate to narrowly ovate, entire or serrulate, all sessile or the lowermost ones short-petioled; (introd.).
   2   Corolla to 5 cm long (about 3 cm broad when pressed); calyx-lobes, floral-axis, and pedicels copiously glandular-pubescent; leaves oblanceolate to lance-ovate, with several strong pairs of lateral veins beneath . . . . *D. ambigua*
   2   Corolla at most 2 cm long (less than 1.5 cm broad when pressed); lateral veins of leaves less distinct.
      3   Raceme rather open, its axis and pedicels glabrous or minutely glandular-puberulent, its bracts greatly reduced upwardly; calyx-lobes minutely glandular-ciliate, otherwise essentially glabrous; corolla to about 1.5 cm long, its lobes subequal; leaves lanceolate to oblanceolate (mostly broadest near or above the middle), sparsely pubescent along the veins beneath or glabrate . . . . *D. lutea*
      3   Raceme dense, its axis and pedicels copiously woolly, its bracts only gradually reduced upwardly; calyx-lobes woolly; corolla to about 2.5 cm long, the lower half strongly reticulated with purple or purple-brown, the middle lobe of the lower lip nearly white and about twice as long as the lateral ones; leaves lance-attenuate (broadest below the middle), glabrous beneath . . . . *D. lanata*

D. ambigua Murray   Yellow Foxglove
Eurasian; an occasional garden-escape in N. America, as in sw B.C. (Aldergrove, near Vancouver; V) and s Ont. (near Aurora, York Co.; CAN; TRT). [*D. grandiflora* Lam.].

D. lanata Ehrh.   Grecian Foxglove
European; an occasional garden-escape to roadsides, open woods, and waste places in N. America, as in sw B.C. (Sooke, Vancouver Is.; Herb. V).

D. lutea L.   Straw Foxglove
European; an occasional garden-escape in N. America, as in sw Que. (at the foot of cliffs below Mt. Royal, Montreal, where taken by Frère Cléonique-Joseph in 1942; MT).

D. purpurea L.   Common Foxglove
European; a garden-escape (sometimes locally abundant) to roadsides, old fields, and waste places in N. America, as in se Alaska (Hultén 1949), w B.C. (Queen Charlotte Is.; Vancouver Is. and adjacent islands and mainland e to Manning Provincial Park, about 30 mi se of Hope), Ont. (n to the Muskoka Dist., Georgian Bay, L. Huron), sw Que. (Rouleau 1947), w Nfld. (a large colony in St. Georges Bay; CAN; GH), and N.S. (Sydney, Cape Breton Is.; CAN). MAP: Hultén 1968*b*:806.

EUPHRASIA L.   [7638]   Eyebright

(Ref.: Fernald and Wiegand 1915; Fernald 1933; Sell and Yeo 1962; Clapham, Tutin, and Warburg 1962. This is an extremely critical genus, accorded widely different treatments by North American and British authors. The following survey of the genus in Canada can only be regarded as tentative.)

1   Floral bracts averaging over twice as long as broad, with distant, acute to awn-tipped teeth; capsules glabrous (very rarely with a few weak marginal bristles); corolla white, to 7 mm long; (introd. in Nfld.) .... [*E. salisburgensis*]
1   Floral bracts averaging less than twice as long as broad; capsules ciliate with long straight hairs.
  2   Corolla rarely over 4 mm long, the 2 terminal lobes of the upper lip very short, rounded, entire, usually revolute-margined, the lower lip scarcely surpassing the upper one, only obscurely fan-shaped; calyx-lobes at most 2 mm long; seeds less than 1.5 mm long; (Que. to Labrador, Nfld., and N.S.) .... *E. oakesii*
  2   Corolla usually at least 4 mm long, the 2 terminal lobes of the upper lip prominent, undulate-truncate, commonly reflexed from near base, the lower lip often fan-shaped, much larger than the upper one; calyx-lobes commonly over 2 mm long; seeds to 2 mm long.
    3   Teeth of floral bracts obtuse to acute but not bristle-tipped; lower corolla-lip white with lilac, violet, or purplish veins, the upper lip often bluish-tinged.
      4   Inflorescence subcapitate, only the lower 1–3 pairs of bracts more or less remote in maturity; teeth of bracts obtuse; leaves very pubescent; corolla to about 7 mm long; (Alaska–Yukon–n ?B.C.) .... *E. mollis*
      4   Inflorescence spicate, becoming loose and elongate; teeth of bracts obtusish to acute; (transcontinental) .... *E. arctica*
    3   Teeth of floral bracts subulate or bristle-tipped; upper corolla-lip purple-tinged, the lower lip whitish with lilac to dark-purple veins.
      5   Flowers borne only along the upper half of the stem and branches; corolla to 1 cm long, the lower lip veined with dark purple, fan-shaped, its lateral lobes wide-spreading; (e Que. to Nfld. and N.S.) .... *E. americana*
      5   Flowers commonly borne nearly to the base of the stem and branches.
        6   Corolla to 1 cm long, its fan-shaped lower lip with dark-purple veins and wide-spreading lateral lobes; (introd.).
          7   Bracts glabrous, all ascending, tapering at base, their teeth bristle-tipped; calyx glabrous, its sharp lobes surpassing the capsule; corolla at most about 8 mm long .... *E. rigidula*
          7   Bracts copiously pubescent, the lower ones spreading-ascending,

rounded at base, the teeth subulate-tipped; calyx densely pubescent, its sharp lobes about equalling the capsule; corolla to 1 cm long . . . .
. . . . *E. tatarica*
6 Corolla averaging less than 8 mm long, with paler veins; lower corolla-lip only weakly fan-shaped, its lateral lobes not wide-spreading.
  8 Bracts broadly oval to ovate or rotund, the blade about as broad as long, glabrous; corolla to 7 mm long, the upper lip violet-tinged, the lower lip white with lilac veins; branches arched-ascending; (Que. to N.S.) . . . . *E. canadensis*
  8 Bracts narrower, distinctly longer than broad; branches commonly more strongly ascending.
    9 Bracts and leaves glabrous or sparingly pubescent beneath, the latter with up to 5 pairs of sharp teeth; corolla to 8 mm long; (introd.) . . . . [*E. condensata*]
    9 Bracts and leaves pubescent, the latter with a few pairs of coarse acute teeth; corolla at most about 6 mm long; (?Alta.; Man. to E Que.) . . . . *E. hudsoniana*

E. americana Wettst.
/T/E/ (T) Fields, pastures, sea-cliffs, and roadsides and waste places from E Que. (Gaspé Pen. and Magdalen Is.) to Nfld., St-Pierre and Miquelon, N.B., P.E.I., N.S., and coastal Maine; introd. in SW B.C. (Vancouver Is.; Langley Prairie). [*E. officinalis* of Canadian reports in large part, not L.].

Hybrids with *E. arctica* (× *E. villosa* Callen), *E. canadensis* (× *E. aequalis* Callen), and *E. pennellii* (× *E. vestita* Callen; *E. pennellii* here included in *E. arctica* var. *submollis*) are reported by E.O. Callen (J. Bot. 78:215–16. 1940) from their type localities in the Gaspé Pen., E Que., at Cloridorme, Cap-des-Rosiers, and Douglastown, respectively.

Sell and Yeo (1962) have referred most of our material of this species and of *E. canadensis*, perhaps all collections that they have seen, to the Old World *E. brevipila* Burnat & Gremli, *E. nemorosa* (Pers.) Mart., and *E. tetraquetra* (Breb.) Arrond. Fernald and Wiegand (1915) had already noted, "It is possible, then, that *E. canadensis* and *E. americana* are derivatives of *E. nemorosa* and of *E. stricta* (doubtfully indigenous in America) or of closely related European species introduced into eastern Canada and eastern Maine by the earliest European colonists, in the 16th and 17th centuries; and, being annuals, the plants have, during hundreds of generations, departed sufficiently from their ancestors now to stand as true American species."

E. arctica Lange
/aST/X/GEA/ (T) Open ground from Alaska (N to near the Arctic Circle) to S-cent. Yukon, Great Bear L., Great Slave L., NW Sask. (L. Athabasca), S Dist. Keewatin (Porsild and Cody 1968), NE Man. (S to Churchill), N Ont. (coasts of Hudson Bay–James Bay), Baffin Is. (N to near the Arctic Circle), Que. (coasts of Hudson Strait and Hudson Bay–James Bay; Côte-Nord; Anticosti Is.; Gaspé Pen.), Labrador (type locality), and Nfld. (reports from N.B. require confirmation; not known from P.E.I. or N.S.), S through B.C.–Alta. to Mont., N Minn., N Mich., and Maine; W Greenland N to 71°25′N, E Greenland N to ca. 74°30′N; Iceland; Scandinavia; NW Siberia. [Incl. vars. *inundata* and *obtusata* (Joerg.) Callen, *E. disjuncta* Fern. & Wieg. and its var. *dolosa* Boivin, *E. frigida* Pugsl., and *E. subarctica* Raup; *E. mollis* var. *?laurentiana* Boivin; *E. latifolia* Pursh in part, not L. nor Schur]. MAPS: Porsild 1957: map 289, p. 197 (incomplete westwards according to the present concept; the map by Porsild 1966: map 126, p. 82, for *E. subarctica* also applies here); Hultén 1968b:814 (N. American range, as *E. disjuncta*), and 1958, map 32 (*E. frigida*; incomplete for W N. America), p. 51.

A hybrid with *E. canadensis* (× *E. aspera* Callen, J. Bot. 78:216. 1940) is known from several localities in the Gaspé Pen., E Que. (type from Rivière-Marsouri). Many collections from E Que., Labrador, and Nfld. distributed as *E. arctica* have been referred by Sell and Yeo (1962) to the Old World *E. curta* (Fries) Wettst., a few to *E. brevipila* Burnat & Gremli, *E. suborbicularis* Sell & Yeo, and *E. vinacea* Sell & Yeo. E.O. Callen (Rhodora 54(642):153. 1952) refers collections from Baffin Is. and northernmost Ungava to var. *submollis* (Joerg.) Callen (var. *minutissima* Polunin; *E. frigida* var. *pusilla* Pugsl.; *E. ?pennellii* Callen; leaves and calyces with an admixture of short-stalked glands in addition to the strong white bristles).

E. canadensis Townsend
/T/E/ (T) Open sterile fields and roadsides from Que. (N to the Côte-Nord and Gaspé Pen.; type from near Quebec City) to P.E.I. (Malpeque, Prince Co.; Fernald and Wiegand 1915), N.S., Maine, N.H., and Mass. [*E. americana* var. *can.* (Townsend) Rob.; *E. officinalis* of Canadian reports in part, not L.; *see* note under *E. americana*].

[*E. condensata* Jord.]
[European; according to Pennell (1935), this species is identical with *E. stricta* Host (not HBK.) and has been introd. in N. America from Nfld. to N.Y. and Maine. Fernald *in* Gray (1950) includes *E. stricta* Host in the synonymy of *E. rigidula* Jord. but Pennell notes Pugsley's opinion that the latter is a distinct species. For the present, reports of *E. condensata* and *E. stricta* from N. America are here included in the treatment of *E. rigidula.*]

E. hudsoniana Fern. & Wieg.
/ST/(X)/ (T) Open ground and shores from ?Alta. (Fernald *in* Gray 1950) to Man. (between Grand Rapids, near the NW end of L. Winnipeg, and Churchill; not known from Sask.), Ont. (N shore of L. Superior; Cochrane, 50°N; shores of James Bay–Hudson Bay N to ca. 56°N), Que. (Chimo, S Ungava Bay; coasts of Hudson Bay–James Bay; Gaspé Pen.; Anticosti Is.; type from the Koksoak R. S of Ungava Bay, where taken by Spreadborough in 1896), and N Labrador (Ryan's Bay at 59°37′N; E.C. Abbe, Rhodora 38(448):158. 1936).
The above statement of range of this critical species is based largely upon collections in CAN and GH, many of them determined or verified by Sell and Yeo (1962). Its range is included in that of *E. (arctica* var.*) disjuncta* by Hultén (1968b:814).

E. mollis (Ledeb.) Wettst.
/Ss/W/eA/ (T) Subalpine meadows of the Aleutian Is., S Alaska (N to ca. 61°N), SW Yukon (St. Elias Mts.; CAN, detd. A.E. Porsild), and N ?B.C. (Rydberg 1922); E Asia. [*E. officinalis* var. *mollis* Ledeb.]. MAP: Hultén 1968b:814.

E. oakesii Wettst.
/sT/E/ (T) Turfy or gravelly slopes, calcareous cliffs, and brackish shores from Que. (St. Lawrence R. estuary from near Quebec City to the Côte-Nord, Anticosti Is., Gaspé Pen., and Magdalen Is.) to Labrador (N to Indian Harbour, 54°27′N), Nfld., N.B., P.E.I., N.S., Maine, and N.H. [Incl. *E. randii* Rob. and its vars. *farlowii* Rob. and *reeksii* Fern. (*E. purpurea* Reeks, not Desf.) and *E. williamsii* Rob. and its var. *vestita* Fern. & Wieg.].
Forma *lilacina* Fern. & Wieg. (corolla-lobes deep lilac rather than whitish with violet lines) is known from the type locality, Blanc-Sablon, Côte-Nord, E Que.

E. rigidula Jord.
European; introd. into dry fields and sterile grasslands in E N. America, as in Que. (Montreal dist.; Gaspé Pen.; Magdalen Is.), Nfld. (GH; MT), ?N.B. (a collection in CAN from Salt Springs, 15 mi SW of Sussex, Kings Co., has been referred by Yeo to a possible hybrid between *E. nemorosa* and *E. stricta*), and N.S. [*E. ?stricta* Host, not HBK. (see *E. condensata*); *E. borealis sensu* M.L. Fernald, Rhodora 9(105):163. 1907, not Wettst.].

[*E. salisburgensis* Funck]
[European; the only record of this species in N. America is a collection in CAN (var. *hibernia* Pugsl.; detd. Yeo) from Daniel's Harbour, W Nfld., about 100 mi N of Cornerbrook, where taken by James Richardson in 1861.]

E. tatarica Fisch.
European; known in N. America from calcareous cliffs and gravels of E Que. (Côte-Nord, Anticosti Is., and Gaspé Pen.) and Labrador (near Nain, 56°32′N; CAN, detd. as *E. curta* by Yeo). [*E. stricta* var. *tat.* (Fisch.) F. & W.; *E. ?curta* (Fries) Wettst.].
Concerning the possibility of this species being native in E N. America (as proposed by Fernald *in* Gray 1950), *see* note under *Luzula campestris.*

GERARDIA L.   [7604]   Gerardia

1   Calyx-tube distinctly reticulate-veiny; corolla pink, less than 2 cm long, its lobes widely spreading; seeds yellowish; plants yellow-green, rarely darkening in drying.

  2   Teeth of calyx-tube subulate, thickened, minute, the sinuses between them nearly flat; stem smoothish, with spreading branches; leaves commonly linear-oblanceolate; (s ?Ont.) .... [G. obtusifolia]

  2   Teeth of calyx-tube triangular, thin, the sinuses between them deeply concave; leaves linear, acuminate; (s Ont.).

    3   Stem smoothish, scarcely angled, very abundantly branched, most of the spreading branches bearing a single apparently terminal flower, a terminal raceme scarcely developed on the main axis; calyx-teeth to nearly 2 mm long; stigmas to 3 mm long .... [G. gattingeri]

    3   Stem usually somewhat scabrous on the 4 narrow wings, simple or with short ascending branches, the main axis terminating in a normal raceme; calyx-teeth less than 1 mm long; stigmas at most 2 mm long .... G. skinneriana

1   Calyx-tube scarcely reticulate-veiny; corolla pink or roseate to rose-purple, to over 3.5 cm long; seeds blackish or dark brown; plants deep green to purple-tinged, mostly blackening on drying.

  4   Leaves linear, sharp-tipped, harshly scabrous above, with axillary clusters of smaller leaves; stem scabrous; calyx-teeth lanceolate, at least 1.5 mm long; corolla to 2.5 cm long, its 2 upper lobes ascending; capsule distinctly longer than thick; (s Man.) .... G. aspera

  4   Leaves and stem glabrous or somewhat scabrous; capsule subglobose.

    5   Corolla glabrous within, at most 1.5 cm long, the upper lobes arching forward; pedicels spreading, to over 2.5 cm long, mostly longer than the flowers; (s Man. to sw Que.) .... G. tenuifolia

    5   Corolla pubescent on both surfaces, the lobes all spreading; pedicels ascending, shorter than the flowers; leaves linear.

      6   Leaves obtuse or subacute, rather fleshy; calyx-teeth obtuse; pedicels to 12 mm long; (coastal saline marshes of N.B. and N.S.) .... G. maritima

      6   Leaves acute, not fleshy; calyx-teeth acute; pedicels at most about 5 mm long; (se Man. to N.B. and N.S.) .... G. purpurea

G. aspera Dougl.

/T/EE/   (T)   Dry prairies and sandy or rocky slopes from s Man. (type from the Red R.; also known from Portage la Prairie, Morden, Emerson, and Stony Mountain; reports from Sask. require confirmation) to Minn., Wisc., and Ill., s to N.Dak. and Okla. [*Agalinis* Britt.]. MAP: Pennell 1935: map 112, p. 429.

[*G. gattingeri* Small]

[This species of the E U.S.A. (N to Nebr., Minn., and Mich.) is reported from s Ont. by Pennell (1935; near L. St. Clair, presumably in Kent Co., this station indicated on his MAP 140, p. 474). The relevant collection has not been seen and the species is probably now extinct in Canada. (*Agalinis* Small; *G. tenuifolia* var. *asperula* Gray, the report of which from w Ont. by John Macoun (1886; Lonely L.) is based upon *G. tenuifolia* var. *parviflora* Nutt., the relevant collection in CAN (Macoun's report from Stony Mountain, s Man., is referable to *G. aspera*)).]

G. maritima Raf.

/T/EE/   (T)   Coastal salt marshes from E N.B. (Kent Co., where taken by Fowler in 1865; MTMG; not known from P.E.I.) and sw N.S. (Argyle Head and Wedgeport, Yarmouth Co.; ACAD; CAN; GH; NSPM) to Fla. and Tex. [*Agalinis* Raf.]. MAPS: Pennell 1935: map 113, p. 429, and 1929: map 21, p. 153.

    Forma *alba* Erskine (*A. mar.* f. *candida* Boivin; flowers white rather than purplish) is known from sw N.S. (type from Wedgeport, Yarmouth Co.).

[G. obtusifolia (Raf.) Pennell]
[The report of this species of the E U.S.A. (Coastal Plain from Md. and Del. to Fla. and La.) from S Ont. by Dodge (1915; Squirrel Is., Lambton Co.) requires confirmation, perhaps being based upon *G. skinneriana* (*see* Gaiser and Moore 1966:106). (*Agalinis* Raf.; *G. parvifolia* (Hook.) Chapm.). The MAP by Pennell (1935: map 137, p. 472) indicates no Canadian stations.]

G. purpurea L.
/T/EE/ (T) Damp open ground, shores, and bogs, the aggregate species from SE Man. (Stony Mountain, near Winnipeg, and Lake of the Woods; DAO; MTMG) to Ont. (N to the Ottawa dist.), Que. (N to l'Islet, about 45 mi NE of Quebec City; MT), N.B. (St. John R. mouth; not known from P.E.I.), and N.S., S to E Tex. and Fla. MAPS and synonymy: *see* below.
1 Corolla at least 2 cm long; calyx-teeth at most 2 mm long; [*Agalinis* Pennell; reported
    from Niagara, S Ont., by Boivin 1966*b*; MAP: Pennell 1935: map 117, p. 438] . . . . . var. *purpurea*
1 Corolla not over 2 cm long.
    2 Calyx-teeth to 8 mm long; [*G. (Agalinis) neoscotica* Greene; N.S. (Annapolis (type
       from Middleton), Digby, Yarmouth, Shelburne, and Queens counties and Sable Is.);
       MAPS: Roland 1947: map 399, p. 535; Pennell 1929: map 23, p. 161] . . . .
       . . . . var. *neoscotica* (Greene) Gleason
    2 Calyx-teeth at most 3.5 mm long . . . . var. *parviflora* Benth.
       3 Corolla white; [*G. paupercula* var. *borealis* f. *albiflora* Vict. & Rousseau, the type
          from Cap-Rouge, near Quebec City, Que.; *Agalinis purp.* var. *parv.* f. *kucyniakii*
          Boivin] . . . . f. *albiflora* (Vict. & Rousseau) Scoggan
       3 Corolla pink to rose-purple; [var. *paupercula* Gray; *G. (Agalinis) paupercula*
          (Gray) Britt. and its var. *borealis* (Pennell) Deam; SE Man. to N.B.; MAPS: Pennell
          1935: map 114, p. 434, and 1929: map 22, p. 157; McLaughlin 1932: fig. 9,
          p. 345] . . . . f. *parviflora*

G. skinneriana Wood
/t/EE/ (T) Dry sandy prairies, hillsides, and dunes from S Wisc. to S Ont. (Squirrel Is., Lambton Co.; OAC; reported from Kent Co. by Pennell 1935), S to Okla., Ark., and Ohio. [*Agalinis* Britt.]. MAP: Pennell 1935: map 135, p. 469.

G. tenuifolia Vahl
/T/EE/ (T) Prairies and open woods, the aggregate species from SE Man. (Winnipeg dist.) to Ont. (N to the Ottawa dist.), Que. (N to near Oka and Montreal), Vt., and Conn., S to Wyo., Colo., Tex., Ala., and Ga. MAP and synonymy: *see* below.
1 Calyx-teeth subulate, at most 1 mm long; capsule at most 5 mm long; anthers densely
    villous; leaves linear, to 3.5 mm broad; [*Agalinis* Raf.; S Ont. and SW Que.; MAP: Pennell
    1935: map 131, p. 459] . . . . var. *tenuifolia*
1 Calyx-teeth broadly triangular, to 2 mm long; capsule at least 5 mm long.
    2 Anthers sparsely pilose to nearly glabrous; leaves linear, at most 3.5 mm broad, often
       with conspicuous axillary fascicles of smaller leaves; [*Agalinis ten.* var. *parv.* (Nutt.)
       Pennell; SE Man. (near Winnipeg) to SW Que. (N to near Oka and Montreal); MAP: on
       the above-noted map by Pennell] . . . . var. *parviflora* Nutt.
    2 Anthers densely villous; leaves linear to linear-lanceolate, to 6 mm broad, sometimes
       with axillary fascicles; [*G. besseyana* Britt.; S Ont. (Essex, Kent, Elgin, and Norfolk
       counties) and SW Que.; MAP: on the above-noted map by Pennell] . . . .
       . . . . var. *macrophylla* Benth.

GRATIOLA L. [7542] Hedge-hyssop

1 Corolla golden yellow, to 18 mm long; a pair of filiform capitate-tipped anther-filaments
    present; fruit about 3 mm long, shorter than the calyx-lobes; leaves linear to ovate, mostly
    less than 3 cm long; perennial with fleshy rhizomes and purplish stolons; (Ont. to Nfld.
    and N.S.) . . . . *G. aurea*
1 Corolla at most about 1 cm long; sterile filaments minute or none; leaves to about 5 cm
    long; fibrous-rooted annuals.

2   Pedicels lacking bracteoles (minute bracts) at the summit, the sepals evidently 5, elongate and pointed, often well over 1 cm long; corolla to 7 mm long; capsule subglobose, not pointed, 4 or 5 mm long; plants glabrous or only obscurely glandular above; (s B.C.) . . . . *G. ebracteata*

2   Pedicels bearing a pair of bracteoles below the calyx, the sepals thus apparently 7, less pointed, to 7 mm long; corolla to 1 cm long; capsule broadly ovoid, pointed, to 7 mm long; plant typically more or less glandular-viscid; (B.C. to N.S.) . . . . *G. neglecta*

G. aurea Muhl.   Golden-pert

/T/EE/   (Hpr)   Open swamps and sandy, gravelly, or peaty shores from E N.Dak. to Ont. (N to Mattawa, Renfrew Co., and the Ottawa dist.), Que. (N to L. St. Peter in St-Maurice Co.; MT), Nfld. (Whitbourne, near St. John's; GH), and N.S. (not known from N.B. or P.E.I.), s to Ill., N.Y., and Fla. [*G. lutea* of auth. in part, perhaps not Raf.]. MAP: Pennell 1935: map 9 (*G. lutea*), p. 70.

Forma *leucantha* Bartlett (corolla milk-white rather than golden yellow) is known from N.S. (Queens Co.; CAN; NSPM). Forma *pusilla* Fassett (the dwarf submersed sterile phase with leaves mostly less than 5 mm long rather than to 3 cm long) is known from Ont. (L. Nipissing and Mattawa) and Que. (Rupert R. SE of James Bay at ca. 51°20′N).

G. ebracteata Benth.

/t/W/   (T)   Wet meadows, muddy shores, and shallow water from SW B.C. (Vancouver Is. and adjacent islands; the report from Kamloops by John Macoun 1890, taken up by Henry 1915, is based upon *G. neglecta,* the relevant collection in CAN) and w Mont. to Calif.

G. neglecta Torr.

/T/X/   (T)   Wet places, muddy shores, and shallow water from s B.C. (N to Kamloops and Sicamous; CAN) to s Alta. (near Hanna, about 100 mi NE of Calgary; CAN), s Sask. (Battleford, Yorkton, and Moose Jaw; CAN), s Man. (N to Foxwarren, about 70 mi NW of Brandon), Ont. (N to Horton, Renfrew Co., and the Ottawa dist.), Que. (N to Beauport, near Quebec City), and N.S. (Middle Stewiacke, Colchester Co.; ACAD; CAN; not known from N.B. or P.E.I.), s to Calif., Tex., and Ga. [*G. virginiana* of auth., not L.; *G. lutea* Raf. in part, the name of doubtful application; *Steironema quadriflorum sensu* Lowe 1943, not (Sims) Hitchc., the relevant Foxwarren collection in WIN]. MAP: Pennell 1935: map 12, facing p. 80.

Var. *glaberrima* Fern. (leaves rounded at base rather than tapering at both ends, they and the upper stem-internodes glabrous rather than more or less viscid-pubescent; corolla milk-white except at base rather than creamy white and with a yellowish tube) is known from Que. (type from Anse-St-Vallier, Bellechasse Co.).

KICKXIA Dumort.   [7479]   Fluellin

1   Leaves broadly triangular-ovate, the middle ones hastate at base, the upper ones sagittate; pedicels to 3 cm long, glabrous throughout or minutely villous near the base and summit only; corolla to 9 mm long, the spur straight; (introd. in s B.C.) . . . . *K. elatine*

1   Leaves ovate or suborbicular, rounded or subcordate at base; pedicels to 2 cm long, villous throughout; corolla to 11 mm long, the spur curved; (introd. in s B.C. and s Ont.) . . . . *K. spuria*

K. elatine (L.) Dumort.   Canker-root

European; occasionally introd. in N. America, particularly on wharf-ballast. Reported from SW B.C. by Eastham (1947; Saanich, Vancouver Is.) and there is a collection in CAN from Victoria, where taken on a street by Miss M.C. Melburn in 1967. [*Antirrhinum* L.; *Linaria* Mill.].

K. spuria (L.) Dumort.

European; introd. along roadsides and gravelly shores and in dry fields in N. America, as in SE B.C. (Duncan, Vancouver Is.; Eastham 1947) and s Ont. (Walkerton, Bruce Co., and Woodville, Victoria Co.; OAC). [*Antirrhinum* L.; *Elatinoides* Wettst.; *Linaria* Mill.].

LAGOTIS Gaertn. [7581]

L. glauca Gaertn.

/aSs/W/EA/ (Hs) Rocky tundra of the Aleutian Is., Alaska–Yukon (N to the arctic coast; *see* Hultén 1949: map 1033b, p. 1470), and NW Dist. Mackenzie; NE Europe; N Asia. [*Bartsia* Poir.; *B. gymnandra* L. f.; *Gymnandra gmelinii* C. & S.]. MAPS (aggregate species): combine the maps by Hultén 1968b:804 and 805 (ssp. *minor*; *see* map by W.J. Cody, Nat. Can. (Que.) 98(2): fig. 6, p. 148. 1971).

Forma *candida* Lepage (flowers white rather than bluish, the floral bracts relatively pale) is known from the type locality in the Talkeetna Mts., Alaska. Most of our material is referable to ssp. *minor* (Willd.) Hult. (*Gymnandra minor* Willd.; *G. (L.) stelleri* C. & S.; *L. ?hultenii* Polunin; stems relatively erect, the basal leaves tending to be lanceolate and serrate rather than ovate to suborbicular and with blunt or rounded teeth, the stamens mostly with relatively long filaments and short anthers); MAPS: Hultén 1968b:805; *Atlas of Canada* 1957: map 6 (*L. stell.*), sheet 38.

LIMOSELLA L. [7558] Mudwort

1 Leaves with elliptic to oblong blades; corolla pink; calyx regular; capsule-valves not thickened at the margins; pedicels not strongly arching; (transcontinental) . . . . *L. aquatica*
1 Leaves filiform or subterete to tip; corolla white; calyx regular or the sepals somewhat united in pairs; capsule-valves thickened at the margins; pedicels soon recurving; (Que. eastwards) . . . . *L. subulata*

L. aquatica L.

/aST/X/GEA/ (T) Fresh to brackish shores and wet sands from the W Aleutian Is. and Alaska (N to the Seward Pen.) to S-cent. Yukon, Great Bear L., N Alta. (L. Athabasca), Sask. (N to Leacross, 53°03′N; Breitung 1957a), Man. (N to Churchill), Ont. (N to the Severn R. at ca. 55°50′N), Que. (N to Chimo, S Ungava Bay, and the Côte-Nord), Labrador (N to the Hamilton R. basin), and Nfld. (not known from the Maritime Provinces), S to Calif., N.Mex., Minn., and S James Bay; W Greenland N to ca. 71°N, E Greenland N to 63°35′N; Iceland; Eurasia. MAPS: Hultén 1968b:797, and 1958: map 187, p. 207; Pennell 1935: map 32 (incomplete northwards), p. 165; Meusel 1943: fig. 28c (incomplete for N. America).

L. subulata Ives

/T/EE/ (T) Brackish or saline coastal sands and muds from Que. (St. Lawrence R. estuary from near Quebec City to the Côte-Nord, Gaspé Pen., and Magdalen Is.; Chicoutimi, near L. St. John; not known from Anticosti Is.) to Nfld., N.B., P.E.I., and N.S., S along the Atlantic coast to Va. [*L. aquatica* of reports from the Maritime Provinces, not L., relevant collections in several herbaria; *L. aquat.* var. *tenuifolia* of most or all Canadian reports, not *L. tenuifolia* Wolf]. MAPS: Hultén 1958: map 187, p. 207; Pennell 1935: map 34, p. 168.

Collections in CAN from SW B.C. (Alberni, Vancouver Is.) and Alta. (Oliver, near Edmonton) have been placed here but probably belong to the bladeless form of *L. aquatica,* (var. *tenuifolia* (Wolf) Schübler & Martens; *see* M.L. Fernald, Rhodora 20(237):160–64. 1918, and Pennell 1935:631).

LINARIA Mill. [7480] Toadflax

1 Stem-leaves lance-ovate to broadly ovate, acute or somewhat acuminate, clasping at the cordate-auriculate base; corolla yellow, to over 4 cm long (including the spur); seeds wingless; perennial; (introd.) . . . . *L. dalmatica*
1 Stem-leaves linear to linear-lanceolate; flowers mostly smaller.
  2 Corolla bright yellow with an orange palate, the body to about 1.5 cm long, the spur about 1 cm long; capsules about 1 cm long, the seeds with a circular wing; perennial; (introd.) . . . . *L. vulgaris*
  2 Corolla various shades of blue, violet, or purple (if yellowish, striped with violet lines); capsules shorter.
    3 Corolla pale blue, whitish, or creamy, striped with violet lines, with a prominent

palate closing the throat, the spur much shorter than the body; stem relatively leafy; perennials; (introd.).

    4   Corolla not much over 1 cm long, creamy to bluish, with a conical spur to 5 mm long; seeds wingless .... *L. repens*

    4   Corolla to 2 cm long, usually deeper yellow, the compressed spur to 7 mm long .... × *L. sepium*

3   Corolla violet, blue-violet, or purple, the slender spur sometimes longer than the corolla-body; leaves usually more remote.

    5   Perennial to 9 dm tall, glabrous and glaucous, the stems branched above; flowers numerous in dense terminal racemes, violet (rarely bright pink), about 8 mm long, the incurved spur more than half as long as the corolla; seeds wingless; (introd.) .... [*L. purpurea*]

    5   Annuals or biennials, more or less viscid or glandular in the inflorescence.

        6   Spur to 9 mm long, strongly curved, placed transversely or obliquely; corolla (including spur) to nearly 2.5 cm long, the lower lip with merely a pair of whitish rounded ridges at base rather than a well-formed palate; seeds wingless; leaves to 3 cm long; plants with short trailing basal offshoots (with opposite leaves) forming winter rosettes; stems to 6 dm tall; (s B.C.; Sask.; s Ont. to N.B. and N.S.) .... *L. canadensis*

        6   Spur nearly straight, vertical; lower corolla-lip with a full yellow or orange palate; leaves to about 4 cm long; plants lacking basal offshoots; (introd.).

           7   Corolla (including spur) about 3.5 cm long, violet-purple with a small whitish or paler yellow patch on the yellow palate; seeds with up to 6 ring-like wings; stems to 5 dm tall .... [*L. maroccana*]

           7   Corolla (including spur) usually less than 2 cm long, purple, the yellow palate reticulate with purple veins; seeds minutely rugose, wingless; stems to over 1 m tall .... [*L. pinifolia*]

L. canadensis (L.) Dumont   Old-field-Toadflax
/T/X/ (T (Hs, bien.)) Dry sandy or sterile soil (often weedy in sandy loam) from sw B.C. (Vancouver Is. and adjacent islands and mainland; CAN; V; not known from Alta.) to Sask. (Alsask, about 125 mi NW of Swift Current; Breitung 1957a; not known from Man.), Ont. (Durham and Welland counties; OAC; TRT), Que. (N to Papineau Co. and the Montreal dist.; reported from Rivière-du-Loup, Temiscouata Co., by Saint-Cyr 1887), N.B., and N.S. (not known from P.E.I.), s to Calif., Mexico, Tex., and Fla. [*Antirrhinum* L.]. MAP (aggregate species): combine the maps by Pennell 1935: map 83, p. 307, and map 82 (*L. texana*), p. 303.

The plant of B.C. and Sask. is referable to var. *texana* (Scheele) Pennell (*L. tex.* Scheele; corolla-body to about 1.5 cm long rather than usually less than 1 cm, the spur to 9 mm long rather than at most 6 mm, the seeds densely tuberculate rather than smooth or nearly so; *see* the above-noted map by Pennell).

L. dalmatica (L.) Mill.
European; introd. or a garden-escape to roadsides and waste places in N. America, as in B.C. (N to Fort St. John, ca. 56°10′N), Alta. (N to near Edmonton), Sask. (N to near Prince Albert), Man. (N to near The Pas), Ont. (N to the N shore of L. Superior at Thunder Bay and Michipicoten), Que. (N to St-Fidèle, Charlevoix Co.), and N.S. (Halifax and Victoria counties; ACAD; GH). [*Antirrhinum* L.]. MAP: J.F. Alex, Can. J. Bot. 40(2): fig. 5, p. 305. 1962.

Some of our material is referable to var. *macedonica* (Griseb.) Vandas (*L. mac.* Griseb.; floral bracts much reduced, at most about half the length of the pedicels rather than equalling or surpassing them, these to 3 cm long rather than mostly less than 1.5 cm; calyx-segments much shorter than the corolla-tube rather than subequal to it; spur about equalling the corolla-body rather than often shorter than it).

[L. maroccana Hook. f.]
[A native of North Africa; reported as a casual introduction into a garden in P.E.I. by D.S. Erskine (1960; Brackley Beach, Queens Co., a 1927 collection being the basis of the report of *L. reticulata* from there by Groh and Frankton 1949b).]

[L. pinifolia (Poir.) Thell.]
[European; an occasional garden-escape in N. America but scarcely established, as in E-cent. B.C. (Fort St. John, ca. 56°10′N) and w-cent. Alta. (Beaverlodge, 55°13′N), these reports by Groh and Frankton (1949*b*; as *L. reticulata*). (*Antirrhinum* Poir.; *L. reticulata* (Sm.) Desf.).]

[L. purpurea (L.) Mill.]   Purple Toadflax
[European; an occasional garden-escape in N. America but scarcely established; reported from SW B.C. by Boivin (1966*b*). (*Antirrhinum* L.).]

L. repens (L.) Mill.   Striped Toadflax
European; introd. along roadsides and in thickets, fields, and waste places in N. America, as in Nfld. (CAN; GH), N.B. (near Chatham and Newcastle, Northumberland Co.), and N.S. (Wolfville, Kings Co.; ACAD). [*Antirrhinum* L.; *L. striata* DC.].
    A probable hybrid between *L. repens* and *L. vulgaris* (× *L. sepium* Allman; keyed out above) is known from Nfld. (roadsides and railway embankments between St. John's and Waterford Bridge; GH; CAN).

L. vulgaris Hill   Butter-and-eggs, Common Toadflax.   Gueule de lion
Eurasian; a common weed of fields, roadsides, and waste places in N. America, as in Alaska (Fairbanks; Hultén 1949), S Dist. Mackenzie (Fort Smith, ca. 60°N), and all the provinces (in Man., N to Churchill); sw Greenland. [*Antirrhinum* (*Linaria*) *linaria* L.]. MAP: Hultén 1968*b*:793.
    Forma *leucantha* Fern. (corolla whitish rather than bright yellow) is known from Ont. (Boivin 1966*b*), SW Que. (Napierville Co. and St. Helen's Is., Montreal), and N.S. (Cumberland and Colchester counties; type from Amherst, Cumberland Co.). Forma *peloria* (L.) Rouleau (*Peloria peloria* L., the floral characters so different from those of the typical form that Linnaeus placed the plant in a separate genus; the corolla regular, spurless or with 3 or 5 spurs at base, the flowers usually sterile, rather than corolla very irregular (as in the snapdragon), 1-spurred at base) is known from Ont. (Ottawa, where taken by James Fletcher in 1879; CAN; reported from Toronto by Pennell 1935), Que. (St. Helen's Is., Montreal; Rouleau 1945), and N.S. (John Macoun 1884).

LINDERNIA All.   [7562]   False Pimpernel

1   Leaves mostly rounded at base, to about 2 cm long, usually distinctly shorter than the subtended stem-internodes; pedicels usually conspicuously surpassing their subtending leaves; seeds brownish yellow; (S B.C. and S ?Ont.) . . . . *L. anagallidea*
1   Leaves mostly cuneate at base, to about 3 cm long, usually equalling or longer than the subtended internodes; pedicels shorter than or only slightly surpassing their subtending leaves; seeds very pale yellow; (Ont. to N.S.) . . . . *L. dubia*

L. anagallidea (Michx.) Pennell
/t/X/   (T)   Moist shores, sands, and banks from S B.C. (Boivin 1966*b*; collections in OAC from Norfolk and Waterloo counties, S Ont., have also been placed here but require confirmation) and Wash. to N.Dak., Wisc., N.Y., and N.H., S to Mexico, Tex., and Fla.; S. America. [*Gratiola* Michx.; *Ilysanthes* Raf.]. MAP: Pennell 1935: map 31 (indicating no Canadian stations), facing p. 160.

L. dubia (L.) Pennell
/T/(X)/   (T)   Shores, damp ground, and disturbed soil, the main area of the aggregate species from Ont. to N.B. and N.S., S to E Tex., La., Ala., and Ga., with isolated stations (?introd.) in S B.C., Wash., Oreg., N Calif., and N Mexico; S. America. MAP and synonymy: *see* below.
1   Bracteal leaves rounded at tip, little reduced; flowers all cleistogamous; [E Que.: fresh tidal shores of the St. Lawrence R. estuary at St-Vallier, Bellechasse Co., and Beauport, Québec Co.] . . . . var. *inundata* Pennell
1   Bracteal leaves blunt to acutish; pedicels to 2 cm long.
    2   Bracts conspicuously smaller than the foliage-leaves, the upper ones at most 6 mm broad; later flowers often cleistogamous; [*Ilysanthes rip.* Raf.; S Ont. (Welland, Peel, and Waterloo counties) and Que. (Venise, Missisquoi Co.; Batiscan, Champlain Co.)] . . . . . .
    . . . . var. *riparia* (Raf.) Fern.

2 Bracts about equalling the foliage-leaves, the upper ones to 1 cm broad; corollas all expanding; [ssp. *major* Pennell; *Gratiola* L.; *Ilysanthes* Barn.; *I. gratioloides* Benth.; s B.C. (New Westminster and South Kootenay, where probably introd.); Ont. (N to Carleton and Russell counties), Que. (N to 71 mi NW of Mont-Laurier and Batiscan, Champlain Co.), N.B., and N.S.; MAP: Pennell 1935: map 27, p. 143] . . . . var. *dubia*

MELAMPYRUM L. [7635]

M. lineare Desr. Cow-wheat
/sT/X/ (T) Mossy coniferous forest, bogs, heaths, and peaty or rocky barrens, the aggregate species from B.C. (N to Kispiox, about 125 mi NE of Prince Rupert at ca. 55°N; Eastham 1947) to L. Athabasca (Alta. and Sask.), Man. (N to Wekusko L., about 90 mi NE of The Pas), Ont. (N to Big Trout L. at ca. 53°45′N, 90°W), Que. (N to the E James Bay watershed at ca. 53°45′N, L. Mistassini, the Côte-Nord, Anticosti Is., and Gaspé Pen.; the report from s Labrador by Fernald *in* Gray 1950, may refer to the E Que. side of the Blanc-Sablon R.), Nfld., N.B., P.E.I., and N.S., s to N Wash.–Idaho–Mont., Minn., Wisc., Ind., Tenn., Ga., and S.C. MAP and synonymy: *see* below.
1 Bracteal leaves essentially entire or the uppermost ones with a few short basal teeth.
   2 Foliage-leaves generally linear and rarely over 5 mm broad; stem simple or loosely few-branched; [*M. pratense* of Canadian reports, not L.; *M. sylvaticum sensu* Hooker 1838, not L.; transcontinental; MAP: Pennell 1935: map 148, p. 508] . . . . var. *lineare*
   2 Foliage-leaves lanceolate to ovate, to over 1 cm broad; stem commonly bushy-branched; [Ont. to N.B., P.E.I., and N.S.; MAP: on the above-noted map by Pennell] . . . .
   . . . . var. *latifolium* Bart.
1 Bracteal leaves with several sharp slender teeth; leaves linear to lanceolate; stem commonly bushy-branched.
   3 Leaves to 1 cm broad; blade of bracts (excluding teeth) to 2 cm broad, the lower bracts to 6 cm long; teeth of middle and upper bracts shorter than the blade-width; [*M. americanum* Michx.; Ont. to N.B., P.E.I., and N.S.] . . . .
   . . . . var. *americanum* (Michx.) Beauverd
   3 Leaves (and blades of bracts, excluding the teeth) less than 1 cm broad, the lower bracts usually not over 3.5 cm long; teeth of middle and upper bracts about as long as the blade-width; [*M. lineare pectinatum* Pennell; E U.S.A. only, but to be searched for, particularly in s Ont.; MAP: on the above-noted map by Pennell] . . . .
   . . . . [var. *pectinatum* (Pennell) Fern.]

[MIMETANTHE Greene] [7547]

[*M. pilosa* (Benth.) Greene]
[The report of this species of the w U.S.A. (N to Wash. and Idaho) from sw B.C. by J.M. Macoun (1913; Swan L., Vancouver Is., as *Mimulus pil.*) is probably based upon a heretofore unnamed collection in CAN taken at that locality by John Macoun in 1908, referable to *Mimulus guttatus*. (*Herpestes* Benth.; *Mimulus* Wats.).]

MIMULUS L. [7524] Monkey-flower. Mimule

(Ref.: A.L. Grant 1924)
1 Corolla various shades of pink, red, blue-violet, or purple (often marked with yellow).
   2 Low slender copiously glandular-stipitate annual to about 1.5 dm tall; corolla reddish to light purple, at most 1 cm long; calyx to 6 mm long, regular, the short teeth nearly equal; leaves linear to linear-elliptic or linear-oblanceolate, sessile, entire, to 2 cm long and 4 mm broad, obscurely 3-nerved; (s B.C.) . . . . *M. breweri*
   2 Taller, stouter, rhizomatous perennials to over 1 m tall; corolla commonly at least 2 cm long; calyx at least 1 cm long; leaves lanceolate or oblanceolate to ovate or obovate, entire or more commonly serrate.
      3 Leaves palmately veined from near the base, sessile, entire or irregularly callous-dentate, viscid-villous; corolla pink-purple, commonly over 3 cm long; calyx to 2.5 cm long; (B.C. and Alta.) . . . . *M. lewisii*

      3   Leaves pinnately veined; corolla violet-purple; calyx to 1.5 cm long.
          4   Leaves petioled, coarsely toothed; corolla to 2.5 cm long; calyx-lobes to 2 mm
              long, bristle-tipped; pedicels mostly less than 2 cm long; angles of stem
              slightly winged; (S Ont.; ?extinct) .... [*M. alatus*]
          4   Leaves sessile, strongly rounded to cordate-clasping at base, very shallowly
              toothed; corolla to 4 cm long; calyx-lobes to 8 mm long; pedicels to over 6 cm
              long; stem not winged; (Sask. to N.S.) .... *M. ringens*
1   Corolla yellow, commonly lined, dotted, or blotched with maroon or red-brown.
    5   Annuals, lacking rhizomes or stolons; leaves palmately or subpalmately veined from
        near the base (or often subpinnately veined in *M. floribundus*).
      6   Corolla strongly 2-lipped (the lower lip longer than the upper one and deflexed
          from it).
          7   Upper calyx-tooth much the largest; calyx to over 1.5 cm long in anthesis,
              accrescent, the 2 small lower teeth tending to fold upwards in fruit; corolla to 4
              cm long; leaves ovate to rotund or reniform-cordate, 3–7-nerved, the
              uppermost ones sessile and tending to be connate; plant soft and often
              somewhat succulent, glabrous or pubescent, very variable in stature; (B.C. to
              sw Sask.) .... *M. guttatus*
          7   Upper calyx-tooth the same size and shape as the two lateral acute upper
              teeth (the lower pair of teeth rounded and slightly longer); calyx to 7 mm long;
              corolla to about 1.5 cm long; leaves elliptic to deltoid or subrhombic,
              3–5-nerved, the blade to 2 cm long; plant glandular-pubescent to partly
              glabrous, to 3 dm tall; (S B.C.) .... *M. alsinoides*
      6   Corolla only slightly 2-lipped (the lower lip only slightly longer than the upper one
          and not much deflexed); calyx-teeth subequal, short.
          8   Leaves abruptly contracted to the petiole, the blades to 3 cm long, mostly
              deltoid-ovate to subcordate, callous-toothed; calyx to 8 mm long, its teeth
              acute; corolla to 14 mm long, the tube much surpassing the calyx; stems to
              over 2.5 dm long; plant erect to subprostrate, conspicuously to sometimes
              obscurely viscid-pubescent and glandular (often tending to be clammy); (B.C.
              and ?Alta.) .... *M. floribundus*
          8   Leaves tapering to the short-petioled or sessile base, the blades mostly not
              over 2 cm long, entire or minutely toothed; calyx to 5 mm long at anthesis;
              corolla to 8 mm long, the tube slightly surpassing the calyx; plants finely
              glandular-puberulent.
             9   Leaves linear to narrowly oblong or oblanceolate, sessile or the lowest
                 ones short-petioled; calyx-teeth mostly rounded and mucronate-tipped;
                 fruiting pedicels tending to be widely spreading but with suberect tips;
                 stems to about 1 dm long; (?B.C.) .... [*M. suksdorfii*]
             9   Leaves mostly relatively broader, narrowly elliptic to rhombic-elliptic,
                 commonly short-petioled; calyx-teeth more or less acute; fruiting pedicels
                 generally more ascending; stems to about 2 dm long; (S B.C.) ... *M. breviflorus*
    5   Perennials with rhizomes; leaves lance-ovate to rotund or reniform-cordate, sessile or
        short-petioled.
      10 Leaves pinnately veined, the blade lance- to elliptic-ovate or ovate, remotely and
          sometimes obscurely callous-dentate; calyx to 13 mm long, the upper tooth often
          a little larger than the others; corolla obscurely 2-lipped, to 3 cm long; plant
          clammy-villous, the stem often prostrate at base and rooting at the nodes; (S B.C.;
          introd. eastwards) .... *M. moschatus*
      10 Leaves palmately (longitudinally) veined from near the base, the blade elliptic to
          ovate, rotund, or reniform-cordate, entire or irregularly toothed; upper calyx-tooth
          conspicuously larger than the others; corolla strongly 2-lipped; plants glabrous or
          sparingly pubescent but not at all clammy-villous.
          11 Corolla mostly 1 or 2 cm long, sparingly if at all red-dotted, its throat open;
              lateral and lower pairs of calyx-teeth blunt and mostly very short; plant
              glabrous or inconspicuously hairy, the weak stems decumbent to creeping and
              rooting at the nodes (sometimes floating); (S Sask. to Que.) .... *M. glabratus*

11 Corolla to 4 cm long, strongly dotted or blotched with red-brown on or about the prominent palate (which nearly closes the throat); lateral and lower calyx-teeth more or less acute, the lower ones tending to fold upwards in fruit.

   12 Stems very variable in size, from dwarf to robust and up to nearly 1 m tall, with stolons but only rarely with definite creeping rhizomes; flowers often more than 5 (when few, commonly less than 2 cm long); (B.C. to sw Sask.) . . . . *M. guttatus*

   12 Stems rarely over 2 dm tall, from well-developed (often sod-forming) creeping rhizomes, and also often stoloniferous; flowers often over 2 cm long, solitary or commonly not more than 5; (mts. of s B.C. and sw Alta.) . . . .
. . . . *M. tilingii*

[M. alatus Ait.]
[This species of the E U.S.A. (N to Nebr., Mich., and Conn.) is apparently known from Canada only through two collections in CAN and TRT from Rondeau Harbour, Kent Co., s Ont. (an 1897 collection by A.J. Stevenson and an undated collection by Stevenson and J. Dearness, both verified by Pennell). The plant may have been introduced, the Stevenson label indicating the habitat as "wet places in a ditch". If once native in s Ont., it can now be considered extinct, no later collections having been reported.]

M. alsinoides Dougl.
/t/W/ (T) Moist shady places and mossy cliffs from s B.C. (Vancouver Is. and adjacent islands and mainland N to Yale, E to Similkameen, s of Penticton) to N Calif.

M. breviflorus Piper
/t/W/ (T) Moist open places in the valleys and plains from s B.C. (Adams Lake, about 35 mi NE of Kamloops, and Newgate, about 40 mi SE of Cranbrook; Herb. V, both detd. J.A. Calder) and Idaho to N Calif.

M. breweri (Greene) Rydb.
/T/W/ (T) Moist to dryish meadows and slopes at low to moderate elevations from s B.C. (near Rossland; Sproat, SE of Revelstoke; CAN, verified by Calder) to s Calif. [*Eunanus* Greene; *M. rubellus* var. *br.* (Greene) Jeps.].

M. floribundus Lindl.
/T/W/ (T) Moist open places at low to fairly high elevations from s B.C. (N to Chilcotin, Lytton, and Armstrong) and sw ?Alta. (Moss 1959) to Calif. and N Mexico. [*M. peduncularis* Dougl.].

M. glabratus HBK.
/T/(X)/ (Hpr) Swampy places, shores, and shallow water from Mont. to s Sask. (Whitewood, about 100 mi E of Regina; Breitung 1957a), s Man. (Aweme, about 20 mi SE of Brandon; CAN; reported from Notre Dame de Lourdes, about 60 mi SE of Brandon, by Lowe 1943), Ont. (N to Matheson, 48°32'N; CAN), and Que. (62 mi N of Amos, Abitibi-East Co.; C. Rousseau, S. Payette, and A. Asselin, Nat. can. (Que.) 97(2):177. 1970), s to Nev., Ariz., Mexico, Tex., Ill., Mich., and s Ont.; S. America. [Incl. var. *fremontii* (Benth.) Gray (*M. jamesii* var. *fre.* Benth., not *M. fre.* (Benth.) Gray); *M. geyeri* Torr.]. MAP: Pennell 1935: map 23, p. 117.

M. guttatus DC.
/ST/WW/ (Hpr (T)) Wet places at low to moderate elevations from the Aleutian Is. (type material grown from seeds collected at Unalaska), Alaska (N to ca. 65°N), and s Yukon to B.C., Alta. (Crowsnest Pass and Waterton Lakes; CAN), and sw Sask. (Cypress Hills; CAN), s to Calif., N.Mex., and N Mexico; introd. elsewhere, as in the E U.S.A. and Europe. [*M. grandiflorus* Howell; *M. langsdorfii* Donn; *M. rivularis* Nutt.; *M. luteus* of auth., not L.; incl. the reduced extreme, generally lacking stolons, var. *depauperatus* (Gray) Grant (*M. lang.* var. *dep.* Gray; *M. microphyllus* Benth.; *M. minimus* Henry; *M. nasutus* Greene)]. MAP: Hultén 1968b:796.

   Ssp. *haidensis* Calder & Taylor (peduncles puberulent but nonglandular rather than glandular-pubescent; leaves acute rather than blunt-tipped) is known from the type locality on Mt. Moresby, Moresby Is., Queen Charlotte Is., B.C.

M. lewisii Pursh
/T/W/ (Hpr) Wet places at moderate to high elevations from the southernmost Alaska Panhandle (Hyder; *see* Hultén 1949: map 1024, p. 1469) through B.C. and w Alta. (N to the Smoky R. at ca. 55°N; John Macoun 1884) to Calif., Utah, and Wyo. [*M. roseus* Dougl.; *Penstemon ?frutescens sensu* Hooker 1838, and Macoun 1884, not Lamb.]. MAP: Hultén 1968*b*:796.

Forma *alba* (Henry) Boivin (corolla white rather than pink-purple) is known from SW B.C. (Mt. Cheam, near Chilliwack (type locality of var. *alb.* Henry, 1915, as the first place cited) and North Vancouver).

M. moschatus Dougl.   Muskflower
/T/W/ (Hpr) Moist places at low to moderate elevations from S B.C. (N to Revelstoke; CAN) to Calif. and Colo.; perhaps introd. elsewhere in E N. America, as in Ont. (N to the Ottawa dist.), Que. (N to Grindstone Is., Magdalen Is.; CAN; GH), Nfld., N.B., P.E.I., and N.S.; introd. in S. America and Europe. [Incl. var. *sessilifolius* Gray, with relatively large sessile leaves].

According to Pennell (1935) and Fernald *in* Gray (1950), this species is apparently native in Nfld., Magdalen Is., and parts of the E U.S.A. (*See* note under *Luzula campestris*). The closely related *M. dentatus* Nutt. is reported as possibly occurring in SW B.C. by Boivin (1967*a*). It should be searched for there, having a presently accepted range from Wash. to N Calif. It differs from *M. moschatus* in being somewhat hirsute but scarcely viscid (even slimy), the calyx-tube hirsute only along the 5 ribs rather than viscid-villous over the surface as well as along the ribs, the usually longer corolla more strongly 2-lipped and with a more expanded throat.

M. ringens L.
/T/EE/ (Hpr) Shores, meadows, and wet places from E Sask. (near Hudson Bay Junction, 52°52'N; CAN) to Man. (N to Hill L., N of L. Winnipeg; CAN), Ont. (N to the SW James Bay watershed at 52°11'N), Que. (N to the Gaspé Pen.), N.B., P.E.I., and N.S., S to NE Tex. (an isolated station in Colo.), La., and Ga. [Incl. the reduced extreme, var. *colpophilus* Fern.]. MAP: Pennell 1935: map 24, p. 125.

Forma *peckii* House (flowers white rather than pinkish to blue-violet) is known from S Ont. (Seymour, Northumberland Co., where taken by John Macoun in 1877; CAN).

[M. suksdorfii Gray]
[The inclusion of B.C. in the range of this species of the w U.S.A. (Wash. to Calif. and Colo.) by Rydberg (1922) requires clarification.]

M. tilingii Regel
/T/W/ (Hpr) Wet places at moderate to high elevations from B.C. (collections in CAN and V from Vancouver Is., Garibaldi, N of Vancouver, mts. along the Skagit and Chilliwack rivers, Manning Provincial Park, about 30 mi SE of Hope, Kokanee, near Nelson, Flathead, South Kootenay Pass, the Selkirks at Rogers Pass between Revelstoke and Golden, Chilcotin, Terrace, E of Prince Rupert, and the junction of the Dease and Liard rivers at ca. 59°40'N; reports from Alaska are probably erroneous) and SW Alta. (Waterton Lakes; Breitung 1957*b*) to Baja Calif. and N.Mex. [*M. alpinus* (Gray) Piper; incl. *M. caespitosus* Greene].

ODONTITES Ludwig   [7644]

O. verna (Bell.) Dum.
Eurasian; introd. along roadsides and in fields in N. America, as in Alta. (Edson, about 120 mi W of Edmonton; Groh 1947), Man. (near Gimli, about 50 mi N of Winnipeg), Ont. (near Thunder Bay), Que. (N to Magdalen Is. and the Gaspé Pen. at Grande-Rivière; QSA; CAN), ?Nfld. (Boivin 1966*b*), N.B. (Westmorland Co. and Grand Manan Is.; CAN; MT), P.E.I. (Prince and Queens counties; CAN; GH), and N.S. [*Euphrasia (Bartsia) odontites* L.; *O. rubra* Gilib.; incl. *O. serotina* (Wettst.) Dum.].

ORTHOCARPUS Nutt. [7633] Owl-clover

1 Anthers 1-locular; leaves pinnately divided to near the midrib into linear or filiform segments; (s B.C.).
 2 Corolla red-purple (sometimes yellow), at most 6 mm long and scarcely surpassing the calyx; stamens in anthesis exserted from the galea (hooded upper corolla-lip); leaves to 3 cm long, minutely spreading-hispid; plant to 2 dm tall, very slender, the spike elongate and often extending to near the base of the stem, the lower flowers remote . . . . O. pusillus
 2 Corolla predominantly sulphur-yellow, to 2.5 cm long, the filiform tube at least twice as long as the calyx; stamens in anthesis not exserted from the galea; leaves glabrous, or puberulent above; (introd. on Vancouver Is.).
  3 Galea dark purple; leaves more or less purplish . . . . O. erianthus
  3 Galea pure white; leaves greenish . . . . O. faucibarbatus
1 Anthers 2-locular.
 4 Corolla crimson or pink-purple, the galea hooked at apex; (Vancouver Is.).
  5 Lower lip of corolla (to 2 cm long) simply saccate or nearly so; leaves short-pubescent with spreading or appressed hairs, the lower ones entire, the upper ones 3-cleft and passing into the divergently 3-lobed bracts . . . . . . O. bracteosus
  5 Lower lip of corolla (to about 3 cm long) more or less trisaccate; leaves and bracts villous-pubescent, pinnately parted into many filiform or narrowly linear lobes; (introd.) . . . . [O. purpurascens]
 4 Corolla commonly yellow (sometimes whitish or pinkish); lower leaves mostly entire, the upper ones commonly 3-cleft (sometimes 5-cleft).
  6 Lower lip of the yellow corolla simply saccate or nearly so; galea incurved or hooked at tip; plants to 3 or 4 dm tall.
   7 Bracts and calyces glandular-pubescent, the bracts gradually differentiated from the leaves; corolla yellow, less than 1.5 cm long, the galea and lower lip subequal; leaves or their lobes linear; stem spreading-hairy (or finally glabrate below); (B.C. to w Ont.) . . . . O. luteus
   7 Bracts and calyces scarcely or not at all glandular, the broad bracts abruptly differentiated from the leaves, entire or the lower ones with a pair of slender, commonly hispid-ciliate, lateral lobes; upper bracts with conspicuous pink-purple petaloid tips; corolla yellow (or purplish at tip), to about 2 cm long, the galea about 1 mm longer than the lower lip; leaves or their lobes narrowly linear; stem puberulent; (s B.C.) . . . . O. tenuifolius
  6 Lower lip of corolla more or less trisaccate; galea nearly straight, relatively slender.
   8 Bracts green throughout (rarely slightly purple-tinged but not at all showy); corolla to 2 cm long, white or light yellow, the teeth of the lower lip inconspicuous; leaves and their lobes linear; plant spreading-hairy throughout; (s B.C.) . . . . O. hispidus
   8 Upper bracts with white, yellow, or purplish, more or less petaloid tips or lobes; corolla to 2.5 cm long, the lower lip with slender terminal teeth to 3 mm long; plants short-spreading-pubescent throughout; (sw B.C.).
    9 Corolla linear (lower lip scarcely expanded), whitish or pink-tinged, the lower lip more or less yellowish and with some purple spots; bracts petaloid only at the tip, the slender spike scarcely showy; leaves and their segments lance-linear, long-acuminate . . . . O. attenuatus
    9 Corolla clavate (lower lip somewhat inflated), yellow with some purple markings; bracts and their lobes more evidently petaloid, the stout spike rather showy; leaves and their segments lanceolate to ovate or oblong . . . .
    . . . . O. castillejoides

O. attenuatus Gray Valley-tassels
/t/W/ (T) Meadows, pastures, and grassy slopes from sw B.C. (Vancouver Is. and Saltspring Is.; CAN; V) to Calif.

O. bracteosus Benth.   Pink Owl's-clover
/t/W/   (T)   Meadows at low elevations from SW B.C. (several localities on Vancouver Is.; CAN) to N Calif.

O. castillejoides Benth.   Johnny-nip
/t/W/   (T)   Salt marshes and other saline soils along the coast from SW B.C. (several localities on S Vancouver Is.; CAN; V) to Calif.

O. erianthus Benth.   Johnny-tuck
A native of California; introd. in SW B.C. (Victoria, Vancouver Is., where taken by John Macoun in 1908; CAN).

O. faucibarbatus Gray
Native in the W U.S.A. from Oreg. to Calif.; known in Canada only from SW B.C. (Vancouver Is.; Carter and Newcombe 1921; collection in CAN from Ten Mile Point, Victoria, where taken by Eastham in 1942).
    The Vancouver Is. plant is referable to ssp. *albidus* Keck (lower corolla-lip pure white rather than yellow, often fading to pink). Of this, D.D. Keck (Madroño 5:165. 1940) notes, "This is the subspecies that was introduced toward the end of the last century on Vancouver Island near Victoria."

O. hispidus Benth.
/t/W/   (T)   Moist places at low to moderate elevations from SW B.C. (Vancouver Is. and L. Osoyoos, near the U.S.A. boundary S of Penticton; CAN; introd. at Skagway, SE Alaska), Idaho, and Mont. to S Calif. [*Triphysaria* Rydb.]. MAP: Hultén 1968*b*:813.

O. luteus Nutt.
/T/WW/   (T)   Dry prairies and plains from B.C. (N to Endako, ca. 54°N; Eastham 1947) to Alta. (N to Peace Point, 59°07′N; CAN), Sask. (N to N of Prince Albert; CAN), Man. (N to Norway House, off the NE end of L. Winnipeg), and W-cent. Ont. (probably introd.; collections in CAN from the N shore of L. Superior at Silver Islet, near Thunder Bay, where taken in a gravelly field, and from Peninsula, near Marathon, where taken in a sandy burn near the railway), S to Calif., N.Mex., Nebr., and Minn. [*O. strictus* Benth.]. MAP: Pennell 1935: map 149 (not indicating any Ontario stations), p. 517.

[O. purpurascens Benth.]
[The report of this Californian species from SW B.C. by Eastham (1947; Mt. Finlayson, Vancouver Is.) requires confirmation.]

O. pusillus Benth.
/t/W/   (T)   Moist places near the sea from SW B.C. (Vancouver Is., Saltspring Is., and the adjacent mainland at Douglas; CAN; V) to Calif.

O. tenuifolius (Pursh) Benth.
/t/W/   (T)   Moist or dry plains and valleys at low to moderate elevations from S B.C. (Carson, SW of Grand Forks; CAN; reported from Cranbrook, Fort Steele, and Flagstone by Eastham 1947, and from L. Osoyoos, S of Penticton, by John Macoun 1884) to Oreg. and Mont. [*Bartsia* Pursh].

PARENTUCELLIA Viviani   [7642]

P. viscosa (L.) Car.
European; a weed in low moist ground of W N. America, as in SW B.C. (Vancouver Is. and adjacent islands and mainland at New Westminster; V). [*Bartsia* L.].

PEDICULARIS L.   [7648]   Lousewort, Wood-betony

1   Stem-leaves opposite or whorled or the plant scapose and the leaves all or chiefly basal; perennials.

2　Stems leafy, the leaves opposite or whorled, their teeth forming a whitish-cartilaginous margin; galea (the upper helmet-shaped corolla-lip) lacking a pair of subapical marginal teeth.

　　3　Stem-leaves opposite, coarsely lobed or more or less pinnatifid to about halfway to the midrib, to about 1 dm long; basal leaves somewhat reduced, not forming a rosette; calyx 2-lobed, otherwise entire except for an ovate or oblong, foliaceous, toothed appendage on each lateral half; corolla pale yellow, to 2.5 cm long; stem glabrous, to about 9 dm tall; (s Man. and s Ont.) .... *P. lanceolata*

　　3　Stem-leaves mostly in whorls of 3 or more, the principal ones pinnately parted nearly or quite to the midrib; calyx shortly 5-toothed; corolla reddish- or bluish-purple.

　　　　4　Corolla to 2.5 cm long, the galea acuminate-beaked; stem-leaves to about 1 dm long; rosette-leaves few or none, smaller than the stem-leaves; plant glabrous (or very sparingly pubescent in the inflorescence), to 6 dm tall; (Aleutian Is. and Alaska) .... *P. chamissonis*

　　　　4　Corolla about 1 cm long, the galea truncate at apex; stem-leaves mostly more or less reduced and short-petioled, the principal leaves long-petioled in a basal rosette; plant copiously pubescent; (Alaska to Dist. Mackenzie and N B.C.) .... *P. verticillata*

2　Stem scapose (lacking leaves or with only 1 or 2 reduced ones), the very deeply lobed slender-petioled leaves all or nearly all in a basal rosette; calyx unevenly 5-toothed or 5-lobed.

　　5　Galea with a straight horizontally spreading beak to 4 mm long, lacking a pair of subapical marginal teeth; corolla purple, to about 1.5 cm long; inflorescence capitate (1 or 2 small lower clusters sometimes present); plant to 3 dm tall, glabrous below the more or less villous inflorescence; (SE Alaska–B.C.) .... *P. ornithorhyncha*

　　5　Galea nearly or quite beakless; (transcontinental).

　　　　6　Corolla creamy yellow, often reddish-tinged, to about 4 cm long, the 2–4(6) flowers in a capitate cluster; galea essentially entire at apex, lacking an obvious pair of subapical marginal teeth; calyx 5-lobed, the lobes up to about the length of the tube; capsule scarcely surpassing the calyx; stem glabrous or short-pubescent, at most about 1.5 dm tall .... *P. capitata*

　　　　6　Corolla dark reddish-purple, less than 2.5 cm long; inflorescence at first subcapitate, elongating in fruit; galea typically with a pair of broadly deltoid subapical marginal teeth; calyx with 5 lanceolate serrulate teeth; capsule somewhat surpassing the calyx; stem densely woolly above and into the inflorescence, to about 5 dm tall .... *P. sudetica*

1　Stem-leaves alternate.

　　7　Inflorescence copiously white-woolly; galea beakless; perennials.

　　　　8　Corolla predominantly yellow but the galea deep purple; (B.C. and SW Alta.) .... *P. oederi*

　　　　8　Corolla more uniformly coloured, pale or bright pink to rose-purple; (arctic and subarctic regions).

　　　　　　9　Corolla bright pink to rose-purple, about 2 cm long; galea lacking a pair of subapical marginal teeth; capsule about 2 cm long; seeds reticulate; stem and inflorescence densely woolly; taproot stout, bright lemon-yellow; (transcontinental) .... *P. lanata*

　　　　　　9　Corolla pale pink, about 1.5 cm long; galea sometimes with a pair of low blunt teeth near the apex; capsule rarely over 1.5 cm long; seeds smooth; stem and inflorescence rather sparingly woolly; taproot pale and more slender; (eastern arctic and subarctic regions) .... *P. hirsuta*

　　7　Inflorescence essentially glabrous or rather sparingly short-pubescent (moderately woolly in *P. langsdorfii*).

　　　　10　Corolla typically pink to reddish or purple (at least as to the galea; partly or even wholly white or yellow in varieties).

11 Galea with a long slender curving beak, lacking subapical marginal teeth; corolla to 1.5 cm long, typically pink to purplish (white or ochroleucous in *P. racemosa* var. *alba*); essentially glabrous perennials.

   12 Leaves deeply pinnatifid, the largest ones (to about 2 dm long) in a basal rosette; calyx with 5 short entire subequal lobes; beak of galea upcurved after an initial downcurving at base; inflorescence elongate but fairly dense, the bracts mostly much shorter than the flowers; (transcontinental) . . . . *P. groenlandica*

   12 Leaves merely doubly serrate (the secondary teeth often inconspicuous), commonly not over 1 dm long and 1.5 cm broad, the lower ones much reduced; calyx deeply cleft on the lower (outer) side into 2 broad-based oblique acuminate segments; beak of galea downcurved; inflorescence mostly lax and elongate, the flowers or peduncled flower-clusters subtended and surpassed by scarcely reduced leaves; (B.C.) . . . . *P. racemosa*

11 Galea nearly or quite beakless.

   13 Perennials with simple stems, these commonly subtended by a rosette of long-petioled leaves; calyx 5-lobed.

      14 Bracts sharply differentiated from the leaves, mostly shorter than the flowers; leaf-blades to over 1.5 dm long, their narrowly lanceolate to linear-oblong segments to 7 cm long; corolla red or purple (occasionally yellow), to about 2 cm long, the galea lacking a pair of subapical marginal teeth; plant pubescent below the inflorescence, to about 1 m tall, coarsely fibrous-rooted, some of the roots tuberous-thickened; (B.C. and Alta.) . . . . *P. bracteosa*

      14 Bracts scarcely differentiated from the leaves, longer than the flowers; leaf-blades to about 7 cm long, their ovate segments commonly 3 or 4 mm long; corolla bright pink, to 2.5 cm long, the galea with a pair of subapical marginal teeth; plant moderately woolly-villous (particularly in the inflorescence), the tufted stems to 3 dm tall, from a taproot; (B.C. and sw Alta.) . . . . *P. langsdorfii*

   13 Mostly low-branching annuals or biennials lacking a rosette of basal long-petioled leaves; flowers in capitate clusters terminating the branches.

      15 Corolla less than 1.5 cm long, purple; galea usually lacking prominent subapical marginal teeth but with a pair of short blunt glandular-margined appendages less than 1 mm long 3 or 4 mm below the apex; calyx with 2 irregularly lacerate lobes; plant glabrous except for the ciliate-fringed lower corolla-lip; (Alaska–B.C. to E James Bay) . . . . *P. parviflora*

      15 Corolla usually 2 or 3 cm long; galea with a pair of slender subapical marginal teeth; plants sparingly pubescent at least in the inflorescence.

         16 Corolla rose-purple, its violet galea with a pair of short blunt glandular-margined appendages about 1 mm long 5 or 6 mm below the apex in addition to the pair of subapical marginal teeth; calyx with 2 low broad crested-toothed lobes, much surpassed by the capsule; central raceme less than half the height of the plant; stem to over 8 dm tall, with stiffly ascending branches; (Que. to Nfld. and N.S.) . . . . *P. palustris*

         16 Corolla flesh-pink; galea lacking appendages other than the pair of subapical marginal teeth; calyx with 4 or 5 subequal foliaceous teeth, these equalling or surpassing the capsule; central raceme often more than half the height of the plant; stem at most about 2 dm tall, simple or diffusely branched; (introd. in Nfld.) . . . . . . *P. sylvatica*

10 Corolla white, ochroleucous, or yellow; plants perennial (or *P. labradorica* perhaps biennial), commonly with a basal rosette of long-petioled leaves.

   17 Galea distinctly beaked, lacking a pair of subapical marginal teeth.

18 Corolla usually uniformly pale yellow, the galea-beak short-conical, about 2 mm long; inflorescence subcapitate; calyx split in front, otherwise merely shallowly undulate at summit; leaf-segments rather crowded, lanceolate to ovate, commonly less than 5 mm long; plant more or less puberulent, to 2.5 dm tall; (transcontinental) .... *P. lapponica*

18 Corolla white or ochroleucous (often finely marked with purple), the elongate beak lunately downcurved and usually hidden within the lower lip; calyx tipped with 5 narrow teeth, the upper tooth the shortest; leaf-segments relatively narrow and remote, to over 1 cm long; plant glabrous throughout, to 6 dm tall; (mts. of B.C. and Alta.) .... *P. contorta*

17 Galea nearly or quite beakless.

19 Calyx split in front, otherwise subentire or merely shallowly undulate at summit; galea with a pair of slender subapical marginal teeth; capsule usually at least twice as long as the mature calyx.

20 Stem freely branched from near base; corolla about 1.5 cm long, the galea purplish or purple-tipped; fruiting spike less than 1 dm long; leaves with only the tips of the teeth white and cartilaginous; (transcontinental) .... *P. labradorica*

20 Stem simple; corolla about 2 cm long, usually uniformly yellow; fruiting spike to 2 dm long; leaves with white cartilaginous margins; (Man. to Que.) .... *P. canadensis*

19 Calyx distinctly 5-lobed, the lobes subequal.

21 Corolla commonly about 1 cm long, the galea strongly tipped with brownish red or reddish purple, lacking a pair of subapical marginal teeth; pedicels to over 8 mm long; capsule about twice as long as the ovate-lobed mature calyx; roots fusiform or tapering; plant essentially glabrous, commonly less than 1.5 dm tall; (transcontinental) ... *P. flammea*

21 Corolla to over 2 cm long, the galea not strongly purple-tipped; pedicels relatively short.

22 Galea with a pair of slender subapical marginal teeth; capsule little surpassing the linear-lanceolate to narrowly oblong calyx-teeth; principal leaves 2-pinnatifid; roots fibrous; plant to over 9 dm tall, pubescent at least in the inflorescence or glabrate; (N.B.) ....
.... *P. furbishiae*

22 Galea lacking a pair of subapical marginal teeth, its tip commonly purple-tinged; capsule much surpassing the ovate calyx-teeth; leaves 1-pinnatifid, their narrowly triangular to ovate segments crenate-dentate; roots fusiform; plants usually sparingly woolly at least in the inflorescence; (B.C. and sw Alta.) .... *P. oederi*

P. bracteosa Benth.
/T/W/ (Grt) Woods, meadows, and moist open montane slopes from B.C. (N to Mt. Selwyn, ca. 56°N) and sw Alta. (N to Jasper; the type was collected by Drummond in the Rocky Mountains, probably in Alta.) to N Calif. and Colo. [*P. montanensis* Rydb.; *P. recutita* Pursh].

Material from the Marble Range, NW of Clinton, B.C., is referable to var. *latifolia* (Pennell) Cronq. (*P. lat.* Pennell; free tips of the lateral sepals mostly shorter than the connate portion above the dorsal sinus rather than very slender and elongate, very finely, if at all, glandular rather than evidently glandular).

P. canadensis L. Wood-betony, Common Lousewort
/T/EE/ (Hs) Sandy or loamy soil of open woods and clearings, the aggregate species from Man. (N to Moosehorn, about 90 mi N of Portage la Prairie) to Ont. (N to Chalk River, Ottawa, and Kapuskasing), Que. (N to Quebec City; John Macoun 1886), and Maine, s to N Mexico, Tex., and Fla. MAP (aggregate species): Pennell 1935: map 147, p. 499.

The report from Matane, Gaspé Pen., E Que., by d'Urban (*in* R. Bell, Geol. Surv. Can., Report of Progress for the year 1858, pages 243–63. 1859) may be based upon *P. palustris,* as, also, the reports from Nfld. by Reeks (1873) and from N.S. by Lindsay (1878) and John Macoun (1884). The report from N.B. by Fowler (1885; Grand Falls, Victoria Co.) is probably based upon *P. furbishiae,* known from that locality; a collection in NBM from near Fredericton may also prove to be that species. Var. *dobbsii* Fern. (stems mostly solitary, scarcely clustered, the basal offsets prolonged, often creeping and rooting) is known from SE Man. (near Otterburne, about 30 mi S of Winnipeg; Löve and Bernard 1959) and S Ont. (Tobermory, Bruce Co.; GH, detd. Fernald). The typical form has uniformly yellow or yellowish corollas. Forma *bicolor* Farw. (corolla crimson on the back, otherwise yellow or yellowish) is known from S Ont. (Lambton Co.; OAC). Forma *praeclara* Moore (the corolla crimson throughout) is known from S Ont. (Lambton Co.; OAC) and SW Que. (Boivin 1966*b*).

P. capitata Adams
/AST/X/GA/ (Grh) Calcareous tundra and rocky slopes at low to fairly high elevations from the Aleutian Is. and coasts of Alaska–Yukon–Dist. Mackenzie–Dist. Keewatin to Banks Is., Melville Is., and Ellesmere Is. (N to ca. 81°N), S in the West to the mts. of B.C. (S to Mt. Selwyn, ca. 56°N; CAN) and SW Alta. (Jasper dist.; CAN), farther eastwards S to Great Bear L., Southampton Is., and S Baffin Is.; NW Greenland between ca. 76° and 81°30′N; N Asia. [*P. nelsonii* R. Br.; *P. verticillata* Pursh, not L.]. MAPS: Hultén 1968*b*:825; Porsild 1957: map 293, p. 197; Raup 1947: pl. 34; Fernald 1925: map 59, p. 325; Tolmachev 1932: map 7, p. 53.

P. chamissonis Stev.
/s/W/eA/ (Hs) Subalpine meadows of the Aleutian Is. and SW Alaska (Pavlov Bay and St. Paul Is.; described from material from Unalaska Is. and Siberia); E Asia. MAPS: Hultén 1968*b*:817, and 1949: map 1051, p. 1472.

P. contorta Benth.
/T/W/ (Hs) Wooded or open slopes and drier meadows at moderate to high elevations from SE B.C. (N to near the Alta. boundary W of Banff; CAN) and SW Alta. (N to Banff; CAN) to N Calif. and Wyo.

P. flammea L.
/aST/(X)/nE/ (Hs) Moist peats, gravels, and tundra from Great Bear L. (the N Alaska map by Wiggins and Thomas 1962:403, indicates an isolated station in N-cent. Alaska but *P. oederi* may be the species involved, an early Yukon report being referred to it by Porsild 1951*a*) and the coasts of Dist. Mackenzie–Dist. Keewatin to E Devon Is., Baffin Is., and northernmost Ungava–Labrador, S to NE Man. (Churchill), N Ont. (coasts of Hudson Bay–James Bay S to ca. 55°N), Que. (coasts of Hudson Strait and Hudson Bay–James Bay S to ca. 53°N; Shickshock Mts. of the Gaspé Pen.), and Nfld.; Iceland; N Norway. MAPS: Eric Hultén 1968*b*:826; 1958: map 163, p. 183; and Sven. Bot. Tidskr. 55(1): fig. 1, p. 194. 1961; Porsild 1957: map 294, p. 197; Gjaerevoll 1963: fig. 3, p. 264.

Porsild's map indicates an isolated station in the mts. on the S B.C.–Alta. boundary, this based upon old reports considered erroneous by Hultén. Forma *flavescens* Polunin (corolla uniformly yellow rather than with a purple-tipped galea) is known from the type locality, Lake Harbour, S Baffin Is.

P. furbishiae Wats.
/T/E/ (Hs) Known only from banks of the St. John R. system in N.B. (between Andover and Grand Falls, Victoria Co.; CAN; GH; NBM) and N Maine.

P. groenlandica Retz. Elephant's-head, Little Red Elephants
/aST/X/G/ (Hs) Wet meadows from southernmost E Yukon (about 130 mi E of Teslin at ca. 60°N; CAN) and the mts. of B.C.–Alta. to Sask. (N to Turtleford, about 130 mi NW of Saskatoon at 53°23′N; Breitung 1957*a*), N Man. (Hayes and Nelson rivers from about 160 mi S of Churchill to York Factory and Churchill), N Ont. (coasts of Hudson Bay–James Bay and watershed S to SW James Bay), N Que. (coasts of James Bay–Hudson Bay N to ca. 58°30′N; Timiskaming R. at 51°11′N; Knob Lake dist. at ca. 55°N; S Ungava Bay), and Labrador (S to the Hamilton R. basin), S in the West to Calif. and N.Mex.; W Greenland (a 1941 collection in CAN by Porsild at ca. 64°N confirming the occurrence in W Greenland, the presumed type locality, following a long period of

search for it there since publication of the species by Retzius in 1795). [*Elephantella* Rydb.; incl. the slender narrow-leaved extreme, f. *gracilis* Lepage, and var. *surrecta* (Benth.) Gray (*P. sur.* Benth.), with a relatively long-beaked galea]. MAP: Hultén 1968b:818.

Forma *pallida* Lepage (corolla predominantly white, drying yellowish, purplish only at base, rather than uniformly pink-purple to almost red) is known from the type locality, Fort George, E James Bay, Que.

P. hirsuta L.   Hairy Lousewort

/AS/EE/GEA/   (Hs)   Moist rocky or sandy tundra and shores from the coast of W Dist. Keewatin to Bathurst Is., Axel Heiberg Is., and Ellesmere Is. (N to ca. 80°N), S to SE Dist. Keewatin, N Que. (S to S Ungava Bay), and N Labrador (Crater L., ca. 58°N; DAO; the report from Ford Harbour, 56°27′N, by John Macoun 1886, is based upon *P. labradorica,* the relevant collection in CAN); W and E Greenland S to ca. 65°N; Spitsbergen; N Norway; arctic Asia. MAPS: Hultén 1958: map 9, p. 29 (dots indicating stations in NE Man. require confirmation); Porsild 1957: map 297 (a dot should be added for the 58°N station in Labrador), p. 198.

Forma *albiflora* Abrom. (flowers white rather than pale pink) is known from Greenland, the type locality.

P. labradorica Wirsing

/aST/X/GEA/   (Hs)   Dryish peaty or rocky tundra and slopes from the coasts of Alaska–Yukon–Dist. Mackenzie–Dist. Keewatin to S Baffin Is. and northernmost Ungava–Labrador (type from Labrador), S to N B.C. (S to Mt. Selwyn, ca. 56°N; CAN), Alta. (S to Jasper and near Edmonton; CAN), northernmost Sask. (S to McKeever L. at ca. 59°55′N; G.W. Argus, Can. Field-Nat. 80(3):139. 1966), N Man. (S to Churchill), N Ont. (Hudson Bay–James Bay watershed S to 54°22′N), islands in James Bay, Que. (S to E James Bay at ca. 54°N, Knob Lake, 54°48′N, and the Côte-Nord at St-Augustin), and S Labrador; W Greenland N to ca. 68°N; NE Europe; N Asia. [*P. euphrasioides* Steph. and its var. *simplex* Hult.]. MAPS: Hultén 1968b:820; Porsild 1957: map 292, p. 197; Böcher 1954: fig. 70, p. 268; Raup 1947: pl. 34.

Var. *sulphurea* Hult. (the corolla uniformly sulphur-yellow and relatively long) is reported from the type locality along the Blackstone R., the Yukon, by Hultén (1968a).

P. lanata C. & S.   Woolly Lousewort

/ASs/X/GEA/   (Hs)   Stony or gravelly tundra and slopes at low to fairly high elevations from the Aleutian Is. and coasts of Alaska–Yukon–Dist. Mackenzie–Dist. Keewatin to N Banks Is., Ellesmere Is. (N to ca. 80°N), and Baffin Is., S to B.C. (S to Queen Charlotte Is. and Smithers, ca. 54°40′N; Herb. V; reports from Alta. require confirmation), S-cent. Dist. Mackenzie–Dist. Keewatin, and northernmost Que. (Hudson Strait and Akpatok Is., Ungava Bay); W Greenland between ca. 65° and 80°N; Spitsbergen; NE Europe; N Asia. [*P. langsdorfii* var. *lan.* (C. & S.) Gray; incl. *P. adamsii* Hult. and *P. kanei* Durand]. MAPS: Hultén 1968b:827 (*P. kanei*); Porsild 1957: map 296, p. 197; *Atlas of Canada* 1957: map 2, sheet 38; Böcher 1954: fig. 32 (map 2), p. 134; Raup 1947: pl. 34.

Forma *alba* Cody (flowers white rather than bright pink to rose-purple) is reported from NW Dist. Mackenzie by Boivin (1966b) and from the type locality, Southampton Is., N Hudson Bay, by W.J. Cody (Can. Field-Nat. 65(4):143. 1951). According to Hultén (1968a), *P. kanei* Dur. is the correct name for the species, *P. lanata* C. & S. being a later name than *P. lanata* Pallas.

P. lanceolata Michx.

/T/EE/   (Hs)   Moist meadows and shores (often calcareous) from S Man. (N to Ochre River, about 85 mi N of Brandon; CAN; reports from Sask. require confirmation) to S Ont. (Bruce and York counties; collection in TRT from near Thunder Bay, where perhaps introd.; not known from Que. or the Atlantic Provinces) and Mass., S to Nebr., Mo., Tenn., and N.C. MAP: Pennell 1935: map 146, p. 495.

P. langsdorfii Fisch.

/AST/X/GA/   (Hs)   Meadows, rocky ridges, and slopes at moderate to high elevations from the Aleutian Is. and N coast of Alaska (type material from Unalaska and St. Lawrence Is.) to N Banks Is., Ellesmere Is. (N to ca. 81°N), and Baffin Is., S in the West to B.C. (S to Rainbow Mt., near Bella Coola, Mt. McLean, NW of Lillooet, and the Marble Range, NW of Clinton; CAN; V) and SW Alta.

(Jasper), farther eastwards s to the coasts of the Yukon–Dist. Mackenzie–Dist. Keewatin and s Baffin Is.; NW Greenland between ca. 76° and 80°N; N Asia. [*P. purpurascens* Cham.]. MAP (aggregate species): combine the maps by Hultén 1968*b*:822 (ssp. *langs.* and ssp. *arctica*).

According to Hultén's map, the typical form is confined in N. America to the Aleutian Is. and the Seward Pen., Alaska, the rest of our material being referable to ssp. *arctica* (R. Br.) Pennell (*P. arctica* R. Br., the type from Melville Is.; *P. hians* Eastw.; galea shorter than or at most about equalling the lower corolla-lip and often toothless rather than distinctly surpassing the lower lip and commonly with a slender tooth on each side of the lower margins just below the summit). MAPS: on the above-noted map by Hultén; Porsild 1957: map 295 (*P. arct.*), p. 197.

P. lapponica L.
/aSs/X/GEA/ (Hs) Tundra and meadows at low to moderate elevations from N Alaska–Yukon and the coast of Dist. Mackenzie to Baffin Is. (N to near the Arctic Circle) and northernmost Ungava–Labrador, s to Great Bear L., SE Dist. Keewatin, N Man. (from 65 mi s of Churchill to Churchill), N Ont. (coast of Hudson Bay), N Que. (coasts of James Bay–Hudson Bay to s Ungava Bay), and N Labrador (s to ca. 57°N); w Greenland N to ca. 72°30′N, E Greenland between ca. 68° and 74°30′N; N Eurasia. MAPS: Hultén 1968*b*:819; Porsild 1957: map 291, p. 197.

P. oederi Vahl
/ST/W/EA/ (Hs) Meadows and rocky slopes at moderate to high elevations from the E Aleutian Is., N Alaska–Yukon (N to ca. 69°N), and NW Dist. Mackenzie (Porsild and Cody 1968) through B.C. and SW Alta. (N to ca. 51°30′N) to the Beartooth Mts. of Wyo.; Eurasia. [*P. versicolor* Wahl.; *P. flammea* of Alaska–Yukon reports, not L.]. MAPS: Hultén 1968*b*:826; Porsild 1966: map 127, p. 82.

Material from Snow Creek Pass, about 65 mi NW of Banff, Alta., is referable to var. *albertae* (Hult.) Boivin (*P. alb.* Hult., the type from that locality; inflorescence copiously white-woolly rather than merely more or less villous). MAP: Eric Hultén, Sven. Bot. Tidskr. 55: fig. 1, p. 194. 1961.

P. ornithorhyncha Benth.
/T/W/ (Hs) Subalpine and alpine meadows and open slopes from SE Alaska (N to ca. 58°N) and w B.C. (N to Tuya L. at ca. 59°N; V) to Mt. Rainier, Wash. [*P. nasuta* Bong.; *P. pedicellata* Bunge; *P. subnuda* Benth.]. MAP: Hultén 1968*b*:819.

P. palustris L.   Swamp-Lousewort
/T/E/E/ (T) Marshes and wet meadows from Que. (St. Lawrence R. estuary from near Quebec City to Anticosti Is., the Gaspé Pen., and Magdalen Is.; the report from Holton Harbour, Labrador, by Waghorne 1898, is based upon *P. labradorica*, the relevant collection in CAN) to Nfld. and NE N.S. (St. Lawrence Bay, Victoria Co.; ACAD; CAN; not known from N.B., P.E.I., or the U.S.A.); Europe. MAP: Hultén 1958: map 141, p. 161.

Forma *laurentiana* Vict. & Rousseau, the relatively pubescent extreme, is known from Que. (type from St-Tite, Charlevoix Co.).

P. parviflora Sm.
/ST/(X)/A/ (T) Marshes and wet meadows from the E Aleutian Is. and Alaska (N to ca. 70°15′N; the type is a Menzies collection from the coast of w N. America, probably in s Alaska or N B.C.) to southernmost Yukon (Porsild 1951a; L. Nares), L. Athabasca (Sask.; not known from Dist. Mackenzie), and SE Dist. Keewatin (mouth of the McConnell R. at ca. 60°50′N; CAN), s to s-cent. B.C. (s to Barkerville, about 75 mi SE of Prince George; Hitchcock et al. 1959, noting a report from Oreg.), Alta. (s to Ma-Me-O Beach, s of Edmonton at ca. 54°N; CAN), Sask. (s to McKague, 52°37′N), cent. Man. (s to Kettle Rapids on the Nelson R. about 160 mi s of Churchill), N Ont. (James Bay–Hudson Bay N to ca. 56°N), and w-cent. Que. (E James Bay N to ca. 54°N; see Que. map by Dutilly and Lepage 1947: fig. 12, p. 258; a report from L. Mistassini noted by Hitchcock et al. 1959); N Asia. [*P. pennellii* spp. *insularis* Calder & Taylor; incl. *P. macrodonta* Richards., with relatively deeply lobed floral bracts]. MAP: combine the maps by Hultén 1968*b*:821 (ssp. *parv.* and ssp. *penn.*) and p. 820 (*P. macro.*).

Some of the Alaskan material is referable to ssp. *pennellii* Hult. (*P. penn.* Hult., the type from King Cove, Alaska; galea with a pair of long marginal teeth near the apex rather than lacking these). MAP: Hultén 1968*b*:821.

P. racemosa Dougl.

/T/W/   (Hp)   Coniferous montane woods, dry meadows, and open slopes from s B.C. (Mt. Mark, Vancouver Is., and the mainland N to Sicamous and Revelstoke, E to Kicking Horse Pass and the Flathead R. near the Alta. boundary; reports from sw Alta. require confirmation, possibly being based upon SE B.C. material) to Calif. and N.Mex.

Much of our material is referable to var. *alba* (Pennell) Cronq. (corolla white or ochroleucous rather than pink to purplish; leaves averaging narrower than those of the typical form).

P. sudetica Willd.

/ASs/X/EA/   (Hs (Hr))   Meadows, tundra, and rocky slopes at low to fairly high elevations, the aggregate species from the Aleutian Is. and coasts of Alaska–Yukon–Dist. Mackenzie–Dist. Keewatin to Banks Is., Melville Is., Ellesmere Is. (N to ca. 80°N), and Baffin Is., s to B.C. (s to mts. at 52°43′N; V), Great Bear L., NE Man. (s to Warkworth Creek, s of Churchill), N Ont. (coasts of James Bay–Hudson Bay N to ca. 56°50′N), islands in James Bay, and w-cent. Que. (E James Bay); Eurasia. MAPS and synonymy: *see* below.

1   Bracts not markedly dilated at base, commonly unlobed (or the basal ones sometimes lobed); flowers purplish.
   2   Inflorescence flat-topped at anthesis; corolla-tube relatively short; leaves with a broad rachis; [Europe only; MAPS: Eric Hultén, Sven. Bot. Tidskr. 55(1): fig. 4, p. 199. 1961; Porsild 1957: map 298 (aggregate species), p. 198; Raup 1947: pl. 34 (aggregate species)] . . . . [ssp. *sudetica*]
   2   Inflorescence pyramidal at anthesis, often already prolonged; corolla-tube relatively long; leaves with a narrow rachis; [*P. scopulorum sensu* J.M. Macoun 1896, and Henry 1915, not Gray; Alaska and the Yukon (type from Whitehorse) to s B.C.; MAPS: Hultén, loc. cit., fig. 4, p. 199, and 1968b:823] . . . . ssp. *interior* Hult.
1   Bracts strongly dilated (often hyaline) at base, the basal ones with apical or lateral lobes.
   3   Flowers comparatively small, purple; lower bracts with small basal lobes, the middle and upper ones entire; [Alaska to E Hudson Bay–James Bay; N Asia; MAPS: Hultén, loc. cit., fig. 4, p. 199, and 1968b:824] . . . . ssp. *interiorioides* Hult.
   3   Flowers larger, with a thick galea; middle (as well as lower) bracts usually lobed.
      4   Spike copiously lanate; all except the lowest bracts short and lacking a prolonged apical lobe; flowers with a purple-tipped pink galea and a white or pinkish, spotted lip; [*P. sud.* f. ?*alba* Cody; Victoria Is. (type locality) to Ellesmere Is. and Baffin Is.; E Asia; MAPS: Hultén, loc. cit., fig. 5, p. 200, and 1968b:824] . . . . . ssp. *albolabiata* Hult.
      4   Spike less pubescent; bracts with prolonged apical lobes; flowers relatively large, purplish; [Aleutian Is. and Alaska (type from St. Paul Is.); E Asia; MAPS: Hultén, loc. cit., fig. 5, p. 200, and 1968b:825] . . . . ssp. *pacifica* Hult.

P. sylvatica L.   Small Lousewort

European; known in N. America only in moist ground of the Avalon Pen., SE Nfld. (CAN; DAO; GH), where considered native by Fernald *in* Gray (1950) but more likely introd. (*see* D.B. Savile, Can. J. Bot. 45(7):1101. 1967, and note under *Luzula campestris*). MAPS: Hultén 1958: map 134, p. 153; Fernald 1929: map 27, p. 1502.

P. verticillata L.

/aST/W/EA/   (Hs)   Meadows, tundra, and rocky slopes at low to fairly high elevations from the Aleutian Is., the coasts of Alaska–Yukon, and NW Dist. Mackenzie to northernmost B.C. (Dease Lake, ca. 58°30′N; CAN). MAP: Hultén 1968b:817.

PENSTEMON Mitchell   [7508]   Beard-tongue

(Ref.: Keck 1945)
1   Anthers more or less densely long-woolly with tangled hairs; corolla lavender to pink-purple or purple-violet, the tube flaring; inflorescence glandular, generally somewhat paniculate; leaves glabrous (or occasionally rough-puberulent in *P. lyallii*); (B.C. and Alta.).

2 Staminode (sterile anther-filament) not bearded; corolla lavender, to 4.5 cm long, glabrous externally, conspicuously woolly-villous along the prominent ventral ridges within; leaves all cauline, entire or remotely serrulate, nearly or quite sessile, the lower ones reduced, the others narrow and elongate, to over 1 dm long; stems to 8 dm tall, glabrous below the inflorescence or hairy in lines; (B.C. and Alta.) .... *P. lyallii*

2 Staminode bearded.

   3 Inflorescence generally more or less paniculate; corolla pink-purple, less than 3.5 cm long, glandular-hairy outside, glabrous inside; leaves all cauline, the lower ones reduced, the others distinctly petioled, conspicuously serrate, lanceolate to ovate, the blades to about 1 dm long and 4 cm broad; stems to 8 dm tall, glabrous or finely puberulent, herbaceous nearly or quite to base; (S B.C.) .... *P. nemorosus*

   3 Inflorescence generally simple and essentially racemose; corolla glabrous outside, more or less pubescent inside toward the base of the lower lip; stems usually more or less distinctly woody toward base, at most 4 dm tall, the subsessile or short-petioled leaves tending to be clustered near its base, usually on short sterile shoots; (B.C. and Alta.).

      4 Stems to 4 dm tall, essentially glabrous below the inflorescence or those of the season often finely puberulent, erect or more or less ascending, commonly bushy-branched above the base (the clusters of basal leaves not forming mats); corolla blue-lavender to light purplish .... *P. fruticosus*

      4 Stems at most about 1.5 dm tall, minutely strigose or short-spreading hairy, the lower portions creeping and forming dense mats on the ground ....
      .... *P. davidsonii*

1 Anthers glabrous or more or less short-hairy, but not at all woolly; staminode more or less bearded at least near tip (usually glabrous in *P. deustus*).

   5 Corolla predominantly white to ochroleucus or yellow (pale violet in *P. laevigatus*); pollen-sacs opening throughout their length (or remaining indehiscent at apex), horizontally spreading after dehiscence; leaves of the flowering stems sessile or short-petioled; basal tufts of petioled leaves commonly present in addition to the stem-leaves (sometimes poorly developed in *P. confertus*).

      6 Plants distinctly woody toward the much-branched shrubby base, the flowering stems erect, glabrous below the inflorescence to finely puberulent or somewhat glandular, to 6 dm tall; leaves essentially glabrous, usually more or less sharply toothed, to about 6 cm long and 2.5 cm broad, those of the flowering stems mostly ovate and sessile or subsessile, those of the sterile shoots mostly elliptic to obovate and short-petioled; corolla commonly dull whitish with some purplish lines within (or sometimes faintly ochroleucous or washed with lavender), to 2 cm long, glandular-hairy inside and out, the tube cylindrical; calyx to 6 mm long; inflorescence usually more or less glandular, consisting of several simple whorls (the lower whorls sometimes remote); (S ?B.C.) .... [*P. deustus*]

      6 Plants herbaceous nearly or quite to base; leaves entire or with a few scattered low teeth.

         7 Corolla ochroleucous to sulphur-yellow, rarely over 12 mm long, often declined, glabrous outside, bearded on the palate within, the slender tube cylindrical, the limb not strongly 2-lipped; calyx to about 5 mm long; inflorescence nonglandular, consisting of a group of simple compact whorls (the lower ones usually remote), its bracts (as also the calyx-segments) with broad scarious erose margins; leaves glabrous, to 1 dm long and 2.5 cm broad; stem glabrous, or minutely puberulent in the inflorescence, to about 6 dm tall; (B.C. and Alta.; introd. in Sask.) .... *P. confertus*

         7 Corolla predominantly white (often violet tinged; rather uniformly pale violet in *P. laevigatus*); inflorescence glandular, often markedly paniculate (usually of simple whorls in *P. albidus*).

            8 Corolla copiously glandular-puberulent within near summit (as well as short-stipitate-glandular outside), white or violet-tinged.

               9 Stem to about 4 dm tall, its internodes densely puberulent; leaves finely cinereous-puberulent or glabrate, linear to lance-oblong, to about

8 cm long; corolla to 2.5 cm long, rather abruptly enlarged above base; calyx at anthesis about 1 cm long, very densely glandular, its lobes narrowly triangular; (s Alta. to s Man.) .... *P. albidus*

    9 Stem to about 1 m tall, its internodes glabrous; leaves glabrous, oblong to oblong-ovate, to over 1 dm long; corolla to 2 cm long, gradually dilated from base to summit; calyx at anthesis 3 or 4 mm long, glabrous or sparingly glandular, its lobes ovate; (introd. in s Ont.) .... *P. tubaeflorus*

8 Corolla not glandular within; (introd.).

    10 Leaves soft-pubescent on both surfaces, firm and coriaceous, pale, entire or remotely toothed, lanceolate to lance-oblong, the principal ones to about 1 dm long and rarely as much as 2 cm broad; corolla white outside, lined with purple inside, to about 2 cm long, the throat not strongly inflated; inflorescence copiously glandular; staminode densely long-bearded above the middle; stem to about 1 m tall, copiously and evenly short-pubescent throughout; (s ?Ont.) .... [*P. pallidus*]

    10 Leaves glabrous, somewhat leathery in texture, to about 1.5 dm long and 4 cm broad; corolla-throat strongly inflated beyond the tubular part; inflorescence rather sparingly glandular; beard of staminode relatively sparse and short; stem sometimes glabrous below the inflorescence, more commonly minutely puberulent (usually in lines); (introd.).

        11 Corolla pale violet externally, usually less than 2 cm long; anthers usually glabrous; leaves lanceolate to narrowly oblong; stem dull, to 12 dm tall .... [*P. laevigatus*]

        11 Corolla white or very faintly violet-tinged (usually marked with purple lines within), to 3 cm long; anthers usually with a few stiff hairs along the back; leaves oblong-lanceolate to narrowly oblong or narrowly triangular; stem sublustrous, slightly glaucous, to 1.5 m tall .... *P. digitalis*

5 Corolla pale lavender or violet (*P. laevigatus* may be sought here; see contrasting lead) to deep blue or various shades of blue or purple.

  12 Pollen-sacs opening across their confluent apices, their free indehiscent tips remaining vertically reflexed and nearly parallel, the anther permanently horseshoe-shaped; corolla-throat flaring; leaves all cauline (the lower ones reduced), sessile or short-petioled, the principal ones coarsely serrate to laciniate-pinnatifid; stems to 7 or 8 dm tall; (B.C.).

    13 Plants not at all glandular (even in the inflorescence); corolla glabrous outside as well as inside (except for ciliate-margined lobes in *P. venustus*); leaves lanceolate to broadly ovate or oblong, glabrous, sharply serrate; stems glabrous or puberulent.

        14 Corolla bright lavender to purple-violet or purple, to over 3.5 cm long, its lobes ciliate; calyx to 6.5 mm long, its lobes sometimes ciliolate, otherwise glabrous; inflorescence commonly forming a narrow thyrsoid panicle; fertile filaments more or less pubescent; leaves to over 1 dm long but commonly less than 3 cm broad .... [*P. venustus*]

        14 Corolla deep blue to dark purple, usually not over 2.5 cm long, its lobes not ciliate; calyx to 9 mm long, its lobes ciliolate and sometimes short-hairy on the back; inflorescence often a single rather compact terminal whorl, sometimes several whorls (these occasionally open and branched to form a more paniculate inflorescence); fertile filaments glabrous; leaves commonly not over 8 cm long but up to 3.5 cm broad .... *P. serrulatus*

    13 Plant distinctly glandular in the inflorescence, the stem commonly puberulent below the inflorescence; leaves glabrous or puberulent; corolla glandular-puberulent outside, glabrous inside (or with a few long hairs); fertile filaments glabrous; inflorescence commonly loosely paniculate; (B.C.).

        15 Corolla to about 3 cm long, bright lavender (the lower lip striped within),

the upper lip cleft less than half its length; calyx to 8 mm long;
leaves sharply toothed to laciniate-pinnatifid, to 7 cm long and 3 cm broad . . . . . .
. . . . *P. richardsonii*

15 Corolla less than 2 cm long, blue-lavender to light purple-violet, the upper
lip cleft more than half its length; calyx to 6 mm long; leaves rather sharply
toothed, to 5 cm long and 1 cm broad, sometimes in whorls of 3 or even 4 . . . . . .
. . . . [*P. triphyllus*]

12 Pollen-sacs opening their whole length (or remaining indehiscent at apex),
horizontally spreading after dehiscence.

16 Plants (including inflorescence and corolla) not glandular (sometimes
glutinous in *P. acuminatus*); inflorescence consisting of usually several more
or less remote dense whorls; leaves entire or sometimes remotely low-
toothed, glabrous, the basal ones tufted (sometimes poorly developed or
wanting in *P. procerus*); plants rarely over 6 dm tall, usually glabrous
throughout (except for the staminode-beard; corolla-palate bearded and the
calyx sometimes puberulent in *P. procerus*).

17 Corolla rarely over 1 cm long, often more or less deflexed, the tube
scarcely flaring, the palate bearded; pollen-sacs at most 0.7 mm long;
capsule 4 or 5 mm long, the seeds about 1 mm long; basal leaves (when
present) to about 1 dm long and 1.5 cm broad; (B.C. to w Man.) . . . . *P. procerus*

17 Corolla bright blue, to about 2 cm long, mostly spreading-ascending, the
tube gradually flaring toward the mouth, the palate glabrous; calyx to 9 mm
long; pollen-sacs commonly about 1 mm long; capsule to 12 mm long
(excluding beak), the seeds commonly about 3 mm long; leaves thick and
very coriaceous, they and the stem very glaucous; stem-leaves lanceolate
to ovate, the bracteal leaves relatively broader.

18 Bearded portion of the staminode at least 2 mm long, the hairs often
well over 0.5 mm long (sometimes more than 1 mm long); basal leaves
to about 1 dm long and 2.5 cm broad; plant commonly not over 3 dm
tall; (s B.C. to sw Man.) . . . . *P. nitidus*

18 Bearded portion of staminode usually less than 1.5 mm long, the hairs
rarely over 0.5 mm long; basal leaves to 1.5 dm long and 2 cm broad;
plant to 6 dm tall . . . . [*P. acuminatus*]

16 Plants more or less glandular in the inflorescence (and on the outside of the
corolla); corolla distinctly flaring toward mouth (except in *P. hirsutus*), the
palate bearded; basal tufts of leaves commonly present (often poorly
developed or soon withering in *P. eriantherus, P. gracilis,* and *P. hirsutus*).

19 Principal stem-leaves relatively coarsely toothed; whorls rather loose and
often few-flowered.

20 Principal stem-leaves relatively narrow (mostly at least 3 times as long
as broad), at first pubescent, becoming glabrate; calyx to 7 or 8 mm
long; corolla scarcely flaring, dull violet or purplish with white lobes, to
about 2.5 cm long, the base of the lower lip arched upwards and nearly
or quite closing the throat; anther-sacs as broad as long, saucer-like at
maturity; capsules to 9 mm long; stem to about 9 dm tall, copiously
spreading-pubescent; (Ont. and Que.) . . . . *P. hirsutus*

20 Principal stem-leaves relatively broad (mostly elliptic to ovate,
deltoid-ovate, or subcordate and commonly less than 3 times as long
as broad); calyx to 6 mm long; corolla distinctly flaring, blue (or the
tube purplish), the lower lip much surpassing the upper one and
downwardly arched; anther-sacs longer than broad, cup-like at
maturity; capsules to 7 mm long, the seeds to 1.5 mm long; (B.C.).

21 Corolla commonly not over 1.5 cm long; leaves and stems
glandular-hirsute to merely puberulent or even essentially glabrous, the stems commonly less than 4 dm tall, the leaves mostly
not over 2 cm broad . . . . *P. pruinosus*

21 Corolla to over 2 cm long; stems and leaves (at least on the midrib beneath) spreading-hirsute to essentially glabrous, the stems to about 1 m tall, the leaves to about 5 cm broad . . . . *P. ovatus*

19 Principal stem-leaves mostly entire or with only a few low teeth, relatively narrow (mostly at least 3 times as long as broad); corolla flaring toward mouth.

22 Ovary and capsule (to 12 mm long) glandular-puberulent near the summit; calyx to 13 mm long, with herbaceous and entire lobes; corolla pale lavender to red-purple or deep blue-purple, to 4 cm long, the staminode somewhat exserted from its orifice; seeds 2 or 3 mm long; stem to 4 dm tall, it and the leaves cinereous-puberulent to villous-hirsute or partly glabrate, the basal leaves often poorly developed; (B.C. to s Sask.) . . . . *P. eriantherus*

22 Ovary and capsule (at most 8 or 9 mm long) glabrous; calyx to 7 mm long; seeds less than 2 mm long; leaves glabrous; stems commonly more or less pubescent (often in lines) below the inflorescence, or glabrate.

23 Staminode bearded with long yellow hairs for more than half its length; corolla to 2.5 cm long.

24 Corolla blue-purple, the staminode somewhat exserted; leaves entire or sometimes shallowly serrulate toward the apex; stems to 3 dm tall; (Alaska–Yukon–Dist. Mackenzie–N B.C.) . . . . *P. gormanii*

24 Corolla pale lilac or pale violet, its lower lip violet-blue, the staminode not exserted; leaves mostly with minute sharp teeth; stems to 5 dm tall; (s B.C. to w Ont.) . . . . *P. gracilis*

23 Staminode bearded for at most about 1/3 of its length, not exserted beyond the orifice of the corolla; corolla rarely over 2 cm long.

25 Corolla various shades of lavender, blue, or purple (sometimes pale yellow or nearly white), neither paler within nor obviously marked with deeper coloured lines; calyx to 7 mm long; stems to 9 dm tall . . . . [*P. attenuatus*]

25 Corolla blue, with paler throat and evident lines of deeper colour inside; calyx to 5 mm long; stems rarely over 4 dm tall.

26 Corolla light blue (sometimes pink), to 2 cm long; (s B.C. and s Alta.) . . . . *P. albertinus*

26 Corolla bright blue with purplish or violet throat, at most 18 mm long . . . . [*P. virens*]

[P. acuminatus Dougl.]
[The report of this species of the w U.S.A. (Wash. and Oreg.) from Austin, sw Man., by Lowe (1943; also Rossburn, but the collection not seen) is based upon *P. albidus,* the relevant collection in WIN. The reports from Man., Sask., and Alta. by John Macoun (1884; 1886) are chiefly or wholly referable to *P. nitidus,* most of the relevant collections in CAN.]

P. albertinus Greene
/T/W/ (Hs (Ch)) Dry open rocky places from the foothills to fairly high elevations from s B.C. (N to Kinbasket L., about 65 mi N of Revelstoke; Keck 1945) and sw Alta. (Coleman and the type locality, Sheep Mt., Waterton Lakes; CAN) to Idaho and Mont. [*P. glaucus sensu* John Macoun 1884, as to the B.C. plant, not Graham, the relevant collection in CAN, his Fort Selkirk, the Yukon, citation probably referring to the type locality of *P. gormanii; P. ?humilis sensu* Macoun 1884, and Henry 1915, not Nutt.; *P. ?pseudohumilis sensu* Rydberg 1922, as to the B.C.–Alta. area, not as to type, which is *P. attenuatus* var. *pseudoprocerus; P. ?virens sensu* Rydberg 1922 (as to the Alta. plant, this taken up by Moss 1959), Ulke 1935, and Eastham 1947, not Pennell]. MAP: Keck 1945: fig. 16, p. 193.

P. albidus Nutt.
/T/WW/ (Hs) Open plains, prairies, and hillsides from Mont. and s Alta. (N to Carmangay, about

35 mi NW of Lethbridge; CAN) to S Sask. (N to Swift Current; Breitung 1957a) and S Man. (N to Duck Mt.; DAO; collection in TRT from a fencerow at Oriole, York Co., S Ont., where presumably introd.), S to N.Mex. and Tex. MAP: Pennell 1935: map 65, p. 254.

[P. attenuatus Dougl.]
[The report of this species of the W U.S.A. (Wash. and Mont. to Oreg. and Wyo.) from B.C.-Alta. by Rydberg (1922; *P. pseudohumilis*) may be partly based upon a so-named collection in CAN from Crowsnest L., near the B.C.-Alta. boundary, where taken by John Macoun in 1897, referable to *P. albertinus*. The MAP by Keck (1945: fig. 11, p. 170) indicates no Canadian stations. (Incl. var. *pseudoprocerus* (Rydb.) Cronq. (*P. pseud.* Rydb.; *P. pseudohumilis* Rydb., not Jones)).]

P. confertus Dougl.
/T/W/ (Hs (Ch)) Meadows and fairly moist open or wooded places at low to moderate elevations from SE B.C. (Keremeos, about 20 mi SW of Penticton, to the Alta. boundary, N to Canal Flats, about 45 mi N of Cranbrook; the report from Fort Selkirk, the Yukon, by John Macoun 1884, is referable to *P. procerus* according to Hultén 1949) and SW Alta. (N to Banff; CAN; reported from along railway tracks at Swift Current, Sask., by Breitung 1957a, noting it as probably introd.) to Oreg. and Mont. MAP: Keck 1945: fig. 12, p. 175.

P. davidsonii Greene
/T/W/ (Ch) Ledges and talus slopes at moderate to high elevations, the aggregate species from S B.C. (N to near Kamloops and Revelstoke) and SW Alta. (N to Jasper) to Calif.
1 Leaves of the erect flowering-stems relatively well developed and not greatly reduced upward, rarely less than 1 cm long; calyx to 1.5 cm long; corolla deep lavender, to 4 cm long; [*P. ellipticus* C. & F.; S B.C. (N to Griffin L., near Kamloops, and Revelstoke) and SW Alta. (N to Jasper National Park)] . . . . var. *ellipticus* (Coult. & Fisch.) Boivin
1 Leaves of the erect flowering-stems small and often bract-like, commonly less than 1 cm long; calyx to 1 cm long; corolla blue-lavender to purple-violet, to 3.5 cm long.
  2 Leaves often distinctly serrulate, tending to be broadest near, or even below, the middle, sometimes acutish at apex; [*P. menziesii* Keck, the type from Nootka, W Vancouver Is.; W B.C.: N to Mt. Waddington, ca. 51°20'N, and Bella Coola, ca. 52°20'N; reports from E B.C. and SW Alta. probably refer to var. *ellipticus*] . . . . var. *menziesii* (Keck) Cronq.
  2 Leaves entire, tending to be broadest above the middle, usually obtuse or rounded at apex; [SW B.C.: Vancouver Is.; Chilliwack; Hope; Manning Provincial Park, SE of Hope; Lillooet, about 70 mi W of Kamloops] . . . . var. *davidsonii*

[P. deustus Dougl.]
[The rather ambiguous reports of this species of the W U.S.A. (N to Wash. and Mont.) from the B.C.-U.S.A. boundary by John Macoun (1884), Rydberg (1922), and Henry (1915) require confirmation. The MAP by D.D. Keck (Am. Midl. Nat. 23: fig. 1, p. 601. 1940) indicates no Canadian stations.]

P. digitalis Nutt.
Originally native largely in the Mississippi Basin of the E U.S.A. but now spread to fields and clearings over a large area outside that region and known in Canada from Ont. (N to the Ottawa dist.), Que. (N to L. Nominingue, about 80 mi NW of Montreal; Frère Lucien Lévesque, Ann. ACFAS 13:90. 1947), N.B. (Charlotte and Carleton counties; CAN), and N.S. (Cape Blomidon, Kings Co.; D.S. Erskine 1951). [*P. laevigatus* var. *dig.* (Nutt.) Gray]. MAP: Pennell 1935: map 41, p. 207.

P. eriantherus Pursh
/T/WW/ (Hs (Ch)) Dry open places at low to moderate elevations from SE B.C. (Cranbrook, Newgate, Waldo, Fairmont Hot Springs, Briscoe, Windermere, and Fort Steele; CAN; V), SW Alta. (N to Castle Mt., NW of Banff), and S Sask. (Estevan, about 80 mi SE of Regina; Breitung 1957a; reports of *P. cristatus* from Sask. and Man. by John Macoun 1884, are otherwise based upon *P. albidus* (corolla-mouth copiously glandular-puberulent within rather than glabrous) according to relevant collections in CAN) to Oreg., Colo., and Nebr. [*P. cristatus* Nutt.; *P. saliens* Rydb.].

P. fruticosus (Pursh) Greene
/T/W/ (N (Ch)) Rocky, open or wooded places from the foothills to rather high elevations from s B.C. and sw Alta. to Oreg., Mont., and Wyo.
1 Leaves entire or more or less toothed, to 5 or 6 cm long.
  2 Leaves mostly elliptic or oblanceolate, to 1.5 cm broad; corolla to 4 cm long; [*Gerardia* Pursh; U.S.A. ?only, reports from B.C. and Alta. all probably referable to var. *scouleri*] . . . . .
  . . . . [var. *fruticosus*]
  2 Leaves relatively very narrow, linear-elliptic or -oblanceolate, mostly not over 5 mm broad; corolla to 5 cm long; [*P. scouleri* Lindl.; *P. menziesii* var. *scouleri* (Lindl.) Gray; s B.C. (N to Kamloops) and sw Alta. (N to Jasper)] . . . . var. *scouleri* (Lindl.) Cronq.
1 Leaves prominently toothed, relatively small, the blade mostly not over 2.5 cm long; corolla to 4 cm long; [ssp. *serratus* Keck; s B.C.: collection from Manning Park, SE of Hope, in the herbarium of the Manning Provincial Park museum] . . . . var. *serratus* (Keck) Cronq.

P. gormanii Greene
/Ss/W/ (Hs) Dry slopes at moderate elevations in Alaska (N to ca. 68°N), the Yukon (N to ca. 65°N; type from near Fort Selkirk), sw-cent. Dist. Mackenzie (Tsichu R. at 63°20′N; CAN), and northernmost B.C. (Liard Crossing and near Cassiar, both at ca. 59°10′N; CAN; V). [*P. glaucus* of the Yukon reports by Watson, Science 3:253. 1884, and John Macoun 1884, not Graham; *P. cristatus sensu* John Macoun, Ottawa Naturalist 13(9):215. 1899, not Nutt.]. MAP: Hultén 1968b:794.
  Forma *albiflora* Porsild (corolla white rather than deep blue, fading to light purple) is known from s Yukon (Alsek L.; type from along the Lapie R.).

P. gracilis Nutt.
/T/WW/ (Hs) Dry prairies and sandy or rocky places from ?B.C. (Boivin 1966b) to Alta. (N to Peace River, 56°14′N; CAN), Sask. (N to N of Prince Albert; CAN), Man. (N to Steeprock, about 100 mi N of Portage la Prairie; CAN), and w Ont. (Kenora; WIN), s to N.Mex., S.Dak., Iowa, and Wisc. MAP: Pennell 1935: map 56, p. 235.

P. hirsutus (L.) Willd.
/T/EE/ (Hs) Dry or rocky ground from Wisc. to Ont. (N to Renfrew, Frontenac, Leeds, and Stormont counties), sw Que. (N to Maniwaki, about 65 mi N of Hull; MT; CAN; reports from P.E.I. require confirmation), and s New Eng., s to Tenn., Ky., and Va. [*Chelone* L.; *P. pubescens* Ait.]. MAP: Pennell 1935: map 59, p. 240.

[*P. laevigatus* Ait.]
[This species of the E U.S.A. (N to Pa. and N.J.; very similar to *P. digitalis* but noted by Fernald *in* Gray 1950, as lacking the weedy and aggressive nature of that species) is reported from s Ont. by Montgomery (1945; Waterloo Co.) and from Mt. Royal, Montreal, Que., by Frère Cléonique-Joseph (Contrib. Inst. Bot. Univ. Montréal 27:39. 1936), and collections in MT from Jacques-Cartier and Iberville counties, sw Que., have been referred to it. However, a so-named collection in GH from Brighton, Northumberland Co., s Ont., has been referred to *P. digitalis* by Shumovich and further studies are necessary to confirm the above reports and collections. The MAP by Pennell (1935: map 44, p. 211) indicates no Canadian stations.]

P. lyallii Gray
/T/W/ (Hp (Ch)) Gravel bars, rocky slopes, and cliffs from the foothills to moderate elevations from SE B.C. (Okanagan, Fernie, Elko, and Flathead; CAN; V) and sw Alta. (Crowsnest Pass and Waterton Lakes; CAN) to N Idaho and Oreg. [*P. venustus sensu* John Macoun 1884, as to the Crowsnest Pass report, not Dougl., the relevant collection in CAN].

P. nemorosus (Dougl.) Trautv. ·
/T/W/ (Hp) Woodlands and moist rocky slopes at low to high elevations from s B.C. (Vancouver Is. and South Kootenay Pass, on the B.C.–Alta. boundary, where taken by Dawson in 1881; CAN) to NW Calif. [*Chelone* Dougl.; *Nothochelone* Straw].

P. nitidus Dougl.
/T/WW/ (Hs) Grassy hillsides, prairies, and plains at low to moderate elevations from SE B.C. (Crowsnest Pass, on the B.C.–Alta. boundary) to S Sask. (N to Moose Jaw and part of the type locality along the Saskatchewan R., this including the Saskatchewan, Assiniboine, and Red rivers) and S Man. (N to St. Lazare, about 75 mi NW of Brandon), S to Wash., Wyo., and N.Dak. [*P. acuminatus* var. *minor* Hook.].

P. ovatus Dougl.
/T/W/ (Hs) Open woods at low to moderate elevations from S B.C. (Vancouver Is. and mainland N to Bella Coola, ca. 52°20′N, and the Big Bend, a northern route between Revelstoke and Golden) to Oreg., Idaho, and Mont. MAP: Keck 1945: fig. 15, p. 187.
Much of the B.C. material appears referable to the glabrous extreme, var. *pinetorum* Piper (*P. pinetorum* Piper; *P. wilcoxii* Rydb.) but the map for *P. wilcoxii* by Keck (1945: fig. 16, p. 193) indicates no Canadian stations.

[P. pallidus Small]
["Sandy or loamy woods and openings, taking to fields and roadsides, N.E. to Mich. and Ia., S to Ga., Tenn., Ark. and Kans., eastw. as a natzd. plant only." (Fernald *in* Gray 1950). The report from S Ont. by Landon (1960; Middleton Township, Norfolk Co.) requires confirmation, being undoubtedly based upon an introd. plant if correctly identified. The MAP by Pennell (1935: map 49, p. 226) indicates no Canadian stations.]

P. procerus Dougl.
/sT/WW/ (Hs (Ch)) Dry plains, prairies, and open or wooded slopes at low to fairly high elevations, the aggregate species from S Yukon (N to ca. 61°30′N; the report from Nome, Alaska, by Hultén 1949 (indicated on his map 1022, p. 1469) is not included in his 1968*b* map) and B.C. to Alta. (N to the Peace River dist. at ca. 56°N; John Macoun 1884), Sask. (N to Scott and Saskatoon), and SW Man. (N to Fort Ellice, about 75 mi NW of Brandon), S to Calif. and Colo. [*P. micranthus* Nutt.; *P. confertus* var. *caeruleopurpureus* Gray]. MAPS (aggregate species): Keck 1945: fig. 6, p. 145; Hultén 1968*b*:795.
Var. *formosus* (Nels.) Cronq. (*P. form.* Nels.; *P. pulchellus* Greene; the dwarf alpine extreme with obtuse to short-cuspidate calyces at most 3 mm long rather than to 6 mm, the inflorescence commonly reduced to a single whorl) is known from S B.C. (Tamihy Mts., near Chilliwack; CAN), where taken by J.M. Macoun in 1901. Some of our B.C. material is referable to another reduced alpine phase (but the calyces more or less strongly caudate-tipped and to 6 mm long), var. *tolmiei* (Hook.) Cronq. (*P. tol.* Hook.; differing from the typical form in having well-developed rosettes of basal leaves). The typical form has deep blue-purple flowers. Forma *albescens* Boivin (flowers white) is known from the type locality, Watson L., the Yukon. Forma *jenkensii* Boivin (flowers pink) is known from the type locality, Hoosier, Sask.

P. pruinosus Dougl.
/T/W/ (Hs (Ch)) Open rocky places at low to moderate elevations from S B.C. (between Princeton and 18 mi W of Cranbrook; CAN) to Wash. MAP: Keck 1945: fig. 15, p. 187.

P. richardsonii Dougl.
/t/W/ (Hp) Dry rocky places and crevices at lower elevations from S B.C. (Dry Interior in the Okanagan Valley N to Kelowna; Eastham 1947; the inclusion of Alta. in the range by Rydberg 1922, requires clarification) to Oreg.

P. serrulatus Menzies
/T/W/ (Hp (Ch)) Moist or wet places at low to high elevations from the southernmost Alaska Panhandle (near Hyder; Hultén 1949, and his map 1020 for *P. diffusus*, p. 1469) through B.C. (E to Revelstoke; Eastham 1947) to Oreg. [*P. diffusus* Dougl.]. MAP: Hultén 1968*b*:795.

[P. triphyllus Dougl.]
[The inclusion of B.C. in the range of this species of Wash., Idaho, and Oreg. by Rydberg (1922) is

probably based upon the report by Gray noted by John Macoun (1884; "On rocks from Oregon to British Columbia."), this requiring clarification.]

P. tubaeflorus Nutt.
Native in the E U.S.A. ("Open woods, fields and roadsides, Neb. to e. Tex., e. to Miss., Tenn., Ind. and Wisc., partly adv. eastw.; adv. locally to Atl. states."; Fernald *in* Gray 1950). It is reported from S Ont. by Pennell (1935; Forest, Lambton Co.), where probably introd. [Incl. var. *achoreus* Fern., this reported from Ont. by Fernald (*in* Gray 1950), probably on the basis of Pennell's citation]. MAP: Pennell 1935: map 60, p. 244.

[P. venustus Dougl.]
[The report of this species of Wash., Idaho, and Oreg. from Crowsnest Pass, on the B.C.–Alta. boundary, by John Macoun (1884) is based upon *P. lyallii,* the relevant collection in CAN. Macoun (1886) also refers his South Kootenay Pass and Wigwam R. citations to *P. lyallii* and his Kicking Horse Pass citation to *P. (davidsonii* var.) *menziesii. (P. ?dasyphyllus* Gray).]

[P. virens Pennell]
[See *P. albertinus.* The MAP by Keck (1945: fig. 14, p. 183) indicates no Canadian stations.]

RHINANTHUS L.   [7647]   Yellow Rattle

1   Floral bracts with at least the lower teeth lance-attenuate and ending in slender bristle-tips;
(transcontinental, introd.) . . . . *R. crista-galli*
1   Floral bracts with the deltoid teeth blunt or merely acute, lacking bristle-tips.
  2   Stem green; leaves oblong, crenate-dentate; corolla uniformly yellow; (transcon-
tinental) . . . . *R. borealis*
  2   Stem bronze-tinged, marked with black lines above; leaves lanceolate, serrate-dentate;
corolla mostly yellow but the lower lip mottled with brown; (introd. in E Canada) . . . .
. . . . *R. stenophyllus*

R. borealis (Sterneck) Druce
/aST/X/GEeA/   (T)   Meadows and shores from the Aleutian Is., Alaska (N to ca. 62°N), the Yukon (N to Whitehorse; CAN), and Dist. Mackenzie (N to Fort Norman, ca. 65 °N; CAN) to S Dist. Keewatin, Man. (N to Churchill; not known from Sask.), northernmost Ont., Que. (N to S Ungava Bay, L. Mistassini, the Côte-Nord, Anticosti Is., and Gaspé Pen.), Labrador (N to Hebron, 58°12'N), Nfld., and N.S. (Cape Breton Is. and St. Paul Is.; not known from N.B. or P.E.I.), S to an uncertain limit in the U.S.A. through confusion with *R. crista-galli;* w Greenland N to 65°10'N, E Greenland N to near the Arctic Circle; Iceland; Europe; Copper Is., NE Asia. [*Alectorolophus bor.* Stern., the type from Unalaska, Aleutian Is.; *R. minor* ssp. *bor.* (Stern.) Löve; *A. pacificus* Stern.; *A. (R.) arcticus* Stern.; incl. *R. groenlandicus* Chab. and *R. oblongifolius* Fern.]. MAPS: Hultén 1958: map 172 (*R. groenl.*), p. 191, and 1968b:815 (*R. minor* ssp. *bor.;* w N. America and Copper Is. only, excluding the eastern *R. groenl.* from the complex, the treatment tentative).
  Some of our material is referable to ssp. *kyrollae* (Chab.) Pennell (*R. kry.* Chab., the type from "Annapolis, evidently not "United States" but likely Nova Scotia" (Abrams 1951); *R. rigidus* Chab.; fruiting calyces essentially glabrous rather than finely pubescent).

R. crista-galli L.   Common Yellow Rattle.   Claquette or Sonnette
Eurasian; introd. along roadsides and in old fields and waste places in N. America, as in S Dist. Mackenzie (Great Slave L.) and all the provinces. [Incl. var. *fallax* (Wimm. & Grab.) Druce; *R. minor* L.]. MAP: Hultén 1958: map 119, p. 139.
  Fernald *in* Gray (1950) believes the species to be native southwards in N. America and introd. northwards. Both Hultén (1958) and Pennell (1935), however, consider it native only in the Old World. *See* note under *Luzula campestris.*

R. stenophyllus (Schur) Druce
European; apparently known in N. America only from boggy meadows and shores of E Que.

(Gaspé Pen.), Nfld., P.E.I. (Souris, Kings Co.; CAN), and N.S. [*Alectorolophus* Schur; *R. minor* ssp. *sten.* (Schur) Schwarz].

This species is considered by Fernald *in* Gray (1950) to be native in E Que. (and presumably elsewhere in E N. America as noted above). *See* note under *Luzula campestris*.

SCROPHULARIA L.   [7505]   Figwort.   Scrophulaire

1   Rudimentary stamen greenish yellow, flabellate to subreniform, to 1.8 mm broad, mostly broader than long; corolla greenish brown (or greenish yellow with a light-maroon overcast especially above), lustrous, to 14 mm long; fruit dull, to 9 mm long, slenderly ovoid, acuminate; panicle rather strict, rarely over 8 cm thick; leaves cuneate to broadly rounded at base, sharply serrate or incised (or doubly serrate), mostly more than 3 times as long as their narrowly wing-margined petioles; stem with rounded angles and flat sides; (essentially transcontinental) .... *S. lanceolata*
1   Rudimentary stamen brown to dull purple, spatulate to obovate, at most 1 mm broad, commonly slightly longer than broad; corolla more distinctly brown; leaves mostly rather coarsely simply or doubly serrate or incised, their petioles scarcely winged.
   2   Principal leaves commonly more than 4 times as long as their petioles (these mostly averaging not over 2 or 3 cm long), scarcely attenuate at tip; corolla to about 8 mm long; panicle rather strict, its short branches stiffly ascending; leaves rather finely to coarsely (but mostly simply) serrate; stem glabrous below the inflorescence, with acute angles and flat sides; (introd. in Que. and Nfld.) .... *S. nodosa*
   2   Principal leaves commonly 2 or 3 times as long as their slender petioles; leaves commonly more coarsely (often doubly) serrate or incised; stems with more rounded angles.
      3   Panicle loose and irregular (with spreading-ascending branches), to about 1.5 dm thick; corolla rarely over 8 mm long; leaves acuminate; stem glabrous below the inflorescence; (Ont. and Que.) .... *S. marilandica*
      3   Panicle strict (the ascending or erect-ascending branches relatively short), commonly not over 5 or 6 cm thick; corolla to about 12 mm long; leaves scarcely acuminate; stem commonly glandular-pubescent nearly or quite to base; (sw B.C.) .... *S. californica*

S. californica C. & S.
/t/W/   (Hp)   Moist low ground from sw B.C. (Vancouver Is. and Triangle Is.; CAN; reported from Moodyville and Griffin L., Kamloops dist., by Henry 1915, Moodyville perhaps being the present-day Port Moody, near New Westminster) to Calif. and Nev. [*S. oregana* Pennell].

S. lanceolata Pursh
/T/X/   (Hp)   Open woods, thickets, and clearings from s B.C. (N to Kamloops and Revelstoke; not known from Alta.) to s Sask. (Mortlach, about 50 mi w of Regina; Breitung 1957a; reports from Man. require confirmation), Ont. (N to near Thunder Bay; TRT), Que. (N to St-Léon, Rimouski Co., and the Gaspé Pen. at Port Daniel, Bonaventure Co.; not known from P.E.I.), N.B., and N.S., s to N Calif., N N.Mex., Okla., Ill., and S.C. [*S. leporella* Bickn.; *S. occidentalis* (Rydb.) Bickn.]. MAP: Pennell 1935: map 78 (somewhat incomplete northwards), p. 278.

Some of our B.C. material (Griffin L. and Revelstoke; CAN) is referable to the apparently commoner western phase, f. *velutina* Pennell (leaves soft-pubescent beneath rather than essentially glabrous).

S. marilandica L.   Carpenter's-square
/T/EE/   (Hp)   Rich woods, thickets, and clearings from Minn. to Ont. (N to Middlesex, York, and s Grenville counties; the report from Casselman, near Ottawa, by John Macoun 1884, is based upon *S. lanceolata*, the relevant collection in CAN; the report from SE Man. by Lowe 1943, may also refer to *S. lanceolata*, whose occurrence there, however, requires confirmation), sw Que. (Philipsburg, Missisquoi Co.; GH; reports farther northwards as far as Quebec City by John Macoun 1884, may refer to *S. lanceolata;* the report from P.E.I. by McSwain and Bain 1891, may also refer to *S. lanceolata,* now evidently extinct there), and sw Maine, s to Okla., La., Ala., and Ga. [*S. glauca* Raf.]. MAP: Pennell 1935: map 79, p. 285.

S. nodosa L.

Eurasian; known in N. America from ballast and waste places in New Eng., N.J., and sw Que. (Montreal; MT) and rocky or gravelly woods and thickets of Nfld. (where considered native by Fernald *in* Gray 1950, but more likely introd.; *see* note under *Luzula campestris*). [*S. marilandica sensu* Waghorne 1898, not L., the relevant collection in GH]. MAP: Hultén 1958: map 143, p. 163.

SYNTHYRIS Benth. [7583]

S. borealis Pennell

/Ss/W/ (Hs) Ridges and solifluction areas in Alaska-Yukon (N to ca. 65°N; type from Double Mt., McKinley Park, Alaska) and NW Dist. Mackenzie (Porsild and Cody 1968). MAPS (the occurrence in NW Dist. Mackenzie should be indicated): Hultén 1968*b*:805, and 1949: map 1034, p. 1470.

[TONELLA Nutt.] [7504]

[T. tenella (Benth.) Heller]

[Native in the w U.S.A. from Wash. to Calif. and to be searched for in s B.C. (*Collinsia tenella* Benth., not (Pursh) Piper, the report of which from sw B.C. by J.M. Macoun (1913; as "*Collinsia tenella,* (Pursh)") is based upon *Collinsia parviflora,* the relevant collection in CAN).]

VERBASCUM L. [7460] Mullein. Molène

1 Plants green and glabrous or somewhat glandular-pubescent above with simple hairs; anther-filaments all violet-villous; fruit subglobose, glandular-pubescent; (introd.).
  2 Pedicels solitary at the nodes, mostly at least 1 cm long, longer than the subglobose fruits; pubescence wholly of simple glandular hairs .... *V. blattaria*
  2 Pedicels up to 5 per node, usually less than 5 mm long and shorter than the fruits; simple and branched glandless hairs present in addition to glandular hairs .... *V. virgatum*
1 Plants with simple or branched eglandular pubescence; leaves tomentose at least beneath, entire or very shallowly toothed; flowers sessile or short-pedicelled; fruit ovoid or cylindric, densely tomentose; (introd.).
  3 Hairs on the anther-filaments violet or purple; flowers to about 2 cm broad; sepals linear, acute; leaves thinly pubescent and dark green above, pale and more conspicuously stellate-pubescent beneath .... *V. nigrum*
  3 Hairs on the anther-filaments whitish or yellowish.
    4 Leaves green above, ashy-pubescent beneath; flowers at most 2 cm broad, in numerous loose open racemes of a freely branching panicle; sepals linear, acute; pedicels to about 1 cm long; fruit 4 or 5 mm long .... [*V. lychnitis*]
    4 Leaves drab-tomentose on both surfaces; flowers broader, subsessile in a spike-like raceme; sepals triangular-ovate, acuminate; fruit to 1 cm long.
      5 Raceme loose, often branching at base; flowers to 4 cm broad; leaves sessile, not decurrent on the stem or only slightly so; plant loosely tomentose ....
      .... *V. phlomoides*
      5 Raceme dense, simple; flowers rarely over 2.5 cm broad; leaves decurrent to the next leaf below, the lower ones petioled; plant densely woolly throughout ....
      .... *V. thapsus*

V. blattaria L. Moth-Mullein

Eurasian; introd. along roadsides and in old fields in N. America, as in s B.C. (N to Armstrong, about 15 mi N of Vernon, and Revelstoke; CAN; V), Ont. (N to the Ottawa dist.), sw Que. (N to Trois-Rivières, St-Maurice Co.; MT), and N.B. (Woodstock, Carleton Co.; CAN; reported from St. Andrews, Charlotte Co., by Fowler 1885; the report from N.S. by Cochran 1829, may refer to *V. virgatum*).

The typical form has uniformly yellow corollas. Forma *albiflora* (Don) House (corolla whitish, with a purplish base) is known from s Ont. (Pelee Is., Essex Co., and Port Ryerse, Norfolk Co.; DAO; TRT). Forma *erubescens* Brüg. (corolla uniformly reddish) is reported from Ont. and N.B. by Boivin (1966*b*).

[V. lychnitis L.]   White Mullein
[Eurasian; locally introd. along roadsides and in old fields in N. America but apparently known from Canada only through an early collection near "Lake Erie", s Ont. (MTMG; Boivin, personal communication). The report from Prescott, Ont., by Montgomery (1957) requires confirmation. The citations from West Augusta, Ont., by John Macoun (1884) and from Sandwich, Essex Co., by J.M. Macoun (1906; taken up by Dodge 1914) are based upon *V. nigrum* and *V. virgatum*, respectively, the relevant collections in CAN, revised by Boivin.]

V. nigrum L.   Black Mullein
Eurasian; apparently known in N. America only from Alta. (Fort Saskatchewan; DAO) and s Ont. (Augusta, Grenville Co., where taken by P. Byrne in 1861; CAN; Boivin 1966*b*). [*V. lychnitis sensu* John Macoun 1884, not L.].

V. phlomoides L.   Clasping Mullein
Eurasian; introd. along roadsides and in fields in N. America, as in sw B.C. (Fort Langley, near Vancouver; V), Alta. (Fort Saskatchewan; Groh 1946), Ont. (Lambton, Norfolk, York, and Glengarry counties; OAC; TRT), Que. (Oka; MT), and P.E.I. (Boivin 1966*b*).

V. thapsus L.   Common Mullein, Flannel-plant.   Tabac du diable or Bouillon blanc
Eurasian; introd. along roadsides and in old fields and waste places in N. America, as in s Alaska (Hultén 1949; Juneau, where "unable to maintain itself"), B.C. (N to Revelstoke; CAN), Alta. (Fort Saskatchewan and Crowsnest Pass; not known from Sask.), Man. (N to Bowsman, N of Duck Mt.), Ont. (N to Kapuskasing, 49°24′N), Que. (N to the Gaspé Pen.), Nfld., N.B., P.E.I., and N.S.

V. virgatum Stokes
European; locally introd. along roadsides and in waste places in N. America, as in s Ont. (Sandwich, Essex Co., where taken by John Macoun in 1901; CAN) and N.S. (Kings and Cape Breton counties; ACAD; CAN). [*V. lychnitis sensu* J.M. Macoun 1906, not L., the relevant collection in CAN].

VERONICA L.   [7579]   Speedwell.   Véronique

1   Leaves linear to linear-lanceolate, acuminate, sessile, entire or obscurely toothed; racemes in the axils of the opposite leaves; corolla lilac or pale violet; sepals ovate-oblong, much shorter than the capsule; capsule strongly flattened, broadly notched at summit, much broader than long; style about as long as the capsule; pedicels filiform, reflexed in fruit, much longer than the flower or fruit; perennial with filiform rhizomes; (transcontinental) . . . . *V. scutellata*
1   Leaves broader.
  2   Flowers long-pedicelled in the axils of scarcely reduced, mostly alternate foliage-leaves (lower leaves often opposite); capsule pubescent, broader than long, somewhat flattened, deeply notched; seeds cup-shaped, rugose, to 3 mm long; leaves ovate to reniform, serrate or dentate, their petioles to 5 mm long; stem weak, decumbent and often creeping at base; pubescent annuals; (introd.).
    3   Corolla about 1 cm broad, much surpassing the calyx; fruiting pedicels to over 3 cm long; style much surpassing the calyx-lobes.
      4   Leaves broadly ovate, obtuse; stems scarcely matted . . . . *V. persica*
      4   Leaves reniform; stems densely matted . . . . *V. filiformis*
    3   Corolla smaller, only slightly surpassing the calyx; fruiting pedicels at most 1 cm long; style shorter; leaves ovate to suborbicular.
      5   Sepals acute, slightly surpassing the sparingly pubescent capsule; style not much surpassing the capsule-lobes; corolla pale blue, the lower 1 or 3 lobes white or very pale . . . . *V. agrestis*
      5   Sepals blunt, equalling or barely surpassing the densely pubescent capsule; style distinctly surpassing the capsule-lobes; corolla usually bright blue, rarely the lower lobe paler . . . . *V. polita*
  2   Flowers sessile or on pedicels at most about twice as long as their subtending bracts or much reduced foliage-leaves.

6 Flowers sessile or subsessile in the axils of much reduced alternate upper leaves; lower leaves opposite; seeds flat, rarely over 1 mm long; stem ascending; annuals.

 7 Corolla whitish; sepals oblanceolate, subequal; capsule broader than long, shallowly notched, the minute style shorter than the capsule-lobes; leaves narrowly oblong to oblanceolate, rather fleshy, entire or obscurely toothed; (B.C. to N.S.) .... *V. peregrina*

 7 Corolla violet-blue; lower pair of sepals longer than the upper pair; capsule about as broad as long, more deeply notched, pilose; style longer; leaves relatively broad, the lower ones oval to suborbicular, scarcely fleshy; plants more or less pilose; (introd.).

  8 Leaves entire or low-toothed; style equalling or surpassing the capsule-lobes .... *V. arvensis*

  8 Median leaves deeply cleft into linear lobes; style much shorter than the capsule-lobes .... *V. verna*

6 Flowers short-pedicelled in the axils of small narrow alternate bracts, the inflorescence loosely or densely subspicate or racemose; foliage-leaves all opposite; perennials.

 9 Racemes solitary and terminal on the main axis or forming the branches of a terminal panicle.

  10 Leaves coarsely toothed, relatively long-petioled, the upper ones some-times in 3's; racemes dense; corolla blue-violet; style 2 or 3 times surpassing the plump barely notched capsule, this longer than broad; stem stoutish, to about 1.5 m tall, from a woody base; (introd.).

   11 Leaves lanceolate, acuminate, tapering or rounded at base, to 1.5 dm long, very sharply and doubly serrate; raceme attenuate; fruiting pedicels nearly as long as the calyx; sepals ciliate, otherwise essentially glabrous; plant minutely tomentose .... *V. longifolia*

   11 Leaves cordate-ovate, blunt, to about 6 cm long, coarsely and simply or somewhat doubly crenate-dentate; fruiting pedicels shorter than the calyx; raceme scarcely attenuate; whole plant (including sepals) minutely glandular-cinereous .... *V. grandis*

  10 Leaves entire to shallowly crenate, sessile or short-petioled, none in 3's; stem relatively slender, rarely over 5 or 6 dm tall.

   12 Leaves to about 8 cm long, lanceolate to narrowly ovate, shallowly crenate; corolla blue or blue-violet; style much surpassing the plump barely notched capsule; raceme dense, attenuate; stem woody at base, to 6 or 7 dm tall; (introd.).

    13 Plant copiously white-woolly, with many sterile matted basal offshoots .... [*V. incana*]

    13 Plant minutely glandular-cinereous and with an admixture of long slender multicellular hairs .... *V. spicata*

   12 Leaves mostly not over 3 cm long, elliptic to suborbicular, entire or only obscurely toothed; stem from a slender creeping base.

    14 Capsule obcordate, broader than long, deeply notched; sepals (and leaves) eciliate; raceme narrow and elongate, loosely flowered; sterile leafy basal offshoots present and mat-forming; (transcontinental) .... *V. serpyllifolia*

    14 Capsule broadly ovoid, tending to be slightly longer than broad, shallowly notched; corolla deep-blue to blue-violet; sterile leafy offshoots few or none.

     15 Style elongate, commonly at least 6 mm long and usually longer than the glandular-pubescent capsule; anther-filaments to 8 mm long; sepals glandular-pubescent on the back but not conspicuously ciliate; leaves completely glabrous, entire; (mts. of s B.C.) .... *V. cusickii*

     15 Style less than 4 mm long, shorter than the capsule; anther-filaments at most 4 mm long; sepals and upper leaves

conspicuously ciliate with long white multicellular hairs; (transcontinental) .... *V. alpina*

9 Racemes lateral, borne in the axils of opposite foliage-leaves (the main axis
sometimes little developed at anthesis, the inflorescence thus pseudo-
terminal).

16 Stem and at least the young leaves pilose; capsule pubescent.

17 Leaves subentire or only obscurely serrate, copiously long-pilose at
least when young with reddish flattened multicellular hairs, the stems
similarly pilose, less than 1 dm tall; (Aleutian Is.) .... *V. grandiflora*

17 Leaves rather coarsely toothed; (introd.).

18 Pedicels shorter than the subequal obtuse sepals, these 2 or 3 mm
long; leaves obovate-elliptic or cuneate-oblong, short-petioled;
corolla about 5 mm broad; capsule glandular-pubescent, obcordate, as broad as or slightly broader than long, much surpassing
the calyx; stem extensively creeping, it and the branches
mat-forming .... *V. officinalis*

18 Pedicels about equalling to much surpassing the acutish sepals;
corolla about 1 cm broad; pubescence of capsule nonglandular;
stems not mat-forming; (introd.).

19 Leaves sessile or nearly so, oblong; sepals 5, narrowly
lanceolate, a short one about 1 mm long, 2 about 2.5 mm long,
and 2 about 3.5 mm long; capsule obovoid, slightly surpassing
the calyx; style about 7 mm long; stem stiff and erect ..... *V. latifolia*

19 Leaves short-petioled, ovate or cordate; sepals 4, broadly
lanceolate, only slightly unequal; capsule obcordate, broader
than long, shorter than or slightly surpassing the calyx; style 4
or 5 mm long; stem weak, from a creeping base .... *V. chamaedrys*

16 Stem glabrous or minutely glandular, decumbent or creeping at base;
capsule glabrous or glandular, reniform or suborbicular, turgid, shorter
than or slightly surpassing the calyx; style to 3 mm long.

20 Leaves distinctly petioled and toothed; capsule very shallowly notched.

21 Leaves mostly lance-ovate, broadest toward base, acute or
acutish, shallowly serrate; style about 3 mm long; lower fruiting
pedicels to over 1 cm long; stem decumbent or short-creeping;
(transcontinental) .... *V. americana*

21 Leaves elliptic to obovate, broadest near or above the middle,
rounded at tip, shallowly crenate; style about 2 mm long; fruiting
pedicels mostly 4 or 5 mm long; stem strongly creeping; (introd.).........
.... *V. beccabunga*

20 Leaves sessile, at least the upper ones cordate-clasping.

22 Capsule ovate to subglobose, barely notched, about as long as
broad or a little longer, not surpassing the acute or acuminate
sepals, on a usually rather strongly ascending pedicel; corolla pale
lavender-violet, 4 or 5 mm broad; principal leaves lance- to
obovate-oblong, commonly less than 3 times as long as broad;
(introd.) .... *V. anagallis-aquatica*

22 Capsule broadly obcordate to round-reniform, deeply notched,
mostly a little broader than long, longer than the obtuse or
short-pointed sepals, on a usually widely spreading pedicel; corolla
whitish or pale pink to roseate, smaller; principal leaves lanceolate
to lance-oblong, to about 5 times as long as broad; (B.C. to SW
Que.) .... *V. catenata*

V. agrestis L. Field-Speedwell

Eurasian; introd. along roadsides and in cult. and waste ground in N. America, as in SE B.C.
(Okanagan; Henry 1915), Alta. (N to Beaverlodge, 55°13′N), Ont. (reported N to Ottawa by Groh
and Frankton 1949*b*, but not listed by Gillett 1958), Que. (N to the Gaspé Pen. near

Grande-Rivière; GH), Nfld., N.B., ?P.E.I. (the report from Charlottetown by Herbert Groh, Sci. Agric. 7(10):394. 1927, is considered by D.S. Erskine 1960, perhaps referable to *V. persica*), and N.S.

V. alpina L.   Alpine Speedwell
/aST/X/GEA/   (Hpr)   Moist meadows, bogs, and open slopes at moderate to high elevations, the aggregate species from the Aleutian Is., Alaska (N to ca. 67°30′N), the Yukon (N to ca. 65°N), Great Bear L., s Dist. Keewatin, s Baffin Is., and northernmost Ungava–Labrador, s in the West through B.C.–Alta. (not known from Sask., Man., or Ont.) to Calif. and N.Mex. and in the East to Que. (s to SE James Bay, Mollie T Lake at ca. 55°N, and the Shickshock Mts. of the Gaspé Pen.), NW Nfld. (not known from the Maritime Provinces), and the mts. of Maine and N.H.; W and E Greenland N to ca. 72°N; Iceland; Europe; widely disjunct stations in Asia. MAPS and synonymy (together with distinguishing keys to the closely related *V. fruticans* of Greenland and *V. stelleri* of Alaska): *see below.*

1   Stems slightly woody, much branched, decumbent-based and often somewhat matted, to
    2 dm tall; inflorescence short, downy with curled, eglandular, short hairs; corolla-lobes
    mostly broader than long; [s half of Greenland; MAPS: Hultén 1958: map 70, p. 89; Böcher
    1938: fig. 100, p. 178] . . . . *V. fruticans* Jacq.
1   Stems from a slender creeping base, simple, erect or barely decumbent-based, often
    taller; inflorescence sparsely to densely villous or pilose with spreading or incurved (also
    often viscid or glandular) hairs; corolla-lobes mostly distinctly longer than broad.
    2   Fruiting pedicels to 11 mm long; leaves sharply serrate; [incl. the glabrescent ex-
        treme, var. *glabrescens* Hult.; Aleutian Is. and s Alaska; MAP: Hultén 1968b:803] . . . .
        . . . . *V. stelleri* Pallas
    2   Fruiting pedicels rarely over 5 mm long; leaves entire or shallowly crenate . . . . *V. alpina*
        3   Backs of sepals essentially glabrous except for marginal ciliation; capsules
            glabrous; fruiting raceme dense, most of the fruits overlapping; stems to about 2
            dm tall; [northernmost Ungava–Labrador and s Baffin Is.; MAPS: Hultén 1958: map
            35, p. 55; Meusel 1943: fig. 13b (aggregate species)] . . . . var. *alpina*
        3   Backs of sepals and the capsules more or less pubescent.
            4   Fruiting raceme dense, to about 6.5 cm long and 1.5 cm thick, the fruits
                moderately crowded to strongly overlapping (except sometimes the lowermost
                ones), villous with multicellular hairs, their styles less than 2 mm long.
                5   Sepals and blackish capsules long-villous with gland-tipped hairs; leaves
                    blackened on drying; stems to 3 dm tall, spreading-villous above a shorter
                    glandular pubescence . . . . var. *unalaschcensis* C. & S.
                    6   Corolla deep blue-violet; [incl. vars. *australis* Wahl., *lasiocarpa* Hartm.,
                        villosa (Wormskj.) Lange (*V. vill.* Wormskj.), and *wormskjoldii* (R. & S.)
                        Hook. (*V. worm.* R. & S.); *V. nutans* Bong. (*V. worm. nutans* (Bong.)
                        Pennell); *V. pumila* All.; transcontinental, the type from Unalaska,
                        Aleutian Is.; MAPS: Hultén 1968b:802, and 1958: map 36 (*V. worm.*
                        and its var. *nutans*), p. 55; Porsild 1957: map 286, p. 196; Raup 1947:
                        pl. 34] . . . . f. *unalaschcensis*
                    6   Corolla white; [SE Yukon, the type from the upper Rose R.] . . . .
                        . . . . f. *albiflora* Porsild
                5   Sepals and pale-brown capsules sparingly villous; leaves at most only
                    slghtly blackening; stems to 2 dm tall, incurved-villous above, scarcely
                    glandular; [type from the Highlands of St. John, NW Nfld.; MAP: Hultén
                    1958: map 35, p. 55] . . . . var. *terra-novae* Fern.
            4   Fruiting raceme lax, to 1.5 dm long, the distinctly pedicelled fruits or pairs of
                fruits mostly becoming distant; [western varieties].
                7   Upper leaves alternate, with blunt or rounded tips, commonly drying green;
                    flowers and fruits mostly alternate; lower bracts mostly linear to lanceolate,
                    the upper ones inconspicuous; stems to about 3.5 dm tall; [*V. wormskjoldii*
                    ssp. *alt.* (Fern.) Pennell] . . . . var. *alterniflora* Fern.
                    8   Corolla deep blue-violet; [Alaska–s Yukon–sw Dist. Mackenzie–B.C.;
                        MAP: Hultén 1968b:803] . . . . f. *alterniflora*

8   Corolla white; [SE Yukon, the type from Rose-Lapie Pass] . . . .
. . . . f. *albiflora* Porsild

7   Leaves, flowers, and fruits mostly opposite, the leaves drying black.

9   Stem-leaves (excluding floral bracts) up to 8 pairs, the upper ones ovate and acutish; lower floral bracts mostly lanceolate to ovate, the upper ones similar but narrower and shorter; calyx to 7 mm long; style to 1.5 mm long; stems to 4 dm tall; [mts. of S B.C. and SW Alta. (Lake Louise)] . . . . var. *geminiflora* Fern.

9   Stem-leaves at most 6 pairs, elliptic or elliptic-ovate, with blunt or rounded tips; lower floral bracts mostly linear to lanceolate, the upper ones inconspicuous; calyx to 4 mm long; style to 3 mm long; stems usually less than 3 dm tall; [S Alaska (near Juneau) and mts. of B.C.; MAP: Hultén 1958: map 36 (*V. worm.* var. *cas.*), p. 55] . . . .
. . . . var. *cascadensis* Fern.

V. americana Schwein.   American Brooklime

/ST/X/eA/   (Ch)   Swampy ground and shallow water from the Aleutian Is., Alaska (N to ca. 65°N), the Yukon (N to ca. 63°30′N), and SW Dist. Mackenzie to B.C.–Alta., Sask. (N to Waskesiu Lake, ca. 54°N), Man. (N to Riding Mt.), Ont. (N to the SW James Bay watershed at 52°11′N), Que. (N to SE James Bay at 51°29′N, L. Mistassini, Anticosti Is., and the Gaspé Pen.), Nfld., N.B., P.E.I., and N.S., S to S Calif., Mexico, Nebr., and N.C.; E Asia. [*V. beccabunga* var. *amer.* Raf.]. MAPS: Hultén 1968*b*:798, and 1958: map 126, p. 145; Porsild 1966: map 128, p. 82; Pennell 1935: map 88 (somewhat incomplete northwards), p. 254.

Forma *rosea* Henry (corolla pink rather than blue) is known from SW B.C. (Vancouver Is. and adjacent islands; type, as first collection cited, from Alberni, Vancouver Is.). Concerning the authorship of the species, *see* Hultén (1968*a*).

V. anagallis-aquatica L.   Water-Speedwell, Brook-Pimpernel

Eurasian; wet places, ditches, and shores of the U.S.A. (where considered apparently both native and introd. by Fernald *in* Gray 1950). Introd. in SE Alaska (Craig; ?Sitka), S Yukon (Whitehorse; CAN), B.C. (N to Prince Rupert, ca. 54°20′N; CAN), and Ont. (Boivin 1966*b*; area uncertain through confusion with *V. catenata*). MAPS: Hultén 1968*b*:798, and 1958: map 126, p. 145.

See Pennell (1935:363) concerning the probability that the N. American plant is entirely introd. Reports and so-named collections from Que. and the Maritime Provinces require further study, perhaps being referable largely to *V. catenata,* as also the report from Alta. by John Macoun (1884; as *V. anagallis,* omitting the terminating inverted-triangle symbol indicating *aquatica* used by Linnaeus). An 1894 collection in CAN by Macoun from the Cypress Hills of SW Sask. has been referred to *V. catenata* by Pennell. A collection in TRT from Bradford, York Co., S Ont., has been referred to f. *anagalliformis* (Boreau) Beck (stem and axis of inflorescence more or less glandular rather than glabrous) but may prove referable to *V. catenata* ssp. *glandulosa.*

V. arvensis L.   Corn-Speedwell

Eurasian; introd. in pastures, open woods, and waste places in N. America, as in S Alaska (Sitka and Juneau; Hultén 1949), B.C. (N to Prince Rupert and Queen Charlotte Is.), Ont. (N to the SE shore of L. Superior and the Ottawa dist.), Que. (N to Ste-Anne-de-la-Pocatière, Kamouraska Co., and Magdalen Is.), ?Labrador (Boivin 1966*b*), Nfld., N.B. (the report from P.E.I. by Hurst 1952, requires confirmation), and N.S.; SW Greenland. MAP: Hultén 1968*b*:800.

V. beccabunga L.   Brooklime

Eurasian; introd. in wet ditches and shallow water in N. America, as in ?Ont. (tentatively reported from Stormont Co. by Dore and Gillett 1955) and Que. (N to La Malbaie, about 80 mi NE of Quebec City; Groh and Frankton 1949*b*; the report from N.S. by Cochran 1829, requires clarification).

V. catenata Pennell

/T/X/   (HH (Hpr))   Ditches and slow-flowing streams from SW B.C. (Vancouver Is. and adjacent mainland E to Manning Provincial Park, SE of Hope; CAN; V) to Alta. (Waterton Lakes; Breitung 1957*b*), Sask. (N to Hudson Bay Junction, 52°52′N; Breitung 1957*a*), S Man. (N to Dropmore, about

100 mi NW of Brandon), Ont. (N to the Ottawa dist.), SW Que. (N to the Montreal dist.), and Vt., S to Calif., N.Mex., Okla., Mo., Ohio, and Pa. [*V. connata (comosa)* var. *glaberrima* Pennell; *V. aquatica (salina)* f. *laevipes* Beck]. MAP: Pennell 1935: map 89 (*V. connata* and its ssp. *glaberrima*), p. 366.

Some of our material is referable to ssp. *glandulosa* (Farw.) Pennell (*V. anagallis-aquatica (comosa)* ssp. *gland.* Farw.; *V. connata* Raf.; *V. salina* Schur; leaf-rachises, pedicels, and upper part of stem finely stipitate-glandular rather than glabrous).

V. chamaedrys L.   Bird's-eye
Eurasian; introd. along roadsides and borders of woods and in fields in N. America, as in S Alaska (Sitka; Hultén 1949), B.C. (Vancouver Is.; Abbotsford), Ont. (N to Ottawa and the Timagami Forest Reserve N of Sudbury), Que. (N to Kenogami, near L. St. John), Nfld., N.B., P.E.I., and N.S. MAP: Hultén 1968b:799.

V. cusickii Gray
/T/W/   (Hpr)   Moist meadows and rocky slopes at rather high elevations from SW B.C. (collections in CAN and in the park herbarium from Manning Provincial Park, SE of Hope) and W Mont. to Calif.

V. filiformis Sm.
Asiatic; introd. in lawns and waste places in N. America, as in B.C. (Queen Charlotte Is.; New Westminster), Ont. (Ottawa; Grimsby, Lincoln Co.; OAC), and SW Que.

V. fruticans Jacq.
/aST/-/GE/   (Ch)   Calcareous rocks and slopes in W and E Greenland N to ca. 71°N; Iceland; Europe. (*V. saxatilis* Scop.). MAPS: Hultén 1958: map 70, p. 89; Böcher 1938: fig. 100 (Greenland map), p. 178. (Keyed out under *V. alpina*).

V. grandiflora Gaertn.
/s/W/eA/   (Ch)   Rocky places in the Aleutian Is. (*see* Hultén 1949: map 1027, p. 1469); E Asia. [Var. *minor* Hult.; *V. kamtschatica* L. f.]. MAP: Hultén 1968b:804.

V. grandis Fisch.
Eurasian; reported by Fernald *in* Gray (1950) as spreading locally to roadsides in Que. (Grondines, about 35 mi SW of Quebec City; GH). [*V. bachofenii* Heuff.].

[V. incana L.]   Woolly Speedwell
[Eurasian; reported from S Ont. by Boivin (1966b; Grimsby Beach, Lincoln Co.), where probably a casual garden-escape and scarcely established.]

V. latifolia L.
Asiatic; a garden-escape in N. America, as in Sask. (Boivin 1966b), Ont. (N to Frontenac, Hastings, and Grenville counties), and Nfld. (Rouleau 1956). [*V. teucrium* L.].

V. longifolia L.
Eurasian; introd. along roadsides and in thickets and fields in N. America, as in Alta. (Moss 1959), Sask. (Raymore, about 60 mi N of Regina; Breitung 1957a), S Ont. (N to Bruce and Leeds counties), Que. (N to Cacouna, Temiscouata Co., and Magdalen Is.), Nfld. (Pennell 1935), N.B., P.E.I., and N.S. [*V. maritima* L.; *Verbena hastata sensu* Fowler 1885, as to the Kouchibouguac, N.B., plant, not L., the relevant collection in NBM].

Forma *glabra* (Schrad.) Aschers. & Graebn. (plant essentially glabrous rather than ashy-puberulent) is known from N.B. (Rothesay, Kings Co., and Perth, Victoria Co.; CAN) and P.E.I. (Summerside, Prince Co.; CAN; GH).

V. officinalis L.   Common Speedwell, Gypsyweed.   Thé d'Europe
Eurasian; a common weed of fields, waste places, and open woods in N. America (ranges of Canadian taxa outlined below).
1   Leaves usually not over 3 cm long and about 1.5 cm broad, mostly much shorter than the
    filiform peduncles; racemes lax and flexuous; corolla blue-violet; capsules often broader

than long; [*V. tournefortii* Vill., not Gmel.; Nfld., P.E.I., and N.S.] . . . .
. . . . var. *tournefortii* (Vill.) Reichenb.

1   Leaves to about 6 cm long and 3 cm broad, longer or shorter than the stoutish peduncles;
racemes densely many-flowered; corolla lilac-blue or lavender; capsules about as broad
as long . . . . var. *officinalis*
   2   Corolla white; [sw Que.: Philipsburg, Missisquoi Co.; MT] . . . . f. *albiflora* (Don) House
   2   Corolla lavender to lilac-blue; [s B.C. (Vancouver Is. and adjacent islands; Hope;
   Rossland; Ainsworth, NE of Nelson), Ont. (N to the Ottawa dist.), Que. (N to the Gaspé
   Pen.), Nfld., N.B., P.E.I., and N.S.; MAP (aggregate species): Hultén 1958: map 120,
   p. 139]. Fernald *in* Gray (1950) believes that this species is both native and introd. in
   N. America . . . . f. *officinalis*

V. peregrina L.   Neckweed, Purslane-Speedwell
/T/X/ (T) Marshes, estuaries, and other damp or wet places (introd. northwards; ranges of
Canadian taxa outlined below), s to Baja Calif., Mexico, Tex., and Fla.; introd. in Eurasia. MAPS and
synonymy: *see* below.

1   Plant glabrous; [sw B.C. (Vancouver Is.; ?introd.); s Ont. (N to Huron, Wellington, Peel,
York, and Frontenac counties), Que. (N to Sorel, Richelieu Co.), N.B. (Fowler 1885; F.W.
Pennell, Rhodora 23(265):18. 1921), and P.E.I. (in disturbed soil in a nursery at Bunbury,
Queens Co., where doubtless introd.; ACAD); MAP: Pennell 1935: map 87, p. 338] . . . .
. . . . var. *peregrina*
1   Capsules and upper part of stem more or less short-stipitate-glandular; [incl. var.
laurentiana Vict. & Rousseau; *V. xalapensis* HBK.; cent. Alaska–Yukon (introd.); Great
Slave L. (?introd.) and B.C. to Alta–Sask. (N to L. Athabasca), Man. (N to York Factory,
Hudson Bay, ca. 57°10'N), Ont. (N to Ingolf, near the Manitoba boundary at ca. 49°50'N,
and the Nipigon R. N of L. Superior), Que. (N to Montmagny Co.), N.B., and N.S.; MAPS:
on the above-noted map by Pennell; Hultén 1968b:801] . . . .
. . . . var. *xalapensis* (HBK.) St. John & Warren

V. persica Poir.   Bird's-eye
Eurasian; introd. along roadsides and in waste places in N. America, as in SE Alaska (Hultén 1949;
Juneau), s B.C. (Vancouver Is. and mainland E to Nelson), Alta. (N to Fort Saskatchewan), Sask.
(Swift Current; Breitung 1957a), Man. (N to Dauphin, N of Riding Mt.), Ont. (N to Sault Ste. Marie
and Ottawa), Que. (N to the Gaspé Pen. at Gaspé Basin; GH), Nfld., N.B., P.E.I., and N.S. [*V.
buxbaumii* Tenore, not Schmidt; *V. tournefortii* of auth., not Gmel. nor Vill.]. MAP: Hultén 1968b:800.
   According to Boivin (1966b), most of our material is referable to var. *aschersoniana* (Lehm.)
Boivin (*V. tournifortii* ssp. *asch.* Lehm.; lower corolla-lip white rather than pale blue, the other lobes
deep blue; *see* Bernard Boivin, Nat. can. (Que.) 79:174. 1952). He also reports var. *corrensiana*
(Lehm.) Boivin (*V. tourn. corr.* Lehm; all the corolla-lobes uniformly deep blue) from Ont., Que.,
Nfld., N.B., and N.S.

V. polita Fries   Wayside Speedwell
Eurasian; introd. along roadsides and in lawns and waste places in N. America, as in s Man.
(Cartwright, about 75 mi sw of Portage la Prairie; DAO, detd. Boivin) and s Ont. (Vineland, Lincoln
Co.; DAO). [*V. didyma* Tenore 1830, not 1811].

V. scutellata L.   Marsh-Speedwell
/ST/X/EA/ (Hpr) Swamps, shores, and wet places from Alaska (N to ca. 65°N), the Yukon (N to
ca. 63°N), Great Slave L., s Dist. Keewatin, and B.C.–Alta.–Sask.–Man. to Ont. (N to the Severn R.
at ca. 55°40'N), Que. (N to s Ungava Bay, the Côte-Nord, Anticosti Is., and Gaspé Pen.), Labrador
(N to Tikkoatokok Bay at ca. 57°N; CAN), Nfld., N.B., P.E.I., and N.S., s to Calif., Colo., Iowa, and
Va.; Iceland; Eurasia. MAPS: Hultén 1968b:799, and 1958: map 45, p. 65; Porsild 1966: map 129,
p. 83; Pennell 1935: map 90, p. 371.
   Forma *alba* Boivin (flowers white rather than lilac or bluish) is known from the type locality,
Sasaginnigak L., about 125 mi NE of Winnipeg, Man. Forma *villosa* (Schum.) Pennell (var. *pilosa*
Vahl; var. *pubescens* Macoun; leaves and at least the upper part of the stem more or less villous
rather than glabrous) occurs throughout the range.

V. serpyllifolia L.   Thyme-leaved Speedwell

/sT/X/EA/   (Hpr)   Grasslands, clearings, roadsides, and waste places (the typical form introd. in N. America, var. *humifusa* native; ranges of Canadian taxa outlined below), s to Calif., Mexico, Minn., Mich., N.Y., and New Eng.; s Greenland (introd.); Eurasia. MAPS and synonymy: *see* below.

1   Rachis and pedicels incurved-puberulent, nonglandular; corolla at most 5 mm broad, whitish or pale blue, the veins darker blue; anther-filaments less than 3 mm long; [introd. in SE Alaska (Hyder and Haines; *see* Hultén 1949: map 1028, p. 1469), the Yukon (Boivin 1966*b*), E B.C. (Glacier and Yoho National Parks), ?Man. (the report from Warren Landing, near the NE end of L. Winnipeg, by Lowe 1943, requires confirmation), Ont. (N to Schreiber, N shore of L. Superior), Que. (N to L. St. John, the Côte-Nord, Gaspé Pen., and Magdalen Is.), ?Labrador (Boivin 1966*b*), St-Pierre and Miquelon, Nfld., N.B., P.E.I., and N.S.; MAP: Hultén 1968*b*:801] .... var. *serpyllifolia*

1   Rachis and pedicels pubescent with spreading glandular hairs; corolla to 8 mm broad, pale blue to deep blue-violet; anther-filaments to 4 mm long; [var. *borealis* Laest.; var. *decipiens* Boivin; *V. humifusa* Dickson; *V. tenella* All.; Aleutian Is.–s Alaska (*see* Hultén 1949: map 1031 (*V. tenella*), p. 1470) and B.C.–Alta. to Sask. (Cypress Hills; not known from Man.), Ont. (N to the Nipigon R. N of L. Superior), Que. (N to E James Bay at 52°37′N, the Côte-Nord, and Gaspé Pen.), Labrador (N to Melville, Hamilton R. basin), Nfld., N.B., and N.S.; MAP: Hultén 1968*b*:802] .... var. *humifusa* (Dickson) Vahl

V. spicata L.

Eurasian; introd. along roadsides and in rocky places in N. America, as in s Ont. (Maple, York Co.; TRT) and sw Que. (Missisquoi, Chambly, Vaudreuil, and Terrebonne counties; MT).

V. verna L.

Eurasian; introd. in waste ground in N. America, as in B.C. (Boivin 1966*b*) and s Ont. (Stokes Bay, L. Huron, Bruce Co.; GH; TRT).

VERONICASTRUM Fabricius   [7579]

V. virginicum (L.) Farw.   Culver's-root

/T/EE/?eA/   (Grh (Hpr))   Open woods and meadows from SE Man. (N to Arnaud, about 40 mi s of Winnipeg; DAO; reports N to Winnipeg require confirmation) to Ont. (N to Savanne, about 45 mi NW of Thunder Bay; CAN; not known from Que. or the Atlantic Provinces, reports from N.S. possibly referring to *V. longifolia*), Vt., and Mass., s to E Tex. and Fla.; the scarcely separable *V. sibiricum* (L.) Pennell in E Asia. [*Veronica* L.; *Leptandra* Nutt.]. MAP: Pennell 1935: map 86, p. 324.

Forma *villosum* (Raf.) Pennell (plant copiously villous rather than glabrous or minutely pubescent) is known from s Ont. (Walpole Is., Lambton Co.; TRT, detd. Pennell).

[NEMESIA Vent.]   [7476]

[N. strumosa Benth.]

[This South African species (not keyed out above) is reported as an ephemeral from Que. by C. Rousseau (Nat. can. (Que.) 98(4):721. 1971; Ste-Foy, near Quebec City), where probably a garden-escape. It is an annual to about 6 dm tall with linear to lanceolate, dentate, sessile, opposite leaves, the flowers to about 2.5 cm broad, in terminal racemes, white or variously yellow or purplish, the bearded throat with a pouch at the base.]

BIGNONIACEAE (Bignonia Family)

Woody vines or trees with opposite, simple or pinnately compound leaves. Flowers large and showy, perfect, gamosepalous and gamopetalous, hypogynous, somewhat irregular or 2-lipped, the lower corolla-lobe slightly larger than the other 4 lobes. Fertile stamens 2 or 4, the other 3 or 1 sterile and rudimentary. Style 1, the stigma 2-lipped. Ovary superior. Fruit an elongate 2-locular capsule, the flat seeds broadly 2-winged.

1   Leaves pinnate with usually 9 or 11 ovate, sharply and coarsely serrate, acuminate
      leaflets to 8 cm long; calyx 5-toothed; flowers in corymbs; corolla tubular-funnelform,
      orange and scarlet, to 8 cm long; fertile stamens 4, didynamous (2 long, 2 short); capsule
      flattened, oblanceolate, to about 2 dm long; wings of seeds merely erose; woody vine
      trailing or climbing by aerial rootlets; (S Ont.) .... *Campsis*
1   Leaves simple, entire or shallowly lobed, ovate or cordate-ovate, acuminate, to about 3
      dm long; flowers in panicles; calyx commonly splitting at anthesis into 2 unequal lobes;
      fertile stamens commonly 2; flowers in panicles; capsule linear-cylindric, terete, to over 5
      dm long; wings of seeds hairy-fringed at tip; trees; (introd. in Ont.) .... [*Catalpa*]

CAMPSIS Lour.   [7714]   Trumpet-flower

C. radicans (L.) Seem.   Trumpet-creeper, Cow-itch
/t/EE/   (MM (vine))   Moist woods and thickets from Iowa to Ill., southernmost Ont. (shores of L. Erie at Amherstburg, Pelee Point, and Pelee Is. and other islands of the Erie Archipelago in Essex Co.; formerly near Chatham, Kent Co.; *see* S Ont. map by Soper and Heimburger 1961:34), and N.J., S to Tex. and Fla. [*Bignonia* L.; *Tecoma* Juss.].

According to Fernald *in* Gray (1950), the plant is natzd. N to Mich. and Conn. John Macoun (1884) writes that, "This species is either indigenous on Pelee Island and Pelee Point, or it has become so naturalized as to run wild, and appear to be native. It is quite hardy eastward as far as Belleville and Prince Edward Co., and bears the winter cold without being taken off the trellis." Soper and Heimburger believe that it is native in S Essex (and, formerly, Kent) Co., noting that it is "Frequently planted as an ornamental vine as far north as Georgian Bay but barely hardy in the Ottawa District." Dodge (1915) notes that it is cult. but not spreading in Lambton Co., adjacent to Kent Co.

[CATALPA Scop.]   [7727]   Catalpa, Catawba, Indian-bean

1   Leaves glabrous or soon glabrate beneath, often with sharp lobes; corolla yellow,
      orange-striped and purple-spotted, the limb 1 or 2 cm broad; capsule at most 8 mm thick ......
      .... [*C. ovata*]
1   Leaves soft-pubescent beneath; corolla white, more or less spotted with yellow and
      purple-brown; capsules at least 8 mm thick.
      2   Leaves long-acuminate; corolla-limb to 6 cm broad, it and the tube inconspicuously
            spotted, its lower lobe notched at apex .... [*C. speciosa*]
      2   Leaves abruptly short-acuminate; corolla-limb to about 4 cm broad, it and the tube
            conspicuously spotted, its lower lobe entire .... [*C. bignonioides*]

[*C. bignonioides* Walt.]   Common Catalpa
[Native in the S U.S.A. from Miss. to Ga.; cult. elsewhere (N to the Montreal dist., Que.) and reported by Dodge (1914) as planted and escaping along the L. Erie shore in Essex Co., S Ont., but (1915), in Lambton Co., "Cult. as a street and lawn tree but not spreading." (*Bignonia (C.) catalpa* L.). MAP (native area): Hough 1947:405.]

[*C. ovata* Don]   Chinese Catalpa
[Asiatic; cult. in N. America and, according to Fernald *in* Gray (1950), escaped and natzd. N to S Ont. (collection in CAN from Ottawa, lacking information as to whether or not escaped).]

[C. speciosa Warder]   Catawba-tree, Cigar-tree
[Native in damp woods and swamps of the E U.S.A. from Iowa and Ind. to Tex. and Tenn. and escaped and often natzd. elsewhere. Dodge (1914) reports it as often planted and apparently spreading near Kingsville, Essex Co., s Ont., but (1915) not spreading in Lambton Co., somewhat farther north. A collection in OAC from Guelph, Wellington Co., lacks information as to whether or not escaped. MAPS (native area): Preston 1961:360; Hough 1947:407.]

MARTYNIACEAE (Martynia Family)

PROBOSCIDEA Schmidel [7785] Unicorn-plant

Clammy-pubescent annual with large round-cordate, entire or somewhat undulate, long-petioled leaves, the lower ones opposite, the upper ones alternate. Flowers perfect, gamopetalous, hypogynous, in racemes. Calyx unequally 5-cleft. Corolla to 5 cm long, gibbous, 5-lobed and somewhat 2-lipped, dull whitish or yellowish, mottled with various shades of purple. Fertile stamens 4, didynamous (2 long, 2 short), with an additional sterile one (staminodium). Style 1. Ovary superior. Fruit drupaceous, the flesh folding away in 2 valves, the upper part woody and terminating in 2 long upwardly curved hooked horns. Seeds several, with a thick roughish coat, wingless. (Introd. in s Ont.).

P. louisianica (Mill.) Thell.   Ram's-horn, Proboscis-flower
Native in the U.S.A. from Minn. to Va., s to N Mexico and Ga.; cult. for pickles elsewhere and occasionally escaping to streambanks and waste places, as in Sask. (Boivin 1966b) and s Ont. (collection in OAC from a poultry-yard at Guelph, Wellington Co.; collections in CAN from Niagara Falls, Welland Co., and Hamilton, Wentworth Co., lacking information as to whether or not escaped; reported from the Ottawa dist. by Gillett 1958). [*Martynia* Mill.; *M. proboscidea* Gloxin].

OROBANCHACEAE (Broom-rape Family)

Root-parasitic, yellowish to brownish or purplish low herbs with scales in place of green leaves. Flowers irregular, gamopetalous, hypogynous, solitary or in spikes. Calyx variously lobed. Corolla 2-lipped, the limb oblique, the upper lip notched or 2-lobed, the lower lip 3-lobed. Stamens 4, didynamous (2 long, 2 short), inserted on the corolla-tube. Stigma capitate or 2-lobed, terminating an elongate style. Ovary superior. Fruit a l-locular, 2-valved, many-seeded capsule.

1 Stem slender, usually abundantly branched, the subsessile flowers scattered along the branches; corollas of upper sterile flowers white, commonly with 2 brown-purple stripes, about 1 cm long; lower cleistogamous fertile flowers about 5 mm long; plant parasitic on roots of the beech (*Fagus grandifolia*); (Ont. to N.S.) . . . . *Epifagus*
1 Stem stouter, rarely forking; flowers all fertile, in dense or loose spikes or racemes.
  2 Corolla to 2.5 cm long; calyx with 5 subequal lobes; stamens included; flowers either in spikes or solitary on long naked pedicels . . . . *Orobanche*
  2 Corolla rarely over 1.5 cm long; stamens more or less exserted; flowers numerous in dense spikes, subtended by 1 or 2 minute bractlets in addition to the floral bract, the calyx deeply cleft on the lower side only.
    3 Plant pale brown or yellowish throughout, essentially glabrous, to about 2 dm tall; scale-leaves to about 2 cm long, ovate; (Man to N.S.) . . . . *Conopholis*
    3 Plant yellowish, brownish red, or purple throughout; lower scale-leaves triangular and sharp-pointed, the floral bracts blunter and broadest above the middle; (Alaska, the Yukon, Dist. Mackenzie, and B.C.) . . . . *Boschniakia*

BOSCHNIAKIA C.A. Meyer [7796]

1 Stems commonly not more than about 1 dm tall; bracts all essentially glabrous, their margins eciliate; orifice of corolla with short-ciliate margins; (Vancouver Is.) . . . . *B. hookeri*
1 Stems to about 4 dm tall; at least the upper bracts copiously ciliate; orifice of corolla more densely ciliate; (Alaska–Yukon–Dist. Mackenzie–N B.C.) . . . . *B. rossica*

B. hookeri Walpers Ground-cone
/t/W/ (Gp (root-parasite)) Parasitic on salal (*Gaultheria shallon*) on or near the coast from SW B.C. (S Vancouver Is.; Herb. V) to N Calif. According to Henry (1915), "Eaten by the Indians, who called it Poque." [*Orobanche (B; Kopsiopsis) tuberosa* Hook., not Vell.; *B. strobilacea* of auth., not Gray].

B. rossica (C. & S.) Fedtsch.
/Ss/W/A/ (Gp (root-parasite)) Parasitic chiefly on *Alnus* and *Picea* in Alaska–Yukon–Dist. Mackenzie (N to ca. 69°30′N, E to Great Slave L.) and northernmost B.C. (Mucho L. at ca. 58°N; CAN); Asia. [*Orobanche* C. & S.; *O. (B.) glabra* Hook.]. MAPS: Hultén 1968b:828; Porsild 1966: map 130, p. 83; Raup 1947: pl. 34.

CONOPHOLIS Wallr. [7790]

C. americana (L.) Wallr. Squawroot
/T/EE/ (Gp (root-parasite)) Rich woods (often hidden by fallen leaves) from Man. (Boivin 1966b, also reporting it from Alaska; not listed by Hultén 1949 and 1968b) to Ont. (N to Carp L., near Sault Ste. Marie, and Ottawa), SW Que. (St-Césaire and Wakefield, Gatineau Co.; Rougemont and Mt-St-Hilaire, NE of Montreal; the reports from Nfld. by Reeks 1873, and Waghorne 1898, require confirmation), and N.S. (Annapolis, Kings, Queens, and Lunenburg counties; not known from N.B. or P.E.I.), S to Ala. and Fla. [*Orobanche* L.].

EPIFAGUS Nutt. [7792]

E. virginiana (L.) Bart. Beech-drops
/T/EE/ (Gp (root-parasite)) Parasitic or saprophytic on beech (*Fagus grandifolia*) from Ont. (N to

the Muskoka Lakes E of Georgian Bay and Ottawa) to Que. (N to Kamouraska Co.; CAN), N.B., P.E.I., and N.S., S to La., Miss., Ala., and Fla. [*Orobanche* L.; *E. americana* Nutt.].

OROBANCHE L. [7791] Broom-rape

1 Flowers solitary on long naked pedicels much longer than the calyces, these not subtended by bracts; corolla with a long curved tube and a spreading, subequally 5-lobed and scarcely 2-lipped limb; calyx 5-lobed.

  2 Pedicels mostly at least 4, to about 2 dm long, about equalling the more or less elongate stem, the lower ones often longer than the upper, resulting in a loose, flat-topped corymb; corolla purple (sometimes yellowish); calyx-lobes rather narrowly triangular, shorter than to equalling the tube; scale-leaves pubescent, at least the upper ones acuminate; (B.C. to S Ont.) . . . . *O. fasciculata*

  2 Pedicels rarely more than 3, much longer than the usually very short stem; corolla creamy-white to lilac; scale-leaves glabrous, blunt or short-pointed; (transcontinental) . . . . . .
. . . . *O. uniflora*

1 Flowers several to many, sessile or on pedicels to about 3 cm long, the calyces subtended by a large bract and usually a pair of smaller ones; corolla rather distinctly 2-lipped.

  3 Calyx 4-lobed, the lanceolate acute lobes usually shorter than the tube; corolla to 3 cm long, dull bluish-purple, suffused with yellow at base; inflorescence a lax spike; (introd.) . . . . [*O. purpurea*]

  3 Calyx 5-lobed.

    4 Calyx less than 1 cm long, the triangular lobes about equalling or a little shorter than the tube; inflorescence loosely paniculate; corolla to 2 cm long, yellowish, marked with purplish brown, the lobes acute; (S B.C.) . . . . *O. pinorum*

    4 Calyx over 1 cm long, the linear lobes much longer than the tube; corolla pink or purplish, its lobes rounded or somewhat pointed.

      5 Inflorescence spicate, the flowers sessile or nearly so; calyx-lobes less than twice as long as the well-developed tube; corolla to about 2 cm long; anthers glabrous; (B.C. to Man.) . . . . *O. ludoviciana*

      5 Inflorescence corymbose, the flowers (especially the lower ones) distinctly pedicelled, the pedicels to about 3 cm long; calyx-lobes several times longer than the short tube; corolla to 3 cm long; anthers woolly; (S B.C.).

        6 Lower lip of corolla continuous with the line of the tube (or slightly arched toward tip), at most 6 mm long . . . . [*O. californica*]

        6 Lower lip of corolla strongly spreading, to 1.5 cm long . . . . *O. grayana*

[O. californica C. & S.]
[The report of this species of the W U.S.A. (Wash. and Mont. to S Calif. and Utah) from B.C. by Rydberg (1922; taken up by Hitchcock et al. 1959) requires confirmation. (*Myzorrhiza (O.) corymbosa* Rydb.).]

O. fasciculata Nutt. Clustered Broom-rape
/sT/(X)/ (Gp (root-parasite)) Parasitic on a variety of hosts (particularly *Artemisia* and *Eriogonum*) in dry plains and prairies from S Alaska, S-cent. Yukon (N to ca. 62°30′N), and B.C.–Alta. to Sask. (N to Saskatoon), S Man. (N to Neepawa, about 30 mi NE of Brandon), and S Ont. (Cloche Is., N L. Huron; OAC), S to Calif., N Mexico, Nebr., and Ind. [*Aphyllon* Gray; *Anoplanthus* Walp.; *Thalesia* Britt.; *T. lutea* (Parry) Rydb.]. MAP: Hultén 1968b:828; (Hultén (1949) also notes a map in Pflanzenareale 1:7. 1927).

Forma *lutea* (Parry) Beck (the flowers, and whole plant, yellowish rather than purplish) is reported from SW Alta. by D.M. Achey (Bull. Torrey Bot. Club 60(6):449. 1933; Banff and Rosedale).

O. grayana Beck
/t/W/ (Gp (root-parasite)) Meadows and open slopes, parasitic chiefly on various species of Compositae, from SW B.C. (Vancouver Is. and adjacent islands and mainland) to Baja Calif. and Nev. [*O. (Aphyllon) comosa* Hook.].

O. ludoviciana Nutt.

/T/WW/ (Gp (root-parasite)) Dry prairies and plains, parasitic on Compositae (particularly *Ambrosia* and *Artemisia*), from s B.C. (Vancouver Is.; Okanagan; Yoho) to s Alta. (Milk River; Hand Hills; Medicine Hat), s Sask. (N to Wakaw, about 50 mi NE of Saskatoon), and sw Man. (N to St. Lazare, about 75 mi NW of Brandon), s to s Calif., Mexico, Tex., and Ind. [*Aphyllon* Gray; *Myzorrhiza* Rydb.].

Forma *albinea* Boivin (flowers whitish rather than purplish) is known from the type locality, Val Marie, Sask.

O. pinorum Geyer

/t/W/ (Gp (root-parasite)) Mostly in coniferous woods, where parasitic on various conifers, from s B.C. (Cowichan, Vancouver Is.; V; reported from Okanagan by Henry 1915) to NW Calif. and Idaho.

[O. purpurea Jacq.]

[Eurasian; locally introd. into grasslands in N. America, as in s Ont. (Wingham, Huron Co., where taken by J.A. Morton in 1895 but apparently not established, no later collections being known; TRT).]

O. uniflora L.   One-flowered Cancer-root

/T/X/ (Gp (root-parasite)) Parasitic on various plants in damp woods and thickets, the aggregate species from s B.C. (evidently confined to s of 50°N; reports from Alaska (and probably the Yukon) are based upon *O. fasciculata,* according to Hultén 1968a) to sw Alta. (Waterton Lakes; Banff), Sask. (Boivin 1966b; not known from Man.), Ont. (N to Georgian Bay, L. Huron, and Ottawa), Que. (N to Anticosti Is. and the Gaspé Pen.), Nfld., N.B., P.E.I., and N.S., s to s Calif., Tex., and Fla. (*See* D.M. Achey, Bull. Torrey Bot. Club 60(6):442–47. 1933).

1   Calyx-lobes narrowly lanceolate, tapering gradually from base to apex, only slightly longer
    than the tube; [*Anoplanthus* Endl.; *Aphyllon* Gray; *Thalesia* Britt.; *O. (Anoplon;*
    Phelipaea) biflora Nutt.; incl. *O. terrae-novae* Fern.; Ont. to Nfld. and N.S.] . . . . var. *uniflora*
1   Calyx-lobes narrowly subulate from a broad base, to about twice as long as the tube;
    [western taxa].
    2   Anthers usually more or less woolly; corolla purple, to 3.5 cm long, the throat strongly
        flaring, to 8 mm broad at the throat; [*Thalesia purp.* Heller, not *O. purp.* Jacq.; s B.C.:
        Jardo; Manning Provincial Park, SE of Hope] . . . . var. *purpurea* (Heller) Achey
    2   Anthers glabrous; corolla to about 2.5 cm long, the tube at most about 5 mm broad at
        the throat.
        3   Corolla purple; [*Aphyllon (Thalesia) min.* Suksd.; s B.C. (Vancouver Is., Selkirk
            Mts., and Trail; Achey, loc. cit., noting it as parasitic on Saxifragaceae) and sw
            Alta. (Waterton Lakes; Banff)] . . . . var. *minuta* (Suksd.) Beck
        3   Corolla yellow or tinged with lavender; [*Aphyllon (O.; Thalesia) sedi* Suksd.; s
            B.C. (Mt. Finlayson, Vancouver Is., and Sproat, s of Revelstoke; Achey, loc. cit.,
            noting it as parasitic on *Sedum* and Compositae)] . . . . var. *sedi* (Suksd.) Achey

LENTIBULARIACEAE (Bladderwort Family)

Small herbs with entire or filiform-dissected leaves, the plants insectivorous by greasy-viscid leaves in *Pinguicula* or bladder-traps in *Utricularia*. Flowers zygomorphic (bilaterally symmetrical), gamopetalous, perfect, hypogynous, solitary on naked scapes or in bracted racemes. Corolla 2-lipped, 5-lobed, the lower 3-lobed lip with a prominent (usually bearded) palate and spurred at base. Stamens 2. Ovary superior. Fruit a 1-locular capsule.

1 Flowers violet, solitary on naked scapes; upper lip of calyx deeply 3-cleft, the lower lip
   2-cleft; corolla open at throat, the lobes spreading; leaves elliptic to ovate, entire,
   greasy-viscid (thus trapping insects), borne in a basal rosette; (transcontinental) . . . . *Pinguicula*
1 Flowers yellow (sometimes purple), commonly racemose on bracted scapes; calyx parted
   to base into 2 entire segments (upper segment often the broader); corolla-throat closed by
   the palate, the upper lip usually ascending; leaves linear-filiform and simple or (more
   commonly) finely dissected into linear bladder-bearing segments, submersed or buried in
   mud . . . . *Utricularia*

PINGUICULA L.   [7898]   Butterwort.   Grassette

1 Scapes densely hairy, commonly about 2 or 3 cm tall; corolla pale violet, about 1 cm long . . . .
   . . . . *P. villosa*
1 Scapes glabrous, commonly 1 dm tall or more; corolla violet-purple, to 2 cm long . . . . . *P. vulgaris*

P. villosa L.
/aST/X/EA/   (Hr)   Damp mossy tundra and peats (often on or between hummocks) from Alaska (N to ca. 69°N), the Yukon (N to ca. 68°N), and the coast of NW Dist. Mackenzie to Great Bear L., Great Slave L., S-cent. Dist. Keewatin, Ont. (N to W Hudson Bay at ca. 56°N), Que. (N to S Ungava Bay; not known from E Que. or the Atlantic Provinces), and Labrador (N to Hebron, 58°12′N; Hustich and Pettersson 1943), S to B.C. (Queen Charlotte Is.; CAN; a sterile collection in CAN from Mt. Arrowsmith, Vancouver Is., may also belong here), L. Athabasca (Alta. and Sask.), Man. (S to Gillam, about 165 mi S of Churchill), cent. Que. (L. Mistassini), and Labrador (S to Indian Harbour, 54°27′N; CAN); Eurasia. [*P. acutifolia* Michx.; *P. involuta* Schrank; *P. ?alpina sensu* Gray 1886, not L.]. MAPS: Hultén 1968*b*:830; Raup 1947: pl. 34 (the occurrence in N Man.–Ont. should be indicated).

P. vulgaris L.   Common Butterwort
/aST/X/GEA/   (Hr)   Wet rocks and moist places (chiefly in calcareous areas) from the Aleutian Is., Alaska (N to ca. 69°N), the Yukon (N to ca. 68°N), and the coast of Dist. Mackenzie to S Baffin Is. and N Ungava–Labrador (N to ca. 60°30′N), S through B.C. and SW Alta. (Waterton Lakes; Breitung 1957*b*) to Oreg., Mont., Sask. (Hasbala L., ca. 59°30′N; Prince Albert; Indian Head), Man. (S to Gilbert Plains, N of Riding Mt.; J.L. Parker, Can. Field-Nat. 76(2):125. 1962; CAN), Ont., N Minn., N Mich., Que. (S to S James Bay, L. Mistassini, the Côte-Nord, Anticosti Is., and Gaspé Pen.), Nfld., N N.B. (Restigouche Co.; not known from P.E.I.), N.S. (Inverness Co., Cape Breton Is., and St. Paul Is.), N.Y., and Vt.; W and E Greenland N to ca. 74°N; Iceland; Eurasia. MAPS: Hultén 1968*b*:829, and 1958: map 211, p. 231; Porsild 1957: map 299, p. 198.

   Some of our western material is referable to var. *macroceras* (Link) Herder (*P. mac.* Link; *P. arctica* Eastw.; *P. microceras* Cham.; flowers relatively large, the corolla-lobes often overlapping, the relatively long spur blunt rather than acute).

UTRICULARIA L.   [7901]   Bladderwort

1 Leaves simple or slightly forking, they and the minute bladder-traps mostly hidden beneath
   the surface of the moist soil on slender basal branches; scapes filiform.
   2 Bract at base of pedicel centrally peltate; traps borne on separate branches; flowers
      long-pedicelled, whitish, yellow, or purplish, the blunt spur appressed; capsule much
      surpassing the calyx; (N.S.) . . . . *U. subulata*
   2 Bract at base of pedicel basally attached; traps borne on leafy branches; flowers with a
      divergent spur.

   3   Flowers purple, solitary, with a short curved spur, the erect pedicel subtended by a truncate or notched cup-like bract; sepals all obtuse; capsule exserted; (Ont. to N.S.) . . . . *U. resupinata*

   3   Flowers often 3 or more, with a subulate spur, subsessile or very short-pedicelled, each subtended by a free bract and 2 smaller bractlets; longer sepal acuminate; capsule not exserted; (Ont. to Labrador, Nfld., and N.S.) . . . . *U. cornuta*

1 Leaves more or less copiously dissected into elongate segments; plants aquatic or amphibious.

   4   Leaves all whorled and uniform, long-petioled, the beakless non-flagellate traps borne at the tips of the segments; corolla purple, the lower lip about twice as long as the appressed spur; (Ont. to N.S.) . . . . *U. purpurea*

   4   Leaves mostly alternate, sessile or short-petioled, the beaked traps borne laterally on the segments or on separate branches, with long flagellae projecting from the orifice; flowers yellow or yellowish.

      5 Scape bearing a whorl of leaves with inflated petioles; submersed leaves 4–6-forked into capillary segments; (N.S.) . . . . *U. inflata*

      5 Scape naked or merely with small scarious bracts; submersed leaves less divided.

         6 Leaf-segments flat; (transcontinental).

            7 Traps borne on the leaves; terminal leaf-segments with entire margins . . . . *U. minor*

            7 Traps borne on separate elongate branches; terminal leaf-segments minutely serrulate.

               8 Traps borne on leafless branches . . . . *U. intermedia*

               8 Traps borne on branches with small leaves subtending many of the pedicels . . . . [*U. ochroleuca*]

         6 Leaf-segments terete or capillary; traps not on separate branches.

            9 Plant bearing small apetalous cleistogamous flowers in addition to the normal ones; lower lip of corolla somewhat longer than the thick blunt spur; (Que., Nfld., N.B., and N.S.) . . . . *U. geminiscapa*

            9 Plants lacking cleistogamous flowers.

              10 Stems at least 0.5 mm thick, free-floating; scape stout, usually with at least 6 flowers; spur slightly longer than the lower lip of the corolla; fruiting pedicels arched-recurving; (transcontinental) . . . . *U. vulgaris*

              10 Stems less than 0.5 mm thick, creeping; scape filiform, 1–6-flowered; fruiting pedicels erect.

                  11 Mature leaves mostly with only 2 segments; corolla at most 12 mm long, its spur shorter than the lower lip; (s Ont. to N.B. and N.S.) . . . . *U. gibba*

                  11 Mature leaves with at least 3 segments; corolla to 17 mm long, its spur one-half to nearly as long as the lower lip; (s ?Ont.) . . . . . [*U. biflora*]

[*U. biflora* Lam.]
[The listing of this U.S.A. species (N to Okla. and s New Eng.) for s Ont. by Soper (1949) is probably based upon a collection in OAC from Puslinch, Wellington Co., referable, according to Bernard Boivin (personal communication) to *U. intermedia*.]

U. cornuta Michx.
/sT/EE/ (Hel (HH)) Bogs and muddy or sandy shores from ?Man. (the report from Riding Mt. by Lowe 1943, requires confirmation) to Ont. (N to the Attawapiskat R. at 53°44′N), Que. (N to the E James Bay watershed at 53°40′N, L. Mistassini, the Côte-Nord, Anticosti Is., and Gaspé Pen.), Labrador (N to Goose Bay, 53°20′N; DAO; RIM), Nfld., N.B., P.E.I., and N.S., s to E Tex., Minn., Ohio, Pa., Del., and Fla. [*Stomoisia* Raf.]. MAP: McLaughlin 1932: fig. 18 (incomplete northwards), p. 349.

U. geminiscapa Benj.
/T/EE/ (HH) Ponds and sluggish streams from Wisc. and Mich. to Que. (Pontiac, Missisquoi, and Kamouraska counties and Magdalen Is.; not known from Ont.), Nfld. (CAN; GH), N.B. (Kent

Co. and Grand Manan Is.; not known from P.E.I.), and N.S., s to N.Y. and Va. [*U. clandestina* Nutt.]. MAP: Fernald 1933: map 6 (somewhat incomplete northwards), p. 85.

U. gibba L.
/T/X/ (HH) Shallow ponds and quaking bogs from Minn. to Ont. (collections in CAN from the Muskoka Lakes dist. E of Georgian Bay, L. Huron; the report from Man. noted by Lowe 1943, requires confirmation), Que. (N to St-Tite, about 30 mi NE of Quebec City; MT), N.B. (Grand Manan Is.; GH; not known from P.E.I.), and N.S., s to Calif., Mexico, Tex., Okla., and Fla.; W.I.; Central America.

U. inflata Walt.
/T/EE/ (HH) Ditches, ponds, and sluggish streams from Ind. to Pa. and N.S. (Yarmouth, Queens, Lunenburg, and Halifax counties; ACAD; CAN; GH; s to Tex. and Fla.; S. America.
    The N.S. plant is referable to var. *minor* Chapm. (*U. radiata* Small; the whole plant reduced and few-flowered, the flowers smaller and the pedicels shorter than those of the typical form).

U. intermedia Hayne
/aST/X/GEA/ (HH) Shallow ponds and sluggish streams from N Alaska–Yukon–Dist. Mackenzie. (N to ca. 69°N) to Great Bear L., Great Slave L., s Dist. Keewatin, northernmost Man.–Ont., Que. (N to E James Bay at 54°25'N, L. Mistassini, the Côte-Nord, Anticosti Is., and Gaspé Pen.), Labrador (Goose Bay, 53°20'N), Nfld., N.B. (not known from P.E.I.), and N.S., s to Calif., Iowa, Ohio, and Del.; W Greenland at ca. 66°N; Iceland; Eurasia. MAP: Hultén 1968*b*:831.

U. minor L.
/aST/X/GEA/ (HH) Shallow pools, wet meadows, bogs, and shores from the E Aleutian Is. and Alaska (N to ca. 68°30'N) to the Yukon (N to ca. 64°N), Great Bear L., Great Slave L., L. Athabasca (Sask.), Man. (N to Churchill), Ont. (N to Hawley L., 54°34'N), Que. (N to Ungava Bay, the Côte-Nord, Anticosti Is., and Gaspé Pen.), Labrador (N to Makkovik, 55°05'N; Hustich and Pettersson 1943), Nfld., P.E.I. (Mt. Stewart and Watervale, Queens Co.; CAN; PEI; not known from N.B.), and N.S., s to Calif., Colo., N.Dak., Pa., and N.J.; W Greenland N to ca. 71°N; Iceland; Eurasia. MAP: Hultén 1968*b*:831.

[U. ochroleuca Hartm.]
[Usually regarded as a hybrid between *U. intermedia* and *U. minor* and occurring essentially throughout the range of those species.]

U. purpurea Walt.
/T/EE/ (HH) Ponds and sluggish streams from Wisc. and Mich. to Ont. (N to the Sudbury dist.; TRT), Que. (N to Mont-Laurier, about 80 mi N of Hull), N Nfld. (near Colinet, Avalon Pen.; Hilda Smith, Can. Field-Nat. 80(3):182. 1966), N.B. (Grand Manan Is. and near St. Andrews, Charlotte Co.; CAN; GH; not known from P.E.I.), and N.S., s to La. and Fla.; W.I.; Central America. [*Vesiculina* Raf.].

U. resupinata Greene
/T/EE/ (HH (Hel)) Local in shallow waters and along shores from Wisc. to Ont. (N to near Thunder Bay and Sault Ste. Marie), Que. (N to Timiskaming and Nominingue), N.B. (Phipp's L., Kings Co., where taken by Livingstone in 1886; CAN; not known from P.E.I.), and N.S., s to Ill., Pa., and Fla. [*Lenticula* Barnh.]. MAP: McLaughlin 1932: fig. 17 (incomplete northwards), p. 349.

U. subulata L.
/T/EE/ (Hel) Wet peats, sands, and shores from W N.S. (Annapolis, Digby, Yarmouth, Shelburne, and Queens counties; the listing for s Ont. by Soper 1949, presumably based upon a collection in TRT from Long Point, Norfolk Co., requires confirmation) to Fla., Ark., and Tex.; W.I. [*Setiscapella* Barnh.]. MAPS: Fernald 1921: map 4, pl. 130, facing p. 120, and 1929: map 25, p. 1499.
    Forma *cleistogama* (Gray) Fern. (flowers usually whitish and only 1 or 2 mm long rather than yellow and up to 12 mm long) is known from N.S. (Fernald 1921).

U. vulgaris L.   Common Bladderwort

/ST/X/EA/   (HH)   Deep or shallow quiet waters from Alaska (N to ca. 70°N), the Yukon (N to ca. 64°30′N), and NW Dist. Mackenzie to Great Bear L., Great Slave L., L. Athabasca (Alta. and Sask.), Man. (N to Churchill), Ont. (N to Hawley L., 54°34′N), Que. (N to the Wiachouan R. at 56°10′N, L. Mistassini, the Côte-Nord, Anticosti Is., and Gaspé Pen.), Labrador (N to the Hamilton R. basin), Nfld., N.B., and N.S. (not known from P.E.I.), S to Calif., Mexico, Tex., Mo., and Va.; Eurasia. [Incl. var. *americana* Gray (*U. macrorhiza* Le Conte), the corolla-spur somewhat more slender and pointed than that of the typical form, to which the N. American plant is sometimes referred]. MAP: Hultén 1968b:830 (*U. vulg.* ssp. *mac.*).

ACANTHACEAE (Acanthus Family)

JUSTICIA L.   [8094]   Water-willow.   Dianthère

Subaquatic glabrous perennial herb from a stout base with numerous cord-like stolons and rhizomes, the usually simple stem to about 1 m tall. Leaves linear to narrowly lanceolate or narrowly oblong, simple, entire, opposite, to about 1.5 dm long and 2.5 cm broad, tapering about equally to the blunt apex and the sessile or short-petioled cuneate base. Flowers gamopetalous, perfect, hypogynous, opposite in dense axillary long-peduncled short spikes or heads. Calyx nearly regular, deeply 5-cleft. Corolla 2-lipped (the upper lip erect or ascending, concave, entire or emarginate, the lower lip deflexed or spreading, 3-lobed), the lips about equalling the tube, pale violet to nearly white, marked with purple at the base of the lower lip. Stamens 2, the terminal anther-sac horizontal. Style 1. Ovary superior. Fruit a short-stalked capsule with usually 4 warty-rugose seeds.

J. americana (L.) Vahl   Water-willow
/T/EE/   (Hel)   Shallow water and muddy shores from Kans. to Mo., Wisc., s Ont. (Pelee Point, Essex Co., and Dufferin Is., Niagara Falls, Welland Co.), sw Que. (Montreal dist.; *see* s Que. map by Robert Joyal, Nat. can. (Que.) 97(5): map H, fig. 2, p. 564. 1970), N.Y., and Vt., s to Tex. and Ga. [*Dianthera* L.; *Dicliptera* Wood; *J. pedunculosa* Michx.].

PHRYMACEAE (Lopseed Family)

PHRYMA L.   [8115]   Lopseed

Perennial glabrous or slightly pubescent herb to about 1 m tall. Leaves simple, opposite, ovate, coarsely serrate, to about 1.5 dm long, the lower ones on petioles to 5 cm long, the upper ones commonly sessile. Flowers gamopetalous, perfect, hypogynous, white to pale purple, about 8 mm long, opposite and horizontal in long-peduncled, interrupted, terminal and axillary spike-like racemes. Calyx zygomorphic, the 3 upper subulate lobes about equalling the tube, hooked at tip, the 2 lower broadly triangular lobes very short. Corolla cylindric, the upper lip erect and emarginate, the lower lip much longer, spreading, 3-lobed. Stamens 4, didynamous (2 long, 2 short), inserted on the corolla-tube. Style 1. Ovary superior. Fruit a dry 1-seeded achene.

P. leptostachya L.   Lopseed
/T/EE/   (Hp (Hpr))   Rich woods and thickets from s Man. (Portage la Prairie, Graysville, Carman, and Morden, sw of Winnipeg) to Ont. (N to Ottawa), Que. (N to l'Ange-Gardien, about 10 mi NE of Quebec City; *see* Que. map by Doyon and Lavoie 1966: fig. 13, p. 818), and N.B. (St. John R. system; not known from P.E.I. or N.S.), s to E Tex. and Fla.

PLANTAGINACEAE (Plantain Family)

Herbs with entire or subentire leaves, these commonly in a basal rosette. Flowers small, whitish or pale, gamopetalous, hypogynous, commonly regular and 4-merous. Stamens usually 4 (sometimes 2). Ovary superior. Fruit a usually circumscissile capsule (the top falling off like a lid) or an achene.

1 Flowers unisexual, the staminate ones solitary at the top of naked or 1-bracted scapes to about 4 cm long; pistillate flowers usually 2, sessile at the base of the scape; fruit a single blackish achene about 2 mm long; leaves linear-subulate, arching, to about 6 cm long, in a basal rosette; plant abundantly stoloniferous and forming turf in shallow water; (Ont. to Nfld. and N.S.) .... *Littorella*
1 Flowers mostly perfect, several or many in a spike at the top of the scape (or terminating numerous peduncles in the leafy-stemmed *P. psyllium*); fruit a usually circumscissile capsule with at least 2 seeds; plants terrestrial, nonstoloniferous .... *Plantago*

LITTORELLA Bergius [8117]

L. uniflora (L.) Aschers.   Shore-weed
/T/EE/E/   (Hel)   Sandy, gravelly, or muddy shores and margins of ponds and lakes from N Minn. to Ont. (Gull L., Peterborough Co.; L. Nipissing; shore of L. Superior near Thunder Bay), Que. (N to the Côte-Nord and Gaspé Pen.; not known from Anticosti Is.), Nfld., N.B. (L. Utopia, Charlotte Co.; not known from P.E.I.), N.S., N.Y., Maine, and Vt.; Europe. [*L. lacustris* var. *uni.* L.].

The N. American plant may be distinguished as var. *americana* (Fern.) Gl. (*L. amer.* Fern.; achenes blackish, smooth or barely rugulose, rather than pale brown and coarsely rugose; calyces at most 4 mm long rather than to 7 mm, their lobes oblong rather than lanceolate; anthers relatively small, their filaments at most 12 mm long rather than to 4 cm; peduncles of staminate flowers at most about 4 cm long rather than to 6 cm; leaves flattish, to about 6 cm long, rather than subterete and to 1.5 dm long).

PLANTAGO L.   [8116]   Plantain, Ribgrass, Ribwort.   Plantain or Queue de rat

1 Stems elongate, leafy; leaves opposite, linear-attenuate, chartaceous, to about 1 dm long, the lower ones with short very leafy shoots in their axils, the upper ones subtending stiff peduncles to about 8 cm long; spikes 1 or 2 cm long, ellipsoid or subglobose, their lower bracts with prolonged green tips; annual; (introd.) .... *P. psyllium*
1 Stems very short, the leaves forming a basal rosette.
　2 Leaves mostly deeply pinnatifid and 1-nerved, linear, to about 6 cm long; bracts ovate, often long-acuminate and with spreading tips, sometimes obtuse and appressed; biennial, usually pubescent; (introd.) .... *P. coronopus*
　2 Leaves entire to shallowly lobed but scarcely pinnatifid.
　　3 Leaves narrowly to broadly linear, the blade obscurely differentiated from the petiole.
　　　4 Bracts of spike very conspicuous, awn-tipped and much exserted from the spike; corolla-tube glabrous; stamens 4; seeds normally 2; leaves glabrous or loosely villous; annual; (B.C. to S Man.) .... *P. patagonica*
　　　4 Bracts shorter than the flowers or the lower ones somewhat exserted from the spike.
　　　　5 Plants perennial from a deep root and 1 to many crowns; leaves thick and fleshy, obscurely ribbed; corolla-tube pilose, its lobes to 1.5 mm long; stamens 4; seeds 2–6, plump; plants of alkaline or saline flats, brackish shores, and coastal rocks; (transcontinental) .... *P. maritima*
　　　　5 Plants annual, from a slender well-developed taproot; corolla-tube glabrous.
　　　　　6 Leaves and spikes more or less densely white-woolly (becoming tawny); flowers uniformly fertile; corolla-lobes to 2 mm long; stamens 4; seeds 2; (B.C. to S Man.) .... *P. patagonica*
　　　　　6 Leaves and spikes glabrous or the leaves merely rough-puberulent;

flowers both fertile and sterile, some with reduced stamens, others with reduced pistils; corolla-lobes to 1 mm long; stamens 2; seeds mostly at least 4.

    7   Corolla-lobes mostly erect in age and forming a beak; seeds 4, to 1.8 mm long .... [*P. pusilla*]

    7   Corolla-lobes spreading or reflexed in age, rarely forming a beak; seeds often more numerous.

        8   Scape and leaves mostly erect, the plants to 1.5 dm tall; seeds commonly 4 or 5, rugose-pitted, dark brown, to 2.5 mm long; (s B.C. to sw Man.) .... *P. elongata*

        8   Scapes and leaves mostly decumbent to semierect, less than 1 dm tall; seeds up to 9 or more, irregularly and coarsely pitted, dark brown to black, to 2 mm long; (sw B.C.) .... *P. bigelovii*

3   Leaves broader, the blade lanceolate or oblanceolate to elliptic or ovate, well differentiated from the petiole; corolla-tube glabrous; stamens 4.

    9   Flowers both fertile and sterile, the fertile ones with anthers included and with the corolla closed over the maturing capsule and forming a beak, the sterile ones with exserted anthers and spreading corolla-lobes; seeds 2; plant annual or biennial, permanently hoary-villous throughout with a pubescence of long multicellular hairs; (introd.) .... *P. virginica*

    9   Flowers uniformly fertile, the corolla-lobes spreading or reflexed; plants commonly perennial (sometimes also annual), the pubescence less obvious (plants sometimes villous at base).

      10 Leaves relatively narrow, the blade narrowly to broadly lanceolate, elliptic, or oblanceolate, mostly at least 5 times as long as broad, entire or remotely denticulate.

        11 Capsules 6 or 7 mm long, indehiscent and falling entire; seeds 2 in number, 4 or 5 mm long; corolla-lobes to 2 mm long; leaves (including petiole) to 4.5 dm long and 3.5 cm broad; plant essentially glabrous throughout; (s Alaska–w B.C.) .... *P. macrocarpa*

        11 Capsules at most 3 or 4 mm long, circumscissile below the middle; seeds about 2 mm long.

          12 Outer sepals (the two adjacent to the bract) united; bracts acuminate or caudate-acuminate, their tips often exserted; corolla-lobes 2 or 3 mm long; seeds commonly 2; leaves to about 4 dm long, villous to glabrate; crown (base of plant) more or less densely tan-woolly; (introd.) .... *P. lanceolata*

          12 Sepals all free; bracts obtuse to acute; seeds commonly 3 or 4; leaves at most about 2 dm long; crown less conspicuously woolly; (Alaska to w Dist. Mackenzie; mts. of sw Alta.) .... *P. canescens*

      10 Leaves relatively broad, mostly narrowly to broadly ovate, entire to undulate or coarsely dentate.

        13 Seeds commonly more than 4 (up to 30), smooth, plump; corolla-lobes at most 1 mm long; scape solid, not brown-woolly at base.

          14 Capsule circumscissile near base; seeds less than 10; leaves thin; (Ont. to N.S.) .... *P. rugelii*

          14 Capsule circumscissile near middle, with up to 30 reticulate seeds, these less than 2 mm long; (transcontinental, partly or wholly introd.) .... *P. major*

        13 Seeds at most 4 in number; corolla-lobes over 1 mm long.

          15 Scape hollow; spike loosely flowered; capsule commonly 5 or 6 mm long, circumscissile near the middle; seeds plump; (s Ont. and s ?Que.) .... *P. cordata*

          15 Scape solid; capsule commonly 3 or 4 mm long.

            16 Capsule circumscissile near middle; seeds flat or slightly concave on the inner face; spike dense, to about 1 dm long; plant not brown-woolly at base; (introd.) .... *P. media*

16 Capsule circumscissile near base; seeds plump, shining,
reddish brown; spike interrupted toward base, to about 2 dm
long; plant copiously and generally conspicuously brown-woolly
at base; (B.C. to Man.; E Que.) .... *P. eriopoda*

P. bigelovii Gray
/t/W/ (T) Wet rocky coastal bluffs and brackish shores from SW B.C. (Vancouver Is. and
adjacent islands) to Baja Calif., NW Mexico, and Ariz. [*P. elongata* of B.C. reports in part, not
Pursh]. MAP: I.J. Bassett, Can. J. Bot. 44(4): fig. 3, p. 473. 1966.

P. canescens Adams
/aST/W/A/ (Hr) Grassy or gravelly slopes at low to moderate elevations from Alaska (N to ca.
69°N), the Yukon (N to ca. 67°N), and the coast of Dist. Mackenzie to S Banks Is. and N Victoria Is.;
isolated in the mts. of SW Alta. (W of Pincher Creek; near Jasper) and Mont.; Asia. [Incl. vars.
cylindrica (J.M. Macoun) Boivin (*P. eriopoda* var. *cyl.* Macoun) and *glabrata* Pilger; *P. richardsonii*
Dcne.; *P. septata* Morris]. MAPS: Hultén 1968*b*:833; I.J. Bassett, Can. J. Bot. 45(5): fig. 6, p. 572.
1967; Porsild 1957: map 301 (*P. sept.*), p. 198.

P. cordata Lam.
/t/EE/ (Hr) Swampy woods, margins of streams, and ditches from Mo. to Wisc., Mich., S Ont.
(collections in CAN from near Amherstburg, Essex Co., where taken by John Macoun in 1882; near
Lucan, Middlesex Co., where taken by Dearness in 1894; and near Thedford, Lambton Co., where
taken by Voss in 1967; collection in TRT from Chatham, Kent Co.; the report from Beauharnois,
near Montreal, Que., by R. Campbell, Can. Rec. Sci. 6(6):342–51. 1895, requires confirmation), and
N.Y., S to La., Ala., and N Fla. MAP: I.J. Bassett, Can. J. Bot. 45(5): fig. 1, p. 568. 1967.

P. coronopus L. Buck's-horn Plantain, Crowfoot
Eurasian; locally introd. in N. America, as in SW B.C. (Ladysmith, Vancouver Is.; Herb. V) and S
Man. (Brandon; G.A. Stevenson, Can. Field-Nat. 79(3):176. 1965); W-cent. Greenland.

P. elongata Pursh
/T/WW/ (T) Dry to moist alkaline places from S B.C. (Vancouver Is. and adjacent islands and
mainland N to Tranquille and Kamloops) to S Alta. (N to near Red Deer), S Sask. (N to Nokomis,
51°30′N), and SW Man. (Melita; Miniota; Brenda), S to Calif. and Tex. MAP: I.J. Bassett, Can. J. Bot.
44(4): fig. 1, p. 469. 1966.
   Some of the material from SW B.C. (Vancouver Is. and adjacent islands) is referable to ssp.
pentasperma Bassett (capsules mostly 5-seeded rather than 4-seeded, one of the seeds smaller
than the others and irregular in shape; spikes relatively densely flowered; leaves and scapes
usually essentially glabrous rather than pubescent with long multicellular hairs). MAP: Bassett, loc.
cit., fig. 2, p. 473.

P. eriopoda Torr.
/aST/(X)/ (Hr) Saline or alkaline soils from E Alaska (N to ca. 63°N), S Yukon, and the coast of
Dist. Mackenzie to B.C.–Alta.–Sask. and Man. (N to Norway House, off the NE end of L. Winnipeg),
S to Calif., Mexico, and Nebr., with an isolated area in E Que. (St. Lawrence R. estuary from
St-Roch-des-Aulnets, l'Islet Co., to Anticosti Is. and the Gaspé Pen.). [*P. ?tweedyi* of Alta.–Sask.
reports, not Gray]. MAPS: Hultén 1968*b*:834; I.J. Bassett, Can. J. Bot. 45(5): fig. 3, p. 568. 1967;
Porsild 1966: map 131, p. 83.

P. lanceolata L. Ribgrass, Ripplegrass, Buckhorn, English plantain
Eurasian; a common weed of grasslands, fields, and waste places in N. America, known from S
Alaska (N to ca. 61°N) and all the provinces (in Sask., N to Waskesiu Lake, ca. 54°N). MAP: Hultén
1968*b*:835.
   Forma *composita* Farw. (a common monstrosity, the spike very compound, much branched and
lobed) probably occurs throughout the area and has been taken in Que. (Montreal and Quebec City
districts; MT). Plants with strongly villous leaves may be distinguished as var. *angustifolia* Poir.
Some of our material may also be separated as var. *sphaerostachya* Mert. & Koch (spikes ovoid or

subglobose, less than 2.5 cm long, rather than cylindric in fruit and up to over 8 cm long; perhaps merely the reflection of a sterile habitat). Of this phase, f. *eriophora* (Hoffmgg. & Link) Beck (upper leaf-surface copiously greyish-pubescent rather than glabrous or only sparingly pubescent) is known from Que., Nfld., and N.S. and f. *vernalis* Béguinot (leaves essentially glabrous as in the typical form but elliptic rather than lanceolate) is known from Que. and Nfld.

P. macrocarpa C. & S.

/sT/W/eA/ (Hr) Sphagnum bogs and wet places near the coast from the Aleutian Is. (type from Unalaska) and s Alaska (N to ca. 60°30'N) through w B.C. to Oreg.; E Asia (Commander Is.; ?Karaginsk Is.). MAPS: Hultén 1968*b*:832; I.J. Bassett, Can. J. Bot. 45(5): fig. 5, p. 572. 1967.

P. major L. Common Plantain, Whiteman's-foot. Grand Plantain

Eurasian; a very common weed of roadsides, dooryards, and other waste places in N. America (some taxa considered native by Fernald *in* Gray 1950), as in Alaska (N to ca. 68°N), the Yukon (N to ca. 64°N), Dist. Mackenzie (N to Norman Wells, 65°17'N; W.J. Cody, Can. Field-Nat. 74(2):96. 1960), and all the provinces (in Sask., N to L. Athabasca; in Man., N to Churchill); w Greenland. MAPS and synonymy: *see* below.

1 Leaves relatively thin, essentially glabrous, tapering to slender ascending petioles; capsules conic above, circumscissile near the tips of the sepals; [var. *asiatica* of American auth., not *P. asiatica* L.; incl. var. *dumanii* Lepage; native, according to Fernald *in* Gray 1950, on "River-gravels, damp ledges, etc., Nfld. to n. B.C., s. to N.B., n. N.E., L. Sup., Ont., N.D. and Ariz. (Eu)"; MAP: Hultén 1968*b*:836] . . . . var. *pilgeri* Domin
1 Leaves relatively thick, more or less pubescent, tapering or broadly rounded to a relatively short broad petiole.
  2 Leaves decumbent or slightly ascending; scape arched-ascending; capsules with rather broadly rounded summits; [incl. var. *ungavensis* Lepage; ssp. *pleiosperma* Pilger; *P. halophila* Bickn.; native, according to Fernald *in* Gray 1950, on "Brackish shores, rarely inland, P.E.I. and C.B. to Del.; James Bay; Wash. to Calif. . . . (Eu.)"] . . . .
. . . . var. *scopulorum* Fries & Broberg
  2 Leaves and scape ascending to erect; capsules broadly conic at summit; [transcontinental, introd.] . . . . var. *major*
    3 Summit of scape below inflorescence bearing a rosette of small broad-bladed leaves; [monstr. *bracteata* (Moench) Pilger (*P. bracteata* Moench); s Ont. (Pelee Point, Essex Co.; Kaladar, Lennox-Addington Co.) and sw Que. (Missisquoi, Chambly, Labelle, and Charlevoix counties)] . . . . f. *rosea* (Dcne.) Prahl
    3 Summit of scape naked.
      4 Spike branching and panicle-like; [f. *?ramosa* Beckh.; SE Man. (Löve and Bernard 1959) and s Ont. (Stamford, Welland Co.)] . . . . f. *paniculata* Domin
      4 Spike normal, unbranched.
        5 Spike at most 3 cm long; leaf-blades at most 4 cm long; [var. *vulgaris* subvar. *mic.* (Hayne) Pilger; var. *?minima* Dcne.; SE Man.: Löve and Bernard 1959] . . : . . . . f. *microstachya* (Hayne) Pilger
        5 Spike to 5 dm long; leaf-blades mostly longer.
          6 Leaves essentially glabrous, smooth to the touch; [var. *?pachyphylla* Pilger (*P. nitrophila* Nels.); ssp. *eumajor* var. *intermedia* (Gilib.) Dcne.; SE Man. (Löve and Bernard 1959), E Que. (Gaspé Pen.; Anticosti Is.; Magdalen Is.), Nfld., N.B., P.E.I., and N.S.] . . . . . f. *intermedia* (Gilib.) Pilger
          6 Leaves roughish on one or both surfaces with minute hairs; [transcontinental, introd.; MAP: Hultén 1968*b*:835] . . . . f. *major*

P. maritima L. Seaside-Plantain

/aST/X/GEA/ (Hr) Salt marshes and coastal sands and ledges (inland around salt springs or saline marshes): Pacific coast from the Aleutian Is. and Alaska (N to ca. 65°N; isolated stations in the Mackenzie R. Delta and at Great Bear L.) through coastal B.C. to Calif. (isolated stations in saline soil along the Red Deer R., Alta., where taken by John Macoun in 1881, and in Wood Buffalo National Park, N Alta.; CAN; reported from Great Salt L., Utah); Man. (coast of Hudson Bay N to Churchill; salt springs at Dawson Bay, N L. Winnipegosis) to N Ont. (coasts of James Bay–Hudson

Bay N to ca. 56°50′N), Que. (coasts of James Bay–Hudson Bay N to ca. 55°N; Ungava Bay; St. Lawrence R. estuary from near Quebec City to the Côte-Nord, Anticosti Is., and Gaspé Pen.), S Baffin Is., northernmost Labrador, Nfld., N.B., P.E.I., N.S., and N.J.; S. America; W Greenland N to ca. 71°N, E Greenland N to ca. 65°30′N; Iceland; Eurasia. [*P. juncoides* Lam. and its var. *californica* Fern., var. *decipiens* (Barn.) Fern. (*P. dec.* Barn.), var. *glauca* (Hornem.) Fern. (*P. borealis* Lange and its f. *pygmaea* Lange), and var. *laurentiana* Fern.; incl. *P. oligosanthes* R. & S. (*P. ?pauciflora* Pursh) and its var. *fallax* Fern.; see M.L. Fernald, Rhodora 27(318):93–104. 1925]. MAPS: Hultén 1968*b*:833; Porsild 1957: map 300 (E area; as *P. junc.* var. *gl.*), p. 198; Potter 1932: map 8 (E area; as *P. junc.* var. *dec.*), p. 75.

Forma *vivipara* (Vict. & Rousseau) Boivin (most or all of the flowers replaced by bulblets) is known from the type locality, Berthier-en-Bas, Montmagny Co., Que.

P. media L.   Hoary Plantain.   Plantain bâtard
Eurasian; introd. into lawns, fields, and waste places in N. America, as in B.C. (Chilliwack; Revelstoke), S Man. (Brandon), Ont. (N to Ottawa), Que. (N to the Gaspé Pen. at Gaspé Basin; GH), N.B. (Bathurst; DAO), and ?N.S. (Cochran 1829). [*P. ?cucullata* Lam.]. MAP: I.J. Bassett, Can. J. Bot. 45(5): fig. 7, p. 572. 1967.

P. patagonica Jacq.   Patagonia Indian-wheat
/T/(X)/   (T)   Dry open places (ranges of Canadian taxa outlined below), S to Calif., Tex., La., and Ind. (natzd. eastwards to New Eng. and the Atlantic states and probably in part of the Canadian area, particularly at Halifax, N.S., where taken by Mackay in 1896 but apparently not since that date; CAN).
1   Bracts of the spike short, hidden by or barely projecting from the dense silky-villous pubescence; [var. *gnaphaloides* (Nutt.) Gray (*P. gnaph.* Nutt.); *P. purshii* R. & S.; apparently native in B.C. (N to Lillooet and Kamloops), S Alta. (Red Deer Lakes; Wood Mt.; Three Buttes; Cypress Hills), S Sask. (Cypress Hills; Skull Creek; Clearwater; Elbow; Saskatchewan Landing), and S Man. (Emerson, about 55 mi S of Winnipeg)] . . . var. *patagonica*
1   Bracts very conspicuous, awn-tipped and much exserted.
    2   Bracts smooth or sparingly pilose, the longer ones several times longer than the flowers (at least in larger plants), their short-pilose linear-attenuate awns to 5 cm long; [*P. aristata* Michx.; *P. purshii* var. *ar.* (Michx.) Jones; apparently introd. in the Yukon (Dawson; Porsild 1951*a*); B.C. (N to Revelstoke) and Alta. (Walsh; Manyberries); introd. eastwards, as in S Ont. (Essex, Lambton, Middlesex, Waterloo, Lincoln, and Carleton counties) and N.S. (Halifax); MAP (*P. aristata*): Hultén 1968*b*:834] . . . .
. . . . var. *aristata* (Michx.) Gray
    2   Bracts long-villous at base, less than 1.5 cm long, at most 4 times longer than the flowers, their stiff bristleform long-villous awns at most 1 cm long; [*P. spinulosa* Dcne.; *P. aristata* of Sask. reports, not Michx.; apparently native in S B.C., S Alta., and S Sask. (N to Saskatoon)] . . . . var. *spinulosa* (Dcne.) Gray

P. psyllium L.   Flaxseed Plantain
Eurasian; introd. along roadsides and in waste places in N. America, as in S B.C. (Locarno Park, Vancouver, and Sicamous, about 60 mi E of Kamloops; Eastham 1947), S Man. (Brandon; I.J. Bassett and C.W. Crompton, Can. J. Bot. 46(4):351. 1968), Ont. (N to Ottawa), SW Que. (Cowansville; Quebec City; Montreal dist.), and N.S. (Halifax; D.S. Erskine 1951). [*P. arenaria* Waldst. & Kit.; *P. indica* L. (*see* Hitchcock et al. 1959:443); *P. ramosa* Asch.].

[**P. pusilla Nutt.**]
[The Sask. citations of this species of the E U.S.A. (*see* MAP by I.J. Bassett, Can. J. Bot. 44(4): fig. 5, p. 476. 1966) by John Macoun (1884) are based upon *P. elongata,* the relevant collections in CAN, revised by E.L. Morris.]

P. rugelii Dcne.
/T/EE/   (Hr)   Damp shores, roadsides, and waste places (perhaps largely or possibly wholly introd. in our area) from Ont. (N to Ottawa; the report from Winnipeg, Man., by Lowe 1943, requires confirmation) to Que. (N to Cap-Rouge, near Quebec City, according to Groh 1946; reported from

Anticosti Is. by John Adams, Can. Field-Nat. 48(4):65. 1934), N.B. (Woodstock and St. John; CAN), ?P.E.I. (*see* D.S. Erskine 1960), and N.S., s to Tex. and Fla.

P. virginica L.   Hoary or Pale-seed-Plantain
Native in the U.S.A. (N to Oreg. and Maine); probably introd. in s Canada, as in s Ont. (reported by Dodge 1915, as occasional in poor and dry open ground in Lambton Co.; personal communication by Roland Beschel, noting its occurrence in 1968 near Kaladar, about 40 mi NW of Kingston) and N.B. (St. John, where taken on wharf-ballast by G.U. Hay in 1877; ACAD).

RUBIACEAE (Madder Family)

Herbs (*Cephalanthus* a shrub) with square or terete stems and simple entire leaves, these either in whorls and lacking evident stipules or opposite and connected by interposed stipules. Flowers regular, gamopetalous, epigynous, commonly 4-merous (sometimes 3-merous in *Galium*). Stamens 4 (sometimes 3 in *Galium*). Style 1. Ovary inferior. Fruit various.

1 Leaves in whorls; fruit a pair of dry or leathery 1-seeded carpels separating at maturity.
   2 Corolla rotate, commonly 4-parted (sometimes 3-parted), white, greenish white, yellow, or purple; calyx-teeth obsolete; inflorescence cymose; leaves in whorls of 4–8; stem square in cross-section . . . . *Galium*
   2 Corolla funnelform to campanulate, with a slender tube, the limb usually 4-lobed; leaves mostly in whorls of 8; (introd.).
      3 Calyx-teeth lanceolate; corolla pink or blue, about 3 mm long, commonly 4-lobed (sometimes 5-lobed); fruit pubescent, about 2 mm long; flowers in terminal heads subtended by an involucre of lanceolate ciliate leaves connate at base; stem square, more or less pubescent . . . . *Sherardia*
      3 Calyx-teeth obsolete; corolla white, blue, purple, or red, 4-lobed; stem square or terete . . . .*[Asperula]*
1 Leaves opposite (occasionally in whorls of 3 or 4 in *Cephalanthus*); corollas funnelform or campanulate, usually 4-lobed.
   4 Shrub to over 3 m tall; leaves lance- to ovate-oblong, tapering at both ends, to about 1.5 dm long; flowers white, they and the obconic fruits densely crowded in globose heads to 3 cm thick; (Ont. to N.B. and N.S.) . . . . *Cephalanthus*
   4 Herbs at most about 3 dm tall; flowers solitary, twinned, or in cymose clusters.
      5 Stem trailing; leaves round-ovate, shining, petioled, evergreen, often variegated with whitish lines, 1 or 2 cm long; flowers twinned, white, mostly terminal, producing a scarlet berry-like double drupe; (Ont. to Nfld. and N.S.) . . . . *Mitchella*
      5 Stem erect; leaves broadly linear to narrowly oblong, deciduous, to about 3 cm long; flowers solitary or in cymose clusters; fruit a 2-locular capsule with several or many seeds . . . . *Houstonia*

[ASPERULA L.] [8485] Woodruff

1 Flowers blue, purple, or red, subsessile in terminal heads subtended by a whorl of leaves; fruit glabrous or pubescent; leaves ciliate, in whorls of 4–8, the upper ones linear to lanceolate, blunt; stem to 3 dm tall, retrorse-scabrous on the angles (at least below) . . . .
. . . . *[A. arvensis]*
1 Flowers white, in peduncled branching cymes; leaves mucronate, mostly in 8's, 2 or 3 cm long.
   2 Leaves linear, erect or ascending; peduncles terminal and from the upper axils, 2-ternate; flowers 2 or 3 mm long; fruit glabrous, about 2 mm long; stem terete, to 8 dm tall . . . . *[A. glauca]*
   2 Leaves oblanceolate, spreading; peduncles 1–3, terminal, each bearing an umbel-like cluster of flowers, these to 5 mm long; fruit densely covered with hooked bristles, 3 or 4 mm long; stem square, to about 2 dm tall; (plant strongly resembling *Galium asprellum* and *G. triflorum* in habit) . . . . *[A. odorata]*

[A. arvensis L.]
[Eurasian; occasionally introd. into N. America but not established, as in sw B.C. (Essondale, near Vancouver; Groh and Frankton 1949*b*), s Ont. (near Hamilton, where taken along the edge of a marsh by Dickson in 1895; CAN), and s N.Y.]

[A. glauca (L.) Bess.]
[Eurasian; locally introd. into waste places or a garden-escape in N. America but scarcely established, as in Ont. (Ottawa dist.; Gillett 1958), Que. (St-Adrien, Megantic Co.; Montreal; Rivière-du-Loup, Temiscouata Co.; MT), New Eng., and N.J. (*Galium* L.; *A. galioides* Bieb.).]

[A. odorata L.]   Sweet Woodruff
[Eurasian; an occasional garden-escape in N. America but scarcely established, as in sw B.C. (Victoria, Vancouver Is.; Eastham 1947), s Ont. (Dorcas Bay, Bruce Co.; TRT), and sw Que. (Rouleau 1947; ?escaped).]

CEPHALANTHUS L.   [8230]

C. occidentalis L.   Buttonbush.   Bois noir
/T/(X)/   (N (Mc))   Swamps and margins of ponds and streams from cent. Calif., s N.Mex., Tex., Nebr., and Minn. to Ont. (N to Renfrew, Carleton, and Russell counties), sw Que. (N to L. St. Peter in St-Maurice Co.; John Macoun 1884; MT), N.B., ?P.E.I. (McSwain and Bain 1891; probably now extinct), and N.S., s to Baja Calif., Mexico, and Fla.; W.I. MAPS: Preston 1961:362, and 1947:274; Hough 1947:409.
Var. *pubescens* Raf. (twigs and at least the lower leaf-surfaces soft-pubescent rather than glabrous, the upper leaf-surfaces relatively pale) is reported from sw Que. by Raymond (1950b; L. Champlain, Missisquoi Co., where growing with the typical form but in separate colonies).

GALIUM L.   [8486]   Bedstraw, Cleavers.   Gaillet

1   Fruit (or ovary) bristly, hairy, or tuberculate.
  2   Fruits tuberculate, 3 or 4 mm long, on strongly recurved pedicels, the pedicels, peduncles, and stem harshly retrorse-scabrous on the angles; peduncles mostly 3-flowered, the inflorescences scarcely surpassing the linear-lanceolate mucronate leaves, these mostly glabrous above except for the ciliate margins; annual; (introd.) . . . . *[G. tricornutum]*
  2   Fruits bristly or hairy.
    3   Stem harsh with downward-pointing bristles on the angles; principal leaves mostly 8 in a whorl, narrowly oblanceolate, retrorsely hispid above and on the margins and midvein beneath, bristle-tipped; fruit to 4 mm long; annual; (transcontinental) . . . . *G. aparine*
    3   Stem smooth (or sparingly retrorse-scabrous in *G. triflorum*).
      4   Principal leaves mostly 6 in a whorl, narrowly elliptic, cuspidate, their margins minutely upwardly ciliate; fruit about 2 mm long; stem usually retrorse-scabrous on the angles at least below; perennial; (transcontinental) . . . . *G. triflorum*
      4   Principal leaves in whorls of 4.
        5   Flowers solitary in the leaf-axils; peduncles at first short, later elongating to as much as 3 cm; corolla white, 3-lobed; fruit nodding, pubescent with short hooked hairs; leaves linear-elliptic or a little broader, mostly 1 or 2 cm long (one pair often smaller than the other), they and the erect, simple or moderately branched stem glabrous; annual; (s B.C. and sw Alta.) . . . . *G. bifolium*
        5   Flowers commonly numerous in terminal cymose panicles; perennials.
          6   At least some of the flowers sessile or subsessile along the branches of the inflorescence; leaves 3–5-nerved; (Ont. and sw Que.).
            7   Middle and upper leaves lance-acuminate, to 8 cm long and 2.5 cm broad, glabrous or sparsely short-hispid on the nerves; corolla glabrous, yellowish, turning dull purple . . . . *G. lanceolatum*
            7   Middle and upper leaves ovate-oblong to oval, blunt, at most about 5 cm long and 1.5 cm broad, upwardly ciliate; corolla greenish, its lobes usually hairy outside . . . . *G. circaezans*
          6   Flowers all pedicelled; leaves 3-nerved.
            8   Leaves linear to linear-lanceolate, broadest near the base, to 5 cm long and 8 mm broad, nearly uniform in size except for the lowermost reduced ones; flowers numerous in dense ascending panicles, bright white; stems commonly short-bearded just below the nodes, otherwise glabrous or minutely scabrous; (transcontinental) . . . . *G. boreale*

8 Leaves broader, elliptic to broadly oval or obovate, broadest near or slightly above the middle, the upper ones gradually reduced; flowers greenish- to yellowish-white or purplish, relatively few.

  9 Leaves firm, dull, subequal, more or less spreading-pilose at least beneath, oval, mostly more than twice as long as broad, in numerous whorls, their lateral nerves obscure; stem hirsute to glabrate; (s Ont.) .... *G. pilosum*

  9 Leaves thin, lustrous, essentially glabrous except for the upwardly ciliate margins (or the veins sparingly hairy), broadly elliptic to obovate, less than twice as long as broad, distinctly 3-nerved and increasing in size toward the top of the stem, the latter glabrous or nearly so.

    10 Stem with at most 4 or 5 whorls of leaves; leaves more or less cuneate at base; inflorescence few-flowered (flowers commonly 2 or 3 on each of the 1–3 terminal peduncles); (widespread but localized) .... *G. kamtschaticum*

    10 Stem with up to 8 (sometimes 9) whorls of leaves; leaf-margins tending to be convexly rounded; flowers relatively numerous, each primary peduncle tending to be cymosely branched and several-flowered; (?Vancouver Is.) .....
.... [*G. oreganum*]

1 Fruit (or ovary) typically smooth and glabrous; perennials.

  11 Stems relatively short, erect or ascending, smooth or merely more or less pubescent; flowers numerous in panicles.

    12 Principal leaves in whorls of 4, lanceolate to ovate-lanceolate; flowers purple, in loose simple or compound cymes .... [*G. latifolium*]

    12 Principal leaves in whorls of 6–8; flowers white or yellow; (introd.).

      13 Flowers yellow, the whole panicle elongate; leaves linear-acicular; stems pubescent, at least in the inflorescence.

        14 Lower branches of panicle much surpassing the adjacent internodes, the panicle thus rather dense; flowers less than 3 mm broad .... *G. verum*

        14 Lower branches of panicle at anthesis shorter than or barely surpassing the adjacent internodes, the panicle slender and interrupted; flowers about 3 mm broad; (introd. in Que.) .... *G. wirtgenii*

      13 Flowers white, in loose leafy panicles; leaves flat; stems glabrous throughout.

        15 Leaves lanceolate, thin, to 5 cm long, broadest near the middle, pale beneath; (introd. in s Ont.) .... *G. sylvaticum*

        15 Leaves firm, narrowly oblanceolate, rarely up to 2.5 cm long; (introd., transcontinental) .... *G. mollugo*

  11 Stems weak, matted, reclining or loosely ascending.

    16 Leaves sharply cuspidate or mucronate, the principal ones in whorls of 6–8; corollas 4-lobed, white.

      17 Leaves retrorse-scabrous on the margins, narrowly elliptic to oblanceolate, the principal ones in whorls of 6; stems retrorse-scabrous on the angles.

        18 Leaves oval-lanceolate; (Ont. to Nfld. and N.S.) .... *G. asprellum*

        18 Leaves narrowly lanceolate or oblanceolate; (introd. in w Greenland) ....
.... *G. uliginosum*

      17 Leaves smooth or minutely upwardly spinulose on the margins.

        19 Principal leaves in 6's, linear to linear-lanceolate, at most minutely upwardly scabrous on the margins; branches and stem smooth or minutely scabrous, the stem lacking conspicuous leafy basal offshoots; panicles very lax; (s Ont.) .... *G. concinnum*

        19 Principal leaves 6–8 in a whorl, spatulate to narrowly obovate, upwardly spinulose-margined; branches smooth; stem producing prostrate leafy basal offshoots; panicle open-cylindric; (introd. in Que., St-Pierre and Miquelon, and s Nfld.) .... *G. saxatile*

    16 Leaves blunt or rounded at tip.

20 Flowers numerous in terminal cymes, the pedicels widely divergent; corollas 4-parted; leaves chiefly in whorls of 4-6; (transcontinental) . . . . *G. palustre*
20 Flowers solitary or in mostly simple few-flowered cymes (if numerous, with ascending pedicels).
21 Corollas white, to 2.5 mm broad, with 4 acutish lobes; stems erect or ascending, lacking matted basal offshoots; principal leaves in 4's.
22 Leaves soon reflexed; fruits less than 2 mm thick; (transcontinental) . . . .
. . . . *G. labradoricum*
22 Leaves spreading or ascending; fruits at least 2.5 mm thick; (Ont. to N.S.) . . . . *G. obtusum*
21 Corollas greenish white, not over 1.5 mm broad, mostly with 3 obtuse lobes; stems reclining, developing matted basal autumnal offshoots.
23 Corolla commonly 2 or 3 rnm broad; inflorescence irregularly cymose and several-flowered; leaves to about 2.5 cm long; (s ?B.C.) . . . .
. . . . [*G cymosum*]
23 Corolla rarely over 1.5 mm broad; flowers 2 or 3 at the ends of terminal or axillary peduncles which may themselves be borne in 2's or 3's; leaves mostly not over 2 cm long; (transcontinental) . . . . *G. trifidum*

G. aparine L.   Cleavers, Goosegrass
/sT/X/EA/   (T)   Woods, thickets, shores, and waste places (probably both native and introd.) from the Aleutian Is. and s Alaska (N to ca. 61°N) to B.C., Alta. (N to Athabasca, 54°43′N; introd. at Kelvington, Sask., according to Fraser and Russell 1944), Man. (N to The Pas; ?introd.), Ont. (N to the SW James Bay watershed at ca. 52°10′N), Que. (N to the E James Bay watershed at 51°29′N, Anticosti Is., and the Gaspé Pen.), Nfld. (Boivin 1966b; not known from P.E.I.), N.B., and N.S., s to Calif., Tex., and Fla.; introd. in s Greenland; Eurasia; introd. in S. America, Africa, Australia, and s Asia. [Incl. vars. *intermedium* (Mérat) Briq. (*G. ?spurium* L.) and *minor* Hook.]. MAP: Hultén 1968b:838.
Some of our material, at least from B.C.-Alta., appears referable to the small-fruited extreme, var. *echinospermum* (Wallr.) Farw. (*G. vaillantii* DC.; *G. ?micranthum* Pursh; fruits at most 3 mm long rather than to 4 or 5 mm).

G. asprellum Michx.   Rough Bedstraw
/T/EE/   (Hp)   Damp woods, thickets, and low ground from Ont. (NW to the Kaministikwia R. about 20 mi w of Thunder Bay, N to the Moose R. s of James Bay at ca. 51°N) to Que. (N to the SE James Bay watershed at 52°37′N, the Côte-Nord, and Gaspé Pen.; not known from Anticosti Is.), Nfld., N.B., P.E.I., and N.S., s to Nebr., Ohio, and N.C.

G. bifolium Wats.
/T/W/   (T)   Moist or dryish places from the foothills to high elevations from s B.C. (collection in CAN from the Dewdney Trail, SW of Rossland, where taken by J.M. Macoun in 1902; collections in V from Nelson and the Columbia Valley) and SW Alta. (Jasper; Herb. V) to s Calif. and Colo.

G. boreale L.   Northern Bedstraw
/aST/X/GEA/   (Hpr)   Meadows, prairies, open woods, and shores, the aggregate species from Alaska (N to ca. 69°N), the Yukon (N to ca. 65°N), and the Mackenzie R. Delta to Great Bear L., Great Slave L., L. Athabasca (Alta. and Sask.), Man. (N to Churchill), northernmost Ont., Que. (N to Bagotville, Chicoutimi Co., and the Gaspé Pen.), N.B., and N.S. (reports from Nfld. by Reeks 1873, and from P.E.I. by McSwain and Bain 1891, require confirmation), s to Calif., Tex., Mo., Ohio, and Del.; Greenland; Iceland; Eurasia. MAPS and synonymy: see below.
1 Ovary and fruit glabrous; [*G. hyssopifolium* Hoffm.; Ont. (N to Bruce, York, and Glengarry counties), Que. (N to the Gaspé Pen.), and N.B. (Restigouche and Charlotte counties)] . . . .
. . . . var. *hyssopifolium* (Hoffm.) DC.
1 Ovary and fruit bristly-hairy.
2 Fruit thinly to densely pubescent with short, strongly incurved bristles; [Ont. (N to Kenora, Thunder Bay, and w James Bay at 54°12′N), Que. (N to Bagotville, Chicoutimi Co.), N.B., and N.S.] . . . . var. *intermedium* DC.

2 Fruit densely hirsute with straight, spreading or ascending bristles; [var. *linearifolium*
Rydb.; *G. septentrionale* R. & S.; *G. rubioides sensu* John Macoun 1884, not L., as to
the Belleville, Ont., plant (relevant collection in CAN) and perhaps as to his reports from
Alaska–B.C.; transcontinental; MAPS (aggregate species): Hultén 1968*b*:837, and
1958: map 86, p. 105; A. Löve and D. Löve, Am. Midl. Nat. 52(1): fig. 1, p. 95. 1954] ....
.... var. *boreale*

G. circaezans Michx.   Wild Licorice
/T/EE/ (Hp (Hpr)) Rich woods from Minn. to Ont. (N to the Ottawa dist.) and SW Que. (N to
Chelsea and Kingsmere, N of Hull, and the Montreal dist.; MT; the report from near Quebec City by
John Macoun 1884, requires confirmation), s to Tex., Mo., Ky., and N.C. [Var. *glabrum* Britt.; incl.
the coarser and more densely pubescent extreme, var. *hypomalacum* Fern.].

G. concinnum T. & G.
/t/EE/ (Hp) Woods and thickets from Minn. to s Ont. (tentatively admitted to our flora on the
basis of the report from Pelee Is., Essex Co., by Core 1948, and so-named collections in TRT from
Summerville, Peel Co., and Vineland, Lincoln Co.), s to Kans., Ark., Ky., and Va.

[G. cymosum Wieg.]
[Reported from SW B.C. by Henry (1915; South Westminster) and otherwise known from ?Mont.
and Oreg. Scarcely separable from *G. trifidum*.]

G. kamtschaticum Steller
/sT/D/eA/ (Hpr) Moist woods and mossy places from the Aleutian Is. and s Alaska (N to ca.
59°N) through coastal B.C. (Queen Charlotte Is.; Alice Arm, ca. 55°20′N; reported from Vancouver
Is. by Carter and Newcombe 1921) to N Wash.; isolated stations in NW Dist. Mackenzie (Porsild and
Cody 1968) and on E L. Superior, Ont. (Mamainse Point, about 40 mi NW of Sault Ste- Marie; CAN);
Que. (Brome, Portneuf, Quebec, and Temiscouata counties; Shickshock Mts. of the Gaspé Pen.),
Nfld., N.B. (Summit Depot, Restigouche Co.; ACAD; CAN; not known from P.E.I.), and N.S. (Cape
Breton Is.: Inverness and Victoria counties) to the mts. of N N.Y. and N New Eng.; E Asia. MAPS:
Hultén 1968*b*:839; *Atlas of Canada* 1957: map 18, sheet 38; Fernald 1933: map 25, p. 309.

G. labradoricum Wieg.
/sT/X/ (Hp) Mossy woods, thickets, and bogs from s Dist. Mackenzie–Dist. Keewatin (N to ca.
61°N; CAN; not known from B.C.) to Alta.-Sask.-Man., northernmost Ont., Que. (N to E Hudson Bay
at ca. 56°10′N, Ungava Bay, the Côte-Nord, Anticosti Is., and Gaspé Pen.), Labrador (N to the
Hamilton R. basin), Nfld., N.B. (St. John; St. Andrews), P.E.I., and N.S. (Bay St. Lawrence, Victoria
Co.; ACAD; not listed by Roland 1947), s to Minn., Ill., Ohio, Pa., and N.J.

G. lanceolatum Torr.   Wild Licorice
/T/EE/ (Hp) Dry woods from Minn. to Ont. (N to the Ottawa dist.; CAN), SW Que. (N to L. St.
Peter in St-Maurice Co.; MT; the report from near Quebec City, Que., by John Macoun 1884,
requires confirmation), and Maine, s to Tenn. and N.C.

[G. latifolium Michx.]
[The report of this species of the E U.S.A. (N to W.Va. and Pa.) from s Ont. by Soper (1949) may be
based upon a collection in OAC from Kitchener, Waterloo Co., this perhaps referable to *G.
lanceolatum*. The report from P.E.I. by McSwain and Bain (1891) also requires confirmation.]

G. mollugo L.   White Bedstraw.   Gaillet mollugine
Eurasian; introd. in fields and along roadsides in N. America, as in B.C. (Vancouver Is. and
adjacent mainland; Eastham 1947), Ont. (N to Manitoulin Is., N L. Huron, and Renfrew Co.), Que. (N
to the Gaspé Pen.), Nfld., N.B., P.E.I., and N.S.; SW Greenland.
    Material from E Que. (Gaspé Pen. at Matapédia), P.E.I. (Southport and Charlottetown), and N.S.
(Windsor, Hants Co.) is referable to ssp. *erectum* (Huds.) Syme (*G. erectum* Huds., not Don nor
Hoffm.; stem and branches relatively erect; leaves mostly linear-lanceolate rather than oblanceolate

or obovate; flowers to 4 mm broad rather than 3 mm; fruits to 2 mm thick rather than 1 mm; panicle-branches ascending rather than spreading).

G. obtusum Bigel.
/T/(X)/ (Hp) Swampy places and wet shores from Nebr. to Minn., Ont. (N to the Ottawa dist.; Gillett 1958), SW Que. (N to Oka and the Montreal dist.; MT; N to Montmagny and Kamouraska counties if so-named collections in MT prove to be correctly identified), N.B. (Youghall, Gloucester Co.; CAN; not known from P.E.I.), and N.S. (Queens and Yarmouth counties), S to Ariz. and Fla.

Var. *ramosum* Gl. (leaves relatively thin, hispid-ciliate, mostly about 4 times as long as broad, rather than scabrous-margined and to about 7 times as long as broad; stems diffusely-branched throughout rather than branched chiefly from the base) is reported from SW Que. by Gleason (1958).

[G. oreganum Britt.]
[The report of this species of Wash. and Oreg. from SW B.C. by J.M. Macoun (1913; Vancouver Is.) requires confirmation. (*G. kamt.* ssp. *oreg.* (Britt.) Piper).]

G. palustre L.   Marsh Bedstraw
/T/EE/EA/ (Hp) Wet meadows, thickets, and shores from Ont. (N to Cochrane, 49°04′N; introd. in S Yukon; so-named collections from Alta. and Man. require further study) to Que. (N to the Rupert R. S of James Bay at ca. 51°25′N and the Côte-Nord), Nfld., N.B., P.E.I., and N.S., S to Wisc., Mich., Pa., and New Eng.; Eurasia. MAPS: Hultén 1968b:839, and 1958: map 151, p. 171; Fernald 1925: map 48, p. 319. (Hultén's maps indicate a station near Hamilton Inlet, Labrador, and Fernald's map indicates stations in S Greenland, these perhaps referable to *G. brandegei*, included below in the *G. trifidum* complex).

G. pilosum Ait.
/t/EE/ (Hp) Dry woods and thickets from Mich. to S Ont. (N to Huron and Lincoln counties; the report from Montreal, Que., by R. Campbell, Can. Rec. Sci. 6(6):342–51. 1895, requires confirmation) and S N.H., S to Tex. and Fla.

G. saxatile L.   Heath-Bedstraw
European; apparently known in N. America only from Que. (St-Flavien, Lotbinière Co.) and St-Pierre and Miquelon (Boivin 1966b) and from along a roadside and the borders of a woods in peaty barrens in SE Nfld. (M.L. Fernald, Rhodora 28(347):236. 1926; Trepassey, where considered native by Fernald but more likely introd.; *see* discussion under *Luzula campestris*). [G. hercynium Weigel]. MAP: Hultén 1958: map 133, p. 153.

G. sylvaticum L.   Scotch-mist, Baby's-breath
European; a garden-escape to fields and roadsides in N. America, as in S Ont. (Lambton, Lincoln, and York counties; OAC; TRT).

[G. tricornutum Dandy]
[European; occasionally introd. into waste and cult. ground in N. America. Reported by John Macoun (1886) from gardens at London, Middlesex Co., S Ont., where scarcely established, apparently no other collections having been made since that date. The report from Anticosti Is., E Que., by John Adams (Can. Field-Nat. 50(7):117. 1936) is based upon *G. trifidum*, the relevant collection in DAO. (*G. tricorne* Stokes in part).]

G. trifidum L.
/aST/X/GEA/ (Hpr) Moist places at low to high elevations, the aggregate species from the Aleutian Is. and N Alaska–Yukon–Dist. Mackenzie (N to ca. 69°30′N) to Great Bear L.; Great Slave L., L. Athabasca (Alta. and Sask.), S Dist. Keewatin (60°16′N; CAN), northernmost Man.–Ont., Que. (N to S Ungava Bay, the Côte-Nord, Anticosti Is., and Gaspé Pen.), Labrador (N to ca. 58°N), Nfld., N.B., P.E.I., and N.S., S to Calif., Mexico, Tex., and Ga.; W Greenland N to 64°20′N, E Greenland N to 65°37′N; Iceland; Eurasia. MAPS and synonymy: *see* below.
1   Leaves, pedicels, and usually the stem smooth, the leaves mostly 4 in a whorl; [var. *halophilum* Fern. & Wieg.; incl. *G. brandegei* Gray (*G. palustre* var. *minus* Lange);

transcontinental; MAPS (*G. brand.*): Hultén 1968*b*:841; A. Löve 1950: fig. 19, p. 53] . . . .
. . . . var. *pusillum* Gray
1 Leaves with backwardly-scabrous margins and midvein; upper internodes of stem often retrorse-scabrous.
  2 Principal leaves 5 or 6 in a whorl, oblanceolate to oblong-spatulate; pedicels smooth, relatively stiff and straight.
    3 Pedicels mostly 3 terminating a peduncle; [var. *latifolium* Torr.; *G. claytonii* Michx.; *G. tinctorium* L.; Ont. to Nfld. and N.S.] . . . . var. *tinctorium* (L.) T. & G.
    3 Pedicels usually single, at most 2 terminating a peduncle; [var. *subbiflorum* Wieg.; *G. claytonii (tinctorium)* var. *sub.* Wieg.; *G. sub.* (Wieg.) Rydb.; *G. columbianum* Rydb.; the MAP for *G. tri.* ssp. *columb.* by Hultén 1968*b*:840, indicates a restriction in N. America to the West in spite of the fact that he includes the transcontinental *G. tinct.* var. *sub.* in the synonymy; *see* Hiroshi Hara, Rhodora 41(489):387–88. 1939] . . . . var. *pacificum* Wieg.
  2 Principal leaves 4 in a whorl, linear to linear-oblanceolate; pedicels scabrous, arching, single or 2 or 3 terminating a peduncle; [incl. *G. brevipes* Fern. & Wieg.; transcontinental; MAP: Hultén 1968*b*:840] . . . . var. *trifidum*

G. triflorum Michx. Sweet-scented Bedstraw
/aST/X/GEA/ (Hp) Woods and thickets from the E Aleutian Is., Alaska (N to ca. 68°N), the Yukon (N to ca. 63°N), and SW Dist. Mackenzie to L. Athabasca (Alta. and Sask.), Man. (N to Wekusko L., about 90 mi NE of The Pas), Ont. (N to Big Trout L. at ca. 53°45′N, 90°W), Que. (N to E Hudson Bay at ca. 56°20′N and the Côte-Nord), Labrador (N to the Hamilton R. basin), Nfld., N.B., P.E.I., and N.S., S to Calif., Mexico, and Fla.; W Greenland N to 64°10′N, E Greenland N to 62°40′N; Eurasia. MAPS: Hultén 1968*b*:838, and 1958: map 241, p. 261; Porsild 1966: map 132, p. 83.

Forma *rollandii* Vict., the leaves nearly reduced to merely the ciliate midrib, is known from the type locality, Longueuil, near Montreal, Que. Some of the Ont. and Que. material is referable to var. *asprelliforme* Fern. (inflorescence a diffuse many-flowered panicle rather than essentially simple, the peduncles freely forking and bearing whorls of reduced leaves at their nodes subtending lateral flowering branches).

G. uliginosum L.
European; reported by J. Groentved (Bot. Tidsskr. 44(2):253. 1937) as apparently well established in two localities in W Greenland (CAN).

G. verum L. Yellow or Our Lady's Bedstraw
Eurasian; introd. in dry fields and along roadsides in N. America, as in SW B.C. (Victoria, Vancouver Is.; Eastham 1947), Alta. (Calgary), S Man. (Holland; High Bluff; Altamont), Ont. (N to Moose Factory, near SW James Bay), Que. (N to Cap-à-l'Aigle, Charlevoix Co., and Anticosti Is.), Nfld., and N.S. (Kings Co.; ACAD). MAP: Hultén 1968*b*:837.

G. wirtgenii Schultz
European; locally introd. into fields and meadows in E N. America, as in SW Que. (in a field near Georgeville, Stanstead Co., where taken by A.S. Pease in 1903 and distributed as *G. verum*, revised by Fernald; GH).

HOUSTONIA L. [8141]

1 Peduncles 1-flowered; corolla lilac to bluish, with a yellow eye, its lobes glabrous; stamens included; capsule flattened, much broader than long; stem rarely over 2 dm tall; (?Ont. to N.S. and St-Pierre and Miquelon) . . . . *H. caerulea*
1 Peduncles several-flowered, the pedicelled flowers in bracted cymose clusters at their tips; corolla funnelform, purplish to white, to 9 mm long, its lobes pubescent within; stamens exserted; capsule globose, not flattened; stems firm, to over 3 dm tall.
  2 Basal leaves numerous at anthesis, they and the stem-leaves usually distinctly ciliate; corolla purple, the lobes more than half as long as the tube; calyx-lobes to 3 mm long; (S Ont.) . . . . *H. canadensis*

2   Basal leaves usually none at anthesis (if present, they and the stem-leaves not ciliate); corolla purple or white, the lobes about half as long as the tube; calyx-lobes 1 or 2 mm long; (Alta. to s Que.) . . . . *H. longifolia*

H. caerulea L.   Bluets, Innocence, Quaker-ladies
/T/EE/   (Hsr)   Meadows and moist places from Wisc. to s ?Ont. (Fernald *in* Gray 1950), Que. (N to Cap-à-l'Orignal, near Bic, Rimouski Co., and the Gaspé Pen. at Métis; MTMG), St-Pierre and Miquelon, N.B., and N.S. (not known from P.E.I.), s to Mo., Ala., and Ga. [*Hedyotis* Hook.]. MAP: W.H. Lewis and E.E. Terrell, Rhodora 64(760): fig. 2, p. 320. 1962.
Forma *albiflora* Millsp. (flowers white rather than pale lilac to bluish) is reported from Que. by Boivin (1966b). The St-Pierre and Miquelon plant is referable to var. *faxonorum* Pease & Moore (*H. serpyllifolia* Graham; upper leaves scarcely reduced rather than much reduced; corolla-tube short rather than to 1 cm long; stem fleshy, drying blackish).

H. canadensis Willd.
/T/EE/   (Hs)   Rocky or gravelly shores from Minn. to s Ont. (N to Manitoulin Is., N L. Huron, and Northumberland and Hastings counties), s to Ill., Tenn., Pa., and N.Y. [*H. ciliolata* Torr.; *H. purpurea* var. *cil.* (Torr.) Gray]. MAP: E.E. Terrell, Rhodora 61(726): map 2, p. 175. 1959.

H. longifolia Gaertn.
/T/(X)/   (Hs)   Rocky or gravelly places from Alta. (near Bruderheim, about 25 mi NE of Edmonton; CAN) to Sask. (N to Meadow Lake, 54°08′N; Breitung 1957a), Man. (N to Riding Mt.; introd. at Churchill), Ont. (N to the mouth of the Rainy R. about 65 mi s of Kenora and Parry Sound, Georgian Bay, L. Huron), and s Que. (Boivin 1966b), s to Okla., Miss., and Ga. [*Hedyotis* Hook.; incl. vars. *musci* and *soperi* Boivin; *H. ?tenuifolia sensu* Dawson 1875, not Nutt.]. MAP: E.E. Terrell, Rhodora 61(727): map 3, p. 194. 1959).

MITCHELLA L.   [8451]

M. repens L.   Partridge-berry.   Pain de perdrix
/T/EE/   (Ch (evergreen))   Dry or moist woods from Minn. to Ont. (N to Batchawana Falls, NW of Sault Ste. Marie, and Renfrew, Carleton, and Russell counties), Que. (N to Ste-Anne-de-la-Pocatière, Kamouraska Co.; QSA), Nfld., N.B., P.E.I., and N.S., s to Tex. and Fla. [*Perdicesca* Prov.].

SHERARDIA L.   [8482]

S. arvensis L.   Field-Madder
European; introd. into fields, orchards, and waste places in N. America, as in w B.C. (Queen Charlotte Is.; Vancouver Is. and adjacent islands and mainland), Ont. (N to Ottawa), sw Que. (Mt. Royal, Montreal), and N.S. (Tatamagouche, Colchester Co.; Lindsay 1878).

CAPRIFOLIACEAE (Honeysuckle Family)

Mostly shrubs (*Triosteum* a coarse herb) with opposite, commonly simple (compound in *Sambucus*), usually exstipulate leaves. Flowers regular or slightly irregular, gamopetalous, epigynous, perfect, usually 5-merous and commonly in cymes. Calyx subentire to deeply cleft nearly or quite to base. Corolla rotate or tubular. Stamens commonly 5 (rarely fewer), inserted on the corolla-tube. Style 1. Ovary inferior. Fruit a capsule or drupe.

1 Leaves pinnately compound, the lanceolate to ovate leaflets closely serrate; inflorescence a terminal compound cyme; corolla regular, rotate or saucer-shaped, white, yellowish white, or sometimes pink; style short, 3-lobed; fruit berry-like, with 3 (sometimes 4 or 5) small nutlets . . . . *Sambucus*
1 Leaves simple.
  2 Stems extensively creeping, slender; leaves 1 or 2 cm long, sparingly low-crenate, broadly oval to obovate; flowers mostly paired, drooping from the tips of long erect axillary peduncles, the funnelform corolla whitish, striped with rose-purple; style slender, exserted; fruit dry, 1-seeded; (transcontinental) . . . . *Linnaea*
  2 Stems erect or climbing, relatively stout.
    3 Coarse herbs to over 1 m tall; flowers sessile in clusters of up to 4 in each leaf-axil, 1 or 2 cm long; corolla narrowly campanulate, somewhat unequally lobed; style elongate, included or exserted; fruit a dry drupe with 3 bony nutlets; leaves obovate to subrhombic, to over 2 dm long; (eastern species) . . . . *Triosteum*
    3 Shrubs.
      4 Leaves entire or merely undulate or somewhat lobed (the lobes entire); style elongate.
        5 Corolla campanulate, pink (sometimes white), regular or nearly so; fruit a 2-seeded white berry-like drupe . . . . *Symphoricarpos*
        5 Corolla funnelform to tubular, often gibbous at base, irregularly lobed; fruit a several-seeded berry . . . . *Lonicera*
      4 Leaves toothed.
        6 Corolla broadly campanulate to rotate; cymes compound, terminal; style short, 3-lobed; fruit a soft-pulpy drupe with a single flat stone . . . . *Viburnum*
        6 Corolla funnelform, yellow (becoming reddish), to 2 cm long; cymes axillary and terminal, few-flowered; style elongate, exserted, with a capitate stigma; fruit a slender long-beaked capsule with persistent slender calyx-lobes; (Sask. to Nfld. and N.S.) . . . . *Diervilla*

DIERVILLA Duham. [8524]

D. lonicera Mill.   Bush-Honeysuckle.   Herbe bleue
/T/EE/ (N) Dry woods, clearings, and rocky thickets from E Sask. (Bjorkdale and McKague, both ca. 52°40′N) to Man. (N to The Pas), Ont. (N to Sandy L. at ca. 53°N, 93°W), Que. (N to L. Mistassini, the Côte-Nord, Anticosti Is., and Gaspé Pen.), Nfld., N.B., P.E.I., and N.S., S to Iowa, Ohio, and N.C. [*Lonicera (D.) diervilla* L.; *D. canadensis* Bartr.; *D. canadensis* Willd.; *D. acadiensis* Duham.; *D. humilis* Pers.; *D. lutea* Pursh; *D. tournefortii* Michx.; *D. trifida* Moench].

Some of the Ont. material is referable to var. *hypomalaca* Fern. (leaves densely pilose beneath rather than essentially glabrous; type from the Timagami Forest Reserve; MAP: N.C. Fassett, Bull. Torrey Bot. Club 69(4): fig. 1, p. 317. 1942).

LINNAEA Gronov. [8520]

L. borealis L.   Twinflower
/aST/X/GEA/ (Ch (evergreen)) Open or dense woods and mossy openings, the aggregate species from N Alaska–Yukon–Dist. Mackenzie (N to ca. 69°30′N) to Great Bear L., Great Slave L., L. Athabasca (Alta. and Sask.), S Dist. Keewatin, northernmost Ont., Que. (N to E Hudson Bay at ca. 57°N, Ungava Bay, and the Côte-Nord), Labrador (N to Okak, 57°33′N), Nfld., N.B., P.E.I., and

N.S., s to Calif., N.Mex., S.Dak., Ind., and Md.; w Greenland N to 69°33'N, E Greenland N to 63°28'N; Eurasia. MAPS and synonymy: see below.

1   Corolla relatively small, at most about 11 mm long, narrowly campanulate, flaring from within the calyx, the tube proper (as opposed to the flaring throat) very short or even wanting; leaves more consistently subrotund than in the following taxa; [*L. serpyllifolia* Rydb.; Aleutian Is. and Alaska (N to ca. 68°N; *see* Hultén 1949: map 1083a, p. 1475) and s Yukon (lower Lapie R.; CAN); MAPS (the last two of the aggregate species): Hultén 1968*b*:843; Raup 1947: pl. 35; Meusel 1943: fig. 20a] .... var. *borealis*

1   Corolla to 1.5 cm long, more funnelform, the slender tube about equalling or slightly surpassing the calyx; leaves broadly elliptic to obovate or subrotund ...... var. *longiflora* Torr.

   2   Flowers uniformly white; [B.C., Alta. (Banff), and Ont.; Boivin 1966*b*] ... f. *candicans* House

   2   Flowers basically white but tinged and striped with rose-purple; [incl. f. *insularis* Wittr. and var. *americana* (Forbes) Rehd. (*L. amer.* Forbes); *L. long.* (Torr.) Howell; range of the species; MAPS: combine the maps by Hultén 1968*b*:843 (ssp. *amer.*) and 844; the above-noted map by Raup and the N. American area of the above-noted map by Meusel are also applicable here] .... f. *longiflora*

LONICERA L.   [8523]   Honeysuckle.   Chèvrefeuille

1   Some of the upper leaves connate, the flowers in sessile whorled clusters from their axils or in a terminal more or less peduncled cluster; berries orange-red; stems usually twining.

   2   Some of the leaves with well-developed, broadly ovate or roundish, connate stipules, the leaves firm, glaucous beneath, variously hirsute or puberulent to glabrous, often subcordate at base; corolla pink, or yellow tinged with pink, to about 2 cm long, the tube deeply 2-lipped, densely hairy within, sparingly hairy outside at base; anther-filaments attached nearly at the orifice of the corolla; (sw B.C.) .... *L. hispidula*

   2   Leaves not subtended by stipules.

      3   Leaves long-ciliate; corolla yellow to orange-red, hairy within.

         4   Corolla yellow to orange, clammy-pubescent outside, to 2.5 cm long, deeply 2-lipped; anther-filaments attached nearly at the orifice of the corolla; leaves dull green, more or less appressed-long-strigose above, downy-hairy and somewhat paler beneath; branchlets glandular-villous; (Ont. and Que.) ... *L. hirsuta*

         4   Corolla orange-yellow to -red, glabrous outside, to 4 cm long, only weakly 2-lipped; anther-filaments attached well down into the corolla-tube; leaves glabrous except for the more or less ciliate margins, strongly glaucous beneath; branchlets glabrous; (s B.C.) .... *L. ciliosa*

      3   Leaves not ciliate.

         5   Corolla nearly regular (scarcely 2-lipped), hairy within toward base, usually deep red outside and yellow inside, commonly 4 or 5 cm long; anther-filaments attached well down into the corolla-tube; leaves green above, glabrous to villous beneath, the united involucral pairs forming rhombic-elliptic disks; (introd. in s Ont. and Que.) .... *L. sempervirens*

         5   Corolla distinctly irregular (2-lipped, the lips commonly about equalling the tube); anther-filaments attached near the orifice of the corolla-tube.

            6   Corolla to 5 cm long, pale yellow (often purple-tinged), the tube glabrous within; (introd.).

               7   Whorls of flowers usually in long-peduncled heads; corolla-tube usually distinctly longer than the 2-lipped limb, usually glabrous outside; berry red; leaves often somewhat pubescent .... *L. etrusca*

               7   Whorls usually sessile in the axils of connate leaves; corolla-tube only slightly longer than the 2-lipped limb, often hairy outside; berry orange-coloured; leaves glabrous, glaucous beneath .... [*L. caprifolium*]

            6   Corolla at most about 3 cm long, the tube hairy within.

               8   United involucral leaves glaucous above, forming a suborbicular disk rounded or often retuse at the ends; lower leaves obovate to suborbicular; flowers pale yellow, only slighly gibbous at base, glabrous outside; (introd.) .... *L. prolifera*

8   United involucral leaves green above, forming rhombic-elliptic disks acutish to rounded and bluntly pointed at the ends; lower leaves elliptic or oblong; flowers greenish yellow to brick-red or purplish, gibbous on one side at base; (B.C. to Que.) .... *L. dioica*

1 Leaves all distinct.

9   Flowers in opposite sessile 3-flowered clusters composing 6-flowered whorls, up to 5 whorls closely crowded into a capitate inflorescence; corolla pale yellow, sometimes purple-tinged, 3–5 cm long, the tube glandular outside; berries red; plant twining or trailing; (introd.) .... *L. periclymenum*

9   Flowers in pairs terminating solitary axillary peduncles, at most about 2 cm long; plants not climbing.

10 Involucral bracts 4, green to dark purple, broadly oval and foliaceous, reflexed in fruit; leaves ovate to obovate, acuminate, tapering at base to a short petiole, strongly ascending, to 1.5 dm long; corolla pale yellow, nearly regular, glandular outside; berries purple-black; (B.C. to E Que.) .... *L. involucrata*

10 Involucral bracts 2, linear to lance-oblong; leaves mostly smaller.

11 Peduncles shorter than to about equalling the flowers, at most about 1.5 cm long.

12 Berries blue, solitary, consisting of 2 ovaries surrounded by a fleshy cup; corolla yellowish white or straw-colour; leaves oblong to oblong-oblanceolate, blunt or rounded at apex, nearly sessile; (Alta. to Labrador, Nfld., and N.S.) .... *L. villosa*

12 Berries normally red (sometimes yellow), in distinct pairs.

13 Corolla glabrous outside, ochroleucous or light yellow, 1 or 2 cm long, obscurely 2-lipped, the slightly unequal lobes much shorter than the tube; berries bright red; leaves elliptic to somewhat ovate or oblong, to about 8 cm long and 4 cm broad, glabrous above, glabrous or often hirsute beneath; (S B.C. and S Alta.) .... *L. utahensis*

13 Corolla pubescent outside, the lobes equalling to twice as long as the tube; (introd.).

14 Corolla whitish, yellowish, or pinkish, about 1 cm long, deeply 2-lipped, the upper lip relatively shallowly lobed; berries red; peduncles pubescent; leaves broadly oval, usually less than twice as long as broad, pale and pubescent beneath, petioled . . . *L. xylosteum*

14 Corolla white, turning yellow, about 1.5 cm long, obscurely 2-lipped, the upper lip deeply 4-lobed; berries red or yellow; peduncles densely pilose; leaves narrowly elliptic to oblong, grey-tomentose beneath, short-petioled .... *L. morrowii*

11 Peduncles mostly longer than the flowers, to about 4 cm long.

15 Branchlets hollow at centre; ovaries subtended by distinct broad bractlets; (introd.).

16 Corolla about 1 cm long, whitish, yellowish, or pinkish, deeply 2-lipped, the upper lip relatively shallowly lobed, the tube pubescent outside; berries red; leaves broadly oval, rounded to short petioles, pubescent .... *L. xylosteum*

16 Corolla to 2 cm long, pink, nearly regular, very deeply 5-lobed, the short tube glabrous outside; berries red or yellow; leaves ovate or oblong, rounded or subcordate above the short petiole, glabrous .... *L. tatarica*

15 Branchlets filled by pith; bractlets minute or obsolete.

17 Leaves oblong-ovate, often cordate, definitely petioled, ciliate, otherwise glabrate; corolla about 2 cm long, yellowish green or straw-colour, glabrous, nearly regular, rather shallowly 5-lobed; berries red; (Ont. to N.S.) .... *L. canadensis*

17 Leaves oblong to narrowly obovate, tapering or obtuse above the short petiole; corolla at most 1.5 cm long, yellowish white, deeply 2-lipped, the lower lip divergent; berries orange-yellow to deep red; (Sask. to N.S.) ...., ... *L. oblongifolia*

L. canadensis Bartr.   Fly-Honeysuckle
/T/EE/   (N)   Cool woodlands from Ont. (N to Renison, S of James Bay at ca. 51°N; Hustich 1955; the report of *Xylosteon canadense* from Sask. by Fraser and Russell 1944, is referred to *Symphoricarpos albus* by Breitung 1957a; its report from Man. by Lowe 1943, requires clarification) to Que. (N to the Côte-Nord and Gaspé Pen.), N.B., P.E.I., and N.S., S to Iowa, Wisc., Ohio, and N.C. [*L. (Caprifolium; Xylosteum) ciliata* Muhl.].

[L. caprifolium L.]   Italian Woodbine
[Eurasian; cult. in N. America and Fernald *in* Gray (1950) notes that it "has long been reported but seems not to be a true member of our flora." Boivin (1966b) reports it from Porter's Point, N.S., and there is a collection in CAN from Barrington Passage, Shelburne Co., N.S., where taken by John Macoun in 1910 but lacking data as to whether or not escaped.]

L. ciliosa (Pursh) DC.   Western Trumpet Honeysuckle
/t/W/   (Mc (vine))   Woods and thickets at low to moderate elevations from S B.C. (Vancouver Is. and adjacent islands; mainland N to Lillooet, about 70 mi W of Kamloops, E to Creston) and Mont. to Calif. [*Caprifolium* Pursh; *C. (L.) occidentale* Lindl.].

L. dioica L.   Limber Honeysuckle
/sT/(X)/   (Mc (vine))   Dry woods, thickets, and rocky slopes (ranges of Canadian taxa outlined below; not known from the Atlantic Provinces), S to SE B.C. (Field), Okla., Mo., Ky., and Ga.
1   Leaves sparsely to densely villous beneath; corolla-tube glandular and villous externally; style hirsute.
  2   Ovary densely glandular; [S Ont.; Gleason 1958] . . . . var. *orientalis* Gleason
  2   Ovary glabrous; [*L. glaucescens* Rydb.; *L. douglasii* Hook.; *Caprifolium parviflorum sensu* Richardson 1823, perhaps not *L. parviflora* L.; SW Dist. Mackenzie (N to near Fort Simpson at ca. 62°N) and B.C. to Alta.–Sask. (N to L. Athabasca), Man. (N to the Hayes R. at ca. 56°N, about 100 mi SW of York Factory), Ont. (N to the Fawn R. at ca. 55°30'N, 88°W), and Que. (N to N of Mont-Laurier, this about 80 mi N of Hull)] . . . .
    . . . . var. *glaucescens* (Rydb.) Butters
1   Leaves glabrous; corolla-tube externally and style glabrous or sparsely hairy; ovary glabrous . . . . var. *dioica*
  3   Upper foliage leaves in whorls of 3, the involucral leaves 3, connate; [type from Mt. Royal, Montreal, Que.] . . . . f. *trifolia* Vict. & Rousseau
  3   Leaves all in opposite pairs (including the involucral set); [*L. glauca* Hill; Ont. (N to Renfrew and Carleton counties; reports from farther westwards all appear referable to var. *glaucescens*) and SW Que. (N to the Montreal dist.)] . . . . f. *dioica*

L. etrusca Santi   Etruscan Honeysuckle
European; noted by Hitchcock et al. (1959) as "now established in thickets along the coast of Oreg. . . . and on S. Vancouver I." Reported from Queen Charlotte Is., B.C., by Calder and Taylor (1968).

L. hirsuta Eat.   Hairy Honeysuckle
/sT/EE/   (Mc (vine))   Moist woods, thickets, and shores from Ont. (N to Fort Hope, SW James Bay, 51°34'N; Dutilly, Lepage, and Duman 1954) to Que. (N to Moose Factory, SE of James Bay at ca. 51°15'N; CAN) and New Eng., S to Nebr., Minn., Ohio, and Pa. [*Caprifolium pubescens* Goldie].
    The reports from Man. by John Macoun (1886; Fort Ellice and Doghead, L. Winnipeg) are based upon *L. dioica* var. *glaucescens* (the relevant collections in CAN), as also, probably, the inclusion of Sask. in the range by Fernald *in* Gray (1950). Var. *schindleri* Boivin is reported from SE Man. by Boivin (1968; type from Falcon L., Whiteshell Forest Reserve, E of Winnipeg) but, from its description, appears scarcely separable from *L. dioica* var. *glaucescens.* Most or all of our material is evidently referable to var. *interior* Gl. (corolla-tube at most 18 mm long rather than to 22 mm; hypanthium glabrous or with a few scattered glands rather than densely glandular), the accrediting of which to Man. by Gleason (1958) also requires clarification.

L. hispidula (Lindl.) Dougl.   Purple Honeysuckle
/t/W/   (Mc (vine))   Woods and thickets from sw B.C. (Vancouver Is. and adjacent islands; CAN; V) to s Calif. [*Caprifolium* Lindl.; *L. microphylla* Hook., not Willd.].

L. involucrata (Richards.) Banks   Black Twinberry
/sT/X/   (N)   Cool moist woods and thickets at low to moderate elevations from SE Alaska (N to ca. 59°30′N) and B.C. to Alta. (N to Waterways, 56°42′N), Sask. (N to Mistatim, 52°52′N; Breitung 1957a), Man. (N to Gillam, about 165 mi s of Churchill), Ont. (N to Fort Severn, w Hudson Bay, ca. 56°N), and Que. (E James Bay watershed N to ca. 52°N; L. Mistassini; Taschereau; Gaspé Pen.; the report from NE N.B. by Fowler 1885, requires confirmation; not known from P.E.I. or N.S.), s to Calif., N Mexico, N.Mex., Wisc., and Mich. [*Xylosteum* Rich.; *Distegia* Cock.; *L. flavescens* Dippel]. MAPS: Hultén 1968b:844; Dansereau 1957: map 1C, p. 33; Raymond 1950b: fig. 19, p. 34.

L. morrowii Gray
Asiatic; a garden-escape to roadsides and thickets in N. America, as in Sask. (Saskatoon; Boivin 1968), Ont. (N to the Ottawa dist.; Gillett 1958), and Que. (Boivin 1966b).
    A hybrid with *L. tatarica* ·(× *L. bella* Zabel) is reported from Sask. and from Ont. to N.B. by Boivin (1966b).

L. oblongifolia (Goldie) Hook.   Swamp-Fly-Honeysuckle
/sT/EE/   (N)   Wet woods and thickets and bogs from Sask. (N to Sikip, 54°21′N; Breitung 1957a) to Man. (N to Gypsumville, about 125 mi N of Portage la Prairie; the report from Flin Flon by Lowe 1943, requires confirmation), Ont. (N to L. Nipigon and the sw James Bay watershed at 52°11′N), Que. (N to Ellen L. at ca. 52°35′N and the Gaspé Pen.), N.B. (St. Leonard, Charlotte Co.; CAN; not known from P.E.I.), and N.S. (William L., Halifax Co.; GH; not listed by Roland 1947), s to Minn., Ohio, and Pa. [*Xylosteum oblongifolium* Goldie, the type from near Montreal, Que.].
    Var. *altissima* (Jennings) Rehd. (*L. alt.* Jennings; plant essentially glabrous from the first rather than minutely downy beneath, finally glabrate) is reported from s Ont. by Fernald *in* Gray (1950).

L. periclymenum L.   Woodbine Honeysuckle
European; a garden-escape to roadsides and thickets in N. America, as in B.C. (Kamloops; Groh and Frankton 1949b), s Ont. (Niagara Falls, Welland Co.; Groh and Frankton 1949b), Nfld. (St. John's; GH), and N.S. (Yarmouth and Cape Breton counties).

L. prolifera (Kirchn.) Rehd.   Grape-Honeysuckle
Native in the E U.S.A. from Wisc. to N.Y., s to Kans., Ark., and Tenn.; introd. elsewhere, as in sw Que. (Boivin 1966b; Pierreville, Yamaska Co.) and N.S. (Grand Pré, Kings Co.; DAO). The reports of *L. sullivantii* from Fort Ellice and Manitoba House, Man., by John Macoun (1884) are based upon *L. dioica* var. *glaucescens* (relevant collections in CAN; his other Man. reports and the Pic R., L. Superior, Ont., one may also refer to that taxon). Its reports from s Ont. by John Macoun (1886; Hatchley, Oxford Co., and Saugeen, Bruce Co.; probable basis of the listing of *L. prolifera* for s Ont. by Soper 1949) require confirmation. [*L. sullivantii* Gray].

L. sempervirens L.   Trumpet- or Coral-Honeysuckle
Native in the E U.S.A. (N to Nebr., N.Y., and Maine); a garden-escape elsewhere, as in s Ont. (Soper 1949) and Que. (Boivin 1966b). [*Phenianthus* Raf.].

L. tatarica L.   Tatarian Honeysuckle
Eurasian; a garden-escape to thickets, borders of woods, and shores in N. America, as in Alta. (Lethbridge; Groh and Frankton 1948), Sask. (N to Nipawin, 53°22′N; Breitung 1957a), Man. (N to The Pas), Ont. (N to near Thunder Bay and Ottawa), Que. (N to Ste-Anne-de-la-Pocatière, Kamouraska Co.; RIM), N.B. (cemetery at St. John, where taken by Fowler in 1863; NBM), and N.S. (Wolfville, Kings Co.; ACAD).
    Forma *albiflora* (DC.) House (flowers white rather than pink) is known from s Ont. (Lambton Co.; Gaiser and Moore 1966) and N.B. (Fredericton; CAN). A hybrid with the Asiatic *L. ruprechtiana* Regel (× *L. notha* Zabel) is reported from NE B.C. by Boivin (1966b).

L. utahensis Wats.   Red Twinberry
/T/W/   (N)   Moist wooded or open slopes at moderate to rather high elevations from s B.C. (N to the Dean R. at ca. 52°N and the Big Bend, a northern route between Revelstoke and Golden; CAN) and sw Alta. (Crowsnest Pass and Waterton Lakes; CAN) to N Calif., Utah, and Wyo. [*Xylosteon* Howell; *L. ebractulata* Rydb.; *L. caerulea* and *L. ciliata sensu* John Macoun 1884, as to the B.C. plants, not L. nor Muhl., respectively, the relevant collections in CAN].

L. villosa (Michx.) R. & S.   Mountain-Fly-Honeysuckle
/ST/X/   (N)   Swampy, peaty, or rocky places at low to moderate elevations, the aggregate species from Wash. and ?B.C. (Boivin 1966*b*) to Labrador and the Atlantic Provinces, s to Calif., Nev., Wyo., Minn., and Pa. (Incl. the red-fruited *L. cauriana* Fern., to which the w N. American plant was referred by M.L. Fernald, Rhodora 27(313):10. 1925).

1   Leaves densely short-villous on both surfaces; branchlets tomentose beneath a long pilosity; calyx-limb ciliate at anthesis; corolla villous or pilose; [*Xylosteum villosum* Michx., the type locality given by Michaux as "a sinu Hudsonis ad Canadam"; *L. caerulea* var. *vill.* (Michx.) T. & G.; Que. (N to SE James Bay at ca. 52°15′N) to Labrador (N to Goose Bay), Nfld., and N.S.] .... var. *villosa*
1   Leaves glabrous to hirsute on both surfaces; calyx-limb and corolla glabrous.
  2   Ends of branchlets glabrous; leaves sparingly pilose or glabrate; [Ont. (N to Big Trout L. at ca. 53°45′N) to Que. (N to E James Bay at 53°50′N and L. Mistassini), Nfld., and N.S.] .... var. *tonsa* Fern.
  2   Ends of branchlets puberulent.
    3   Ends of branchlets both puberulent and more or less short-pilose; leaves pilose beneath, strigose to glabrate above; [*Xylosteum sol.* Eat.; Alta. (near Edmonton) to Que. (N to Ungava Bay at ca. 58°N), Nfld., and N.S.] .... var. *solonis* (Eat.) Fern.
    3   Ends of branchlets merely puberulent; leaves finely pilose to glabrate beneath; [*L. caerulea* var. *cal.* F. & W., the type from Goose Pond, Nfld.; Ont. (N to w James Bay at ca. 53°N) to Que. (N to Ungava Bay at 58°50′N), Nfld., and N.S.] ....
      .... var. *calvescens* (Fern. & Wieg.) Fern.

L. xylosteum L.   Fly-Honeysuckle
Eurasian; a garden-escape to roadsides and thickets in N. America, as in Ont. (N to the Ottawa dist.; Gillett 1958) and sw Que. (N to Hull and Laprairie, near Montreal).

SAMBUCUS L.   [8515]   Elder, Elderberry.   Sureau

1   Inflorescence pyramidal or strongly rounded, commonly longer than broad, panicle-like, with the main axis extending well beyond the lowest branches; flowers yellowish white, ill-scented; leaflets 5 or 7; pith brown; (transcontinental) .... *S. racemosa*
1   Inflorescence flattish-topped, commonly broader than long, the axis scarcely or not at all produced beyond the mostly 4 or 5 subumbellately clustered principal branches (or a pair of large, opposite or subopposite basal branches often present and the inflorescence then more rounded and appearing 3-rayed); flowers usually fragrant; leaflets often more numerous; pith white.
  2   Stem herbaceous, usually less than 1 m tall; leaflets 7–11, most of the lateral ones strongly asymmetrical at base, the lowest pair symmetrical and stipule-like, close to the stem; flowers white or roseate; (introd.) .... *S. ebulus*
  2   Stem woody, the plant a shrub or small tree; petioles not subtended by a pair of stipule-like leaflets.
    3   Leaflets mostly 5 or 7; fruits lustrous-black, about 7 mm thick; cyme usually 3-rayed from base; (introd.) .... *S. nigra*
    3   Leaflets 5–11; fruit to 5 or 6 mm thick; cyme usually 5-rayed from base.
      4   Fruit bluish black beneath a dense waxy bloom, thus appearing pale powdery blue; leaflets mostly rather finely serrate (with commonly 4–6 teeth per cm); (sw B.C.) .... *S. cerulea*
      4   Fruit purplish black, lustrous, without a bloom; leaflets generally more coarsely toothed (commonly with 3–5 teeth per cm); (SE Man. to N.S.) .... *S. canadensis*

S. canadensis L.   Common Elder.   Sureau blanc
/T/EE/   (N (Mc))   Moist woods, meadows, and fields from SE Man. (Shoal L., near the Ont. boundary E of Winnipeg; WIN; the inclusion of Sask. in the range by Rydberg 1922 and 1932, requires clarification) to Ont. (N to the Rainy R. S of Lake of the Woods and Renfrew and Carleton counties), Que. (N to Ste-Modiste, Temiscouata Co.; RIM; the report from Nfld. by Bachelot de la Pylaie 1823, is probably based upon *S. racemosa* var. *pubens*), N.B., P.E.I., and N.S., S to Okla., La., and Ga.

Forma *chlorocarpa* Rehd. (fruits greenish rather than purple-black; leaves pale) is reported from S Ont. by Soper (1949) and Boivin (1966*b*; Morpeth, Kent Co.).

S. cerulea Raf.
/t/W/   (N (Mc))   Valleys and open slopes at low to moderate elevations from S B.C. (Vancouver Is. and mainland chiefly or wholly S of 50°N; the inclusion of Alta. in the range by Rydberg 1922, (and in the following maps) requires clarification) and w Mont. to Calif. and N.Mex. [*S. glauca* Nutt.].   MAPS (*S. gl.*): Hosie 1969:314; Canada Department of Northern Affairs and Natural Resources 1956:284; Preston 1961:364.

S. ebulus L.   Danewort, Dwarf Elder
Eurasian; spreading from cult. to roadsides and waste places in N. America, as in SW Que. (Montreal dist.: Mt. Royal, Longueuil, and Boucherville), where well-established.

S. nigra L.   European Elder
Eurasian; occasionally spread from cult. to waste ground in N. America, as in S Ont. (Goderich, Huron Co.) and Nfld. (Steady Brook, Humber dist.; CAN; MT).

S. racemosa L.   Red-berried or Stinking Elder.   Sureau rouge
/sT/X/EA/   (Mc)   Woods, thickets, and meadows at low to moderate elevations, the aggregate species from the Aleutian Is. and S Alaska (N to ca. 62°30′N) to B.C.-Alta., Sask. (N to 60 mi NE of Nipawin at ca. 53°30′N; Breitung 1957a), Man. (N to Dawson Bay, N L. Winnipegosis), Ont. (N to Renison, S of James Bay at ca. 51°N; Ilmari Hustich, Acta Geogr. 13(2):47. 1955), Que. (N to L. Mistassini, the Côte-Nord, Anticosti Is., and the Gaspé Pen.), Nfld., N.B., P.E.I., and N.S., S to Calif., N.Mex., S.Dak., Ill., Tenn., and Ga.; Eurasia. MAPS and synonymy: *see* below.

1   Leaves glabrous; stipules relatively well developed and persistent; [Eurasia only] . . . .
. . . .[var. *racemosa*]
1   Leaves more or less pubescent beneath; stipules small and often soon deciduous;
   (N. America).
   2   Fruits black or purplish-black; nutlets slightly rugose or pebbly; leaflets glabrous to
      somewhat pubescent beneath; shrub to 2 m tall; [*S. mel.* Gray; B.C.-Alta.] . . . .
      . . . . var. *melanocarpa* (Gray) McMinn
   2   Fruits typically bright red (varying to yellow or even white); leaflets generally more or
      less pubescent beneath.
      3   Nutlets mostly smooth; shrub to 6 m tall; [*S. pubens* var. *arb.* T. & G.; *S. arb.* (T.
         & G.) Howell; *S. callicarpa* Greene; *S. leiosperma* Leib.; Aleutian Is., coastal and
         subcoastal Alaska, and w B.C.; MAP: Hultén 1968*b*:841] . . . .
         . . . . var. *arborescens* (T. & G.) Gray
      3   Nutlets mostly slightly rugose or pebbly; shrub to about 3 m tall . . . .
         . . . . var. *pubens* (Michx.) Koehne
         4   Flowers roseate; [*S. rosaeflora* Carr.; Que.: Batiscan, Champlain Co., and l'Ile
            d'Orléans, near Quebec City; Pierre Dansereau, Nat. can. (Que.) 72(5/6):143.
            1945] . . . . f. *rosaeflora* (Carr.) Scoggan
         4   Flowers white or cream-colour.
            5   Fruit yellow; [*S. pubens* f. *xan.* (Cock.) Rehd.; S Man.: Delta; Boivin
               1966*b*] . . . . f. *xanthocarpa* Cockerell
            5   Fruit bright red.
               6   Leaflets finely dissected; [*S. pubens dissecta* Britt.; S Man.: introd.
                  near Otterburne, about 30 mi S of Winnipeg; Löve and Bernard 1959] . . . . . .
                  . . . . f. *dissecta* (Britt.) Scoggan

6 Leaves merely sharply serrate.
    7 Branchlets glabrous; [*S. pubens* f. *calva* Fern.; Ont., Que. (type
       from the Gaspé Pen.), N.B., and N.S.] .... f. *calva* (Fern.) Scoggan
    7 Branchlets pubescent; [*S. pubens* Michx.; *S. ?canadensis sensu*
       Bachelot de la Pylaie 1823, not L.; transcontinental; MAP (aggre-
       gate species): Hultén 1968*b*:841] .... f. *pubens*

SYMPHORICARPOS Duhamel [8518] Snowberry

1 Stems trailing and rooting at the nodes, the branches rising less than 5 dm; corolla to 6
  mm long and about as broad, its lobes about as long as the tube; anthers about 1 mm
  long, a little shorter than their filaments; style glabrous, 2 or 3 mm long, included; fruit to 6
  mm long; nutlets to 3 mm long; (S B.C.) .... *S. mollis*
1 Stems erect, more or less branching, to over 1 m tall.
    2 Corolla relatively long and narrow, to about 1 cm long and 6 mm broad, its lobes not
      more than half the length of the tube; anthers mostly 1 or 2 mm long, about as long as
      their filaments; style glabrous, to 4 mm long, included; fruit to 1 cm long; (B.C.) ....
      .... *S. oreophilus*
    2 Corolla about as broad as long, the lobes about equalling the tube.
      3 Flowers sessile, relatively many in a cluster; corolla-lobes 3 or 4 mm long; style
        usually at least 4 mm long, commonly long-hairy near the middle, more or less
        exserted; anthers about half as long as their filaments; leaves becoming
        coriaceous; (B.C. to w Ont.; introd. eastwards) .... *S. occidentalis*
      3 Flowers short-pedicelled, usually few in a cluster; corolla-lobes 2 or 3 mm long;
        style 2 or 3 mm long, glabrous, included in the corolla; anthers about equalling or
        longer than their filaments; leaves relatively thin; (transcontinental) .... *S. albus*

S. albus (L.) Blake   Snowberry.   Belluaine
/sT/X/   (N)   Woods, thickets, and open slopes at low to moderate elevations from SE Alaska (N to
ca. 59°30′N) and S Dist. Mackenzie (Fort Smith, near the Alta. boundary; not known from the
Yukon) to L. Athabasca (Alta. and Sask.), Man. (N to Gillam, about 165 mi S of Churchill), Ont. (N to
Sandy L. at ca. 53°N, 93°W), and Que. (N to L. Mistassini, the Côte-Nord, and Gaspé Pen.; var.
laevigatus introd. in the Atlantic Provinces), S to Calif., Colo., Nebr., and Va. [*Vaccinium album* L.,
the type probably from near Quebec City, Que., according to H.K. Svenson, Rhodora
39(467):461–62. 1937; *Xylosteon* Moldenke; *S. pauciflorus* (Rob.) Britt.; *S. pubescens* Pers.; *S.
(Lonicera; Symphoria) racemosus* Michx.]. MAP: Hultén 1968*b*:842.

    Some (?all) of our Pacific slope material is referable to var. *laevigatus* (Fern.) Blake (*S.
racemosus* var. *laev.* Fern.; *S. rivularis* Suksd.; leaves and branchlets essentially glabrous, the
fruits to 1.5 cm long, rather than leaves more or less pilose beneath, the branchlets minutely
puberulent or glabrate, the fruits mostly not over 1 cm long; introd. in the East, as in S Ont. (Soper
1949), Que. (Tadoussac, Saguenay Co.; GH), Nfld. (Rouleau 1956), N.B., P.E.I., and N.S.). MAP:
Hultén 1968*b*:842.

S. mollis Nutt.   Western Snowberry
/t/W/   (Ch)   Woods and open slopes at low to moderate elevations from SW B.C. (Vancouver Is.
and adjacent islands and mainland E to Yale and Manning Provincial Park, SE of Hope) and N Idaho
to S Calif.
    The B.C. plant is referable to var. *hesperius* (Jones) Cronq. (*S. hesp.* Jones; plant subglabrous
or only moderately hairy rather than densely hairy).

S. occidentalis Hook.   Wolfberry
/sT/WW/   (N)   Open prairies and moist low ground from Great Slave L. and B.C.–Alta. to Sask.
(N to Prince Albert), Man. (N to Flin Flon, ca. 55°N), and w Ont. (probably native E to Kenora,
possibly to Thunder Bay; apparently introd. farther eastwards in Ont. and at Farnham, Missisquoi
Co., SW Que.), S to Wash., Utah, N.Mex., Kans., Mo., and Mich. [*Symphoria* R. Br.].

S. oreophilus Gray   Western Snowberry
/sT/W/   (N)   Dry meadows and open slopes at low to high elevations from s B.C. (Mt. Sophia, sw of Rossland, where taken by J.M. Macoun in 1902; CAN, verified by G.N. Jones) and ?Alta. (a sterile collection in CAN from near Peace River, ca. 56°14'N, has been placed here by Raup) to Calif., N Mexico, and N.Mex. [*S. utahensis* and *S. vaccinioides* Rydb.].

TRIOSTEUM L.   [8517]   Feverwort, Horse-Gentian

1   Stem sparingly glandular-puberulent, usually with an admixture of spreading glandless
     hairs to 1.5 mm long; leaves usually all distinct and tapering to a narrow base, rather
     sparingly pubescent beneath; sepals about 2 mm broad; corolla red-purple; style usually
     included; fruit ellipsoid-ovoid, bright orange-red; (Ont. to N.S.) . . . . *T. aurantiacum*
1   Stem densely glandular-puberulent above; leaves connate-perfoliate at base, soft-
     pubescent beneath; sepals about 1.5 mm broad; corolla dull greenish-yellow to purplish;
     style exserted; fruit subglobose, dull orange-yellow; (Ont. and sw Que.) . . . . *T. perfoliatum*

T. aurantiacum Bickn.   Wild Coffee
/T/EE/   (Hp)   Rich woods and thickets from Ont. (N to Batchawana Falls, N of Sault Ste. Marie, Ottawa, and Prescott) to Que. (N to Grosse-Ile, about 25 mi NE of Quebec City; MT), N.B. (St. John R. system in Carleton and Victoria counties; GH; NBM; not known from P.E.I.), and N.S. (Inverness Co., Cape Breton Is.; ACAD; CAN), s to Iowa and Ga.

T. perfoliatum L.   Tinker's-weed, Wild Coffee
/T/EE/   (Hp)   Rich woods and thickets from Nebr. to Minn., Ont. (N to Ottawa and Hawkesbury; CAN), sw Que. (Kingsmere, N of Hull; CAN), and N.Y., s to E Kans. and Ga.

VIBURNUM L.   [8516]   Viburnum, Arrow-wood.   Viorne

1   Marginal flowers sterile and with greatly enlarged white corollas.
   2   Leaves pinnately veined, finely serrate, broadly ovate-cordate; cymes sessile; drupe
        red, becoming blackish; young growth rusty-stellate-tomentose; (Ont. to N.S.) . . . .
        . . . . *V. alnifolium*
   2   Leaves palmately 3–5-nerved and lobed, more coarsely serrate, obtuse to truncate at
        base; cymes peduncled; pubescence not stellate; (transcontinental) . . . . *V. opulus*
1   Marginal flowers fertile and not differentiated.
   3   Principal leaves 3-lobed toward summit, broadly rounded to subcordate at the
        palmately 3–5-ribbed base; cymes peduncled.
        4   Leaves coarsely toothed, soft-downy and copiously dotted beneath, 3-ribbed at
             base; petioles slender; stipules often present; cymes on peduncles to 5 cm long;
             drupe purple-black; young growth finely stellate-pubescent; (Ont. and sw Que.) . . . .
             . . . . *V. acerifolium*
        4   Leaves more finely toothed, pilose beneath at the 3–5-ribbed base and in the
             vein-axils; petioles stout; stipules none; cymes shorter-peduncled; drupe yellow,
             becoming orange or red; pubescence not stellate; (transcontinental) . . . . *V. edule*
   3   Leaves unlobed, finely to coarsely toothed, pinnately veined; mature drupes dark
        purple to blue-black or black.
        5   Lateral veins of leaves straightish, simple or once or twice forked, ending in the
             coarse teeth; pubescence (when present) consisting of tufts of hairs; winter-buds
             with 2 pairs of outer scales.
             6   Petioles glabrous or stellate-pubescent, to 3 cm long; stipules usually none;
                  leaves glabrous or pilose on the veins beneath; cyme long-peduncled; (Ont.,
                  sw Que., and N.B.) . . . . *V. dentatum*
             6   Petioles pubescent, mostly less than 1 cm long; linear stipules present; cyme
                  sessile to long-peduncled; (Man. to sw Que.) . . . . *V. rafinesquianum*
        5   Lateral veins of leaves more or less curved, freely branching and forming a
             network before reaching the fine teeth; winter-buds naked or with a single pair of
             scurfy outer scales.

7 Cyme sessile; leaves abruptly and sharply long-acuminate, sharply serrulate; winter-buds with a pair of outer scales; (SE Sask. to Que.) .... *V. lentago*
7 Cyme more or less peduncled; leaves obtuse to bluntly short-acuminate.
    8 Winter buds naked; leaves closely serrulate, rounded or subcordate at base; stone with 3 ventral grooves; young growth and lower leaf-surfaces cinereous with minute stellate pubescence; (introd.) .... *V. lantana*
    8 Winter-buds with a pair of outer scales; leaves crenulate or dentate, rounded or tapering at base; stone not grooved; young shoots and lower leaf-surfaces more or less brown-scurfy; (Ont. to Nfld. and N.S.) ....
.... *V. cassinoides*

V. acerifolium L.   Maple-leaved Viburnum, Dockmackie
/T/EE/ (N) Dry or rocky woods from Minn. to Ont. (N to Renfrew and Carleton counties; *see* S Ont. map by Soper and Heimburger 1961:38), SW Que. (N to S Pontiac and S Gatineau counties and the Montreal dist.), and New Eng., S to Tenn. and Ga.
Forma *collinsii* Rouleau (petals pink rather than creamy white) is reported from S Ont. by Gaiser and Moore (1966; Lambton Co.).

V. alnifolium Marsh.   Hobblebush.   Bois d'Orignal
/T/EE/ (N (Mc)) Woods and cool ravines from Ont. (N to L. Nipissing, Chalk River, Renfrew Co., and the Ottawa dist.; *see* S Ont. map by Soper and Heimburger 1961:39) to Que. (N to Ste-Anne-de-la-Pocatière, Kamouraska Co.; QSA), N.B., P.E.I., and N.S., S to Tenn. and Ga. [*V. lantanoides* Michx.].

V. cassinoides L.   Witherod, Wild-raisin.   Alisier or Bleuets sains
/T/EE/ (N (Mc)) Swamps, thickets, borders of woods, and clearings from Ont. (N to L. Abitibi at ca. 48°50′N) to Que. (N to the Côte-Nord, Anticosti Is., and Gaspé Pen.), Nfld., N.B., P.E.I., and N.S., S to Wisc., Ohio, Tenn., and Ala. [*V. nudum* var. *cass.* (L.) T. & G.; *V. nudum* of most or all Canadian reports, not L., relevant collections in CAN, GH, and NSPM; *V. prunifolium sensu* Hooker 1833, not L.; *Cornus paniculata sensu* Lindsay 1878, not L'Hér., the relevant collection in NSPM].

V. dentatum L.   Arrow-wood
/T/EE/ (N (Mc)) Damp thickets from Ont. (N to Ottawa; *see* S Ont. map by Soper and Heimburger 1961:44) to SW Que. (Ste-Agathe, about 45 mi NW of Montreal; Marcel Raymond, Ann. ACFAS 23:96. 1957) and N.B. (Fernald *in* Gray 1950; not known from P.E.I. or N.S.), S to Tex. and Fla. [*V. pubescens* (Ait.) Pursh].
Our material is apparently all referable to the northern, essentially glabrous extreme, var. *lucidum* Ait. (*V. recognitum* Fern.).

V. edule (Michx.) Raf.   Squashberry, Mooseberry.   Pimbina
/ST/X/ (N) Woods and thickets from Alaska–Yukon–Dist. Mackenzie (N to ca. 69°N) to L. Athabasca (Alta. and Sask.), northernmost Man.–Ont., Que. (N to Ungava Bay and the Côte-Nord), Labrador (N to Turnavik, 55°18′N), Nfld., N.B. (St. John and Nipisiguit River systems; not known from P.E.I.), and N.S. (Inverness and Victoria counties, Cape Breton Is.), S to Oreg., Idaho, Colo., Minn., Pa., and New Eng. [*V. opulus* var. *ed.* Michx.; *V. pauciflorum* La Pylaie; *V. acerifolium sensu* Bongard 1833, not L.; *V. oxycoccus sensu* Richardson 1823, and Hooker 1833, in part, not Pursh]. MAPS: Hultén 1968*b*:842; Raup 1947: pl. 34.

V. lantana L.   Wayfaring-tree, Twistwood
Eurasian; spread from cult. to fence-rows and roadsides in N. America, as in Ont. (Rondeau Park, Kent Co.; Guelph, Wellington Co.; Ottawa) and SW Que. (Boivin 1966*b*).

V. lentago L.   Sweet Viburnum, Nannyberry.   Alisier or Bourdaine
/T/(X)/ (Mc (Ms)) Woods and thickets from SE Sask. (Gainsborough, about 160 mi SE of Regina; Breitung 1957*a*) to Man. (N to Duck Mt.), Ont. (N to the Rainy R. N of Lake of the Woods and Renfrew and Carleton counties; *see* S Ont. map by Soper and Heimburger 1961:42), and SW Que. (N to S Portneuf Co.; reports from farther eastwards in Que. and from the Atlantic Provinces are

apparently largely or wholly based upon *V. cassinoides,* relevant collections in CAN, GH, NBM, and NSPM), s to Colo., S.Dak., Mo., and Ga. MAPS: Hosie 1969:312; Preston 1961:366; Hough 1947:417; Braun 1935: fig. 3, p. 355.

V. opulus L.   Guelder-Rose
/T/X/EA/   (Mc)   Moist woods, thickets, and shores (ranges of Canadian taxa outlined below), s to Wash., Wyo., S.Dak., Ill., Pa., and New Eng.; Eurasia. MAP and synonymy: *see* below.
1   Glands on the petioles near the juncture with the leaf-blade sessile or subsessile, concave-topped, thicker than long; stipules attenuate-tipped; [Eurasian; a garden-escape in s Ont. (Rock Glen, Lambton Co.; OAC) and ?N.S. (Boivin 1966*b*). Var. *roseum* L., the commonly cultivated snowball-tree, has a more rounded inflorescence with all of the flowers neutral and enlarged] .... var. *opulus*
1   Glands on the petioles columnar or clavate and thus more or less stipitate, round-topped, as long as or longer than thick; stipules thick-tipped and bluntish; [ssp. *trilobum* (Marsh.) Hult. (*V. tri.* Marsh.); *V. americanum* of auth., not Mill.; *V. oxycoccus* Pursh; *V. acerifolium sensu* Lindsay 1878, not L. (relevant collection in NSPM); B.C. (N to Quesnel, ca. 53°N), Alta. (N to Edmonton), Sask. (N to Mistatim, 52°52′N), Man. (N to Hill L., N of L. Winnipeg), Ont. (N to w James Bay at 52°14′N), Que. (N to L. Waswanipi at 49°39′N, 76°30′W, L. St. John, Anticosti Is., and the Gaspé Pen.), Nfld., N.B., P.E.I., and N.S.; MAP: Hultén 1958: map 55 (*V. tri.*), p. 75] .... var. *americanum* Ait.

V. rafinesquianum Schultes   Downy Arrow-wood
/T/EE/   (N)   Open woods, thickets, and barrens from Man. (N to Duck Mt.) to Ont. (N to North Fowl L., near the Minn. boundary w of Thunder Bay, and the Ottawa dist.; *see* s Ont. map by Soper and Heimburger 1961:43) and sw Que. (N to s Pontiac Co. and the Montreal dist.; not known from the Atlantic Provinces), s to Ark., Ky., and Ga. [*V. affine* var. *hypomalacum* Blake].

Var. *affine* (Bush) House (*V. affine* Bush; leaves softly stellate-pubescent beneath rather than merely pilose on the veins, the petioles often shorter than the stipules rather than usually surpassing them) is known from s Ont. (Lambton Co.; OAC; reported from the Bruce Pen., L. Huron, by Krotkov 1940).

ADOXACEAE (Moschatel Family)

ADOXA L.   [8526]   Moschatel

Low delicate glabrous herb to about 2 dm tall from a scaly rhizome. Basal leaves long-petioled, 1–3-ternate into oblong or obovate divisions, the stem-leaves a single similar opposite pair (but smaller and less divided). Flowers perfect, gamopetalous, semi-epigynous, in clusters of about 5 on a slender peduncle. Sepals 2, 3, or 4. Corolla yellowish or greenish, rotate, about 8 mm broad, with 4 or 5 lobes, each sinus bearing a pair of separate or slightly united stamens. Styles 4 or 5. Ovary partially inferior. Fruit a dry greenish drupe with 4 or 5 nutlets.

A. moschatellina L.   Moschatel
/ST/X/EA/   (Grh)   Cool mossy woods, thickets, and wet rocks (chiefly calcareous) from Alaska (N to Cape Lisburne, ca. 68°N), the Yukon (N to ca. 65°N), and sw Dist. Mackenzie to B.C.–Alta., Sask. (Candle Lake, 53°45′N; Breitung 1957*a*), Man. (Duck Mt.; CAN; reported N to N of Berens R., L. Winnipeg, by Lowe 1943), Ont. (Boivin 1966*b*), Minn., Wisc., Iowa, and the Catskill Mts. of N.Y., S in the West to Colo., N.Mex., and S.Dak.; Eurasia. MAPS: Hultén 1968*b*:845, and 1937: fig. 14, p. 129; Porsild 1966: map 133, p. 83; Meusel 1943: fig. 30f.

VALERIANACEAE (Valerian Family)

Herbs with opposite, simple or pinnately compound, exstipulate leaves. Flowers regular or slightly irregular, perfect or sometimes unisexual, gamopetalous, epigynous, in terminal panicled or capitate cymes. Corolla commonly gibbous on one side or with an unequally 5-lobed limb. Stamens 3, inserted on the corolla-tube. Style 1. Ovary inferior. Fruit dry and indehiscent, 1-seeded, either 1-locular or 3-locular (then with 2 of the locules empty).

1 Principal stem-leaves deeply pinnate; calyx consisting of inrolled bristles expanding to form a conspicuous plumose pappus; flowers white to roseate; fruit 1-locular . . . . *Valeriana*
1 Stem-leaves entire or merely toothed near base; calyx minute or obsolete.
  2 Corolla at most 4 mm long; stigma 3-lobed; fruit 3-locular but only 1-seeded; stem dichotomously branched above, the flowers borne in cymose clusters at the ends of the branches; (s Ont.) . . . . *Valerianella*
  2 Corolla to over 6 mm long; stigma 2(3)-lobed; fruit 1-locular (the two sterile locules obsolete); stem simple or with opposite axillary branches, the inflorescence consisting of spike-like or head-like cymose clusters; (s B.C.) . . . . *Plectritis*

PLECTRITIS DC. [8527]

1 Convex side of the fruit more or less sharply keeled, not grooved; wings of fruit (when developed) tending to be connivent toward base and divergent above; hairs of fruit (when present) more or less pointed; cotyledons transverse to the ventral face of the fruit; corolla to 6 mm long . . . . *P. congesta*
1 Convex side of the fruit broader, scarcely keeled, bearing a narrow groove down the centre; wings of fruit (when present) commonly about equally divergent above and below; hairs of fruit (when present) clavate or long-cylindrical and blunt; cotyledons parallel to the ventral face of the fruit; corolla to 8 mm long . . . . *P. macrocera*

P. congesta (Lindl.) DC.   Sea-blush
/t/W/   (T)   Open slopes and moist meadows from w B.C. (Queen Charlotte Is.; Vancouver Is. and adjacent islands and mainland E to Vernon; CAN; V) to s Calif. [*Valerianella* Lindl.; *V. (P.) anomala* Gray; *Betckea (P.; V.) samolifolia* DC.].

P. macrocera T. & G.
/t/W/   (T)   Streambanks and moist open places from sw B.C. (Vancouver Is. and adjacent islands; mainland N to Oliver, about 60 mi sw of Kamloops) and Mont. to s Calif. and Utah. [*Valerianella* Gray].

VALERIANA L.   [8532]   Valerian

(Ref.: Meyer 1951)
1 Leaves thickish, nearly parallel-ribbed, those of the stem simple or pinnate with narrowly oblanceolate segments, the basal ones entire or sometimes with 1 or 2 pairs of lobes, gradually narrowed to short winged petioles; flowers unisexual, the yellowish-white corolla essentially rotate; inflorescence panicle-like even at anthesis; fruit ovoid; root fusiform; (s B.C.; s Ont.) . . . . *V. edulis*
1 Leaves thin, reticulate-veined, the basal ones abruptly narrowed to long slender petioles; flowers white to roseate, at least some of them perfect; inflorescence corymbiform at anthesis (but often more expanded in fruit); fruit lanceolate to lance-oblong in outline; roots fibrous.
  2 Basal leaves all pinnate with at least 8 pairs of lateral lobes, they and their rachises hirsute; stem-leaves with up to 25 lance-acuminate leaflets, the terminal segment not greatly enlarged; corolla roseate, the tube about 4 mm long, the lobes 1 mm long; (introd.) . . . . *V. officinalis*
  2 Basal leaves undivided or with a few lobes or leaflets; stem-leaves with at most 6

pairs of lateral leaflets and a much larger terminal one; rachises and lower surfaces of leaves glabrous; corolla white or pinkish.

  3 Corolla at most 4 mm long, the spreading lobes about equalling the indistinctly gibbous or straight tube; bractlets of inflorescence eciliate; some of the flowers perfect, others pistillate.

    4 Achenes commonly lanceolate, glabrous; lateral lobes or leaflets of the stem-leaves mostly well under 1 cm broad; plant usually less than 4 dm tall, not very leafy; (transcontinental) .... *V. dioica*

    4 Achenes mostly somewhat broader and short-hairy (occasionally glabrous); lateral lobes or leaflets of stem-leaves often over 1 cm broad; plant to about 9 dm tall, usually more leafy; (s ?B.C.) .... [*V. occidentalis*]

  3 Corolla usually at least 5 mm long (to 9 mm), the ascending lobes much shorter than the distinctly gibbous tube; flowers all perfect.

    5 Upper stem-leaves sessile or nearly so, simple or 3-lobed to 3-foliolate; bracts of inflorescence glabrous or pubescent only at base; (Alaska–Yukon–w Dist. Mackenzie–N B.C.) .... *V. capitata*

    5 Stem-leaves all distinctly petioled, 3–5-foliolate; bracts of inflorescence ciliate; (essentially transcontinental but with a large west-central gap) .... *V. sitchensis*

V. capitata Pallas

/aSs/W/EA/ (Grh) Moist tundra and slopes at low to fairly high elevations from the Aleutian Is. and coasts of Alaska–Yukon–NW Dist. Mackenzie to N B.C. (Ogilvie Mt., Cassiar dist., ca. 58°30′N; reports from farther southwards by John Macoun 1884, are based upon *V. sitchensis*, the relevant collections in CAN); NE Europe; Asia. [*V. bracteosa* Britt., not Philippi]. MAPS: Hultén 1968*b*:845; Porsild 1966: map 134, p. 83; Meyer 1951: fig. 8 (somewhat incomplete), p. 403.

V. dioica L. Marsh Valerian

/ST/X/EA/ (Grh) Moist places and wet meadows at low to high elevations from cent. Yukon (N to ca. 64°N) and Great Bear L. to B.C.–Alta., Sask (N to Prince Albert National Park and Emma Lake, 53°34′N; Breitung 1957a), Man. (N to Churchill), northernmost Ont., Que. (N to E Hudson Bay at ca. 57°N, L. Mistassini, Anticosti Is., and the Gaspé Pen.; not known from the Maritime Provinces, early N.B. reports apparently referring to *V. sitchensis* ssp. *uliginosa*), SW Labrador (Labrador City, ca. 53°N, 67°W; CAN), and Nfld., S to Wash., Idaho, Wyo., S Sask.–Man., and cent. Ont. (S to Longlac, N of L. Superior at 49°48′N); Europe; SW Asia. MAPS: Hultén 1968*b*:846, and 1958: map 50, p. 69; Meyer 1951: fig. 15 (ssp. *sylv.*; somewhat incomplete northwards), p. 418; Raup 1947: pl. 35.

The N. American plant has been distinguished as var. *sylvatica* (Sol.) Gray (*V. sylv.* Sol., not Schmidt.; *V. septentrionalis* Rydb.) on minor and perhaps insignificant characters.

V. edulis Nutt.

/t/(X)/ (Grt) Moist meadows (sometimes saline) and open places from southernmost B.C. (Midway, about 50 mi SE of Penticton, where taken by J.M. Macoun in 1905; CAN, verified by J.A. Calder; reported from SE Kootenay by Henry 1915) and Wash. to Mont., S.Dak., Minn., Mich., and S Ont. (Middlesex, Huron, Brant, and Waterloo counties; the report from Swan River, Man., by Jackson et al. 1922, requires confirmation), S to Oreg., Mexico, N.Mex., S.Dak., Iowa, and Ohio. [*Patrinia* (*V.*) *ceratophylla* Hook.]. MAP (aggregate species): Meyer 1951: fig. 18, p. 425.

The S Ont. plant may be separated as ssp. *ciliata* (T. & G.) Meyer (*V. cil.* T. & G.; leaves more or less short-hairy rather than usually glabrous).

[V. occidentalis Heller] Western Valerian

[This species of the w U.S.A. (N to Oreg. and Mont.) is accredited to B.C. in the range given by Rydberg (1922) and collections in CAN from S B.C. (Vancouver Is.; Cascade) have been placed here. However, the MAP by Meyer (1951: fig. 14, p. 415) indicates no Canadian stations and the B.C. plants may prove referable to *V. dioica*.]

V. officinalis L. Common Valerian

Eurasian; a garden-escape to roadsides and thickets in N. America, as in S B.C. (Yarrow and

Chilliwack; V), Man. (Riding Mt.; Lowe 1943; tentatively reported from Ochre River by Jackson et al. 1922), Ont. (N to Espanola, N of L. Huron, and Ottawa), Que. (N to La Malbaie, Charlevoix Co.; Groh and Frankton 1949b), N.B., P.E.I., and N.S.

V. sitchensis Bong.
/ST/X/ (Grh (Hsr)) Moist open or wooded places at low to fairly high elevations: cent. Alaska–Yukon–w Dist. Mackenzie through B.C. and w Alta. to N Calif. and Idaho; Mich. to s Ont. (N to Grey, Victoria, Peterborough, and Hastings counties), Que. (N to E Hudson Bay at ca. 56°30′N and the Gaspé Pen.), and N.B. (Charlotte, Restigouche, and Gloucester counties; not known from P.E.I. or N.S.), s to Ohio, N.Y., and Vt. MAPS and synonymy: *see* below.
1 Stem-leaves with (3)4–6 pairs of lateral lobes, the terminal lobe lanceolate to elliptic, acute to acuminate; [*V. sylvatica* var. *ulig.* T. & G.; *V. ulig.* (T. & G.) Rydb.; Ont. to N.B., as outlined above; MAP: Meyer 1951: fig. 7, p. 395] .... ssp. *uliginosa* (T. & G.) Meyer
1 Stem-leaves with 1–3(4) pairs of lateral lobes, the terminal lobe obovate to suborbicular, acute or obtuse.
    2 Slender plant to about 7 dm tall; leaves chiefly basal, their segments entire or nearly so, the terminal lobe less than 4 cm broad; corolla to 9 mm long; [*V. scouleri* Rydb.; s B.C. (Vancouver Is.; Chilliwack; Rossland; Yale; Skagit Valley) and sw Alta. (N to Banff); MAP: Meyer 1951: fig. 6, p. 395] .... ssp. *scouleri* (Rydb.) Meyer
    2 Robust plant to about 12 dm tall; leaves chiefly cauline, their segments coarsely crenate or wavy-margined, the terminal lobe to over 6 cm broad; corolla to 7 mm long; [western part of the range, the type from Sitka, SE Alaska; MAPS: Hultén 1968b:846; Porsild 1966: map 135, p. 83; Meyer 1951: fig. 5, p. 395] .... ssp. *sitchensis*

VALERIANELLA Mill.   [8529]   Corn-salad, Lamb's-lettuce

1 Corolla 3 or 4 mm long, white or with pinkish lobes; fruit dorsiventrally compressed, the fertile locule twice as long as the two united empty ones; bractlets subtending cymules glabrous, broadly lanceolate; (s Ont.) .... *V. chenopodifolia*
1 Corolla about 1.5 mm long, white with bluish lobes; fruit laterally compressed, the fertile locule about as broad as the two empty ones together and with a very rigid corky back; bractlets subtending cymules ciliate, the outer ones spatulate, rounded at summit; (introd. on Vancouver Is. and in s Ont.) .... *V. locusta*

V. chenopodifolia (Pursh) DC.
/t/EE/ (T) Meadows and moist places from Ind. and s Ont. (St. Thomas, Elgin Co.; Ridgeway, Welland Co.; Yorkshire Is., near Kingston, Frontenac Co.; CAN; OAC; TRT) to Pa. and N.Y. [*Fedia* Pursh; *see* Sarah Dyal, Rhodora 40(473):195. 1938].

V. locusta (L.) Betcke   Corn-salad, Lamb's-lettuce
Eurasian; a garden-escape to roadsides, old fields, and waste places in N. America, as in sw B.C. (Vancouver Is.; Henry 1915) and s Ont. (St. Catherines, Lincoln Co., where taken by Dearness in 1892; CAN; reported from Essex and Welland counties by John Macoun 1884). [*Valeriana* L.; *Valerianella olitoria* (L.) Poll.].

DIPSACACEAE (Teasel Family)

Herbs with opposite, simple or compound, exstipulate leaves. Flowers gamopetalous, perfect or in part unisexual, more or less irregular, epigynous, in dense heads subtended by a foliaceous involucre. Corolla usually 4-cleft (sometimes 5-cleft). Stamens commonly 4 (sometimes 2), inserted on the corolla-tube, their anthers distinct (in the similar Compositae, stamens 5, their anthers united into a tube). Style 1. Ovary inferior. Fruit achene-like, 1-seeded. (Introd. species).

1   Plant spiny and thistle-like; involucre consisting of several linear-filiform spine-like incurving prickly leaves to over 8 cm long; head ovoid, to about 7 cm long, the receptacular chaff tapering into long rigid awns; calyx 4-toothed or -lobed; leaves lanceolate to lance-oblong, subentire to coarsely pinnatifid, prickly on the midrib beneath and often on the margins; (introd.) .... *Dipsacus*
1   Plant not spiny; (introd.).
   2   Involucre consisting of several series of rigid silky-hairy overlapping bracts; flowers creamy or yellow, in rather flat heads to about 5 cm thick; calyx 4-toothed or -lobed; leaves pinnatifid into linear to oblong segments .... *Cephalaria*
   2   Involucre consisting of herbaceous bracts in 1 or 2 series; heads smaller; calyx consisting of usually 5 bristle-teeth terminating a cup-like base.
      3   Receptacle merely more or less hairy, lacking chaff; calyx consisting of usually 8 ciliate bristle-teeth terminating a cup-like base; corolla lilac to purple; middle and upper stem-leaves 1–2-pinnatifid, the lower leaves often entire or merely shallowly lobed .... *Knautia*
      3   Receptacle chaffy (each individual flower subtended by a bractlet); calyx consisting of usually 5 bristle-teeth terminating a cup-like base.
         4   Leaves entire or shallowly toothed; receptacular chaff about equalling the flowers .... *Succisa*
         4   Basal leaves lyrate or sometimes subentire, most of the stem-leaves divided to base into linear-lanceolate segments (the segments themselves commonly pinnatifid), greyish-pubescent; receptacular chaff much shorter than the yellowish-white flowers; (introd. on Vancouver Is.) .... [*Scabiosa*]

CEPHALARIA Schrad. [8541]

C. alpina (L.) Schrad.
Eurasian; apparently known in N. America only from Que. (railway ballast between Carleton and New Richmond, Bonaventure Co., Gaspé Pen.; GH; OAC; reported by Marcel Raymond and James Kucyniak, Nat. can. (Que.) 74(3/4):63. 1947, as an aggressive weed at the Montreal Botanical Garden). [*Scabiosa* L.].

Most of the Que. material was originally distributed as *C. tatarica* (L.) Schrad. (marginal flowers of the head enlarged and falsely radiate rather than not conspicuously enlarged), this noted by Bailey (1949a) as the garden-plant in N. America often called *C. alpina*. The Que. plant may prove referable to this former species.

DIPSACUS L. [8450]

D. fullonum L.   Teasel.   Cardère
Eurasian; introd. along roadsides and in old fields and pastures in N. America, as in SW B.C. (near Vancouver; Eastham 1947), Ont. (N to Timmins and Ottawa; see S Ont. map by Montgomery 1957: fig. 11, p. 29), and SW Que. (Montreal dist.). [*D. sylvestris* Huds.].

According to Clapham, Tutin, and Warburg (1962), "This is the wild plant whose receptacular bracts are too flexible for use in combing cloth." Our material is chiefly or wholly referable to this phase in spite of the fact that ssp. *sativus* (L.) Thell. (*D. sativus* (L.) Scholler, the Fuller's Teasel with receptacular bracts ending in a stiff recurved spine rather than a long straight spine) was formerly cultivated in N. America for use in wool-mills. Concerning the varying applications of the names used here, *see* Hitchcock et al. (1959:480), Fernald *in* Gray (1950:1347), and Clapham, Tutin, and Warburg (1962:797).

KNAUTIA L.   [8543]

K. arvensis (L.) Coult.   Field Scabious
Eurasian; introd. along roadsides and in dry fields and pastures in N. America, as in B.C. (N to Fort Fraser and Vanderhoof, both ca. 54°N), Alta. (N to Fort Saskatchewan), Sask. (N to Shellbrook, about 25 mi W of Prince Albert), Man. (Bethany, about 35 mi N of Brandon; Winnipeg dist.), Ont. (Guelph, Wellington Co.; OAC), Que. (N to Carleton, S Gaspé Pen.), Nfld. (near St. John's; CAN; GH), and N.B. (Boivin 1966b). [*Scabiosa* L.].

[SCABIOSA L.]   [8546]

[S. ochroleuca L.]   Devil's-bit
[Eurasian; an occasional garden-escape to waste places in N. America but scarcely established, as in SW B.C. (a vacant lot at Victoria, Vancouver Is., where taken by John Macoun in 1908; CAN).]

SUCCISA Moench   [8542]

1   Leaves narrowly lanceolate, only those of the branches much reduced; heads finally subcylindric; calyx shortly 4–5-toothed, awnless; corolla pale blue; fruit glabrous, strongly 8-ribbed; bractlets subtending individual flowers glabrous, 8-ribbed, with a small crenate-lobed spreading border . . . . *S. australis*
1   Leaves mostly in a basal tuft, oblanceolate to ovate, the 1 or 2 pairs of upper ones greatly reduced; heads subglobose; calyx with 4 or 5 short awn-teeth; corolla mauve to dark blue-purple (rarely white); fruit downy, 4-angled; bractlets subtending flowers villous, somewhat 4-angled, 4-toothed . . . . *S. pratensis*

S. australis (Wulf.) Reichenb.
European; introd. into wet fields and ditches in N. America, as in S Ont. (a 1910 collection at Guelph, Wellington Co., noted by Montgomery 1957) and SW Que. (Chambly, Iberville, and St-Jean counties; CAN; GH; MT). [*Scabiosa* Wulf.].

S. pratensis Moench   Devil's-bit Scabious
Eurasian; introd. into fields and waste places in N. America, as in S Ont. (Grey and Wellington counties; TRT), SW Que. (Covey Hill, Huntingdon Co.; MT), and N.S. (Louisbourgh, Cape Breton Co., where taken by John Macoun in 1898 and by later collectors in 1949 and 1951; ACAD; CAN). [*Scabiosa succisa* L.].

CUCURBITACEAE (Cucumber or Gourd Family)

Vines (trailing or climbing by tendrils) with large simple alternate leaves, these commonly lobed and shallowly serrate. Flowers regular, unisexual, epigynous, the 5 or 6 sepals sometimes distinct to base, the 5 or 6 petals more or less united. Stamens usually 3 (one of the anthers 1-locular), sometimes 5, united by their anthers (and sometimes, also, by their filaments). Style 1. Ovary inferior. Fruit a commonly fleshy (sometimes membranaceous) pepo (cucumber, melon, etc.).

1 Leaves deeply pinnately divided into 3 or 4 pairs of broadly obovate lobes (the lobes themselves lobed and toothed), ovate to ovate-oblong in outline, cordate at base, to about 1.5 dm long; tendrils forking; flowers solitary in the leaf-axils; corolla light yellow, rotate, about 4 cm broad, the 5 lobes obovate and obtuse; fruit the familiar watermelon; hairy annual; (introd.) .... [*Citrullus*]
1 Leaves entire or at most lobed about half the width of the blade.
  2 Leaves unlobed, cordate-ovate, acuminate, minutely and regularly but remotely serrulate, harshly scabrous; tendrils forking; plants unisexual, the staminate and pistillate flowers alike, commonly solitary in the leaf-axils, with yellow corollas to about 2.5 cm long, the corolla-lobes recurving; fruit not prickly, about 5 cm long; tuberous-rooted perennial; (introd.) .... *Thladiantha*
  2 Leaves commonly palmately 5-lobed (or merely angled or 3-lobed in *Cucumis*), the lobes pointed.
    3 Fruit a smooth red globose berry to 8 mm thick; plants dioecious, the greenish flowers in axillary cymes, those of the male plants pedicelled in corymbose clusters, to 18 mm broad, those of the female plants subsessile in umbellate clusters, to 12 mm broad; stem from a massive tuberous root, angled, hispid with pustular-based hairs, climbing by simple tendrils; (introd.) .... [*Bryonia*]
    3 Fruit prickly, fleshy or dry, more or less elongate.
      4 Tendrils simple; fruits prickly with sharp elevations, many-seeded; flowers yellowish, about 3 cm broad, the pistillate ones solitary, the staminate ones often several in an axil; rough-hairy annual; (introd.) .... [*Cucumis*]
      4 Tendrils forking; flowers relatively small.
        5 Corolla 6-parted, small, whitish or greenish white; staminate flowers borne in narrow panicles; pistillate flowers (and fruits) solitary or sometimes in pairs; fruits inflated, soft-prickly, 2-locular, finally bursting at summit to expel the 4 seeds explosively; essentially glabrous annual; (B.C. to N.B. and N.S.) .... *Echinocystis*
        5 Corolla commonly 5-parted, whitish or greenish white.
          6 Pistillate flowers (and fruits) in capitate clusters; staminate flowers in corymbs; fruits dry and indehiscent, small, filled by the single seed, covered with readily detached barbed prickly bristles; clammy-hairy annual; (introd.) .... *Sicyos*
          6 Pistillate flowers (and fruits) mostly solitary; staminate flowers in racemes; fruits more or less inflated, weakly spiny, fibrous-netted within, to 8 cm long, finally bursting irregularly at summit, with 1 or 2 seeds in each of the 2–4 locules; leaves scabrous-hispid above, sparsely hairy or subglabrous beneath; perennial from a much enlarged woody root; (s B.C.) .... *Marah*

[BRYONIA L.] [8595]

[B. dioica Jacq.] Bryony
[Eurasian; reported from s Man. by Boivin (1966b; Altona, NW of Emerson), where, however, probably not established.]

[CITRULLUS Neck.]   [8598]

[C. vulgaris Schrad.]   Watermelon.   Pastèque
[A native of Africa; occasionally spontaneous in waste places in N. America but scarcely established, as in s Ont. (Toronto; TRT) and sw Que. (St. Helen's Is., Montreal; Rouleau 1945.]

[CUCUMIS L.]   [8599]

[C. sativus L.]   Cucumber
[Asiatic; sometimes germinating on dumps and in waste places in N. America but not established, as in s Ont. (Toronto, where taken by Scott in 1897; TRT) and Que. (Ste-Anne-de-Beaupré, E of Quebec City; TRT).]

ECHINOCYSTIS T. & G.   [8629]   Wild Balsam-apple

E. lobata (Michx.) T. & G.   Wild or Prickly Cucumber.   Concombre grimpant
/T/X/   (T (vine))   Thickets and moist places from Mont. to Sask. (N to Tisdale, 52°51′N; Breitung 1957a; introd. in s B.C.–Alta.), Man. (N to Wekusko L., about 90 mi NE of The Pas), Ont. (N to Kapuskasing, 49°24′N), Que. (N to the sw Gaspé Pen. at Matapédia), N.B., P.E.I., and N.S., s to Idaho, Ariz., Tex., and Fla. [Sicyos Michx.; Micrampelis Greene; Momordica echinata Willd.].

MARAH Kell.   [8629]

M. oreganus (T. & G.) Howell   Bigroot, Manroot
/t/W/   (Mc (vine))   Moist fields, thickets, and open hillsides from sw B.C. (Vancouver Is. and adjacent islands; CAN; V) to N Calif. [Sicyos T. & G.; Megarrhiza Torr.; Micrampelis Greene].

SICYOS L.   [8637]

S. angulatus L.   Bur-Cucumber
/T/EE/   (T)   Streambanks and damp yards from Minn. to Ont. (N to Ottawa), sw Que. (Montreal dist.; St-Jean), and Maine, s to Tex. and Fla.

THLADIANTHA Bunge   [8558]

T. dubia Bunge   Manchu Tuber-gourd
Asiatic; introd. along roadsides and in thickets and waste places in N. America, as in s Man. (Brandon; DAO), Ont. (N to Ottawa), and sw Que. (Gatineau and Deux-Montagnes counties and the Montreal dist.).

[CUCURBITA L.]   [8622]

[C. pepo L.]   Pumpkin
[Var. condensa Bailey of this commonly cultivated species (the genus not keyed out above) of reputedly North American origin is reported from Que. by Lionel Cinq-Mars et al. (Nat. can (Que.) 98(2):196. 1971; roadside at Cap Rouge, near Quebec City), where undoubtedly ephemeral.]

CAMPANULACEAE (Bluebell or Harebell Family)

Herbs with milky juice and simple alternate exstipulate leaves, these entire or toothed. Flowers white, violet, or blue, gamopetalous, regular, epigynous, the calyx and usually campanulate (rotate in *Specularia*) corolla deeply 5-lobed. Stamens 5, free from the corolla or inserted at the base of the corolla-tube, alternate with the corolla-lobes. Style 1, usually hairy above. Ovary inferior. Fruit a 3-locular many-seeded capsule.

1 Flowers small (about 5 mm long) and numerous, sessile or subsessile in a terminal head; corolla blue (rarely white), 5-parted nearly to base, the lobes very narrow; style long-exserted; leaves narrowly lanceolate or oblong, to about 5 cm long, undulate or crenate, ciliate, the basal ones narrowed to a short petiole; pubescent, usually biennial plant to about 5 dm tall; (introd. in s B.C.) .... [*Jasione*]
1 Flowers mostly larger, solitary or in racemes or panicles (only those of *Campanula glomerata* in a head-like cluster, but the corolla then 2 or 3 cm long).
  2 Flowers commonly in spike-like or open terminal racemes, or solitary (but then terminating the stem; often appearing irregularly scattered in *Githopsis* through elongation of the central axis), all with well-developed corollas (cleistogamous flowers wanting).
    3 Capsules opening only at apex, to 1.5 cm long; filaments scarcely expanded at the non-ciliate base; corolla blue with whitish throat, commonly less than 1 cm long and shorter than the calyx-lobes (but sometimes to 2 cm long); leaves lanceolate to oblong, remotely serrate, all sessile, to about 1.5 cm long and 3 mm broad, the lower ones much reduced; glabrous or minutely spreading-hairy annuals to about 3 dm tall; (introd. on Vancouver Is.) .... [*Githopsis*]
    3 Capsules opening by vertical slits; filaments expanded and ciliate at base; leaves mostly larger, the lower ones commonly the largest and often long-petioled; perennials (except *C. americana* and *C. medium*) .... *Campanula*
  2 Flowers solitary and sessile or subsessile in or opposite the leaf-like rotund bracts (the plant often floriferous nearly or quite to base); lower flowers cleistogamous, with reduced or abortive corollas; leaves (and bracts) rotund-ovate or broader, serrate, all sessile and more or less cordate-clasping (or a few of the lowermost ones obovate and obscurely petioled); glabrous or more or less scabrous or spreading-hispid annuals with simple or sparingly branched stems.
    4 Upper flowers blue, their corollas to 6 mm long, their lobes shorter than the tube; calyx-lobes ovate or broader, foliaceous and veiny, to 4 mm long; bracts inserted opposite the flowers; capsule opening near base; leaves commonly less than 1 cm long, sharply serrate, obscurely veined; stem lax and very slender, commonly not over 3 dm tall; (s B.C.) .... *Heterocodon*
    4 Upper flowers pale lavender to deep purple, their corollas to 13 mm long, cleft nearly to base; calyx-lobes narrowly triangular-acuminate, to 8 mm long; flowers in the bract-axils; capsule opening near the summit; leaves to 3 cm long, palmately veined; stem commonly stouter, to about 6 dm tall; (s B.C.; s Ont. and sw Que.) ....
    .... *Triodanis*

CAMPANULA L. [8644] Harebell, Bellflower. Campanule

1 Corolla rotate or shallowly campanulate-spreading, cut to well below the middle into lanceolate to deltoid-ovate lobes; style long-exserted, often recurved; leaves usually all cauline.
  2 Inflorescence spike-like, to about 6 dm long, the lower bracts leaf-like; leaves lanceolate to ovate-oblong, acuminate at both ends, coarsely serrate, the blade to about 1 dm long and 4 cm broad, commonly hirsute on the veins beneath; annual to over 1 m tall; (s Ont.) .... *C. americana*
  2 Flowers solitary at the ends of the stems; leaves narrowly lanceolate or oblanceolate, remotely denticulate, commonly less than 3 cm long and 1 cm broad; glabrous perennial to about 3 dm tall; (Alaska–Yukon–Dist. Mackenzie–N B.C.) .... *C. aurita*

1   Corolla shallowly to deeply campanulate or funnelform, seldom cut below the middle; style usually included or barely exserted (long-exserted in *C. scouleri*).

    3   Stems weak, filiform, somewhat 3-angled, downwardly scabrous on the angles and thus supported on other plants; leaves linear to narrowly lanceolate, often scabrous on the margins and on the midrib beneath; flowers solitary on long slender pedicels.

        4   Corolla nearly white, at most 1 cm long; calyx-lobes mostly less than 3 mm long; pedicels ascending, mostly less than 5 cm long; (Sask. to N.S.) .... *C. aparinoides*

        4   Corolla pale blue, to 12 mm long; calyx-lobes to 4 mm long, often slightly toothed at base; pedicels to 1 dm long, strongly divergent; (Sask. to Que. and N.B.) .... *C. uliginosa*

    3   Stems erect or ascending, terete or nearly so; corolla blue to blue-violet.

        5   Flowers short-pedicelled in racemes or sessile in a terminal cluster; corollas to over 3.5 cm long; (introd.).

            6   Flowers sessile in a terminal cluster subtended by an involucre of reduced leaves; leaves minutely dentate, the lower ones oblong-lanceolate, petioled, the upper ones lanceolate to oblong or ovate, sessile and more or less clasping .... *C. glomerata*

            6   Flowers short-pedicelled in loose racemes.

                7   Hispid-hairy biennial; corolla inflated-campanulate or urceolate, to 5 cm long, the short lobes flaring or somewhat reflexed; calyx bearing large reflexed cordate appendages at base; stem-leaves lance-oblong, sessile and more or less clasping, crenulate and undulate, to about 5 cm long .... *C. medium*

                7   Glabrous or sparingly pubescent perennials; corolla campanulate, blue to blue-violet, to about 3 cm long; calyx unappendaged.

                    8   Leaves linear to linear-lanceolate, minutely crenate, the lower stem-leaves more or less spatulate, nearly sessile; flowers erect; plant glabrous .... *C. persicifolia*

                    8   Leaves lanceolate to ovate, irregularly serrate, usually sparsely pubescent beneath, the upper ones short-petioled or sessile, the lower ones long-petioled; flowers nodding .... *C. rapunculoides*

    5   Flowers solitary at the ends of the stems or short-pedicelled in loose open clusters; perennials.

        9   Calyx and unexpanded corolla bristly with long pale hairs; corolla to 4 cm long; flowers in loose open clusters; leaves short-bristly, coarsely and irregularly serrate, the lower ones triangular or cordate, long-petioled, the upper ones lanceolate, short-petioled; stem often bristly above; (introd.) .... *C. trachelium*

        9   Calyx and corolla not bristly; flowers solitary to several.

          10   Corolla to about 12 mm long; calyx-lobes to about 5 mm long.

              11   Flower solitary, the style not exserted from the corolla; calyx-tube glabrous or often sparingly villous, the plant otherwise glabrous; leaves entire or obscurely callous-toothed, rarely over 3 cm long (including the obscure petiole), the larger (lower) ones lanceolate or oblanceolate; stems commonly less than 2 dm tall; (transcontinental) .... *C. uniflora*

              11   Flowers commonly several in a loose open raceme, the style long-exserted from the corolla; calyx glabrous; leaves sharply serrate with callous teeth, the larger (lower) ones with ovate or rotund-ovate blades to 4 cm long on petioles of about the same length or longer; plants glabrous or inconspicuously short-hairy, to about 4 dm tall; (SE Alaska–w B.C.) .... *C. scouleri*

          10   Corolla to about 3 cm long; calyx-lobes to over 1 cm long.

              12   Calyx-tube (and petioles) glabrous, the lobes entire; flowers often several; larger (basal) leaves long-petioled (often deciduous), sometimes with broadly ovate to subrotund or cordate-rotund, angular-toothed blades to 2 cm long, sometimes merely oblanceolate; (transcontinental) .... *C. rotundifolia*

12 Calyx-tube moderately to densely woolly-villous; flowers usually solitary (rarely 2); (alpine and low- to high-Arctic species).

13 Calyx with reflexed appendages between the entire (rarely crenate-serrate) lobes; lower leaves broadly spatulate to obovate, crenate; capsule drooping, opening near the base; (Aleutian Is.) . . . .
. . . . *C. chamissonis*

13 Calyx lacking appendages, the lobes with one or more sharp teeth or laciniations; lower leaves broadly oblanceolate, remotely sharp-serrate or laciniate; (Alaska to Dist. Mackenzie; mts. of B.C. and Alta.) . . . . *C. lasiocarpa*

C. americana L.   Tall Bellflower

/t/EE/   (T)   Rich moist soil from Minn. to s Ont. (N to Lambton, Middlesex, Wellington, Wentworth, and York counties) and N.Y., s to Mo., Ala., and Fla. [*Campanulastrum* Small]. MAP: M.L. Fernald, Rhodora 43(514): map 7, p. 503. 1941.

C. aparinoides Pursh   Marsh-Bellflower

/T/(X)/   (Hpr)   Wet meadows and shores from Sask. (N to Montreal River, 54°03′N) to Man. (N to the Hayes R. at ca. 55°N; CAN), Ont. (N limit uncertain; the report from Renison, sw James Bay at ca. 51°N, by Hustich 1955, may be based upon *C. uliginosa*), Que. (N to Grosse-Ile, about 20 mi NE of Quebec City), N.B. (Charlotte, Queens, and Kings counties; CAN; NBM; not known from P.E.I.), and N.S., s to Colo., Iowa, and Ga.

C. aurita Greene

/Ss/W/   (Hpr)   Rock crevices in Alaska–Yukon (N to ca. 68°N; type from the Yukon R., Alaska), w Dist. Mackenzie (N to ca. 64°30′N), and northernmost B.C. (s to Summit Pass at 58°31′N). MAPS: Hultén 1968*b*:849.

C. chamissonis Federov

/s/W/eA/   (Hsr)   Stony tundra in the Aleutian Is.; s Kamchatka, the Kuril Is., and N Japan. [*C. dasyantha* of N. America reports, not Bieb.]. MAPS: Hultén 1968*b*:848, and 1949: map 1091 (*C. dasy.*), p. 1475.

C. glomerata L.   Clustered Bellflower

Eurasian; introd. along roadsides and in old fields and pastures in N. America, as in s Ont. (Frontenac, York, and Halton counties; CAN; TRT), Que. (N to Rivière-du-Loup, Temiscouata Co.; CAN), and N.S. (Hants and Kings counties).

C. lasiocarpa Cham.

/aST/W/eA/   (Hsr)   Sandy tundra and alpine heaths from the Aleutian Is. (type from Unalaska), Alaska (N to the N coast), and the Yukon–w Dist. Mackenzie (N to ca. 65°N) through B.C. and sw Alta. (N to Jasper; CAN) to N Wash.; NE Siberia, Kamchatka, and Japan. MAPS: Hultén 1968*b*:847; *Atlas of Canada* 1957: map 6, sheet 38; Raup 1947: pl. 35.

Some of the Aleutian Is. material is referable to var. *latisepala* Hult. (not *C. latisepala* Hult., a member of the *C. rotundifolia* complex; calyx-lobes broadly triangular rather than linear or lanceolate, more laciniate than those of the typical form). MAP: Hultén 1968*b*:848.

C. medium L.   Canterbury-bells

European; an occasional garden-escape to roadsides and waste places in N. America, as in sw B.C. (a large colony observed by the writer in 1964 at Nanaimo, Vancouver Is., the flowers ranging in colour from white to roseate or deep blue-purple; CAN), Ont. (Owen Sound, Grey Co.), and N.B. (in a salt marsh near St. Andrews, Charlotte Co.; CAN).

C. persicifolia L.

Eurasian; an occasional garden-escape or persisting in old gardens in N. America, as in sw B.C. (Nanaimo dist., Vancouver Is.; CAN), s Ont. (Guelph, Wellington Co.; OAC), Que. (Oka, near Montreal; Berthier, Montmagny Co.), and N.B. (Lakeside).

C. rapunculoides L.
Eurasian; a garden-escape to roadsides, thickets, and waste places in N. America, as in SE B.C. (Cranbrook; CAN), Alta. (Moss 1959), S Man. (Brandon; Winnipeg), Ont. (N to Timmins and Ottawa), Que. (N to Baie-Comeau, Saguenay Co., and the Gaspé Pen.), Nfld., N.B., P.E.I., and N.S.

Var. *ucranica* (Bess.) Koch, the subglabrous extreme, is reported from Que. by Boivin (1966*b*; Grondines, about 40 mi SW of Quebec City).

C. rotundifolia L.   Harebell, Bluebell
/aST/X/GEA/   (Hsr)   Fields, rocky shores, and grassy slopes at low to fairly high elevations from cent. Alaska–Yukon–Dist. Mackenzie to Great Bear L., Great Slave L., L. Athabasca (Alta. and Sask.), Man. (N to the Nelson R. about 150 mi S of Churchill), Ont. (N to NW James Bay at ca. 55°N), Baffin Is. (N to ca. 71°N), northernmost Ungava–Labrador, Nfld., N.B., and N.S. (a garden-escape in P.E.I.), S to N Calif., N Mexico, Tex., Nebr., Pa., and N.J.; W and E Greenland N to ca. 75°N; Iceland; Spitsbergen; Eurasia. MAPS: Hultén 1968*b*:850; Porsild 1957: map 303, p. 198.

Forma *albiflora* Rand. & Redf. (flowers white rather than blue-purple) apparently occurs locally throughout the range. Forma *laciniata* Rousseau & Raymond (corolla deeply cleft rather than shallowly lobed) is known from the type locality along the George R., N Que., at ca. 57°30′N.

In a cytological study of the *C. rotundifolia* complex, T.W. Böcher (Ann. Bot. Fenn. 3(3):287–98. 1966) concludes that it is "owing to the high degree of intercrossability and high degree of cytological instability perhaps the most intricate polyploid complex we know." Material from Greenland is referred by him to *C. gieseckiana* Vest and its ssp. *groenlandica* (Berl.) Böcher (*C. gr.* Berl.), with 2n chromosomes-counts of 34 and 68, respectively. He also reports a 2n count of 68 for typical *C. rotundifolia* from Europe and for *C. intercedens* Witasek from Nfld., believing that the E N. American plant is referable to the latter species. He includes in *C. intercedens* a series of plants having stem-leaves ranging from linear to lanceolate (*C. dubia* DC.; type from Nfld.) to rather broadly ovate (*C. pratensis* DC.; type from Nfld.). Reports of the following taxa from our area, whether strictly identical or not, are included in the present concept: f. *pygmaea* Hartz; vars. *alaskana* Gray, *arctica* Lange, *hirsuta* Macoun, *intercedens* (Wit.) Farw., *lancifolia* Mert. & Koch, and *petiolata* (DC.) Henry; *C. canadensis* Prov.; *C. dubia* DC.; *C. gieseckiana* Vest; *C. groenlandica* Berl.; *C. heterodoxa* Bong.; *C. heterodoxa* Vest; *C. intercedens* Witasek; *C. langsdorffiana* Fisch.; *C. latisepala* Hult. and its var. *dubia* Hult.; *C. linifolia* Lam.; *C. petiolata* DC.; *C. pratensis* DC.; *C. scheuchzeri* Vill.; *C. stylocampa* Eastw.

C. scouleri Hook.
/T/W/   (Hpr)   Open or dense woods and rock outcrops at low to moderate elevations from the S Alaska Panhandle through W B.C. to N Calif. MAP: Hultén 1968*b*:849.

C. trachelium L.   Nettle-leaved Bellflower, Bats-in-the-Belfry
Eurasian; a garden-escape to roadsides, thickets, and waste places in N. America, as in SE Man. (near Otterburne, about 30 mi S of Winnipeg; Löve and Bernard 1959), Ont. (Essex, Norfolk, Waterloo, and Wellington counties), and Que. (N to Bic, Rimouski Co.).

C. uliginosa Rydb.   Marsh-Bellflower
/ST/EE/   (Hpr)   Wet meadows, thickets, and shores from Sask. (N to Montreal Lake, 54°03′N) to Man. (N to Churchill), Ont. (N to Big Trout L. at ca. 53°45′N, 93°W), Que. (N to the E James Bay watershed at 52°37′N and Cabano, Temiscouata Co.), and N.B. (not known from P.E.I. or N.S.), S to Nebr., Ohio, Pa., and Mass. [*C. aparinoides* var. *ulig.* (Rydb.) Gl. and var. *grandiflora* Holz].

C. uniflora L.
/AST/X/GEA/   (Hs)   Stony ridges on tundra and rocky or grassy places at low to high elevations from the E Aleutian Is. and coasts of Alaska–Dist. Mackenzie–Dist. Keewatin (in the Yukon, N to ca. 68°N) to Banks Is., Ellesmere Is. (N to ca. 80°N), and northernmost Ungava–Labrador, S in the West through the mts. of B.C. and SW Alta. (Columbia Icefield, about 50 mi SE of Jasper; Herb. V) to Mont. and Colo., farther eastwards S to NE Man. (Churchill; not known from Sask. or Ont.) and Que. (coasts of Hudson Strait and E Hudson Bay S to Cape Jones, 54°37′N; isolated in the Shickshock

Mts. of the Gaspé Pen.); w and E Greenland N to ca. 79°N; Iceland; Spitsbergen; arctic Europe; NE Siberia. MAPS: Hultén 1968b:851, and 1958: map 173, p. 193; Porsild 1957: map 302, p. 198; Fernald 1929: map 20, p. 1498.

[GITHOPSIS Nutt.]   [8666]

[G. specularioides Nutt.]
[Native in the w U.S.A. from s Wash. to s Calif. Eastham (1947) notes a 1926 report from Sooke, Vancouver Is., SW B.C., where either a casual waif and evidently not established or, if once native, apparently now extinct.]

HETEROCODON Nutt.   [8650]

H. rariflorum Nutt.
/t/W/   (T)   Moist open places in the valleys and foothills from s B.C. (Vancouver Is., Sproat, SW of Revelstoke, and Nelson; CAN) to Calif., Nev., and ?Wyo. [*Specularia* McVaugh].

[JASIONE L.]   [8674]

[J. montana L.]   Sheep's-bit
[European; according to Eastham (1947), "European and commonly cult. Golf-links, Qualicum, V.I. (PM); spontaneous and persisting for several years on site of Jericho Air Station, Vancouver, but now probably destroyed (DA)."]

TRIODANIS Raf.   [8649]

T. perfoliata (L.) Nieuwl.   Venus's Looking-glass
/T/(X)/   (T)   Open ground from s B.C. (Vancouver Is. and mainland N to Sproat, SW of Revelstoke, E to Cranbrook) and Mont. to S.Dak., Minn., Ont. (N to the Ottawa dist.; Gillett 1958), SW Que. (Granby, Shefford Co.; MT), N.Y., and Maine, s to Calif., N Mexico, Tex., and Fla.; tropical America (?introd.); introd. in Europe. [*Campanula* L.; *Specularia* DC.].

LOBELIACEAE (Lobelia Family)

Herbs with milky juice and simple alternate exstipulate leaves. Flowers irregular and somewhat 2-lipped, gamopetalous, epigynous. Calyx 5-lobed. Upper corolla-lip erect and 2-lobed, the lower lip spreading and 3-lobed. Stamens 5, inserted at the base of the corolla-tube, united by their anthers (and also sometimes by their filaments). Style 1. Ovary inferior. Fruit a many-seeded capsule.

1 Flowers sessile in the axils of lanceolate to ovate leaf-like bracts to about 2.5 cm long (but appearing long-stalked because of the much elongated linear or subulate hypanthium); corolla white to pink, blue, or purplish, marked with yellow or white; capsules to 5 cm long, elastically but tardily dehiscent by long slits on the sides; foliage-leaves generally smaller than the bracteal leaves and soon deciduous, subulate to lanceolate (or the uppermost ones broader), sessile, entire or with a few minute teeth; (B.C. to sw Sask.) . . . . *Downingia*
1 Flowers distinctly pedicelled in terminal bracted racemes, the bracts much smaller than the foliage-leaves; capsules 2-locular, dehiscent by apical valves . . . . *Lobelia*

DOWNINGIA Torr. [8706]

1 Corolla showy and well surpassing the calyx, to over 1.5 cm long (seldom less than 7 mm); anther-column more or less strongly incurved (commonly at about a right angle to the filament-column, this to about 1 cm long); capsule 1-locular, with parietal placentation (seeds borne in 2 rows along the inner wall); stem to about 5 dm tall, erect or curved-ascending; (se B.C.) . . . . *D. elegans*
1 Corolla at most 7 mm long, about equalling or barely surpassing the calyx; anther-column nearly erect on the filament-column (this to 2.5 mm long); capsule 2-locular, with axile placentation (seeds borne along the central axis); stem to 2 dm tall, often decumbent at base and sometimes rooting at the lowermost nodes; (s Alta. and sw Sask.) . . . . *D. laeta*

D. elegans (Dougl.) Torr.
/t/W/ (T) Wet meadows, borders of ponds, and shallow vernal pools from se B.C. (Leech L., Kootenay Flats, near Creston; CAN, detd. Porsild) to n Calif. and n Nev. [*Clintonia* Dougl.; *Bolelia* Greene; *C. (D.) corymbosa* DC.; *D. laeta sensu* A.E. Porsild, Can. Field-Nat. 63(3):116. 1949, not Greene].

D. laeta Greene
/T/W/ (T) Wet (often alkaline) meadows, prairies, and borders of ponds from w Mont., s Alta. (Foremost, sw of Medicine Hat; Moss 1959), and sw Sask. (Crane Lake, sw of Swift Current, where taken by John Macoun in 1894; CAN, detd. McVaugh; ?extinct) to n Calif., Utah, and Wyo. [*Bolelia* Greene; *B. brachyantha* Rydb.].

LOBELIA L. [8694] Lobelia. Lobélie

(Ref.: Rogers McVaugh 1936, and N. Am. Flora 32A:35–99. 1943)
1 Corolla-tube fenestrate (with 2 lateral longitudinal openings near base).
  2 Flowers to 4.5 cm long; corolla scarlet; filament-tube usually over 2.5 cm long; leaves narrowly to broadly lanceolate, irregularly serrate; (s Ont. to Que. and N.B.) . . . . *L. cardinalis*
  2 Flowers at most about 3 cm long; corolla blue to purple, rarely white; filament-tube at most 1.5 cm long.
    3 Calyx-lobes with foliaceous basal auricles up to 5 mm long; pedicels bracted near the middle; flowers usually over 2.5 cm long; inflorescence scarcely 1-sided; leaves acute at both ends; (s Man. and s Ont.) . . . . *L. siphilitica*
    3 Calyx-lobes with basal auricles either minute or wanting; pedicels bracted at or near base; flowers less than 2.5 cm long; inflorescence more or less 1-sided; leaves lanceolate to narrowly obovate, subentire or with callous-tipped teeth, obtuse at apex, cuneate at base, usually more or less pubescent on both sides; (sw ?Que.) . . . . [*L. puberula*]

1 Corolla-tube lacking longitudinal basal openings.
  4 Plant scapose, the hollow scape usually partly submersed, naked or with a few small bracts; leaves in a basal rosette, lance-linear, fleshy, hollow; flowers usually about 1.5 cm long, white to pale violet; (transcontinental) . . . . *L. dortmanna*
  4 Plant leafy-stemmed; flowers usually less than 1.5 cm long.
    5 Leaves entire or the larger ones sometimes obscurely toothed, those of the stem linear, the basal ones spatulate to narrowly obovate, pubescent; flowers to about 1.5 cm long, in lax 1-sided racemes; corolla blue, with a large white eye; (transcontinental) . . . . *L. kalmii*
    5 Leaves lanceolate to oblong-obovate, dentate or serrate; flowers at most 12 mm long, in denser non-secund racemes.
      6 Mature capsule inflated, not exserted; corolla whitish to pinkish or pale violet; racemes terminal and at the tips of branches; stem-leaves ovate to obovate; stem long-hirsute at least below; (s Ont. to N.S.; introd. in B.C.) . . . . *L. inflata*
      6 Mature capsule not inflated, partly exserted; corolla normally white to pale blue; raceme terminal; stem-leaves lanceolate to obovate; (s Alta. to N.S.) . . . . *L. spicata*

L. cardinalis L.   Cardinal-flower
/T/EE/   (Hs)   Damp shores, meadows, and swamps from Minn. to Mich., Ont. (N to Mattawa and the Ottawa dist.; the inclusion of Sask. in the range by Rydberg 1922 and 1932, requires clarification), Que. (N to St-Onésime, Kamouraska Co.; QSA), and N.B. (Charlotte, Kent, and York counties; not known from P.E.I. or N.S.), s to E Tex. and Fla. MAP: McVaugh 1936: fig. 3, p. 276.

A hybrid with *L. siphilitica* is reported from s Ont. by Landon (1960; Norfolk Co.). Forma *alba* (Eat.) St. John (corolla white rather than vermilion to deep red) and f. *rosea* St. John (corolla pink) are reported from Que. by Boivin (1966b; also tentatively reporting f. *rosea* from Ont.).

L. dortmanna L.   Water-Lobelia, Water-gladiole
/T/(X)/E/   (Hel)   Shallow water and sandy or gravelly margins of ponds: sw B.C. (Vancouver Is. and adjacent islands and mainland N to Powell River, about 60 mi NW of Vancouver; Eastham 1947) to Oreg.; N Sask. (between Windrum L., ca. 56°N, and the s shore of L. Athabasca at ca. 59°N; CAN; a notable filling in of the large gap between the two disjunct areas, validating the report from Lac la Loche, Sask., 56°28′N, by Hooker 1829, and lending credence to his report of a Richardson collection from "Slave Lake", the name often used by early authors for Great Slave L., Dist. Mackenzie; the report from Man. by Burman 1909, taken up by later authors, requires confirmation); Minn. to Ont. (N to the E shore of L. Superior and Renfrew and Carleton counties), Que. (N to E James Bay at 53°33′N, L. Mistassini, the Côte-Nord, and Gaspé Pen.), SE Labrador (Battle Harbour, 52°16′N; Hustich and Pettersson 1943), Nfld., N.B., P.E.I., and N.S., s to Pa. and N.J.; N Europe. [*L. lacustris* Salisb.]. MAPS: Hultén 1958: map 192, p. 211, and 1937b: fig. 12, p. 126; McVaugh 1936: fig. 30, p. 358; Fernald 1933: map 8, p. 87; Meusel 1943: fig. 26b.

L. inflata L.   Indian-tobacco
/T/EE/   (T)   Fields, roadsides, waste places, and open woods from Minn. to Ont. (N to Renfrew and Carleton counties), Que. (N to L. St. John and Rimouski, Rimouski Co.), N.B., P.E.I., and N.S., s to E Kans., Ark., and Ga. MAP: McVaugh 1936: fig. 20, p. 323.

The species is reported from s B.C. by W.M. Bowden (Can. J. Genet. Cytol. 1(1):55. 1959; near Agassiz and Chilliwack), where he believes it to have been introd. Reports from Sask. are probably based upon the ambiguous phrase "to the Saskatchawan and Hudson's Bay" used by Hooker 1829; reports from Man. may be based, at least in part, upon a collection in DAO from Winnipeg, referable to *L. spicata* var. *hirtella*.

L. kalmii L.
/sT/X/   (Hs)   Damp or wet soils and ledges (often calcareous) from N B.C. (N to the Liard R. at 59°23′N; CAN) and Alta. to Great Slave L., Sask. (N to Bjorkdale, 52°43′N), Man. (N to York Factory, Hudson Bay, 57°N), Ont. (N to the Fawn R. at ca. 55°30′N, 88°W; CAN), Que. (N to E James Bay at 51°21′N, L. Mistassini, and the Côte-Nord), Nfld., N.B. (reports from P.E.I. require

confirmation; if once there, now evidently extinct), and N.S., s to Wash., Mont., Colo., S.Dak., Ohio, and N.J. [*L. strictiflora* (Rydb.) Lunnell]. MAP: McVaugh 1936: fig. 29 (somewhat incomplete northwards), p. 357.

Forma *leucantha* Rouleau (flowers white rather than blue; type from New Liverpool, Lévis Co., Que.) is known from Ont., Que., Nfld., and N.B. and probably occurs throughout the range.

[L. puberula Michx.]
[The report of this species of the E U.S.A. (N to Ill. and Pa.) from Montreal, Que., by R. Campbell (Can. Rec. Sci. 6(6):342–56. 1895) requires confirmation. The MAPS by McVaugh (1936; combine his fig. 12, p. 294, and fig. 13, p. 296) indicate no Canadian stations.]

L. siphilitica L.   Great Lobelia, Blue Cardinal-flower
/T/EE/   (Hs)   Rich moist woods and swamps, the aggregate species from sw Man. to s Ont. and Maine, s to Tex., La., and N.C. MAPS: *see* below.
1   Leaves lanceolate, subglabrous and often subentire; calyx subglabrous; flowers rarely
    more than 20; [S Man.: near Turtle Mt., about 50 mi s of Brandon, where taken by
    Burgess in 1874; TRT; MAP: McVaugh 1936: fig. 6, p. 283] . . . . var. *ludoviciana* DC.
1   Leaves broadly lanceolate to ovate, strigose above, irregularly serrate; calyx hirsute;
    flowers often more than 20 . . . . var. *siphilitica*
    2   Flowers white; [Ont.; Boivin 1966*b*] . . . . f. *albiflora* Britt.
    2   Flowers blue; [S Ont.: N to Grey, York, Leeds, and Stormont counties; MAP: McVaugh
        1936: fig. 5, p. 280] . . . . f. *siphilitica*

L. spicata Lam.   Pale-spike-Lobelia
/T/(X)/   (Hs)   Moist meadows, fields, and thickets, the aggregate species from s Alta. (Craigmyle, about 50 mi NE of Calgary; McVaugh 1936) to Sask. (N to near the South Saskatchewan R., where taken by John Macoun in 1872, the exact locality uncertain but perhaps represented by the dot near Prince Albert on McVaugh's below-noted map), Man. (N to Eriksdale, about 60 mi N of Portage la Prairie), Ont. (N to Thunder Bay, the Timagami Provincial Forest NE of Sudbury, and Ottawa), Que. (N to the E Gaspé Pen. at the mouth of the Grande-Rivière), N.B. (Charlotte and Westmorland counties; CAN), P.E.I. (Charlottetown; DAO), and N.S. (Yarmouth, Kings, and Cumberland counties), s to Kans., Ark., and Ga. MAPS and synonymy: *see* below.
1   Plant bristly pubescent nearly throughout; [Canadian range as outlined above; MAP:
    McVaugh 1936: fig. 16, p. 314] . . . . var. *hirtella* Gray
1   Plant essentially glabrous except toward base . . . . var. *spicata*
    2   Anthers white; flowers dark purplish-blue, at most about 30 in a raceme; capsules
        globose, often somewhat inflated; [var. *camp.* McVaugh; s Man. (Brandon; DAO), s
        Ont. (Branchton, Waterloo Co.; OAC), and Que. (Boivin 1966*b*); the MAP by McVaugh
        1936: fig. 17, p. 317, indicates no Canadian stations] . . . .
        . . . . f. *campanulata* (McVaugh) Bowden
    2   Anthers blue; flowers light blue; racemes densely many-flowered; capsules short-
        hemispheric; [*L. claytoniana* Michx.; Ont. to the Maritime Provinces, somewhat more
        southern than var. *hirtella*; MAP: McVaugh 1936: fig. 15 (var. *originalis*), p. 309] . . . .
        . . . . f. *spicata*

COMPOSITAE (Composite Family)

(Ref.: P.A. Rydberg, N. Am. Flora 33(pt. 1):1–46. 1922; 34(pts. 1–4):1–360. 1914–1927; H.A. Gleason, N. Am. Flora 33(pt. 1):47–110. 1922; E.E. Sherff, N. Am. Flora, Ser. II(pt. 2):1–149. 1955) Annual, biennial, or perennial herbs (only a few Canadian species with more or less woody stems) with watery or milky juice. Leaves usually simple (sometimes compound), exstipulate, variously arranged. Flowers small, epigynous, in compact heads, these solitary or in spikes, racemes, corymbs, or panicles. Calyx-tube united with the 1-locular ovary, the limb (when present) cup-shaped or consisting of a pappus of bristles, awns, scales, or teeth. Corolla either strap-shaped (ligulate) or tubular and typically 5-lobed, the heads either entirely ligulate, discoid (corollas all tubular), or radiate (central disk-florets tubular, marginal ray-florets with ligulate corollas). Stamens typically 5, inserted on the corolla, their anthers united into a tube. Style usually 2-cleft. Ovary inferior. Fruit a dry seed-like achene. (Including Ambrosiaceae, Carduaceae, and Cichoriaceae).

KEY TO GROUPS

1 Heads discoid (composed entirely of tubular florets, as in the thistle, *Cirsium,* or the marginal florets with inconspicuous ligules scarcely surpassing the plane of the disk; marginal florets sometimes enlarged and ray-like in *Centaurea*).
  2 Leaves all or chiefly alternate.
    3 Leaves entire or nearly so .... *GROUP* 1
    3 Leaves distinctly toothed to deeply lobed .... *GROUP* 2 (p. 1444)
  2 Leaves all or chiefly opposite or basal (sometimes whorled in *Eupatorium*)...............
    .... *GROUP* 3 (p. 1446)
1 Heads normally radiate or entirely ligulate (but discoid forms sometimes occur).
  4 Heads radiate (central disk-florets tubular, marginal ray-florets with ligulate corollas, as in the ox-eye daisy, *Chrysanthemum leucanthemum*).
    5 Leaves all or chiefly alternate.
      6 Leaves entire or nearly so .... *GROUP* 4 (p. 1448)
      6 Leaves distinctly toothed to deeply lobed .... *GROUP* 5 (p. 1449)
    5 Leaves all or chiefly opposite or basal (sometimes whorled in *Actinomeris* and *Silphium*) .... *GROUP* 6 (p. 1452)
  4 Heads entirely ligulate (composed entirely of strap-shaped ligulate florets, as in the dandelion, *Taraxacum*); receptacle almost invariably naked (chaffy in *Hypochaeris*); pappus usually consisting of capillary bristles (minute scales in *Cichorium*; wanting in *Arnoseris* and *Lapsana*); juice often milky.
    7 Leaves all or chiefly alternate.
      8 Leaves entire or nearly so .... *GROUP* 7 (p. 1455)
      8 Leaves distinctly toothed to deeply lobed .... *GROUP* 8 (p. 1456)
    7 Leaves all or chiefly basal, stem-leaves, when present, much reduced ....
    .... *GROUP* 9 (p. 1456)

KEY TO GENERA

GROUP 1
(Heads discoid; leaves all or mostly alternate, entire or nearly so)

1 Phyllaries with spreading, subulate, inwardly hooked tips in several unequal rows (forming the characteristic bur of the burdock); heads pink or purplish; receptacle bristly; achenes with a short pappus of numerous somewhat chaffy rough deciduous bristles; coarse biennial weeds with large ovate to roundish, mostly cordate leaves, these more or less floccose-tomentose beneath; (introd.) .... *Arctium*
1 Phyllaries lacking hooked tips.
  2 Shrubs with linear or narrowly oblanceolate to oblong leaves; heads in terminal cymose clusters; involucres narrow, with commonly not more than 4 or 5 yellow flowers; receptacle naked; pappus of capillary bristles.
    3 Phyllaries (and flowers) 4, equal; primary leaves to 3 cm long and 4 mm broad

(their axils sometimes bearing fascicles of shorter and proportionately broader leaves); plant finely and densely white-tomentose; (s B.C.) .... *Tetradymia*

3   Phyllaries commonly at least 15, in several series of unequal length; (B.C. to sw Sask.) .... *Chrysothamnus*

2 Herbs (at most somewhat woody-based).

4   Receptacle bristly or chaffy (sometimes so only near the marginal florets; sometimes naked or nearly so in *Saussurea*).

5   Pappus present at the summit of the achenes.

6   Pappus of elongate plumose bristles; heads violet-purple, few in corymbiform or capitate clusters; (western arctic and alpine regions) .... *Saussurea*

6   Pappus of simple bristles or scales (sometimes none in *Centaurea*); (introd.).

7   Florets small, yellowish, in heads 4 or 5 mm high, these in small clusters of 2 or more; phyllaries linear, blunt, spreading like a star when mature, with woolly hairs extending to the tip; achenes about 0.6 mm long, basally attached; leaves linear-oblanceolate, commonly about 1 cm long, they and the stem (commonly simple below but with short ascending lateral branches above) copiously white-woolly; (introd.) .... *Filago*

7   Florets often showy, white, pink, blue, or purplish (rarely yellow), the marginal ones often enlarged and ray-like; phyllaries with a membranous or scarious terminal appendage which is usually toothed, pectinate, or spiny, rarely entire; achenes longer, obliquely or laterally attached near base to the receptacle .... *Centaurea*

5   Pappus none or rudimentary.

8   Heads completely discoid; receptacle chaffy throughout; phyllaries few, roundish; anthers nearly separate; leaves at most about 3 cm long; (*I. axillaris*; s B.C. to s Man.) .... *Iva*

8   Heads with short yellow marginal ray-florets; receptacle chaffy only near the margin; phyllaries elongate; anthers united nearly their whole length; leaves often longer .... *Madia*

4   Receptacle naked; pappus present, of capillary bristles; phyllaries in several series of unequal length (except in *Luina* and sometimes in *Aster* and *Erigeron*).

9   Phyllaries equal, the thinly tomentose involucre to 7 (sometimes 8) mm high; heads dull yellowish, slender-peduncled in a short corymbiform (sometimes umbelliform) inflorescence; leaves elliptic to broadly ovate, sessile, green and thinly tomentose or glabrate above, white-tomentose beneath, to about 6 cm long and 3.5 cm broad, the lower ones reduced to mere bracts; stems to about 4 dm tall, from a stout branched woody caudex; (s B.C.) .... *Luina*

9   Phyllaries in several series of markedly unequal length (except sometimes in *Aster* and *Erigeron*).

10 Flowers all perfect.

11 Heads large, in spikes or racemes; pappus single, plumose or barbellate; leaves more or less punctate, at least the lower ones petioled; stems mostly from a corm-like base; (s Alta. to s Ont.) ..... *Liatris*

11 Heads terminating the branches or subcorymbosely clustered; leaves sessile or subsessile, not punctate; stem not from a corm-like base.

12 Heads relatively small (involucres less than 1 cm high, their phyllaries mostly blunt to merely mucronate, shorter than the florets), subcorymbosely clustered, purple or rose-purple; pappus in 2 unequal series, the outer row of short bristles or scales, the inner of capillary bristles; plants to 2 m tall; (SE Sask. to s Ont.) .... *Vernonia*

12 Heads relatively large (involucres commonly more than 1 cm high, strongly striate, their acute or acuminate phyllaries equalling the florets), terminating the branches or subcorymbosely clustered,

white or creamy to pink-purple; pappus-bristles 1-rowed; plants glandular-puberulent, to about 6 dm tall, generally with many stems from a woody caudex; (*B. oblongifolia*; s B.C.) . . . . *Brickellia*

10 Outer flowers (or all the flowers of some heads) pistillate.

  13 Plants glabrous or more or less pubescent, but not white-woolly; leaves narrowly linear to oblanceolate.

    14 Heads on leafy-bracted peduncles; phyllaries subequal but in more than 1 series, their tips green; hairy tips of the style-branches acutish . . . . *Aster*

    14 Heads on naked or scaly-bracted peduncles; phyllaries in essentially 1 series, their tips not green; hairy tips of the style-branches obtuse . . . . *Erigeron*

  13 Plants more or less white-woolly, at least on the lower leaf-surfaces; leaves linear to broadly obovate.

    15 Plants bisexual, the outer (pistillate) florets of each head thread-like to summit, the central (perfect) florets coarser and distinctly flaring at summit; phyllaries whitish to brown or purplish, often with a dark spot near tip . . . . *Gnaphalium*

    15 Plants unisexual (or female plants of *Anaphalis* commonly with a few central hermaphrodite sterile florets in the heads).

      16 Basal leaves conspicuous, tufted, persistent, the stem-leaves usually few and much reduced; inflorescence relatively small; phyllaries white to yellow-tinged or roseate . . . . *Antennaria*

      16 Basal leaves soon deciduous, the stem-leaves scarcely reduced; inflorescence many-headed and relatively large; phyllaries pearly-white; (transcontinental) . . . . *Anaphalis*

GROUP 2 (see p. 1442)
(Heads discoid; leaves all or mostly alternate, toothed to deeply lobed)

1 At least some of the phyllaries spine-tipped (except in *Cirsium muticum* and some species of *Centaurea*); receptacle distinctly bristly or chaffy (except in *Onopordum*); leaves often spiny- or prickly-toothed.

  2 Heads unisexual, the phyllaries of the pistillate ones united into a tubercled or prickly bur (this often only obscurely tuberculate in *Ambrosia psilostachya*); pappus none.

    3 Burs armed with numerous hooked prickles, 2-flowered, 2-seeded; phyllaries of staminate involucres distinct; heads 1–few in the leaf-axils; leaves at most moderately deeply lobed, the lobes broad . . . . *Xanthium*

    3 Burs armed with 1–several series of tubercles or straight spines; phyllaries of staminate involucres united; staminate heads in spikes or racemes; leaves normally deeply lobed or dissected.

      4 Bur with several series of flat lance-attenuate spines, 1–4-flowered and 1–4-seeded; principal leaves 2-pinnatifid and petioled; (B.C. to sw Man.) . . . . *Franseria*

      4 Bur with a single row of tubercles near summit (except sometimes in *A. psilostachya*), 1-flowered, 1-seeded; (essentially transcontinental species) . . . . *Ambrosia*

  2 Heads normally composed of many perfect florets, their phyllaries distinct; pappus usually present (sometimes wanting in *Centaurea*).

    5 Leaves not prickly or spiny; (introd.).

      6 Phyllaries hooked at the spiny tip; achenes basally attached . . . . *Arctium*

      6 Phyllaries not hooked, either terminated by a straight spine or more commonly by a broad, erose to lacerate or pectinate appendage; achenes obliquely or laterally attached . . . . *Centaurea*

    5 Leaves prickly or spiny on the margins.

      7 Heads 1-flowered, whitish to blue, very numerous in a terminal globose cluster

to about 6 cm thick; pappus of short narrow scales or awns; leaves white-tomentose beneath, pinnatifid into lanceolate to oblong-triangular lobes, to about 4 dm long and 2 dm broad, at least the middle and upper ones clasping; (introd.) .... *Echinops*

  7 Heads many-flowered, simple, hemispheric to campanulate or cylindrical.

      8 Heads yellow, 3 or 4 cm high; phyllaries broad, the spines of the inner ones pinnatisect; pappus double, consisting of 10 outer smooth long awns and 10 much shorter sparsely pectinate inner ones, the obliquely attached achene also with a crown of 10 short horny teeth; plant spreading-short-villous; (introd.) .... *Cnicus*

      8 Heads white to violet or purple (creamy in some species of *Cirsium*); achenes basally attached; pappus single, consisting of naked or plumose capillary bristles.

         9 Stem wingless or essentially so (except in *Cirsium palustre* and *C. vulgare*).

            10 Phyllaries broad-based, sharply serrate, at least the outer ones considerably surpassing the disk; pappus-bristles not plumose; leaves glabrous or slightly tomentose, more or less white-mottled along the main veins; (introd.) .... *Silybum*

            10 Phyllaries narrower, mostly entire, shorter than the disk; pappus-bristles plumose to the middle; leaves mostly not white-mottled .... *Cirsium*

         9 Stem conspicuously spiny-winged by the decurrent leaf-bases; heads usually red-purple (sometimes white in *Carduus*); pappus not plumose; (introd.).

            11 Receptacle densely bristly; wings of stem toothed to base .... *Carduus*

            11 Receptacle deeply honeycomb-pitted, often with short bristles on the intervening ridges but not densely bristly; wings of stem with a fairly broad uncut portion adjoining the white-woolly stem, this to about 1.5 m tall .... *Onopordum*

1 Phyllaries not spine-tipped; receptacle naked or nearly so (hairy in *Artemisia absinthium* and *A. frigida* and chaffy in some species of *Saussurea*); leaves scarcely spiny-toothed.

  12 Pappus consisting of elongate bristles.

      13 Pappus-bristles plumose; heads violet-purple, few in corymbiform or capitate clusters; (western arctic and alpine regions) .... *Saussurea*

      13 Pappus-bristles capillary, naked or merely barbed.

         14 At least the lower and middle leaves narrowly to broadly deltoid or subcordate, the blades to about 2 dm long and about as broad at base.

            15 Phyllaries in several series of unequal length, striate; heads greenish white or creamy; plant pubescent; (*B. grandiflora*; B.C. and Alta.) .... *Brickellia*

            15 Phyllaries in a single series; heads white or flesh-colour; plant minutely puberulent or glabrous, often glaucous .... *Cacalia*

         14 Leaves mostly narrower in outline.

            16 Heads purple; pappus purplish (often tawny in *V. missurica*), double, the inner of long bristles, the outer of very short bristles; phyllaries in several series of unequal length; leaves sessile or subsessile, finely to coarsely serrate; (SE Sask. to S Ont.) .... *Vernonia*

            16 Heads whitish, yellow, or orange-yellow; pappus white.

               17 Phyllaries in 4 or 5 series of unequal length, puberulent; heads yellow; (*H. carthamoides* and *H. nuttallii*; S B.C. to S Sask.) .... *Haplopappus*

               17 Phyllaries subequal (in *Senecio*, occasionally subtended by a row of short bractlets).

                  18 Heads whitish; achenes with 10–12 pale ribs, the intervening brown furrows usually strigose; leaves coarsely and irregularly serrate with callous-tipped teeth; stems grooved, to over 3 m tall; (Ont. to N.S.) .... *Erechtites*

        18 Heads yellow or orange-yellow; achenes 5–10-nerved, glabrous or
            strigose; leaves often deeply lobed, their teeth not callous-tipped;
            stems not grooved, rarely over 1 m tall .... *Senecio*

12 Pappus none or rudimentary, or consisting of small scales.
    19 Leaves triangular-subcordate, undulate-toothed, sub-basal, on winged petioles;
        heads whitish; phyllaries green, subequal, only about 2 mm long, finally
        deciduous, they, the branches of the inflorescence, and the achenes more or less
        stipitate-glandular; (s B.C. and sw Alta.; s Ont.) .... *Adenocaulon*
    19 Leaves narrower or deeply dissected, more uniformly distributed on the stem.
      20 Phyllaries subequal, herbaceous or subherbaceous, to about 1.5 cm long,
          stipitate-glandular or viscous; heads creamy white to sometimes pink, usually
          several in a corymbiform flat-topped inflorescence (or the lateral branches
          overtopping the main stem); leaves 1–3-pinnatifid, to 12 cm long, their thickish
          segments characteristically curled; plant sparingly to densely tomentose or
          arachnoid, to 6 dm tall; (s B.C.) .... *Chaenactis*
      20 Phyllaries usually in 2 or more series of unequal length, commonly more or
          less scarious-margined, they and the inflorescence not stipitate-glandular;
          heads yellow.
        21 Inflorescence spike-like, racemose, or paniculate; heads relatively small
            and numerous .... *Artemisia*
        21 Inflorescence corymbiform or heads solitary at the ends of the branches.
          22 Leaves merely crenate to coarsely incised; (introd.).
            23 Heads numerous in a corymbiform inflorescence; achenes
               10-ribbed, sessile; plant fragrant; (*C. balsamita*) ...... *Chrysanthemum*
            23 Heads solitary at the ends of the branches; achenes of the outer
               flowers broadly winged, long-stipitate, the others merely 2-nerved,
               short-stipitate; plant not fragrant; (*C. coronopifolia*) .... *Cotula*
          22 Leaves deeply cut-toothed to finely pinnate or dissected (*Cotula
            australis* may key out here).
            24 Leaves pinnately divided to midrib into remote linear-filiform
               segments (these either entire or themselves pinnate), the actual
               leaf-surface relatively small; phyllaries obovate-oblong, tomentose;
               achenes densely villous, the long hairs hiding the short pappus;
               leaves, branches, and stem more or less tomentose, the stem to
               about 3.5 dm tall; (s Alta. to s Sask.) .... *Hymenopappus*
            24 Leaves deeply cut-toothed or compoundly dissected into closer
               segments, the total leaf-surface much larger; achenes glabrous or
               merely glandular.
              25 Mature receptacle strongly conical; achenes glabrous; leaves
                 deeply dissected into narrow segments; bruised plant with
                 odour of pineapple; (*M. matricarioides*; introd.) .... *Matricaria*
              25 Mature receptacle flat or merely somewhat convex; achenes
                 commonly glandular; leaves 1–3-pinnatifid; plant with a charac-
                 teristic strong scent .... *Tanacetum*

GROUP 3 (see p. 1442)
(Heads discoid (or disciform, inconspicuous rays sometimes present);
leaves all or mostly opposite or basal)

1 Leaves basal, toothed or lobed; receptacle naked or nearly so; pappus consisting of
  capillary bristles.
   2 Leaves dissected into 3–5 deeply cleft small divisions, the few reduced stem-leaves
      linear and mostly uncleft; heads yellow, solitary on the scapes, these less than 1.5 dm
      tall; (*E. compositus*) .... *Erigeron*
   2 Leaves merely dentate to deeply once 5–7-lobed, the reduced stem-leaves subentire;
      heads creamy white, racemose or corymbose; plants more or less white-woolly ....
      .... *Petasites*

1 Leaves opposite on the stem (sometimes whorled in *Eupatorium* and *Psilocarphus*).
  3 Leaves entire or obscurely serrate; receptacle bristly or chaffy.
    4 Pappus consisting of elongate awns, these and the angles of the achenes usually minutely barbed; disk commonly more than 5 mm broad; rays yellow or orange, minute; plant glabrous or more or less hispid .... *Bidens*
    4 Pappus none or an obscure toothed crown; disk rarely over 5 mm broad; erect to prostrate and often matted low annuals.
      5 Achenes rugose-warty, slightly hairy at the summit; receptacle flat; phyllaries in 2 rows; rays present but minute, whitish; leaves lance-linear to oblong, acute at each end, sessile or short-petioled, entire or remotely serrulate, to 1 dm long and 2.5 cm broad; plant strigose; (introd. in s Ont.) .... [*Eclipta*]
      5 Achenes smooth; receptacle truncately obovoid to subglobose; true involucre none, but the heads commonly subtended by several foliage-leaves; each pistillate flower loosely enclosed by a hood-shaped woolly receptacular bract bearing a hyaline appendage laterally (at the top of the open side); plants copiously white-woolly; (B.C. and Alta.) .... *Psilocarphus*
  3 Leaves distinctly toothed to deeply lobed (stem-leaves divided into separate leaflets in *Eupatorium cannabinum*).
    6 Receptacle bristly or chaffy.
      7 Pappus present, consisting of elongate awns, these and the angles of the achenes usually minutely barbed; disk yellow to orange, usually about 1 or 2 cm broad .... *Bidens*
      7 Pappus none or rudimentary.
        8 Phyllaries united, at least at base.
          9 Heads unisexual, numerous and small; mature phyllaries usually with a single series of tubercles or short erect spines near the pointed apex ....
          .... *Ambrosia*
          9 Heads perfect, at most 3, to 9 mm long; phyllaries unarmed; leaves crowded near base, irregularly pinnatifid to nearly 2-pinnatifid into relatively few lobes (these to 5 cm long and 3 mm broad), the upper ones reduced to simple (often alternate) bracts; essentially glabrous plant to about 2 dm tall, from a creeping rootstock; (s Alta.) ... *Thelesperma*
        8 Phyllaries free to base, unarmed.
          10 Heads heterogamous, the few marginal florets pistillate, the many central florets staminate; anthers only slightly united .... *Iva*
          10 Heads essentially homogamous, all but the obscurely ligulate pistillate marginal florets perfect; rays whitish, usually minute; anthers united nearly their whole length; (s Ont.).
            11 Leaves commonly about 1 dm long, lance-linear to oblong, acute at each end, sessile or short-petioled, obscurely serrate; achenes rugose-warty; strigose annual, often rooting at the lower stem-nodes; (introd. in s Ont.) .... [*Eclipta*]
            11 Leaves to over 3 dm long, broadly ovate to oblong in outline, pinnately few-lobed and also toothed, petioled; achenes 3-ribbed, not roughened; clammy-pubescent perennial .... *Polymnia*
    6 Receptacle naked or nearly so (short-bristly in *Dyssodia*); pappus present.
      12 Stem climbing by twining; leaves broadly triangular-cordate, acuminate or caudate-tipped, long-petioled; heads white or pinkish, small and numerous in corymbiform clusters on long axillary peduncles; pappus consisting of capillary bristles; principal phyllaries 4, subequal, occasionally a few short outer ones present; (s ?Ont.) .... [*Mikania*]
      12 Stem not climbing.
        13 Leaves at most about 5 cm long, 1–2-pinnatifid into linear or filiform, bristly-toothed segments, they and the phyllaries dotted with large translucent glands; rays few, yellow or orange, scarcely surpassing the disk; pappus a row of chaffy scales dissected into rough bristles; ill-scented annual; (introd. in s Ont.) .... [*Dyssodia*]

13 Leaves longer, merely serrate, perfoliate or subsessile to distinctly petioled, sometimes whorled; rays none; pappus a row of slender capillary bristles; perennials .... *Eupatorium*

GROUP 4 (see p. 1442)
(Heads radiate, with a central disk of tubular florets and marginal ligulate florets; leaves all or mostly alternate, entire or nearly so)

1 Receptacle bristly or chaffy.
  2 Receptacle with a single series of chaff-like bracts near the margin (between the ray- and disk-florets), otherwise naked; phyllaries essentially 1-rowed and equal, enfolding and usually completely enclosing the ray-achenes (the involucre thus usually appearing deeply grooved); pappus none or a short crown or a few scales; rays yellow, broad and 3-cleft; leaves linear to linear-oblong; more or less glandular and typically tar-scented annuals .... *Madia*
  2 Receptacle chaffy throughout; phyllaries not enfolding the achenes.
    3 Leaves linear, to 2.5 cm long and 1.5 mm broad; involucre to 7 mm high, glandular and sometimes hairy, the phyllaries in 2 subequal series; rays broad, 3-lobed, to 1 cm long, they and the disk-flowers white; achenes silky-villous; pappus of narrow fringed scales or sometimes reduced or wanting; slender, scabrous-puberulent and often spreading-hairy annual to 3 dm tall; (s ?B.C.) .... [*Blepharipappus*]
    3 Leaves relatively broader; involucre usually considerably larger; rays commonly longer and entire (sometimes shallowly toothed at apex); plants taller, chiefly perennial (sometimes biennial).
      4 Receptacle flat or merely convex; rays yellow.
        5 Stem-leaves all alternate; pappus a crown of short scales; taprooted perennials; (s ?B.C.) .... [*Wyethia*]
        5 Lower stem-leaves often opposite; pappus consisting of 2 readily deciduous awned scales (rarely with some additional short scales); perennials from rhizomes or tuberous roots (rarely annuals) .... *Helianthus*
      4 Receptacle conical.
        6 Receptacular bracts with firm spiny tips conspicuously surpassing the disk-corollas; rays pink or purple .... *Echinacea*
        6 Receptacular bracts not spine-tipped but sometimes shortly awn-pointed; rays yellow to orange, sometimes purplish at base .... *Rudbeckia*
1 Receptacle naked or nearly so.
  7 Rays white, pink, blue, or purple (yellowish in *Aster ptarmicoides* var. *lutescens*).
    8 Pappus consisting of several minute bristles and usually 2–4 awns up to 2 mm long; achenes glabrous, very flat, rather broadly callous-winged; phyllaries in several series of unequal length; heads in leafy-bracted corymbs; rays white to lilac; disk yellow; leaves firm, broadly linear to lance-elliptic, to about 1.5 dm long, barely narrowed at the subsessile base; plant to 1.5 m tall; (s Sask. and s Man.) .... *Boltonia*
    8 Pappus consisting of capillary bristles; achenes glabrous or pubescent.
      9 Peduncles commonly leafy or bracteate; rays commonly broad and relatively few; achenes several-nerved in most species, glabrous or pubescent; pappus copious and firm, double only in *A. linariifolius* and *A. umbellatus*; hairy tips of the style-branches relatively long, generally acute or acuminate; phyllaries relatively broad (the usually expanded tip commonly remaining green), commonly in 2 or more series of unequal length, sometimes subequal and then usually green throughout or some of the outer ones expanded and somewhat foliaceous; plants mostly tall and leafy, mostly flowering in late summer and fall .... *Aster*
      9 Peduncles commonly naked or with few and much reduced leaves; rays mostly narrow and numerous; achenes usually 2-nerved, commonly more or less pubescent; pappus usually scanty and fragile, usually double (with an

outer series of small bristles); hairy tips of the style-branches short, mostly obtuse or merely acutish; phyllaries usually relatively long and narrow, mostly equal or subequal and without green tips; plants commonly low and with relatively narrow, chiefly basal leaves, mostly beginning to flower in spring and early summer .... *Erigeron*

7 Rays cream-colour to yellow or deep orange.
  10 Pappus none; heads solitary on terminal peduncles, to over 5 cm broad; phyllaries subequal, 1–3-seriate; (introd.).
    11 Leaves linear-oblong, tapering to base; achenes of two kinds, those of the ray-florets 3-angled, those of the disk-florets compressed and with thickened or winged borders .... [*Dimorphotheca*]
    11 Leaves thickish, oblong to oblong-obovate, more or less clasping; achenes uniform, produced only by the ray-florets .... [*Calendula*]
  10 Pappus present; heads few to many, much smaller; phyllaries usually in several series of unequal length.
    12 Pappus consisting of several awns or scales; heads crowded in small clusters, cylindrical; ray- and disk-florets each 3 or 4; phyllaries coriaceous, with green tips; leaves linear to narrowly oblanceolate, to about 3 cm long; stems woody at base, to about 3 dm tall, from a taproot; plants glabrous and often glutinous; (s Alta. to s Man.) .... *Gutierrezia*
    12 Pappus (at least an inner series) consisting of capillary bristles.
      13 Disk rarely over 5 or 6 mm broad; heads numerous, in racemes, corymbs, panicles, or axillary clusters .... *Solidago*
      13 Disk to 2.5 cm broad; heads solitary to few.
        14 Leaves white-woolly beneath, to 3 or 4 dm long, the cauline ones cordate-clasping; stems erect; pappus simple; (*I. britannica*; introd. in s Ont.) .... *Inula*
        14 Leaves not woolly or felted, mostly much smaller, scarcely clasping; stems usually several from a branched caudex, commonly arched-ascending; (western species).
          15 Pappus double (the outer series much shorter than the inner) ....
          .... *Chrysopsis*
          15 Pappus simple (the bristles generally unequal but not divided into 2 series) .... *Haplopappus*

GROUP 5 (see p. 1442)
(Heads radiate, with a central disk of tubular florets and marginal ligulate florets; leaves all or mostly alternate, distinctly toothed to deeply lobed)

1 Receptacle more or less chaffy (or merely bristly in *Gaillardia*); pappus consisting of chaff, scales, or awns, a minute crown, or none.
  2 Rays normally white.
    3 Lower leaves toothed to rather irregularly and coarsely pinnatifid, to 6 cm long and about 1.5 cm broad, the upper leaves commonly entire; rays broad and 3-cleft, to 1.5 cm long; receptacular chaff confined to a ring between the ray- and disk-flowers; ray-achenes lacking a pappus; disk-achenes usually with a pappus of slender scales; phyllaries herbaceous, with abruptly dilated thin margins enfolding the achene; spreading-hairy and more or less stipitate-glandular branching annual; (s ?B.C.) .... [*Layia*]
    3 Leaves mostly finely dissected and "fern"-like (merely finely serrate in *Achillea ptarmica* and closely pectinate in *A. sibirica*); receptacle usually chaffy throughout (only toward the middle in *Anthemis cotula*); pappus none or a minute crown; phyllaries dry, papery and scarious-margined.
      4 Heads relatively large, solitary at the ends of the branches, the disk to 12 mm broad, the rays to about 13 mm long; receptacle hemispherical to conical; achenes quadrangular or subterete; (introd.) .... *Anthemis*

      4   Heads small, several to numerous in a terminal corymb; disk to about 5 mm broad; rays to about 5 mm long; receptacle flattish; achenes strongly-flattened, callous-margined .... *Achillea*

  2 Rays yellow to orange (sometimes partly or wholly purple in *Helenium, Gaillardia,* and *Ratibida*).

      5   Phyllaries dry, papery and scarious-margined; heads solitary at the ends of the branches, the disk about 1.5 cm broad, the rays to 1.5 cm long; receptacle hemispherical; pappus none or a minute crown; leaves deeply pinnately divided into narrow toothed or pinnatifid segments; (*A. tinctoria*; introd.) .... *Anthemis*

      5   Phyllaries not papery; leaves less divided.

          6   Stem distinctly square in cross-section, the angles usually more or less winged; leaves lanceolate to oblong or oblong-lanceolate, sharply serrate to subentire, to about 2.5 dm long; heads corymbed, the 2–8 irregular rays to about 3 cm long; achenes spreading in all directions (forming a globose head), mostly broadly winged, the pappus consisting of 2 or 3 smooth persistent awns; (s Ont.) .... *Actinomeris*

          6   Stems terete or subterete; fruiting head not globose, the achenes uniformly disposed.

              7   Disk-florets sterile, with undivided style, producing sterile stalk-like achenes; ray-achenes strongly flattened and overlapping in 2 or 3 marginal series; pappus none or consisting of 2 awns confluent with the broad achene-wings; receptacle flat; tall, coarse perennials with few large corymbed heads and copious resinous juice; (s Ont.) .... *Silphium*

              7   Disk-florets fertile, with divided styles; ray-florets either fertile or sterile.

                  8   Receptacle merely bristly-hairy, convex to hemispheric; achenes partly or wholly covered with long ascending hairs; pappus consisting of up to about 10 awned scales; rays purple-brown at base or sometimes purplish throughout; disk-corollas purple-tipped, woolly-villous near summit; phyllaries in 2 or 3 series, reflexed in fruit; (B.C. to Man.; introd. eastwards) .... *Gaillardia*

                  8   Receptacle chaffy; achenes commonly glabrous except for the sometimes ciliate margins; phyllaries not reflexed.

                      9   Receptacle flat to merely convex; pappus consisting of 2 or 4 deciduous scales .... *Helianthus*

                      9   Receptacle conical or columnar; pappus none or a short crown, with or without 1 or 2 awn-teeth.

                          10 Receptacle columnar, its truncate chaff bearded at tip; rays often tinged with brown-purple at base, subtended by receptacular chaff; achenes flat, margined; (s B.C. to s Ont.) .... *Ratibida*

                          10 Receptacle hemispheric or ovoid-conical, its chaff short and subtending only disk-florets; achenes 4-angled, not margined ....
                            .... *Rudbeckia*

1 Receptacle naked or merely short-hairy around the achene-pits.

  11 Rays white, pink, bluish, or purplish (yellowish in *Aster ptarmicoides* var. *lutescens*).

    12 Pappus a short crown or none; phyllaries dry, papery and scarious-margined; rays white.

      13 Leaves deeply pinnately dissected into narrow segments; mature receptacle conical .... *Matricaria*

      13 Leaves coarsely toothed to deeply broad-lobed; receptacle flat or merely convex .... *Chrysanthemum*

    12 Pappus consisting of capillary bristles; phyllaries not papery; leaves merely toothed.

      14 Peduncles commonly leafy or bracteate; rays commonly broad and relatively few; achenes several-nerved in most species, glabrous or pubescent; pappus copious and firm, simple; hairy tips of the style-branches relatively long, generally acute or acuminate; phyllaries relatively broad (the usually expanded

tip commonly remaining green), usually in 2 or more series of unequal length, sometimes subequal and then usually green throughout, or some of the outer ones expanded and leafy; plants mostly tall, flowering in late summer and fall .... *Aster*

    14 Peduncles commonly naked or with few and much reduced leaves; rays mostly relatively narrow and numerous; achenes usually 2-nerved; pappus usually scanty and fragile, commonly double (with an outer series of short setae); hairy tips of the style-branches short, commonly obtuse or merely acutish; phyllaries usually relatively long and narrow, equal or subequal and lacking green tips; plants mostly low and with relatively narrow, chiefly basal leaves, beginning to flower in spring and early summer .... *Erigeron*

11 Rays yellow to orange.
  15 Plant very glutinous with resinous glands; leaves linear- to ovate- or narrowly obovate-oblong, serrate or crenate, to about 7 cm long; phyllaries in 3 or 4 series of unequal length, their tips strongly recurving; (w Canada) .... *Grindelia*
  15 Plant rarely strongly glutinous (glandular-viscid in a few species of *Senecio*, with subequal principal phyllaries).
    16 Pappus consisting of scales, a short crown, or wanting.
      17 Achenes of 2 kinds, those of the ray-florets 3-angled, those of the disk-florets compressed and with thickened or winged borders; leaves linear-oblong, tapering to base, to about 9 cm long; (introd.) .... [*Dimorphotheca*]
      17 Achenes uniform.
        18 Phyllaries dotted with resinous particles, firm, in 2 series of unequal length, those of the lower series united to middle; involucre commonly about 8 mm high, woolly; achenes densely hirsute with ascending hairs; pappus consisting of 5 hyaline scales; leaves to about 1 dm long, pinnatifid into linear divisions; stem to about 3 dm tall; (*H. richardsonii*; s Alta. and Sask.) .... *Hymenoxys*
        18 Phyllaries not resinous-dotted, distinct.
          19 Phyllaries not papery, subequal; pappus of chaffy scales.
            20 Leaves shallowly toothed, glandular-punctate; phyllaries mostly at least 20, herbaceous or subherbaceous, not individually subtending the ray-achenes, they and the rays deflexed; minutely puberulent to glabrate perennials to over 1 m tall .... *Helenium*
            20 Leaves mostly ternate or coarsely pinnatifid (the upper ones often entire), to 8 cm long; phyllaries usually less than 15, firm, partly embracing the ray-achenes, they and the rays not deflexed; tomentose perennial to 6 dm tall; (s B.C.) .... *Eriophyllum*
          19 Phyllaries dry and papery, scarious-margined, in 2 or more series of unequal length; pappus a crown of short scales or wanting.
            21 Principal leaves merely coarsely toothed to moderately deeply lobed; (*C. segetum*; introd.) .... *Chrysanthemum*
            21 Principal leaves 1–2-pinnate-pinnatifid.
              22 Ray-achenes narrowly 3-winged; annual; (*C. coronarium*; introd.) .... *Chrysanthemum*
              22 Achenes 5-ribbed or 5-angled; rhizomatous perennials.......... .... *Tanacetum*
    16 Pappus of capillary bristles.
      23 Phyllaries subequal and essentially 1-rowed (sometimes with a series of bractlets at base); heads solitary or in corymbs .... *Senecio*
      23 Phyllaries in 2 or more series of unequal length.
        24 At least the upper ovate stem-leaves cordate-clasping; lower leaves long-petioled; (introd.).
          25 Leaves densely white-woolly beneath, the lower ones elliptic; (*I. helenium*) .... *Inula*

25 Leaves more or less pubescent but not woolly, the lower ones
   deeply cordate .... *Doronicum*

24 None of the leaves cordate-clasping.

26 Leaves (or their teeth or lobes) tipped with a short whitish spine;
   achenes copiously appressed-villous.

27 Upper leaves entire, the lower ones toothed or sometimes
   pinnatifid, to about 8 cm long; phyllaries broadly ovate, with
   whitish margins; annual or biennial to over 1 m tall; (introd. in
   s Ont.) .... *[Xanthisma]*

27 Leaves all conspicuously toothed or lobed, rarely over 6 cm
   long; phyllaries lanceolate; perennials from branching woody
   caudices; (*H. nuttallii* and *H. spinulosus*; w Canada) ....
   .... *Haplopappus*

26 Leaves (or their teeth or lobes) not spinulose-tipped.

28 Heads relatively large, solitary or in corymbs; pappus double,
   the outer series much shorter than the inner; stems several
   from a heavy crown, mostly arched-ascending; (B.C. to s Ont.) ......
   .... *Chrysopsis*

28 Heads small and numerous, in racemes, corymbs, panicles, or
   axillary clusters; pappus single; stems mostly solitary and erect .....
   .... *Solidago*

GROUP 6 (see p. 1442)

(Heads radiate, with a central disk of tubular florets and marginal ligulate florets;
leaves all or mostly opposite or basal, rarely whorled)

1 Leaves all or mostly basal (stem-leaves, when present, much reduced).

2 Leaves entire to obscurely undulate-crenate.

3 Pappus consisting of capillary bristles; achenes densely appressed-hirsute;
   receptacle naked; phyllaries in several series of unequal length; leaves linear to
   linear-oblanceolate, at most about 8 cm long, tapering to obscure margined
   petioles; low plants with woody branching caudices; (s B.C. to s Man).

4 Heads sessile among the leaves, these silky-strigose, mostly less than 5 cm
   long, not evidently nerved; rays pinkish or purplish; phyllaries narrowly
   lanceolate .... *Townsendia*

4 Heads solitary or occasionally in pairs at the top of a scape to about 1.5 dm
   tall bearing 1 or 2 leafy bracts; leaves rather strongly triple-nerved; rays
   yellow; phyllaries lanceolate to oval or oblong .... *Haplopappus*

3 Pappus consisting of hyaline scales or lacking; rays yellow, with darker veins;
   heads solitary on naked or leafy-bracted scapes.

5 Receptacle chaffy throughout; pappus usually none; achenes usually glabrous
   or nearly so; rays to over 4 cm long; leaf-blades to about 2 dm long,
   long-petioled; scapes to about 1 m tall; (s B.C. and s Alta.) .... *Balsamorhiza*

5 Receptacle naked; pappus consisting of about 5 hyaline scales; achenes
   densely hairy; rays about 1.5 cm long; leaves linear-oblanceolate, more or
   less appressed-silky, resinous-dotted, less than 1 dm long, tapering to an
   obscure petiole; scapes commonly less than 1.5 dm tall; (*H. acaulis*; s Alta., s
   Sask., and s Ont.) .... *Hymenoxys*

2 Leaves distinctly toothed to deeply lobed (sometimes entire in *Crocidium*); phyllaries
   subequal.

6 Heads white or purplish, numerous in a corymb or racemose panicle terminating a
   scape bearing broad sheathing scaly bracts; receptacle naked; pappus consisting
   of capillary bristles .... *Petasites*

6 Head solitary at the top of the scape or scapose stem.

7 Receptacle chaffy throughout; rays yellow, to 4.5 cm long; leaves to 4 dm
   long, long-petioled, deeply parted nearly or quite to the midrib into narrow,

toothed to deeply lobed segments up to 1 dm long; plant with a carrot-like taproot and simple crown; (*B. hirsuta*; s ?B.C.) . . . . *Balsamorhiza*

7   Receptacle naked.

    8   Scape naked; receptacle conical; rays white to pink or purple; pappus none; leaves elliptic to obovate, narrowed to margined petioles; (introd.) . . . .
. . . . *Bellis*

    8   Scape abundantly scaly- or leafy-bracted; rays yellow; pappus consisting of capillary bristles.

        9   Basal leaves cordate-rotund, with a deep narrow sinus, callous-denticulate and shallowly lobed, glabrous above, persistently white-tomentose beneath, to about 2 dm long and broad, long-petioled; receptacle flat; stems thinly tomentose, to about 5 dm tall, from creeping rhizomes; (introd.) . . . . *Tussilago*

        9   Basal leaves oblanceolate to narrowly obovate, often coarsely few-toothed (sometimes entire), slightly fleshy, to about 2.5 cm long (including the petiolar base); receptacle strongly conic; delicate annual rarely over 2 dm tall, bearing loose tufts of wool in the leaf-axils, otherwise glabrous; (Vancouver Is.) . . . . *Crocidium*

1   Leaves opposite on the stem (sometimes also in basal tufts or the upper leaves alternate).

  10 Leaves entire or nearly so; rays mostly yellow (white to roseate in *Coreopsis rosea*).

    11 Pappus of capillary barbed (rarely subplumose) bristles; receptacle merely minutely fringed around the pits; heads solitary or few in a corymb . . . . *Arnica*

    11 Pappus none or of teeth, scales, or awns.

      12 Receptacle naked, conic; heads mostly solitary; phyllaries in several series of unequal length; rays at most 5 mm long; pappus none; leaves linear-oblanceolate to narrowly oblong, to 6 cm long and 6 mm broad; succulent lax glabrous rhizomatous perennial to about 3 dm tall; (Vancouver Is.) . . . . *Jaumea*

      12 Receptacle chaffy.

        13 Achenes laterally compressed at right angles to the phyllaries, embraced by the receptacular chaff; phyllaries usually green and more or less herbaceous, subequal or in 2 or more series of unequal length; leaves simple.

          14 Achenes strongly compressed, thin-edged, their persistent pappus consisting of several confluent fringed short scales and commonly 2 slender awns; leaves lanceolate to elliptic (or the lowermost ones oblanceolate and smaller), to 1.5 dm long and 5 cm broad, scabrous; stems harshly puberulent to spreading-hirsute (or glabrous below), to 1 m tall, clustered from a branching caudex; (s B.C.) . . . . *Helianthella*

          14 Achenes 4-sided and only slightly or moderately compressed, their readily deciduous pappus consisting of 2 large awned or awnless scales, rarely with some additional short scales . . . . *Helianthus*

        13 Achenes compressed parallel to the phyllaries, free from the receptacular chaff; phyllaries dimorphic, 2-rowed, the outer series more or less herbaceous; leaves simple or compound.

          15 Pappus consisting of up to 6 mostly downwardly-barbed (rarely smooth) awns; achenes not wing-margined . . . . *Bidens*

          15 Pappus none or consisting of 2 short upwardly-barbed teeth or a few minute bristles or both; achenes narrowly to broadly winged . . . . . *Coreopsis*

  10 Leaves mostly distinctly toothed to deeply lobed.

    16 Receptacle naked (*Baeria* and *Bahia*) or the pits merely minutely fringed (*Arnica*); rays yellow or orange-yellow.

      17 Leaves deeply ternate or 5-lobed into linear to oblong segments (or the upper ones entire), to about 5 cm long, impressed-punctate and minutely strigose; receptacle flat or nearly so; rays 3 or 4 mm long; pappus a crown of 8 lanceolate to ovate scales; achenes glandular; stems to 2.5 dm tall, from a creeping rhizome; (introd. in s Alta. and s Sask.) . . . . *Bahia*

17 Leaves merely serrate or dentate.
  18 Lax, somewhat succulent annual to about 5 dm tall, more or less
    woolly-villous above when young; leaves narrowly oblong, to 5 cm long;
    involucre to 6 mm high; phyllaries few; receptacle conic, with a broad
    stalk-like base for each flower; pappus consisting of up to 5 awns and
    about as many alternating fringed shorter scales; (Vancouver Is. and
    adjacent islands) .... *Baeria*
  18 Lax to erect perennials from a rhizome or caudex; leaves and involucres
    commonly longer; phyllaries rather numerous; receptacle convex; pappus
    consisting of barbed or subplumose capillary bristles .... *Arnica*
16 Receptacle chaffy; pappus none or consisting of scales, teeth, or barbed awns.
  19 Stem square in cross-section, to over 2.5 m tall; achenes winged; rays yellow.
    20 Leaves lanceolate to ovate-lanceolate or oblong, pointed at both ends;
      stem usually winged above, somewhat hairy; rays 2–8, irregular, to about
      3 cm long; achenes spreading in all directions, forming a globose head,
      the pappus consisting of 2 or 3 smooth persistent awns; (s Ont.) ....
      .... *Actinomeris*
    20 Leaves ovate, scabrous, the upper ones united into a cup-shaped connate
      base; stem wingless, glabrous; rays numerous, to 3.5 cm long; achenes in
      a hemispheric head, the pappus wanting or consisting of 2 awns confluent
      with the achene-wings; (*S. perfoliatum*; s Ont.) .... *Silphium*
  19 Stem terete, subterete, or only obscurely 4-angled.
    21 Rays white to lilac or roseate.
      22 Leaves at most rather coarsely toothed, lance-ovate to ovate, petioled;
        rays 4 or 5, white, small, roundish; disk 3 or 4 mm broad; receptacle
        conical; pappus consisting of fringed scales; annual; (introd.) .... *Galinsoga*
      22 Leaves deeply lobed; rays mostly longer; disk at least 6 mm broad;
        receptacle flat.
        23 Leaves pinnately dissected into entire linear or linear-filiform
          segments; rays lilac to roseate; achenes with a beak 1–6 mm long
          tipped by a pappus of 2 or 3 short barbed awns, or the pappus
          wanting; glabrous or minutely scabrous annual; (garden-escape) ....
          .... *[Cosmos]*
        23 Upper leaves more or less deeply 3–5-lobed, the lower ones
          deeply 1-pinnatifid, all of them also irregularly dentate and
          glandular-punctate; rays whitish; achenes 3-ribbed and 3-angled,
          beakless, lacking a pappus; clammy-hairy perennial to about 1.5 m
          tall; (s Ont.) .... *Polymnia*
    21 Rays yellow to orange-yellow.
      24 Rays remaining attached to the 4-sided marginal achenes and
        becoming papery; receptacle conical; phyllaries in 2 or 3 subequal
        series; pappus none or a small crown; leaves ovate, coarsely serrate,
        petioled; perennial to about 1.5 m tall; (B.C. to Ont.; introd. eastwards) ......
        .... *Heliopsis*
      24 Rays soft, deciduous from the achenes at maturity.
        25 Rays few and inconspicuous, not much surpassing the disk;
          involucre a single row of phyllaries united into a firm cup, with a
          few loose bracts at base; pappus consisting of numerous scales
          (each divided to near base into 5 or more bristles); leaves dotted
          with large pellucid glands, 1–2-pinnatifid into filiform or linear
          segments, at most about 5 cm long; strong-scented, diffusely
          branched, essentially glabrous annual to about 5 dm tall; (introd. in
          s Ont.) .... *[Dyssodia]*
        25 Rays commonly numerous and showy (often more or less reduced
          in *Bidens*); phyllaries in 2 or more subequal or unequal series.
          26 Disk-florets sterile, with undivided style, producing sterile
            stalk-like achenes; ray-achenes strongly flattened and overlap-

ping in 2 or 3 marginal series; receptacle flat; pappus none or consisting of 2 awns confluent with the achene-wings; tall perennials; (S Ont.) .... *Silphium*

    26 Disk-florets fertile, with divided styles; ray-florets either fertile or sterile .

        27 Phyllaries mostly in 2 or more series of unequal length, not dimorphic; receptacular chaff concave, partly enclosing the achenes, these slightly compressed at right angles to the phyllaries; pappus readily deciduous, consisting of 2 thin scales on the principal achene-angles .... *Helianthus*

        27 Phyllaries 2-rowed and dimorphic, the outer ones more or less herbaceous; receptacular chaff flattish, scarcely enveloping the achenes, these flattened parallel to the phyllaries.

            28 Pappus consisting of up to 6 mostly downwardly-barbed (rarely smooth) awns; achenes wingless; leaves simple and subentire to coarsely toothed or deeply lobed, or ternately or pinnately compound with serrate leaflets (submersed leaves filiform-dissected in *B. beckii*) ... *Bidens*

            28 Pappus consisting of 2 short smooth or upwardly-barbed teeth, or a few minute bristles, or both, or pappus none; achenes usually narrowly to broadly winged (wingless in *C. tinctoria*) .... *Coreopsis*

GROUP 7 (see p. 1442)
(Heads ligulate; leaves all or mostly alternate, entire or nearly so; juice often milky)

1   Pappus double, consisting of an inner row of numerous fragile capillary bristles subtended by a row of short scales; heads 2–7, orange; phyllaries 1-rowed; stem-leaves 1–3, oblong or oval, clasping; roots fibrous; (*K. biflora*; S Man. and S Ont.) .... *Krigia*

1   Pappus a single row of capillary bristles.

    2   Pappus-bristles densely plumose; achenes all or mostly long-beaked.

        3   Plants glabrous; leaves somewhat grass-like, clasping; heads solitary, large, yellow or purple; involucre a single row of subequal phyllaries; (introd.) .... *Tragopogon*

        3   Plants more or less spiny-hispid; leaves lanceolate to oblanceolate, the lower ones petioled, the upper ones somewhat clasping; heads usually at least 2, yellow; involucre consisting of subequal spine-tipped phyllaries in 2 rows; (*P. echioides*; introd.) .... *Picris*

    2   Pappus not plumose; heads smaller.

        4   Achenes flat or flattish, beaked or tapering to summit; heads blue, purple, or yellow; phyllaries in 2 or more series of unequal length .... *Lactuca*

        4   Achenes terete or nearly so.

            5   Branches rush-like; heads pink to reddish-purple; involucre of usually 5–9 linear phyllaries subtended by a ring of small bractlets; achenes linear-filiform; (S B.C. to S Man.) .... *Lygodesmia*

            5   Branches not rush-like; heads yellow or orange (white or cream-colour in *Hieracium albiflorum*; sometimes pink in *Crepis*); achenes narrowed at base.

                6   Head solitary on the scape.

                    7   Achenes distinctly beaked; pappus white; head yellow to orange (often drying pinkish or purplish); plant from a taproot .... *Agoseris*

                    7   Achenes beakless; pappus more or less sordid or tawny; head yellow; plant abundantly stoloniferous, lacking a taproot .... *Hieracium*

                6   Heads few to numerous; achenes beakless or nearly so.

                    8   Plants with short to elongate rhizomes and fibrous roots; phyllaries in more or less distinct series of unequal length (or at least with a basal series of smaller bractlets); pappus more or less sordid or tawny ....
.... *Hieracium*

8 Plants from a taproot or several strong roots, lacking rhizomes; phyllaries commonly 1-rowed; pappus white or nearly so .... *Crepis*

GROUP 8 (see p. 1442)
(Heads ligulate; leaves all or mostly alternate, distinctly toothed to deeply lobed; juice often milky)

1 Heads pink or blue (occasionally white); pappus present.
  2 Heads mostly pink, terminating the branches, mostly 5-flowered; involucre to about 11 mm long, with mostly 5 principal phyllaries; pappus consisting of plumose bristles; achenes to 6 mm long; leaves filiform or linear, entire or sparingly toothed, to 8 cm long and 3 mm broad, the upper ones often scale-like; plant glabrous or puberulent, with several or many stems from a stout taproot and branching caudex; (s B.C. to sw Sask.) .... *Stephanomeria*
  2 Heads mostly blue.
    3 Heads to over 3 cm broad, sessile or short-peduncled; achenes obscurely 5-angled; pappus consisting of minute scales; phyllaries 2-rowed, the outer ones at most half as long as the inner; (introd.) .... *Cichorium*
    3 Heads smaller, panicled; achenes flat or flattish; pappus consisting of capillary bristles; phyllaries in 2 or more series of unequal length .... *Lactuca*
1 Heads yellow to orange (rarely whitish, pinkish, or purplish); pappus (wanting in *Lapsana*) consisting of capillary bristles.
  4 Pappus-bristles plumose (*Picris*) or pappus wanting (*Lapsana*).
    5 Pappus none; heads corymbed or panicled; phyllaries subequal, with a subtending series of minute bractlets; leaves ovate to subrotund; plant hirsute to subglabrous; (introd.) .... *Lapsana*
    5 Pappus-bristles plumose (or the outer short ones merely barbed); heads terminating leafy stems; outer phyllaries loose and spreading; leaves lanceolate or somewhat broader; (introd.) .... *Picris*
  4 Pappus-bristles not plumose.
    6 Pappus double, an inner row of numerous fragile capillary bristles subtended by a row of short scales; roots fibrous; (SE Man. and s Ont.) .... *Krigia*
    6 Pappus a single row of capillary bristles.
      7 Achenes more or less strongly flattened, beakless; pappus white; phyllaries in 2 or more series of unequal length; (introd.) .... *Sonchus*
      7 Achenes terete or nearly so.
        8 Head solitary, yellow or orange (often drying pinkish or purplish); achenes usually distinctly beaked; pappus white .... *Agoseris*
        8 Heads usually 2 or more; achenes beakless.
          9 Phyllaries in a single series; heads yellow or orange; pappus white or whitish; plant from a taproot or several strong roots, lacking rhizomes ....
           .... *Crepis*
          9 Phyllaries in 2 or more series.
            10 Inflorescence a branching raceme or panicle of slender, drooping, whitish, creamy, or pink heads; pappus white to reddish brown; lower leaves often deeply cleft; root tuberous .... *Prenanthes*
            10 Inflorescence a corymb or panicle of erect or ascending, yellow (often white in *H. albiflorum*), broad heads; pappus more or less sordid or tawny; leaves merely toothed; root not tuberous .... *Hieracium*

GROUP 9 (see p. 1442)
(Heads ligulate; leaves all or mostly basal, stem-leaves, when present, much reduced; juice often milky)

1 Leaves entire or essentially so (sometimes remotely runcinate-toothed); receptacle naked.
  2 Heads few to numerous; achenes beakless or nearly so.

3 Plants with short to elongate rhizomes and fibrous roots; phyllaries weakly to strongly overlapping, or at least with a basal series of bractlets; pappus-bristles more or less sordid or tawny .... *Hieracium*

3 Plants from a taproot or several strong roots, lacking rhizomes; involucre commonly a single row of equal phyllaries; pappus-bristles white or whitish ..... *Crepis*

2 Head usually solitary on the scape (except in robust individuals of *Microseris nutans*).

 4 Achenes typically more or less strongly beaked; pappus-bristles white; heads yellow or orange (often drying pinkish or purplish) .... *Agoseris*

 4 Achenes beakless or nearly so; heads yellow.

  5 Plants fibrous-rooted and with short to elongate rhizomes; pappus-bristles usually more or less sordid or tawny; involucre not calyculate .... *Hieracium*

  5 Plants with a taproot; involucre often calyculate (subtended by bractlets).

   6 Pappus consisting of brownish barbed capillary bristles more or less united at base and tending to fall in a ring; plant essentially glabrous; (B.C.)........... .... *Apargidium*

   6 Pappus various, the segments distinct and falling separately; (B.C. to Man.) .... *Microseris*

1 Leaves distinctly toothed to deeply lobed; heads yellow (sometimes pink or red-orange).

 7 Pappus none; achenes strongly ribbed, shining; heads 1–few, on conspicuously upwardly-thickened peduncles; receptacle naked; leaves oblanceolate or spatulate; plant glabrous or minutely puberulent; (introd.) .... *Arnoseris*

 7 Pappus present.

  8 Pappus double, an inner row of numerous fragile capillary bristles subtended by a row of short scales; phyllaries subequal; receptacle naked; roots fibrous; (Man. and s Ont.) .... *Krigia*

  8 Pappus a single series of capillary bristles.

   9 Pappus-bristles plumose; (introd.).

    10 Inner achenes slender-beaked; receptacle chaffy; leaves hirsute .... *Hypochaeris*

    10 Achenes not slender-beaked; receptacle naked .... *Leontodon*

   9 Pappus-bristles not plumose; receptacle naked.

    11 Heads few to numerous; achenes beakless or nearly so.

     12 Plants with short to elongate rhizomes and fibrous roots; phyllaries weakly to strongly overlapping, or at least with a basal series of small bractlets; pappus more or less sordid or tawny .... *Hieracium*

     12 Plants from a taproot or several strong roots, lacking rhizomes; involucre commonly a single row of subequal phyllaries; pappus white or whitish .... *Crepis*

    11 Head solitary on the scape.

     13 Achenes strongly tuberculate at least above, long-beaked; head yellow to orange-yellow; scapes hollow, weak .... *Taraxacum*

     13 Achenes not tuberculate; scapes usually solid and firm.

      14 Achenes typically more or less strongly beaked; pappus white; heads yellow or orange, often drying pinkish or purplish; plants with a taproot (except *A. heterophylla*) .... *Agoseris*

      14 Achenes beakless; pappus more or less sordid or tawny; heads yellow; plants with short to elongate rhizomes and fibrous roots .... *Hieracium*

ACHILLEA L. [9332] Yarrow. Achillée

1 Flowers yellow, the heads in dense convex compound corymbs; leaves pinnate-pinnatifid; (introd.) .... [*A. filipendulina*]

1 Flowers white (atypically pink to purplish).

 2 Leaves lanceolate to broadly ovate in outline, much dissected and "fern"-like in appearance (2-pinnate-pinnatifid to 3-pinnate); (transcontinental) .... *A. millefolium*

 2 Leaves lance-linear, not 2-pinnately dissected.

3 Leaves merely appressed-serrulate; rays surpassing the involucre by about 4 or 5 mm; stems glabrous or slightly pubescent; (introd.) .... *A. ptarmica*
3 Leaves pectinately divided to about 3/4 their width into numerous serrulate segments about 1 mm broad; rays surpassing the involucre by about 1 mm; stems villous or becoming glabrate; (B.C. to E Que.) .... *A. sibirica*

[A. filipendulina Lam.] Fernleaf Yarrow
[Asiatic; apparently recorded for N. America only from Ont. (Galetta, Carleton Co.; Boivin 1966*b*; presumably a garden-escape but lacking information as to whether or not established). Description: Bailey (1949*a*:991).]

A. millefolium L. Common Yarrow, Milfoil. Herbe à dindes
/aST/X/EA/ (Hsr) Meadows and gravelly or sandy slopes and shores at low to moderate elevations (the typical Eurasian form introd. along roadsides and in waste places in the S part of the N. American area), the aggregate species from the Aleutian Is. and coasts of Alaska–Yukon–Dist. Mackenzie to N Sask. (L. Athabasca), Man. (N to Churchill), northernmost Ont., Que. (N to E Hudson Bay at ca. 59°N, Ungava Bay, the Côte-Nord, Anticosti Is., and Gaspé Pen.), Labrador (N to Hebron, 58°12′N), Nfld., ?P.E.I. (not known from N.B.), and N.S., S to Calif., Mexico, Tex., and Fla.; introd. in S Greenland; Eurasia. MAPS and synonymy: *see* below.
1 Phyllaries with dark-brown to blackish margins; rays surpassing the involucre by 3 or 4 mm; stem more or less lanate; [transcontinental] .... var. *borealis* (Bong.) Farw.
  2 Ray-ligules white; [var. *nigrescens* Mey. (*A. nig.* (Mey.) Rydb.); ssp. and var. *atrotegula* Boivin; *A. borealis* Bong., the type from Sitka, Alaska; the common form northwards; MAPS: Hultén 1968*b*:888; the map by G. A. Mulligan and I.J. Bassett, Can. J. Bot. 37: fig. 4 (solid dots), p. 77. 1959, for plants with "hexaploid size pollen" is presumably largely referable here] .... f. *borealis*
  2 Ray-ligules pink or reddish; [ssp. and var. *atrotegula* f. *rhodantha* Lepage, the type from Cape Henrietta Maria, NW James Bay, Ont.; other pink-flowered forms have been named ssp. *atrotegula* vars. *parvula* f. *discolor* Boivin and *fulva* f. *roseiflora* Boivin, var. *parvula* Boivin being the small-rayed extreme and var. *fulva* Boivin the plant with rust-tinged pubescence] .... f. *rhodantha* (Lepage) Scoggan
1 Leaves lanceolate to broadly ovate in outline, 2-pinnate-pinnatifid to 3-pinnate.
  3 Corymb more or less round-topped, usually less than 1 dm broad; leaf-lobes generally rather strongly incurved; stem densely lanate; [transcontinental] ....
.... var. *lanulosa* (Nutt.) Piper
    4 Ray-ligules white; [ssp. and var. *pallidotegula* Boivin; var. *russeolata* Boivin; var. *occidentalis* DC. (*A. occidentalis* (DC.) Raf.); *A. lanulosa* Nutt.; *A. alpicola* Rydb.; *A. megacephala* Raup; *A. subalpina* Greene; *A. tomentosa* Pursh, not L.; MAPS: Hultén 1968*b*:889 (*A. lan.*); the maps by Mulligan and Bassett, loc. cit., fig. 1, p. 76, and fig. 4 (open rings), p. 77, are largely applicable here] .... f. *lanulosa*
    4 Ray-ligules pink or reddish .... f. *roseoides* Breitung
  3 Corymb flattish-topped, to 3 dm broad; leaf-lobes horizontally spreading; stem sparingly cobwebby to essentially glabrous; [introd., transcontinental] ..... var. *millefolium*
    5 Ray-ligules white; [*A. arenicola* Heller; *A. dentifera* DC.; *A. ligustica* All.; *A. pannonica* Scheele; *A. setacea* Waldst. & Kit.; common throughout the southern part of the area; MAP: Hultén 1968*b*:888] .... f. *millefolium*
    5 Ray-ligules pink or purple.
      6 Ray-ligules pink; [more or less throughout the range of f. *millefolium*] ....
.... f. *rosea* Rand & Redf.
      6 Ray-ligules deep rose-purple; [*A. ?asplenifolia* Vent.; Man. to N.B. and N.S.] ....
.... f. *purpureum* (Gouan) Schinz & Thell.

A. ptarmica L. Sneezeweed. Herbe à éternuer
Eurasian; originally cult. in N. America, where now found along roadsides and in fields, thickets, clearings, and ditches, as in SE Alaska (Petersburg and Juneau; Hultén 1950), S B.C. (Langley Prairie, near Vancouver; V), Alta. (N to High Prairie, W of Lesser Slave L.; Groh and Frankton

1949b), SE Man. (Otterburne, about 30 mi S of Winnipeg; Löve and Bernard 1959), Ont. (N to W James Bay at 51°15′N), Que. (N to the Côte-Nord and Gaspé Pen.), Nfld., N.B., P.E.I., and N.S.; SW Greenland. [*Ptarmica vulgaris* DC.]. MAP: Hultén 1968b:887.

Forma *multiplex* (Reynier) Heimerl (*A. mult.* Reynier; the commonly cult. "pearls" or bachelors'-buttons, the florets of the head all ligulate) is the common form in our area.

A. sibirica Ledeb.
/ST/D/eA/ (Hsr) Damp open woods, thickets, and shores from Alaska (N to ca. 68°N), the Yukon (N to ca. 65°N), and NW Dist. Mackenzie to Great Slave L., L. Athabasca (Alta.), Sask. (N to Newnham L. at 59°06′N), and Man. (N to the Hayes R. at Knee L., ca. 55°N), S to cent. B.C. (Clayhurst, near the Alta. boundary at ca. 55°30′N; Herb. V), S-cent. Alta.–Sask., S Man., and N.Dak., with isolated stations in cent. Ont. (near Thunder Bay, where probably introd. along the railway; TRT) and E Que. (a meadow in the valley of the Ste-Anne-des-Monts R., Gaspé Pen., where taken by K.P. Jansson in 1928 and apparently native; *see* M.L. Fernald, Rhodora 31(370):219–20. 1929); E Asia. [*A. multiflora* Hook.; *A. ptarmica sensu* Richardson 1823, not L.]. MAP: Hultén 1968b:887.

ACTINOMERIS Nutt. [9215]

A. alternifolia (L.) DC. Wing-stem
/t/EE/ (Hp) Rich woods and borders of thickets from Iowa to S Ont. (Essex and Kent counties; *see* S Ont. map by Soper 1962: fig. 32, p. 49) and N.Y., S to La. and Fla. [*Coreopsis* L.; *Verbesina* Britt.; *A. squarrosa* Nutt.].

ADENOCAULON Hook. [9082]

A. bicolor Hook. Trail-plant, Silver-green
/T/(X)/ (Hs) Moist shady woods from S B.C. (Vancouver Is. and adjacent islands; mainland N to near Hope and Nelson) and SW Alta. (Waterton Lakes; Breitung 1957b) to Calif. and Mont.; Black Hills of S.Dak.; N Minn., N Mich., and S Ont. (Cape Croker, Bruce Pen., L. Huron, where taken by Massey in 1895 and apparently now extinct; CAN). MAP: Fernald 1935: map 9, p. 213.

AGOSERIS Raf. [9601] False Dandelion, Mountain-dandelion

1 Slender crisp-hairy to subglabrate annual to 4 dm tall, the scapes commonly several; leaves oblanceolate, to 1.5 dm long and 1.5 cm broad, entire to toothed or pinnatifid; uppermost internodes sometimes developed, the leaves then not all strictly basal; body of achene to 5 mm long, the beak 2 or 3 times as long as the body; (S B.C.) ..... *A. heterophylla*
1 Perennials from a taproot; achene-body often over 5 mm long; leaves to over 2.5 dm long and 3 cm broad; plants glabrous or somewhat villous.
  2 Achene-beak stout and more or less striate, commonly to about half as long as the body (sometimes longer in var. *laciniata;* sometimes wanting in var. *dasycephala*); flowers yellow, often drying pinkish; outer phyllaries not strongly ciliate; leaves to 3 cm broad; (B.C. to Ont.) .... *A. glauca*
  2 Achene-beak long and slender, scarcely or not at all striate, over half as long as the achene-body; outer phyllaries usually strongly ciliate.
    3 Achene-beak at least twice the length of the body; flowers yellow, often drying pinkish; leaves entire to deeply pinnatifid; (B.C. and Alta.) .... *A. grandiflora*
    3 Achene-beak not much longer than the body of the achene.
      4 Flowers burnt-orange, often drying purplish; leaves to 3 cm broad, entire or with a few divergent teeth or small lobes; (B.C. and Alta.; Que.) ..... *A. aurantiaca*
      4 Flowers yellow, sometimes drying pinkish; leaves to 7 cm broad, entire or more or less pinnatifid; (S ?B.C.) .... [*A. elata*]

A. aurantiaca (Hook.) Greene
/sT/D/ (Hr) Meadows and woodlands at moderate to high elevations: SE Alaska and the Yukon

(N to ca. 62°30′N) through B.C. and SW Alta. (N to Jasper) to Calif. and N.Mex.; isolated in Que. (Otish Mts., cent. Que., 52°20′N, 70°35′W, the type locality of *A. naskapensis*; Tabletop Mt. and Mt. McNab of the Shickshock Mts., Gaspé Pen., the type of *A. gaspensis* from Mt. McNab; CAN; GH; MT). [*Troximon* Hook.; *T. (A.) gracilens* Gray; *A. carnea* Rydb.; *A. gaspensis* Fern.; *A. gracilenta* and *A. graminifolia* Greene; *A. greenei* (Gray) Rydb.; *A. naskapensis* Rousseau & Raymond]. MAPS (W N. America; Fernald's map also indicates the occurrence in E Que.): Hultén 1968*b*:953; Porsild 1966: map 136, p. 83; Fernald 1925: map 12 (*A. gracilens* and *A. gaspensis*), p. 253.

According to Henry (1915), var. *purpurea* (Gray) Cronq. (*A. purp.* (Gray) Greene; phyllaries relatively broad and blunt, conspicuously imbricate, strongly mottled or blotched with purple) may occur in S B.C.

[A. elata (Nutt.) Greene]
[Collections from S B.C. (Herb. V and Herb. Manning Provincial Park) have been referred to this species of the W U.S,A. (Wash. to Calif.) but require confirmation. (*Stylopappus* Nutt.).]

A. glauca (Pursh) Raf.
/ST/(X)/ (Hr) Prairies and meadows, the aggregate species from SE Alaska, southernmost Yukon, and NW Dist. Mackenzie (N to ca. 69°30′N) to Great Slave L., Sask. (N to Tisdale, 52°51′N; CAN), Man. (N to Grand Rapids, near the NW end of L. Winnipeg), and N Ont. (between the Ekwan R. at 53°27′N and the coast of Hudson Bay at ca. 56°40′N), S through B.C.–Alta. to Calif., Ariz., Colo., S.Dak., and Minn. MAP and synonymy: see below.
1 Plant glabrous, or merely sparingly ciliate at the base of the leaf-margins and on the petioles; leaves acute or acuminate, entire or sometimes with a few irregular teeth; [*Troximon* Pursh; *Macrorhynchus* Eat.; *A. lapathifolia* and *A. procera* Greene; Alaska-Yukon-Dist. Mackenzie-B.C. to Ont. (N to W Hudson Bay at ca. 56°40′N; see N Ont. map by Lepage 1966: map 17, p. 236); MAP: Hultén 1968*b*:953] . . . . var. *glauca*
1 Plant more or less pubescent at least on the involucre or just below it.
  2 Leaves mostly oblanceolate and more or less obtuse, entire or sometimes weakly laciniate toward base; plant seldom over 2.5 dm tall, tending to be rather densely pubescent; achenes sometimes beakless; [*Troximon (A.) glaucum* var. *dasy.* T. & G.; *T. (A.) pumilum* Nutt.; *A. aspera* and *A. villosa* Rydb.; *A. leontodon* vars. *aspera* and *pygmaea* Rydb.; *A. eisenhoweri* Boivin; *Ammogeton (Agos.) scorzoneraefolius* Schrad.; B.C. (N to the Halfway R. at ca. 56°30′N), Alta., and S Man.] . . . .
  . . . . var. *dasycephala* (T. & G.) Jeps.
  2 Leaves mostly lanceolate, acute or acuminate, usually laciniate, often nearly or quite glabrous.
    3 Plant usually over 2.5 dm tall; outer phyllaries tending to be partly pinkish; [*A. agrestis* Osterh.; *A. altissima* and *A. ?turbinata* Rydb.; Alta. and S Sask.] . . . .
    . . . . var. *agrestis* (Osterh.) Jones
    3 Plant usually less than 2.5 dm tall; phyllaries generally not pinkish; [incl. var. *monticola* (Nutt.) Greene (*A. ?laciniata* (Nutt.) Greene); *A. ?tenuifolia* (Gray) Rydb.; *Macrorhynchus (Troximon) glaucus* var. *lac.* Eat.; *Troximon (A.) parviflorum* and *T. taraxacifolium* Nutt.; *T. (A.) glaucum* var. *parviflorum* Gray; *A. ?pubescens* Rydb.; B.C.] . . . . var. *laciniata* (Eat.) Smiley

A. grandiflora (Nutt.) Greene
/T/W/ (Hr) Meadows and open ground at low to moderate elevations from S B.C. (Vancouver Is. and adjacent islands; mainland N to Spences Bridge, about 45 mi SW of Kamloops; CAN; V; also reported from W-cent. Alta. by Boivin 1966*b*) to Calif. and Utah. [*Stylopappus* Nutt.; *Troximon* Gray].

A. heterophylla (Nutt.) Greene
/t/W/ (T) Dry open places in the lowlands and foothills from S B.C. (Vancouver Is.; Osoyoos; Armstrong; Trail; Creston) to Calif. and Ariz. (*Macrorhynchus* Nutt.; *Troximon* Greene; *T. humile sensu* John Macoun 1884, perhaps not Gray, the relevant collection in CAN). MAP: K.L. Chambers, Quart. Rev. Biol. 38(2): fig. 1b, p. 126. 1963.

AMBROSIA L.   [9146]   Ragweed

1   Leaves mostly long-petioled, palmately 3–5-lobed or undivided, serrate, all opposite;
staminate involucres 3-ribbed on the outer side; fruit (including beak) to about 12 mm
long, the beak to 4 mm long; annual; (B.C. to N.S.) .... *A. trifida*
1   Leaves sessile or relatively short-petioled, pinnately lobed, the lower ones opposite, the
upper ones alternate; staminate involucres indistinctly radiate-veined; fruit 4 or 5 mm
long.
2   Staminate involucres strigose-hispid; fruit nearly beakless, unarmed or with about 4
short blunt tubercles; leaves scabrous above and somewhat hoary with stiffish short
hairs, mostly 1-pinnatifid; perennial from a creeping rhizome; (B.C. to N.S.) ....
.... *A. psilostachya*
2   Staminate involucres glabrous or pilose; fruit with usually 4–7 sharp tubercles
encircling the base of the subulate beak, this 1 or 2 mm long; leaves smoothish above
and relatively thin; annual; (B.C. to Nfld. and N.S.) .... *A. artemisiifolia*

A. artemisiifolia L.   Common Ragweed.   Petite herbe à poux
/T/X/   (T)   Beaches, roadsides, dooryards, and waste or cult. land (the native area uncertain
because of its extremely weedy nature) from southernmost Dist. Mackenzie (Fort Smith, ca. 60°N,
where undoubtedly introd) and s B.C. (Vancouver Is.; Fernie) to Alta. (N to High Prairie, w of Lesser
Slave L.), Sask. (N to ca. 55°N), Man. (N to Dawson Bay, L. Winnipegosis), Ont. (N to Kenora and
the N shore of L. Superior), Que. (N to the Côte-Nord and Gaspé Pen.), Nfld., N.B., P.E.I., and N.S.,
s to Mexico, Tex., and Fla. MAP: I.J. Bassett and J. Terasmae, Can. J. Bot. 40(1): fig. 1, p. 143.
1962.
The typical form (leaves simple or 1-pinnatifid (or the lower ones 2-pinnatifid); staminate
involucres to 7 mm broad) is reported from Nfld. and E Que. (Magdalen Is.) by Fernald *in* Gray
(1950). Most of our material is referable to var. *elatior* (L.) Desc. (*A. elatior* L.; *A. diversifolia* (Piper)
Rydb.; leaves 2–3-pinnatifid; staminate involucres at most 5 mm broad) and its spreading-villous
extreme, f. *villosa* Fern. & Grisc. A hybrid with *A. trifida* (× *A. helenae* Rouleau) is known from the
type locality, St. Helen's Is., Montreal, Que.

A. psilostachya DC.   Perennial Ragweed.   Herbe à poux vivace
/T/X/   (Grh (Hpr))   Dry prairies, roadsides, and waste places (the native area uncertain because
of its weedy nature) from s B.C. (Matsqui, near Vancouver; Nakusp, about 50 mi SE of Revelstoke)
to Alta. (N to ca. 51°N), Sask. (N to ca. 54°N), Man. (N to Dauphin, N of Riding Mt.), Ont. (N to
Kenora and the N shore of L. Superior), Que. (N to the Côte-Nord and Gaspé Pen.), P.E.I., and N.S.
(not known from N.B.), s to Calif., Mexico, Tex., and La. MAP: I.J. Bassett and J. Terasmae, Can. J.
Bot. 40(1): fig. 3 (*A. cor.*), p. 144. 1962.
Our material is referable to var. *coronopifolia* (T. & G.) Farw. (*A. cor.* T. & G.; staminate
involucres strigose-hispid to pilose with long slender hairs rather than merely minutely
scabrous-hirtellous or puberulent).

A. trifida L.   Giant or Great Ragweed.   Grande herbe à poux
/T/X/   (T)   Moist soil and waste places (the native area uncertain because of its weedy nature)
from s B.C. (Vancouver Is.; Revelstoke) to Alta. (N to Peace River, 56°14′N), Sask. (N to ca. 54°N),
Man. (N to Wekusko L., about 90 mi NE of The Pas), Ont. (N to Kenora and the N shore of L.
Superior), Que. (N to L. St. John, the Côte-Nord, and Gaspé Pen.), N.B., P.E.I., and N.S., s to N
Mexico and Fla.; introd. in Europe. [*A. striata* Rydb.]. MAP: I.J. Bassett and J. Terasmae, Can. J.
Bot. 40(1): fig. 2, p. 143. 1962.
Forma *integrifolia* (Muhl.) Fern. (*A. integrifolia* Muhl.; leaves unlobed rather than deeply and
palmately 3(5)-cleft into ovate-lanceolate serrate lobes) occurs essentially throughout the range.

ANAPHALIS DC.   [8983]

A. margaritacea (L.) Clarke   Pearly Everlasting.   Immortelle
/sT/X/A/   (Hpr)   Meadows, dry fields, roadsides, open woods, and thickets at low to moderate
elevations, the aggregate species from the Aleutian Is., s Alaska (N to ca. 61°30′N; not known from

the Yukon), and sw Dist. Mackenzie to B.C.–Alta., Sask. (N to Cut Knife, SE of Prince Albert), Man. (N to Bissett, about 110 mi NE of Winnipeg), Ont. (N to Big Trout L. at ca. 53°45′N, 90°W), Que. (N to the E James Bay watershed at ca. 53°N, L. St. John, and the Côte-Nord), Labrador (N to the Hamilton R. basin), Nfld., N.B., P.E.I., and N.S., s to s Calif., N.Mex., S.Dak., and New Eng.; introd. in Europe; Asia. MAP and synonymy: *see* below.

1 Leaves broadly linear to narrowly oblong, bluntish to slightly attenuate, scarcely reduced above.
  2 Leaves bright green and soon glabrate above, the upper ones to 12 cm long and 2 cm broad; [var. *occidentalis* Greene (*A. occ.* (Greene) Heller); *A. angustifolia* Rydb.; *Gnaphalium* L.; *Antennaria* R. Br.; this and the following varieties all transcontinental; MAP (aggregate species): Hultén 1968*b*:882] . . . . var. *margaritacea*
  2 Leaves ashy-tomentose above, the upper ones rarely over 7 cm long and 1.5 cm broad; [*A. subalpina* (Gray) Rydb.] . . . . var. *subalpina* Gray
1 Leaves linear to linear-lanceolate, attenuate, the upper ones usually much reduced.
  3 Leaves bright green and essentially glabrous above; [var. *revoluta* Suksd.; f. *anochlora* Fern.; *Antennaria cinnamomea* var. *ang.* Miquel] . . . . var. *angustior* (Miquel) Nakai
  3 Leaves tomentose or cobwebby above . . . . var. *intercedens* Hara

ANTENNARIA Gaertn. [8978] Pussy-toes, Everlasting, Ladies'-tobacco. Immortelle

(Ref.: Porsild 1950, 1965)
1 Basal leaves erect, linear to lanceolate or oblanceolate, acute or short-acuminate, commonly over 8 times as long as broad, similar to the stem-leaves, not rosette-forming; plants lacking stolons, scarcely mat-forming (but often with several or many stems from a branched rhizome or caudex); achenes glabrous; plants persistently more or less silky-tomentose; (chiefly western species; *A. pulcherrima* transcontinental) . . . . GROUP 1
1 Basal leaves spreading and forming depressed rosettes, distinctly broader in outline than the stem-leaves (these often terminated by a scarious appendage), obtuse or rounded at summit, with or without a terminal mucro; plants mat-forming, usually with numerous leafy stolons but sometimes densely tufted and lacking stolons.
  2 Rosette-leaves (at least some of them) over 5 mm broad; heads commonly 2 or more (*A. spathulata* and *A. appendiculata* sometimes with solitary heads); achenes usually more or less papillate at least when young . . . . GROUP 2 (p. 1463)
  2 Rosette-leaves rarely over 5 mm broad; heads solitary to several . . . . GROUP 3 (p. 1464)

GROUP 1

1 Involucres glabrous and scarious to the base, to 5 or 6 mm high; leaves to about 8 cm long.
  2 Involucres dark brownish; inflorescence very compact, usually not over 1.5 cm broad; leaves linear, mostly 1 or 2 mm broad, commonly only the midrib prominent; (?B.C.) . . . . [*A. stenophylla*]
  2 Involucres pale or nearly white; inflorescence commonly corymbiform and often much more than 1.5 cm broad; basal leaves mostly linear-lanceolate, to 8 mm broad, often several-nerved; (s B.C. and sw Alta.) . . . . A. *luzuloides*
1 Involucres densely pubescent toward the non-scarious base; leaves commonly 3-nerved.
  3 Involucre white or whitish, to 8 mm high; phyllaries obtuse or acutish, in 3 or 4 unequal series, with or without a dark spot at base; stems to about 5 dm tall; (B.C. to sw Sask.) . . . . A. *anaphaloides*
  3 Involucre deep brown to blackish; phyllaries in up to 7 series, at least the inner ones acuminate.
    4 Involucres to 8 mm high, the phyllaries in 5 or 6 series; leaves to 1 dm long and 1 cm broad; plants commonly 1 or 2 dm tall; (B.C. and Alta.) . . . . A. *lanata*
    4 Involucres to 1 cm high, the phyllaries in 6 or 7 series; leaves to over 1.5 dm long and nearly 2 cm broad; plants to about 5 dm tall; (transcontinental) . . . . A. *pulcherrima*

GROUP 2 (see p. 1462)

1 Rosette-leaves to over 4 cm broad, with 3 or 5(7) nerves more or less prominent and prolonged beneath; middle and upper stem-leaves tapering to a dark subulate tip.

  2 Inflorescence loose and open, becoming very elongate (to about half the height of the plant), the peduncles commonly to over 4 cm long; basal leaves to 8 cm long and 5 cm broad, persistently cottony-tomentose beneath, green and glabrous or subglabrous above; (B.C. and Alta.) . . . . *A. racemosa*

  2 Inflorescence a crowded or even subcapitate cyme, the peduncles relatively short.

    3 Pistillate involucres at most 7 mm high; central corollas less than 4.5 mm long; achenes to 1.5 mm long; pappus to 5.5 mm long; rosette-leaves obovate to suborbicular, minutely canescent above, to 4 cm broad, their rounded summits with a minute terminal mucro; (SE Man. to N.S.) . . . . *A. plantaginifolia*

    3 Pistillate involucres commonly 7 mm long or more; central corollas 4.5 mm long or more; achenes mostly over 1.5 mm long; pappus to 9 mm long.

      4 Rosette-leaves mostly subtruncate, ashy-tomentose above, to about 2 cm broad, concave-arching from above the middle to the slenderly cuneate base; pappus to 6 mm long; (S Ont.) . . . . *A. farwellii*

      4 Rosette-leaves acutish or gradually rounded at summit, more gradually narrowed to base; mature pappus at least 6 mm long.

        5 Blades of rosette-leaves to about 3.5 cm long and 2.5 cm broad, loosely tomentose above; stems stipitate-glandular above and in the corymb; involucres to 8.5 mm high; corollas to 5.5 mm long; pappus to 7 mm long; (transcontinental) . . . . *A. neglecta*

        5 Blades of rosette-leaves to 8 cm long and 5 cm broad, densely ashy-tomentose above; stem glandless; involucres to 11 mm high; corollas to 7 mm long; pappus to 9 mm long; (Ont. and ?Que.) . . . . *A. munda*

1 Rosette-leaves at most about 2 cm broad, only the midrib prominent to tip, the lateral nerves short and obscure.

  6 Middle and upper stem-leaves of pistillate plants with a terminal flattish or twisted scarious tip.

    7 Rosette-leaves more or less persistently tomentose above; (transcontinental) . . . . *A. neglecta*

    7 Rosette-leaves bright green and essentially glabrous above from the first.

      8 Rosette-leaves oblanceolate to narrowly obovate, acutish and tipped with a mucro at least 0.5 mm long; heads often more than 10; (transcontinental) . . . . *A. neglecta*

      8 Rosette-leaves spatulate to cuneate-oblanceolate, the terminal mucro less than 0.5 mm long; heads solitary or at most 6; (chiefly Que. and Nfld.).

        9 Rosette-leaves broadly rounded at summit; lower and median stem-leaves obtuse and merely mucronate-tipped, only usually the upper 1–3 with scarious appendages . . . . *A. spathulata*

        9 Rosette-leaves subacute to round-tipped; all but the very lowermost stem-leaves scarious-appendaged . . . . *A. appendiculata*

  6 Middle and upper stem-leaves merely mucronate or subulate-tipped (only the leaves of the corymb sometimes appendaged).

    10 Stolons slender and elongate, cord-like, tardily developing terminal rosettes; rosette-leaves cuneate-oblanceolate to spatulate-obovate, canescent above when young; middle and upper stem-leaves with a coloured awn-like tip; pistillate involucres to 11 mm high; (transcontinental) . . . . *A. neglecta*

    10 Stolons and basal offshoots short, their ascending tips terminated by rosette-leaves, these relatively broader and often evidently petioled.

      11 Stem-leaves all or nearly all blunt; rosette-leaves rounded to an obscure point, densely grey-tomentose, at most about 1 cm broad; pistillate involucres to 1 cm high, their phyllaries whitish; (B.C. to Ont.) . . . . *A. parviflora*

11 Stem-leaves and rosette-leaves subulate-tipped; the latter to 2 cm broad; pistillate involucres to 9 mm high, their phyllaries tinged with green, red, purple, or brown; (Ont. to Nfld. and N.S.) .... *A. neodioica*

GROUP 3 (see p. 1462)

1 Heads usually solitary (additional lateral ones, when present, smaller and long-peduncled).
  2 Pappus-bristles of staminate florets upwardly barbellate but scarcely clavate; staminate involucres to about 7 mm high, the phyllaries colourless and hyaline at the margins and apex, otherwise dingy blackish green or brownish; pistillate involucres to 1.5 cm high, their narrow slenderly pointed phyllaries at least partly tinged with brown or reddish brown; achenes puberulent; leaves numerous, persistently silky-tomentose, linear or oblanceolate, to about 3 cm long; dwarf plant commonly less than 5 cm tall; (s B.C. to s Sask.) .... *A. dimorpha*
  2 Pappus-bristles of staminate florets with clavate or scarious-dilated tips; achenes glabrous; (arctic, subarctic, and alpine regions).
    3 Inner phyllaries with prominent broad creamy or straw-coloured tips; stem-leaves lacking brown scarious tips, not much reduced, they and the rosette-leaves usually thinly but persistently tomentose beneath, glabrous or glabrate and fresh green above; (the Yukon to Labrador) .... *A. pygmaea*
    3 Inner phyllaries olivaceous or light brown, their long-attenuate tips more or less erose.
      4 Plant fresh green, the leaves glabrous or nearly so; (Baffin Is. and Greenland) ......
        .... *A. glabrata*
      4 Plant grey-green, thinly tomentose, the basal leaves commonly green and glabrate above; (transcontinental) .... *A. angustata*
1 Heads commonly 2 or more.
  5 Inner phyllaries dark rose or pink, fading in age; involucres 4 or 5 mm high, their phyllaries 3-seriate; stems to about 3 dm tall, with up to 10 leaves, these lacking terminal appendages; rosette-leaves densely appressed-tomentose, from strongly developed freely branching stolons; (transcontinental) .... *A. rosea*
  5 Phyllaries white to pale brown or blackish.
    6 Young flowering heads nodding; pistillate involucres 5–7 mm high, white to pale sulphur-yellow, the staminate ones more deeply sulphur-yellow; achenes glabrous; upper stem-leaves with attenuate brownish scarious tips; rosette-leaves silky-tomentose and more or less silvery-lustrous on both surfaces; (var. *nitida*; transcontinental) .... *A. rosea*
    6 Young flowering heads scarcely nodding; involucres whitish to straw-coloured or light brown, sometimes greenish or pink-tinged, but not sulphur-yellow.
      7 Involucre 4 or 5 mm high, the phyllaries with a large dark-brown or blackish spot at base; leaves thinly but persistently tomentose, those of the short ascending basal offshoots narrowly oblanceolate, mucronate, tapering gradually to base; flowering-stems slender, to about 3 dm tall, thinly tomentose, the several heads short-peduncled in a compact cluster; (se B.C. to Sask.) ....
        .... *A. corymbosa*
      7 Plants with not all of the above characters.
        8 Inner phyllaries with blunt and entire, light straw- or cream-coloured to pink tips.
          9 Stolons procumbent, the stems branching below and more or less mat-forming; phyllaries normally obtuse or rounded at tip; achenes papillate; (Ungava–Labrador) .... *A. rousseauii*
          9 Stolons mostly ascending, the stems often more or less tufted; phyllaries erose at tip; achenes glabrous; (transcontinental) ....
            .... *A. umbrinella*
        8 Inner phyllaries with attenuate, more or less erose tips; at least the

uppermost stem-leaves with prominent brownish scarious tips; achenes glabrous (or minutely papillate below the middle in *A. friesiana*).
  10 Plants tufted or caespitose, not mat-forming (the crowded, sessile or subsessile rosettes from a rhizomatous base); (transcontinental) . . . .
  . . . . *A. friesiana*
  10 Plants mat-forming, with prostrate or ascending leafy stolons.
    11 Plant thinly tomentose, green and completely glabrous (or at least the upper surface of the basal leaves glabrate); (mts. of w Alta.; Dist. Keewatin–Ungava–Labrador) . . . . *A. ungavensis*
    11 Plant densely white-tomentose; (sw Dist. Mackenzie; B.C.–Alta.) . . . .
    . . . . *A. media*

NOTE

The present treatment of *Antennaria* is tentative. The genus includes several clear-cut species but, also, several species-complexes many of whose entities have been separated as distinct species in spite of the apomictic type of reproduction that characterizes them. An attempt has been made to steer a middle course between extreme "lumping" into relatively few species or species-complexes and extreme "splitting" into numerous, scarcely separable "microspecies". *Antennaria* is a top-priority candidate for a thorough study based upon the "Species-standard Method" advocated by Rollins (*see* Introduction, p. 4).

A. anaphaloides Rydb.
/T/W/ (Hs) Open woods and grassy foothills up to moderate elevations in the mts. from s B.C. (N to Chilcotin and Kamloops; CAN; V), sw Alta. (N to near Jasper; CAN), and sw Sask. (Cypress Hills; Breitung 1957a) to Oreg., Nev., and Colo. [Incl. var. *straminea* Boivin].

A. angustata Greene
/aST/X/G/ (Ch) Gravelly or rocky places from the coasts of Alaska–Yukon–Dist. Mackenzie-Dist. Keewatin to Banks Is., Ellesmere Is. (N to ca. 80°N), Baffin Is., and northernmost Ungava-Labrador (type from Cape Chidley), s in the West to the mts. of s B.C. (Skagit R. and Columbia R.; CAN) and sw Alta. (Lake Louise; CAN), farther eastwards s to Great Bear L., cent. Dist. Keewatin, Labrador (s to ca. 55°30′N), and Nfld.; w Greenland between ca. 67° and 79°30′N. [*A. monocephala* ssp. *ang.* (Greene) Hult.; incl. *A. nitens* Greene, *A. burwellensis*, *A. congesta*, and *A. hudsonica* Malte, and *A. fernaldiana* and *A. tansleyi* Polunin; *A. ?columnaris* Fern.]. MAPS: Hultén 1968b:875 (*A. mono.* ssp. *ang.*); Porsild 1950: map 6, p. 8, 1957: map 312, p. 199, and 1965: map 2, p. 54.

A. appendiculata Fern.
/sT/E/ (Ch) Peaty or calcareous soil from James Bay (Charlton Is.; Dutilly, Lepage, and Duman 1958) to Que. (Côte-Nord, Anticosti Is., and Gaspé Pen.; type from banks of the Grande-Rivière, Gaspé Pen.) and Nfld.

A. corymbosa Nels.
/T/W/ (Ch) Meadows and moist open woods up to fairly high elevations from Mont. to s B.C. (Glacier National Park; CAN), Alta. (Boivin 1966b), and sw Sask. (Cypress Hills; CAN; Breitung 1957a), s to N Calif. and Colo.

A. dimorpha (Nutt.) T. & G.
/T/WW/ (Ch) Dry open places from s B.C. (N to Cache Creek, about 40 mi NW of Kamloops), s Alta. (Medicine Hat; J.M. Macoun 1895), and s Sask. (Climax and Divide; DAO; not listed by Breitung 1957a) to s Calif., Colo., and Nebr. [*Gnaphalium* Nutt.].

A. farwellii Greene
/T/EE/ (Ch) Dry gravelly banks and rocky bluffs from N Mich. (Keweenaw Co.) to s Ont. (Jordan Harbour, Lincoln Co.; OAC; reported from the Bruce Pen., L. Huron, by Krotkov 1940, and from fields at Ottawa by M.L. Fernald, Rhodora 1(8):152. 1899). [*A. parlinii* var. *far.* (Greene) Boivin].

A. friesiana (Trautv.) Ekman

/ASs/(X)/GeA/ (Ch) Meadows and dry slopes up to fairly high elevations from the coasts of Alaska–Yukon–Dist. Mackenzie–Dist. Keewatin to Ellesmere Is. (N to ca. 79°30′N) and Baffin Is., S to S Alaska–Yukon, cent. Dist. Keewatin, and N Ungava–Labrador (S to Cape Mugford, 57°48′N); W Greenland between ca. 67° and 79°30′N, E Greenland at ca. 73°N; N Siberia. MAPS and synonymy (together with distinguishing keys to several closely related taxa (?microspecies)): *see* below.

1   Head solitary . . . . [*A. monocephala* DC.]
    2   Plant with horizontal stolons; stem slender; [*A. philonipha* Porsild, the type from the
        arctic coast of NW Dist. Mackenzie; incl. var. *latisquamea* Hult.; Alaska–Yukon–NW
        Dist. Mackenzie; MAPS: Porsild 1950: map 16 (*A. phil.*), p. 9; Hultén 1968*b*:874] . . . .
        . . . . ssp. *philonipha* (Porsild) Hult.
    2   Plant caespitose, forming dense tufts, lacking horizontal stolons; stem relatively stout . . . . . .
        . . . . ssp. *monocephala*
        3   Leaves pubescent above; [*A. exilis* Greene, the type from St. Paul Is., Alaska; E
            Aleutian Is. and S Alaska–Yukon; MAP: Hultén 1968*b*:874] . . . . var. *exilis* (Greene) Hult.
        3   Leaves glabrous above; [*A. alpina* var. *mon.* (DC.) T. & G.; *A. shumaginensis*
            Porsild; Aleutian Is. (type from Unalaska), Alaska–Yukon–W Dist. Mackenzie, and
            NE B.C. (S to Mt. Selwyn and Redfern L., both ca. 56°N; the report from Waterton
            Lakes, SW Alta., by Breitung 1957*a*, requires confirmation); NE Siberia; MAPS:
            Hultén 1968*b*:873; Porsild 1950: map 12, p. 9; Raup 1947: pl. 36] . . . var. *monocephala*
1   Heads normally 2–4 (sometimes very densely condensed and simulating a single head).
    4   Plants with more or less horizontal stolons, forming mats; leaves pubescent on both
        sides at least when young.
        5   Inner phyllaries obtuse, brownish green to brown; [Aleutian Is., Alaska (type from
            Disenchantment Bay), and N B.C. (S to ca. 57°N); MAPS: Hultén 1968*b*:875;
            Porsild 1950: map 14, p. 9, and 1966: map 140, p. 84] . . . . [*A. pallida* Nels]
        5   Inner phyllaries acute, dark brown to blackish.
            6   Basal shoots slender, their oblanceolate glabrescent leaves acute; [S Alaska,
                Great Slave L., and N B.C. (type from Mt. Selwyn, ca. 56°N); MAPS: Hultén
                1968*b*:878; Porsild 1950: map 7, p. 8] . . . . [*A. atriceps* Fern.]
            6   Basal shoots with dense, blunt, ligulate, densely tomentose leaves; [Alaska–
                Yukon N to ca. 64°30′N, the type from the MacMillan R., the Yukon; MAPS:
                Hultén 1968*b*:879; Porsild 1950: map 17, p. 9] . . . . [*A. stolonifera* Porsild]
    4   Plants caespitose, lacking horizontal stolons . . . . *A. friesiana*
        7   Basal leaves short, densely crowded, oblanceolate to obovate, permanently
            tomentose on both sides; [*A. compacta* Malte, the type from Bernard Harbour,
            coast of Dist. Mackenzie; *A. crymophila, A. densifolia*, and *A. ?neoalaskana*
            Porsild; *A. alpina* f. *latifolia* Ekm.; essentially the N. American range of the
            species; MAPS: Hultén 1968*b*:877; Porsild 1965: map 5 (*A. comp.*), p. 54; Savile
            1961: map L, p. 929] . . . . ssp. *compacta* (Malte) Hult.
        7   Basal leaves petioled, averaging narrower.
            8   Basal leaves more or less rounded and mucronate at apex, often glabrescent
                above; both staminate and pistillate plants known; [*A. alaskana* Malte, the type
                from near Port Clarence, Alaska; incl. var. *beringensis* Hult.; Alaska–N Yukon;
                MAPS: Hultén 1968*b*:877; Porsild 1950: map 5 (*A. alask.*), p. 8] . . . .
                . . . . ssp. *alaskana* (Malte) Hult.
            8   Basal leaves relatively narrow and acute, permanently tomentose on both
                sides; staminate plant unknown; [*A. alpina* var. *fr.* Trautv.; *A. megacephala*
                Fern.; *A. ekmaniana* and *A. ?pedunculata* Porsild; *A. ?subcanescens*
                Ostenf.; *A. angustifolia* Ekm., not Rydb.; *A. labradorica* of auth., not Nutt.;
                range of the species; MAPS: Hultén 1968*b*:876; Porsild 1965: map 6, p. 54,
                1957: map 313, p. 200, and 1950: map 10, p. 8 (all as *A. ekm.*); Böcher 1954:
                fig. 33 (map 3; *A. ekm.*), p. 135] . . . . ssp. *friesiana*

A. glabrata (Vahl) Greene

/a/E/G/ (Ch) Grassy tundra and snowbeds on calcareous soil: E Baffin Is. near the Arctic

Circle; w Greenland between ca. 65° and 73°N. [*A. alpina* var. *glab*. Vahl, the type from NW Greenland; incl. f. *ramosa* Porsild and the pubescent extreme, f. *tomentosa* Ekm.]. MAP: Porsild 1957: map 311, p. 199.

A. lanata (Hook.) Greene
/T/W/ (Hs) Ledges and cliffs at subalpine to alpine elevations from s B.C. (N to Rogers Pass and Revelstoke; a collection in V from Atlin, ca. 59°N, may also belong here) and sw Alta. (N to Jasper; CAN) to Oreg., Idaho, and Wyo. [*A. carpathica* var. *lan*. Hook., the type a Drummond collection in the Rocky Mountains of B.C.–Alta. at ca. 52°N].

A. luzuloides T. & G.
/T/W/ (Hs) Moist or dryish gravels and ledges in the foothills and at moderate elevations in the mts. from s B.C. (N to Kamloops; CAN) and sw Alta. (Waterton Lakes; Breitung 1957b) to N Calif. and Colo. [*A. oblanceolata* Rydb.].

A. media Greene
/sT/W/ (Ch) Dryish meadows and slopes up to high elevations from sw Dist. Mackenzie (Colonel Mt., ca. 62°N; Raup 1947) through the mts. of B.C.–Alta. to Calif. and Colo. [*A. acuta* Rydb.; *A. candida, A. chlorantha, A. macounii, A. modesta,* and *A. pulvinata* Greene; *A. reflexa* Nels.; *A. alpina* var. *media* (Greene) Jeps.; *A. cana* (Fern. & Wieg.) Fern. (*A. alpina* var. *cana* F. & W.), in part]. MAP: Hultén 1968b:878.

A. munda Fern.
/T/EE/ (Ch) Dry sands, gravels, and rocks from w Ont. (N to near Thunder Bay; CAN; TRT) and ?Que. (collections in CAN and GH from Bic, Rimouski Co., have been referred to *A. occidentalis* by Fernald and Collins but Que. is not included in the range of *A. munda* by Fernald *in* Gray 1950) to Ind. and Va. [*A. occidentalis* of Canadian reports, not Greene].

A. neglecta Greene
/sT/X/ (Ch) Sterile fields, pastures, rocky barrens, and open woods from s Yukon and Great Slave L. to B.C.–Alta., Sask. (N to L. Athabasca), Man. (N to Gillam, about 165 mi s of Churchill), Ont. (N to Swan L. at ca. 54°30′N, 91°W), Que. (N to E Hudson Bay at ca. 56°10′N and the Côte-Nord), Nfld., N.B., P.E.I., and N.S., s to Calif., Ariz., and Va. [A very plastic complex, here taken to include *A. campestris* and *A. ?solitaria* Rydb., *A. denikeana* Boivin, *A. obovata* Nels., *A. manicouagana* Landry, *A. brainerdii, A. gaspensis, A. glabrifolia, A. petaloidea,* and *A. rupicola* Fern., and *A. athabascensis, A. callilepis, A. canadensis, A. eximia, A. howellii, A. lunellii, A. ?petasites,* and *A. stenolepis* Greene; *A racemosa sensu* Fraser and Russell 1944, not Hook.]. MAP: Hultén 1968b: 881.

A. neodioica Greene
/T/EE/ (Ch) Dry fields, pastures, rocky barrens, and open woods from Ont. (N to the N shore of L. Superior) to Que. (N to the Gaspé Pen.), Nfld., N.B., P.E.I., and N.S., s to Minn., Ind., and Va. [Incl. vars. *chlorophylla, grandis,* and *interjecta* Fern.; scarcely separable from *A. neglecta,* with whose var. *attenuata* (Fern.) Cronq. it is merged by Arthur Cronquist, Rhodora 47(557):184. 1949].

A. parvifolia Nutt.
/sT/WW/ (Ch) Dry open places from s B.C. (N to Clinton, Chilcotin, and Cariboo; CAN) to L. Athabasca (Alta. and Sask.) and Man. (N to Hill L., N of L. Winnipeg; CAN; reports from Alaska, Dist. Keewatin, and Ont. by Boivin 1966b, require clarification), s to Wash., Nev., and Ariz. [*A. aprica* Greene; *A. bracteosa* Rydb.; *A. minuscula* Boivin].
The typical form has white phyllaries. Forma *brunnea* (Boivin) Breitung (phyllaries brownish-tipped) is known from sw Sask. (Cypress Hills; type from Swift Current). Forma *roseoides* (Boivin) Breitung (phyllaries pink-tipped) is known from B.C., Alta., and Sask. (type from Swift Current).

A. plantaginifolia (L.) Hook.
/T/EE/ (Ch) Fields, clearings, and open woods from SE Man. (N to the Winnipeg dist.) to Ont. (N

to the Ottawa dist.), Que. (N to Aylmer, Buckingham, and Montreal), and N.S. (reports from N.B. and P.E.I. require confirmation), s to E Tex., Tenn., and Va. [Incl. *A. arnoglossa* and *A. fallax* Greene and *A. parlinii* Fern.].

A. pulcherrima (Hook.) Greene

/ST/(X)/ (Hs) Moist places at low to high elevations from Alaska (N to ca. 68°N) to the Yukon (N to ca. 62°N), the Mackenzie R. Delta, Great Bear L., northernmost Alta.-Sask.-Man.-Ont., Que. (N to E Hudson Bay at ca. 56°10'N; also known from Anticosti Is.), and Nfld. (not known from the Maritime Provinces), s in the West to Wash., Utah, and Colo., farther eastwards s to s-cent. Sask., N Man., and James Bay (Ont. and Que.). [*A. carpathica* var. *pulch.* Hook., the type a Drummond collection from the Rocky Mountains, probably of Alta.; *A. carp.* var. *humilis* Hook.; *A. eucosma* Fern. & Wieg.; incl. vars. *angustisquamata* Porsild and *sordida* Boivin]. MAPS: Hultén 1968*b*:872; Porsild 1950: map 3, p. 8, and 1965: map 1, p. 54; Marie-Victorin 1938: fig. 64 (requiring considerable expansion), p. 551.

A. pygmaea Fern.

/aS/(X)/ (Ch) Gravels, turfs, and mossy ledges of the Yukon (N to ca. 65°N; CAN), Dist. Mackenzie (Mackenzie R. Delta; Great Bear L.), sw-cent. Dist. Keewatin, Southampton Is., s Baffin Is., and northernmost Ungava-Labrador (s to ca. 56°N; type from Okak, Labrador, 57°33'N). [Incl. *A. tweedsmuirii* Polunin]. MAPS: Porsild 1950: map 29, p. 20, and 1965: map 17, p. 55.

A. racemosa Hook.

/T/W/ (Ch) Cool montane woods from B.C. (N to Hazelton, Stuart Lake, Prince George, Clinton, and Revelstoke; CAN) and sw Alta. (N to Jasper; CAN) to Oreg., Idaho, and Wyo. [*A. oblancifolia* Nels.].

A. rosea Greene

/ST/X/ (Ch) Prairies, meadows, and open woods at low to moderate elevations from Alaska (N to ca. 69°N) to cent. Yukon, the Mackenzie R. Delta, Great Bear L., s Dist. Keewatin, Man. (N to Churchill), Ont. (Neaka, w James Bay, 53°16'N; reported from Fort William (Thunder Bay) by P.A. Hyypio, Rhodora 54(647):291. 1952, where taken by McMorine in 1879), James Bay (Manawanan Is., ca. 53°N), N Que. (Hudson Bay–James Bay s to ca. 52°N), and N Labrador (s to ca. 56°N; not known from the Atlantic Provinces), s in the West to Calif. and N.Mex. [*A. dioica* var. *rosea* (Greene) Eat.; *A. arida, A. concinna,* and *A. imbricata* Nels.; *A. acuminata, A. erigeroides,* and *A. oxyphylla* Greene; *A. breitungii, A. elegans,* and *A. incarnata* Porsild; *A. leontopodioides* Cody]. MAPS and key to var. *nitida* and three closely related species (?microspecies) of w N. America: *see* below.

1 At least the inner phyllaries paper-white or brownish (except for the green base; in *A. leuchippi,* with purple spots visible under a lens).
  2 Upper stem-leaves with broad flat brownish scarious tips; [SE Yukon, the type from Whitehorse; MAPS: Hultén 1968*b*:881; Porsild 1950: map 26, p. 20] ....
  .... [*A. leuchippii* M.P. Porsild]
  2 Upper stem-leaves with subulate tips; [*A. nitida* Greene, the type from Charlton Is., James Bay; *A. laingii* Porsild; *A. viscidula* Neils.; *A. microphylla* Rydb., not Gand.; *A. ?isolepis* Greene; *A. subviscosa* of w N. America reports, not Fern.; some of the synonyms under *A. rosea* should probably be placed here; MAPS: Raup 1947: pl. 35 (*A. isolepis*); combine the maps by Porsild 1950: map 27 (*A. nitida*) and map 24 (*A. isolepis*), p. 20, and 1965: map 16 (*A. isolepis*), p. 55; Hultén 1968*b*:876 (*A. isolepis*)] .....
  .... *A. rosea* var. *nitida* (Greene) Breitung
1 Inner phyllaries pink to roseate.
  3 Leaves obovate, rounded (often mucronate) at apex, glabrous above.
    4 Stem thinly lanate, distinctly glandular; [cent. Alaska–Yukon–w Dist. Mackenzie to northernmost B.C.; isolated in sw B.C.; type from the Pelly Range, the Yukon; MAPS: Hultén 1968*b*:880; Porsild 1950: map 19, p. 9, and 1966: map 137, p. 84] ....
    .... [*A. alborosea* A.E. Porsild]
    4 Stem densely lanate, not glandular; [*Gnaphalium* L.; *A. insularis* Greene; Aleutian Is.; Eurasia; MAP: Hultén 1968*b*:879] .... *A. dioica* (L.) Gaertn.

3   Leaves oblanceolate, acutish, pubescent above; [MAPS: Porsild 1950: map 30, p. 20; Hultén 1968b:880] .... *A. rosea*

A. rousseauii Porsild

/Ss/E/ (Ch) An endemic of N Ungava–Labrador in gravelly and rocky places between ca. 52° and 60°N; type from the Payne R. w of Ungava Bay at 59°17′N; MAP: Porsild 1965: map 18, p. 55.

    Keyed out below are three closely related taxa (?microspecies) endemic to Greenland.

1   Achenes smooth; [known only from Greenland, the type locality].

    2   Phyllaries with olive-brown tips, becoming cream- or straw-coloured in age; [*A. alpina* var. *intermedia* Rosenv.; MAPS: Porsild 1965: map 15, p. 55; T.W. Böcher, Medd. Groenland 148(3): fig. 35, p. 48. 1963] .... [*A. intermedia* (Rosenv.) M.P. Porsild]

    2   Phyllaries with pink tips, becoming cream-white in age, the outer ones with prominent dark spots near the base; [*A. dioica* var. *hyperborea* Lange; *A. groenlandica* M.P. Porsild; MAPS: Porsild 1965: map 14, p. 55; Böcher, loc. cit., fig. 29, p. 47] ....
.... [*A. hansii* Kerner]

1   Achenes strongly papillate even when immature.

    3   Plant dwarf, densely white-tomentose; [known only from w Greenland, the type locality; MAPS: Porsild 1965: map 13, p. 55; Böcher, loc. cit., fig. 42, p. 51] ....
.... [*A. affinis* Fern.]

    3   Plant taller (the flowering stems to 2.5 dm tall), thinly tomentose .... *A. rousseauii*

A. spathulata Fern.

/T/E/ (Ch) Turfy and peaty places in E Que. (L. Mistassini; Côte-Nord; Anticosti Is.), St. Pierre and Miquelon, and Nfld. (type from Rushy Pond; GH). [*A. canadensis* var. *spath.* Fern.; incl. var. *continentis* Fern. & St. John and *A. ?wiegandii* Fern.; scarcely separable from the *A. neglecta* complex].

[*A. stenophylla* Gray]

[The report of this species of the w U.S.A. (Wash. to Nev. and Idaho) from SE B.C. by Eastham (1947; collection in V from Creston) requires confirmation, perhaps being based upon *A. luzuloides*, with which it probably intergrades.]

A. umbrinella Rydb.

/ST/X/ (Ch) Gravels, ledges, and turfs from the Yukon (N to ca. 64°N), SW Dist. Mackenzie, and N Sask. (L. Athabasca) through B.C.–Alta. to Calif., Ariz., and Colo.; isolated in cent. Ont. (N shore of L. Superior) and E Que. (Bic and St-Donat, Rimouski Co.; calcareous cliffs and talus, N Gaspé Pen.). [A very plastic species, here taken to include *A. albescens* and *A. flavescens* Rydb., *A. peasei* and *A. subviscosa* Fern. (E Que. endemics), *A. albicans*, *A. bayardii*, *A. brunnescens*, *A. confusa*, *A. foggii*, *A. longii*, and *A. straminea* Fern. (these last seven Nfld. endemics), *A. mucronata* Nels., and *A. aizoides*, *A. lanulosa*, *A. maculata*, *A. sansonii*, and *A. sedoides* Greene]. MAPS (*A. subviscosa*): Porsild 1950: map 31, p. 20, and 1966: map 141, p. 84.

A. ungavensis (Fern.) Malte

/ST/(X)/ (Ch) Moist sandy places, tundra, and wooded slopes: mts. of SW Alta. (Porsild's map); s-cent. Dist. Keewatin (Chesterfield Inlet; CAN); N Que. (coasts of Hudson Strait and Hudson Bay; Ungava Bay and watershed; Knob Lake dist. at ca. 54°45′N) and northernmost Labrador (S to Saglek, 58°35′N). [*A. alpina* var. *ung.* Fern., the type from the Stillwater R., N Que.; *A. arenicola* Malte]. MAP: Porsild 1965: map 11, p. 55.

    Keyed out below are four closely related taxa (?microspecies):

1   Plants densely white-tomentose.

    2   Plant densely matted, with creeping and spreading leafy stolons; flowering stems to 2 dm tall, weak and flexuous; style much exserted in anthesis, distinctly bifid; [*A. alpina* var. *canescens* Lange and var. *cana* Fern. & Wieg., in part; *A. vexillifera* Fern; Great Slave L. to S Baffin Is., Labrador, and Greenland; Iceland; MAPS: Porsild 1957: map 310, p. 199, and 1965: map 4, p. 54] .... [*A. canescens* (Lange) Malte]

    2   Plant rhizomatous, with short leafy branches; flowering stems rarely over 12 cm tall, stiffly erect; styles at most only slightly exserted in anthesis; [*A. brevistyla* Fern.; N

Ungava, N Labrador (type from Ramah), S Baffin Is., and Greenland; MAPS: Porsild
1957 (1964 revision): map 342, p. 203, and 1965: map 9, p. 54] ..... [*A. sornborgeri* Fern.]
1 Plants glabrous to thinly tomentose (at least the upper surface of the basal leaves
   glabrous and green in age).
   3 Plant glabrous and green; basal leaves oblanceolate; flowering stems to 1 dm tall,
      stiff; [incl. f. *roseola* Ekm.; W and E Greenland and N Scandinavia; MAPS: Hultén 1958:
      map 165, p. 185; Böcher 1938: fig. 105 (Greenland), p. 188; Porsild 1965: map 8
      (Greenland), p. 54] .... [*A. porsildii* Ekman]
   3 Basal leaves glabrous only on the upper surface.
      4 Flowering stems 5 or 6 cm tall, stiff; basal leaves oblanceolate, glabrous or only
         glabrate in age; [*A. canescens* var. *pseudoporsildii* Böcher, basionym, the type
         from Greenland; also known from S Baffin Is. and N Ungava; MAP: Porsild 1965:
         map 3, p. 54] .... [*A. boecheriana* Porsild]
      4 Flowering stems to 2.5 dm tall, weak and flexuous; basal leaves obovate,
         distinctly mucronate .... *A. ungavensis*

ANTHEMIS L.   [9330]   Dogfennel, Chamomile.   Camomille

1 Rays yellow; disk to 2 cm broad, its corolla-tubes compressed; receptacle chaffy
   throughout; achenes striate; leaves 1-pinnatifid, the segments toothed; whitish-pubescent
   stoloniferous perennial; (introd.) .... *A. tinctoria*
1 Rays white; disk to 12 mm broad, its corolla-tubes cylindric; leaves 2–3-pinnatifid;
   annuals; (introd.).
   2 Receptacle chaffy only toward the middle; ray-florets sterile; achenes ribbed and
      rough-tuberculate; leaves essentially glabrous; plant ill-scented .... *A. cotula*
   2 Receptacle chaffy throughout; ray-florets fertile; achenes smooth except for the 10
      nerves; leaves hairy or even somewhat woolly beneath; plants lacking a distinctive
      odour .... *A. arvensis*

A. arvensis L.   Corn-Chamomille
Eurasian; introd. along roadsides and in waste places in N. America, as in SW B.C. (Vancouver Is.
and adjacent islands and mainland), Ont. (N to Ottawa), Que. (N to the Gaspé Pen.), Nfld. (St.
John's; DAO), N.B., P.E.I., and N.S.; SW Greenland.
   Most or all of our material appears referable to var. *agrestis* (Wallr.) DC. (*A. agr.* Wallr.;
receptacular chaff shorter than the disk-florets rather than surpassing them).

A. cotula L.   Mayweed, Dog-Fennel.   Maroute or Camomille des chiens
Eurasian; introd. along roadsides and in waste places in N. America, as in Alaska–Yukon (N to ca.
65°N) and B.C., Alta. (N to McMurray, 56°44′N), Sask. (Troy, where taken by J.M. Macoun in 1883;
CAN), Man. (N to The Pas), Ont. (N to near Thunder Bay), Que. (N to the Gaspé Pen. at York; GH),
Nfld., N.B., P.E.I., and N.S. [*Maruta* DC.]. MAP: Hultén 1968b:886.

A. tinctoria L.   Yellow Chamomille
Eurasian; introd. or a garden-escape to roadsides and fields in N. America, as in SE Alaska (Sitka)
and B.C., Alta. (N to Peace River, 56°14′N), Sask. (near Moose Jaw), Man. (Winnipeg), Ont. (N to
Kapuskasing and Ottawa), Que. (N to the Gaspé Pen. at Tourelle, Matane Co.), Nfld. (St. John's;
DAO), N.B., and N.S. MAP: Hultén 1968b:885.

APARGIDIUM T. & G.   [9597]

A. boreale (Bong.) T. & G.
/sT/W/   (Hr)   Sphagnous bogs and wet meadows at low to fairly high elevations from S Alaska (N
to ca. 60°30′N) and coastal B.C. (Queen Charlotte Is.; Vancouver Is. and adjacent islands; CAN;
DAO; V) to NW Calif. [*Apargia borealis* Bong., the type from Sitka, Alaska; *Leontodon* DC.;
Microseris and *Scorzonella* Greene]. MAP: Hultén 1968b:943.

ARCTIUM L.   [9452]   Burdock.   Bardane

1   Heads more or less corymbosely arranged, mostly long-peduncled; larger leaf-blades
    rounded at apex; petioles strongly angled; (introd.).
    2   Petioles mostly solid; heads commonly at least 3 cm broad; involucre glabrous, its
        middle and upper phyllaries subequal and surpassing the corollas .... *A. lappa*
    2   Petioles usually hollow; heads less than 3 cm broad; involucre cobwebby, its middle
        and upper phyllaries successively longer, mostly shorter than the corollas ....
        .... *A. tomentosum*
1   Heads racemose or racemosely clustered, sessile to long-peduncled; involucre glabrous
    or somewhat tomentose; larger leaf-blades tapering at apex; petioles usually hollow, only
    slightly angled; (introd.).
    3   Heads at most 2.5 cm broad; achenes to 6 mm long; phyllaries shorter than the
        corollas .... *A. minus*
    3   Heads to 4 cm broad; achenes to 1 cm long; some phyllaries equalling or surpassing
        the corollas .... *A. nemorosum*

A. lappa L.   Great Burdock.   Grande Bardane, Artichaut, etc.
Eurasian; introd. along roadsides and in thickets, fields, and waste places in N. America, as in SW
B.C. (Vancouver Is. and adjacent mainland), Man. (N to Warren Landing, near the NE end of L.
Winnipeg; CAN), Ont. (N to Moosonee, SW James Bay, 51°17′N), Que. (N to La-Malbaie, Charlevoix
Co.; reported from Anticosti Is. by Schmitt 1904), N.B., and N.S. [*A. majus* Bernh.].
    A hybrid with *A. minus* (× *A. nothum* (Rühm.) Weiss) is reported from SW Que. by Boivin (1966*b*;
Buckingham).

A. minus (Hill) Bernh.   Common Burdock.   Bardane, Tabac du Diable, etc.
Eurasian; introd. along roadsides and in thickets, fields, and waste places in N. America, as in B.C.
(N to Queen Charlotte Is. and Williams Lake, ca. 52°N), Alta. (N to Edmonton), Sask. (N to Tisdale,
52°51′N), Man. (N to Brandon and Winnipeg), Ont. (N to Cochrane and Ottawa), Que. (N to Anticosti
Is. and the Gaspé Pen.), Nfld., N.B., P.E.I., and N.S. [*Lappa* Hill; *A. lappa* var. *minus* (Hill) Gray].
    The typical form has entire to shallowly toothed leaves and pale-pink to roseate flowers. Forma
laciniatum Clute (leaves laciniate or reduced to narrow blades, the flowers sterile) is known from S
Ont. (Simcoe, Norfolk Co.; OAC), Que. (Montreal and Quebec City; MT; OAC), and P.E.I.
(Charlottetown and Kensington; ACAD). Forma *pallidum* Farw. (corolla white) is known from P.E.I.
(near Kensington, Prince Co.; MT). Forma *purpureum* (Blytt) Evans (corolla deep purple) is known
from SE Man. (Otterburne, about 30 mi S of Winnipeg; Löve and Bernard 1959).

A. nemorosum Lej.
European; introd. along roadsides and in waste places in N. America, as in B.C.–Alta. (Boivin
1966*b*), Man. (near Otterburne, about 30 mi S of Winnipeg; Löve and Bernard 1959), Ont. (N to the
Timagami Forest Reserve, NE of Sudbury), Que. (N to the Gaspé Pen. at Mont-Louis; GH), Nfld.,
P.E.I., and N.S. [*A. minus* var. *corymbosum* Wieg. and ssp. *nem.* (Lej.) Syme; *A. vulgare sensu*
Evans, not *Lappa vulgaris* Hill].

A. tomentosum Mill.
Eurasian; introd. along roadsides and in waste places in N. America, as in Alta. (Edmonton; Groh
and Frankton 1948), Sask. (Rosthern, about 40 mi NE of Saskatoon; Breitung 1957*a*), Man. (N to
Pine River, E of Duck Mt.), Ont. (Simcoe Co.; Montgomery 1957), Que. (N to the Gaspé Pen. at
Tourelle, Matane Co.; CAN), Nfld. (St. John's; CAN), N.B., P.E.I. (Bonshaw, Queens Co.; NSPM),
and N.S. [*A. lappa* var. *tom.* (Mill.) Gray].

ARNICA L.   [9396]   Arnica

(Ref.: Maguire 1943; P.A. Rydberg, N. Am. Flora 34:321–57. 1927)
1   Well-developed stem-leaves mostly 5–12 pairs, the upper ones not greatly reduced;
    heads generally several to rather numerous; anthers yellow.
    2   Phyllaries obtuse or merely acutish, bearing a tuft of long white hairs at apex; leaves

lanceolate to oblanceolate, entire or slightly toothed, to 2 or 3 dm long; stems solitary, from long rhizomes; plants variously hairy to subtomentose; (B.C. to Que.) . . . .
. . . . *A. chamissonis*

2   Phyllaries more sharply acute, their tips not markedly long-hairy; leaves rarely much over 1 dm long; (B.C. and Alta.).

     3   Leaves narrowly lanceolate to lance-elliptic, entire or nearly so; plants more or less scabrid-puberulent at least above, densely tufted (the rhizome commonly shortened into a branching caudex), no well-developed basal leaves present either on the stem or on the separate short sterile leafy shoots; pappus stramineous to tawny, barbed or subplumose . . . . *A. longifolia*

     3   Leaves narrowly lance-elliptic to ovate, at least the basal ones usually rather coarsely toothed; plants more or less hairy especially above, or subglabrous, seldom much tufted (the rhizome mostly more elongate); tufted basal leaves commonly present; pappus tawny, subplumose . . . . *A. amplexicaulis*

1 Well-developed stem-leaves rarely more than 3 or 4 (sometimes none in *A. lessingii* and *A. louiseana*), at least the upper 1 or 2 pairs usually greatly reduced; rhizomes mostly well developed (except in *A. alpina*, the solitary stems from a short, mostly ascending rhizome or mere caudex).

    4   Heads typically discoid and usually lacking rays (but the marginal corollas sometimes enlarged), nodding in youth, commonly several (rarely solitary); pappus tawny; anthers yellow; principal leaves mostly lanceolate to ovate, to 2 dm long and 5 cm broad, entire or denticulate; stems solitary from freely rooting to nearly naked rhizomes; (B.C. and Alta.) . . . . *A. parryi*

    4   Heads typically radiate.

       5   Anthers purple; heads solitary; pappus tawny; stems commonly not over 2.5 dm tall.

          6   Head usually nodding; rays to 2 cm long and 8 mm broad; pappus barbed (rarely subplumose); phyllaries purplish, they and the peduncle densely pilose with multicellular hairs with purple cross-walls; leaves mostly basal, entire or denticulate (rarely sharply dentate); rhizome naked; (Alaska to Dist. Mackenzie; N B.C.) . . . . *A. lessingii*

          6   Head erect; rays to 17 mm long and 7 mm broad; pappus subplumose; phyllaries greenish, they and the peduncle clothed with moniliform (necklace-like) hairs with whitish-translucent cross-walls; stems with up to 4 or 5 pairs of serrulate to coarsely serrate leaves; rhizome densely clothed with fibrous leaf-bases; (Aleutian Is.) . . . . *A. unalaschcensis*

       5   Anthers yellow; heads solitary to several.

         7   Pappus plumose or subplumose, more or less tawny; phyllaries acute or acuminate; flowering stems rarely with evident tufts of basal leaves, solitary or in loose tufts from freely rooting rhizomes.

           8   Involucres hemispheric-campanulate or broader, the disk-flowers relatively numerous; heads few or solitary; stem-leaves lanceolate or oblanceolate to elliptic, ovate, or obovate, entire or irregularly denticulate (or dentate); (B.C. and Alta.; E Que. and N.B.) . . . . *A. mollis*

           8   Involucres narrower, more or less top-shaped, the disk-flowers fewer; heads generally several; stem-leaves narrowly or broadly elliptic to deltoid or ovate, irregularly low-dentate; (B.C. and Alta.) . . . . *A. diversifolia*

        7   Pappus merely barbellate, usually white or nearly so (sometimes somewhat tawny in *A. fulgens*); basal leaves often conspicuous (sometimes small or wanting).

          9   Leaf-blades narrowly to broadly ovate, commonly not over twice as long as broad, usually more or less coarsely toothed, distinctly pinnate-veined; basal leaves frequently present but mostly on separate short shoots.

            10 Achenes generally glabrous at least toward base; heads solitary to several; basal leaves sometimes present on short shoots, rarely cordate, the stem-leaves even more rarely cordate, the middle ones commonly larger than those below; (B.C. and Alta.) . . . . *A. latifolia*

    10 Achenes mostly uniformly short-hairy or glandular (or both); heads
        solitary or 2 or 3; basal leaves commonly present on short shoots, they
        and the lower stem-leaves usually strongly cordate and larger than the
        middle ones; (B.C. to w Sask.; introd. in Man.) .... *A. cordifolia*
  9 Leaf-blades mostly distinctly narrower in outline (commonly 3 or 4 times as
    long as broad), tapering or rounded to base (never cordate), entire or
    denticulate (rarely dentate).
    11 Stems rarely over 2 or 3 (commonly about 1 or 1.5) dm tall, often
        subscapose; leaves elliptic-lanceolate to elliptic or oblanceolate, the
        lower ones regularly denticulate or dentate, their faint lateral veins
        joining the midrib well above the leaf-base; heads solitary (rarely 2 or
        3); achenes glabrous below the middle or sometimes throughout; (B.C.
        and Alta.; E Que. and Nfld.) .... *A. louiseana*
    11 Stems commonly taller and more leafy; lateral veins of leaves
        generally very conspicuous and extending nearly or quite to base;
        achenes uniformly short-hairy throughout.
        12 Lower leaves mostly rather abundantly denticulate or dentate,
            lanceolate to lance-elliptic (sometimes ovate); heads commonly 4
            or 5 (up to 10 or 11); phyllaries lanceolate; (essentially transconti-
            nental) .... *A. lonchophylla*
        12 Lower (and upper) leaves entire or obscurely denticulate; heads 1
            to few.
            13 Base of stem with conspicuous dense tufts of long brown wool
                in the axils of the old leaves; leaves lanceolate; phyllaries
                lanceolate to elliptic-oblong; (B.C. to Man.) .... *A. fulgens*
            13 Base of stem lacking conspicuous tufts of brown wool (or the
                wool scanty and white when present in *A. sororia*).
                14 Plants more or less densely long-villous-tomentose; leaves
                    lanceolate, the stem-leaves rarely more than 2 or 3 pairs;
                    heads solitary to few; phyllaries lanceolate; (trans-
                    continental) .... *A. alpina*
                14 Plants not tomentose; stem-leaves commonly 3 or 4 pairs.
                    15 Phyllaries oblanceolate, their tips short or only acutish;
                        leaves lanceolate to elliptic or oblanceolate (sometimes
                        broader); (Ungava–Labrador) .... *A. plantaginea*
                    15 Phyllaries lanceolate to elliptic-oblong, acute or acutish;
                      (mts. of B.C. and Alta.).
                        16 Leaves lanceolate to oblanceolate or spatulate
                          (sometimes broader), the lower stem-leaves tending
                          to be sessile; involucre turbinate-campanulate; rays
                          mostly about 8 .... *A. rydbergii*
                      16 Leaves lanceolate, the lower stem-leaves generally
                        petioled; involucre nearly hemispheric; rays usually
                        more than 10 .... *A. sororia*

A. alpina (L.) Olin
/AST/X/GEA/ (Hs) Open rocky slopes up to high elevations, the aggregate species from the coasts of Alaska–Yukon–Dist. Mackenzie to Ellesmere Is. (N to ca. 82°N), Baffin Is., and northernmost Ungava–Labrador, S in the West through the mts. of B.C. and SW Alta. (N to Jasper) to Mont., farther eastwards S to L. Athabasca (Sask.), N Man. (S to Gillam, about 165 mi S of Churchill), N Ont. (Hudson Bay–James Bay watershed S to the Missinaibi R. at 50°03′N), Que. (S to E Hudson Bay at ca. 56°10′N; isolated at Knob Lake, ca. 54°48′N, 68°W), and Nfld.; W and E Greenland N to ca. 78°N; Spitsbergen; N Eurasia. MAPS and synonymy: *see* below.
1 Pilosity on the peduncles and at the base of the involucres with little or no admixture of stipitate glands; stem-leaves 2 or 3 pairs; plant commonly less than 3 dm tall.
  2 Teeth of the ray-ligules mostly 2.5–4 mm long; stipitate glands lacking or very short and

obscure; [*A. montana* var. *alpina* L.; N Scandinavia; MAPS: Eric Hultén 1958: map 183, p. 203, and Sven. Bot. Tidskr. 43(2/3): fig. 11, p. 405. 1949; Maguire 1943: fig. 1, p. 409] . . .
. . . . [ssp. *alpina*]

2 Teeth of ray-ligules 0.5–2(2.5) mm long; peduncles and base of involucres more or less evidently stipitate-glandular; [*A. angust.* Vahl; *A. alp.* f. *inudata* M.P. Porsild; transcontinental; MAPS: Porsild 1957: map 323, p. 201; also on the above-noted maps by Hultén and Maguire; Hultén 1968*b*:923] . . . . ssp. *angustifolia* (Vahl) Maguire

1 Pilosity on the peduncles and at the base of the involucres with an obvious admixture of long-stipitate glands; stem-leaves commonly 4 or 5 pairs.

3 Leaves and stem conspicuously villous-tomentose; achenes often over 6 mm long; plants to about 2 dm tall; [*A. tomentosa* J.M. Macoun, the type from Waterton Lakes, Alta.; *A. pulchella* Fern.; *A. alpina* ssp. *attenuata* var. *?vestita* Hult.; SE Alaska–S Yukon–W Dist. Mackenzie and mts. of B.C. and Alta.; Table Mt., Nfld.; MAPS: on the above-noted maps by Hultén (1958) and Maguire (only Hultén's 1958 map indicating the Nfld. locality but this reported by Maguire); Hultén 1968*b*:924] . . . .
. . . . ssp. *tomentosa* (J.M. Macoun) Maguire

3 Leaves and stem inconspicuously pilose; achenes at most about 6 mm long.

4 Heads solitary or 2 or 3 (rarely more) on a stem; disk-corollas to 1 cm long; [*A. sornborgeri* and *A. terrae-novae* Fern.; *A. sorn.* var. *ungavensis* Boivin; *A. plantaginea sensu* Fernald 1933, not Pursh, the relevant collections in GH; N Que. (between E Hudson Bay at ca. 56°10′N and Ungava Bay), N Labrador (S to ca. 54°30′N; type from Ramah), and Nfld.; MAPS: on the above-noted maps by Hultén and Maguire] . . . . ssp. *sornborgeri* (Fern.) Maguire

4 Heads mostly 3 or more; disk-corollas at most about 8.5 mm long . . . .
. . . . ssp. *attenuata* (Greene) Maguire

5 Radical leaves linear, little broader than the stem-leaves; [Alaska, the type from Fort Yukon] . . . . var. *linearis* Hult.

5 Radical leaves linear-lanceolate to lanceolate, to about 18 mm broad and distinctly broader than the stem-leaves; [var. *?vahliana* Boivin; *A. attenuata* Greene; *A. lowii* Holm; transcontinental; MAPS: on the above-noted maps by Hultén (1958) and Maguire; Hultén 1968*b*:923] . . . . var. *attenuata*

A. amplexicaulis Nutt.

/ST/W/ (Hpr) Streambanks and moist woods from Alaska–Yukon–W Dist. Mackenzie (N to ca. 63°N) through B.C. and the mts. of SW Alta. (N to Jasper) to N Calif. and W Mont. [*A. amplexifolia, A. borealis,* and *A. elongata* Rydb.; *A. aspera, A. filipes,* and *A. macounii* Greene]. MAPS: Hultén 1968*b*:919; Porsild 1966: map 142, p. 84; Maguire 1943: fig. 16, p. 469.

Some of the more northern material is referable to ssp. *prima* Maguire (stem-leaves at most 7 pairs, the lower 2 or 3 pairs distinctly petioled, rather than up to 12 pairs and usually all sessile, less conspicuously toothed than those of the typical form). MAPS: Hultén 1968*b*:919; on the above-noted map (squares) by Maguire.

A. chamissonis Less.

/ST/(X)/ (Hpr) Mostly in meadows and wet places, the aggregate species from the Aleutian Is. (type from Unalaska) and Alaska–Yukon (N to ca. 63°N) to Great Slave L., L. Athabasca (Alta.), Sask. (N to Meadow Lake, 54°08′N), Man. (N to the Churchill R. at ca. 57°15′N), northernmost Ont., and W-cent. Que. (E James Bay at ca. 52°N; Dutilly, Lepage, and Duman 1958), S in the West through B.C.-Alta. to Calif. and N.Mex., farther eastwards S to S Sask.-Man. and cent. Ont. (W James Bay watershed at ca. 51°N; a collection in DAO from Thunder Bay (?introd.) has also been placed here). MAPS and synonymy: *see* below.

1 Pappus tawny, subplumose; hairs at base of involucre with very prominent cross-walls; leaves usually distinctly toothed; [var. *interior* Maguire; *A. kodiakensis* Rydb.; Aleutian Is. (type from Unalaska), Alaska, B.C., and Alta.; MAPS: Maguire 1943: fig. 14, p. 462 (ssp. *genuina*; the report from Man. by Maguire, not indicated on his map, relates to ssp. *foliosa*); Porsild 1966: map 143, p. 84; Hultén 1968*b*:920] . . . . ssp. *chamissonis*

1 Pappus whitish to stramineous, merely barbed; hairs at base of involucre with less
   prominent cross-walls; leaves commonly entire or remotely denticulate . . . .
   . . . . ssp. *foliosa* (Nutt.) Maguire
    2 Leaves conspicuously silvery-tomentose; [*A. foliosa* var. *incana* Gray (*A. incana*
      (Gray) Greene); *A. cana* Greene; s Yukon–sw Dist. Mackenzie; MAPS: on the
      above-noted maps by Maguire and Porsild, a B.C. report by Maguire not indicated on
      either map] . . . . var. *incana* (Gray) Hult.
    2 Leaves less densely hairy and scarcely silvery; [var. *angustifolia* Herder; *A. foliosa*
      Nutt.; *A. columbiana* Greene, not Nels.; *A. maguirei* and *A. rhizomata* Nels.; *A.*
      rubricaulis Greene; *A. wilsonii* Rydb.; s Alaska–B.C. to s James Bay, Ont.; reported
      from James Bay, Que., by Dutilly, Lepage, and Duman 1958; MAPS: on the
      above-noted maps by Maguire and Porsild; Hultén 1968*b*:921] . . . . var. *foliosa*

A. cordifolia Hook.
/ST/D/ (Hsr) Woods and meadows at low to high elevations from SE ?Alaska (according to
Boivin, personal communication, the Muir Glacier report by Hultén 1950, is based upon *A. latifolia*;
the other collections by Krause, being Berlin war casualties, cannot be confirmed), the Yukon (N to
ca. 64°N), and sw Dist. Mackenzie (between Fort Liard and Fort Simpson; CAN) to B.C., w Alta. (N
to Dunvegan, ca. 56°N), and w Sask. (Cypress Hills and Waskesiu Lake, ca. 54°N; Breitung 1957*a*;
an isolated station on Riding Mt., Man., where taken by D.A. Blood in 1962, evidently introd., being
annotated "roadside weed, common in patches"), s to Calif., N.Mex., and S.Dak.; ssp. *whitneyi*
(Fern.) Maguire (*A. whitneyi* Fern.; leaves relatively broad, the basal sinus deep and narrow)
isolated in Keweenaw Co., Mich. [*A. andersonii* Piper; *A. evermannii* and *A. subcordata* Greene; *A.*
humilis and *A. pumila* Rydb.]. MAPS: Hultén 1968*b*:918; *Atlas of Canada* 1957: map 9, sheet 38;
Maguire 1943: fig. 10, p. 450.
    Var. *eradiata* Gray (*A. discoidea* Benth. and its var. *erad.* (Gray) Cronq.; *A. grayii* Heller;
ray-ligules wanting, the heads discoid) is reported from B.C. by Henry (1915; *A. disc.*, "East of the
Cascades") and collections in Herb. V from Little Shuswap L., E of Kamloops, and Deer Park, about
30 mi NW of Trail, have been referred to it. The MAPS by Maguire (1943: fig. 11 (*A. grayii*) and fig. 13
(*A. disc.*), p. 447, indicate no Canadian stations.

A. diversifolia Greene
/sT/W/ (Hsr) Rocky places at moderate to high elevations from SE Alaska (Copper R. region at
ca. 61°30'N; the report from s Yukon by P.A. Rydberg, N. Am. Flora 34:355. 1927, requires
confirmation) through B.C. and sw Alta. (N to Jasper; CAN; V) to Calif. and Utah. [*A. latifolia* var.
viscidula Gray]. MAPS: Hultén 1968*b*:921; Maguire 1943: fig. 15, p. 469.

A. fulgens Pursh
/T/WW/ (Hsr) Meadows and slopes up to moderate elevations from SE B.C. (N to Kamloops; V;
collections in V from farther north at Chezacut, Cariboo, and Sifton Pass, ca. 58°N, require
confirmation) to Alta. (N to Fairview, ca. 56°N; CAN), s Sask. (N to Tramping Lake, 52°08'N; CAN),
and s Man. (N to Wheatland, about 20 mi NW of Brandon; CAN), s to Calif. and Colo. [*A.*
pedunculata Rydb.; *A. alpina sensu* John Macoun 1884, as to the Man. plant, not (L.) Olin, relevant
collections in CAN]. MAP: Maguire 1943: fig. 4, p. 424.

A. latifolia Bong.
/ST/W/ (Hsr) Moist woods, meadows, and moist open places at moderate to high elevations,
the aggregate species from Alaska-Yukon (N to ca. 63°N; type from Sitka, Alaska) and w Dist.
Mackenzie (Porsild and Cody 1968) through B.C. and sw Alta. (N to Jasper; CAN) to Calif. and
Colo. MAPS and synonymy: *see* below.
1 Stems slender and mostly not over 3 dm tall, usually several from a rhizome that is
   commonly shortened into a slender, loosely branched, scaly caudex; leaves seldom over
   2.5 cm broad; heads 3–9, the involucres to 13 mm high; [*A. gracilis, A. leptocaulis,* and
   A. puberula Rydb.; *A. betonicaefolia* and *A. lactucina* Greene; *A. columbiana* Nels., not
   Greene; s B.C. (N to Revelstoke) and sw Alta. (N to Jasper); MAP: Maguire 1943: fig. 7 (*A.*
   gracilis), p. 440] . . . . var. *gracilis* (Rydb.) Cronq.

1   Stems to about 6 dm tall, solitary or few together from elongate, mostly naked rhizomes; leaves to about 8 cm broad; heads mostly 1–3, the involucres to over 15 mm high; [*A. aprica, A. grandifolia, A. laevigata,* and *A. ventorum* Greene; *A. granulifera* and *A. oligolepis* Rydb.; *A. menziesii* Hook.; range of the species; MAPS: Hultén 1968b:918 (aggregate species); on the above-noted map by Maguire] .... var. *latifolia*

A. lessingii Greene
/Ss/W/eA/   (Hsr)   Alpine and subalpine meadows in the E Aleutian Is., Alaska (N to ca. 70°N; type from "Alaskan shores and islands"), the Yukon (N to ca. 67°N), W Dist. Mackenzie (N to ca. 68°N), and N B.C. (S to Robb L. at ca. 56°30'N); Kamchatka. [*A. montana* var. *angustifolia* Hook. in part; *A. obtusifolia* var. *acuta* Raup; *A. porsildiorum* Boivin; not *A. angustifolia lessingii* T. & G.; incl. the relatively leafy high-grown extreme, ssp. *norbergii* Hult. & Maguire]. MAPS: Hultén 1968b:916; Raup 1947: pl. 36; Maguire 1943: fig. 20, p. 487.

A. lonchophylla Greene
/ST/(X)/   (Hsr)   Dry open places and calcareous gravels, ledges, and cliffs (ranges of taxa in the two disjunct areas outlined below, the species confined to Alaska–Canada except for isolated stations in the Black Hills of S.Dak. and in NE Minn.). MAPS and synonymy: *see* below.
1   Leaves 3–5-nerved, to about 1.5 cm broad, subentire or inconspicuously dentate; [Alaska–Yukon–w Dist. Mackenzie (N to ca. 65°30'N) to Great Bear L., Great Slave L., L. Athabasca (Sask.), S Dist Keewatin, and northernmost Ont., S to SW Alta. (type from the Athabasca R., probably near Jasper), cent. Sask. (Meridian L., 54°32'N; Breitung 1957a), Man. (S to The Pas), and cent. Ont. (Fawn R. at ca. 54°15'N; CAN); MAPS: Hultén 1968b:922; Maguire 1943: fig. 6 (ssp. *genuina*), p. 429] .... ssp. *lonchophylla*
1   Leaves mostly 5–7-nerved, to about 3.5 cm broad, more strongly dentate; [*A. chionopappa* Fern., the type from the banks of the Grand R., Gaspé Pen., E Que.; *A. gaspensis* Fern.; *A. fernaldii* Rydb.; *A. arnoglossa* of Canadian reports, not Greene; E Que. (Anticosti Is.; Gaspé Pen.) and W Nfld.; MAPS: Fernald 1933: map 29 (*A. chion.*), p. 367; on the above-noted maps by Hultén and Maguire] .... ssp. *chionopappa* (Fern.) Maguire

A. longifolia Eat.
/T/W/   (Hp)   Open ground and cliffs at moderate to high elevations from w-cent. B.C. (collection in V from Prince George, ca. 54°N; collection in CAN from near Pine Pass, ca. 55°20'N) and SW Alta. (Waterton Lakes; Banff) to Calif. and Colo. [*A. myriadena* Piper]. MAP: Maguire 1943: fig. 15, p. 469.

A. louiseana Farr
/aST/D/eA/   (Hsr)   Rock slides, slopes, and ridges up to high elevations (ranges of taxa in the western and eastern disjunct areas outlined below, the species confined in N. America to Alaska–Canada); NE Siberia. MAPS and synonymy: *see* below.
1   Phyllaries uniformly short-stipitate-glandular; achenes mostly glandular as well as sparingly hirsute toward summit; [*A. frigida* var. *?glandulosa* Boivin; mts. of Alta. near L. Louise, the type locality; MAPS (both as ssp. *genuina*): Maguire 1943: fig. 2, p. 418; Porsild 1966: map 144, p. 84] .... ssp. *louiseana*
1   Phyllaries becoming glabrate above, they and the achenes scarcely glandular.
    2   Base of involucre densely yellowish-lanate-pilose; [*A. frigida* Mey. and its var. *pilosa* Maguire; *A. brevifolia, A. illiamnae, A. mendenhallii,* and *A. nutans* Rydb.; coasts of Alaska–Yukon–Dist. Mackenzie (E to Coronation Gulf) and northernmost B.C. (S to Summit Pass, 58°31'N); E Asia; MAPS: on the above-noted maps by Maguire and Porsild; Hultén 1968b:917 (*A. frigida*)] .... ssp. *frigida* (Mey.) Maguire
    2   Base of involucre whitish-pilose (sometimes scantily so); [*A. griscomii* Fern.; E Que. (type from Mt. Mattaouisse, Gaspé Pen.) and Nfld.; MAPS: on the above-noted maps by Maguire and Porsild] .... ssp. *griscomii* (Fern.) Maguire

A. mollis Hook.
/ST/D/   (Hsr)   Ledges or gravelly shores and wet cliffs at low to high elevations: S-cent. Alaska (a single station at ca. 63°N indicated in Hultén's map), S Yukon (upper Rose R.; CAN), and SW Dist.

Mackenzie (Brintnell L., ca. 62°N; CAN) through B.C. and sw Alta. (N to Jasper; CAN) to Calif. and Colo.; Que. (Megantic and Rimouski counties and Gaspé Pen.) and N N.B. (Aroostook R., Victoria Co.; Bathurst and the Nepisiguit R., Gloucester Co.; MTMG; GH; Fowler 1879; not known from P.E.I. or N.S.) to the mts. of Maine and N.H. [*A. confinis, A. crocea, A. crocina,* and *A. rivularis* Greene; *A. lanceolata* Nutt.]. MAPS: Hultén 1968*b*:920; combine the maps by Maguire 1943: fig. 18 (w area; *A. mollis*), p. 478, and fig. 17 (E area; *A. lanc.*), p. 469.

Some of our eastern material is referable to var. *petiolaris* Fern. (*A. pet.* (Fern.) Rydb.; all but the upper pairs of leaves narrowed to slender petioles rather than only the lowest leaves petioled).

A. parryi Gray
/sT/W/ (Hsr) Open woods, meadows, and moist slopes at low to moderate elevations from S Yukon (two stations at ca. 62°N; CAN) through B.C. and the mts. of sw Alta. (N to Jasper; CAN) to Calif. and Colo. [*A. angustifolia* var. *eradiata* Gray (*A. erad.* (Gray) Heller), not *A. cordifolia* var. *erad.* Gray]. MAPS: Hultén 1968*b*:922; Maguire 1943: fig. 15 (incomplete northwards), p. 469.

A. plantaginea Pursh
/Ss/E/ (Hs) Brooksides and cool slopes of N Que. (Ungava Bay watershed S to L. Marymac at ca. 57°N; CAN) and N Labrador (S to Indian Harbour, 54°27′N; type from Labrador; according to Maguire 1943, the citations from Nfld. by Fernald 1933, are based upon *A. alpina* ssp. *sornborgeri*). [*A. alpina* var. *lessingii sensu* Fernald and Sornborger 1899, not T. & G., the relevant Ramah, Labrador, collection in GH]. MAP: Maguire 1943: fig. 3, p. 418.

A. rydbergii Greene
/T/W/ (Hsr) Dry meadows and open slopes, mostly at high elevations, from B.C. (N to Hazelton, ca. 55°15′N; Herb. V) and sw Alta. (N to Jasper; CAN) to N Calif. and Colo. [*A. aurantiaca* Greene; *A. cascadensis* St. John; *A. ovalis* Rydb.]. MAP: Maguire 1943: fig. 3 (solid dots), p. 418.

A. sororia Greene
/T/W/ (Hsr) Open dryish places at low to moderate elevations from S B.C. (N to Cariboo, Chilcotin, and Kamloops; CAN; V; type from Cascade, near the U.S.A. boundary S of Grand Forks) and sw Alta. (N to Banff; CAN) to Calif. and Utah. [*A. stricta* Greene, not Nels.; *A. trinervata* Rydb.]. MAP: Maguire 1943: fig. 5, p. 424.

A. unalaschcensis Less.
/sT/W/eA/ (Hsr) Meadows in the Aleutian Is. (type from Unalaska); S Kamchatka, the Kuril Is., and N Japan. MAPS: Hultén 1968*b*:917, and 1950: map 1189, p. 1682; Maguire 1943: fig. 20 (crosses), p. 487.

ARNOSERIS Gaertn. [9564]

A. minima (L.) Schweig. & Koerte   Lamb-Succory
Eurasian; locally introd. along roadsides and in dry fields in N. America, as in ?Ont. (an early report by Britton and Brown noted by Groh and Frankton 1949*b*), ?N.B. (Boivin 1966*b*), P.E.I. (wharf-ballast at Charlottetown, where taken by Hurst in 1936), and N.S. (railway ballast at Belleville, Yarmouth Co.; CAN; GH). [*Hyoseris* L.; *A. pusilla* Gaertn.].

ARTEMISIA L. [9358] Sagebrush, Wormwood. Armoise

(Ref.: P.A. Rydberg, N. Am. Flora 34:244–85. 1916)
1 Principal leaves entire to coarsely toothed or with relatively few and shallow entire lobes (the lower, generally deciduous leaves of *A. dracunculus* often with 1 or 2 elongate linear lobes; some varieties of *A. ludoviciana* with much dissected blades), usually all cauline (basal tufts wanting or poorly developed).
  2 Shrubs with woody stems and branches; leaves to 5 cm long, finely and closely canescent; involucres to 5 mm high; flowers all perfect.
    3 Leaves linear to linear-lanceolate, usually acute, entire or sometimes with 1 or 2

teeth or lobes (or the lower ones sometimes rather deeply 3-parted near apex);
plant to about 1 m tall; (B.C. to sw Man.) .... *A. cana*

3 Leaves narrowly cuneate, the principal ones 3-toothed at apex; plant to over 2 m
tall; (s B.C. and s ?Alta.) .... *A. tridentata*

2 Herbs (or the stem sometimes distinctly woody at base); marginal flowers pistillate,
the disk-flowers perfect.

4 Leaves glabrous to silky-canescent or villous-puberulent (but not at all tomentose), linear, commonly 3 or 4 cm long and 2 or 3 mm broad (but up to 8 cm long
and 1 cm broad), the basal ones often deeply 3-cleft; involucres to 4 mm high;
disk-flowers sterile, the ovary abortive; (B.C. to Ont.) .... *A. dracunculus*

4 Leaves white-tomentose beneath, lanceolate to elliptic or sometimes broader;
disk-flowers fertile, with normal ovary.

5 Principal leaves relatively narrow, commonly not over 1 cm broad (exclusive of
the lobes, when present); plants rarely over 1 m tall.

6 Stems more or less woody at base (not dying back entirely to the ground
each year), tending to be taprooted and lacking well-developed rhizomes;
leaves usually green and glabrous or glabrate above, to about 5 cm long
but rarely as much as 1 cm broad, commonly with axillary fascicles which
may develop into short shoots; involucre about 3 mm high; inflorescence
narrow, consisting of short axillary few-headed spikes; (s ?B.C.; plants
evidently confined to the shores of lakes and rivers below the high-water
mark) .... [*A. lindleyana*]

6 Stems herbaceous to base, not taprooted, often with well-developed
rhizomes; leaves often thinly tomentose above (as well as more densely
so beneath), to over 1 dm long.

7 Stems clustered from stout woody crowns to 2 cm thick, lacking
stolons; leaves linear-attenuate, entire or the lowest ones rarely
toothed, revolute-margined; inflorescence usually narrow and compact;
(Alta. to s Man.) .... *A. longifolia*

7 Stems scattered or few together from slender cord-like rhizomes, soon
producing slender elongate stolons; leaves mostly lanceolate or
lance-elliptic, entire or with a few coarse teeth or entire forward-
directed lobes, flat-margined; inflorescence narrow to ample; (B.C. to
Man.; introd. eastwards) .... *A. ludoviciana*

5 Principal leaves commonly over 1.5 cm broad (exclusive of the lobes, when
present), usually green and glabrous or glabrate above, the larger ones often
with a few forward-directed teeth or small lobes; inflorescence ample and
paniculate; plants often well over 1 m tall.

8 Involucres nearly or quite glabrous, yellow-green and shining, subcylindric
or oblong, less than 2 mm broad, with at most 15 flowers; heads erect or
nearly so in anthesis; (s B.C.) .... *A. suksdorfii*

8 Involucres persistently more or less tomentose (usually densely so).

9 Involucres oblong or subcylindric, to 3 mm broad, with at most 15
flowers; heads erect or nearly so at anthesis; leaves to 1.5 cm broad,
exclusive of the lobes .... [*A. herriottii*]

9 Involucres broad-campanulate, with up to about 35 flowers; heads
nodding at anthesis; leaves to 5 cm broad, exclusive of the lobes ....
.... [*A. douglasiana*]

1 Principal leaves deeply dissected and often "fern"-like in appearance, the primary
segments themselves often lobed.

10 Heads mostly in a dense terminal globular cluster (or sometimes also in smaller
peduncled lateral clusters); marginal flowers pistillate, the disk-flowers perfect; leaves
chiefly in a dense basal tuft, those of the stem wanting or few and reduced and mostly
entire; plants densely tufted or cushion-forming, the stems terminating the branches of
a woody caudex, rarely over 1 dm tall.

11 Corolla more or less pilose; involucres 5 or 6 mm broad; principal leaves
2–3-ternate, silky-canescent; (Alaska–Yukon) .... *A. glomerata*

11 Corolla glabrous, or merely glandular-granuliferous; leaves rarely twice ternate; (Alaska).

    12 Plant densely long-hirsute; involucres 4 or 5 mm broad .... *A. senjavinensis*

    12 Plant sparingly silky; involucres to 7 mm broad .... *A. globularia*

10 Heads in spikes, racemes, or panicles; stems mostly relatively leafy and usually taller.

  13 Shrubs with more or less woody stems and branches; receptacle naked.

    14 Inflorescence narrow, spicate-racemose, the subsessile heads erect; involucres canescent, campanulate, about 3.5 mm high and 2 mm broad, with not more than 8 flowers, these all perfect; leaves canescent, the principal ones divided nearly to base into linear or linear-oblanceolate segments that are themselves commonly deeply 3-cleft; shrub to about 6 dm tall; (S B.C.) ....
.... *A. tripartita*

    14 Inflorescence ample, open paniculate, the short-stalked heads nodding; involucres subglabrous or thinly cobwebby, hemispheric, to 3 mm high and 5 mm broad, with up to 20 perfect disk-flowers and about 10 pistillate marginal flowers; leaves glabrous above, usually thinly tomentose beneath, 2-pinnately dissected into linear-filiform divisions with revolute margins; plant to about 2 m tall; (introd. from Alta. to Que. and Nfld.) .... *A. abrotanum*

  13 Herbs (or the stem sometimes distinctly woody at base); marginal flowers pistillate, the disk-flowers perfect (but sterile in *A. campestris*).

    15 Leaves persistently more or less white-tomentose (sometimes grey-tomentose) beneath; disk-flowers fertile; receptacle naked; perennials with tough, often somewhat woody bases or crowns.

      16 Leaf-lobes lanceolate to ovate (relatively broad and mostly over 2 mm broad), directed forward.

        17 Involucre 6 or 7 mm high and up to 9 mm broad; inflorescence narrow and often dense, elongate; leaves 1-pinnatifid into oblong, obtuse, entire or few-toothed lobes, tomentose above; plant densely white-woolly, the decumbent stems matted from extensively creeping and forking stout rhizomes; (introd. on sandy beaches and dunes in B.C. (Boivin 1966*b*), S Ont. (L. Erie), E Que., and the Maritime Provinces) ....
.... *A. stelleriana*

        17 Involucres at most 5 mm high, narrower; stems to about 1.5 m tall.

          18 Leaves ordinarily lacking stipule-like lobes at base, the principal ones moderately to deeply divided into a few lanceolate, often acuminate, entire segments; inflorescence narrow and almost thyrsoid to more often narrowly but rather openly panicle-like; (Alaska–B.C. to N Man. and W James Bay, Ont.) .... *A. tilesii*

          18 Leaves ordinarily with 1 or 2 pairs of stipule-like lobes at base; divisions of the principal leaves again toothed, cleft, or lobed; inflorescence an ample leafy panicle with ascending spike-like branches; stem to 1.5 m tall; (introd., transcontinental) ...... *A. vulgaris*

      16 Leaf-lobes relatively narrow (mostly less than 2 mm broad), the blade commonly more copiously and finely dissected and often "fern"-like in appearance, usually more or less tomentose above, the lobes generally more spreading than forward-directed.

        19 Upper (as well as lower) leaves mostly 2-pinnatifid into linear, often narrow-toothed blunt segments; leaves to 3 cm long, often with a pair of stipule-like lobes at base; involucre densely pubescent, to 3 mm high and 4 mm broad, the slender-pedicelled nodding heads scattered along the racemes or racemose slender branches of the panicle; (introd. from Man. to Que. and N.S.) .... *A. pontica*

        19 Upper (commonly bracteal) leaves markedly less dissected than the lower leaves; involucre to about 4.5 mm high.

          20 Leaves typically bright green and glabrous above, bipinnately parted into narrow, often saliently toothed segments; inflorescence commonly spike-like or thyrsoid; (B.C. and Alta.) .... *A. michauxiana*

20 Leaves commonly more or less tomentose above; inflorescence narrow or ample; (B.C. to Man.; introd. eastwards) ...... *A. ludoviciana*

15 Leaves glabrous or variously pubescent but not tomentose.

  21 Receptacle long-hairy between the flowers, these all fertile; involucres 3 or 4 mm high; leaves silvery-silky.

    22 Ultimate segments of the 2–3-pinnatifid leaves mostly oblong and obtuse, to 4 mm broad, the stem-leaves to over 8 cm long; stems coarse, herbaceous, erect, to about 1.5 m tall; (introd., transcontinental) .... *A. absinthium*

    22 Ultimate segments of the 2–3-ternate leaves linear-filiform, at most about 1 mm broad, the stem-leaves to about 12 mm long; stems slender, often decumbent, from a branching woody base, to about 5 dm tall; (B.C. to Man.; introd. eastwards) .... *A. frigida*

  21 Receptacle naked.

    23 Glabrous annuals or biennials from a taproot; stems leafy, to 3 m tall; (introd.).

      24 Involucres 2 or 3 mm high; inflorescence dense and strict, leafy, the erect heads obscurely peduncled; leaves deeply pinnatifid, their narrow lobes usually sharply toothed, or the lower leaves 2-pinnatifid; achenes 4–5-nerved; plant not sweet-scented; (introd., transcontinental) .... *A. biennis*

      24 Involucres 1 or 2 mm high; inflorescence paniculate, the nodding heads distinctly peduncled; leaves mostly 2–3-pinnatifid; achenes obscurely or scarcely nerved; plant sweet-scented; (introd. from Ont. to the Maritime Provinces) .... *A. annua*

    23 Perennials (rarely biennials) with tough, more or less woody bases or crowns, the principal leaves in a basal tuft, mostly 2–3-pinnately parted, the upper stem-leaves often entire.

      25 Plant glabrous; inflorescence narrow, racemose-spicate; disk-flowers fertile; involucres to 1 cm broad; stems to 3 dm tall, reddish and striate; (Alta.) .... [*A. laevigata*]

      25 Plants usually more or less pubescent.

        26 Involucres to 7 mm high and over 1 cm broad, their prominently dark-margined phyllaries glabrous or moderately woolly-villous; disk-flowers fertile; achenes commonly glabrous; inflorescence spike-like to narrowly spicate-racemose; leaves loosely and copiously villous to essentially glabrous; stems to 6 dm tall; (Alaska–Yukon–Dist. Mackenzie–B.C.–Alta.) .... *A. arctica*

        26 Involucre less than 5 mm high.

          27 Disk-flowers fertile, the ovary developing into an achene; inflorescence commonly spicate (sometimes spicate-racemose); phyllaries more or less dark-margined; leaves canescent- or silvery-silky, the principal ones pinnately divided, their segments again mostly deeply 3-cleft; stems to about 3 dm tall; (Alaska–Yukon–B.C.) .... *A. furcata*

          27 Disk-flowers sterile, the ovary abortive; leaves silky or villous to glabrous, 2–3-pinnatifid or ternate; (transcontinental) .... *A. campestris*

A. abrotanum L.   Southernwood, Lad's Love
European; locally introd. or a garden-escape to roadsides and waste places in N. America, as in Alta. (Medicine Hat and Grande Prairie; CAN), Sask. (McKague, 52°37′N; Breitung 1957a), s Man. (Portage la Prairie; Macgregor; Winnipeg), Ont. (N to Carleton and Russell counties), Que. (N to L. Timiskaming at ca. 47°30′N), and Nfld. (Rouleau 1956).

A. absinthium L.   Absinthe, Absinthium, Common Wormwood.   Armoise
Eurasian; introd. along roadsides and in pastures and waste places in N. America, as in B.C. (N to

near Quesnel, ca. 53°N; Herb. V), Alta. (N to Grande Prairie, 55°10′N; CAN), Sask. (N to Montreal Lake, 54°03′N; Breitung 1957a), Man. (N to Wekusko L., about 90 mi NE of The Pas), Ont. (N to Moose Factory, SW James Bay, 51°16′N), Que. (N to the Gaspé Pen.), Nfld., N.B., P.E.I., and N.S. [Incl. var. *insipida* Stechmann].

A. annua L.   Sweet Wormwood
Eurasian; introd. along roadsides and in fields and waste places in N. America, as in Ont. (N to North Bay and Ottawa), Que. (Roberval, L. St. John; Iberville, SE of Montreal; DAO; MT), N.B. (Chatham, Northumberland Co.; Groh and Frankton 1949b), and P.E.I. (Charlottetown; CAN; GH).

A. arctica Less.
/aST/W/eA/   (Hs)   Meadows and rocky slopes at low to fairly high elevations, the aggregate species from the Aleutian Is. and coasts of Alaska–Yukon to the Richardson Mts. of NW Dist. Mackenzie, S through B.C. and W Alta. (N to Sheep Creek, ca. 54°N; CAN; see *A. laevigata*) to Calif. and Colo.; E Asia. MAPS and synonymy (together with a distinguishing key to the Eurasian *A. norvegica* Fries, from which it is scarcely separable): *see* below.
1   Heads 1.3–1.5(2) cm broad when pressed, all long-peduncled, relatively few in a
    corymbose-racemose inflorescence; leaves mostly more or less subpalmately divided;
    [Eurasia only; MAP: Eric Hultén, Nytt Mag. Bot. (Oslo) 3: fig. 8, p. 78. 1954] . . . .
    . . . . [*A. norvegica* Fries]
1   Heads averaging about 1 cm broad when pressed, mostly short-peduncled or sessile
    and more numerous in an erect, simple or compound raceme; leaves distinctly pinnate . . . .
    . . . . *A. arctica*
    2   Inflorescence dense, thick and spike-like, the heads normally all short-peduncled or
        sessile; plant low-grown.
        3   Pubescence of the raceme cinnamon-colour or brown; leaves permanently villous,
            their rachises narrow; [var. *villosa* Hult., not (Koidz.) Tatew.; Aleutian Is. and
            islands in the Bering Sea, the type from St. Paul Is., Alaska; MAPS: on the
            above-noted 1954 map by Hultén, and Hultén 1950: map 1162b, p. 1978, and
            1968b:906] . . . . ssp. *beringensis* Hult.
        3   Pubescence of the raceme greyish; leaves long-hairy, glabrate in age, their
            rachises broad; [*A. comata* Rydb.; coasts of Alaska (type of *A. comata* from
            Collison Point), the Yukon, and W Dist. Mackenzie; Mt. Selwyn, B.C.; cent. Alta.;
            MAP: on the above-noted 1954 map by Hultén, and Hultén 1950: map 1162c, p.
            1679, and 1968b:907] . . . . ssp. *comata* (Rydb.) Hult.
    2   Inflorescence open, at least the lower peduncles or raceme-branches elongate; plant
        relatively tall; [*A. chamissoniana* Bess. and its var. *saxatilis* Bess.; *A. ?laciniatiformis*
        Kom.; *A. laciniata* of American auth., perhaps not Willd.; *A. ?macrobotrys* Ledeb.; *A.
        norvegica* var. *pacifica* Gray; *A. saxicola* Rydb.; Aleutian Is. (type from Unalaska),
        Alaska, the Yukon, and W Dist. Mackenzie, and mts. of B.C. and SW Alta.; MAPS: on
        the above-noted 1954 map by Hultén, and Hultén 1968b:906; Raup 1947: pl. 36
        (aggregate species)] . . . . ssp. *arctica*

A. biennis Willd.   Biennial Wormwood
Apparently native in the W U.S.A. but described from New Zealand material; introd. in Europe and elsewhere outside its uncertain native limits, as in Alaska (N to Fairbanks, ca. 65°N), W Dist. Mackenzie (N to Norman Wells, 65°17′N; W.J. Cody, Can. Field-Nat. 74(2):98. 1960), and all the provinces except Nfld. (in Man., N to Churchill). MAP: Hultén 1968b:909.

A. campestris L.   Sagewort Wormwood
/aST/X/GEA/   (Hs (Ch))   Open sandy soil and dry slopes at low to fairly high elevations, the aggregate species from the coasts of Alaska–Dist. Mackenzie–Dist. Keewatin (in the Yukon, N to ca. 65°N) to Victoria Is., S Baffin Is. (N to near the Arctic Circle), northernmost Ungava–Labrador, and Nfld., S through all the provinces (probably extinct in P.E.I.) to N ?Calif., Ariz., Colo., Mich., S Ont., Vt., and Maine; W Greenland between ca. 64° and 74°N; NE Europe; Asia. MAPS and synonymy: *see* below.

1 Heads orange or reddish; disk-corollas to 3 mm long; involucres to 4 mm high and 6 mm broad, green or purplish, with up to over 45 florets, the phyllaries with narrow scarious whitish margins; panicle commonly narrow and spicate-thyrsoid; perennial, usually not over 3 dm tall, with well-developed basal rosettes, commonly with several decumbent-based stems from a short branching caudex; [the aggregate subspecies, as indicated by the following maps for *A. borealis*, transcontinental in chiefly arctic, subarctic, and alpine regions; MAPS: Hultén 1968*b*:910; Porsild 1957: map 318, p. 200; Böcher 1954: fig. 51 (top), p. 189; Fernald 1925: map 66, p. 339] . . . . ssp. *borealis* (Pall.) Hall & Clements

 2 Leaves, stems, and involucres minutely silky to glabrate or glabrous.

 3 Basal leaves with narrowly lanceolate or oblanceolate ultimate segments mostly not over 2 mm broad; [*A. borealis* Pall. and its f. *adamsii* Bess.; range of the subspecies] . . . . var. *borealis*

 3 Basal leaves relatively thick and with oblanceolate or oblong ultimate segments often 3 or 4 mm broad; [E Que. (Anticosti Is.), Labrador, and w Nfld.: Fernald *in* Gray 1950] . . . . var. *latisecta* Fern.

 2 Leaves and stems densely and more or less permanently silky; involucres and floral-axis usually villous; [*A. borealis* vars. *purshii* and *wormskioldii* Bess. and *besseri* T. & G.; *A. aleutica* Hult.; *A. groenlandica* Wormsk.; *A. spithamaea* Pursh; *A. desertorum* var. *hookeriana* Bess. in part (not *A. hookeriana* Bess.); range of the subspecies] . . . . var. *purshii* (Bess.) Cronq.

1 Heads yellow; involucres glabrous, their green phyllaries with whitish scarious margins; panicle more open, with strongly ascending elongate branches; basal leaves glabrous or sparsely silky and glabrate; stems usually taller.

 4 Involucres hemispheric, 3 or 4 mm high and up to 5 mm broad, with up to over 45 florets, the phyllaries with a narrow green centre and broad white scarious margins; disk-corollas to 3 mm long; perennial, commonly with several decumbent-based stems from a short branching caudex, well-developed basal rosettes usually present, the leaves of these with ultimate segments to about 2 mm broad; [*A. canadensis* Michx. (type from "Hudson Bay", probably from near L. Mistassini, Que., according to Dutilly and Lepage 1945) and its f. *dutillyanus* Rousseau, f. *pumila* and f. *rupestris* Vict. & Rousseau, and f. *peucedanifolia* (Juss.) Vict. & Rousseau; *A. caudata* vars. ?*majuscula* Vict. & Rousseau, *richardsoniana* (Bess.) Boivin (*A. desertorum* var. *rich.* Bess., not *A. rich.* Bess.), and *rydbergiana* Boivin; *A. maccalliae* Rydb.; essentially transcontinental but not yet known from Labrador, P.E.I., or N.S.; MAP (*A. can.*): Hultén 1968*b*:909] . . . . ssp. *canadensis* (Michx.) Scoggan

 4 Involucres subglobose, to 3 mm high and broad, with up to about 25 florets, their green phyllaries with narrow whitish scarious margins; disk-corollas to 2 mm long; biennial or short-lived perennial, the stems commonly solitary and lacking well-developed basal rosettes, the ultimate leaf-segments about 0.5 mm broad; [*A. caudata* Michx. and its vars. *calvens* Lunell and *douglasiana* (Bess.) Boivin (*A. desertorum* var. *doug.* Bess., not *A. douglasiana* Bess.); *A. bourgeauana* and *A. camporum* Rydb.; *A. forwoodii* Wats. and its var. *calvens* Lunell; *A. pacifica* Nutt.; *A. scouleriana* Bess.; *A. desertorum* var. *hookeriana* Bess. in part (not *A. hookeriana* Bess.); apparently essentially the range of ssp. *canadensis* but more southern] . . . . ssp. *caudata* (Michx.) Hall & Clements

A. cana Pursh Silvery Sagebrush
/T/WW/ (N) Plains, prairies, and slopes at low to fairly high elevations from Wash. to s Alta. (N to Castor, 52°13′N; CAN; reports from B.C. require confirmation; an isolated station in sw Yukon on a sandy flat at the s end of Kluane L., ca. 61°N, where taken by J.P. Anderson in 1944 and probably introd., the collection distributed as *A. bigelovii*), s Sask. (N to Whiteshore Lake, 52°08′N; CAN), and sw Man. (N to Millwood, about 85 mi NW of Brandon; also known from Melita, Oak Lake, and St. Lazare), s to N Calif., Utah, N.Mex., and Kans. [*A. columbiensis* Nutt.; *A. bigelovii sensu* Hultén 1950, not Gray, according to Hultén 1968*b*, the relevant above-noted Kluane L., the Yukon, collection in CAN]. MAPS: Hultén 1968*b*:900; G.H. Ward, Contrib. Dudley Herb. 4(6): fig. 9 (incomplete northwards), p. 190. 1953.

[A. douglasiana Bess.]
[The reports of this species of the w U.S.A. (Wash. to Calif.) from B.C. by John Macoun (1884; Vancouver Is., as *A. vulgaris* var. *californica*) and J.M. Macoun (1897), taken up by Henry (1915; *A. heterophylla*), are based upon *A. suksdorfii,* relevant collections in CAN. The report from s of Battleford, Sask., by John Macoun (1884; *A. lud.* var. *dougl.*) is based upon *A. herriotii* according to a revision by Rydberg of the collection in CAN, this species, however, perhaps best merged with *A. tilesii.* [*A. ludoviciana (vulgaris)* var. *dougl.* (Bess.) Eat. (not *A. desertorum* var. *dougl.* Bess., which is *A. campestris* ssp. *caudata* of the present treatment); *A. heterophylla* Nutt., not Bess. (which is *A. furcata* Bieb.)]. The MAP by Keck (1946: fig. 17, p. 455) indicates no Canadian stations.]

A. dracunculus L.   Tarragon
/sT/WW/   (Hp)   Prairies, plains, and rocky slopes at low to moderate elevations from s Alaska–Yukon (N to ca. 61°30′N; ?introd.) and B.C. to Alta. (N to Wood Buffalo National Park at 59°07′N), Sask. (N to Saskatoon), and Man. (N to Dawson Bay, N L. Winnipegosis; an isolated station near Toronto, Ont., indicated on Hultén's map, where perhaps introd.), s to Baja Calif., N Mexico, Tex., and Mo.; introd. eastwards to New Eng. [*A. aromatica* Nels.; *A. dracunculoides* Pursh; *A. glauca* Pall. and its var. *megacephala* Boivin; *A. nuttalliana* Bess.]. MAPS: Hultén 1968*b*:899; Porsild 1966: map 146, p. 85.

A. frigida Willd.   Prairie-Sagewort
/aST/WW/EA/   (Ch)   Dry plains, prairies, foothills, and roadsides from Alaska (N to ca. 69°30′N), the Yukon (N to ca. 65°N), and the coast of NW Dist. Mackenzie (Richards Is. and Liverpool Bay; CAN) to L. Athabasca (Alta. and Sask.) and Man. (N to Wekusko L., about 90 mi NE of The Pas), s to Ariz., Tex., Kans., Minn., and Wisc.; introd. eastwards, as in Ont. (N to Ingolf, near the Man. boundary w of Kenora), Que. (Aylmer, Gatineau Co.; Sacré-Coeur and Rimouski, Rimouski Co.; Gaspé Pen. at Métis, Méchins, Cap-Chat, Mont-St-Pierre, and Mont-Louis), and N.B. (Fairville, St. John Co.; GH); cent. E Europe; Asia. MAP: Hultén 1968*b*:904.

   A. rupestris L. of Eurasia, habitally similar to depauperate individuals of *A. frigida* (and keying out to it by reason of its long-hairy involucres and filiform-dissected leaves; differing in its dark-green rather than silvery-silky leaves, these merely glandular-punctate or with a few long hairs along the midrib, the stem-leaves 1-pinnate rather than mostly 2–3-ternate, the inflorescence racemose rather than paniculate, the whole plant markedly viscid-glandular), is reported from sw Yukon by J.A. Neilson (Can. Field-Nat. 82(2):116. 1968, with MAP, fig. 1, p. 115; Sheep Mt., near Kluane L., ca. 61°N; CAN). MAP: Hultén 1968*b*:908 (ssp. *woodii* Neilson).

A. furcata Bieb.
/ST/W/A/   (Hs)   Ledges and rocky or sandy slopes at low to high elevations from Alaska (N to ca. 69°N), sw Yukon (N to ca. 61°N), the coast of Dist. Mackenzie (E to Coronation Gulf), and w Victoria Is. through B.C. to Wash.; Asia. [*A. alaskana, A. hyperborea, A. minuta,* and *A. tyrrellii* Rydb.; *A. trifurcata* Steph.; *A. kruhsiana* Bess.; *A. ?caespitosa sensu* Hooker 1833, not Ledeb.; incl. var. *heterophylla* (Bess.) Hult. (*A. het.* Bess., not Nutt., which is *A. douglasiana* Bess.), the form with leaves more dissected but the segments relatively short]. MAPS (all but Hultén's as *A. hyperborea*): Hultén 1968*b*:910; Porsild 1966: map 147, p. 85, and 1957: map 317, p. 200; *Atlas of Canada* 1957: map 17, sheet 38.

A. globularia Bess.
/Ss/W/eA/   (Ch)   Rocky slopes of the Aleutian Is. and Alaska (N to ca. 69°30′N); NE Siberia (Chukch Pen.). MAPS: Hultén 1968*b*:897, and 1950: map 1168, p. 1679.

   Forma *lutea* (Hult.) Boivin (var. *lutea* Hult.; flowers yellow rather than white) is known from the type locality, Hall Is., Alaska.

A. glomerata Ledeb.
/aST/W/eA/   (Ch)   Sandy slopes of Alaska (N to the N coast) and N Yukon (a single station at ca. 69°N); E Asia. MAPS: Hultén 1968*b*:898, and 1950: map 1169, p. 1679.

   The glabrous extreme may be distinguished as var. *subglabra* Hult. (type from Alaska).

[A. herriotii Rydb.]
[This obscure species is reported from the type locality, Edmonton, Alta., and from Sask. by P.A. Rydberg (Bull. Torrey Bot. Club 37(9):455. 1910). However, Keck (1946) believes that it is merely a large-leaved form of *A. ludoviciana* whereas Breitung (1957a) merges it with *A. tilesii* ssp. *unalaschcensis*.]

[A. laevigata Standl.]
[Known only from the type locality along the Smoky R., cent. Alta., at ca. 56°10′N (P.C. Standley, Smithson. Misc. Collect. 56:2. 1912). The type has not been seen but may be merely the glabrous extreme of *A. arctica.*]

[A. lindleyana Bess.]
[This species of the w U.S.A. (Wash., Idaho, and Mont.) is tentatively reported from B.C. by John Macoun (1884; Chilcotin R.) and a collection from "New Westminster: Fort Hope" is placed here by Keck (1946; also, doubtfully, a collection from Sicamous, about 35 mi sw of Revelstoke). There are other so-named B.C. collections in Herb. V from Hell's Gate in the Fraser Canyon and from Fort Steele, near Cranbrook, but it is felt that further confirmation is required. MAP: Keck 1946: fig. 17, p. 455.]

A. longifolia Nutt.
/T/WW/ (Ch (Hp)) Dry open places (often alkaline) from Mont. to s Alta. (N to ca. 50°N; CAN), s Sask. (N to Yellow Grass, ca. 50°N; Breitung 1957a), and s Man. (N to Miniota, 50°10′N; CAN; doubtfully reported from w Ont. by Boivin 1966b, where possibly introd.), s to Wyo., Colo., and S.Dak. [*A. vulgaris* ssp. *long.* (Nutt.) Hall & Clements; *A. falcata* Rydb.; *A. integrifolia sensu* Pursh 1814, not L.]. MAP: Keck 1946: fig. 12 (incomplete northwards), p. 436.

A. ludoviciana Nutt.   Western Mugwort, White Sage
/T/WW/ (Hpr) Dry open places (ranges of Canadian taxa outlined below), s to Calif., Mexico, Tex., Ill., ?Mich., and s ?Ont. (possibly native; introd. northwards); introd. eastwards, as in Que., N.B., and P.E.I. MAPS and synonymy: *see* below.
1 Principal leaves entire or at most coarsely toothed or shallowly lobed.
   2 Leaves becoming glabrate and bright green at least above, entire or some of the
      lower ones falcate-lobed; [*A. vulgaris* ssp. *lud.* (Nutt.) Hall & Clements; reported by
      Keck 1946, from SE B.C. (Kootenay dist.), Alta. (N to Peace River and Grande Prairie),
      Sask. (Indian Head; Moose Jaw), and w Ont. (near Fort William); MAP: Keck 1946: fig.
      14, p. 441] .... var. *ludoviciana*
   2 Leaves permanently tomentose or lanate on both sides.
      3 Principal leaves broadly oblong (at most 4 times as long as broad), with a few
         coarse teeth around the summit, firm and thick; axillary branches short and
         suppressed; [*A. brittonii* Rydb.; s Ont.: Lambton Co.; Gaiser and Moore 1966] ....
         .... var. *brittonii* (Rydb.) Fern.
      3 Principal leaves narrower in outline (many times longer than broad) and softer,
         entire or with marginal falcate teeth.
         4 Leaves mostly folded lengthwise, widely spreading or recurving, often twisted;
            stem or its erect basal branches with suppressed axillary fascicles; [*A.
            rhizomata* var. *pab.* Nels. (*A. pab.* (Nels.) Rydb.); s B.C. (near Fairmont Hot
            Springs) to Alta. (Moss 1959), Sask. (N to North Battleford, Saskatoon, and
            Hudson Bay Junction), and Man. (N to Grand Beach); introd. in Ont. (Quetico
            Park; Thunder Bay) and sw Que. (Buckingham, Papineau Co.)] ....
            .... var. *pabularis* (Nels.) Fern.
         4 Leaves flat, mostly straight, ascending or spreading.
            5 Leaves lanceolate, acute or attenuate, ascending; stems commonly simple
               except for short axillary branches; [*A. gnaphalodes* Nutt.; *A. purshiana*
               Bess.; B.C. (N to Kamloops; CAN; introd. at Bennett, near the Yukon
               boundary, and in sw Dist. Mackenzie sw of Great Slave L.) to Alta., Sask.
               (N to Prince Albert and Rosthern), and Man. (apparently native N to
               Porcupine Mt.; introd. at Churchill); apparently introd. in Ont. (N to Quetico

Park, Thunder Bay, Sioux Lookout, Haileybury, and Ottawa), Que. (N to L. St. John, Baie-St-Paul, Charlevoix Co., Bic, Rimouski Co., and Matane, Matane Co.), N.B. (Bathurst; Sackville; Fredericton), and P.E.I. (Montague, Queens Co.); MAP: Hultén 1968*b*:899] .... var. *gnaphalodes* (Nutt.) T. & G.

  5 Leaves oblong to oblong-oblanceolate, blunt or merely acutish, loosely ascending or spreading; stems often with loosely spreading elongate branches; [*A. purshiana* vars. *?angustifolia* and *latifolia* Bess.; B.C. (Kootenay Valley; John Macoun 1884), ?Alta., ?Sask. (reported from several localities by Macoun 1884; not listed by Breitung 1957*a*), and Man. (N to Gillam, about 165 mi S of Churchill, where apparently introd. along a railway); introd. in Que. (Lac Deschênes, near Hull; l'Annonciation, Labelle Co.; near Montreal; L. St. John) and in N.B. (railway yard at Fairville, St. John Co.; GH)] .... var. *latifolia* (Bess.) T. & G.

1 Principal leaves more or less parted or divided; [B.C. and Alta.].
  6 Heads relatively large, the involucre 4 or 5 mm high and up to 7 mm broad; leaves white-tomentose on both sides; [ssp. *candicans* (Rydb.) Keck (*A. candicans* Rydb.); *A. latiloba* (Nutt.) Rydb.; *A. diversifolia* Rydb.; B.C. (Trail; Lytton; Spences Bridge) and Alta. (Coalhurst, near Calgary)] .... var. *latiloba* Nutt.
  6 Heads smaller, the involucre to about 3.5 mm high and broad.
    7 Panicle spike-like or racemose; disk with up to 45 florets; leaves green above or sometimes tomentose on both sides; [*A. atomifera* Piper; *A. incompta* Nutt.; *A. potens* Nels.; B.C. (Kootenay L.; CAN) and SW Alta. (Waterton Lakes; Breitung 1957*b*); MAP: Keck 1946: fig. 16 (ssp. *candicans* and ssp. *incompta*; not indicating any Canadian stations), p. 448] .... var. *incompta* (Nutt.) Cronq.
    7 Panicle broader, branching and leafy; disk with at most about 18 florets; leaves green above; [*A. mexicana* Willd.; *A. indica* var. *mex.* (Willd.) Bess.; *A. underwoodii* Rydb., this reported from B.C. by Rydberg 1922] ....
    .... var. *mexicana* (Willd.) Fern.

A. michauxiana Bess.
/sT/W/ (Ch) Rocky places, usually at rather high elevations, from S Yukon (Little Atlin L., near the B.C. boundary at ca. 60°22′N; CAN) through B.C. and SW Alta. (N to the Athabasca R. NE of Jasper at ca. 53°30′N; Keck 1946; there is a collection in CAN from Saskatoon, Sask., where reputedly taken by Malte in 1917 and probably introd., unless a confusion of labels is involved) to Calif., Utah, and Wyo. [*A. discolor* Dougl.; *A. graveolens* Rydb.]. MAPS: Hultén 1968*b*:905; Keck 1946: fig. 11, p. 434.

A. pontica L.   Roman Wormwood
European; introd. along roadsides and in old pastures and waste places in N. America, as in S Man. (Dauphin, N of Riding Mt.; DAO), S Ont. (N to Bruce, Simcoe, and Frontenac counties), SW Que. (N to the Montreal dist.), and N.S. (Dartmouth; Sydney).

A. senjavinensis Bess.
/S/W/eA/ (Ch) Coastal rocks of the Seward Pen., W Alaska; NE Siberia. [*A. androsacea* Seem.]. MAPS: Hultén 1968*b*:898, and 1950: map 1173, p. 1680.

A. stelleriana Bess.   Dusty Miller, Beach-Wormwood
Asiatic (but probably native at Shemya Is., W Aleutian Is.; *see* Hultén 1968*a*); according to Fernald *in* Gray (1950), originally spread from cult. in N. America about 1880. It occurs along sandy shores of the Great Lakes in S Ont. (inclined to escape and persist in Lambton Co.; Dodge 1915), Minn., and N.Y., and of the Atlantic Seaboard from E Que. (St. Lawrence R. estuary from St-Roch-des-Aulnets, l'Islet Co., to the Gaspé Pen. and Magdalen Is.), N.B., P.E.I., and N.S. to Va. Boivin (1966*b*) also reports it from B.C. MAP: Hultén 1968*b*:901.

A. suksdorfii Piper
/t/W/ (Hp) Bluffs and rocky or sandy beaches from SW B.C. (collections in CAN, verified by Rydberg, from Vancouver Is. and the adjacent mainland E to the Chilliwack R.) to Calif. [*A. vulgaris*

var. *litoralis* Suksd.; *A. vulgaris* var. *californica sensu* John Macoun 1884, not Bess., which is *A. douglasiana* Bess., not definitely known from B.C.; *A. heterophylla* Nutt. in part, not Bess., which is *A. furcata*]. MAP: Keck 1946: fig. 17, p. 455.

A. tilesii Ledeb.
/aST/(X)/EA/ (Hpr) Open rocky or gravelly places at low to rather high elevations, the aggregate species from the Aleutian Is. and coasts of Alaska–Yukon–Dist. Mackenzie to N Banks Is., Dist. Keewatin (N to ca. 67°N; CAN), NE Man. (Nelson R. about 20 mi SW of York Factory to Churchill; CAN), and NE Ont. (James Bay coast N to ca. 53°N; Dutilly, Lepage, and Duman 1954; collection in CAN, verified by Rydberg, from the mouth of the Moose R. at ca. 51°20′N, where taken by Spreadborough in 1904; the report from w-cent. Que. by Boivin 1966*b*, may be based upon the report by John Macoun 1884, of a collection by Bell on the "East coast of Hudson Bay", the relevant 1879 collection in CAN, however, bearing the annotation by Malte, "Dr. Bell was on the West coast of Hudson Bay in 1879."), S in the West through B.C. and N Alta. (near Carcajou, ca. 57°50′N; Athabasca R. at ca. 56°N; not known from Sask.) to Oreg. and Mont.; NE Europe; N Asia. MAPS and synonymy: see below.

1   Upper leaves lobed; [var. *arctica* Bess.; *A. vulgaris* var. *til.* Ledeb.; *A. gormanii* Rydb.; N part of the N. American area; MAP: combine the maps by Hultén 1968*b*:901 and 902 (ssp. *gormanii*)] . . . . var. *tilesii*
1   Upper leaves entire; [*A. el.* (T. & G.) Rydb.; *A. obtusa* and *A. unalaskensis* Rydb. (the latter not based on var. *unalaschcensis* Bess.); *A. unalaskensis* var. *aleutica* Hult., not *A. aleutica* Hult.; *A. hookeriana* Bess.; *A. ludoviciana* var. *americana* (Bess.) Fern., at least as to the type of *A. vulgaris* var. *amer.* Bess.; *A. ?diversifolia sensu* Dutilly, Lepage, and Duman 1958, not Rydb.; essentially the N. American area of the typical form but slightly less northern; MAPS: combine the maps by Hultén 1968*b*:902 (ssp. *elatior*) and 903; Keck 1946: fig. 18; W.J. Cody, Nat. can. (Que.) 98(2): fig. 15, p. 149. 1971] . . . . . . var. *elatior* T. & G.

A. tridentata Nutt.   Common Sagebrush
/t/WW/ (N) Dry plains and hills from S B.C. (Dry Interior N to Kamloops and Alkali L., about 110 mi NW of Kamloops, E to the Flathead R. near the Alta. boundary) and SW ?Alta. (the report by Moss 1959, requires confirmation) to Baja Calif., N.Mex., and N.Dak. [Var. *angustifolia* Gray; *A. angusta* Rydb.]. MAP: G.H. Ward, Contrib. Dudley Herb. 4(6): fig. 6, p. 173. 1953.

Most of the B.C. material extending above the 4,000-ft elevation is referred by L.S. Marchand, Alastair McLean, and E.W. Tisdale (Can. J. Bot. 44(12): fig. 1, p. 1625. 1966) to the montane ecotype, ssp. *vaseyana* (Rydb.) Vasey (*A. vas.* Rydb.; involucres to 5 mm high and 4 mm broad rather than at most about 4 mm high and 2 mm broad, with about 12 phyllaries rather than about 15, the plant with a spreading-decumbent rather than erect growth-form and flowering relatively early). The species "has appeared casually in an old field in e. Mass." (Fernald *in* Gray 1950).

A. tripartita Rydb.
/t/W/ (N) Dry plains and hills, often in somewhat moister or more favourable sites than the similar *A. tridentata,* from S B.C. (chiefly in the Dry Interior N to Lillooet and Cache Creek, W of Kamloops) and Mont. to Calif. and Colo. [*A. trifida* Nutt., not Turcz.].

A. vulgaris L.   Common Mugwort.   Herbe Saint-Jean
Eurasian; introd. along roadsides and in thickets and waste places in N. America, as in B.C. (N to Smithers, ca. 54°30′N), Alta. (Waterton Lakes; Groh 1944*a*), Sask. (Hoosier, about 120 mi N of Swift Current; Breitung 1957*a*), S Man. (Brandon, where taken by John Macoun in 1896), Ont. (N to Matheson, 48°42′N), Que. (N to Anticosti Is. and the Gaspé Pen.), Nfld., N.B., P.E.I., and N.S.; SW Greenland.

Var. *glabra* Ledeb. (the narrowly lance-acuminate divisions of the principal leaves entire rather than cut-toothed or lacerate) is known from Ont. (Ottawa dist.) and N.S. (Amherst, Cumberland Co.). Var. *latiloba* Ledeb. (leaves relatively thin, with a broadly obovate or rhombic terminal segment and oblong or oblanceolate lateral ones, the segments and their few teeth blunt or merely acutish, the leaves of the typical form cleft nearly to midrib into lance-acuminate segments) is reported from Que. by Fernald *in* Gray (1950).

ASTER L.   [8900]   Aster

1 At least the basal or lower stem-leaves cordate or subcordate and distinctly petioled; perennials with strong caudices, crowns, or creeping rhizomes.
   2 Involucres and peduncles copiously glandular; inflorescence a roundish-topped corymb with relatively few bracts, these large and leaf-like; phyllaries greenish, firm, well imbricated, the outer ones to 2.5 mm broad; rays violet or pale blue; achenes linear, scarcely compressed; rhizomes elongate; (s Man. to N.S.) . . . . *A. macrophyllus*
   2 Plants nonglandular.
      3 Inflorescence corymbiform, roundish-topped, with relatively few bracts, these large and leaf-like; phyllaries in several unequal series; achenes linear, scarcely compressed; rays white; rhizomes commonly elongate.
         4 Large tufted basal leaves abundant on separate short shoots, tending to have broad, more or less rectangular sinuses; involucre slenderly cylindric, to 1 cm high, its phyllaries loosely ascending, the inner ones much prolonged; (s ?Ont.) . . . . [*A. schreberi*]
         4 Large tufted basal leaves on separate shoots rarely produced; leaf-sinuses narrower; involucre ovoid-campanulate, to about 8 mm high, its phyllaries appressed and more gradually increasing in length from row to row; (s Ont.) . . . . *A. divaricatus*
      3 Inflorescence paniculate or racemose, its bracts often more numerous but either narrow or very small or both; achenes flattened, linear to obovate-oblong; rays often coloured; plants rarely colonial (except *A. ciliolatus*).
         5 Principal leaves cordate-clasping or with downwardly-dilated clasping petioles, entire or low-toothed, loosely pilose beneath; rays commonly lilac or blue; achenes minutely pubescent at least above; plant pale with dense minute pubescence; (s Ont. to N.S.) . . . . *A. undulatus*
         5 Leaves not clasping; achenes nearly or quite glabrous.
           6 Leaves entire or occasionally very shallowly serrate, the basal ones much longer than broad; phyllaries with short broad green tips; (s Ont.).
              7 Phyllaries glabrous or merely ciliate; rays commonly deep blue; leaves strongly scabrous on both surfaces, only the lowermost ones definitely cordate . . . . *A. azureus*
              7 Phyllaries minutely pubescent on the back; rays commonly pale violet or sometimes roseate; leaves smooth or slightly scabrous above, glabrous or pubescent beneath, nearly all of the principal ones cordate . . . . *A. shortii*
           6 Principal leaves distinctly toothed.
              8 Inflorescence with relatively few heads, the very unequal peduncles naked or with only 1 or 2 bracts; phyllaries with long narrow green tips; rays commonly pale blue; leaves glabrous, the lower ones broadly lanceolate to narrowly ovate, on winged petioles; (transcontinental) . . . . *A. ciliolatus*
              8 Inflorescence often with more than 100 heads, each well-developed peduncle commonly with several bracts; rays white or pinkish to pale blue-violet.
                9 Phyllaries loosely ascending, linear-attenuate or subulate, about 0.5 mm broad, with an obscure slender green median line; leaves commonly scabrous above, otherwise glabrous, the lower ones lance-ovate, on winged petioles; (se ?Man. and Ont.) . . . . *A. sagittifolius*
                9 Phyllaries closely appressed, the outermost ones linear-oblong and up to 1 mm broad, blunt or short-tipped, with a distinct dilated green median band.
                  10 Leaves narrowly ovate, very smooth, rather fleshy, very shallowly toothed, pale beneath when fresh, the principal ones on broadly winged petioles; (Ont. and Que.) . . . . *A. lowrieanus*

          10 Leaves broadly ovate, glabrous or sparingly pubescent, sharply
              serrate, on wingless petioles; (Man. to N.S.) .... *A. cordifolius*

1 None of the leaves both cordate and distinctly petioled.
  11 Head usually solitary on the stem.
     12 Leaves (at least the principal ones) usually distinctly toothed, lanceolate to
       oblong-elliptic or oblanceolate, nearly or quite sessile (or the lowermost ones
       short-petioled); stems from slender creeping rhizomes and stolons.
       13 Phyllaries (and top of peduncle) more or less white-villous or -woolly, loose,
         acute or acuminate, chartaceous below, commonly purple-tinged; rays to
         about 12 mm long; achenes pubescent; stems commonly 1 or 2 dm tall (to
         about 4 dm); (mts. of B.C. and Alta.) .... *A. sibiricus*
       13 Phyllaries puberulent or subglabrous (or their margins ciliate) but neither
         villous nor woolly; stems commonly taller; (Que. eastwards).
         14 Phyllaries firm, obtuse, with a broad green band at the summit above a
           pale coriaceous base; rays about 1 cm long; achenes glabrous ...... *A. radula*
         14 Phyllaries thin and flexible, acute or acuminate, often purple-tinged; rays to
           over 1.5 cm long; achenes glandular-puberulent .... *A. blakei*
     12 Leaves nearly or quite entire (*A. sibiricus* may key out here); achenes pubescent.
       15 Stems arising singly from the tips of long slender rhizomes and stolons,
         slender and very leafy, to about 9 dm tall; leaves linear to narrowly lanceolate,
         sessile, their scabrous margins revolute; phyllaries very narrowly linear to
         linear-lanceolate, sharp-pointed, often purple-tinged, in several series of
         unequal length; rays lilac-purple; achenes glandular-puberulent; (Ont. to s
         Labrador, Nfld., and N.S.) .... *A. nemoralis*
       15 Stems arising from a woody caudex or short rhizome, usually less than 5 dm
         tall; achenes hairy but not glandular; (B.C. and Alta.).
         16 Principal leaves long-petioled in basal tufts, the sessile stem-leaves much
           reduced; phyllaries oblong-linear, in subequal series; rays commonly white
           (sometimes lavender or violet); achenes flattened, 2-nerved; stems to
           about 3 dm tall; (mts. of B.C. and Alta.) .... *A. alpinus*
         16 Leaves all or chiefly cauline; phyllaries linear or lance-linear; achenes
           mostly several-nerved; (s B.C.).
          17 Rays white; phyllaries in subequal series; pappus simple or with some
            short outer bristles; leaves pinnately veined, to 4 cm long and 13 mm
            broad; plant rather copiously glandular-puberulent, to about 5 dm tall;
            (sw B.C.) .... *A. paucicapitatus*
          17 Rays lilac to violet or blue; phyllaries in several series of unequal
            length; pappus distinctly double (with an outer series of very short
            bristles); leaves with a prominent midrib, otherwise nerveless,
            scabrous-puberulent.
           18 Leaves very firm, commonly not over 1.5 cm long and 3 mm broad;
             achenes densely silky; stems usually not over about 1 dm tall;
             (?B.C.) .... [*A. scopulorum*]
           18 Leaves to about 4 cm long and 4 mm broad, laxer; achenes less
             densely silky; stems to about 3 dm tall; (s B.C.) .... *A. stenomeres*
  11 Heads usually 2 or more on each stem.
     19 Heads more or less discoid in appearance, the white or pinkish rays at most 3 mm
       long; achenes pubescent; plants essentially glabrous except for the commonly
       ciliate leaf-margins.
       20 Perennial from slender creeping rhizomes, the stem (to about 3 dm tall)
         usually simple and unbranched up to the compact terminal cluster of heads;
         rays white, to 3 mm long, rarely more than 3 in each head; leaves
         oblanceolate, to 3.5 cm long and 9 mm broad; (sw B.C.) .... *A. curtus*
       20 Annuals or biennials with fibrous roots; inflorescence commonly open and
         ample; rays to 2 mm long, white or pinkish, more numerous than the
         disk-florets.

21 Rays about 2 mm long, surpassing the short style; phyllaries obtuse or acutish; leaves seldom over 6 cm long, the upper ones linear and sessile, the lower ones oblanceolate and short-petioled; stems to over 1 m tall; (s ?B.C.) .... [*A. frondosus*]

21 Rays less than 1 mm long or wanting, the tubular corolla of the marginal flowers shorter than the style; stems usually less than 7 dm tall; (transcontinental) .... *A. laurentianus*

19 Heads distinctly radiate, the rays usually at least 4 mm long, the ray-florets mostly fewer than the disk-florets.

22 Annuals or biennials (sometimes short-lived perennials) with a short taproot or an erect taproot-like caudex, lacking rhizomes or numerous fibrous roots; rays blue or purplish; puberulent appendages terminating the style-branches equalling or longer than the stigmatic portion.

23 Leaves pinnately incised to 3-pinnatifid, glandular and hairy; heads terminating leafy branches, the disk relatively large (at least 1 or 2 cm broad); (s Alta.) .... *A. tanacetifolius*

23 Leaves entire or remotely salient-toothed, spatulate-oblanceolate or the upper ones linear.

24 Plant glabrous; phyllaries subequal, linear-subulate or attenuate, thin and scarious-margined, the outer ones herbaceous; rays 4 or 5 mm long, rolled outwards; leaves entire, to 1.5 dm long; stem to 1 m tall; (s Ont.; NE N.B.) .... *A. subulatus*

24 Plant cinereous-puberulent; phyllaries in 5 or 6 unequal series, lanceolate, obtuse to short-acuminate, canescent or glandular (var. *viscosus*), thick and firm, their green tips spreading or reflexed; rays to 1 cm long; leaves entire to remotely salient-toothed (the teeth spinulose-tipped), commonly not over 3 cm long; stem usually less than 5 dm tall; (s B.C. to s Sask.) .... *A. canescens*

22 Perennials, fibrous-rooted and often with creeping rhizomes or branching caudices; puberulent style-appendages in most species shorter than the stigmatic portion.

25 Involucres and peduncles glandular; achenes pubescent; inflorescence corymbiform, roundish-topped.

26 Rays white (or fading to pinkish), commonly not more than 13 in each head; phyllaries tending to be keeled; lower leaves reduced and bract-like; plants fibrous-rooted from a caudex or stout rhizome, lacking well-developed creeping rhizomes.

27 Leaves to over 1 dm long, up to 6 times as long as broad; rays to about 2.5 cm long; phyllaries in several series of unequal length; heads often rather numerous; plant to about 1.5 m tall; (B.C. and Alta.) .... *A. engelmannii*

27 Leaves mostly less than 5 cm long and 1.5 cm broad; rays to 1.5 cm long; phyllaries in subequal series; heads few (or solitary); plants to about 5 dm tall; (sw B.C.) .... *A. paucicapitatus*

26 Rays normally violet to blue-purple, commonly at least 15; phyllaries scarcely keeled; lower leaves, if reduced, scarcely bract-like and often deciduous; stems commonly from creeping rhizomes.

28 Leaves coarsely and sharply serrate, thick and firm, elliptic to ovate (or the lower ones obovate), to over 1.5 dm long and about 8 cm broad; phyllaries in several series of unequal length; heads few to many in an open-corymbiform inflorescence; plants to about 1 m tall; (B.C. to Sask.) .... *A. conspicuus*

28 Leaves entire (or sometimes remotely serrulate in *A. modestus*).

29 Leaves linear to narrowly oblong, to about 1 dm long and commonly not over 5 or 6 mm broad; involucre to 8 mm high, its phyllaries in 2 or 3 series of more or less unequal length.

30 Leaves linear to linear-oblanceolate, rather fleshy, the upper ones greatly reduced; rays blue or whitish, to about 6 mm long; heads few; plant glabrous; (saline or alkaline soils from s Dist. Mackenzie and Alta. to Man.) . . . . *A. pauciflorus*

30 Leaves linear to narrowly oblong, firm, the upper ones only gradually reduced; rays violet or purplish, to 12 mm long; heads commonly rather numerous; plants usually copiously but minutely glandular-scabrous; (s B.C. and s Alta.) . . . . *A. campestris*

29 Leaves mostly lanceolate to oblong or oval, commonly cordate-clasping; involucre often over 8 mm high.

31 Phyllaries in several series of markedly unequal length, firm, scarious, with a conspicuous green mid-band; heads on elongate leafy-bracted peduncles; leaves thick and firm, oblong to oval, blunt or merely mucronate-tipped, scabrous-puberulent; (s ?Ont.; ?N.B.) . . . . [*A. patens*]

31 Phyllaries in subequal (or at least not strongly unequal) series; leaves lanceolate to oblong.

32 Leaves strongly cordate-auriculate-clasping, lanceolate, acuminate, scabrous or stiffly appressed-hairy above, more softly hairy beneath, dryish in texture; heads on short naked or few-bracted peduncles; stems spreading-hirsute; (s Man. to N.S.) . . . . *A. novae-angliae*

32 Leaves only half-clasping; heads on relatively long leafy-bracted peduncles.

33 Leaves thin, narrowly to broadly lanceolate, acuminate, essentially glabrous, the lower ones often remotely but sharply low-serrate; stem villous to glabrate; (B.C. to Ont.) . . . . *A. modestus*

33 Leaves hard and harshly scabrous, nearly linear to oblong, entire; stems commonly bushy-branched, cinereous-pilose or -hispid; (s Ont; ?introd.) . . . . *A. oblongifolius*

25 Involucres and peduncles (as well as the rest of the plant) lacking obvious glands.

34 Phyllaries firm, most of them (and most of the firm leaves) tipped with a short whitish spine distinctly differentiated from the blade in colour and texture.

35 Leaves silvery-silky on both surfaces, hard, lanceolate to elliptic or oblong (the oblanceolate basal ones soon deciduous), entire, sessile, to about 4 cm long and 1 cm broad; involucre to 1 cm high, its lance-acuminate silky phyllaries subequal, passing insensibly into the reduced upper leaves; heads several or numerous, corymbose-paniculate, often clustered at the ends of the branches; rays purple-violet, becoming blue, to 1.5 cm long; achenes glabrous; pappus tawny, becoming reddish brown; (SE Man. to w Ont.) . . . . *A. sericeus*

35 Leaves not silvery-silky, usually relatively narrower, many of them commonly subtending reduced tufts or reduced sterile branches; involucres to 8 mm high (often not over 4 or 5 mm); rays rarely over 1 cm long; achenes more or less pubescent.

36 Phyllaries (at least the outer) and leaves bristly-ciliate and commonly with copiously hispid surfaces; primary leaves scarcely narrowed to the sessile base; stems copiously hirsute to hoary-puberulent; plant often slenderly stoloniferous; (B.C. to Que.) . . . . *A. ericoides*

36 Phyllaries glabrous, acute, subequal (or with an outer much shorter series); primary and upper leaves gradually tapering to a petiolar base; stem glabrous or merely pilose; plants nonstoloniferous.

    37 Involucre commonly about 4 mm high and 3 mm broad; disk-flowers rarely more than 12; rays white, 4 or 5 mm long; plant pilose; (s Ont.; introd. in N.S.) . . . . *A. parviceps*

    37 Involucre to about 8 mm high and 6 or 7 mm broad; disk-flowers commonly at least 20; rays white, sometimes purple, to 1 cm long; (s Ont to N.S.) . . . . *A. pilosus*

34 Phyllaries and leaves sometimes subulate-tipped but lacking a distinctly differentiated whitish short terminal spine.

    38 Phyllaries hard or firm, usually pale-coriaceous at base, mostly green-tipped and often with a broad green median band, usually in several markedly unequal series; inflorescence commonly corymbiform; leaves usually firm.

        39 Phyllaries (and tops of peduncles) more or less white-villous or -woolly, loose, acute or acuminate, chartaceous below, commonly purple-margined; rays purple, to 12 mm long; leaves to about 8 cm long and 2.5 cm broad, entire or serrate; stems usually less than 3 dm tall; (mts. of B.C. and Alta.) . . . . . . *A. sibiricus*

        39 Phyllaries puberulent or subglabrous (but their margins ciliate or erose-fimbriate), usually not purple-margined (except commonly in *A. linariifolius* and *A. radulinus*).

            40 Leaves linear to narrowly oblong (the lower ones slightly broader), entire; stems commonly less than 5 dm tall.

                41 Achenes glabrous; pappus bright white; involucre to 7 mm high, its lance-subulate to lance-oblong coriaceous acutish phyllaries with a relatively long green area on the back of the thickened midrib; rays white or yellowish, to 8 mm long; peduncles naked or with 1 or 2 scale-like bracts; (s Sask. to N.B.) . . . . *A. ptarmicoides*

                41 Achenes copiously appressed-silky; leaves linear or the lowest ones very narrowly oblanceolate.

                    42 Phyllaries cartilaginous, rather broadly oblong, yellowish except for a small green spot at the rounded erose-fringed tip; involucre to 7 mm high; rays white, to 1 cm long, not more than 6; pappus white, a single series of capillary bristles; heads numerous, mostly clustered at the ends of the floral-branches, the leaves of these reduced to more or less scale-like bracts; leaves moderately scabrous-margined; (sw ?Que.) . . . . [*A. solidagineus*]

                    42 Phyllaries firm, with a relatively long green area on the back of the thickened midrib, ciliate-fringed on the commonly purplish margins; rays normally blue-violet, to 1.5 cm long, numerous; pappus tawny, double, the outer series of bristles about 1 mm long; heads solitary or few on leafy-bracted peduncles; leaves minutely scabrous and strongly scabrous-margined, nerveless except for the prominent midrib; (s Que. and N.B.) . . . . *A. linariifolius*

            40 Leaves lanceolate to oblanceolate, ovate, or oblong, entire or serrate; rays commonly at least 1 cm long; pappus usually tawny and becoming reddish brown; stems commonly taller.

                43 Achenes nearly or quite glabrous.

44 Leaves thick and rigid, tending to be glaucous, entire or obscurely and remotely serrate, narrowly lanceolate to ovate, their margins smooth or scabrous, the middle and upper leaves auriculate-clasping, the lower ones on long winged petioles; phyllaries acute or acuminate, closely appressed; plant nearly or quite glabrous, from a short stout rhizome or branching caudex, occasionally also with short creeping red rhizomes; (transcontinental) . . . . *A. laevis*

44 Leaves green, firm but relatively thin, sharply serrate or crenate in the middle, scabrous on both faces, rugose-veiny, scabrous-margined, oblong-lanceolate, acute, all sessile but scarcely clasping, the lower ones commonly deciduous; tips of phyllaries often spreading; plant more or less minutely hispid, from a cord-like rhizome, producing long subterranean stolons; (Que. eastwards) . . . . . . *A. radula*

43 Achenes copiously hairy.

45 Phyllaries tending to be keeled, their tips purple-margined around a narrow green mid-band; rays white (or fading pinkish), commonly less than 15, to 2.5 cm long; leaves entire or nearly so, not clasping, subglabrous, the lower ones usually reduced and bract-like; stems from a fibrous-rooted caudex or a stout rhizome; (B.C. and Alta.) . . . . *A. engelmannii*

45 Phyllaries green or with a broad green mid-band near summit; rays commonly more than 15; lower leaves not bract-like; stems usually from slender creeping rhizomes.

46 Leaves mostly strongly serrate, firm, not clasping, sparsely to copiously hispid-puberulent; phyllaries often strongly purple-tinged, with a prominent midrib, in several unequal series; rays white to purple, to 12 mm long . . . [*A. radulinus*]

46 Leaves entire or remotely low-serrate, relatively thin, the upper ones usually more or less clasping; phyllaries less commonly purple-tinged; rays rose-purple to blue or violet.

47 Phyllaries narrow or broad, often in strongly unequal series, the outer ones scarious-margined near the chartaceous yellowish or brownish base; rays to about 2 cm long; leaves glabrous or pubescent, commonly low-serrate; (transcontinental) . . . . *A. subspicatus*

47 Phyllaries linear or oblong, in scarcely to evidently unequal series, often green nearly throughout, the outer ones scarcely scarious-margined; leaves mostly glabrous and entire; (Dist. Mackenzie, B.C., and w Alta.) . . . . *A. occidentalis*

38 Phyllaries thin and scarious or foliaceous, pliable and not markedly coriaceous toward base; leaves mostly relatively thin and soft.

48 Involucres copiously white-villous or -woolly, to about 1 cm high, the loose purplish phyllaries broadly linear, acute or acuminate; rays lilac, blue, or purple; leaves rather firm; stem

spreading-pubescent, from a creeping rhizome; (western
subarctic, arctic, and alpine regions) . . . . *A. sibiricus*
48 Involucres glabrous or only sparingly pubescent, not at all
woolly, the phyllaries normally not markedly purplish (margins
and tips often purplish in *A. borealis*).
49 Inflorescence distinctly corymbiform (roundish- or flattish-
topped; or the head sometimes solitary); phyllaries lacking
herbaceous tips . . . . *GROUP 1*
49 Inflorescence paniculate or corymbiform-paniculate, not
distinctly corymbiform; achenes glabrous or more or less
pubescent but not glandular; pappus a single series of
capillary bristles . . . . *GROUP 2*

GROUP 1

1 Pappus double, the inner series consisting of capillary bristles, the outer series short;
achenes more or less pubescent but not glandular; heads numerous; rays normally white,
to about 8 mm long; leaves pinnate-veined, narrowly to broadly elliptic or elliptic-ovate, to
over 1.5 dm long; stem to about 2 m tall; (Alta. to Nfld. and N.S.) . . . . *A. umbellatus*
1 Pappus a single series of capillary bristles; achenes minutely glandular, otherwise
glabrous; stem to about 1 m tall.
2 Leaves entire, linear to narrowly lanceolate, blunt or acutish, very numerous, to about
6 cm long and 1 cm broad, their margins commonly revolute; heads solitary or
several; rays lilac-purple; (Ont. to s Labrador, Nfld., and N.S.) . . . . *A. nemoralis*
2 Leaves more or less toothed.
3 Rays lilac-purple; leaves lanceolate to oblong, acute or short-acuminate,
low-serrate, to about 2.5 cm broad; (Que., Nfld., N.B., and N.S.) . . . . *A. blakei*
3 Rays white or purple-tinged; leaves oblong-lanceolate to -oblanceolate or broadly
oval, coarsely serrate, to about 6 cm broad; (Ont. to Nfld. and N.S.) . . . . *A. acuminatus*

GROUP 2

1 Principal stem-leaves (or their petioles) more or less auriculate-clasping at base;
inflorescence corymbiform-paniculate; rays normally blue-violet.
2 Leaves tapering or gradually rounded to the auriculate-clasping base, not abruptly
contracted below; (Dist. Mackenzie and Alta. to Labrador, Nfld., and N.S.) . . . . *A. puniceus*
2 At least the lower leaves abruptly contracted below the middle into a winged clasping
petiole; achenes usually pubescent.
3 Leaves only half-clasping, entire or low-serrate; (transcontinental) . . . . . *A. subspicatus*
3 Leaves strongly auriculate-clasping, appearing almost perfoliate at base, the
blade sharply and deeply serrate, lanceolate to narrowly ovate, long-acuminate,
scabrous above, smooth beneath; heads corymbose; rays blue-violet or white; (s
Ont.) . . . . *A. prenanthoides*
1 Principal stem-leaves scarcely or not at all auriculate-clasping (if half-clasping, not
abruptly contracted to a winged petiole).
4 Rays normally white; involucres less than 7 mm high, their phyllaries in several
unequal series.
5 Leaves pubescent beneath at least along the midrib, linear-lanceolate to
subrhombic; lobes of disk-corollas about half (or more) the length of the free limb;
rays to 7.5 mm long; (SE Man. to N.S.) . . . . *A. lateriflorus*
5 Leaves glabrous beneath; lobes of disk-corollas less than half as long as the free
limb.
6 Heads numerous and often on 1-sided racemose branches; rays to 6 mm
long.
7 Leaves of branches rarely over 1.5 cm long; involucres mostly less than 4
mm high; (Ont. and sw Que.) . . . . *A. vimineus*

       7   Leaves of branches often over 1.5 cm long; involucres mostly over 4 mm
          high; (Sask. to Nfld. and N.S.) . . . . *A. simplex*
    6   Heads seldom in a 1-sided arrangement.
       8   Rays to 8 mm long; stems rarely over 6 dm tall; (Ont. to s Nfld. and N.S.) . . . .
          . . . . *A. tradescantii*
       8   Rays to 6 mm long; stems to about 1.5 m tall; (Sask. to Nfld. and N.S.) . . . .
          . . . . *A. simplex*
4   Rays normally violet, bluish, or purple (usually pink or white in *A. eatonii* and
   sometimes so in *A. hesperius* and *A. borealis*).
    9   Phyllaries commonly subequal (the outer ones, if noticeably shorter, with more or
       less recurving tips).
      10 Outer phyllaries broadly foliaceous, to 5 mm broad, usually minutely reticulate
         like the foliage-leaves, equalling or surpassing the inner ones.
         11 Inflorescence long and narrow, leafy, with many heads; leaves mostly
            more than 7 times as long as broad; rays usually pink or white; (s B.C. to
            sw Sask.) . . . . *A. eatonii*
         11 Inflorescence either few-flowered or shorter and more open, its leaves
            commonly reduced; middle leaves mostly less than 7 times as long as
            broad; rays normally blue or violet; (transcontinental) . . . . *A. subspicatus*
      10 Outer phyllaries narrower, only obscurely reticulate; (Ont. to s Labrador and
         the Atlantic Provinces) . . . . *A. novi-belgii*
    9   Phyllaries in 3 or 4 series of unequal length (sometimes subequal in *A.
       hesperius*); involucres to 8 mm high.
      12 Stems stout, to over 2 m tall; leaves typically lanceolate to lance-elliptic, entire;
         (Man. to Que.) . . . . *A. praealtus*
      12 Stems slender, less than 1 m tall; leaves linear to lance-linear.
         13 Branchlets with several to many crowded, firm, much reduced small
            leaves; primary leaves spreading or reflexed, bluntish or abruptly
            mucronate; heads terminating copiously small-bracted branchlets or
            peduncles; rays pale lavender or bluish; (s Ont.) . . . . *A. dumosus*
         13 Branchlets bearing few and scarcely reduced leaves; rays blue, pink, or
            white.
           14 Plants very slender, chiefly of cold bogs; principal leaves linear-
              attenuate, entire or subentire, harshly scabrous on the margins,
              usually slightly clasping, mostly less than 5 mm broad; inner phyllaries
              linear-attenuate, with a slender green midrib; stem commonly less than
              8 dm tall, glabrous below, puberulent in lines above, solitary from very
              slender elongate stolons seldom over 2 mm thick; (Alaska–B.C. to
              N.S.) . . . . *A. borealis*
           14 Plants stouter; leaves broader and blunter, to over 2 cm broad;
             phyllaries broader and with broader green midribs; stolons thicker.
             15 Stem (to about 1.5 m tall) and branches pubescent in lines
                decurrent from the leaf-bases; leaves glabrous or merely scabrous,
                entire or toothed, linear to broadly lanceolate, to 2.5 cm broad;
                phyllaries subequal or more or less imbricate; (sw Dist.
                Mackenzie–B.C. to Sask.) . . . . *A. hesperius*
             15 Stem (to about 1 m tall) pubescent at least above, the pubescence
                more uniformly distributed; leaves commonly more or less
                pubescent, mostly entire, the principal ones linear, mostly less than
                1 cm broad; phyllaries in distinctly unequal series; (Alta. and Sask.;
                introd. ?eastwards) . . . . *A. chilensis*

A. acuminatus Michx.
/T/EE/ (Hpr) Dry or moist woods and clearings from Ont. (N to the Ottawa dist.) to Que. (N to L.
St. John, the Côte-Nord, Anticosti Is., and Gaspé Pen.), s Nfld. (GH), N.B., P.E.I., and N.S., s to E
Tenn. and Ga. MAPS: Dansereau 1957: map 5A, p. 35; R.B. Pike, Rhodora 72(792): fig. 8, p. 424.
1970.

Forma *discoideus* Ktze. (f. *virescens* Vict. & Rousseau, a presumed pathological phase with the florets modified into tufts of chaff rather than normal, the receptacle naked) is known from SE Que. (Sully, Temiscouata Co.; CAN; MT) and SW N.B. (Wolf Is., Charlotte Co.). Forma *subverticillatus* Fern. (upper leaves much reduced and crowded near the top of the stem rather than only gradually reduced and not crowded) is known from E Que. (Tadoussac, Saguenay Co.) and the Maritime Provinces. Var. *magdalenensis* Fern. (leaves nearly half as broad as long, acute or short-acuminate, rather than usually over twice as long as broad and long-acuminate) is known from the type locality, Grindstone Is., Magdalen Is., E Que.

A. alpinus L.

/ST/W/EA/ (Hs (Ch)) Grassy or rocky slopes and summits from Alaska (a station on the NW coast at ca. 67°N indicated in the map by Johnson and Viereck 1962:30; also known from ca. 62°30′N), the Yukon (N to ca. 62°30′N), and Great Bear L. through B.C. and the mts. of SW Alta. (N to Jasper; CAN; the puzzling report from Anticosti Is., E Que., by Saint-Cyr 1887, requires clarification) to Colo.; Eurasia. [Incl. the pubescent but intergrading extreme, ssp. *vierhapperi* Onno, to which the N. American plant is often referred]. MAPS: Hultén 1968*b*:856; Porsild 1966: map 148, p. 85.

A. azureus Lindl.

/T/EE/ (Hs (Hsr)) Dry prairies, thickets, and open woods from Minn. to S Ont. (N to York and Hastings counties; collection in TRT from along railway ballast at North Bay, L. Nipissing, where doubtless introd.) and N.Y., S to E Tex., La., Ala., and Ga. [*A. shortii* ssp. *az.* (Lindl.) Avers; incl. var. *scabrior* Engelm.].

A. blakei (Porter) House

/T/E/ (Hpr) Damp thickets and woods, shores, and bogs from Que. (N to Abitibi and Portneuf counties) to Nfld. (GH), N.B., N.S. (not known from P.E.I.), N.J., and New Eng. [*A. nemoralis* vars. *blakei* Porter and *major* Peck].

As illustrated in the key to species (*GROUP 1*), this taxon is more or less intermediate between *A. acuminatus* and *A. nemoralis* and is considered by some authors to be a hybrid of this parentage. Fernald *in* Gray (1950), however, notes that it is, "rarely associated with either or both, very fertile and uniform and unknown from much of their coincident area." See R.B. Pike (Rhodora 72(792):401–36. 1970, the shaded area on his MAP, fig. 8, p. 424, indicating the area of × *A. blakei*).

A. borealis (T. & G.) Provancher

/ST/X/ (Hpr) Calcareous bogs, swamps, wet gravels, and shores from Alaska (N to near the Arctic Circle), the Yukon (N to ca. 62°30′N), and Great Bear L. to Great Slave L., L. Athabasca, Man. (N to Churchill), Ont. (N to W James Bay at ca. 54°30′N), Que. (N to E James Bay at 54°25′N, Anticosti Is., and the Gaspé Pen.; not known from Nfld.), N.B., P.E.I., and N.S., S to Idaho, Colo., S.Dak., Minn., and N.J. [*A. laxifolius* var. *bor.* T. & G.; *A. franklinianus* and *A. junciformis* Rydb.; *A. junceus* of Canadian auth., not Ait.; *A. salicifolius* Rich., not Lam.; *A. ?longulus* Sheld.]. MAPS (*A. junciformis*): Hultén 1968*b*:859; Porsild 1966: map 149, p. 85.

A. campestris Nutt.

/T/W/ (Hpr) Open places at low to moderate elevations from S B.C. (N to the Marble Range, NW of Clinton, and Canal Flats, about 30 mi N of Kimberley; CAN; V) and SW Alta. (Crowsnest Forest Reserve and Blairmore) to Calif. and Utah.

The habitally very similar *A. yukonensis* Cronq. is known from SW Yukon (several collections in CAN from the St. Elias Mts. and the mouth of Slim's R., S end of Kluane L., ca. 61°N, the type locality; MAP: Hultén 1968*b*:859). The type collection cited by A.A. Cronquist (Madroño 8:99. 1945) is No. 9384 of J.P. Anderson, taken July 23, 1944. An isotype sheet of this same collection in CAN, however, is the habitally very similar (but nonglandular) *Erigeron yukonensis* Rydb. (*E. glabellus* ssp. *pubescens* var. *yuk.* (Rydb.) Hult.), also known from the Yukon. It seems likely that the collection is a mixture of both species. *A. yukonensis* may be distinguished from *A. campestris*, according to Cronquist, "in its lax, herbaceous, equal phyllaries, as contrasted to the firm, chartaceous-based, more or less imbricate phyllaries of *A. campestris*. It also differs in its short simple caudex, instead of creeping rhizomes, as well as in several other features. Its true affinities, as suggested by the involucre, and by the auriculate-clasping bases of the upper leaves, are with *A. modestus* Lindl. and *A. novae-angliae* L." Collections in CAN also show an abundant admixture of long flattened villous hairs in the glandularity of the involucre and upper part of stem, the pubescence of *A. campestris,* at least of collections in CAN, appearing to be entirely short-glandular.

A. canescens Pursh   Hoary Aster
/T/W/   (Hs)   Dry open places at low to fairly high elevations from s B.C. (N to Summerland, about 10 mi NW of Penticton, and Cranbrook; CAN; V), s Alta. (N to Red Deer; CAN), and SW Sask. (N to Swift Current; Breitung 1957a) to s Calif., Ariz., and Colo. [*Dieteria* Nutt.; *Machaeranthera* Gray].

Var. *viscosus* (Nutt.) Gray *(D. (M.) viscosa* and *D. (M.) puberulenta* Nutt.; phyllaries glandular rather than merely canescent) is known from SW Sask. (Crichton and Val Marie; Breitung 1957a).

A. chilensis Nees
/T/W/   (Hpr)   In a wide variety of habitats and elevations from SW ?B.C. (Victoria and Crescent, Vancouver Is.; Henry 1915), Alta. (N to Edmonton; John Macoun 1884), and Sask. (N to Warman, about 10 mi N of Saskatoon; Breitung 1957a) to Calif. and N.Mex. [*A. adscendens* and *A. menziesii* Lindl.; *A. subgriseus* Rydb.].

Hitchcock et al. (1955) report this species as introd. eastwards, probably on the basis of its report from s Ont. (Manitoulin dist., N L. Huron), W Nfld., and the Gaspé Pen., E Que., by Fernald *in* Gray (1950). Further confirmation of these E Canadian reports are required. They may prove referable to the very plastic *A. novi-belgii* complex or to *A. foliaceus* (see *A. subspicatus*), under which names most of the relevant collections in CAN and GH were originally distributed.

A. ciliolatus Lindl.
/sT/X/   (Hpr)   Woodlands and clearings from northernmost B.C. (Liard Hot Springs, 59°25′N) and Great Bear L. to Great Slave L. (type locality), L. Athabasca (Alta. and Sask.), Man. (N to Wekusko L., about 90 mi NE of The Pas; a Gardner collection in CAN purportedly from Churchill was probably taken at Flin Flon), northernmost Ont., Que. (N to E James Bay at 52°17′N, L. Mistassini, and the Côte-Nord), N.B., and N.S. (not known from Nfld. or P.E.I.), s to Mont., Wyo., S.Dak., Minn., and New Eng. [*A. lindleyanus* T. & G. and its var. *cil.* (Lindl.) Gray; *A. maccallai* and *A. wilsonii* Rydb.; incl. the reduced northern extreme, var. *borealis* (Rousseau) Dutilly & Lepage]. MAP: A.E. Porsild and H. Crum, Nat. Mus. Can. Bull. 171: fig. 2, p. 146. 1961.

Forma *comatus* Fern. (petioles, lower leaf-surfaces, and the upper part of the stem pilose rather than glabrous) occurs essentially throughout the range.

A. conspicuus Lindl.   Showy Aster
/sT/W/   (Hpr)   Open woods and clearings from B.C. (N to Fort St. John, ca. 56°10′N; reports from the Yukon require confirmation), Alta. (N to the Caribou Mts. at 58°54′N; CAN), and Sask. (N to Carlton, about 35 mi SW of Prince Albert; the type locality was given as "Carlton House on the Saskatchewan River to the Rocky Mountains"; the report from St. Lazare, SW Man., in the undated supplement to Lowe's 1943 checklist requires confirmation; there is a collection in TRT from Niagara Falls, Welland Co., s Ont., where taken by Scott in 1900 and probably introd., unless a misidentification or confusion of labels is involved) to Oreg., Idaho, and Wyo.

A. cordifolius L.
/T/EE/   (Hsr)   Open woods, thickets, and clearings, the aggregate species from Man. (N to Grand Rapids, near the NW end of L. Winnipeg) to Ont. (N to Renison, near James Bay at ca. 51°N; Hustich 1955), Que. (N to Anticosti Is. and the Gaspé Pen.), N.B., P.E.I., and N.S., s to Mo. and Ga.

A hybrid with *A. paniculatus* is reported from SW Que. by Frère Marie-Victorin (Contrib. Inst. Bot. Univ. Montréal 8:466. 1926; Longueuil, near Montreal). One with *A. simplex* var. *ramosissima* is reported from SW Que. by Père Louis-Marie (Rev. Oka Agron. Inst. Agric. 34:4. 1960; La Trappe, Deux-Montagnes Co.). According to Louis-Marie, this is the plant commonly passing as *A. tardiflorus* L., this, however, considered by A.S. Pease (Rhodora 19(221):88–90. 1917) to be a hybrid between *A. cordifolius* and *A. puniceus.* It is apparently common throughout the range of *A. cordifolius.*

1 Primary panicle-branches 1–several times forked; involucres at most 5 mm high.
  2 Upper stem-leaves cordate; [Man. (N to Grand Rapids, near the NW end of L.
    Winnipeg) to N.B. and N.S.] .... var. *cordifolius*
  2 Upper stem-leaves truncate or tapering at base; [Que.] .... var. *polycephalus* Porter
1 Primary panicle-branches simple or short-forking; involucres to 6.5 mm high; leaves all
  cordate.
  3 Stem and leaves glabrous or nearly so; [Que., N.B., P.E.I. (type from Malpeque), and
    N.S.] .... var. *racemiflorus* Fern.
  3 Stem and petioles densely white-pubescent; lower leaf-surfaces sparingly pilose; [E
    Que. and N.B.] .... var. *furbishiae* Fern.

A. curtus Cronquist
/t/W/ (Hpr) Plains and rocky hillsides from SW B.C. (collections in CAN from Vancouver Is.,
where taken by John Macoun in 1887 and 1893; collection in V from Wellington, near Vancouver) to
Oreg. [*Seriocarpus rigidus* Lindl., not *A. rigidus* L.].

A. divaricatus L. White Wood-Aster
/T/EE/ (Hsr) Dry woods and clearings from Ohio and S Ont. (N to Toronto; TRT; reports from
elsewhere in our area, as from Man. by Lowe 1943, from cent. Ont. by John Macoun 1884, and
from N.S. by Cochran 1829, probably refer to other species) to Maine, S to Tenn., Ala., and Ga. [*A.
(Eurybia) corymbosus* Ait.].

A. dumosus L. Bushy Aster
/T/EE/ (Hpr) Open ground, thickets, and shores from Mich. to S Ont. (N to Simcoe and York
counties), N.Y., and Maine, S to E Tex. and Fla. [Incl. var. *strictior* T. & G.].
  Var. *dodgei* Fern. (stem densely ashy-hispidulous rather than glabrous or more or less
puberulent; leaves harshly scabrous rather than smooth or smoothish) is reported from S Ont. by
Soper (1949).

A. eatonii (Gray) Howell
/T/W/ (Hpr) Streambanks and other moist places at low to moderate elevations from S B.C. (N
to Armstrong, about 15 mi N of Vernon; CAN), SW Alta. (N to Banff; CAN), and SW Sask. (Cypress
Hills; Breitung 1957a) to Calif. and N.Mex. [*A. foliaceus* var. *eat.* Gray; *A. mearnsii* Rydb.; *A.
microlonchus* Greene].

A. engelmannii (Eat.) Gray
/T/W/ (Hpr) Open woods and clearings at moderate to high elevations from S B.C. (N to Lillooet,
about 95 mi W of Kamloops) and SW Alta. (N to Banff; CAN; reported N to Brazeau, about 100 mi SW
of Edmonton, by J.M. Macoun 1899) to Nev. and Colo. [*A. elegans* var. *eng.* Eat.; *Eucephalus*
Greene; *A. ?macounii* Rydb.].

A. ericoides L. Heath Aster
/ST/X/ (Hpr (Hsr)) Dry or moist open places, the aggregate species from Alaska (N to near the
Arctic Circle), the Yukon (N to ca. 62°30′N), and NW Dist. Mackenzie to Great Bear L., N Alta. (L.
Athabasca), Sask. (N to Cumberland House, ca. 54°N), Man. (N to The Pas), S Ont. (N to York and
Prince Edward counties; farther northwards, probably introd.), and Que. (?introd.), S to Calif.,
N.Mex., Tex., Ill., N.J., and Mass. MAP and synonymy (together with a distinguishing key to the
closely related (?hybrid) *A. amethystinus*); see below.
1 Phyllaries acute, in several unequal series, loosely ascending, their tips scarcely
  out-curving; rays usually blue or purplish, to 1 cm long; [plant of moist habitats; possibly a
  recurrent hybrid between *A. ericoides* and *A. novae-angliae*; reported from Essex and
  Lambton counties, S Ont., by Dodge 1914 and 1915] .... [*A. amethystinus* Nutt.]
1 Phyllaries obtuse to broadly rounded at summit, abruptly spinulose-tipped, more or less
  spreading or their tips recurved; rays commonly white, 5 or 6 mm long; [plant of dry
  habitats] .... *A. ericoides*
  2 Plants scarcely stoloniferous, the stems clustered from a short rootstock or caudex;
    involucres commonly 4 or 5 mm high, the phyllaries in 3 or 4 unequal series; heads

numerous in dense, commonly 1-sided racemes; [*A. multiflorus* var. *pansus* Blake; *A. pansus* (Blake) Cronq.; B.C. (Osoyoos and Keremeos; Bernard Boivin, Nat. can. (Que.) 89(2):70. 1962); reports from farther east all probably relate to the following varieties] .... var. *pansus* (Blake) Boivin

2   Plants strongly slender-stoloniferous or rhizomatous, the stems not clustered.

    3   Involucre to over 8 mm high, its phyllaries subequal; heads few or solitary at the ends of the branches; [var. ?*villosus* T. & G.; *A. mult.* vars. *commutatus* T. & G. (*A. comm.* (T. & G.) Gray) and *stricticaulis* T. & G. (*A. str.* (T. & G.) Rydb.); *A. falcatus* and *A. ramulosus* Lindl.; *A. incanopilosus* Sheld.; *A. crassulus* and *A. polycephalus* Rydb.; *A. adsurgens* Greene; *A. elegantulus* Porsild; the Yukon–Alta. to Man.; reported by Boivin, loc. cit., as probably introd. near Thunder Bay and in Renfrew Co., Ont., and by Boivin 1966*b*, from Que., where undoubtedly also introd.; MAP (*A. comm.*): Hultén 1968*b*:860] ..... var. *commutatus* (T. & G.) Boivin

    3   Involucre at most 5 mm high, its phyllaries in 3 or 4 unequal series; heads numerous in dense, commonly 1-sided racemes; [var. *prostratus* (Ktze.) Blake; *A. multiflorus* Ait.; *A. microlonchus* Greene; s Ont. (Kent, Lambton, Elgin, Middlesex, Wentworth, York, and Prince Edward counties)] .... var. *ericoides*

[*A. frondosus* (Nutt.) T. & G.]
[This species of the w U.S.A. (Wash. to Calif., Utah, and Wyo.) is reported as locally abundant in wet soil in s B.C. by Eastham (1947; Penticton and Osoyoos). However, it is so similar to the transcontinental *A. laurentianus* Fern. (*A. brachyactis* Blake) that further material is desirable before the species can be admitted with certainty to our flora. The report from P.E.I. by M.L. Fernald and K.M. Wiegand (Rhodora 12(144):227. 1910) is based upon *A. laurentianus*, relevant collections in CAN and GH. The MAP by Fernald (1925: map 30 (solid dots, not circled), p. 259) indicates no Canadian stations. (*Tripolium* Nutt.).]

A. hesperius Gray
/sT/WW/ (Hpr) Streambanks, ditches, and moist ground at low to moderate elevations from sw Dist. Mackenzie (J.W. Thieret, Can. Field-Nat. 75(3):120. 1961) through B.C.–Alta.–Sask. to s Calif., N.Mex., Mo., and Wisc. [*A. coerulescens* DC.].

    The report eastwards to Que. by Boivin (1966*b*) is probably based upon a broader concept of the species. Some of our material is referable to var. *laetevirens* (Greene) Cronq. (*A. laet.* Greene; *A. osterhoutii* and *A. tweedyi* Rydb.; *A. franklinianus sensu* Fraser and Russell 1944, not Rydb.; *A. longifolius, A. paniculatus,* and *A. salicifolius* of w Canadian reports in part; ray-ligules white or pink rather than blue).

A. laevis L.
/sT/X/ (Hs) Open places at low to moderate elevations from southernmost Yukon (on or near the B.C. boundary; see Hultén 1950: map 1103, p. 1673) and B.C. to Alta. (N to L. Athabasca), Sask. (N to McKague, 52°37′N), Man. (N to Hill L., N of L. Winnipeg), and Ont. (N to near Thunder Bay; reported as introd. at three localities near L. St. John, Que., by R. Cayouette, Nat. can. (Que.) 96(5):743. 1969), s to Oreg., Utah, N.Mex., and Ga. [Var. *cyaneus* (Hoffm.) Gray; *A. rubricaulis* Lam.; *A. laevigatus* of auth., not Willd.].

    The above statement of the Canadian range is tentative. The Yukon plant may have been introd. (or possibly misidentified), as also the Que. plant. The reports from Labrador by Hultén (1950), Nfld. by Waghorne (1898), and N.S. by Lindsay (1878; Halifax) require clarification. The w N. America plant is usually referred to var. *geyeri* Gray (*A. gey.* (Gray) Howell; intergrading with the typical phase but the involucral phyllaries less strongly imbricate and with shorter, diamond-shaped rather than narrow and elongate, green tips). Forma *latifolius* (Porter) Shinners of the typical phase (the relatively broad upper stem-leaves scarcely reduced) is reported from se Man. by Löve and Bernard (1959; near Otterburne, about 30 mi s of Winnipeg).

A. lateriflorus (L.) Britt.   Calico Aster
/T/EE/ (Hsr) Fields, thickets, and clearings from se Man. (N to Victoria Beach, about 55 mi NE of Winnipeg) to Ont. (N to the w James Bay watershed at ca. 51°15′N), Que. (N to Cabano,

Temiscouata Co., and Magdalen Is.), N.B., P.E.I., and N.S., s to Tex. and Fla. [*Solidago* L.; *A. diffusus*, *A. divergens*, and *A. pendulus* Ait.; *A. hirsuticaulis* Lindl.; *A. miser* Nutt.].

A collection in CAN from near Charlottetown, P.E.I., has been named a hybrid with *A. novi-belgii* by Malte. Some of our material is referable to var. *tenuipes* Wieg. (*A. ten.* (Wieg.) Shinners; *A. acadiensis* Shinners; heads on peduncles up to 4 times the length of the involucre rather than subsessile or short-peduncled; leaf-midribs usually glabrous rather than pilose beneath; type from Dundee, Kings Co., P.E.I.).

A. laurentianus Fern.
/sT/X/   (T)   Saline or brackish soil from s Yukon (ca. 60°30′N) and sw Dist. Mackenzie (w of Great Slave L.; CAN) to B.C.–Alta., Sask. (N to Prince Albert), Man. (N to Wabowden, about 135 mi NE of The Pas), Ont. (N to w James Bay at 53°14′N), Que. (N to Cacouna and Ile-Verte, Temiscouata Co., and Magdalen Is.), and P.E.I. (type from Brackley Point, Queens Co.; not known from N.B. or N.S.), s to Wash., Wyo., ?Utah, S.Dak., Mo., Minn., and N.Y.; the closely related *A. ciliatus* (Ledeb.) Fedtsch. in Asia. [*A. brachyactis* Blake; *Crinitaria humilis* Hook.; *Tripolium (A.; Brachyactis) angustum* Lindl., not *A. ang.* Nees; *A. frondosus sensu* M.L. Fernald and K.M. Wiegand, Rhodora 12(144):227 (footnote). 1910, not (Nutt.) T. & G.]. MAPS: Hultén 1968b:860 (*A. br.*); Fernald 1925: map 30 (treating *A. angustus* as a distinct species; incomplete), p. 259.

Depauperate forms with the outer involucral phyllaries barely elongate, the stems less than 1.5 dm tall, may be referred to var. *contiguus* Fern. (outer phyllaries 1 or 2 mm broad, acute or acutish; type from near Tracadie, Gloucester Co., N.B.) and var. *magdalenensis* Fern. (outer phyllaries to 4 mm broad, mostly obtuse; type from Coffin Is., Magdalen Is., E Que.).

A. linariifolius L.
/T/EE/   (Hpr)   Dry soil, ledges, and rocky banks from s Minn. to s Que. (shores of L. St. Peter and the St. Lawrence R. in St-Maurice and Portneuf counties) and N.B. (Nepisiguit and Miramichi valleys; not known from P.E.I. or N.S.; the report from Nfld. by Waghorne 1898, requires confirmation), s to E Tex., Miss., and N Fla. [*Diplopappus* Hook.; *Ionactis* Greene].

Some or all of the Que. material may be referred to var. *victorinii* Fern. (leaves round-tipped rather than subulate-tipped, commonly less than 2 cm long rather than to over 4 cm; type from St-Raymond, Portneuf Co., Que.).

A. lowrieanus Porter
/T/EE/   (Hsr)   Open woods and thickets from s Mich. to Ont. (N to the Ottawa dist.; Gillett 1958; the report from w James Bay at ca. 52°10′N by Dutilly, Lepage, and Duman requires confirmation, perhaps referring to *A. ciliolatus*) and sw Que. (Mt. Johnson, about 20 mi SE of Montreal; Raymond 1950b), s to Ga. and w N.C. [Incl. var. *lanceolatus* Porter].

A. macrophyllus L.
/T/EE/   (Hsr)   Open woods, thickets, and clearings, the aggregate species from s Man. (N to Norway House, off the NE end of L. Winnipeg; John Macoun 1884) to Ont. (N to Longlac, N of L. Superior at 49°47′N), Que. (N to L. St. John and the Gaspé Pen.), N.B., P.E.I., and N.S., s to Ill., Ala., and N.C. MAP and synonymy: *see* below.
1   Glands of the pedicels minute, rarely stipitate; leaves thinnish, smooth or smoothish
    above; [*A. ianthinus* Burgess; Ont. (region N of L. Huron) and Que. (N to l'Ile d'Orléans,
    near Quebec City)] .... var. *ianthinus* (Burgess) Fern.
1   Glands of the pedicels elongate or stipitate; leaves firm.
    2   Stems and petioles villous; leaves pilose beneath.
        3   Most of the stem-leaves truncate to tapering at base; [Ont. to N.B. and N.S.] ....
            .... var. *velutinus* Burgess
        3   Most of the stem-leaves rounded or cordate at base; [Ont. and Que.] ....
            .... var. *sejunctus* Burgess
    2   Summit of stem glabrous or only minutely pubescent.
        4   At least some of the basal leaves smooth or smoothish; pedicels more or less
            glandular; [Ont. (near Thunder Bay) to Que. (L. St. John) and N.S. (Partridge Is.,
            Cumberland Co.)] .... var. *excelsior* Burgess

4   Leaves harshly scabrous; pedicels copiously glandular; [*Biotia* DC.; *Eurybia* Cass.; s Man. (the report as far N as Norway House by John Macoun 1884, requires confirmation) to N.B. and N.S.; MAP (aggregate species; somewhat incomplete northwards): M.L. Fernald, Rhodora 13: map 11, pl. 90 (facing p. 140). 1911] . . . . var. *macrophyllus*

A. modestus Lindl.
/ST/(X)/ (Hpr) Moist open or wooded places from SE Alaska and the Yukon (N to ca. 63°N; *see* Hultén 1950: map 1104, p. 1673) to B.C., Alta. (type from the mouth of the Smoky R. near Peace River at ca. 56°N), Sask. (N to Ile-à-la-Crosse, 55°27′N), SE Man. (Marchand and Sandilands Forest Reserve, SE of Winnipeg), and Ont. (N to w James Bay at 54°12′N; *see* cent. Ont. map by Lepage 1966: map 19, p. 240), s to Oreg., Idaho, and N Mich. [*A. mutatus* T. & G.; *A. sayianus* Nutt.; *A. sayii* Gray; *A. unalaschkensis* var. *major* Hook. (*A. major* (Hook.) Porter)]. MAP: Hultén 1968b:858.

A. nemoralis Ait.   Bog-Aster
/T/EE/ (Hpr) Sphagnous bogs and peaty places from Ont. (N to Schreiber, N shore of L. Superior) to Que. (N to Sims L. at 54°05′N, the Côte-Nord, Anticosti Is., and Gaspé Pen.), s Labrador (N to Goose Bay, Hamilton R. basin), Nfld., N.B., P.E.I., and N.S. (type locality), s to Mich. and N.J. [*Galetta* Nees]. MAPS: M.L. Fernald, Rhodora 22(257): map 3, p. 94. 1920; R.B. Pike, Rhodora 72(792): fig. 8, p. 424. 1970.
Forma *albiflorus* Fern. (ray-ligules white rather than purple) is known from N.S. (Lawson L., Lunenburg Co.; ACAD).

A. novae-angliae L.   New England Aster
/T/EE/ (Hpr) Rich damp ground from s Man. (probably native southwards; perhaps introd. at The Pas) to Ont. (N to L. Nipigon), Que. (N to Cacouna, Temiscouata Co.), and N.S. (probably introd.; not known from N.B. or P.E.I.), s to Colo. (?introd.), Kans., Ark., and Ky. [*A. amplexicaulis* Lam.].
Being much cult. and a garden-escape outside of its native area, its limits as a native member of our flora are impossible to define precisely. The report from Saskatoon, Sask., by Fraser and Russell (1944) is thought by Breitung (1957a) to be probably based upon an introd. plant, as may be the case with the more northern Ont. and Que. material.
Forma *geneseensis* House (ray-ligules white rather than violet-purple) is known from Ont. (Steen Is., Stormont Co.; Ottawa). Forma *roseus* (Desf.) Britt. (ray-ligules pinkish) is known from Ont. (Norfolk Co.; Ottawa) and Que. Boivin (1966b) reports a hybrid with *A. simplex* from sw Que.

A. novi-belgii L.
/sT/EE/ (Hpr) Damp ground and shores (ranges of Canadian taxa outlined below), s along the Atlantic Seaboard to Ga.
1   Stems densely white-villous; leaves densely villous along the midrib beneath . . . .
    . . . . var. *villicaulis* (Gray) Boivin
       2   Ray-ligules white; [*A. gaspensis* f. *alb.* Vict., the type from along the Bonaventure R., Gaspé Pen., E Que.] . . . . f. *albiflorus* (Vict.) Boivin
       2   Ray-ligules blue-violet; [*A. longifolius (johannensis)* var. *vill.* Gray; Ont. to s Labrador and Nfld.] . . . . f. *villicaulis*
1   Stems glabrous or only sparingly pubescent; leaves subglabrous.
       3   Phyllaries relatively broad and straight; [incl. var. *elodes* (T. & G.) Gray (*A. elodes* T. & G.); E Que. (Gaspé Pen.), N.B., and N.S.] . . . . var. *litoreus* Gray
       3   Phyllaries with loosely spreading to recurving (squarrose) tips; [incl. var. *rosaceus* Rousseau; *A. anticostensis* and *A. johannensis* Fern.; *A. gaspensis* Vict.; *A. rolandii* Shinners; *A. laevigatus* Lam., not Willd.; *A. longifolius, A. paniculatus,* and *A. salicifolius* of E Canadian reports in part; Ont. (N to sw James Bay at ca. 52°N), Que. (N to L. St. John, the Côte-Nord, Anticosti Is., and Gaspé Pen.), Labrador (N to Goose Bay), Nfld., N.B., P.E.I., and N.S.] . . . . var. *belgii*

A. oblongifolius Nutt.
Native in the E U.S.A. (N to N.Dak., Minn., and Pa.); reported from near Windsor, Essex Co., s Ont.,

by Dodge (1914) and a collection in TRT from Toronto has been referred to it by Soper and Rao. Boivin (1966*b*) considers the Ont. plant to be introd.

A. occidentalis (Nutt.) T. & G.
/ST/W/ (Hpr) Meadows and slopes at low to fairly high elevations from Great Bear L. (type locality of *A. spathulatus* Lindl.; a report from Alaska requires confirmation) through B.C. and sw Alta. (N to 13 mi N of the Saskatchewan R. Trading Post NW of Banff; CAN) to Calif. and Colo. [*Tripolium* Nutt.; *A. adscendens* var. *fremontii* T. & G. (*A. fre.* (T. & G.) Gray); *A. spathulatus* Lindl., not Lag.; *A. ?ptarmicoides sensu* A.E. Porsild, Nat. Mus. Can. Bull. 101:34. 1945, not (Nees) T. & G., the relevant collection in CAN; incl. the robust extreme, var. *intermedius* Gray].

A. parviceps (Burgess) Mackenz. & Bush
/t/EE/ (Hpr) "Dry open woods and prairies, Ill., Ia. and Mo." (Fernald *in* Gray 1950). There is a collection in Herb. MICH, detd. K.K. Mackenzie, from Point Edward, Lambton Co., s Ont., where taken by Dodge in 1913 and probably at the N limit of its native range. Bernard Boivin (ACFAS, Programme of the 29th Congress 1961:39) reports it as introd. at Windsor, N.S. [*A. ericoides* var. *parv.* Burgess].

[*A. patens* Ait.] Skydrop Aster
[The tentative reports of this species of the E U.S.A. (N to Minn. and Maine) from s Ont. and N.B. by John Macoun (1884) require clarification.]

A. paucicapitatus Rob.
/T/W/ (Hp) Open slopes at moderate to high elevations: sw B.C. (Mt. Arrowsmith, s Vancouver Is.; Herb. V) and NW Wash. (Olympic Mts.). [*A. engelmannii* var. *pauc.* Rob.].

A. pauciflorus Nutt.
/sT/WW/ (Hsr) Salt marshes and alkaline flats from southernmost Dist. Mackenzie (Fort Smith, near the Alta. boundary) and northernmost Alta. (Wood Buffalo National Park) to Sask. (N to Fort Pitt, 53°39′N; Breitung 1957*a*) and Man. (N to Dawson Bay, N L. Winnipegosis; CAN), s to Ariz. and Tex. [*Tripolium* Nees].

A. pilosus Willd.
/T/EE/ (Hpr) Dry thickets, fields, roadsides, and clearings (ranges of Canadian taxa outlined below), s to Ark. and Ga.
1 Stem, branches, and often the leaves pilose or villous.
　2 Plant pilose-hirsute; principal leaves linear to linear-lanceolate; [*A. ericoides* var.
　　pil. (Willd.) Porter; s Ont. (Essex, Lambton, Kent, Elgin, Norfolk, and Haldimand
　　counties) and sw Que. (St-Jean, St-Jean Co., and Rougemont, Rouville Co.; MT)] . . . .
　　. . . . . var. *pilosus*
　2 Stem densely white-villous; principal leaves lanceolate to lance-oblong; [*A. ericoides*
　　var. *plat.* T. & G.; s Ont.: Norfolk Co.; OAC] . . . . var. *platyphyllus* (T. & G.) Blake
1 Stem, branches, and leaves essentially glabrous.
　3 Branches of inflorescence loosely ascending to widely divergent; involucres mostly
　　not over 6 mm high; [*A. ericoides* of auth., not L.; s Ont. (Toronto), sw Que.
　　(Philipsburg, Missisquoi Co.; MT), and ?N.S. (Windsor, Hants Co.; E.C. Smith and
　　J.S. Erskine, Rhodora 56(671):251. 1954; this collection referred to var. *pilosus* by
　　Boivin)] . . . . var. *demotus* Blake
　3 Branches of inflorescence rather short and stiffly ascending; involucres to 8 mm high;
　　[*E. ericoides* var. *pr.* Gray; incl. *A. faxonii* Porter, *A. kentuckiensis* and *A. priceae*
　　Britt., and *A. polyphyllus* Willd., not Moench.; Ont. (N to the region N of L. Huron) and
　　sw Que. (N to the Montreal dist.)] . . . . var. *pringlei* (Gray) Blake

A. praealtus Poir.
/T/EE/ (Hsr) Meadows and thickets from Man. (N to the Churchill R. at ca. 56°10′N; CAN) to Ont. (N to Sioux Lookout, about 175 mi NW of Thunder Bay; CAN) and s Que. (Fernald *in* Gray

1950; var. *angustior* Wieg., the narrow-leaved extreme), s to Tex., Ky., and Md. [*A. salicifolius* Ait., not Lam.].

A. prenanthoides Muhl.
/t/EE/ (Hpr (Hsr)) Damp thickets, rich woods, and shores from Minn. to s Ont. (St. Thomas and Yarmouth Twp., Elgin Co.; TRT) and Mass., s to Iowa, Ky., Va., and Del.

A. ptarmicoides (Nees) T. & G. White Upland Aster
/T/(X)/ (Hsr) Dry prairies and sandy or rocky places from Wyo. to s Sask. (Indian Head; Boulder L.; Souris Plain; Touchwood Hills), Man. (N to the Red Deer R. w of L. Winnipegosis at ca. 53°N and 40 mi s of The Pas), Ont. (N to L. Nipigon and the Fawn R. at ca. 55°N, 88°W), Que. (L. Timiskaming at ca. 47°30'N, Senneterre, 48°24'N, and the Bell R. at 49°43'N; CAN; RIM), and N.B. (near Woodstock, Carleton Co., where taken by John Macoun in 1899; CAN; not known from P.E.I. or N.S.), s to Colo., Mo., and N.Y. [*Doellingeria* Nees; *Solidago* Boivin; *Chrysopsis (Diplopappus; Unamia) alba* Nutt.].
    The tentative report from Normal Wells, w Dist. Mackenzie, by A.E. Porsild (Nat. Mus. Can. Bull. 101:34. 1945) is based upon *A. occidentalis,* the relevant collection in CAN. A hybrid with *Solidago ohioensis* (× *S. krotkovii* Boivin) is known from s Ont. (Bruce Pen., L. Huron; type from Stokes Bay; *see* map by J.P. Bernard, Nat. can. (Que.) 96(2): fig. 1 (bottom map), p. 173. 1969). Var. *lutescens* (Lindl.) Gray (*A. (Solidago; Unamia) lut.* Lindl.; ray-ligules yellowish rather than white) is known from s Sask. (Indian Head; Touchwood Hills) and s Man. (N to the Red Deer R. w of L. Winnipegosis at ca. 53°N). Löve and Bernard (1959) present evidence that this taxon may be of hybrid origin between the typical form and either *Solidago riddellii* or *S. parvirigida* (*S. rigida* var. *humilis* of the present treatment; *see* J.R. Beaudry and D.L. Chabot, Can. J. Bot. 37(2):224. 1959). They report the *S. riddellii* hybrid from s Man. (Kleefeld, about 30 mi s of Winnipeg, its type locality under the name × *S. bernardii* Boivin), *S. riddellii* also being known from the same locality. The distribution of var. *lutescens* is indicated in MAPS by Bernard, loc. cit., fig. 1 (middle map, *A. ptarm.* × *S. ridd.;* top map, *A. ptarm.* × *S. parv.*), p. 173.

A. puniceus L.
/sT/(X)/ (Hpr) Damp thickets, meadows, and shores, the aggregate species from Dist. ?Mackenzie and Alta. (N to Wood Buffalo National Park at 59°31'N) to Sask. (N to McKague, 52°37'N), Man. (N to Churchill), Ont. (N to the Fawn R. at ca. 54°40'N), Que. (N to s Ungava Bay and the Côte-Nord), Labrador (N to Makkovik, 55°05'N), Nfld., N.B., P.E.I., and N.S., s to N.Dak., Ill., Ala., and Ga.
1 Disk-flowers yellow; inflorescence with at most 6 heads; peduncles to 6 cm long; [incl. f. *brachyphyllus* Lepage; *A. calderi* Boivin, the type from Fort Chimo, s Ungava Bay, Que.; also reported from w and E James Bay by Dutilly, Lepage, and Duman 1954 and 1958] . . . .
. . . . var. *calderi* (Boivin) Lepage
1 Disk-flowers red or purple; heads more numerous; peduncles mostly shorter.
   2 Outer phyllaries to 3 mm broad, mostly hiding the inner ones.
      3 Leaves harshly scabrous above; inflorescence compact; [Dist. Mackenzie–Sask.; Ont. (N to w James Bay at ca. 53°N) to Labrador (N to Makkovik, 55°05'N; type from Red Bay), Nfld., and E Que.] . . . . var. *oligocephalus* Fern.
      3 Leaves glabrous and shining; inflorescence open; [known only from wet spots in the coniferous forest at the base of Tabletop Mt., Gaspé Pen., E Que., the type locality] . . . . var. *perlongus* Fern.
   2 Outer phyllaries less than 1 mm broad; leaves scabrous above.
      4 Stem entirely glabrous or sparsely hispid above; leaves glabrous beneath; (note the 3-parted division of lead 5) . . . . var. *firmus* (Nees) T. & G.
         5 Ray-ligules white; [near Otterburne, Man.; Löve and Bernard 1959] . . . .
. . . . f. *etiamalbus* Venard
         5 Ray-ligules reddish; [sw Que.: Beauceville, Beauce Co.; Raymond 1950*b*] . . . .
. . . . f. *rufescens* Fassett
         5 Ray-ligules blue-violet; [incl. vars. *laevicaulis* and *lucidulus* Gray; Ont. to Labrador (N to the Hamilton R. basin; Abbe 1955), Nfld., and N.S.] . . . . f. *firmus*

    4   Stem hirsute with often tuberculate-based hairs; leaves harshly scabrous above,
        minutely hispid on the midrib beneath .... var. *puniceus*
      6   Ray-ligules white; [S Man.; Boivin 1966*b*] .... f. *candidus* Fern.
      6   Ray-ligules blue-violet.
          7   Pedicels and branches of the inflorescence densely hairy with incurved or
            entangled soft hairs; [*A. blandus* Pursh; E Que. (Gaspé Pen.), N.B., and
            N.S.] .... f. *blandus* (Pursh) Fern.
          7   Pedicels and branches of the inflorescence with loosely spreading stiffish
            hairs.
            8   Panicle narrow and thyrsiform, its branches much shorter than the
               subtending leafy bracts; [var. *demissus* Lindl.; reported from Churchill,
               Man., by Gardner 1937, and also known from the Gaspé Pen., E Que.] .....
               .... f. *demissus* (Lindl.) Fern.
            8   Panicle relatively broad, its branches mostly longer than the sub-
               tending leafy bracts; [Alta. (N to the Slave R. at 59°31′N) to Labrador
               (N to ca. 55°N), Nfld., and N.S.] .... f. *puniceus*

A. radula Ait.
/sT/E/ (Hpr) Low woods, swamps, and bogs from Que. (N to the Great Whale R., SE Hudson Bay, ca. 55°20′N, L. Mistassini, the Côte-Nord, Anticosti Is., and Gaspé Pen.) to Labrador (N to Nain, 56°32′N), Nfld., N.B., P.E.I., and N.S. (type locality), S to W.Va. and Va.

    Much of the material from the northern part of our area is referable to var. *strictus* (Pursh) Gray (*A. strictus* Pursh, the type from Labrador; *A. biflorus* Michx.; heads relatively few, the phyllaries subequal, acute or acutish, not markedly scarious-margined, rather than in several unequal series, obtuse, and scarious-margined).

[*A. radulinus* Gray]
[The tentative report of this species of the W U.S.A. (Wash. to Calif.) from S B.C. by John Macoun (1884; taken up by J.M. Macoun 1901, as *Eucephalus macounii*) is based upon *A. conspicuus*, relevant collections in CAN. A so-named collection in CAN from Laggan, Alta., is referable to *A. sibiricus*. (*Eucephalus macounii* Greene).]

A. sagittifolius Wedemeyer
/T/EE/ (Hsr) Dry open woods, thickets, and clearings from N.Dak. and SE ?Man. (Otterburne, about 30 mi S of Winnipeg, somewhat atypical; Löve and Bernard 1959) to Ont. (N to the Ottawa dist.; Gillett 1958) and Vt., S to Mo., Tenn., Ala., and Ga.

    The above Man. citation may prove referable to *A. ciliolatus*, the specimen illustrated by Löve and Bernard lacking the numerous small bracts on the peduncles characteristic of *A. sagittifolius*. Much or all of our material appears referable to var. *drummondii* (Lindl.) Shinners (*A. dr.* Lindl.), the lower leaf-surfaces and at least the upper part of the stem closely ashy-puberulent rather than essentially glabrous or the stem merely pubescent in lines. Forma *hirtella* (Lindl.) Shinners (leaves markedly scabrous above rather than smooth) is known from S Ont. (Elgin and Kent counties).

[*A. schreberi* Nees]
[The report of this species of the E U.S.A. (N to Ill. and New Eng.) from S Ont. by Soper (1949) is probably based upon a collection in TRT from Brighton, York Co., where taken by John Macoun in 1900, this, however, proving referable to *A. macrophyllus*.]

[*A. scopulorum* Gray]
[The report of this species of the W U.S.A. (N to Oreg. and Mont.) from SE B.C. by Eastham (1947; L. McArthur, Yoho) requires confirmation. (*Chrysopsis (Ionactis) alpina* Nutt.).]

A. sericeus Vent.
/T/EE/ (Hp (Hs)) Dry open woods and prairies from SE Man. (Winnipeg; Arnaud; Birds Hill; Stony Mountain) to W Ont. (mouth of the Rainy R. and islands in Lake of the Woods; CAN; John Macoun 1884) and N Mich., S to Tex., Mo., and Tenn.

A. shortii Lindl.
/t/EE/ (Hs (Hsr)) Open woods, thickets, and rocky slopes from E Iowa and Wisc. to S Ont. (Pelee Point and islands of the Erie Archipelago, Essex Co.; TRT; MICH; Core 1948), S to Tenn., Ala., and Ga.

A. sibiricus L.
/aST/W/EA/ (Hpr) Open rocky places and meadows at low to high elevations, the aggregate species from the Aleutian Is. and coasts of Alaska–Yukon–Dist. Mackenzie (E to Coronation Gulf) to SW Victoria Is., S through the mts. of B.C. and Alta. to Oreg., Idaho, and Wyo.; N Eurasia. MAPS and synonymy: *see* below.

1   Phyllaries rather distinctly imbricate in several series of more or less unequal length, relatively firm and straight; [*A. meritus* Nels.; SE B.C. (Kootenay dist.; Ulke 1935) and SW Alta. (Waterton Lakes; Breitung 1957*b*)] . . . . var. *meritus* (Nels.) Raup
1   Phyllaries less distinctly imbricate, rather herbaceous and sometimes a little squarrose.
   2   Heads solitary; leaves oblong, obtuse, entire, glabrous (or woolly along the margins), to about 2.5 cm long, all sessile and commonly more or less clasping; stems mostly not over 1 dm tall; [*A. pygmaeus* Lindl.; S Banks Is., SW Victoria Is., the coast of Dist. Mackenzie between ca. 107° and 117°W, and N Great Bear L.; MAPS (*A. pyg.*): Porsild 1957: map 305, p. 199; *Atlas of Canada* 1957: map 15, sheet 38] . . . .
    . . . . var. *pygmaeus* (Lindl.) Cody
   2   Heads commonly several; leaves oblanceolate to oblong-elliptic, entire or distinctly low-serrate, glabrous above, often somewhat pubescent beneath, to about 7 cm long, the lower ones short-petioled, the upper ones sessile but scarcely clasping; stems to about 4 dm tall . . . . var. *sibiricus*
      3   Ray-ligules white; phyllaries and stem not purple-tinged; [Alaska (type from 20 mi S of Fairbanks) and Alta. (Boivin 1966*b*)] . . . . f. *albinus* Lepage
      3   Ray-ligules violet or purple; phyllaries and stem more or less purplish; [*A. arcticus* Eastw.; *A. richardsonii* Spreng.; *A. (Erigeron) salsuginosus* Rich.; *A. montanus* Rich., not Nutt.; Aleutian Is.–Alaska–Yukon (*see* Hultén 1950: map 1105, p. 1673) and NW Dist. Mackenzie–B.C.–Alta.; MAPS (aggregate species): Hultén 1968*b*:857; Raup 1947:35] . . . . f. *sibiricus*

A. simplex Willd.
/ST/EE/ (Hsr) Damp thickets, meadows, and shores, the aggregate species from Sask. (N to Prince Albert; CAN; not listed by Breitung 1957*a*) to Man. (N to Gillam, about 165 mi S of Churchill), Ont. (N to the Severn R., Hudson Bay, at ca. 56°N), Que. (N to Rupert House, SE James Bay, 51°29′N, and the Gaspé Pen.), Nfld., N.B. (reports from P.E.I. require confirmation), and N.S., S to Kans., Mo., and N.C.

1   Involucres averaging less than 4 mm high; [incl. var. *estuarinus* Boivin; *A. interior* Wieg.; Ont. (N to New Liskeard, 47°31′N) and Que. (N to L. Timiskaming at ca. 47°30′N and St. Augustin, Portneuf Co.)] . . . . var. *interior* (Wieg.) Cronq.
1   Involucres averaging over 4 mm high.
   2   Leaves averaging not much over 1 cm broad, mostly over 10 times as long as broad; [*A. tenuifolius* vars. *bellidiflorus* (Willd.) T. & G. and *ramosissimus* T. & G.; *A. lamarckianus* Nees; *A. longifolius* of Canadian reports in part, not Lam.; *A. paniculatus* Lam. (not Mill.) and its var. *acutidens* Burgess; S Man. (Löve and Bernard 1959) to N.B. and N.S.] . . . . var. *ramosissimus* (T. &. G.) Cronq.
   2   Leaves to about 3.5 cm broad, mostly less than 10 times as long as broad; [*A. panic.* var. *simplex* (Willd.) Burgess; *A. ?bellidiflorus* Willd. (not Nees nor Steud.) in part; Sask. (N to Prince Albert) to Nfld. and N.S.] . . . . var. *simplex*

[*A. solidagineus* Michx.]
[Hooker (1834) reports a Cleghorn collection of this species of the E U.S.A. (Ohio to N.H., S to La. and Ga.) from near Montreal, Que., but John Macoun (1884) writes, "We have no other record of this species and consider its occurrence within our limits very doubtful." The Montreal report may possibly be based upon the habitally similar *Solidago graminifolia. (Seriocarpus* Nees; *Aster (S.) linifolius* L.).]

A. stenomeres Gray
/t/W/ (Hp (Ch)) Open hillsides and clearings at low to moderate elevations from SE B.C. (Dry Interior in the Columbia and Kootenay valleys N to Sproat, S of Revelstoke; CAN) to Wash., Idaho, and Mont. [*Ionactis* Greene].

A. subspicatus Nees
/sT/(X)/ (Hpr) Moist open or wooded places from the E Aleutian Is. and S Alaska (N to ca. 61°N; type from Yakutat Bay) to B.C., Alta. (N to Jasper), Sask. (Boivin 1966*b*; not known from Man.), Ont. (Kenora dist.; CAN), Que. (N to the Côte-Nord, Gaspé Pen., and Anticosti Is.), Labrador (N to Makkovik, 55°05′N), Nfld., N.B. (not known from P.E.I.), and N.S., S in the West to Calif., N.Mex., Idaho, and Mont., and in the East to New Eng. [This is an extremely plastic species, the range still uncertain, the present treatment including the following taxa in the complex: *A. amplus, A. douglasii,* and *A. foliaceus* Lindl.; *A. burkei* (Gray) Howell; *A. butleri, A. ciliomarginatus,* and *A. subcaudatus* Rydb.; *A. carteriana* Henry; *A. cusickii* Gray; *A. oregonus* (Nutt.) T. & G.; *A. robynsianus* Rousseau; *A. adscendens (fol.)* var. *parryi* Eat.; *A. foliaceus* vars. *arcuans, crenifolius, subgeminatus,* and *subpetiolatus* Fern. and vars. *burkei* and *frondeus* Gray; *A. elegans* (*A. perelegans* Nels. & Macbr.) *sensu* John Macoun 1884, not (Nutt.) T. & G., according to Macoun 1886]. MAP: Hultén 1968*b*:857 (conservative treatment).

Some of the B.C.–Alta. material is referable to var. *apricus* (Gray) Boivin (*A. foliaceus* var. *ap.* Gray; heads lacking the broad foliaceous phyllaries characteristic of the typical form).

A. subulatus Michx.
/T/EE/ (T) Fresh to brackish or saline marshes, borders of woods, and clearings from S Mich. to S Ont. (Windsor, Essex Co.; Boivin 1967*a*), N.Y., and N.B. (Bathurst and Tetagouche R., Gloucester Co.; CAN; GH; not known from N.S.; the report from P.E.I. by John Macoun 1890, is based upon *A. laurentianus,* the relevant collection in CAN), S along the coast to Fla. and La. [*Tripolium* Nees].

The above Bathurst, N.B., collection is the type of var. *obtusifolius* Fern., differing from the typical form in its relatively broad, obtuse or round-tipped rather than long-attenuate leaves, its subequal rather than distinctly imbricated phyllaries, and its 1-rowed ray-ligules much surpassing the pappus rather than 2-rowed and only slightly surpassing it.

A. tanacetifolius HBK.
/T/WW/ (T) Plains, open hillsides, and disturbed ground from Mont. and S Alta. (near the U.S.A. boundary along the Belly R. and Milk R.; CAN; John Macoun 1884) to Ariz., Mexico, Tex., and Nebr. [*Machaeranthera* Nees].

A. tradescantii L.
/T/EE/ (Hsr) Damp ground and rocky or gravelly shores from Ont. (N to the Rainy R., Lake of the Woods) to Que. (N to L. St. John), S Nfld. (Murray's Pond, near St. John's; GH), N.B., and N.S. (not known from P.E.I.), S to S Ont., N.Y., and New Eng. [*A. parviflorus* Nees; *A. vimineus* var. *saxatilis* (Blanch.) Fern. (*A. sax.* Blanch.); *A. ontarionis* Wieg.; *A. diffusus* var. *?thyrsoideus* Gray].

A. umbellatus Mill.
/T/(X)/ (Hpr) Thickets, meadows, and swampy places (ranges of Canadian taxa outlined below), S to Nebr., Iowa, Mich., Ky., and Ga.
1 Involucre puberulent, slightly turbinate, with up to about 22 florets (4–7 of these being ray-florets), the relatively few phyllaries acutish; leaves sparingly to usually rather densely tomentulose-puberulent beneath, usually scabrous above; stem puberulent; [*Doellingeria pubens* (Gray) Rydb.; *A. pubentior* Cronq.; Alta. (Moss 1959), Sask. (Spy Hill, about 125 mi E of Regina; A.J. Breitung, Am. Midl. Nat. 61(2):512. 1959), Man. (N to Swan River, N of Duck Mt.), Ont. (N to the Fawn R. at ca. 54°40′N, 88°W), and Que. (N to SE James Bay at 51°21′N)] . . . . var. *pubens* Gray
1 Involucre glabrous, campanulate, with up to over 50 florets . . . . var. *umbellatus*
  2 Heads discoid, ray-florets wanting; [type from the Rupert R. system S of James Bay, Que.] . . . . f. *discoideus* Vict.
  2 Heads radiate, with 7–14 ray-florets.
    3 Leaves more or less pilose over the surface (or at least along the midrib) beneath;

[Que., Nfld. (type from Bonne Bay), N.B. (Wolf Is., Charlotte Co.), and N.S. (Cape Breton Is. and St. Paul Is.)] .... f. *intercedens* Fern.
3 Leaves glabrous; [incl. var. *latifolius* (Gray) T. & G.; *Doellingeria* Nees; *Diplopappus* T. & G.; *A. (Chrysopsis; Diplostephium; Diplopappus; Doell.) amygdalinus* Lam.; *A. humilis* Willd.; Ont. (N to the region N of L. Huron and the Ottawa dist.; collections in CAN from the N shore of L. Superior and Cochrane, ca. 49°N, appear intermediate between this and var. *pubens*), Que. (N to the Côte-Nord and the Marten R. S of James Bay), Nfld., N.B., P.E.I., and N.S.] ....
.... f. *umbellatus*

A. undulatus L.
/T/EE/ (Hsr) Dry open woods, thickets, and clearings from Minn. to S Ont. (N to Middlesex, York, and Hastings counties), ?Que. (Rouleau 1947), and W N.S. (Queens, Kings, and Lunenburg counties; not known from P.E.I.; reports from N.B. by John Macoun 1884, and Fowler 1885, require confirmation, perhaps being based upon *A. macrophyllus*, as is an 1861 Fowler collection in NBM from Kouchibouguac, Kent Co.), S to Ark., La., and Fla.

A. vimineus Lam.
/T/EE/ (Hsr) Fields, meadows, and shores from Mich. to Ont. (N to Casselman, about 30 mi E of Ottawa; John Macoun 1886), SW Que. (near Lacolle, St-Jean Co.; MT), and Maine, S to E Tex. and Ga.
   Some or all of our material is referable to var. *subdumosus* Wieg. (*A. ?foliolosus* Ait.; inflorescence relatively open, the scattered heads on diffuse branchlets or pedicels to 3 cm long rather than mostly less than 7 mm long).

BAERIA F. & M. [9279]

B. maritima Gray   Goldfields
/t/W/ (T) Along the coast from SW B.C. (S Vancouver Is. and adjacent islands; CAN; V) to Calif. [*Burrielia* Gray].

BAHIA Lag. [9301]

B. oppositifolia (Nutt.) DC.
Native in the W U.S.A. from Mont. to N.Dak., S to ?Ariz., N.Mex., and Tex.; introd. in alkaline flats of S Alta. (Coaldale, NE of Lethbridge; CAN) and SW Sask. (Pambrum, SE of Swift Current; CAN; Breitung 1957a). [*Trichophyllum* Nutt.; *Picradeniopsis* Rydb.; *Helenium* Spreng.].

BALSAMORHIZA Nutt. [9188]   Balsam-root

1 Leaves pinnately divided to the midrib into numerous narrow segments to about 1 dm long, these mostly themselves few-cleft into narrow, sometimes further cleft segments, the whole leaf "fern"-like in appearance; a pair of reduced but still fairly well-developed pinnatifid leaves borne several cm above the base of the otherwise scapose stem; tips of middle and lower phyllaries reflexed; taproot carrot-like, bearing a simple caudex; (S ?B.C.) .... [*B. hirsuta*]
1 Leaves entire or shallowly crenate, rather broadly triangular-hastate or -caudate, several strongly reduced or bract-like ones usually borne along the otherwise scapose stem; tips of phyllaries erect; taproot large and woody, bearing a branched caudex.
   2 Leaves silvery beneath when young with a fine felt-like tomentum, green and often glabrate in age; involucre usually woolly-tomentose; (S B.C. and SW Alta.) ..... *B. sagittata*
   2 Leaves green, relatively sparsely hairy with coarser hairs and sometimes also glandular; involucre only slightly or scarcely woolly.
      3 Ray-ligules tending to persist on the achenes and become somewhat papery on drying; (S ?B.C.) .... [*B. careyana*]
      3 Ray-ligules soon deciduous, not becoming papery; (SW B.C.) .... *B. deltoidea*

[B. careyana Gray]
[Hitchcock et al. (1959) report this species of Wash. and Oreg. from s B.C., referring their northern material to var. *intermedia* Cronq., "which differs sharply in its glabrous achenes, and which approaches the more w. *B. deltoidea* in having usually a larger central head, some evidently enlarged and foliaceous outer involucral bracts, and often some crenate teeth on the leaves." They note that it is difficult to distinguish in herbarium specimens from *B. deltoidea* apart from the tendency of the ray-ligules to persist on the achenes rather than being soon deciduous. The species should be searched for in B.C. in an effort to validate the above report.]

B. deltoidea Nutt.
/t/W/ (Grt) Open places and grassy slopes from sw B.C. (6 collections in CAN from the Victoria dist., Vancouver Is., where taken by John Macoun between 1875 and 1913, and one from Drew Harbour, Valdez Is., where taken by G.M. Dawson in 1876) to s Calif.

[B. hirsuta Nutt.]
[The reports of this species of Wash., Oreg., and Nev. from B.C. by John Macoun (1886; "Utah to British Columbia", on the authority of Gray; taken up by Rydberg 1922) and by Henry (1915; South Kootenay) require confirmation.]

B. sagittata (Pursh) Nutt.
/T/WW/ (Grt) Flats and open hillsides at low to moderate elevations from s B.C. (chiefly valleys of the Dry Interior N to Quesnel, ca. 53°N) and sw Alta. (Crowsnest Pass; Waterton Lakes; Cardston; reports from Sask. require confirmation) to s Calif., Colo., and S.Dak. [*Buphthalum* Pursh].

BELLIS L. [8879]

B. perennis L. English Daisy
Eurasian; introd. or a garden-escape (particularly to lawns) in N. America, as in the E Aleutian Is. (Unalaska), B.C. (N to Queen Charlotte Is. and Prince Rupert), Ont. (N to Ottawa), Que. (N to La-Malbaie, Charlevoix Co.; Groh 1946), St-Pierre and Miquelon, Nfld., N.B., and N.S. MAP: Hultén 1968b:855.

BIDENS L. [9237] Bur-Marigold, Beggar-ticks. Bident

(Ref.: E.E. Sherff 1937, and N. Am. Flora, Ser. II (pt. 2):70–129. 1955)
1 Principal leaves compound (with up to 7 leaflets; or filiform-dissected in *B. beckii*).
  2 Plants aquatic, the submersed leaves filiform-dissected into numerous capillary segments, the emersed ones merely serrate; achenes nearly terete; (s B.C.; Sask. to N.S.) .... *B. beckii*
  2 Plants terrestrial or merely subaquatic; achenes flattened parallel to the phyllaries or slender and 4-sided (rarely subterete).
    3 Ray-ligules none or less than 5 mm long and shorter than the outer phyllaries; leaves with at most 5 merely serrate leaflets.
      4 Outer phyllaries eciliate, at most 5; achenes thickish, less than 7 mm long, their awns less than 2.5 mm long, upwardly barbed; lower and primary leaves 3-foliolate, the upper ones often simple; (Ont., Que., and N.S.) .... *B. discoidea*
      4 Outer phyllaries ciliate; achenes flat, to 17 mm long, their awns to about 9 mm long, normally downwardly barbed; principal leaves 3–5-foliolate; (essentially transcontinental).
        5 Outer involucre consisting of at least 10 copiously hispid-ciliate phyllaries; inner achenes to 17 mm long .... *B. vulgata*
        5 Outer involucre consisting of at most 8 sparingly ciliate phyllaries; inner achenes at most 1 cm long .... *B. frondosa*
    3 Ray-ligules well developed, to over 1 cm long, longer than the outer phyllaries; (s Ont.).

     6   Rays white to pale yellow or purplish; achene-awns upwardly barbed; leaves long-petioled, with 3–5 ovate, serrate leaflets; (introd. in s Ont.) .... [*B. pilosa*]

     6   Rays golden yellow; leaves with 3–7 linear to lanceolate, entire to coarsely serrate leaflets.

         7   Achenes at most 2.5 mm broad, marginless, merely strigose-ciliolate, the outer ones cuneate-oblong, to 6 mm long, the inner ones cuneate-linear, to 9 mm long; achene-awns upwardly barbed; petioles at most about 1.5 cm long; (s Ont.) .... *B. coronata*

         7   Achenes to about 5 mm broad, with a thin friable hispid-ciliate margin, the outer ones to 7 mm long, the inner to about 8 mm long; achene-awns to 4 mm long, upwardly or downwardly barbed; petioles to 3 cm long; (introd. in s Ont.) .... [*B. aristosa*]

1   Principal leaves simple (at most deeply cleft into 3–5 coarse segments).

    8   Achenes with a convex cartilaginous summit, their awns typically downwardly barbed; rays usually present; leaves linear to oblanceolate, sessile or occasionally on short winged petioles, subentire or serrate.

       9   Heads campanulate; disk-corollas 4-toothed; rays pale yellow; anthers included, rather pale; achenes coarsely striate, olive-brown to drab; outer phyllaries erect or ascending; (plants of estuarine habitats from Ont. to N.S.) .... *B. hyperborea*

       9   Heads hemispheric; disk-corollas 5-toothed; anthers exserted, purple-black; achenes delicately or obscurely striate; outer phyllaries reflexed or scarcely ascending; plants only incidentally estuarine.

        10   Achenes straight, wingless, not strongly keeled, deep-brown to purplish, the outer to 8 mm long, with awns to 4.5 mm long, the central to 9.5 mm long; leaves appressed-serrate; stem firm, usually smooth; receptacular chaff reddish-tipped; rays golden yellow, to 3 cm long; (s ?Ont.) .... [*B. laevis*]

        10   Achenes curved, with pale margins and keels, olivaceous, the outer not much over 6 mm long, with awns less than 3 mm long, the central less than 8 mm long; stem soft, usually hispid; receptacular chaff yellow-tipped; rays (when present) bright yellow, rarely more than 1.5 cm long; (B.C. to N.S.) ...... *B. cernua*

    8   Achenes truncate or concave at summit (if convex in *B. infirma,* not cartilaginous); heads often discoid.

      11   Mature central achenes strongly 4-angled, usually retrorsely barbed; heads usually discoid; outer phyllaries at most 6, smooth-margined or slightly ciliate; anthers blackish; principal leaves unlobed or merely with up to about 4 broad basal lobes, coarsely sharp-serrate or dentate, sometimes incised; (Ont. to N.S.) .... *B. connata*

      11   Mature central achenes flat or flattish, with slender midribs; outer phyllaries up to 10 in number.

        12   Leaves sessile or broadly wing-petioled; central achenes often over 2 mm broad, their awns normally downwardly barbed.

          13   Leaves sessile (or the lowermost ones on short, very broadly wing-margined petioles), the principal ones deeply 3-parted, the divisions coarsely toothed or incised; outer phyllaries rather closely short-ciliate, the inner ones to 12 mm long; heads radiate (but the rays short and inconspicuous); (sw B.C. and s Man.) .... *B. amplissima*

          13   Leaves more distinctly wing-petioled, incised-serrate, cleft, or 3–5-lobed; (introd.) .... *B. tripartita*

        12   Leaves slender-petioled (at least the principal ones), usually only coarsely serrate but sometimes deeply cleft toward base; central achenes at most about 2 mm broad; outer phyllaries smooth-margined or sparingly ciliate.

          14   Achenes convex at summit, the outer ones at least 6 mm long, awnless, the central ones to 8 mm long, awnless or with upwardly barbed divergent awns less than 1 mm long; outer phyllaries usually not more than 4; (Que.) .... *B. infirma*

          14   Achenes truncate, with erect or ascending awns at least 2 mm long; outer phyllaries usually more numerous.

15 Heads about as broad as long; outer phyllaries up to 7; inner phyllaries less than 1 cm long; outer achenes 4 or 5 mm long, the central ones to 8 mm long; anthers blackish; (E Que. and P.E.I.) . . . . *B. heterodoxa*
15 Heads distinctly longer than broad; outer phyllaries at most 5; inner phyllaries to 13 mm long; outer achenes over 5 mm long, the central ones to 11 mm long; (Que.) . . . . *B. eatonii*

B. amplissima Greene
/T/WW/ (T) Wet places of SW B.C. (several localities on S Vancouver Is., the type from the Somass ("Lomas") R. near Alberni) and S Man. (Boivin 1966b). [*B. cernua* var. *elata* T. & G. (*B. elata* (T. & G.) Sherff); *B. bullata sensu* John Macoun 1890, not L.].
Concerning the possibility of the origin of this species through hybridization between *B. cernua* and *B. frondosa* (or possibly *B. comosa* (*B. tripartita* of the present treatment) or *B. vulgata*), see Sherff (1937, and loc. cit., 1955, p. 94).

[*B. aristosa* (Michx.) Britt.]
[This species of the E U.S.A. (N to Minn. and Del.) is reported as introd. in S Ont. by Soper (1949; an 1893 collection by ?Scott in TRT from Sandwich, Essex Co., has been placed here), where, however, scarcely established.]

B. beckii Torr.   Water-Marigold
/sT/X/ (HH) Ponds and slow streams from SE B.C. (Duck L., Kootenay Flats, near Creston, where taken by J.A. Munro in 1947 and 1949; CAN; V; not known from Alta.) to Sask. (known only from Cumberland House, ca. 54°N; Breitung 1957a), Man. (N to Grand Rapids, near the NW end of L. Winnipeg), Ont. (N to the W James Bay watershed at ca. 53°N; Dutilly, Lepage, and Duman 1954), Que. (N to Duparquet, ca. 48°30'N, in W. Abitibi Co., and between Rivière-du-Loup and Trois-Pistoles, Temiscouata Co.), N.B., and N.S. (not known from P.E.I.), S to Oreg. (probably introd. in S B.C., Wash., and Oreg., according to Hitchcock et al. 1959), Mo., Pa., and N.J. [*Megalodonta* Greene].

B. cernua L.   Stick-tight
/sT/X/EA/ (T) Shores, shallow pools, and wet places, the aggregate species from B.C. (N to Quesnel, ca. 53°N; CAN; introd. in Alaska between ca. 65° and 67°N) to Alta. (N to Wood Buffalo National Park at 59°36'N; CAN), Sask. (N to Prince Albert; Breitung 1957a), Man. (N to Wabowden, about 135 mi NE of The Pas), Ont. (N to the SW James Bay watershed at 52°11'N), Que. (N to the SE James Bay watershed at ca. 52°10'N, L. St. John, and the Gaspé Pen.), N.B., P.E.I., and N.S., S to Calif., Idaho, S.Dak., Mo., and N.C.; Eurasia. MAPS and synonymy: see below.
1 Stem capillary, simple or only slightly forking; leaves petioled; outer involucre with at most 6 phyllaries; [*B. minima* Huds.; Man. (Fernald in Gray 1950), Ont. (N to Cochrane), Que. (N to Magdalen Is.), N.B., and N.S.] . . . . var. *minima* (Huds.) Pursh
1 Stem stoutish, commonly branching; leaves sessile or nearly so; outer involucre with up to 10 phyllaries.
2 Leaves mostly blunt or round-tipped, entire or remotely toothed; outer phyllaries oblong to spatulate, obtuse; heads discoid, lacking ray-ligules; [Que. (N to Magdalen Is., the type locality) and P.E.I.] . . . . var. *oligodonta* Fern. & St. John
2 Leaves long-acuminate; outer phyllaries linear to lanceolate, acutish; heads mostly radiate, the bright-yellow rays to over 1.5 cm long.
3 Leaves elliptic-lanceolate, conspicuously narrowed to base, coarsely toothed; [Que. (N to Rimouski, Rimouski Co.), N.B. (near Sussex, Kings Co.), and P.E.I. (Southport, Queens Co.)] . . . . var. *elliptica* Wieg.
3 Larger leaves with broad sessile or subconnate bases.
4 Leaves relatively broad, finely toothed; [Alta., Sask., Man. (Bernard Boivin, Nat. can. (Que.) 87(2):28. 1960), Ont. (N to SW James Bay at 51°46'N and L. St. John), and P.E.I. (Brackley Point, Queens Co.)] . . . . var. *integra* Wieg.
4 Leaves linear to oblanceolate, coarsely toothed . . . . var. *cernua*
5 Heads discoid, lacking ray-ligules; [N.B. (Grand Manan Is.), P.E.I.

(Dundee, Kings Co.), and N.S. (Cape Breton Is.)] . . . .
. . . . f. *discoidea* (Wimm. & Grab.) Briq. & Cavill.
5 Heads radiate; [*B. dentata* (Nutt.) Wieg.; *B. glaucescens, B. leptopoda,*
and *B. macounii* Greene; range of the species; MAPS (aggregate species):
Hultén 1958: map 260, p. 279, and 1968*b*:884] . . . . f. *cernua*

B. connata Muhl.
/T/EE/ (T) Wet shores and swamps (ranges of Canadian taxa outlined below), S to Kans.,
Tenn., and Va.
1 Principal leaves with up to 4 decurrent or confluent broad basal lobes.
2 Petioles short, broad-margined.
3 Middle leaves and terminal lobes of the divided ones rather closely and sharply
serrate; [often merged with *B. tripartita;* Ont. (N to near Ottawa), Que. (N to near
Quebec City), N.B., P.E.I., and N.S.] . . . . var. *connata*
3 Blades or terminal lobes of middle leaves coarsely dentate; [*B. tripartita* var. *fallax*
Warnst.; Ont. (L. Nipissing) and Que. (tidal flats near Quebec City)] . . . .
. . . . var. *fallax* (Warnst.) Sherff
2 Petioles long and slender; [N.S., the type from Sand Beach, Yarmouth Co.] . . . .
. . . . var. *inundata* Fern.
1 Leaves nearly all unlobed, tapering to slender or narrowly margined petioles.
4 Achenes awnless or with awns at most 1 mm long; [type from near North Bay, L.
Nipissing, Ont.] . . . . var. *submutica* Fassett
4 Achenes with awns to nearly 5 mm long.
5 Awns downwardly barbed; [*B. petiolata* Nutt.; Ont. (N to Thunder Bay and
Ottawa), Que. (N to Hull), N.B., and N.S.] . . . . var. *petiolata* (Nutt.) Farw.
5 Awns upwardly barbed; [*B. tripartita* f. *anom.* (Farw.) Boivin, this tentatively
reported from N.S. by Boivin 1966*b*] . . . . var. *anomala* Farw.

B. coronata (L.) Britt.
/t/EE/ (T) Rich moist ground and prairies from Minn. to S Ont. (Essex, Kent, Lambton, Huron,
and Norfolk counties) and Conn., S to Nebr., Ill., and Va. [*Coreopsis* L.; incl. var. *tenuiloba* (Gray)
Sherff and *B. trichosperma* (Michx.) Britt.].

B. discoidea (T. & G.) Britt.
/T/EE/ (T) Swampy ground and peaty or sandy shores from Minn. to Ont. (N to Ottawa), Que.
(N to the Montreal dist.), and N.S. (beaches at Pictou; Sherff 1937; not known from N.B. or P.E.I.), S
to Tex. and Ala. [*Coreopsis* T. & G.].

B. eatonii Fern.
/T/E/ (T) Tidal shores of Que., N.Y., Maine, Mass., and Conn. MAP and synonymy: *see* below.
1 Awns downwardly barbed.
2 Heads slenderly cylindric; [treated by Boivin 1966*b*, as identical with *B. tripartita* var.
heterodoxa f. *orthodoxa* (Fern. & St. John) Boivin (*B. het.* var. *orth.* Fern. & St. John),
this known from E Que. (Magdalen Is.; GH); MAP (aggregate species): Fassett 1928:
map 3, pl. 12] . . . . var. *eatonii*
2 Heads thick-cylindric to campanulate; [*B. heterodoxa* var. *interstes* Fassett; tidal
shores of the St. Lawrence R., Que., in Québec, Lévis, and Bellechasse counties] . . . .
. . . . var. *interstes* Fassett
1 Awns upwardly barbed; heads slenderly cylindric; [tidal flats of the St. Lawrence R., Que.,
near Quebec City] . . . . var. *fallax* Fern.

B. frondosa L.
/T/(X)/ (T) Damp open places (often a weed in cult. or waste ground; introd. in SE Alaska at ca.
55°N and probably in S B.C.; ranges of Canadian taxa outlined below), S to Calif., La., and Va.;
introd. in Europe. MAP and synonymy: *see* below.
1 Teeth of leaflets broadly triangular; outer phyllaries rarely more than twice as long as the
inner ones . . . . var. *frondosa*

2 Awns of achenes downwardly barbed; [incl. var. *pallida* Wieg., stated by Fernald *in* Gray 1950, to be probably a hybrid between *B. frondosa* and a form of *B. connata*; B.C. (New Westminster and Field, where probably introd.); Man. (N to The Pas), Ont. (N to near North Bay and Ottawa), Que. (N to L. St. John, the Gaspé Pen., and Magdalen Is.), Nfld., N.B., P.E.I., and N.S.; (introd. in Europe; Hultén 1958); MAP: Hultén 1968*b*:884] . . . . f. *frondosa*
2 Awns of achenes upwardly barbed; [Ont., Que., N.B., and N.S.] . . . f. *anomala* (Porter) Fern.
1 Teeth of leaflets lance-attenuate; outer phyllaries up to 4 times as long as the inner ones; [E Que. (Magdalen Is.), Nfld. (type from Whitbourne), and P.E.I.] . . . . var. *stenodonta* Fern. & St. John

B. heterodoxa (Fern.) Fern. & St. John

/T/E/ (T) Fresh, brackish, or saline marshes and borders of brackish ponds in E Que. (Magdalen Is.), P.E.I. (type from Bunbury, Queens Co.), and Conn. [Scarcely separable from *B. tripartita,* of which it is probably better treated as var. *heterodoxa* Fern.].

Some of the Magdalen Is., E Que., material is referable to var. *orthodoxa* Fern. & St. John (awns of the achenes retrorsely barbed rather than upwardly barbed; see *B. eatonii*).

B. hyperborea Greene

/sT/EE/ (T) River estuaries of E Canada (ranges of taxa outlined below), S to Maine, Mass., and N.J. MAPS and synonymy: *see* below.
1 Outer achenes 4 or 5 mm long; inner achenes at most 7 mm long, their awns at most 3 mm long.
   2 Stem simple; leaves blunt, subentire, to 4 cm long; outer phyllaries about 1.5 cm long; [incl. vars. *laurentiana* and *svensonii* Fassett; Ont. (Boivin 1966*b*) and Que. (S James Bay, the type from the mouth of the Rupert R.; Montmagny Co. to the Gaspé Pen.); MAPS (aggregate species): Potter 1932: map 11, p. 77; Fernald 1929: map 26 (not indicating the James Bay station), p. 1499] . . . . var. *hyperborea*
   2 Stem simple or branching; leaves long-acuminate, sharply toothed; outer phyllaries to 4 cm long; [*B. colpophila* Fern. & St. John; N.B. and N.S.] . . . . var. *colpophila* (Fern. & St. John) Fern.
1 Outer achenes to 8.5 mm long; inner achenes to 10 mm long, their awns to 5 mm long.
   3 Leaves thinnish, long-acuminate, with a prominent midrib, the larger ones to over 1.5 dm long and with up to about 12 pairs of sharp teeth; outer phyllaries up to 9 in number, linear-lanceolate, acute or acuminate; [type from the estuary of the Miramichi R. near Newcastle, N.B.] . . . . var. *arcuans* Fern.
   3 Leaves rather fleshy, obtusish, with the midrib less evident, mostly not over 1 dm long, entire or with a few mostly blunt teeth; outer phyllaries at most 6, obtuse to subacute.
      4 Leaves very fleshy, obtuse, at most about 5 cm long, entire or with mostly not more than a single pair of teeth; [Que. (near Quebec City; St. John and Dartmouth rivers, Gaspé Co., the type from the mouth of the Dartmouth R.)] . . . . var. *gaspensis* Fern.
      4 Leaves slightly fleshy, subattenuate, to over 1 dm long, with up to 5 pairs of teeth; (*see* the first lead 2, above) . . . . var. *hyperborea*

B. infirma Fern.

/T/E/ (T) Known only from tidal flats of the St. Lawrence R. estuary, Que., in Lotbinière, Québec, Montmorency, and Bellechasse counties; type from St-Vallier, Bellechasse Co. [Scarcely separable from *B. tripartita* and probably better treated as its var. *heterodoxa* f. *infirma* (Fern.) Boivin].

[B. laevis (L.) BSP.]

[Reports of this species of the U.S.A. (Ind. to N.H., S to Calif., Mexico, Tex., and Fla.) from S Ont. by Dodge (1914: Point Pelee, Essex Co.; 1915: Lambton Co.) and so-named collections in TRT from Cambridge (Galt) and Kitchener, Waterloo Co., require confirmation. (*B. chrysanthemoides* Michx.).]

[B. pilosa L.]
[A native of tropical America; reported as introd. in s Ont. by Boivin (1966b, as var. *radiata* Sch. Bip.; Point Pelee, Essex Co.), where probably not established.]

B. tripartita L.
Eurasian; locally introd. in swampy thickets and waste places, sometimes about ports, in N. America, as in ?B.C.–Alta. (Boivin 1966b) and Que. (Montreal dist.; Napierville; Percé, Gaspé Pen.). [Incl. *B. comosa* (Gray) Wieg.].

The scarcely separable *B. connata* Muhl., *B. eatonii* and *B. infirma* Fern., and *B. heterodoxa* (Fern.) Fern. & St. John should probably be merged with it, in which event the species would have a presumably native area in E N. America.

B. vulgata Greene   Beggar-ticks, Stick-tight
/T/X/   (T)   Moist ground, ditches, roadsides, thickets, and waste places (ranges of Canadian taxa outlined below), s to Calif., Mo., and Ga.
1   Principal leaves 1-pinnate .... var. *vulgata*
    2   Stem glabrous; [B.C. (Kootenay; Henry 1915), s Alta. (Medicine Hat; Seven Persons), Sask. (N to Nokomis, about 80 mi SE of Saskatoon), s Man. (N to Delta), Ont. (N to Thunder Bay, Timmins, and Ottawa), Que. (N to Ste-Anne-de-la-Pocatière, Kamouraska Co.), N.B., and N.S.] .... f. *vulgata*
    2   Stem crisp-puberulent; [*B. frondosa* var. *pub.* Wieg.; *B. pub.* (Wieg.) Rydb.; Sask., Man., Ont., and Que.] .... f. *puberula* (Wieg.) Fern.
1   Principal leaves 2-pinnate, the leaflets themselves sometimes deeply incised; [Sask. (Fernald *in* Gray 1950; not listed by Breitung 1957a) and w Ont. (near Thunder Bay (Fort William); Sherff 1937)] .... var. *schizantha* Lunell

[BLEPHARIPAPPUS Hook.]   [9248]

[B. scaber Hook.]
[The report of this species of the w U.S.A. (Wash. and Idaho to Calif. and Nev.) from SE B.C. by Henry (1915; Kootenay) requires confirmation. (*Ptilonella* Nutt.).]

BOLTONIA L'Hér.   [8892]

B. asteroides (L.) L'Hér.   White Boltonia
/T/EE/   (Hpr)   Shores and wet ground from s Sask.–Man. to N Ohio, N.Y., and N.J., s to Okla., Mo., and N.C.
1   Disk less than 1 cm broad; achenes at most 2 mm long, their awns wanting or rarely over 1 mm long; ligules usually lilac or purplish; phyllaries linear, at most 1 mm broad; [*Matricaria* L.; *B. glastifolia* l'Hér.; E U.S.A. only, reports from Canada being referable chiefly or wholly to var. *recognita*] .... [var. *asteroides*]
1   Disk to 1.5 cm broad; achenes about 2.5 mm long, their awns 1 or 2 mm long; ligules white to lilac.
    2   Phyllaries spatulate-oblong, 1 or 2 mm broad, rounded-obtuse below a short tip; [*B. latisquama* Gray; an escape in the E U.S.A.; collections have been seen from gardens in Simcoe, Norfolk, and Durham counties, s Ont.; probably never taken in the wild state in Canada] .... [var. *latisquama* (Gray) Cronq.]
    2   Phyllaries mostly linear and acute, at most 1.3 mm broad; [var. *occidentalis sensu* Boivin 1966b, perhaps not Gray; *B. latisquama* var. *recog.* F. & G.; s Sask. (*B. asteroides* reported from Weyburn and Torquay by Breitung 1957a) and s Man. (N to Aweme and near Winnipeg)] .... var. *recognita* (Fern. & Grisc.) Cronq.

BRICKELLIA Ell.   [8823]

1   Leaf-blades narrowly to broadly deltoid or subcordate, to 11 cm long and 7 cm broad, rather coarsely serrate, on petioles to 7 cm long; heads terminating the branches of a

corymbiform or subumbelliform inflorescence, this naked except for the leaves subtending a few of the main branches; involucres to 12 mm high, their outer phyllaries tipped with a well-developed slender awn; plant more or less finely puberulent or short-hairy throughout; (s B.C. and sw Alta.) .... *B. grandiflora*
1 Leaf-blades mostly oblong or elliptic-oblong, to 4 cm long and 1.5 cm broad, entire or nearly so, sessile or subsessile; heads terminating leafy-bracted branches of a corymbiform inflorescence; involucres to 2 cm high, their phyllaries acute or acuminate; plant glandular-puberulent; (s B.C.) .... *B. oblongifolia*

B. grandiflora (Hook.) Nutt.   Tassel-flower Brickellia
/T/WW/   (Grt)   Moist or dryish slopes and ledges at low to fairly high elevations from SE B.C. (South Kootenay Pass, near the Alta. boundary; V) and sw Alta. (Waterton Lakes; CAN) to Baja Calif., N.Mex., and Ark. [*Eupatorium* Hook.; *Coleosanthus* Ktze.; *C. (B.) umbellatus* Greene].

B. oblongifolia Nutt.
/t/W/   (Ch (Hp))   Dry, often rocky places, in the lowlands and foothills from s B.C. (collection in CAN from the Skagit R., SE of Hope, where taken by J.M. Macoun in 1905; collection in V, verified by Calder, from Hedley, about 20 mi SE of Princeton) and Mont. to Calif. and N.Mex. [*Coleosanthus* Ktze.; *B. linifolia* Eat.].

CACALIA L.   [9409]   Indian-plantain

1 Heads many-flowered, their involucres with up to about 15 phyllaries.
   2 Leaves triangular, acute, rather coarsely and doubly serrate, hastate at base but otherwise unlobed; involucre subtended by several linear-attenuate calyculate bractlets, with up to 30 flowers .... [*C. suaveolens*]
   2 Leaves cordate-rotund to reniform, palmately cleft, their segments again coarsely toothed or lobed (resembling those of *Petasites palmatus* but lacking divergent callus-tipped teeth); involucre not subtended by calyculate bractlets, with over 30 flowers; (sw B.C.) .... *C. nardosmia*
1 Heads with 5 or 6 florets, 4 or 5 phyllaries, and no calyculate bractlets.
   3 Principal leaves broadly cordate-reniform, to about twice as broad as long, very coarsely toothed and often shallowly lobed, their petioles conspicuously auriculate-clasping; receptacle plane; (Aleutian Is.) .... *C. auriculata*
   3 Principal leaves rarely noticeably broader than long, their petioles not auricled; receptacle commonly with a thickish fringe-like protuberance in the centre.
      4 Leaves green, lance-ovate or oval, entire or shallowly toothed, strongly 5–7-nerved longitudinally; stem angled and grooved; (s Ont.) .... *C. tuberosa*
      4 Leaves pale or glaucous beneath, pinnately veined, the larger ones triangular-ovate to cordate; stem terete; (?Ont) .... [*C. atriplicifolia*]

[*C. atriplicifolia* L.]   Pale Indian-plantain
[The report of this species of the E U.S.A. (N to Minn. and N.Y.) from Canada by Hooker (1834; this perhaps the basis of the w Ont. report by Torrey and Gray noted by John Macoun 1884) is probably based upon some habitally similar species such as *Prenanthes altissima*. (*Mesadenia* Raf.; *Senecio* Hook.).]

C. auriculata DC.
/sT/W/eA/   (Hp)   Thickets and subalpine meadows of the westernmost Aleutian Is. (*see* Hultén 1950: map 1190, p. 1682); E Asia. [Incl. var. *kamtschatica* (Maxim.) Matsum.]. MAP: Hultén 1968b:924.

C. nardosmia Gray
/T/W/   (Grh (Hs))   Meadows and open woods at moderate to high elevations from sw B.C. (Cascade Mts.; Boivin 1966b) to Calif. [*Luina* Cronq.; *Cacaliopsis* Gray; incl. var. *glabrata* (Piper) Boivin].

[C. suaveolens L.]
[The report of this species of the E U.S.A. (N to Iowa and Conn.) from N.S. by Lindsay (1878; Windsor, Hants Co.) is based upon *Erechtites hieracifolia* var. *intermedia,* the relevant collection in NSPM. (*Synosma* Raf.).]

C. tuberosa Nutt.
/t/EE/ (Grt (Gst)) Damp fields, prairies, and marly bogs from Minn. to S Ont. (Lambton, Huron, and Bruce counties; CAN; TRT), S to Tex. and Ala. [*Mesadenia* Britt.].

[CALENDULA L.] [9423]

1   Fruiting heads drooping; outer achenes narrow and not greatly curved, the middle ones boat-shaped, the innermost ones curved into a complete ring; leaves all oblong-lanceolate .... [*C. arvensis*]
1   Fruiting heads erect; achenes all or mostly incurved and boat-shaped; lower leaves oblong-ovate .... [*C. officinalis*]

[C. arvensis L.]
[European; an occasional garden-escape in N. America but scarcely established, as in SW B.C. (Carter and Newcombe 1921; Victoria, Vancouver Is.), Man. (Boivin 1966*b*), and N.B. (St. John, where taken by G.U. Hay in 1882; CAN).]

[C. officinalis L.]   Pot-Marigold
[European· an occasional garden-escape in N. America but scarcely established, as in SW B.C. (Vancouver Is.; Herb. V), Ont. (Boivin 1966*b*), Que. (Rawdon, about 50 mi N of Montreal; Ste-Flavie, SW Gaspé Pen.), Nfld. (Rouleau 1956), and N.S. (Annapolis; CAN, detd. Boivin).]

CARDUUS L.   [9461]   Plumeless Thistle

1   Heads mostly solitary and nodding; involucre to 5 cm thick, its phyllaries to 8 or 9 mm broad, the outer ones abruptly contracted to the strong-spined spreading or reflexed tip; stems usually lacking spiny wings for some distance below the red-purple heads .... *C. nutans*
1   Heads often clustered, not nodding; phyllaries less than 2 mm broad, tapering into a weak spine; heads usually red-purple (sometimes white or yellowish).
   2   Involucre hemispherical, to 2.5 cm thick, its outer phyllaries somewhat herbaceous and spreading; leaves slightly hairy beneath (chiefly along the main veins); plant very strongly spiny, the tough stem spiny-winged to the heads .... *C. acanthoides*
   2   Involucre ovoid, less than 1.5 cm thick, its rigid outer phyllaries scarcely spreading; leaves tomentose beneath; plant weakly spiny; stem brittle, spiny-winged to just below the heads .... *C. crispus*

C. acanthoides L.
European; introd. along roadsides and in fields and waste places in N. America, as in Ont. (N to the Ottawa dist.; Gillett 1958), Que. (St-Armand, Missisquoi Co.; Dundee, Huntingdon Co.), and N.S. (Yarmouth, Yarmouth Co.). MAP: G.A. Mulligan and C. Frankton, Can. Field-Nat. 68(1):33. 1954.

The typical form has purple flowers. Forma *albiflora* (L.) Gross (flowers white) is known from Grey Co., S Ont. Forma *ochranthus* Wallr. (flowers creamy yellow) is known from near Wiarton, Bruce Co., S Ont. A hybrid with *C. nutans* (× *C. orthocephalus* Wallr.) is reported from Grey Co., S Ont., by R.J. Moore and G.A. Mulligan (Can. J. Bot. 34(1):72. 1955) and a white-flowered phase of this purported parentage, f. *mulliganii* Boivin, is reported from the type locality, Flesherton, Grey Co., S Ont., by Boivin (1967*a*).

C. crispus L.   Welted Thistle
Eurasian; introd. along roadsides and in fields and waste places in N. America, as in S Ont. (Inglewood and Snelgrove, Peel Co.; Mulligan and Frankton, loc. cit.), N.B. (near St. John; John Macoun 1884), and N.S. (Cumberland, Pictou, and Cape Breton counties). MAP: Mulligan and Frankton, loc. cit., p. 33.

C. nutans L.   Musk-Thistle

Eurasian; introd. along roadsides and in fields and waste places in N. America, as in B.C. (Chilcotin; CAN, detd. Frankton), Sask. (N to Wilkie, about 30 mi SW of North Battleford; Breitung 1957a), Man. (Haywood, about 20 mi S of Portage la Prairie; DAO), Ont. (N to the Ottawa dist.; Gillett 1958), Que. (N to L. St. John), St-Pierre and Miquelon, Nfld. (Avalon Pen.; DAO), N.B. (Kent and Northumberland counties), and N.S. [Incl. *C. leiophyllus* var. *vestitus* Hal. (*C. nutans* var. *?petrovicii* Arènes) and *C. macrocephalus* Desf.]. MAP: Mulligan and Frankton, loc. cit., p. 33.

Some of our material is referable to the glabrous extreme, var. *leiophyllus* (Petrovic) Arènes (*C. leiophyllus* Petr.). MAP: Mulligan and Frankton, loc. cit., p. 33.

CENTAUREA L.   [9476]   Star-Thistle, Knapweed.   Centaurée

1   At least the middle and outer phyllaries tipped with a rigid spine, this spinose-ciliate along the lower margins; marginal flowers not enlarged; annuals or biennials; (star-thistles; introd.).

2   Stem and branches narrowly winged by the decurrent bases of the progressively reduced leaves, the larger (lower) leaves to about 2 dm long and 5 cm broad, some of them usually deeply lyrate-pinnatifid (middle leaves narrowly lanceolate or oblanceolate and remotely dentate, the upper ones linear or linear-oblong and entire); involucres subglobose, to over 1.5 cm high, not leafy at base; flowers yellow, at least the central ones with a pappus.

3   Larger spines of the phyllaries mostly 1 or 2 cm long; inner phyllaries tipped with a small dilated hyaline appendage; marginal flowers lacking a pappus, that of the inner flowers mostly 3–5 mm long; leaves persistently tomentose (as also the involucres), the basal and lower cauline ones deeply lyrate-pinnatifid; (introd. from Sask. to S Ont.) .... *C. solstitialis*

3   Larger spines of the phyllaries at most 9 mm long; inner phyllaries weakly spinose or merely tapering, not at all enlarged apically, generally purple-tinged; flowers all with a pappus, this mostly 1.5–3 mm long; leaves lightly cobwebby when young (not tomentose), the basal and lower cauline ones dentate to lyrate-pinnatifid; (introd. in S B.C.) .... *C. melitensis*

2   Stem merely angled (not winged); leaves mostly smaller, the principal ones all deeply pinnatifid, only the upper ones becoming irregularly dentate or entire; involucres ovoid, to about 1 cm long; flowers lacking a pappus.

4   Phyllary-spine slender, rarely as much as 5 mm long; involucres not leafy at base; flowers creamy or sometimes purplish; (introd. in S B.C. and S Alta.) .... *C. diffusa*

4   Phyllary-spine very stout, to over 3 cm long, 3 or 4 mm broad at the base; involucre subtended by a whorl of entire or irregularly dentate reduced leaves; flowers purple; (introd. on Vancouver Is. and in S Ont.) .... *C. calcitrapa*

1   Phyllaries not spiny, the terminal appendage at most minutely prickly; (knapweeds; introd.).

5   Principal stem-leaves pinnately divided into linear to lanceolate, entire to pinnatifid segments; flowers purplish, the marginal ones enlarged and falsely radiate; pappus to 2 or 3 mm long.

6   Involucre to 2.5 cm high, its cobwebby middle and outer phyllaries with dark pectinate horseshoe-shaped strongly decurrent tips to about 6 mm long; pappus mostly at least 3 mm long; more or less hirsute perennial from a stout taproot, the summit of the caudex clothed with the elongate fibrous remnants of previous years' leaves; (introd. in SW B.C. and from S Ont. to Nfld. and N.B.) .... *C. scabiosa*

6   Involucre to about 1.5 cm high, its smooth, strongly ribbed, middle and outer phyllaries with only slightly decurrent dark pectinate tips 1 or 2 mm long; pappus to 2 mm long, or rarely wanting; canescent and more or less glandular-punctate biennial with bushy-branched wiry stems; (introd. in S B.C. and from Ont. to N.S.) .... *C. maculosa*

5   Principal stem-leaves simple or at most coarsely toothed or lyrate-based.

7   Flowers yellow, the outer ones not enlarged; heads large, to about 1 dm broad;

leaves ovate-lanceolate, somewhat serrate, scabrous; (introd. in sw Que.) . . . .
. . . . [*C. macrocephala*]
7 Flowers not yellow; heads smaller.
  8 Involucre pale green, to 1.5 cm high, the middle and outer phyllaries with
    subentire broad unappendaged silvery-hyaline tips, the inner with narrower
    plumose-hairy tips; flowers roseate or purplish, the marginal ones not
    enlarged; pappus to over 1 cm long, deciduous; leaves linear-lanceolate to
    narrowly oblong, entire, dentate, or slightly pinnatifid, cobwebby to glabrate;
    coarse bushy-branched perennial from a deep, blackish rhizome, rooting at
    the creeping base; (introd. from B.C. to s Ont.) . . . . *C. repens*
  8 Involucre brownish to nearly black at least at the tips of the phyllaries (or pale
    green in *C. cyanus*); pappus none or at most about 3 mm long.
    9 White-flocculent annual with linear to narrowly lanceolate, essentially
      entire leaves less than 1 cm broad; involucre to about 1.5 cm high; middle
      phyllaries terminated by a white or silvery (often darkened), lacerate or
      pectinate, relatively narrow horseshoe-shaped appendage; flowers blue,
      violet, pink, or white, the outer ones enlarged and falsely radiate; pappus
      usually 2 or 3 mm long; (introd., transcontinental) . . . . *C. cyanus*
    9 Green perennials or biennials with lanceolate to elliptic, entire to toothed
      stem-leaves and lanceolate to broadly oblanceolate or oval, entire to
      sinuate or more or less pinnatifid rosette-leaves.
      10 Leaves entire or merely toothed, broadly lanceolate to obovate-
        oblanceolate, toothed, silvery-white when young, their bases decurrent
        on the stem; flowers blue, the marginal ones enlarged and falsely
        radiate; phyllaries with black, fringed margins; perennial, often
        stoloniferous; (introd. in s B.C. and from Ont. to Nfld. and N.B.) . . . .
        . . . . *C. montana*
      10 At least the basal or rosette-leaves usually sinuate to more or less
        lobed or pinnatifid; flowers roseate to rose-purple; at least the middle
        phyllaries tipped with a conspicuous ovate to orbicular scarious
        appendage.
        11 Phyllary-appendages entire, erose, or irregularly lacerate (not
          regularly toothed or pectinate), tan to dark brown, the inner ones
          often strongly 2-cleft; involucre to 2.5 cm high; stem-leaves hard,
          usually with a pair of short narrow lobes at base; rosette-leaves
          entire or more or less pinnatifid; marginal flowers enlarged and
          falsely radiate; pappus none; plant glabrous or somewhat cob-
          webby; (introd. on Vancouver Is. and in Ont. and Que.) . . . . *C. jacea*
        11 Phyllary-appendages (of at least the middle and lower phyllaries)
          regularly pectinate or toothed, seldom obviously 2-cleft; plants
          usually more or less rough-puberulent.
          12 Middle and outer phyllaries with conspicuous greenish blades
            and blackish pectinate appendages mostly not over 3 mm long;
            pressed involucre at least as high as broad; pappus usually
            none; marginal flowers enlarged and falsely radiate; lower
            leaves mostly oblanceolate, often coarsely and irregularly
            few-lobed; (introd. on Vancouver Is. and in s Ont.) . . . . *C. dubia*
          12 Middle and outer phyllaries with tawny to blackish, pectinate or
            fringed appendages to 6 mm long and hiding the blades;
            pressed involucre usually broader than high; pappus usually
            evident; lower leaves usually merely more or less sinuate.
            13 Marginal flowers not enlarged; involucre deep brown to
              blackish.
              14 Stem-leaves elliptic to oblong, the uppermost ones
                blunt; fringe of appendages only about as long as the
                breadth of the blade; (introd. in St-Pierre and Miquelon
                and N.S.) . . . . *C. nigrescens*

14 Stem-leaves oblong-lanceolate, the uppermost reduced
ones pointed; fringe of median appendages 2 or 3 times
as long as the breadth of the blade; (introd. in s B.C.
and from Ont. to the Atlantic Provinces) .... *C. nigra*
13 Marginal flowers enlarged and falsely radiate; involucre
tawny, rarely blackish; (var. *radiata;* introd. in s Ont., Nfld.,
and N.S.) .... *C. nigra*

C. calcitrapa L.   Star-Thistle, Caltrops
Eurasian; introd. along roadsides and in fields and waste places in N. America, as in sw B.C.
(Victoria and Nanaimo, Vancouver Is.; CAN; V) and s Ont. (Middlesex, Waterloo, York, Grey, and
Perth counties; CAN; OAC; TRT).

C. cyanus L.   Bachelor's-button, Bluebottle, Cornflower.   Bleuet
European; a garden-escape to roadsides, fields, and waste places in N. America, as in B.C.
(Vancouver Is. and adjacent islands; Kootenay L.), Alta. (N to near Notikewin, ca. 57°N; Groh 1949),
Man. (N to Norway House, off the NE end of L. Winnipeg), Ont. (N to North Bay and Ottawa), Que. (N
to the s Gaspé Pen. at Carleton; GH), Nfld., N.B., P.E.I., and N.S.; s Greenland.

C. diffusa Lam.
Eurasian; locally introd. along roadsides and in fields and waste places in N. America and rapidly
spreading, as in s B.C. (valleys of the Dry Interior N to Cache Creek, Kamloops, and Vernon;
particularly abundant near Grand Forks, near the U.S.A. boundary w of Trail) and s Alta. (Boivin
1966*b*).

C. dubia Suter
Eurasian; introd. along roadsides and in fields and waste places in N. America, as in sw B.C.
(Victoria, Vancouver Is.; Herb. V) and s Ont. (Norfolk, Middlesex, Bruce, Waterloo, Wellington, and
Carleton counties).
    At least our Ont. material is referable to ssp. *vochinensis* (Bernh.) Hayek (*C. voch.* Bernh.; upper
stem-leaves rather abruptly reduced rather than passing gradually into the branch-leaves; the short
dark tips of the involucral phyllaries contrasting strongly with the pale body rather than not
conspicuously contrasting).

C. jacea L.
European; introd. along roadsides and in fields and waste places in N. America, as in sw B.C.
(Victoria, Vancouver Is.; Henry 1915), Ont. (N to North Bay and Ottawa), Que. (N to Val-d'Or,
48°06'N; Baldwin 1958); s Greenland. [Incl. var. *lacera* Koch].
    A hybrid with *C. nigra* is reported by Gaiser and Moore (1966) as occupying large stands near
Sarnia, Lambton Co., s Ont.

[C. macrocephala Puschk.]
[A native of Armenia; cult. as an ornamental in N. America and reported from sw Que. by Boivin
(1966*b*; Hull), where presumably a garden-escape but scarcely established.]

C. maculosa Lam.
European; introd. along roadsides and in fields and waste places in N. America and rapidly
spreading, as in sw B.C. (Vancouver Is. and adjacent islands and mainland N to Kamloops and
Revelstoke; CAN; V), Ont. (N to the SE shore of L. Superior at Pancake Bay), Que. (N to Montebello
and the Montreal dist.), N.B. (Boivin 1966*b*), and N.S. (Kings Co.; CAN; ACAD). [*C. paniculata*
Bieb., not L.].

C. melitensis L.
European; occasionally introd. in waste places and on ballast in N. America, as in sw B.C.
(Nanaimo and Esquimault, Vancouver Is.; Henry 1915).

C. montana L.   Mountain-Bluet
European; a garden-escape to roadsides and waste places in N. America, as in sw B.C. (Vancouver Is. and Chilliwack; Herb. V), s Ont. (Guelph, Wellington Co.; OAC), Que. (N to near Quebec City; Marcel Raymond and James Kucyniak, Rhodora 50(595):179. 1948), Nfld. (St. John's; GH), and N.B. (Fredericton and Dalhousie; CAN).

C. nigra L.   Knapweed, Spanish-buttons
European; introd. along roadsides and in fields and waste places in N. America, as in s B.C. (Vancouver Is.; Vancouver; Nakusp, about 50 mi SE of Revelstoke; Westwold, about 25 mi SE of Kamloops), Ont. (N to Ottawa), Que. (N to Anticosti Is. and the Gaspé Pen.), St-Pierre and Miquelon, Nfld., N.B., P.E.I., and N.S.
   Var. *radiata* DC. (marginal florets enlarged, the heads falsely radiate rather than discoid and usually paler than in the typical form) is known from s Ont. (Lambton Co.; OAC), Nfld. (GH), and N.S. (Kings and Hants counties). The typical form has rose-purple flowers. Forma *pallens* Spenn (flowers whitish) is reported from E Que., St-Pierre and Miquelon, and N.S. by Boivin (1966*b*).

C. nigrescens Willd.
European; introd. along roadsides and in fields and waste places in N. America, as in St-Pierre and Miquelon (Fernald *in* Gray 1950) and N.S. (Middleton, Annapolis Co.; CAN; GH).

C. repens L.   Russian Knapweed, Turkestan Thistle
Asiatic; introd. along roadsides and in fields and waste places in N. America and rapidly spreading as a noxious weed, as in B.C. (N to Vanderhoof, w of Prince George at ca. 54°N), Alta. (Lacombe and Blackfalds; Groh 1944*b*), Sask. (N to near Landis and Saskatoon; Breitung 1957*a*), sw Man. (Hartney; Melita; Boissevain), and s Ont. (Brant, York, Simcoe, and Frontenac counties). [*C. (Acroptilon) picris* Pall.].

C. scabiosa L.   Knapweed
Eurasian; introd. or a garden-escape to roadsides, fields, and pastures in N. America, as in sw B.C. (Vancouver Is.; Herb. V), s Ont. (Wellington and Peel counties), Que. (Farnham, Missisquoi Co.; Carleton and St-Omer, s Gaspé Pen.), Nfld. (Rouleau 1956), and N.B. (Boivin 1966*b*).

C. solstitialis L.   Yellow Star-Thistle, Barnaby's Thistle
Eurasian; introd. along roadsides and in fields and waste places in N. America, as in Sask. (Scott, about 30 mi sw of North Battleford; Breitung 1957*a*), s Man. (Shellmouth, about 100 mi NW of Brandon; DAO), and s Ont. (Huron, Middlesex, and Wellington counties).

CHAENACTIS DC.   [9299]

C. douglasii (Hook.) H. & A.
/T/W/ (Hs) Dry, often sandy or rocky places at low to fairly high elevations from s B.C. (chiefly valleys of the Dry Interior N to near Lillooet, Spences Bridge, Nicola, and Vernon, E to Cascade, near the U.S.A. boundary sw of Trail; the report from the Belly R., s Alta., by John Macoun 1884, may refer to some habitally similar species such as *Aster tanacetifolius* or *Hymenopappus filifolius,* these both known from that locality; Macoun's report from Wood Mountain, s Sask., also requires clarification) and Mont. to Calif., Ariz., and Colo. [*Hymenopappus* Hook.; incl. the reduced alpine extremes, var. *alpina* Gray (*C. alpina* (Gray) Jones) and var. *montana* Jones]. MAP: Palmer Stockwell, Contrib. Dudley Herb. 3(4): pl. 41, p. 163. 1940.

CHRYSANTHEMUM L.   [9341] Chrysanthemum.   Chrysanthème

1   Heads discoid or nearly so (rayless or with minute white rays), the disk to 7 mm broad; pappus with a minute crown; leaves elliptic to narrowly ovate, closely crenate, often with a pair of small lobes at base, the radical ones long-petioled, the cauline ones sessile or nearly so; coarse fragrant perennial; (introd.) .... *C. balsamita*
1   Heads radiate; plant not fragrant (but sometimes with a pungent odour).
    2   Rays yellow; annuals; (introd.).

3 Principal leaves rather shallowly pinnatifid; ray-achenes narrowly 2-winged; rays golden yellow .... *C. segetum*
3 Principal leaves pinnate-pinnatifid; ray-achenes narrowly 3-winged; rays yellow or yellowish white .... *C. coronarium*
2 Rays white; perennials.
    4 Leaves linear, entire, to about 2 cm long, villous, chiefly basal or sub-basal, the cauline ones mostly bract-like; stem usually less than 1 dm tall, 1-flowered, woolly above, branching at base; (Alaska and N B.C. to Baffin Is. and N Que.) .... *C. integrifolium*
    4 Leaves broader in outline, more or less toothed or lobed; stem usually taller.
        5 Leaves 2-pinnatifid into ovate, irregularly cut divisions, finely puberulent at least beneath; heads small, several to many in a corymbiform inflorescence; disk not over 1 cm broad; rays less than 1 cm long; plant bushy-branched; (introd.) .... *C. parthenium*
        5 Leaves serrate to deeply lobed or 1-pinnatifid; heads larger.
            6 Stem not over 5 dm tall, with usually a solitary flower to 4 cm broad (including rays); leaves glabrous, rather fleshy, the upper ones linear, the lower ones cuneate-spatulate and with triangular lobes; (Alaska and N B.C. to N Que.) .... *C. arcticum*
            6 Stem to over 1 m tall; heads often broader; (introd.).
                7 Stem-leaves acute to acuminate, deeply and sharply serrate with forward-pointing teeth; heads numerous; stem much branched, to 2 m tall .... *C. uliginosum*
                7 Stem-leaves obtuse or subacute, not deeply toothed.
                    8 Heads relatively small; leaves remotely toothed or notched; stem simple or little branched, rarely as much as 1 m tall; (transcontinental) .... *C. leucanthemum*
                    8 Heads large and showy, often "doubled"; leaves shallowly and closely toothed; stem branched toward the top .... [*C. lacustre*]

[C. alpinum L.]
[European; according to Polunin (1959), this species (not keyed out above) "has been reported to grow in West Greenland, presumably as an ephemerophyte."]

C. arcticum L.
/aSs/(X)/EA/ (Hsr) Coastal salt-marshes and muddy, gravelly, or rocky shores from NW B.C. (Larcom Is., Observatory Inlet, near the S tip of the Alaska Panhandle at ca. 55°N, where taken by McEvoy in 1893; CAN), the Aleutian Is., and coasts of Alaska–Yukon–Dist. Mackenzie–Dist. Keewatin to NE Man. (York Factory and Churchill), N Ont. (Hudson Bay–James Bay S to ca. 52°N), islands in James Bay, and NW Que. (Hudson Strait S to SE James Bay at ca. 51°45′N); NE Europe; NE Asia. [*Leucanthemum* DC.]. MAPS: Eric Hultén 1968*b*:893 (combine the maps for ssp. *arct.* and ssp. *pol.*), and Sven. Bot. Tidskr. 43(2/3): fig. 10, p. 404. 1949.

Our more northern material is referable to the reduced extreme, ssp. *polare* Hult. (*Leucanthemum (Dendranthema) hultenii* Löve & Löve; *Cakile edentula sensu* Lowe 1943, not (Bigel.) Hook., the relevant collection in Herb. Man. Prov. Mus., Winnipeg). MAP: Hultén 1968*b*:893.

C. balsamita L.   Costmary, Mint-geranium
Eurasian; a garden-escape to roadsides and waste places in N. America, as in SW B.C. (Fort Langley; Groh 1946), Sask. (Boivin 1966*b*), Ont. (N to the E shore of L. Superior at Batchawana Bay; Hosie 1938), Que. (N to gravelly shores at Port Daniel, S Gaspé Pen.; MT; GH), and N.S. (Granville, Annapolis Co.; GH). [*Balsamita* Rydb.; *Tanacetum* L.; incl. the completely rayless extreme, var. *tanacetoides* Boiss., to which much of our material appears referable].

C. coronarium L.   Garland-Chrysanthemum
European; an occasional garden-escape in N. America, as in S Ont. (Tillsonburg, Oxford Co., where taken by John Macoun in 1901; CAN) and N.B. (ballast at St. John; Fowler 1885).

C. integrifolium Richards.
/aSs/X/eA/ (Ch (Hr)) Gravelly or rocky calcareous barrens and solifluction areas from the coasts of Alaska–Yukon–Dist. Mackenzie–Dist. Keewatin to Banks Is., Devon Is., and Baffin Is., s to northernmost B.C. (s to Summit Pass at 58°31'N; CAN), Great Bear L., Southampton Is., and northernmost Que. (Akpatok Is., Ungava Bay; Nicholas Polunin, J. Bot. 72:204. 1934); N Kamchatka and NE Siberia. [*Leucanthemum* DC.]. MAPS: Hultén 1968*b*:895; Porsild 1957: map 316, p. 200; *Atlas of Canada* 1957: map 7, sheet 38; Fernald 1925: map 60 (incomplete northwards), p. 325.

The type locality was given as "On the Copper Mountains", probably near Coronation Gulf on the Coppermine R., N Dist. Mackenzie.

[C. lacustre Brot.] Portuguese Daisy
[European; cult. as an ornamental in N. America and reported by Boivin (1966*b*) as introd. at Ottawa, Ont., where, however, probably not established.]

C. leucanthemum L. Ox-eye-Daisy. Marguerite
Eurasian; introd. along roadsides and in fields, meadows, pastures, and waste places in N. America, as in the E Aleutian Is., Alaska (N to ca. 61°30'N), the Yukon (N to ca. 64°N), B.C., Alta. (N to Wood Buffalo National Park at 59°34'N), Sask. (N to Montreal Lake, 54°03'N), Man. (N to Churchill), Ont. (N to w James Bay at ca. 53°N), Que. (N to E James Bay at ca. 51°30'N and the Côte-Nord), Labrador (N to the Hamilton R. basin), Nfld., N.B., P.E.I., and N.S.; sw Greenland. [*Leucanthemum* Rydb.; *L. vulgare* Lam.; incl. var. *subpinnatifidum* Fern.]. MAP: Hultén 1968*b*:894.

Our material includes diploid (2n = 18) and tetraploid (2n = 36) taxa, separated by T.W. Böcher and K. Larsen (Watsonia 4(1):11–16. 1957) as follows:
1 Lower leaves often subpinnatifid, with irregular alternating lobes or teeth; upper leaves linear-lanceolate, with pinnatifid bases but not very regularly toothed; [diploid, the pollen-grains averaging slightly smaller than those of the tetraploid race; "Frequent in north-west Europe"] .... *C. leucanthemum*
1 Lower leaves regularly toothed but not lobed; upper leaves averaging somewhat broader, regularly and mostly closely toothed, not or only slightly pinnatifid at base; [tetraploid; the more widespread Eurasian phase; *C. leuc.* var. *?boecheri* Boivin] .... *C. ircutianum* Turcz.

C. leucanthemum as interpreted above is evidently the widespread N. American plant separated as var. *pinnatifidum* Lec. & Lam. (var. *laciniatum* Vis.; *Leucanthemum ircutianum* var. *pinn.* (Lec. & Lam.) Löve & Bernard). The tetraploid plant, on the other hand, although the common Eurasian phase, is relatively localized in its N. American distribution. For further discussion, *see* M.L. Fernald (Rhodora 5(55):177–81. 1903), Löve and Bernard (1959), and G.A. Mulligan (Rhodora 60(713):122–25. 1958).

C. parthenium (L.) Bernh. Feverfew
Eurasian; a garden-escape to roadsides, fields, and waste places in N. America, as in sw B.C. (Vancouver Is. and adjacent mainland; CAN; V), s Ont. (N to Victoria and Stormont counties), Que. (N to the Gaspé Pen. at York), N.B. (Fowler 1879), P.E.I. (Herbert Groh, Sci. Agric. 7(10):390. 1927), and N.S. (Mill Brook, Pictou Co.; ACAD; CAN). [*Matricaria* L.; *Leucanthemum* Godr.].

C. segetum L. Corn-Marigold
Eurasian; a garden-escape to roadsides, fields, and waste places in N. America, as in the w Aleutian Is. (Kiska), s Ont. (Lambton, Norfolk, and Wellington counties), ?Que. (Boivin 1966*b*), Nfld. (St. Anthony; GH), N.B. (wharf-ballast at St. John and Carleton; CAN), P.E.I. (Hurst 1952), and N.S. (Truro; MT). MAP: Hultén 1968*b*:894.

C. uliginosum Pers. High Daisy
European; a local garden-escape in N. America, as in B.C. (Boivin 1966*b*), Ont. (Ottawa dist.; Morrisburg, Dundas Co.), and sw Que. (Lachine, near Montreal, where first taken by Kucyniak in 1939; GH; MT; reported by James Kucyniak, Rhodora 47(563):389–90. 1945, as still well established at this later date in a moist depression in a pasture). [*Pyrethrum* Waldst. & Kit.].

CHRYSOPSIS Ell. [8844] Golden Aster

1 Leaves linear-attenuate to -oblong, silvery-pubescent, often arching or falcate, with nearly
   parallel nerves or ribs; involucres campanulate; achenes compressed-fusiform; (introd. in
   s Ont.) .... *C. falcata*
1 Leaves typically oblanceolate to oblong, not parallel-nerved; involucres hemispherical;
   achenes obovate, more or less flattened; (B.C. to Man.) .... *C. villosa*

C. falcata (Pursh) Ell.
Native in dry sandy soil of the Atlantic Coastal Plain from Mass. to N.J. There is a collection in DAO
from along railway tracks w of Toronto, Ont., where taken by Sharp in 1949. According to W.J.
Cody (Rhodora 54(648):308. 1952), "The plant had what appeared to be good fruit and, since it is a
perennial, it is quite possible that the species may persist and spread as a weed."

C. villosa (Pursh) Nutt.
/T/WW/ (Hp) Dry prairies, dunes, and roadsides from B.C. (N to Hudson Hope, ca. 56°N; Raup
1934), Alta. (N to Fort Saskatchewan; CAN), Sask. (N to Carlton, about 35 mi sw of Prince Albert),
and Man. (N to Bield, about 120 mi NW of Brandon), s to Calif., N.Mex., Okla., Mo., Ill., and w Ind.
[*Amellus* Pursh; *Diplopappus* Hook.; *C. bakeri* and *C. hirsutissima* Greene; *C. ballardii* Rydb.; incl.
the narrow-leaved extreme, var. *angustifolia* (Rydb.) Cronq. (*C. ang.* Rydb.)].
   Some of our material is referable to var. *hispida* (Hook.) Gray (*Diplopappus (C.) hispidus* Hook.;
C. arida Nels.; *C. barbata* and *C. butleri* Rydb.; *C. columbiana* Greene; pubescence of leaves and
involucre consisting of spreading hairs (these sometimes glandular) rather than appressed and
seldom glandular).

CHRYSOTHAMNUS Nutt. [8855] Rabbit-brush, Rabbit-bush

1 Plant more or less felted-tomentose; leaves not much twisted, at most 3 mm broad and
   with 3 nerves; involucres to 13 mm high, with usually at least 20 phyllaries; disk-corollas
   to about 1 cm long; (B.C. to Sask.) .... *C. nauseosus*
1 Plant glabrous or minutely spreading-puberulent, not at all tomentose; leaves often
   twisted, to 1 cm broad and with up to 5 nerves; involucres to 8 mm high, with about 15
   phyllaries; disk-corollas to 7 mm long; (s B.C.) .... *C. viscidiflorus*

C. nauseosus (Pall.) Britt.
/T/WW/ (N) Dry valleys, plains, and foothills (ranges of Canadian taxa outlined below), s to
Calif., N Mexico, and Tex.
1 Plant woody only at base and seldom over 6 dm tall; [*Chrysocoma* Pall.; *Chrysothamnus*
   frigidus and *C. plattensis* Greene; *C. pulcherrimus* Nels.; s B.C. (N to Kamloops), s Alta.
   (Moss 1959), and sw Sask. (Cypress Hills; Cadillac; Val Marie)] .... var. *nauseosus*
1 Plant more woody, to about 2 m tall.
   2 Phyllaries (at least the outer ones) more or less tomentose-puberulent; tomentum of
      leaves and twigs relatively dense and persistent, greyish or sometimes white; [*C.*
      speciosus var. *alb.* Nutt.; *C. macounii* Greene; s B.C.] .... var. *albicaulis* (Nutt.) Rydb.
   2 Phyllaries nearly or quite glabrous; tomentum of leaves and twigs thin and less persistent, that
      of the twigs close and usually light yellowish-green; [var. *graveolens* (Nutt.) Hall; *C.*
      (Bigelowia; Chrysocoma) graveolens (Nutt.) Greene; *B. grav.* vars. *glab.* and
      hololeuca Gray; s B.C. (Boivin 1966b) and SE Sask. (Estevan; Breitung 1957a)] ....
      .... var. *glabratus* (Gray) Cronq.

C. viscidiflorus (Hook.) Nutt.
/t/WW/ (N) Dry valleys, plains, and foothills from s B.C. (apparently N to Cache Creek, W of
Kamloops; see below) to N.Dak., s to Calif. and N.Mex.
1 Leaves to 4 mm broad, 1–3-nerved, they and the twigs glabrous or the leaves merely with
   marginal ciliation; [*Crinitaria* Hook.; *Bigelowia* DC.; *B. douglasii* Gray; reports from s B.C.
   apparently refer chiefly to var. *lanceolatus*] .... [var. *viscidiflorus*]
1 Leaves and twigs more or less densely puberulent especially in the inflorescence.

2   Leaves to 1 cm broad, 3–5-nerved, seldom much twisted; [*C. lanceolatus* Nutt.; s
    B.C. (the reports of *Bigelowia douglasii* from Cache Creek, w of Kamloops, and from
    the upper valley of the Columbia R. within B.C. by John Macoun 1884, probably
    belong here] .... var. *lanceolatus* (Nutt.) Greene
2   Leaves mostly about 1 mm broad and 1(3)-nerved, more or less twisted; [*Linosyris*
    visc. var. *pub.* Eat. (*C. pub.* (Eat.) Greene); reported from B.C. by Rydberg 1922, and
    a collection in CAN from L. Osoyoos appears referable here] ... var. *puberulus* (Eat.) Jeps.

CICHORIUM L.   [9553]   Chicory, Succory.   Chicorée

1   Leaves glabrous, the basal ones merely sinuate-toothed to more or less deeply lobed;
    heads purple, the stalk of the terminal one conspicuously thickened; phyllaries glabrous;
    pappus up to 1/2 as long as the achenes; bracts commonly surpassing the heads; annual
    or biennial; (garden-escape in Alta.) .... [*C. endivia*]
1   Lower and basal leaves bristly-hairy at least on the midrib beneath, toothed to
    runcinate-pinnatifid; heads blue, pink, or white, the stalk of the terminal one less
    thickened; phyllaries glandular-bristly; pappus at most 1/8 as long as the achenes; bracts
    commonly shorter than the heads; stout deep-rooted perennial; (introd., transcontinental) ......
    .... *C. intybus*

[*C. endivia* L.]   Endive
[Asiatic; cult. in N. America and reported as introd. in Alta. by Boivin (1966*b*: Lethbridge), where
undoubtedly not established.]

C. intybus L.   Blue Sailors
Eurasian; introd. along roadsides and in fields and waste places in N. America, often locally
abundant, as in B.C. (N to Kamloops and Sicamous), Alta. (N to Fort Saskatchewan), Sask., Man. (N
to Winnipeg), Ont. (N to Thunder Bay and Cochrane, ca. 49°N), Que. (N to Anticosti Is. and the
Gaspé Pen.), Labrador (Boivin 1966*b*), St-Pierre and Miquelon, Nfld., N.B., P.E.I., and N.S.
    Forma *album* Neum. (flowers white rather than bright blue (or pink in f. *roseum* Neum.)) is known
from s Ont. (Ajax, Ontario Co.; OAC) and sw Que. (Montreal; MT).

CIRSIUM Mill.   [9462]   Common or Plumed Thistle.   Chardon

1   Flowering stems arising from extensively creeping and freely sprouting perennial roots;
    heads unisexual, the pappus of the pistillate heads surpassing the pink-purple
    (sometimes white) corollas, that of the staminate heads shorter than the corollas;
    involucre 1 or 2 cm high; outer phyllaries tipped with a weak prickle about 1 mm long;
    (introd.) .... *C. arvense*
1   Flowering stems arising from the centre of the previous year's basal rosette, the root
    biennial or perennial; flowers perfect in all heads; involucre mostly higher.
    2   Upper half of stem and branches with prickly elongate wings decurrent from the
        leaf-bases nearly or quite to the node below; flowers purple; (introd.).
        3   Involucre to over 3 cm high, nearly all of its linear to lanceolate phyllaries tipped
            with long spreading prickles; heads mostly scattered; leaves scabrous-hispid and
            green above, thinly white-woolly beneath; stem to about 2 m tall .... *C. vulgare*
        3   Involucre to about 1.5 cm high, the ovate-lanceolate outer phyllaries tipped with a
            prickle about 1 mm long, the inner phyllaries with soft purple linear tips; heads
            clustered; leaves green and hirsute above, paler and pubescent beneath; stem
            rarely over 12 dm tall .... *C. palustre*
    2   Stem and branches not long-winged, the leaves not at all or only slightly decurrent,
        more or less cobwebby-villous, lanate, or tomentose (or glabrous, especially above).
        4   Flowers white or creamy (or often pale purple in *C. foliosum*); outer and middle
            phyllaries prickle-tipped.
            5   Leaves with a very narrowly winged rachis and remote, linear, entire or basally
                few-toothed segments commonly 3 or 4 cm long, densely and persistently

white-tomentose beneath (less so above), the stem also densely white-tomentose; involucre to 3 cm high, the outer ovate phyllaries tipped with prickles to about 2 mm long, the inner ones with a slender weak bristle; (sandy shores of the Great Lakes in s Ont.) .... *C. pitcheri*

5 Leaves with a very broadly winged rachis, the ovate segments commonly less than twice as long as basally broad.

  6 Stem and lower leaf-surfaces densely and persistently white-tomentose, not at all cobwebby, the leaves soon green and subglabrate above; involucre commonly about 2 cm high, the outer and middle phyllaries with short spine-tips and some of them with a dark thickened glutinous dorsal ridge; inflorescence open, the heads terminating the main branches; plant tending to spread by short creeping roots; (?Sask.) .... [*C. brevifolium*]

  6 Stem more or less cobwebby; leaves thinly tomentose or subglabrous beneath, cobwebby or subglabrate above; involucre commonly higher, its phyllaries nearly or quite lacking a thickened glutinous dorsal ridge; inflorescence compact, the heads commonly sessile or short-peduncled in a terminal cluster; creeping roots nearly or quite wanting; (B.C. and Alta.; *C. foliosum* also in E Que.).

    7 Phyllaries relatively broad (middle and outer ones mostly lanceolate to narrowly ovate) and more or less distinctly overlapping below, generally glabrous or nearly so, the outer ones with short spine-tips, some of the inner ones often with dilated fringed tips; stem thick and succulent to summit; (B.C. and Alta.; E Que.) .... *C. foliosum*

    7 Phyllaries of involucre relatively narrow and gradually tapering, mostly narrowly lance-linear, more or less strongly cobwebby (especially along the margins), rarely any of them with dilated fringed tips; stem tapering and becoming slender above.

      8 Plant very strongly spiny; corollas usually less than 2 cm long; phyllaries with terminal spines to about 1 cm long; (?Alta.) .... [*C. polyphyllum*]

      8 Plant only moderately spiny; corollas usually over 2 cm long; phyllaries with short spine-tips; (B.C. and Alta.) .... *C. hookerianum*

4 Flowers normally pink-purple to purplish red.

  9 Leaves densely and persistently white-tomentose beneath; phyllaries rather narrowly lanceolate or lance-ovate, commonly not over 3 or 4 mm broad, the outer and middle ones spine-tipped and often with a dark thickened glutinous dorsal ridge; inflorescence loose, the head or heads terminating long branches.

   10 Upper leaf-surfaces green and glabrous or merely minutely hirsute; involucre 2 or 3 cm high.

    11 Principal stem-leaves divided nearly to midrib into linear-lanceolate to narrowly oblong, remote, simple or forking lobes, the spines strongly developed; inner phyllaries with long-attenuate tips; (SE Man. to Que.) .... *C. discolor*

    11 Principal stem-leaves undivided, the narrowly oblanceolate to oblong-ovate blade merely sinuate-toothed or undulate-pinnatifid, with weakly developed spines; some of the inner phyllaries with a lanceolate or deltoid, usually serrulate, dilated tip .... [*C. altissimum*]

   10 Upper leaf-surfaces persistently more or less grey-tomentose.

    12 Involucre rarely over 2.5 cm high; principal stem-leaves deeply pinnatifid to near midrib into narrowly lanceolate divisions mostly 3–7 mm broad; achenes mostly 3 or 4 mm long; relatively slender plants with a tendency to spread by slender short-lived creeping roots, the individual plants producing deep taproots; (B.C. to w Ont.; introd. eastwards) .... *C. flodmanii*

    12 Involucre usually at least 4 cm high; principal leaves rather shallowly

lobed into broader divisions; achenes to 7 mm long; stouter plants with creeping roots poorly or scarcely developed; (B.C. to SW ?Man.) . . . .
. . . . *C. undulatum*

   9   Leaves not densely white-tomentose beneath; phyllaries mostly lacking a distinct thickened glutinous dorsal ridge (except sometimes in *C. pumilum*).

      13 Outer phyllaries blunt or merely with a sharp-toothed tip; heads commonly several or many; involucre to 3.5 cm high; (Sask. to S Labrador, Nfld., and N.S.) . . . . *C. muticum*

      13 Outer phyllaries distinctly spine-tipped; heads solitary or few in terminal clusters.

         14 Outer phyllaries chartaceous, ovate, to about 7 mm broad; involucre to 4 or 5 cm high; leaves deeply lobed to at least half their breadth, the upper stem-leaves surpassing the heads (plants sometimes nearly or quite stemless); (B.C. to W Ont.) . . . . *C. drummondii*

         14 Outer phyllaries lanceolate to lance-ovate, commonly less than 4 mm broad; involucre to 3 or 4 cm high.

            15 Involucre subglabrous or only sparsely cobwebby, its inner phyllaries dilated below the slender tip into a narrowly elliptic or narrowly oblong, scarious and crisped appendage; (S Ont.) . . . .
  ·   . . . . *C. pumilum*

            15 Involucre more or less strongly cobwebby, its inner phyllaries tapering uniformly and not dilated below the slender tip; (B.C.).

               16 Corolla-tube less than 12 mm long, mostly less than twice as long as the dilated throat, the lobes to 1 cm long; style exserted at least 3 mm beyond the corolla-lobes . . . . *C. edule*

               16 Corolla-tube at least 12 mm long, more than twice as long as the throat, the lobes to 4 mm long; style included or exserted at most 1 mm beyond the corolla-lobes . . . . *C. brevistylum*

[*C. altissimum* (L.) Spreng.]
[Reports of this species of the E U.S.A. (N.Dak. and Minn. to N.Y., S to Tex. and Fla.) from Canada are all apparently based upon the closely related *C. discolor. See* R.A. Davidson (Brittonia 15(3):222–41. 1963) and C. Frankton and R.J. Moore (Can. J. Bot. 41(1):73–84. 1963).

The closely related *C. heterophyllum* (L.) Hill is reported by J. Groentved (Bot. Tidsskr. 44(2):253. 1937) as introd. in one place in W Greenland, where probably not established.]

C. arvense (L.) Scop.   Canada Thistle (but not native).   Chadron or Chaudron
Eurasian; a common weed of roadsides, cult. fields, pastures, and waste places in N. America, as in Alaska (N to ca. 61°30′N), Dist. Mackenzie (Fort Simpson, ca. 62°N; W.J. Cody, Can. Field-Nat. 75(2):68. 1961), and all the provinces (in Man., N to Churchill; in Ont.–Que., N to S James Bay); SW Greenland. MAP and synonymy: *see* below.

   1   Leaves strongly sinuate-pinnatifid and prickly-margined, glabrate in age . . . . var. *arvense*

      2 Flowers pink-purple.

         3 Phyllaries and stem greenish; [*Serratula* L.; *Carduus* Robs.; *Cnicus* Hoffm.; introd., transcontinental; MAP (aggregate species): Hultén 1968*b*:939] . . . . . . f. *arvense*

         3 Phyllaries and stem purplish; [known from the type locality, Pagwa, Ont., and adjacent Moose Factory, S James Bay] . . . . f. *rubricaule* Lepage

      2 Flowers white; [frequent throughout the range of f. *arvense*] . . . .
. . . . f. *albiflorum* (Rand & Redf.) Hoffm.

   1   Leaves entire to merely undulate-lobed.

      4 Stem and lower surface of the entire or merely undulate, nearly prickleless leaves densely white-tomentose; [Sask. (near Saskatoon; Breitung 1957*a*), S Man. (Löve and Bernard 1959), and S Ont. (Soper 1949)] . . . . var. *vestitum* Wimm. & Grab.

      4 Stem and leaves green.

         5 Leaves entire or shallowly crenate, nearly prickleless; [Alta. (Moss 1959), Sask. (Breitung 1957*a*), S Man. (Löve and Bernard 1959), Ont. (Kent, Oxford, and Wellington counties), and Que. (near Montreal)] . . . . . var. *integrifolium* Wimm. & Grab.

5 Leaves undulate-lobed and with a few fine short prickles; [B.C. (Vancouver Is.),
Alta. (Fort Saskatchewan), s Man. (Elkhorn; Virden; Methven; Winnipeg), and Ont.
(N to Ottawa)] .... var. *mite* Wimm. & Grab.

[C. brevifolium Nutt.]
[The inclusion of Sask. in the range of this species of the w U.S.A. (Wash., Idaho, and Oreg.) by Rydberg (1922) requires clarification, possibly being based upon the white-flowered form of either *C. flodmanii* or *C. undulatum*. See R.J. Moore and C. Frankton (Can. J. Bot. 41(11):1559. 1963).]

C. brevistylum Cronq.
/t/W/ (Hs) Meadows and other moist open places at low to moderate elevations from B.C. (s Queen Charlotte Is.; Vancouver Is. and adjacent islands; mainland N to Sicamous, about 35 mi sw of Revelstoke) and Mont. to Calif. MAP: R.J. Moore and C. Frankton, Can. J. Bot. 40(9): fig. 10, p. 1194. 1962.
    A hybrid with *C. edule* (× *C. vancouverense* Moore & Frankton) is known from several localities on s Vancouver Is. (type from Nanaimo), these indicated on the above-noted map.

C. discolor (Muhl.) Spreng.
/T/EE/ (Hs) Moist open places and thickets from SE Man. (Winnipeg; Emerson, about 55 mi s of Winnipeg) to Ont. (N to Stormont Co.; Dore and Gillett 1955), Que. (N to the Montreal dist.), and Maine, s to Mo., Tenn., and Ga. [*Cnicus* and *Carduus* Muhl.; *Cnicus altissimus* var. *dis.* (Muhl.) Gray]. MAP: C. Frankton and R.J. Moore, Can. J. Bot. 41(1): fig. 6, p. 82. 1963.

C. drummondii T. & G.
/ST/WW/ (Hs) Moist open places from w-cent. Dist. Mackenzie (N to the type locality at Fort Franklin, sw end of Great Bear L. at ca. 65°N) and B.C.–Alta. to Sask. (N to Prince Albert; CAN), Man. (N to Grand Rapids, near the NW end of L. Winnipeg; CAN), and w Ont. (Severn R. at ca. 55°N; NW shore of L. Superior), s to Calif., Ariz., and N.Mex. [*Cnicus* Gray; *Carduus pumilus sensu* Hooker 1833, not Nutt.; incl. the nearly or quite stemless *Cnicus dr.* var. *acaulescens* Gray]. MAPS: R.J. Moore and C. Frankton, Can. J. Bot. 42(4): fig. 10, p. 459. 1964, and 45(9): fig. 2, p. 1748. 1967.

C. edule Nutt.
/T/W/ (Hs) Wet meadows and moist open woods at low to rather high elevations from SE Alaska (Hyder, ca. 55°N) through B.C. to Oreg. [*Cnicus* Gray; *Carduus (Cirsium) macounii* Greene]. MAPS: Hultén 1968b:940; R.J. Moore and C. Frankton, Can. J. Bot. 40(9): fig. 9, p. 1193. 1962.
    The closely related *C. kamtschaticum* Ledeb. of E Asia is known from the westernmost Aleutian Is. (*see* Hultén 1950: map 1212, p. 1684). It may be distinguished from *C. edule* as follows:
1 Leaves narrowly to rather broadly oblong, more or less pinnatifid to coarsely toothed,
    sparsely arachnoid-villous on both surfaces .... *C. edule*
1 Leaves oval to oblong-ovate, deeply pinnatifid into broad lobes, nearly glabrous above,
    pubescent on the nerves beneath; [*Cnicus* Maxim.; w Aleutian Is. and NE Asia; MAP:
    Hultén 1968b:941] .... *C. kamtschaticum* Ledeb.

C. flodmanii (Rydb.) Arthur
/T/WW/ (Gr) Moist prairies, fields, and clearings from Alta. (N to Fort Vermilion, 58°24'N) to Sask. (N to Prince Albert), Man. (N to The Pas), and w Ont. (perhaps native at Malachi, about 35 mi NW of Kenora; probably introd. farther eastwards, as near Thunder Bay, the Timagami Provincial Forest, and Port Sydney, E of Georgian Bay, L. Huron), s to Mont., Colo., Iowa, and Minn.; introd. in Vt. and probably so in N.Y. and Que. (St-Jovite, Terrebonne Co.; Nominingue, Labelle Co.). [*Carduus* Rydb.; *Cirsium canescens* of auth., not Nutt.]. MAPS: C. Frankton and R.J. Moore, Can. J. Bot. 39(1): fig. 9, p. 24, and fig. 11, p. 26. 1961.
    Concerning the sporadic eastern distribution, Frankton and Moore write, "Whether these eastern sites are native stations or represent recent migration eastward through human agency is not clear.... The Ontario and Quebec specimens extend in an arc that suggests that they may be relics of a more continuous distribution of early postglacial time." Forma *albiflorum* D. Löve

(flowers white rather than purple) is known from Alta. (Boivin 1966*b*), Sask. (Swift Current; Tyvan), and s Man. (Letellier, about 50 mi s of Winnipeg; type from near Langruth, about 80 mi NW of Winnipeg). Forma *oblanceolatum* (Rydb.) Löve and Bernard (*Carduus (Cirsium) obl.* Rydb.; not only the basal but also the cauline leaves entire or nearly so) is reported from SE Man. by Löve and Bernard (1959; near Otterburne, about 30 mi s of Winnipeg).

C. foliosum (Hook.) DC.

/sT/D/ (Hs) Meadows and other moist places at low to rather high elevations, the main area from the Yukon (N to ca. 62°30'N), sw Dist. Mackenzie (Porsild and Cody 1968), and N Alta. (L. Athabasca) through B.C.–Alta. to Calif., Ariz., and Colo.; isolated along calcareous shores of the Mingan Is., Côte-Nord, E Que., the type locality of *C. minganense* Vict. (var. *ming.* (Vict.) Boivin, this scarcely separable from the typical form). [*Carduus* Hook.; *Cnicus* Gray; incl. *Cirsium minganense* Vict.; *C. ?scariosum* Nutt.; *C. drummondii* of auth., not T. & G.]. MAPS: Hultén 1968*b*:968; Porsild 1966: map 150, p. 85; R.J. Moore and C. Frankton, Can. J. Bot. 42(4): map 9, p. 459. 1964, and 45(9): fig. 2, p. 1748. 1967 (expressing the opinion that the E Que. station of "*C. minganense*" resulted from a chance introduction).

C. hookerianum Nutt.

/T/W/ (Hs) Moist meadows, fields, and open rocky slopes at low to fairly high elevations from s B.C. (N to Cariboo and Williams Lake, ca. 52°N) and sw Alta. (N to Jasper) to N Wash.–Idaho and NW Mont. [*Cnicus* Gray; *Carduus* Heller; *Cnicus ?eriocephalus* Gray]. MAP: R.J. Moore and C. Frankton, Can. J. Bot. 43(5): fig. 5, p. 602. 1965.

According to Hitchcock et al. (1955), "The Canadian specimens seem well characterized, but most of the specimens from s. of the border vary in the direction of *C. foliosum*, as if contaminated by introgression." A hybrid with *C. undulatum* is reported from Merritt, s B.C., by Boivin (1967*a*).

C. muticum Michx.   Swamp-Thistle

/T/EE/ (Hs) Swampy ground, thickets, and wet woods (ranges of Canadian taxa outlined below), s to Tenn., La., and N.C. MAP and synonymy: *see* below.

1   Stem mostly less than 1 m tall; heads approximate or crowded; involucre often glabrous from the first; phyllaries glabrous or nearly so, glutinous; [*Cnicus muticus* var. *mont.* Fern.; Ont. (N to the Ekwan R. at 53°44'N), Que. (N to E James Bay at 52°37'N; type from Mt. Albert, Gaspé Pen.), Labrador (N to the Hamilton R. basin), and Nfld.] . . . .
. . . . f. *monticola* (Fern.) Boivin

1   Stem to 3 m tall; heads remote or clustered; phyllaries cobwebby.
2   Flowers whitish; [Sask. (McKague, 52°37'N; Breitung 1957*a*), Que., and Nfld.] . . . .
. . . . f. *lactiflorum* Fern.

2   Flowers purple.
3   Leaf-margins merely undulate-lobed and with a few short coarse lobes; [E Que. (Anticosti Is.; GH), Nfld., and N.S. (St. Paul Is.; GH)] . . . . f. *subpinnatifidum* (Britt.) Fern.
3   Principal leaves deeply pinnatifid at least 2/3 to the midrib into lanceolate to oblong, often toothed or forking, slightly prickly-margined lobes; [*Cnicus* Pursh; *Carduus* Pers.; Sask. (N to Waskesiu Lake, ca. 54°N) to Man. (N to Cross L., NE of L. Winnipeg), Ont. (N to Renison, s of James Bay at ca. 51°N; Hustich 1955), Que. (N to L. Mistassini and the Côte-Nord), s Labrador, Nfld., N.B., P.E.I., and N.S.; MAP (aggregate species): C. Frankton and R.J. Moore, Can. J. Bot. 41(1): fig. 6, p. 81. 1963] . . . . f. *muticum*

C. palustre (L.) Scop.

Eurasian; introd. into damp clearings and thickets in N. America, as in B.C. (R.J. Moore and C. Frankton, Can. J. Bot. 40(2):288. 1962), ?Ont. (Boivin 1966*b*), Nfld. (Humber Valley; CAN; GH), St-Pierre and Miquelon (Rouleau 1956), and N.S. (Annapolis and Halifax counties; ACAD; DAO). [*Carduus* L.]. MAP: Hultén 1958: map 144, p. 163.

The Nfld. plant is considered apparently native by Fernald but is more likely introd. (*see* note under *Luzula campestris*).

C. pitcheri (Eat.) T. & G.
/T/EE/ (Hs) Sandy shores and dunes of Lakes Michigan, Huron, and Superior in Wisc., Ill., Ind., Mich., and s Ont. (Lambton, Huron, Bruce, and Simcoe counties and Cockburn Is. and Manitoulin Is., N L. Huron; the report from Fort William (Thunder Bay) by John Macoun 1884, requires confirmation). [*Cnicus* Eat.].

[*C. polyphyllum* (Rydb.) Petr.]
[The inclusion of Alta. in the range of this species of the w U.S.A. (Mont., Idaho, and Wyo.) by Rydberg (1922) requires clarification. (*Carduus* Rydb.; *Cirsium tweedyi* Rydb., the correct name through priority according to Hitchcock et al. 1969).]

C. pumilum (Nutt.) Spreng. Pasture- or Bull-Thistle
/T/EE/ (Hs) Dry, often sandy soil from Minn. to Mich., Ohio, s Ont. (Bruce and Simcoe counties; Manitoulin Is., Drummond Is., and Fishing Is., L. Huron; CAN; TRT; John Macoun 1886), Pa., and Maine, s to Ill. and N.C. MAP and synonymy: *see* below.
1   Spines of outer phyllaries stout, to 6 mm long; leaves usually deeply lobed nearly to the
      midvein, bearing strong marginal spines to 7 mm long; plant usually with several long
      branches; root rarely (and then only slightly) tuberous; [*Carduus* Nutt.; *Cnicus* Torr.; E
      U.S.A. only, reports from Canada referring to var. *hillii*; MAP: R.J. Moore and C. Frankton,
      Can. J. Bot. 44(5): fig. 2, p. 590. 1966, and 45(9): fig. 2, p. 1748. 1967] . . . . , [var. *pumilum*]
1   Spines of outer phyllaries slender, to 3 mm long; leaves usually shallowly lobed and with
      fine marginal spines to 6 mm long; plant unbranched or with few and short upper
      branches; root usually strongly tuberous . . . . var. *hillii* (Canby) Boivin
   2   Flowers purple; [*Cnicus (Carduus; Cirsium) hillii* Canby; s Ont. (N to Manitoulin Is.; the
         report from Man. by Lowe 1943, probably refers to *C. drummondii*); MAP: on the
         above-noted map by Moore and Frankton] . . . . f. *hillii*
   2   Flowers white; [f. *candidum* Boivin, the type from the shores of L. Huron, s Ont.,
         where taken by John Macoun in 1874; MTMG; Boivin 1967a] . . . . f. *albiflorum* Scoggan

C. undulatum (Nutt.) Spreng.
/T/WW/ (Hs) Dry open places from s B.C. (N to near Williams Lake, ca. 52°N), s Alta. (N to near Howie, ca. 51°N; the report from Fort Vermilion, 58°24′N, by Groh 1949, is probably based upon *C. flodmanii*), and Sask. (N to near the Alta. boundary at ca. 52°30′N; reports from Man. and s Ont. (Stroud 1941; Wellington Co.) refer chiefly to *C. flodmanii* but *C. undulatum* is reported from SW Man. by Boivin 1966b) to Oreg., Nev., Ariz., N.Mex., Tex., and Mo. [*Carduus* Nutt.; *Cnicus* Gray; *Cirsium engelmannii* Rydb.; incl. the large-headed extreme, var. *megacephalum* (Gray) Fern. (*C. meg.* (Gray) Cock.), the accrediting of which to Man. by Fernald *in* Gray 1950, requires clarification]. MAPS: C. Frankton and R.J. Moore, Can. J. Bot. 39(1): fig. 10, p. 25, and fig. 11, p. 26. 1961.
   Forma *album* Farw. (flowers white rather than deep purple) is reported from SW Sask. by Boivin (1966b).

C. vulgare (Savi) Tenore Bull-Thistle. Piqueux or Gros Chardon
Eurasian; an aggressive weed along roadsides and in pastures and clearings in N. America, as in SE Alaska (Hyder and Sitka; Hultén 1950), B.C. (N to the Cassiar dist. at ca. 59°15′N; V), Alta. (Crowsnest Pass; Waterton Lakes), Sask. (Cypress Hills, Dana, and Tompkins; Breitung 1957a), Man. (Winnipeg Beach, about 40 mi N of Winnipeg; CAN; reported from Winnipeg and Elm Creek by Lowe 1943), Ont. (N to the NW shore of L. Superior near Port Arthur (Thunder Bay); CAN), Que. (N to L. St. John, Anticosti Is., and the Gaspé Pen.), St-Pierre and Miquelon, Nfld. (GH), N.B., P.E.I., and N.S. [*Carduus (Cirsium; Cnicus) lanceolatus* L.]. MAP: Hultén 1968b:939.

CNICUS L. [9479]

C. benedictus L. Blessed Thistle
Eurasian; a garden-escape or introd. along roadsides and in waste places in N. America, as in s B.C. (Port Angelus; Groh and Frankton 1949b), N.B. (Bass River, Kent Co., where taken by Fowler in 1870; GH), and N.S. (West Point R.; John Macoun 1884). [*Centaurea* L.].

COREOPSIS L.   [9227]   Coreopsis, Tickseed

(Ref.: E.E. Sherff, N. Am. Flora, Ser. II (pt. 2):4–40. 1955)
1   Leaves mostly entire (sometimes irregularly lobed); rays distinctly toothed or lobed at summit; perennials.
  2   Leaves chiefly sub-basal, linear to narrowly oblanceolate, to 2 dm long (including the petiole) and over 1.5 cm broad, glabrous to villous or hirsute; rays yellow, to 3 cm long and often over 1 cm broad; style-branches cuspidate; achenes broadly thin-winged, orbicular, to 3 mm long; pappus consisting of 2 short chaffy teeth; stem spreading-villous (especially near base) or glabrous, from a short woody caudex; (s Ont.) . . . .
       . . . . *C. lanceolata*
  2   Leaves uniformly disposed on the stem, linear, to about 5 cm long and 3 mm broad, glabrous; rays white to pink or deep roseate, to 13 mm long; style-branches abruptly and bluntly conic-tipped; achenes wingless, oblong, to 2 mm long; pappus nearly obsolete or a minute crown; stems glabrous, from well-developed creeping rhizomes; (N.S.) . . . . *C. rosea*
1   Leaves (at least some of them) deeply parted nearly or quite to the midrib into narrow segments or distinct leaflets; stems rather uniformly leafy; rays yellow or orange-yellow, sometimes with a reddish-brown base.
  3   Achenes wingless, linear-oblong, to about 4 mm long, the pappus obsolete; rays 3-lobed at summit, to 1.5 cm long; style-branches obtusely conic-tipped; principal leaves once or twice pinnately divided into linear or narrowly lanceolate segments, subsessile or short-petioled; glabrous annual; (B.C. to s Man.; introd. eastwards) . . . .
       . . . . *C. tinctoria*
  3   Achenes wing-margined.
    4   Principal leaves once ternately parted or divided into mostly lanceolate to narrowly oblong divisions; rays to about 2.5 cm long, entire or slightly toothed at summit; style-branches acute; achenes to 6 or 7 mm long; perennials with rhizomes and stolons.
      5   Leaves sessile, the principal ones with 3 linear-oblong elongate lobes arising from near or slightly below the middle (appearing broadly wing-petioled by the undivided basal portion); heads rarely more than 3 (up to 5 or 6); achenes elliptic-oblong; (?Man.) . . . . [*C. palmata*]
      5   Leaves (except the upper entire ones) on petioles to about 3 cm long, divided into 3 (sometimes 5) linear to oblong-lanceolate separate leaflets; heads often more numerous; achenes cuneate-oblong or -obovate; (s Ont.; introd. in sw Que.) . . . . *C. tripteris*
    4   Principal leaves once or twice pinnately or pinnate-ternately divided into linear-filiform to narrowly lanceolate segments; plants normally glabrous.
      6   Heads usually rather numerous; rays to 18 mm long, distinctly toothed or short-lobed at summit; style-branches abruptly subtruncate; achenes oblong-oblanceolate to oblong, to 2.8 mm long; annual or biennial; (s B.C.) . . . .
         . . . . *C. atkinsoniana*
      6   Heads solitary or few; rays to 2.5 cm long; style-branches acute or cuspidate; (introd.).
        7   Achenes orbicular, to about 2.5 mm long, the inner face often with large callosities at top and bottom; leaves divided into linear-filiform to narrowly lanceolate segments, more or less petioled; perennial (rarely annual), lacking rhizomes or stolons . . . . *C. grandiflora*
        7   Achenes oblong-obovate, to 5 mm long, lacking callosities; leaves divided into linear-filiform segments (the basal segments simulating stipules), sessile; perennials with rhizomes and stolons . . . . *C. verticillata*

C. atkinsoniana Dougl.
/t/W/   (T (Hs))   Moist banks of rivers and streams from s B.C. (collection in CAN from Waneta, near the U.S.A. boundary sw of Trail, where taken by J.M. Macoun in 1902; collection in V from L.

Osoyoos, near the U.S.A. boundary s of Penticton; the citation of an early Bourgeau collection from Sask. by E.E. Sherff, Field Mus. Nat. Hist. Publ. Bot. Ser. 11(6):429. 1936, is referred to *C. tinctoria* by Breitung 1957*a*) and Mont. to Oreg. and S.Dak.; introd. eastwards. [*Calliopsis* Hook.].

C. grandiflora Hogg
Native in the U.S.A. (N to Kans., Mo., and Ga.); introd. elsewhere, as in Ont. (N to Wilberforce, Renfrew Co., and the Ottawa dist.), sw Que. (Farnham, Missisquoi Co.; CAN), and N.B. (Boivin 1966*b*).

C. lanceolata L.
/t/EE/ (Hs) Dry sandy, gravelly, or rocky places from Mo. to Wisc., Mich., s Ont. (apparently native in Norfolk and Bruce counties; probably introd. along roadsides and railways in Wellington and York counties and at Gros Cap, w of Sault Ste. Marie), and Va., s to N.Mex., La., and Fla.; a garden-escape eastwards to N.J. and New Eng. and in sw B.C. (near Langley Prairie, E of Vancouver; Herb. V).

Some of our material is referable to var. *villosa* Michx. (stem and leaves pubescent rather than glabrous).

[C. palmata Nutt.]
[Reports of this U.S.A. species (Wisc. to Okla., Mo., Ill., and Ind.) from Man. by Fernald *in* Gray (1950), Jackson et al. (1922), and Lowe (1943) are perhaps based upon a report by Gray from "Winnipeg to Wisconsin" noted by John Macoun (1886), who states, however, "We have never seen Canadian specimens of this."]

C. rosea Nutt.
/T/E/ (Hsr) Damp shores and peaty depressions: N.S. (several localities by lakes in Yarmouth Co., where first taken by Fernald et al. in 1920; CAN; GH); se Pa. and s N.J. to E Md., E Mass., Long Is., and R.I.

C. tinctoria Nutt.
/T/WW/ (T) Moist ground and roadsides from B.C. (Rydberg 1922) to s Alta. (N to near Lethbridge and Medicine Hat), s Sask. (N to Wilcox, about 30 mi s of Regina), and s Man. (Deloraine, about 50 mi s of Brandon; CAN), s to Calif., Tex., and La.; much cult. and a common garden-escape elsewhere (and probably in some of the above Canadian area), as in s Ont. (Lambton Co.; Dodge 1915) and sw Que. (near Aylmer, Gatineau Co.; Chambly, near Montreal; near Trois-Rivières).

C. tripteris L. Tall Coreopsis
/t/EE/ (Hsr) Damp thickets and swampy places from Wisc. to s Ont. (La Salle, Puce, Sandwich, and Windsor, Essex Co.; islands in the Detroit R., Kent Co.; CAN; TRT; John Macoun 1884), s to Kans., La., and Ga.; cult. and a garden-escape elsewhere, as in sw Que. (ditches and pathways at Côteau Landing, Soulanges Co.; Lionel Cinq-Mars, Ann. ACFAS 18:80. 1952).

C. verticillata L.
Native in the E U.S.A. from D.C. and Md. to Ark., Ala., and Fla.; there are Ontario collections in CAN from Rondeau, Kent Co. (where taken by John Macoun in 1885) and from Port Dover, Norfolk Co. (where taken by Macoun in 1890), and in OAC from wet ground at York, Haldimand Co. It may have been introd. in these localities.

[COSMOS Cav.] [9238]

[C. bipinnatus Cav.] Cosmos
[A native of Mexico; a garden-escape to roadsides and waste places in N. America and becoming established in the s U.S.A. There are collections in CAN and MT from sandy flats of the Causapscal R. near a stable at St-Tharsicius, Matapédia Co., Gaspé Pen., E Que., where taken by Bernard Boivin in 1938 but probably not established.]

COTULA L.   [9351]   Brass-buttons

1   Leaves 1–2-pinnately dissected into linear lobes, with scattered soft spreading hairs; heads usually less than 4 mm broad; pistillate flowers in 2 or 3 rows; slender branching annual; (introd. on Vancouver Is.) .... [*C. australis*]
1   Leaves linear to lanceolate or oblong, entire or with a few coarse teeth or narrow lobes, essentially glabrous and somewhat succulent, dilated at base into a short clasping sheath; heads to over 1 cm broad; pistillate flowers in a single row; perennial with usually many clustered decumbent stems; (introd. in Alaska–B.C. and from E Que. to N.S.) ....
.... *C. coronopifolia*

[*C. australis* (Sieb.) Hook. f.]
[A native of Australia and N.Z.; introd., chiefly on wharf-ballast, in Oreg. and Calif. and reported from sw B.C. by J.M. Macoun (1895; ballast-heaps at Nanaimo, Vancouver Is.), where scarcely established. (*Anacyclus* Sieb.; *Lancisia* Rydb.).]

C. coronopifolia L.
A native of South Africa; now thoroughly established in brackish mud of tidal flats in N. America, as in SE Alaska (Wrangell and Gravina Is.; Hultén 1950), B.C. (Queen Charlotte Is.; Vancouver Is. and adjacent islands and mainland), E Que. (Maria, St-Simon, and Bonaventure, Bonaventure Co., S Gaspé Pen.), N.B. (Charlotte, Gloucester, and St. John counties), P.E.I. (Summerside, Prince Co.), and N.S. (Eastern Passage, Halifax Co.; ACAD; not listed by Roland 1947). [*Lancisia* Rydb.]. MAP: Hultén 1968*b*:895.

CREPIS L.   [9605]   Hawk's-beard

(Ref.: Babcock 1947; Babcock and Stebbins 1938)
1   Low glabrous perennials of arctic, subarctic, and arctic-alpine regions; leaves mostly basal, slender-petioled, oblanceolate to obovate, entire or sinuately lyrate-pinnatifid with a few lateral lobes, the blade to about 2.5 cm long; involucre cylindric; midrib of phyllaries scarcely thickened; achenes with an apical disk below the pappus; pappus-bristles promptly deciduous (usually falling together).
   2   Stems to about 2.5 dm tall, the caudex relatively thick and never stoloniferous; involucre to 1 cm long; achenes fusiform, attenuate into a delicate beak about 1/4 as long as the body, the ribs narrow and finely tuberculate; (mts. of B.C. and sw Alta.) ....
   .... *C. elegans*
   2   Stems tufted, commonly less than 1 dm tall, the slender caudex often stoloniferous; involucre to 13 mm long; achenes columnar, at most very short-beaked, the broad ribs smooth or slightly roughened; (transcontinental) .... *C. nana*
1   Taller plants of warmer regions, rarely (if ever) completely glabrous; leaves larger, the principal ones lanceolate or oblanceolate, often strongly toothed or pinnatifid; midrib of phyllaries commonly somewhat keeled or thickened near the base; achenes scarcely beaked (except in *C. vesicaria*), lacking an apical disk; pappus-bristles tardily falling separately.
   3   Achenes (to 5 mm long) all gradually narrowed into a beak about as long as the body; involucres to 12 mm high, tomentose and often glandular; leaves finely pubescent on both sides, the stem-leaves sessile or short-petioled, subentire to pinnatifid (the middle ones clasping), the basal leaves petioled, commonly lyrate- or runcinate-pinnatifid, the lobes very variable in length and width; stems to 8 dm tall, hispid and purplish toward base; plant commonly biennial; (introd. in sw B.C.) .... *C. vesicaria*
   3   Achenes scarcely beaked.
      4   Annuals (occasionally biennials) with leafy stems, the leaves glabrous or minutely hispid; phyllaries and peduncles commonly beset with gland-tipped bristly hairs; (introd.).
         5   Inner phyllaries pubescent within; receptacle ciliate-fringed between the achene-pits; stem-leaves linear to lanceolate; basal leaves minutely toothed to pinnately parted.

6 Achenes dark purplish-brown, 10-ribbed, less than 5 mm long; involucre less than 1 cm high, its inner phyllaries strigose or puberulent within; stem-leaves sagittate-clasping at base .... *C. tectorum*

6 Achenes reddish brown, 13-ribbed, 7 mm long or more; involucre at least 1 cm high, its inner phyllaries downy within; stem-leaves semi-clasping but not or scarcely sagittate at base .... *C. biennis*

5 Inner phyllaries glabrous within; achenes 10-ribbed; stem-leaves lanceolate, sagittate-clasping, the lower ones and the basal leaves runcinate-pinnatifid or pinnately parted.

7 Achenes brownish yellow to dark brown, to 2.5 mm long, their ribs narrow; receptacle glabrous; involucres at most 8 mm high, to 4 mm broad at the middle .... *C. capillaris*

7 Achenes golden brown, to 3.8 mm long, their ribs broader; receptacle ciliate-fringed between the achene-pits; involucres to 1 cm high, to 6 mm broad at the middle .... *C. nicaeensis*

4 Perennials from a taproot and often a simple or branched caudex; involucres to over 1.5 cm high.

8 Stem scapose, merely bracted or with 1 or 2 much-reduced leaves; principal leaves in a basal rosette, entire or remotely dentate to runcinate-dentate; involucre rather broadly campanulate; plant usually more or less hispid but not at all canescent or tomentose; (B.C. to Man.) .... *C. runcinata*

8 Stem distinctly leafy, with generally 2 or 3 well-developed leaves; involucre cylindric to narrowly campanulate; leaves commonly deeply runcinate-pinnatifid halfway or more to the midrib; plants more or less tomentose or canescent-puberulent with appressed crinkled greyish hairs at least when young.

9 Principal leaves with a very narrow rachis and linear of lance-linear, mostly entire segments; heads up to 30 (rarely 40), with up to 30 or more flowers; involucres to 1.5 cm high, with up to 15 inner phyllaries, these usually minutely greyish-tomentose and often with some glandless black setae (rarely glabrous); achenes usually greenish, mostly attenuate at summit; stems to 7 dm tall; (s B.C. to s Sask.) .... *C. atrabarba*

9 Principal leaves mostly with a broad rachis and relatively broad entire or toothed segments.

10 Inner phyllaries glabrous or occasionally minutely tomentose, rarely more than 7; heads up to 100 or more, very narrow (with rarely more than 10 flowers); stem to 7 dm tall, sparingly or not at all setose; (Alta.) .....
.... *C. acuminata*

10 Inner phyllaries commonly beset with setae, otherwise glabrous or minutely tomentose, commonly 10 or more; heads at most about 25, broader (commonly with more than 10 flowers); stems to 3 or 4 dm tall.

11 Involucres to 16 mm high, their setae (and those of the peduncles, when present) glandless; heads less than 10, with up to 60 flowers; achenes greenish black to deep reddish-brown; at least the lower part of the stem conspicuously setose with glandless hairs; (?B.C.) .....
.... [*C. modocensis*]

11 Involucres to about 2 cm high, with up to 40 flowers, their setae (and those of the peduncles, when present) normally gland-tipped (except in ssp. *conjuncta*); heads up to 25, with up to 40 flowers; achenes light to dark brown; (s B.C. to s Sask.) .... *C. occidentalis*

C. acuminata Nutt.

/T/W/ (Hs) Mostly in open places in the foothills from Wash. and Alta. (Waterton Lakes and Crowsnest Lake; CAN, detd. Porsild) to Calif. and N.Mex. [*C. angustata* Rydb.]. MAP: combine the maps by Babcock and Stebbins 1938: fig. 31, p. 169, and fig. 33 (*C. intermedia*), p. 180.

The Alta. material has been referred to var. *intermedia* (Gray) Jeps. (*C. (Hieraciodes) int.* Gray) by Porsild. According to Hitchcock et al. (1955), this appears to be a group of polyploid apomicts involving the features of *C. acuminata* and *C. occidentalis,* as well as sometimes *C. atrabarba* and *C. modocensis.* Reports from B.C. by John Macoun (1884 and 1886), Henry (1915), and Rydberg (1922) are based upon *C. atrabarba,* relevant collections in CAN.

C. atrabarba Heller
/T/W/ (Hs) Dry open places at low to moderate elevations from s B.C. (Dry Interior N to Cariboo and the Marble Mts. NW of Clinton; CAN; V), s Alta. (Waterton Lakes; Bailey Valley), and sw Sask. (Cypress Hills; CAN) to Nev. and Colo. [*C. exilis* Osterh.; *C. occidentalis (intermedia)* var. *gracilis* Eat. (*C. gr.* (Eat.) Rydb.)]. MAP (*C. exilis*): Babcock and Stebbins 1938: fig. 29, p. 160.

C. biennis L.
European; locally introd. on wharf-ballast and in waste places in N. America, as in Nfld. (GH; R.B. Kennedy, Rhodora 32(373):3. 1930). Reports from elsewhere in Canada apparently refer to the closely related *C. capillaris, C. nicaeensis,* and *C. tectorum.*

C. capillaris (L.) Wallr.
European; introd. along roadsides and in waste places in N. America, as in E-cent. Alaska (near Fairbanks, ca. 64°30'N), B.C. (N to Queen Charlotte Is., Cariboo, and Dawson Creek), Ont. (N to Ottawa; Groh 1946), Que. (Montreal dist.; MT), N.B. (Boivin 1966*b*), and N.S. (Shelburne Co.; ACAD; CAN). [*Lapsana* L.; *C. virens* L.]. MAP: Hultén 1968*b*:955.

C. elegans Hook.
/ST/W/ (Hs) Dry plains, bluffs, riverbanks, and sandbars from Alaska-Yukon (N to ca. 68°N) to Great Bear L. and Great Slave L., s through B.C. and sw Alta. (N to Jasper; the report from Sask. by Rydberg 1922, requires confirmation) to Mont. and Wyo. [*Youngia* Rydb.; *Barkhausia* Nutt.; *Hieraciodes* Ktze.]. MAP: Hultén 1968*b*:956.

The type is a Drummond collection purportedly taken on "Battures of the Assisaboyne R." According to John Macoun (1884), "This must be a misprint for "Battures" of the Athabasca, as the former river has no "Battures" (gravel beds extending into the river) while the latter has, near Jasper House where Drummond was collecting."

[C. modocensis Greene]
[The report of this species of the w U.S.A. (Wash. and Mont. to Calif. and Colo.) from B.C. by Hitchcock et al. (1955) requires confirmation. (See *C. acuminata*). (Incl. *C. rostrata* Cov.). The MAP by Babcock and Stebbins (1938: fig. 24, p. 143) indicates no Canadian stations.]

C. nana Richards.
/aST/X/A/ (Hs) Sandy or gravelly banks, talus slopes, and cliffs, chiefly at subalpine and alpine levels, from the coasts of Alaska-Yukon-Dist. Mackenzie (type from the head of the Coppermine R. N of Great Slave L.) to Prince Patrick Is. and N Baffin Is., s in the West through B.C. and the mts. of sw Alta. (N to Jasper; CAN) to Calif. and Utah, farther eastwards s to w-cent. Dist. Keewatin, northernmost Que. (Port Burwell, Hudson Strait), and N Labrador (s to Cape Mugford, 57°48'N; GH); isolated along the Strait of Belle Isle at Pistolet Bay, NW Nfld. (GH); according to M.L. Fernald, Rhodora 49(588):299. 1947, the report from the Gaspé Pen., E Que., by Babcock 1947, is the result of an erroneous interpretation of the discussion by Fernald 1925:252); Asia. [*Barkhausia* DC.; *Hieraciodes* Ktze.; *Youngia* Rydb.]. MAPS: Hultén 1968*b*:955; Porsild 1957: map 332, p. 202; Raup 1947: pl. 36; Tolmatchev 1932: fig. 10, p. 56; Fernald 1929: map 14, p. 1492, and 1925: map 14, p. 253.

Var. *lyratifolia* (Turcz.) Hult. (*Youngia americana* Babcock; leaves more or less pinnatifid or dissected rather than entire or only slightly toothed) is known from Alaska (N to ca. 69°N). MAP: Hultén 1968*b*:956.

C. nicaeensis Balbis
European; sparingly introd. into waste places in N. America, as in sw B.C. (s Vancouver Is., where taken in the Sidney–Victoria dist. by John Macoun in 1887 and 1913; CAN, distributed as *C. biennis,* revised by Boivin).

C. occidentalis Nutt.
/T/WW/ (Hs) Dry open places (ranges of Canadian taxa outlined below), s to s Calif., N.Mex., and S.Dak. MAP and synonymy: *see* below.
1 Involucre with at least some gland-tipped hairs.
    2 Largest heads with at least 18 flowers and 10 inner phyllaries; [var. *glandulosa* Torr.; var. *crinita sensu* John Macoun 1884, not Gray; s B.C. (N to Kamloops) and sw Alta. (near Pincher Creek); MAP: Babcock and Stebbins 1938: fig. 18, p. 120] . . . . ssp. *occidentalis*
    2 Largest heads with at most 14 flowers and 9 inner phyllaries; [var. *costata* Gray; Cypress Hills of sw Sask.; MAP: on the above-noted map] . . . . ssp. *costata* (Gray) Babcock & Stebbins
1 Involucre completely lacking glandular pubescence; [var. *nevadensis* Kellogg; B.C. (N to Kamloops); MAP: on the above-noted map] . . . . ssp. *conjuncta* Babcock & Stebbins

C. runcinata (James) T. & G.
/T/WW/ (Hr) Moist, often alkaline meadows and prairies at low to fairly high elevations from B.C. (N to Revelstoke; CAN) to Alta. (N to Jasper and near Edmonton; CAN), Sask. (N to Waskesiu Lake, ca. 54°N), and Man. (N to Grand Rapids, near the NW end of L. Winnipeg; CAN), s to Calif., N.Mex., Nebr., and Minn. MAP and synonymy: *see* below.
1 Plant distinctly pubescent, the involucre and often the leaves more or less glandular-hairy; basal bractlets up to half as long as the principal phyllaries of the heads.
    2 Involucres sparingly short-pubescent; peduncles and stem usually glabrous; [*Hieracium* James; *H. venosum sensu* Dawson 1875, not L.; *C. glaucella* and *C. perplexans* Rydb.; B.C. to s Man.; MAP: Babcock and Stebbins 1938: fig. 6, p. 91] . . . . ssp. *runcinata*
    2 Involucres, peduncles, and usually also the stem copiously glandular-hairy; [var. *hispidulosa* Howell; *C. pallens, C. platyphylla,* and *C. obtusissima* Greene; Cypress Hills of SE Alta. and sw Sask.; MAP: on the above-noted map] . . . . ssp. *hispidulosa* (Howell) Babcock & Stebbins
1 Plant very sparingly pubescent, not at all glandular; basal bractlets not over 1/4 as long as the principal phyllaries; [*Crepidium (Crepis) glaucum* Nutt.; Alta. (Boivin 1966*b*) to Man.; MAP: on the above-noted map, but incomplete] . . . . ssp. *glauca* (Nutt.) Babcock & Stebbins

C. tectorum L.
Eurasian; introd. along roadsides and in waste places in N. America, as in E-cent. Alaska (near Fairbanks, ca. 65°N), sw Yukon (Mackintosh, ca. 61°N; CAN), sw Dist. Mackenzie (J.W. Thieret, Can. Field-Nat. 75(3):120. 1961), B.C.–Alta., Sask. (N to Prince Albert), Man. (N to Flin Flon), Ont. (N to Thunder Bay), Que. (N to Ste-Flavie, Rimouski Co.), N.B., P.E.I., and N.S.; sw Greenland. MAP: Hultén 1968*b*:954.

C. vesicaria L.
European; there is a collection in Herb. V from Mayne Is., near Vancouver Is., sw B.C., that has been referred to ssp. *taraxacifolia* (Thuill.) Thell. by Piper and this taxon is reported as "Abundant on roadsides, waste ground, hay-fields, etc., in the Nanaimo district of V.I.; Mayne Is." by Eastham (1947). It differs from the typical form in its relatively narrow non-imbricate involucral phyllaries. [Incl. *C. taraxacifolia* Thuill.].

CROCIDIUM Hook. [9398]

C. multicaule Hook.
/t/W/ (T) Sand flats, ledges, and other dry open places at low elevations from sw B.C. (several localities on Vancouver Is.; CAN; V; a collection in V from Queen Charlotte Is. requires confirmation, the species not being listed by Calder and Taylor 1968) to Calif.

[DIMORPHOTHECA Vaill.]   [9425]

[D. sinuata DC.]   Blue-eyed Cape-marigold
[A native of S. Africa; *D. aurantiaca* is reported from Nfld. by Rouleau (1956), where perhaps a garden-escape but certainly not established. (*D. aurantiaca* Hort., not DC.).]

DORONICUM L.   [9400]   Leopard's-bane

1   Heads solitary on nearly naked peduncles; leaves coarsely dentate . . . . *D. caucasicum*
1   Heads usually several, the stem leafy to the top; leaves toothed or entire . . . . *D. pardalianches*

D. caucasicum Bieb.
European; reported from SW B.C. by Boivin (1967a; Moresby Is., Queen Charlotte Is.; not listed by Calder and Taylor 1968) and there is a collection in GH from Murray's Pond, Nfld., where taken as a garden-escape by Agnes Ayre in 1932.

D. pardalianches L.   Great Leopard-bane
European; there is a collection in Herb V from a grassy roadside in the Agassiz dist. E of Vancouver, where taken by Faris in 1954 from apparently well-established clumps.

[DYSSODIA Cav.]   [9312]

[D. papposa (Vent.) Hitchc.]   Fetid Marigold
[Native in the U.S.A. (N to Mont., Minn., and Ohio). *D. chrysanthemoides* is noted by John Macoun (1884) as having been taken by Day as a "railroad weed" at Fort Erie, Welland Co., S Ont., where evidently not established. (*D. chrysanthemoides* Lag.; *Boebera* Rydb.; *Tagetes* Vent.).]

ECHINACEA Moench   [9178]   Purple Coneflower

1   Leaves narrowly linear to lanceolate, entire, 3-nerved, gradually tapering to the petiole; receptacular chaff relatively broad and rigid-tipped; (Sask. and Man.; introd. in S Ont.) . . . .
. . . . *E. pallida*
1   Leaves narrowly to broadly ovate, rounded at base, often toothed, the lowest ones 5-nerved; receptacular chaff linear to narrowly lanceolate, soft-tipped; rays to 8 cm long; (introd. in S Ont.) . . . . *E. purpurea*

E. pallida Nutt.
/T/(X)/ (Hs)   Dry plains and prairies from Mont. to SE Sask. (Carnduff and Estevan; Breitung 1957a), S Man. (N to Brandon and Winnipeg), and Mich., S to Tex., La., and Ala.; introd. elsewhere, as in the Atlantic states and S Ont. (Aylmer and Churchville, Elgin Co.; TRT). [*Brauneria* Britt.].
The native plant of Sask.–Man. is referable to var. *angustifolia* (DC.) Cronq. (*E. (Brauneria) ang.* DC.; ray-ligules spreading, rarely over 4 cm long, rather than strongly reflexed and to 9 cm long; plant rarely over 6 dm tall, with tuberculate-based hairs, rather than to over 1 m tall, the pubescence consisting of slender hairs).

E. purpurea (L.) Moench
Native in the E U.S.A. from Iowa to Va., S to La. and Ga.; introd. elsewhere, as in S Ont. (dry meadow near St. Thomas, Elgin Co., where taken by L.E. James in 1952; TRT). [*Rudbeckia* L.; *Brauneria* Britt].

ECHINOPS L.   [9442]   Globe-Thistle

1   Plant non-viscid; inner phyllaries glabrous on the back; heads bluish; leaves pinnatifid into lanceolate segments . . . . *E. exaltatus*
1   Plant glandular-viscid as well as grey-woolly; inner phyllaries pubescent on the back; heads whitish to blue; leaves sinuate-pinnatifid into oblong-triangular segments . . . .
. . . . *E. sphaerocephalus*

E. exaltatus Schrad.   Russian Globe-Thistle
A native of Siberia; introd., presumably as a garden-escape, in sw B.C. (Boivin 1966*b*; probably Vancouver Is.), Ont. (Ottawa dist.; Gillett 1958), and sw Que. (Boivin 1966*b*).

E. sphaerocephalus L.   Common Globe-Thistle
Eurasian; a garden-escape to fields and waste places in N. America, as in sw B.C. (Victoria, Vancouver Is.; Groh and Frankton 1949*b*), Sask. (Regina and Saskatoon; Breitung 1957*a*), se Man. (Otterburne, about 30 mi s of Winnipeg; Löve and Bernard 1959), Ont. (N to the Ottawa dist.), sw Que. (N to Val-Morin, about 35 mi NW of Montreal), and N.B. (Grand Manan, Charlotte Co.; Weatherby and Adams 1945).

[ECLIPTA L.]   [9166]

[E. alba (L.) Hassk.]   Yerba-de-Tago
[Native in the E U.S.A. (N to Nebr. and Ind.); introd. elsewhere, as in Mass. and s Ont. (Pelee Point, Essex Co., where taken by Klugh near a tobacco field in 1905 but scarcely established; CAN). (*Verbesina* L.)]

ERECHTITES Raf.   [9389]

E. hieracifolia (L.) Raf.   Fireweed.   Crève-à-yeux
/T/EE/   (T)   Damp thickets, clearings, and shores (ranges of Canadian taxa outlined below), s to Tex., La., and Fla.
1   Upper leaves relatively broad at the sessile base.
   2   Upper leaves scarcely reduced; [*Senecio* L.; *Eriophthalmia* Prov.; Ont. (N to near Thunder Bay), Que. (N to L. St. John), and N.S. (Caledonia, Queens Co.); MAP: Fernald 1918*b*: map 19 (aggregate species; incomplete northwards), pl. 14] . . . . var. *hieracifolia*
   2   Upper leaves reduced to bracts below the inflorescence; [*Cacalia suaveolens sensu* Lindsay 1878, not L.; Ont. (N to Ottawa), Que. (N to l'Isle-Verte, Temiscouata Co.), N.B., P.E.I., and N.S.] . . . . var. *intermedia* Fern.
1   Upper leaves tapering to a more or less distinctly petioled base; [*E. praealta* Raf.; s Ont. (Lambton Co.) and sw Que. (Oka)] . . . . var. *praealta* (Raf.) Fern.

ERIGERON L.   [8901]   Fleabane.   Vergerette

(Ref.: Cronquist 1947)
1   Stems scapose or subscapose, from a branching woody caudex, well-developed leaves all or mostly in a basal cluster (depauperate individuals of some other species may key out here); heads often solitary; involucres to 7 or 8 mm high.
   2   Ray-ligules yellow; plants copiously and finely pubescent.
      3   Involucre sparsely to densely woolly-villous (the hairs sometimes with purple cross-walls), its phyllaries sometimes purple-tipped; head solitary; disk to 16 mm broad; rays to 9 mm long; leaf-blades elliptic to obovate or subrotund, to 13 mm broad; plant spreading-pubescent, to 1.5 dm tall; (mts. of B.C.–Alta.) . . . . *E. aureus*
      3   Involucre strigose or strigose-villous and sometimes finely glandular; heads solitary or few; disk to 13 mm broad; rays to 11 mm long; leaves linear, to 9 cm long and 3 mm broad; plant appressed-greyish-strigose, to about 3 dm tall; (s B.C.) . . . . *E. linearis*
   2   Ray-ligules white, pink, or blue.
      4   Basal leaves deeply ternately lobed or dissected, more or less glandular, the few upper reduced leaves mostly entire; head solitary, the disk to 2 cm broad; rays to 12 mm long (sometimes wanting); (Alaska–B.C. to s Sask.; E Que. and Nfld.) . . . . *E. compositus*
      4   Basal (and upper) leaves entire or merely toothed (or some of them apically 3-lobed in *E. lanatus* and *E. pallens* or a few of the earliest ones apically 3-toothed in *E. purpuratus*), with linear-oblanceolate to oblanceolate or spatulate

blades; (mts. of B.C. and Alta.; *E. pumilus* and *E. radicatus* also in SW Sask.).

5 Plants with long trailing sparsely leafy stolons; heads mostly solitary; involucres to about 5 mm high, finely glandular and sparsely to moderately hirsute with usually appressed short hairs; disk to 13 mm broad; rays to 1 cm long and 1 mm broad; leaves entire, the basal ones oblanceolate, to 5 cm long (including the petiole) and 8 mm broad; plants sparsely or moderately appressed-hairy, to about 4 dm tall; (S B.C. and SW Alta.) . . . . *E. flagellaris*

5 Plants lacking stolons, the scapes arising from a branching caudex; involucre commonly higher (at most about 5 mm high only in *E. radicatus*).

6 Involucre to 13 mm high, woolly-villous (the hairs often with purple cross-walls); heads solitary; disk to 23 mm broad; rays to 11 mm long and 2 mm broad; leaves to 3 cm long and 5 mm broad, often apically 3-toothed; plants loosely long-woolly-villous, usually not over 5 cm tall; (the Yukon–B.C. and SW Alta.) . . . . *E. lanatus*

6 Involucre rarely over 1 cm high (but usually at least 5 mm high), variously pubescent or glandular but not woolly-villous; plants often taller.

7 Leaves (at least some) apically 3-lobed, to 2.5 cm long and 4 mm broad, they and the scapes (rarely over 3 cm tall) sparsely to moderately villous and obscurely viscid; head solitary; disk to 13 mm broad; involucres to 8 mm high; rays to about 5 mm long and 0.5 mm broad; (SE B.C. and SW Alta.) . . . . *E. pallens*

7 Leaves entire, they and the scapes usually spreading-pubescent (or the pubescence often appressed or the leaves sometimes merely ciliate in *E. radicatus*).

8 Leaves finely ciliate, otherwise glabrous or sparsely short-pubescent, commonly not over 2 cm long and 2.5 mm broad; scapes commonly less than 1 dm tall, finely pubescent with spreading or sometimes appressed hairs; head solitary; involucre turbinate, to about 5 mm high; disk to 1 cm broad; rays to 8 mm long and 2 mm broad; (SW Alta. and S Sask.) . . . . *E. radicatus*

8 Leaves and scapes more or less spreading-hairy (or the leaves finally more or less glabrate in *E. purpuratus*); involucre hemispheric, to over 7 mm high; disk to over 1.5 cm broad.

9 Involucre viscid-villous (especially near the base) with long multicellular hairs, some of these usually with purplish cross-walls; disk to 1.5 cm broad; rays to 6 mm long and 0.9 mm broad; leaves usually not over 3 cm long and 5 mm broad, villous at least when young; head solitary; scapes to about 1 dm tall; (Alaska–Yukon–Dist. Mackenzie–N B.C.) . . . . . *E. purpuratus*

9 Involucre more or less spreading-hairy and obscurely to conspicuously glandular, but the hairs lacking purplish cross-walls; heads solitary to several; rays to about 1.5 cm long; leaves to about 8 cm long, permanently more or less spreading-hairy; scapes often taller.

10 Achenes very densely long-silky; rays usually over 1.5 mm broad; disk to 2 cm broad; plant more or less spreading-hairy; (SE B.C.) . . . . *E. poliospermus*

10 Achenes not very densely hairy; rays usually less than 1.5 mm broad; disk to 1.5 cm broad; plant copiously spreading-hairy; (S B.C. to S Sask.) . . . . *E. pumilus*

1 Stems distinctly leafy (but the upper leaves often considerably reduced; well-developed individuals of some of the foregoing species, particularly *E. linearis, E. poliospermus, E. pumilus,* and *E. radicatus,* may key out here); rays white, pink, blue, or purplish.

11 Ray-ligules minute or wanting or at most 6 mm long (barely longer than the breadth of the disk) and 1 mm broad, erect, usually very numerous, white or pinkish (sometimes purplish in *E. acris* and *E. uniflorus*); leaves entire or the lowest ones sparingly toothed.

12 Involucres slenderly campanulate, to about 5 mm high; heads few-flowered; leaves with conspicuously bristly-ciliate margins, linear to oblanceolate; stem to 1 or 2 m tall (but often lower in dryish places), remotely spreading-hirsute to summit; annual; (transcontinental) .... *E. canadensis*
12 Involucres hemispheric; heads many-flowered.
   13 Involucres to about 5 mm high; pappus double, the inner series consisting of long capillary bristles, the outer of shorter scales or bristles; rays to 6 mm long and 1 mm broad; plant mostly annual; (transcontinental) .... *E. strigosus*
   13 Involucres to over 1 cm high; pappus a single series of capillary bristles (or sometimes with a few short outer setae).
      14 Involucres and summit of peduncles more or less densely woolly-villous with long multicellular hairs; principal leaves chiefly in a basal cluster; perennial; (transcontinental in arctic, subarctic, and alpine regions) .... *E. uniflorus*
      14 Involucres not woolly-villous; stems abundantly leafy; biennials or short-lived perennials; (transcontinental).
         15 Inflorescence racemiform, the peduncles erect or nearly so (or the head solitary); involucres and summit of peduncles copiously hispid or villous, not glandular-powdery; stem-leaves erect, narrowly linear, bristly-ciliate .... *E. lonchophyllus*
         15 Inflorescence corymbiform, the peduncles arcuate or obliquely ascending (or the head solitary); involucres glabrous or very sparsely hispid at base, they and the peduncles glandular-powdery and viscid; stem-leaves spreading-ascending, lanceolate to linear-oblong or oblanceolate, the ciliation not bristly .... *E. acris*
11 Ray-ligules conspicuous, commonly at least 1 cm long and often much longer than the breadth of the disk, spreading.
   16 Leaves very numerous, linear to linear-oblanceolate, entire, acute, short-ciliate, to about 3 cm long, commonly less than 4 mm broad, not in basal clusters and scarcely reduced up the stem, often with sterile leafy branches in their axils; heads usually solitary on elongate scape-like peduncles terminating a slender stem, the peduncles naked or with 1 or 2 bracts; involucres to 6 mm high; rays to 8 mm long and about 1.5 mm broad; (transcontinental on wet rocks and gravels) .... *E. hyssopifolius*
   16 Leaves less crowded on the stem (the internodes usually less numerous and more remote), the larger ones sub-basal and in basal clusters, the cauline ones more or less reduced; plants of drier, sometimes desert-like habitats.
      17 Involucres more or less densely white-woolly-villous with soft wavy hairs; head solitary; entire-leaved perennials to about 2.5 dm tall.
         18 Ray-ligules yellow, to 9 mm long and 2.5 mm broad; involucre to 8 mm high, sparsely to densely woolly-villous with multicellular hairs (these sometimes with purple cross-walls); leaf-blades elliptic to obovate or subrotund, to 13 mm broad; plant spreading-pubescent; (s B.C. and sw Alta.) .... *E. aureus*
         18 Ray-ligules white, pink, or blue.
            19 Leaves (and stem) densely lanate with slender entangled hairs, their blades to 3 cm long and 7 mm broad, acute or obtuse; (ssp. *muirii*; N Alaska–Yukon) .... *E. grandiflorus*
            19 Leaves ciliate, otherwise more or less hirsute or hirsute-pilose (but not lanate).
               20 Hairs of the involucre with black or purplish-black cross-walls at least near their bases, the involucre to 8 or 9 mm high; rays to about 12 mm long and 2 mm broad.
                  21 Leaves hirsute, acute or sometimes obtuse, the basal ones to about 5 cm long and 4 mm broad; disk-corollas to 4.5 mm long; pappus double, the outer series consisting of a few inconspicuous bristles; (Alaska and w Dist. Mackenzie) .... *E. hyperboreus*

21 Leaves glabrous or only slightly hirsute, the basal ones
rounded or even retuse at the apex, to 6 cm long and nearly 1
cm broad; disk-corollas less than 3.5 mm long; pappus lacking
a well-defined outer series; (mts. of SE ?B.C.) . . . .
. . . . [*E. melanocephalus*]
20 Hairs of the involucre with clear cross-walls (or occasionally the
basal cross-walls bright reddish-purple), the involucre to about 1
cm high; leaves ciliate, otherwise glabrous or sparingly hirsute.
22 Pappus-bristles usually less than 15; outer pappus conspicuous; leaves ciliate, otherwise glabrous or sparingly hirsute, the
basal ones to 8 cm long and 13 mm broad, obtuse or rounded
at apex; (SW ?Alta.) . . . . [*E. simplex*]
22 Pappus bristles 15–20; basal leaves more or less hairy, often
acute.
23 Stem-leaves linear-lanceolate, acuminate; outer pappus
conspicuous; rays usually not more than 75; (the Yukon) . . . .
. . . . [*E. yukonensis*]
23 Stem-leaves lanceolate to ovate, acute; outer pappus
obscure; rays at least 100; (Alaska–Yukon–Dist.
Mackenzie–Victoria Is. and mts. of B.C.–Alta.) . . . .
. . . . *E. grandiflorus*
17 Involucres pubescent with relatively stiff straight hairs; heads solitary to
numerous.
24 Leaves relatively narrow, linear-filiform to rather narrowly oblanceolate or
narrowly spatulate.
25 Stems and leaves appressed-strigose (or sometimes subglabrous);
perennials, the stems from a taproot and crown or a branching woody
caudex.
26 Basal leaves to 8 cm long and 3 mm broad, their bases neither
enlarged nor of different texture than the blades; involucre to 6 mm
high; rays to 13 mm long and 2 mm broad; (S B.C.) . . . . *E. filifolius*
26 Basal leaves to 12 cm long and 5 mm broad, their bases more or
less enlarged, whitish or purplish and somewhat membranous;
involucre to 8 mm high; rays to 12 mm long and 2.4 mm broad;
(Alta. to SW ?Sask.) . . . . *E. ochroleucus*
25 Stem spreading-pubescent, the hairs commonly about 0.5 mm long;
leaves linear-oblanceolate to oblanceolate or spatulate, their pubescence usually more or less spreading.
27 Taprooted biennial or short-lived perennial, the freely branching
stems to about 7 dm tall; leaves 1-nerved, the basal ones often
deciduous, their blades to 2.5 cm long and 1 cm broad, their
petioles to 5 cm long; rays to 1 cm long and 1.2 mm broad;
disk-corollas 2 or 3 mm long; (S B.C.) . . . . *E. divergens*
27 Perennials, the simple or sparingly branched stems terminating the
branches of a woody caudex; leaves distinctly 3-nerved; rays to
over 13 mm long, about 2 mm broad; disk-corollas over 3 mm long.
28 Phyllaries noticeably thickened on the back; basal leaves
usually rounded or obtuse at tip, to 12 cm long and 1.5 cm
broad; stems rarely purplish at base, to 3 dm tall; (B.C. to
Man.) . . . . *E. caespitosus*
28 Phyllaries only slightly or obscurely thickened on the back;
leaves all acute or acuminate, the basal ones to 2.5 dm long
and 1 cm broad; stems usually purplish at base; (S B.C. and W
?Alta.) . . . . *E. corymbosus*
24 Leaves relatively broad, the basal ones rather broadly oblanceolate or
spatulate to elliptic, oval, or obovate, to over 2 cm broad; stems relatively
stout, commonly taller.

29 Involucre to 5 mm high; disk-corollas less than 3 mm long; rays white to lavender, to 1 cm long and 1 mm broad, about as long as the disk; pappus of the ray- and disk-florets dissimilar, that of the ray-florets composed only of some very short setae or scales less than 1 mm long, an inner series of bristles also present in that of the disk-florets; achenes hairy; principal leaves coarsely and sharply toothed; stem sparingly long-spreading-hispid; annual or rarely biennial; (transcontinental) . . . . *E. annuus*

29 Involucre usually over 5 mm high; disk-corollas at least 3.5 mm long; pappus of ray- and disk-florets similar (of bristles, sometimes also with short outer setae or scales).

　30 Stems soft and easily compressed (often hollow), copiously long-spreading-villous; leaves villous, those of the stem cordate-clasping, the lower ones and the basal leaves commonly coarsely toothed; achenes glabrous or sparingly pubescent; biennials or short-lived perennials from a short simple or subsimple caudex.

　　31 Slender whip-like superficial stolons present; heads usually not more than 5, commonly at least 2.5 cm broad and with at least 50 broad, whitish to bluish-purple rays; disk-florets to 4.5 mm long; (s Ont., s Que., and ?N.S.) . . . . *E. pulchellus*

　　31 Slender stolons wanting; heads few to many, not over 2.5 cm broad, with about 100 or more narrow, whitish to pinkish or roseate rays; disk-florets to 3.5 mm long; (transcontinental) . . . . *E. philadelphicus*

　30 Stems firmer, often decumbent-based; stem-leaves sessile or only half-clasping, they and the basal leaves usually entire or rarely slightly toothed; achenes more or less densely hairy; perennials with a short rhizome or a simple or branched somewhat woody caudex.

　　32 Rays at least 2 mm broad, to 2.5 cm long, whitish or pale blue to rich rose-purple or darker; phyllaries glandular or more or less villous (sometimes merely ciliate on the margins and glutinous on the back); pappus mostly simple, of capillary bristles (occasionally with a few short outer setae); leaves soft-pubescent to essentially glabrous; stems moderately villous below to glabrous; (B.C. and Alta.) . . . . *E. peregrinus*

　　32 Rays mostly to about 1 mm broad; pappus mostly distinctly double, an outer series of short setae usually present.

　　　33 Upper stem-leaves only gradually reduced (the stem appearing rather uniformly leafy), the middle leaves often as large as or larger than the often deciduous basal ones; (B.C. and Alta.) . . . . *E. speciosus*

　　　33 Upper stem-leaves strongly reduced, the middle leaves commonly smaller than the mostly persistent ones of the basal cluster.

　　　　34 Stem and involucre glandular or viscid and often also hirsute; heads rarely more than 5 or 6; (sw Alta.) . . . . *E. formosissimus*

　　　　34 Stem and involucre more or less hairy, scarcely glandular or viscid; (Alaska–B.C. to Man.).

　　　　　35 Stem strict; rays usually white, commonly not over 1 cm long; plant rather copiously short-hirsute . . . . *E. asper*

　　　　　35 Stem curved or decumbent at base; rays pink, blue, or purple, rarely white, to 1.5 cm long . . . . . . *E. glabellus*

E. acris L.

/ST/X/EA/ (Hs) Damp thickets, clearings, and shores, the aggregate species from Alaska (N to ca. 69°N) to the Yukon (N to ca. 65°N), the Mackenzie R. Delta, Great Bear L., Great Slave L., SE Dist. Keewatin, northernmost Man.–Ont., Que. (N to Ungava Bay), Labrador (N to Okak, 57°33'N), Nfld., and N.B. (Madawaska, Restigouche, and Kings counties; not known from P.E.I. or N.S.), S to Calif., Utah, Colo., Minn., Mich., and Maine; Eurasia. MAPS and synonymy: *see* below.

1 Peduncles and involucres nearly or quite glabrous; [vars. *arcuans* Fern. and *oligocephalus* Fern. & Wieg.; *E. alpinus* var. *elatus* Hook., the type a Drummond collection from the Rocky Mts., presumably of Alta.; *E. elatus* (Hook.) Greene and its var. *oligocephalus* (F. & W.) Fern.; transcontinental; MAP: Raup 1947: pl. 35 (*E. elatus*)] . . . .
. . . . var. *elatus* (Hook.) Cronq.

1 Peduncles and involucres more or less glandular.

    2 Plant commonly over 3 dm tall and bearing several to many heads; ray-ligules relatively narrow and only slightly if at all surpassing the pappus; [var. *droebachensis* (Muell.) Blytt (*E. droe.* Muell.); *E. asteroides* Andrz.; *E. elongatus* Ledeb., not Moench; *E. kamtschaticus* DC. (*E. angulosus* var. *kamt.* (DC.) Hara); *E. politus* Fries; *E. lapiluteus* and *E. yellowstonensis* Nels.; transcontinental; MAPS: Hultén 1968b:866 (ssp. *politus*) and 867 (ssp. *kamt.*); Raup 1947: pl. 35 (*E. ang.* var. *kamt.*)]. A hybrid with *E. aureus* is reported from SW Alta. by Boivin 1966b, who also reports a hybrid with *E. uniflorus* var. *unalaschkensis* from SE B.C. and SW Alta. . . . . var. *asteroides* (Andrz.) DC.

    2 Plant rarely as much as 3 dm tall, with few or solitary heads; rays relatively broad and more evidently surpassing the pappus; [*E. debilis* (Gray) Rydb.; *E. jucundus* Greene; *E. nivalis* Nutt.; SE Yukon (Porsild 1951a), S Dist. Mackenzie (E end of Great Slave L.), B.C., and SW Alta.; MAPS: Hultén 1968b:867; Raup 1947: pl. 35 (*E. jucundus*)] . . . .
. . . . var. *debilis* Gray

E. annuus (L.) Pers. Daisy-Fleabane. Vergerette anuelle

/T/X/ (T) Pastures, fields, thickets, and waste places from S B.C. (several collections in V from between Vancouver and Hope) to Alta. (Moss 1959), ?Sask.–Man. (Boivin 1966b; not listed for Sask. by Breitung 1957a; reports from Man. by Lowe 1943, require confirmation, perhaps being based upon *E. strigosus*), Ont. (N to Matheson, 48°32'N), Que. (N to the Gaspé Pen.), S Nfld., N.B., P.E.I., and N.S., S through much of the U.S.A.; introd. and thoroughly natzd. in Europe. [*Aster* L.].

Forma *discoideus* Vict. & Rousseau (the heads discoid, lacking ray-ligules) is known from the type locality, Ste-Famille, about 20 mi NE of Quebec City, Que.

E. asper Nutt.

/sT/WW/ (Hs) Meadows and moist open places, the range very uncertain through confusion with *E. glabellus,* with which it is merged by Cronquist (1947). Boivin (1966b) reports it from Alaska–Yukon–B.C. but Hultén (1968b) lists only *E. glabellus.* In Sask., it ranges N to Prince Albert and in Man., N to Gypsumville, about 125 mi N of Portage la Prairie. [*E. glabellus* of Canadian reports in part, not Nutt.].

In its erect (rather than decumbent) habit and white (rather than pink to purple) ray-ligules, the taxon seems fairly distinct from *E. glabellus.* Observations by the writer "in the field" also indicate that it flowers at a considerably earlier date.

E. aureus Greene

/T/W/ (Hs) Rocky places at high elevations from B.C. (N to Redfern L., Penticton, and Kicking Horse Pass, Yoho National Park) and SW Alta. (N to Jasper) to the Cascade Mts. of Wash. [*Haplopappus brandegei* Gray, not *E. brand.* Gray].

Var. *acutifolius* Raup (basal leaves acute rather than obtuse or rounded, somewhat narrower than those of the typical form) is known from the type locality, Redfern L., in the Peace River dist. of E B.C. at ca. 56°N. An apparent hybrid between *E. aureus* and *E. humilis (E. uniflorus* var. *unalaschkensis* of the present treatment) is reported from the Selkirk Mts. of B.C. by Cronquist (1947).

E. caespitosus Nutt.

/ST/WW/ (Hs) Dry, open, often rocky places at low to moderate elevations from Alaska (N to ca. 67°30′N) and the Yukon (N to ca. 63°N) to B.C.–Alta., Sask. (N to Humboldt, about 55 mi E of Saskatoon), and Man. (N to Victor, about 80 mi NW of Brandon; the report from York Factory, Hudson Bay, 57°N, by Jackson et al. 1922, taken up by Lowe 1943, requires confirmation), S to Wash., Utah, Ariz., N.Mex., and Nebr. [Var. *grandiflorus* (Hook.) T. & G. (*Diplopappus gr.* Hook., not *E. gr.* Hook.); *D. (E.) canescens* Hook., not *E. canescens* H. & A.; *E. subcanescens* Rydb.; *E. condensatus* of Sask. reports, not Nels.]. MAP: Hultén 1968b:870.

E. canadensis L.   Horse-weed.   Vergerette du Canada
/sT/X/   (T)   Disturbed ground, cult. fields, waste places, etc. (probably largely or wholly introd. northwards), from S Dist. Mackenzie (Fort Smith, ca. 60°N) and B.C.–Alta. to Sask. (N to the Churchill R. at ca. 56°N; CAN), Man. (N to Wekusko L., about 90 mi NE of The Pas), Ont. (N to the W James Bay watershed at ca. 53°N), Que. (N to L. St. John and the Côte-Nord), Nfld., N.B., P.E.I., and N.S., S to Calif., Mexico, Tex., and Fla.; tropical America; widely introd. in Eurasia (and probably in a large part of the N. American range, particularly northwards and westwards, the native area very uncertain because of its weedy nature). [*Conyza* Cronq.; *Leptilon* Britt.].

E. compositus Pursh
/AST/(X)/GeA/   (Ch)   Dry calcareous rocks, sands, gravels, and ledges, the aggregate species from Alaska (N to ca. 69°N) and the coasts of the Yukon–Dist. Mackenzie to Banks Is., Melville Is., and Ellesmere Is. (N to ca. 79°N), S in the West through B.C.–Alta. and dry hills in the prairie region of S Sask. (Cypress Hills, Elrose, Biggar, Mortlach, Southey, and Moose Mountain Creek; Breitung 1957a) to Calif., Ariz., and S.Dak., farther eastwards S to Great Slave L. and Baffin Is. (S to ca. 70°N; the report of *E. trifidus* from Churchill, Man., by Lowe 1943, requires confirmation); an isolated small eastern area in E Que. (Bic Mt., near St-Fabien, Rimouski Co.; coastal ledges and cliffs of the Gaspé Pen.) and Nfld.; circumgreenlandic (but large gaps); NE Asia (a station on Wrangel Is., off the coast of NE Siberia, indicated in Hultén's below-noted 1968 map, this not Wrangell Is., Alaska). MAPS and synonymy: see below.
1   Leaves mostly 1-ternate; [the Yukon–B.C. to Sask. (the report of *E. trifidus* from Churchill, Man., by Lowe 1943, requires confirmation); isolated in E Que. (Bic, Rimouski Co.; Gaspé Pen.) and Nfld.; see discussion of the range of var. *compositus*] . . . . var. *discoideus* Gray
  2   Ray-ligules wanting; [*E. multifidus* var. *disc.* (Gray) Rydb.; *E. trifidus* var. *disc.* (Gray) Nels.] . . . . f. *discoideus*
  2   Ray-ligules about 1 cm long; [the more usual form; *E. trifidus* Hook.; *E. gormanii* Greene; *Cineraria lewisii* Rich.] . . . . f. *trididus* (Hook.) Fern.
1   Leaves mostly 2–4-ternate.
  3   Best developed leaves mostly regularly 2–3-times ternate, the divisions usually not very long; [var. *multifidus* (Rydb.) Macbr. & Pays. (*E. multifidus* Rydb.); essentially the range of the species] . . . . var. *glabratus* Macoun
  3   Best developed leaves mostly 3–4 times ternate (often irregularly so), with very long linear divisions; [the range of the aggregate species, as indicated in the following maps, is transcontinental in arctic, subarctic, and alpine regions; there is, however, some doubt as to whether this form with extreme leaf-dissection occurs in N. America other than in the U.S.A., maps of the Alaska–Canada area probably being referable to the above varieties; MAPS (aggregate species): Hultén 1968b:862; Porsild 1957: map 306, p. 199; Dansereau 1957: map 2C, p. 33; Marie-Victorin 1938: fig. 32, p. 524, and Contrib. Inst. Bot. Univ. Montréal 5: fig. 3, p. 90. 1925 (both somewhat incomplete); Fernald 1929: map 13, p. 1492, and 1925: map 54, p. 323 (both somewhat incomplete)] . . . . var. *compositus*

E. corymbosus Nutt.
/T/W/   (Hs)   Open dry places (often with sagebrush) from S B.C. (Vancouver Is. and mainland N to Kamloops, Armstrong, and Windermere; CAN; V) and W-cent. ?Alta. (Grande Prairie, 55°10′N; Herb. V; if correctly identified, the plant was probably introd. there) to Oreg., Wyo., and Mont.

E. divergens T. & G.
/t/WW/   (Hs)   Dry and waste places, often sandy, in the valleys and foothills from S B.C. (Grand

Forks, near the U.S.A. boundary sw of Trail; V; reported from Kelowna, L. Okanagan, by Eastham 1947) to Calif., Mexico, and Okla.

E. filifolius Nutt.
/T/W/ (Hs) Dry plains and foothills, often with sagebrush, from s B.C. (chiefly valleys of the Dry Interior N to the Marble Mts. (NW of Clinton), Kamloops, and Invermere; CAN; V) and Mont. to N Calif. and Nev. [*Diplopappus* Hook.].

E. flagellaris Gray
/T/WW/ (Hsr) Dry open or partly shaded places: s B.C. (Dry Interior N to Chilcotin, the Marble Mts. NW of Clinton, Williams Lake, Kamloops, and Armstrong; CAN; V) and sw Alta. (Waterton Lakes; Breitung 1957b); the main area from Wyo. and S.Dak. to Nev., Ariz., and Tex.

E. formosissimus Greene
/T/WW/ (Hs) Meadows and open ground in the mts., often at high elevations: sw Alta. (w of Pincher Creek, sw of Lethbridge; Cronquist 1947) and Red Lodge, Mont.; the main area from Utah, Wyo., and S.Dak. to Ariz. and N.Mex.

E. glabellus Nutt.
/ST/WW/ (Hs) Meadows and moist open ground (ranges of Canadian taxa outlined below), s to Idaho, Utah, Colo., S.Dak., and Wisc. MAP and synonymy: see below.
1 Pubescence of stem appressed or closely ascending; [*Tessenia* Lunell; *E. speciosus sensu* Fraser and Russell 1944, not DC.; Sask. (N to Prince Albert) and Man. (N to Gypsumville, about 125 mi N of Portage la Prairie)] .... var. *glabellus*
1 Pubescence of stem spreading .... var. *pubescens* Hook.
   2 Ray-ligules pink or roseate; [*Tessenia oligodonta* var. *ros.* Lunell; *E. asper* var. *pub.* f. *ros.* (Lunell) Breitung; *E. drummondii* Greene; probably throughout the range of f. *pubescens*] .... f. *roseata* (Lunell) Scoggan
   2 Ray-ligules blue or purple; [*E. asper* var. *pub.* (Hook.) Breitung; *E. anodontus* and *E. oligodontus* Lunell; *E. turneri* Greene; Alaska (N to ca. 67°30′N), the Yukon (N to ca. 64°N), Dist. Mackenzie (N to Great Slave L.), B.C., Alta. (N to L. Athabasca), Sask. (N to Prince Albert), and Man. (N to Rocky Lake, 30 mi N of The Pas); MAP: Hultén 1968b:870] .... f. *pubescens*

E. grandiflorus Hook.
/aST/W/ (Hs (Ch)) Dry tundra and in the mts. to high elevations from the coasts of Alaska–Yukon–w Dist. Mackenzie and w Victoria Is. to (following an apparent gap) the Rocky Mts. of SE B.C. (North Kootenay Pass; John Macoun 1884) and w Alta. (N to ca. 54°N, s to Crowsnest Pass, 49°38′N; type a Drummond collection from "Summits of the Rocky Mountains", probably in Alta.). MAPS: Hultén 1968b:865; J.G. Packer, Nat. can. (Que.) 98(2): fig. 6, p. 134. 1971.

Some of the N Alaska–Yukon material is referable to ssp. *muirii* (Gray) Hult. (*E. (Aster) muirii* Gray, the type from Cape Thompson, Alaska; leaves and involucres densely lanate rather than the leaves hirsute-pilose, the involucres pilose or long-villous). MAP: W.J. Cody, Nat. can. (Que.) 98(2): fig. 19, p. 152. 1971.

E. hyperboreus Greene
/S/W/ (Hs (Ch)) Rocky places and solifluction soils of Alaska (N to ca. 69°N; type from Porcupine River), w ?Yukon (on or very close to the Alaska boundary N of ca. 65°N), and NW Dist. Mackenzie (Tree R.; CAN). [*E. alaskanus* Cronq.]. MAP: Hultén 1968b:865.

E. hyssopifolius Michx.
/ST/X/ (Hpr) Calcareous ledges, talus, and gravelly shores (ranges of Canadian taxa outlined below), s to northernmost ?B.C. (see Hultén's below-noted map), N Alta. (Wood Buffalo National Park), Sask. (s to The Pas; Herb. Walter Krivda), Man. (s to Cowan, NE of Duck Mt.), Ont. (s to Michipicoten, ca. 48°N on the E shore of L. Superior), N Mich., N N.Y., and N Vt. MAP and synonymy: see below.

1   Plant generally less than.1.5 dm tall, the peduncles as long as or longer than the
     leaf-bearing stem, this with relatively crowded leaves .... var. *villicaulis* Fern.
     2   Stem appressed-pubescent; [var. *anticostensis* Vict. & Rousseau; E Que. (Anticosti
          Is., the type locality) and NW Nfld. (Straits of Belle Isle; Ha-Ha Mt.)] ....
          .... f. *appressus* Cronq.
     2   Stem spreading-pubescent; [type from Table Mt., Port-à-Port Bay, Nfld.] .... f. *villicaulis*
1   Plant commonly over 1.5 dm tall, the peduncles usually shorter than the leaf-bearing
     stem .... var. *hyssopifolius*
     3   Stem spreading-pubescent nearly or quite to the top; [type from ledgy banks of the
          Restigouche R. at Matapédia, SW Gaspé Pen., E Que.] .... f. *patens* Cronq.
     3   Stem appressed-pubescent at least above the middle; [*Aster (Galatella) graminifolius*
          Pursh; S Yukon (Hultén 1968a), NW Dist. Mackenzie, and Great Bear L. to N Alta.
          (Wood Buffalo National Park), Sask. (The Pas; Amisk L., near Flin Flon, ca. 55°N;
          Hasbala L. at ca. 59°N), S Dist. Keewatin, Man. (S to Cowan, near the NE edge of
          Duck Mt.), Ont. (N to W Hudson Bay at ca. 56°30'N), Que. (N to L. Mistassini (probable
          type locality), the Côte-Nord, Anticosti Is., and Gaspé Pen.), Nfld., N.B., and N.S.; MAP
          (aggregate species): Hultén 1968b:862] .... f. *hyssopifolius*

E. lanatus Hook.
/ST/W/   (Hr (Ch))   High mts. of SW Yukon (an isolated station in the Kluane Ranges at 60°46'N;
J.A. Neilson, Can. Field-Nat. 82(2):114. 1968), SE B.C. (N to Kootenay National Park, ca. 50°30'N),
SW Alta. (N to Jasper; CAN; type a Drummond collection from "Summits of the Rocky Mountains
between lat. 52° and 56°."), NW Mont., and Colo. [*E. grandiflorus* var. *lan.* (Hook.) Gray]. MAP:
Neilson, loc. cit., fig. 1, p. 115. 1968.

E. linearis (Hook.) Piper
/t/W/   (Hs)   Dry, often rocky soil from the plains and foothills to moderate elevations from S B.C.
(valleys of the Dry Interior N to the Marble Mts. NW of Clinton, E to Midway, near the U.S.A.
boundary about 50 mi SE of Penticton; the report of *E. peucephyllus* from the Cypress Hills of SW
Sask. (SE ?Alta.) by John Macoun 1884, is based upon *E. radicatus,* the relevant collection in CAN,
detd. Cronquist) to Oreg. and Nev. [*Diplopappus* Hook.; *E. peucephyllus* Gray].

E. lonchophyllus Hook.
/ST/(X)/A/   (Hs)   Calcareous meadows, gravels, and shores at low to fairly high elevations, the
main area from Alaska–Yukon (N to ca. 66°30'N) and the Mackenzie R. Delta to Great Slave L.,
Sask. (N to Hasbala L., ca. 59°N; type a Drummond collection from "Saskatchawan"), and Man. (N
to Churchill), S through B.C.–Alta. to S Calif., Utah, N N.Mex., and N.Dak.; isolated areas along the
James Bay coasts of Ont.–Que. (*see* James Bay watershed map by Dutilly, Lepage, and Duman
1958: fig. 13, p. 167) and in E Que. (Mingan Is. of the Côte-Nord; Anticosti Is.); Asia. [*E.
armeriifolius* Turcz.; *E. glabratus* var. *minor* Hook. (*E. minor* (Hook.) Rydb.); *E. kindbergii* Greene;
E. (Tessenia) racemosus Nutt.; *E. politus sensu* Rydb., not Fries; incl. var. *laurentianus* Vict.]. MAPS:
Hultén 1968b:868; Fernald 1925: map 18 (incomplete northwards), p. 255.

[E. melanocephalus Nels.]
[The report of this species of Wyo. from the Horsethief Creek–Purcell Range region near
Windermere, SE B.C., by Ulke (1935) requires confirmation.]

E. ochroleucus Nutt.
/T/WW/   (Hs)   Dry plains and barren places at moderate to high elevations from SW Alta.
(Waterton Lakes, Crowsnest Pass, and North Kootenay Pass; the last two on the B.C.-Alta.
boundary, the species thus to be expected in SE B.C.; reported from the Cypress Hills of SW Sask.
by Cronquist 1947, but the actual locality of the 1894 Macoun collection cited may have been on
the SE Alta. side) to S Wyo. and NW Nebr. [Incl. the reduced alpine phase, var. *scribneri* (Canby)
Cronq. (*E. scr.* Canby; *E. macounii* Greene); *see* note under *E. radicatus*].

E. pallens Cronq.
/sT/W/   (Ch)   Known only from W Dist. Mackenzie (Porsild and Cody 1968) and the mts. of SE

B.C. (known only from the type locality, Glacier, about 35 mi NE of Revelstoke, where taken by Butters and Holway in 1913) and SW Alta. (Lake of the Hanging Glacier, Rocky Mountain National Park, where taken by Sanson in 1928; Shovel Pass and Mt. Whitehorn, Jasper National Park, where taken by Kindle in 1927 and Porsild in 1964, respectively; Waterton Lakes, where taken by Kuijt in 1970).

E. peregrinus (Pursh) Greene
/ST/W/eA/ (Hs) Moist meadows, streambanks, and boggy places at moderate to high elevations (ranges of Canadian taxa outlined below), S to Calif., Utah, and N.Mex.; E Asia (Commander Is.; other closely related taxa in Asia). MAPS and synonymy: *see* below.
1   Phyllaries villous on the back (or sometimes merely ciliate marginally and glutinous on the back), not at all glandular; rays commonly pale or even white; leaves often toothed and soft-pubescent; stem and peduncles usually loosely villous; [MAP: Hultén 1968*b*:869] . . . .
. . . . ssp. *peregrinus*
    2   Upper stem-leaves either relatively large or closely set; [*Aster peregrinus* Pursh, the type from Unalaska, Aleutian Is.; *A. unalaschkensis* Less., not *E. unal.* Vierh.; *E. salsuginosus* var. *unal.* Less.; Aleutian Is.–S Alaska–SW Yukon (*see* Hultén 1950: map 118a, p. 1674) and B.C.] . . . . var. *peregrinus*
    2   Upper stem-leaves reduced and distant; [S Alaska, B.C. (type from Queen Charlotte Is.), and SW Alta. (Banff)] . . . . var. *dawsonii* Greene
1   Phyllaries densely glandular on the back, rarely also with a few long hairs; rays commonly rich rose-purple or darker; leaves usually entire and glabrous; plant usually essentially glabrous except for the closely villous peduncles; [MAP: Hultén 1968*b*:869] . . . .
. . . . ssp. *callianthemus* (Greene) Cronq.
    3   Reduced, often subscapose, alpine plant less than 2 dm tall, with relatively large, obtuse or rounded basal leaves and much smaller stem-leaves; [*E. acutatus* and *E. obtusatus* Greene; *Aster salsuginosus* var. *scaposus* T. & G.; *A. glacialis* Nutt.; SE ?Alaska, B.C., and SW Alta.] . . . . var. *scaposus* (T. & G.) Cronq.
    3   Larger, mostly subalpine plants to 7 dm tall (if smaller, then with narrow and acute basal leaves).
        4   Basal leaves oblanceolate or narrower, those of the stem linear or lanceolate; [*E. angustifolius* Gray; *E. loratus* Greene; mts. of S B.C.] . . . .
. . . . var. *angustifolius* (Gray) Cronq.
        4   Basal leaves oblanceolate or broader, those of the stem mostly ovate or not greatly reduced; [var. *eucallianthemus* Cronq.; *E. callianthemus* Greene; S Alaska–B.C.–SW Alta.] . . . . var. *callianthemus*

E. philadelphicus L.
/ST/X/ (Hs (T)) Moist places (often disturbed) from NW Yukon (an isolated station at ca. 67°30′N), W Dist. Mackenzie (N to ca. 64°N), and B.C. (N to Liard Crossing, ca. 59°25′N) to Alta. (N to Fort Vermilion, 58°24′N, and L. Mamawi, W of L. Athabasca), Sask. (N to Prince Albert), Man. (N to the Churchill R. at ca. 57°20′N), Ont. (N to Sachigo L. at ca. 54°N, 92°W), Que. (N to the E James Bay watershed at 52°37′N and the Côte-Nord), Nfld., N.B., P.E.I., and N.S., S to Calif., Tex., and Fla. [*Tessenia* Lunnell; *E. purpureus* Ait.; incl. the luxuriant extreme, f. *scaturicola* (Fern.) Cronq. (*E. scat.* Fern.); *E. bellidiastrum sensu* John Macoun 1884, not Nutt., the relevant collection in CAN]. MAP: Hultén 1968*b*:871.
    Var. *glabra* Henry (*E. provancheri* Vict. & Rousseau; plant essentially glabrous rather than long-hairy) is known from B.C. (type from Vancouver Is.) and E Que. (type of *E. prov.* from St-Vallier, Bellechasse Co.). Forma *angustatus* Vict. & Rousseau (basal leaves linear-lanceolate, generally less than 1 cm broad, rather than oblanceolate to narrowly obovate and to about 3 cm broad) is known from the type locality, Flower-Pot Is., L. Huron, Bruce Co., S Ont.

E. poliospermus Gray
/t/WW/ (Hs) Dry plains and foothills (often with sagebrush) from SE B.C. (L. Osoyoos, near the U.S.A. boundary S of Penticton, where taken by John Macoun in 1905, distributed as *E. concinnus*, revised by Cronquist; CAN; reported from North Kootenay Pass, on the B.C.–Alta. boundary, by John Macoun 1886) to Oreg.

E. pulchellus Michx.   Robin's-plantain
/T/EE/   (Hsr)   Meadows, thickets, and open woods from Minn. to S Ont. (N to Wellington, Peel, and Ontario counties; CAN; TRT), SW Que. (Stanstead, Richmond, Richelieu, and Terrebonne counties; CAN; MT; reports from N.S. require confirmation), and S Maine, S to Kans., La., Ala., and Fla. [*E. bellidifolius* Muhl.].

There are 3 collections in CAN, detd. Porsild, from the Yukon between ca. 62°30' and 64°N, where almost certainly introd., probably in forage, having been taken in pastures, meadows, and river flats near "Pelly farm" and in a meadowlike spot near the confluence of the Yukon and Pelly rivers.

E. pumilus Nutt.
/T/WW/   (Hs)   Dry plains, valleys, and foothills (often with sagebrush) from S B.C. (valleys of the Dry Interior N to Kamloops and Armstrong), S Alta. (N to Rosedale and Drumheller), and S Sask. (Souris Plain, S of Moose Mt., where taken by J.M. Macoun in 1883; CAN, verified by Cronquist; the tentative report from Sutherland by Breitung 1957a, is referred by A.J. Breitung, Am. Midl. Nat. 61(2):512. 1959, to *E. caespitosus;* reports from S Man. by Lowe 1943, require clarification; the indication of a station at Kluane L. in SW Yukon in Hultén's below-noted map may refer to an introd. plant) to S Calif., N.Mex., and Kans. MAP and synonymy: *see* below.

1   Outer pappus relatively inconspicuous, the inner pappus consisting of up to 27 slender, white, obscurely barbed bristles; rays nearly always white; base of disk-corollas glabrous; [Alta. (reported from Rosedale and Medicine Hat ("Sask."), Alta., by Cronquist 1947; the tentative report of *E. uncialis* var. *conjugans* Blake of the w U.S.A. from SW ?Alta. by Boivin 1966b, may possibly be referable here or to *E. radicatus*) and S Sask.; MAP (aggregate species): Hultén 1968b:863] .... ssp. *pumilus*
1   Outer pappus consisting of coarse bristles, the inner pappus of up to 20 rather coarse, yellowish or sordid, evidently barbed bristles; rays usually pink or blue.
    2   Base of disk-corollas rather copiously puberulent; rays nearly always pink or blue; [*E. concinnus* T. & G.; reported from S B.C. by John Macoun 1884 (Kootenay and Nicola valleys) and Eastham 1947 (Tranquille; Okanagan; Grand Forks) but not cited for Canada by Cronquist 1947, who does, however, note a collection from L. Okanagan under ssp. *intermedius* var. *gracilior*; the Kootenay and Nicola Valley collections also refer to this latter taxon, as probably do the others] .... ssp. *concinnoides* Cronq.
    2   Base of disk-corollas glabrous or slightly puberulent; rays sometimes white ....
    .... ssp. *intermedius* Cronq.
        3   Plant robust, the larger stems over 1.5 mm thick at base and bearing at least 5 heads; [*E. strigosus* var. *hispidissimus* Hook.; S B.C. (Fairmont Hot Springs; Anarchist Mt. near Osoyoos; Midway; Flathead); ?introd. in SW Yukon] ..... var. *intermedius*
        3   Plant slender, the stems at most 1.5 mm thick at base or bearing less than 5 heads, or both; [S B.C.: Kootenay and Nicola valleys; Kamloops; L. Okanagan; L. Osoyoos; Cascade; Trout Creek] .... var. *gracilior* Cronq.

E. purpuratus Greene
/Ss/W/   (Ch (Hrr))   Sandy or gravelly places at moderate to high elevations from Alaska (N to ca. 69°30′N), the Yukon (N to ca. 64°N; type from Fort Selkirk), and w Dist. Mackenzie (Canol Road at ca. 63°N; CAN) to northernmost B.C. (Taku Arm, Tagish L., near Atlin at ca. 59°30′N; CAN; a collection in V from Chilcotin, SW of Williams Lake, requires confirmation). [*E. denalii* Nels.]. MAPS: Hultén 1968b:863; Porsild 1966: map 151, p. 85.

E. radicatus Hook.
/T/W/   (Ch (Hr))   Dry montane slopes and hillsides of SW Alta. (Elbow R. at Moose Mt., SE of Banff, where taken by John Macoun in 1897; CAN; type an early Drummond collection from near Jasper; the report from B.C. by Rydberg 1922, requires confirmation; according to Hultén 1950, Alaskan reports are referable to *E. purpuratus*) and S Sask. (Cypress Hills, where taken by Macoun in 1880 and 1895, and Old Wives Creek and Wood Mountain, near the U.S.A. boundary SW of Moose Jaw, where taken by Macoun in 1895; CAN). [*E. peucephyllus sensu* Macoun 1884, not Gray, the relevant collection in CAN, revised by Cronquist; *E. ?leiomeris sensu* Rydberg 1922, not Gray].

According to Cronquist (1947), "*E. radicatus* is closely allied to *E. ochroleucus* var. *scribneri*, and may intergrade with it. Small forms of the latter may be distinguished by the finer pubescence of their stems, leaves, and involucre, and by usually having at the base a mere stout crown, or only slightly branched caudex, whereas the caudex of *E. radicatus* is well developed and branched, forming something of a cushion."

[E. simplex Greene]
[The report of this species of the w U.S.A. (N to Oreg. and Mont.) from sw Alta. by Moss (1959; Banff) requires clarification, perhaps being based upon *E. uniflorus,* with which, according to Hitchcock et al. (1955), it has often been confused.]

E. speciosus (Lindl.) DC.
/T/WW/ (Hs) Open woods and clearings, mostly in the foothills and at moderate elevations, from s B.C. and w Alta. (ranges of Canadian taxa outlined below) to Baja Calif., Ariz., N.Mex., and the Black Hills of S.Dak.

1   Leaves, stem, and involucres more or less long-hairy; [*E. conspicuus* and *E. subtrinervis* Rydb.; *E. glabellus* var. *?mollis* Gray; s B.C. (Skagit Valley; Anarchist Mt., near Osoyoos; Greenwood, about 50 mi w of Trail; St. Mary R. near Kimberley) and sw Alta. (Waterton Lakes; Belly R. near Fort Macleod)] .... var. *conspicuus* (Rydb.) Breitung
1   Leaves glabrous or nearly so (except for the ciliate margins); stem glabrous or very sparingly hairy below the inflorescence; involucre with few or no long hairs.
   2   Leaves usually glabrous over the surfaces, rarely strongly ciliate, relatively broad (the uppermost ones ovate); stem commonly glabrous except directly below the heads, these with glabrous involucres; [*E. grandiflorus* and *E. macranthus* Nutt., not *E. gr.* Hook.; s B.C. (N to Kamloops and Lac la Hache) and Alta. (Crowsnest Pass; Waterton Lakes; Sweet Grass Hills)] .... var. *macranthus* (Nutt.) Cronq.
   2   Leaves relatively narrow, the uppermost ones lanceolate and often sparingly hairy, tending to be strongly ciliate; stem often sparsely hairy below the heads; involucres commonly sparingly hairy; [*Stenactis* Lindl.; B.C. (N to Chilcotin Plains, ca. 52°20′N) and Alta. (N to Grande Prairie in the Peace River dist. at ca. 55°10′N)] ...... var. *speciosus*

E. strigosus Muhl.   Daisy-Fleabane, White-top.   Vergerette rude
/sT/X/ (T) Dry open places (often in waste or disturbed ground and cult. fields; apparently largely introd. northwards) from B.C. (N to the Pine R. in the Peace River dist. at ca. 56°N; V) to Alta. (N to Wood Buffalo National Park at 59°31′N and L. Athabasca; CAN), Sask. (N to Hudson Bay Junction, 52°52′N), Man. (N to Steeprock, about 100 mi N of Portage la Prairie), Ont. (N to the sw James Bay watershed at ca. 52°N), Que. (N to near L. Waswanipi at ca. 49°30′N and the Gaspé Pen.), Nfld., N.B., P.E.I., and N.S., s to Calif., Idaho, Okla., Tex., and Fla.

1   Larger basal leaves to over 3 cm broad; lower stem-leaves broadly oblanceolate; disks of larger heads to 12 mm broad .... var. *strigosus*
   2   Rays elongate, usually white; [incl. var. *septentrionalis* (Fern. & Wieg.) Fern. (type of *E. ramosus* var. *sept.* F. & W. from Nfld.) which, as noted by Cronquist 1947, bridges the gap between *E. strigosus* and *E. annuus,* and may be the result of hybridization between these two species; *E. ramosus* (Walt.) BSP., not Raf.; transcontinental] ....
   .... f. *strigosus*
   2   Rays very short; [var. *discoideus* Robbins; SE B.C.: Sproat, Columbia Valley; J.M. Macoun 1896] .... f. *discoideus* (Robbins) Fern.
1   Larger basal leaves usually not over 1.5 cm broad; lower stem-leaves nearly linear; disks to about 8 mm broad; rays white or purple; [Otterburne, s Man.; Löve and Bernard 1959] ......
   .... var. *beyrichii* (Fisch. & Mey.) T. & G.

E. uniflorus L.
/AST/X/GEA/ (Hs) Meadows, snow beds, glacial till, and dry slopes (mainly in calcareous soils) at low to high elevations, the ranges of N. American taxa outlined below; circumgreenlandic; N Eurasia. MAPS and synonymy (together with a distinguishing key to the scarcely separable *E. borealis* of Baffin Is. and Greenland): *see* below.

1   Hairs of the involucre with dark blackish-purple cross-walls; phyllaries generally blackish
     purple throughout, usually less than 1 mm broad, attenuate; ray-ligules to 1 mm broad;
     disk to 2 cm broad; [*E. pulchellus* var. *unal.* DC., described from plants taken at
     Unalaska, Aleutian Is., and in E Siberia; *E. unal.* (DC.) Vierh.; *E. humilis* Graham; Aleutian
     Is. and the coasts of Alaska–Yukon–Dist. Mackenzie to Banks Is., Victoria Is., Baffin Is.
     near the Arctic Circle, and northernmost Ungava–Labrador, S to S Alaska–Dist.
     Mackenzie–Dist. Keewatin, NE Man. (known only from Churchill), and the coasts of
     Hudson Bay–James Bay, Que. (S to E James Bay at 53°50′N), and in the mts. through
     B.C. and SW Alta. to Mont.; MAPS: Hultén 1968*b*:864, and 1958: map 175 (*E. hum.*), p.
     195; Porsild 1957: map 309 (*E. unal.*), p. 199; Raup 1947: pl. 35 (*E. unal.*); Böcher 1954:
     fig. 29 (*E. unal.*), p. 115] . . . . var. *unalaschkensis* (DC.) Boivin
1   Hairs of the involucre with clear or bright reddish-purple cross-walls.
     2   Pubescence of the involucre very long and generally very dense, appearing tousled;
          phyllaries rich reddish-purple, their loose tips somewhat attenuate; disk to 3 cm broad;
          involucre to 11 mm high; ray-ligules to about 0.5 mm broad; [*E. eriocephalus* Vahl,
          the type from Greenland; coasts of Alaska–Yukon–Dist. Mackenzie to northernmost
          Ellesmere Is. and northernmost Ungava–Labrador, S to S Dist Mackenzie, NE B.C., S
          Dist. Keewatin, and N Ungava–Labrador at ca. 60°N; nearly circumgreenlandic but
          with extensive gaps; ?Iceland; N Europe; Siberia; MAPS (*E. erio.*): Hultén 1968*b*:864;
          Porsild 1957: map 307, p. 199; Raup 1947: pl. 35] . . . . var. *eriocephalus* (Vahl) Boivin
     2   Pubescence of the involucre shorter, coarser, and generally less dense, not
          appearing tousled; phyllaries generally green, their purplish tips appressed and
          scarcely attenuate; disk to 2 cm broad; involucre to 8 mm high; ray-ligules to 0.8 mm
          broad; [*Trimorphaea* Vierh.; *E. alpiniformis* Cronq.; Baffin Is. (Frobisher Bay); W and E
          Greenland N to ca. 68°N; Iceland; Europe; MAPS: Hultén 1958: map 68, p. 87; Porsild
          1957 (1964 revision): map 343, p. 203; Löve and Löve 1956*b*: fig. 16 (*E. alpin.*), p.
          179] . . . . *E. borealis* (Vierh.) Simm.

[*E. yukonensis* Rydb.]
[This obscure Yukon species (type from Dawson) is included in *E. grandiflorus* by Polunin (1959)
and, according to Hultén (1950), "might as well be regarded as a narrow-leaved race of *E.
glabellus pubescens*.". (*E. glabellus* ssp. *pubescens* var. *yuk.* (Rydb.) Hult.)]

ERIOPHYLLUM Lag.   [9295]

E. lanatum (Pursh) Forbes
/t/W/   (Hp)   Dry open places at low to moderate elevations from SW B.C. (Vancouver Is. and
adjacent islands and mainland E to Hope and Yale; CAN; V) and Mont. to Calif., Utah, and Wyo.
[*Actinella* Pursh; *Bahia* DC.; *Helenium* Spreng.; *Trichophyllum* Nutt.; *B. leucophylla* DC.; *E. (H.)
caespitosum* Dougl. and its var. *leuc.* (DC.) Gray].
     Var. *achillaeoides* (DC.) Jeps. (*Bahia ach.* DC.; *E. ternatum* Greene; *E. cusickii* Eastw. in herb.;
leaves mostly 2-ternate rather than merely deeply pinnatifid or 1-ternate) is reported from B.C. by
P.A. Rydberg (N. Am. Flora 34:93. 1915) and Carter and Newcombe (1921; Vancouver Is.) but
so-named collections in CAN from Vancouver Is. have been referred to the typical form by Lincoln
Constance, his annotation-labels noting that this variety does not occur N of S Oreg. Var.
integrifolium (Hook.) Smiley (*Trichophyllum int.* Hook.; leaves mostly entire or merely 3-lobed at
apex, all or nearly all alternate rather than mostly opposite) is reported from SW B.C. by John
Macoun (1886; Gulf of Georgia, as *E. caesp.* var. *int.* (Hook.) Gray) but the relevant Dawson
collection has not been located.

EUPATORIUM L.   [8816]   Thoroughwort

1   Stem-leaves compound, opposite, sessile or very short-petioled, their 3(–5) coarsely
     serrate leaflets lanceolate to elliptic-acuminate; basal leaves oblanceolate, petioled;
     branch-leaves lanceolate to ovate; receptacle flat; phyllaries 2–3-ranked; flowers whitish
     to reddish-mauve; (introd. in ?B.C.) . . . . [*E. cannabinum*]
1   Leaves all simple.

2 Leaves whorled, lanceolate to ovate or elliptic, petioled (or the uppermost ones subsessile); receptacle flat; phyllaries in several series of unequal length; flowers creamy white to pale pink or purple; (transcontinental) .... *E. purpureum*
2 Leaves opposite (but often with axillary fascicles of reduced leaves).
   3 Leaves sessile (their broad bases united and perfoliate), crenate-serrate, lance-acuminate; receptacle flat; phyllaries 2–3-ranked; flowers whitish or purple-tinged; (SE Man. to the Maritime Provinces) .... *E. perfoliatum*
   3 Leaves mostly long-petioled, ovate; phyllaries subequal.
      4 Base of stem with slender, superficial, creeping stolons, these rooting at the nodes; leaves blunt-toothed; receptacle conical; flowers bluish violet; (S Ont.) ......
          .... *E. coelestinum*
      4 Base of stem lacking stolons; leaves coarsely and often sharply toothed; receptacle flat; flowers bright white; (?Sask.–Man.; Ont. to N.S.) .... *E. rugosum*

[E. cannabinum L.]   Hemp-Agrimony
[Eurasian; reported as introd. in SW B.C. by Henry (1915; Sullivan and Surrey, near Vancouver), where perhaps not established.]

E. coelestinum L.   Mistflower
/t/EE/   (Hpr)   Low woods, damp thickets, clearings, and streambanks from Kans. to Ind., Ohio, S Ont. (known only from Essex Co., where taken by R. Frith in 1965 at "Point Pelee National Park, Tilden subdivision; edge of woods. New to Canada?"; CAN, detd. A.E. Porsild), Pa., Md., and N.J., S to Tex. and Fla.

E. perfoliatum L.   Thoroughwort
/T/EE/   (Hp (Hpr))   Wet thickets, swampy ground, and shores from SE Man. (N to Muskeg Is., L. Winnipeg, ca. 52°N) to Ont. (N to the Nipigon R. N of L. Superior), Que. (N to near Mont-Laurier, about 80 mi N of Hull, and Montmagny Co.; MT), N.B., P.E.I., and N.S., S to Tex. and Fla.
1 Heads purple-tinged; [Ont. (Boivin 1966*b*), Que. (St-Jean, Richelieu Co.), and N.S. (St-Croix, Hants Co.)] .... f. *purpureum* Britt.
1 Heads whitish.
   2 Leaves free at their truncate or broadly rounded bases; [var. *tr.* Gray; Ont. (Curran; Boivin 1966*b*) and N.S. (Havelock, Digby Co.)] .... f. *truncatum* (Gray) Fassett
   2 Leaves united around the stem by their broad bases.
      3 Leaves mostly in whorls of 3; [Ont., Que., and N.S.] .... f. *trifolium* Fassett
      3 Leaves in opposite pairs; [incl. var. *colpophilum* Fern. & Grisc.; *E. connatum* Michx.; S Man. to N.B. and N.S.] .... f. *perfoliatum*

E. purpureum L.   Joe-Pye-weed
/T/X/   (Hp (Hpr))   Damp thickets, meadows, and shores, the aggregate species from S B.C. (several islands adjacent to Vancouver Is.; lower Fraser and Chilliwack valleys) to ?Alta. (reported N to the Clearwater R., ca. 56°45′N, by John Macoun 1884; not listed by Moss 1959), Sask. (N to near Ile-à-la-Crosse, 55°27′N; Breitung 1957*a*), Man. (N to Hill L., N of L. Winnipeg), Ont. (N to the Shamattawa R. at 54°14′N), Que. (N to the E James Bay watershed at 52°15′N, L. Mistassini, and the Côte-Nord), St-Pierre and Miquelon, Nfld., N.B., P.E.I., and N.S., S to Utah, N.Mex., Iowa, Ohio, and N.C.
1 Inflorescence roundish-topped; flowers rarely more than 8 in a head; stem usually unspotted; [*E. dubium* Willd.; *E. falcatum* Michx.; S Ont.: N to Bruce and Hastings counties] .... var. *purpureum*
1 Inflorescence flattish-topped; flowers 8 or more in a head; stem deep purple or purple-spotted .... var. *maculatum* (L.) Darl.
   2 Heads white; [*E. mac.* f. *faxonii* Fern.; Sask. (Boivin 1966*b*), Ont. (SE shore of L. Superior), N.B. (near Edmunston), N.S. (Whycocomagh, Cape Breton Is.), and Nfld.] ......
       .... f. *faxonii* (Fern.) Boivin
   2 Heads pale lilac or pink to purple.
      3 Inflorescence decompound, comprising about the upper 1/3 of the plant, the few-headed corymbs mostly surpassed by the numerous bracteal leaves; [*E. mac.*

f. *anomalum* Vict., the type from Rupert House, James Bay, Que.; also known
from w James Bay, Ont., and from Rimouski and Bonaventure counties, E Que.] . . . .
. . . . f. *anomalum* (Vict.) Boivin

3   Inflorescence relatively short and more regularly branched, the heads usually
more numerous in corymbs mostly surpassing the bracteal leaves.
   4   Floral organs changed into phyllaries; [type from near Otterburne, s Man.] . . . .
   . . . . f. *tegulosum* Boivin
   4   Floral organs normal; [*E. maculatum* L. and its var. *foliosum* (Fern.) Wieg.,
   and f. *erisinatum* Lepage of the latter taxon; *E. bruneri* Gray; *E. fistulosum*
   Barratt; *E. rydbergii* Britt.; transcontinental] . . . . f. *maculatum*

E. rugosum Houtt.   White Snakeroot
/T/EE/   (Hp)   Moist woods, thickets, and clearings from ?Sask.–Man. (*see* Breitung 1957a:64) to
Ont. (N to the Ottawa dist.), Que. (N to L. St. John, the Côte-Nord, and Gaspé Pen.), N.B., and N.S.
(not known from P.E.I.), s to E Tex. and Ga. [*E. ageratoides* L. f.; *E. boreale* Greene; *E.
urticaefolium* Reichard].
   Forma *verticillatum* (Vict.) Scoggan (*E. urticaefolium* f. *vert.* Vict.; most or all of the leaves in 3's
rather in opposite pairs) is known from the type locality, Grosse-Ile, about 40 mi NE of Quebec City,
Que.

FILAGO L.   [8969]   Fluff-weed

1   Heads 2–7; phyllaries spreading in fruit, blunt . . . . *F. arvensis*
1   Heads 20–40; phyllaries erect, the outer shorter ones cuspidate, the inner longer ones
with a shining, yellow, awn-like bristle-tip . . . . [*F. germanica*]

F. arvensis L.
Eurasian; introd. in the w U.S.A., where, according to Hitchcock et al. (1955), becoming common
on overgrazed ranges; in Canada, known from dry roadsides and fields in s B.C. (Saltspring Is.,
near Vancouver Is.; Kootenay dist. at Creston, Erickson, Moyie, Cranbrook, Kitchener, and Elko;
CAN; V) and s Ont. (Manitoulin Is., N L. Huron; DAO). [*Gnaphalium* L.].

[F. germanica (L.) Huds.]   Cudweed, Herba Impia
[Eurasian; reported from B.C. by T.M.C. Taylor (1966b), otherwise locally introd. into dry fields of
the E U.S.A. (*Gnaphalium* L.).]

FRANSERIA Cav.   [9147]   Bur-sage

1   Burs (fruit) glabrous except for minute glands, the spines thin and relatively weak, their
tips straight or slightly curved; staminate heads to 4 mm broad; leaves 1–2-pinnatifid;
coarsely strigose or scabrous-hispid annual; (s Alta., s Sask., and s Man.) . . . . *F. acanthicarpa*
1   Burs more or less hairy, the spines somewhat flattened or subterete, stiffer, their tips
often curved; staminate heads to about 8 mm broad; leaves toothed to pinnate-pinnatifid,
generally appressed-hairy, succulent; perennial, the stems freely branching beneath the
surface; (coastal sands of B.C.) . . . . *F. chamissonis*

F. acanthicarpa (Hook.) Coville   Sandbur
/T/WW/   (T)   Open places and sandy shores from Wash. to s Alta. (Milk River, Medicine Hat,
and Manyberries, s of Medicine Hat; CAN), s Sask. (Cypress Hills, Beverley, Piapot, Beaver Creek,
Mortlach, and Great Sand Hills, NW of Swift Current; Breitung 1957a), and s Man. (near Bernice,
about 35 mi sw of Brandon; DAO; reported from the Red River Valley by Jackson et al. 1922), s to
Calif. and Tex.; introd. eastwards. [*Ambrosia acan.* Hook., the type locality given as "Banks of the
Saskatchewan and Red River"; *F. hookeriana* Nutt.].

F. chamissonis Less.
/t/W/   (Ch (Grh))   Coastal sands from B.C. (Queen Charlotte Is.; Vancouver Is. and adjacent
islands; Vancouver; CAN; V) to s Calif. [*Ambrosia* Greene; *Gaertneria* Ktze.; *F. cuneifolia* Nutt.].

Most of the B.C. material is referable to var. *bipinnatisecta* Less. (*F. bip.* (Less.) Nutt.; leaves mostly 2–3-pinnatifid and less hairy than the usually densely silvery-pubescent and merely toothed leaves of the typical form).

GAILLARDIA Foug.   [9306]   Gaillardia

1   More or less hirsute perennial with entire to sinuate-pinnatifid leaves to about 2 dm long; heads to 1 dm broad, the yellow rays often with a purple base (or sometimes wholly purple); receptacular chaff surpassing the achenes; (B.C. to Man.; introd. eastwards) . . . . *G. aristata*
1   Soft-pubescent annual usually less than 5 dm tall; leaves to about 1 dm long, entire or the lower ones lyrate-pinnatifid; heads to 5 cm broad, the rays yellow at tip, rose-purple at base; receptacular chaff about equalling the achenes; (introd. in Ont.) . . . . [*G. pulchella*]

G. aristata Pursh
/T/WW/   (Hs)   Prairies, plains, and foothills from B.C. (N to near Cariboo and Lac la Hache) to Alta. (N to Jasper; CAN), Sask. (N to Prince Albert), and Man. (N to Grand Rapids, near the NW end of L. Winnipeg), S to Oreg., Utah, Colo., and S.Dak.; occasionally introd. elsewhere, as in the E U.S.A. and in S Dist. Mackenzie (Fort Simpson, ca. 62°N; W.J. Cody, Can. Field-Nat. 75(2):67. 1961). [*G. pulchella* of Canadian reports except those noted under that species; *G. bicolor* Sims, not Lam.].
Some of our material is referable to f. *monochroma* Boivin (ray-ligules uniformly yellow or orange-yellow rather than with a purple base; type from Waldheim, Sask.).

[*G. pulchella* Foug.]
[Native in the S U.S.A. from Colo. and Ariz. to N.C. and Fla.; introd. or a garden-escape elsewhere, as in Ont. (North Gower, Carleton Co.; DAO; reported from Lambton Co. by Gaiser and Moore 1966), where, however, probably not established.]

GALINSOGA R. & P.   [9246]   Quick-weed

1   Peduncles and upper part of stem more or less densely clothed with spreading flexuous simple hairs (at least 0.5 mm long) and glandular hairs; receptacular chaff simple; at least the longer pappus-scales of the disk-achenes narrowed into distinct awns; pappus-scales of the ray-achenes shorter than the corolla; achenes densely hispid on the inner face; leaves rather coarsely serrate . . . . *G. ciliata*
1   Peduncles with short ascending hairs (less than 0.5 mm long); stem glabrous or sparingly appressed-puberulent and glandular; receptacular chaff 3-toothed at apex; pappus-scales not awned, those of the ray-florets none or rudimentary; achenes hispid only at summit; leaves more finely serrate . . . . *G. parviflora*

G. ciliata (Raf.) Blake
A native of Cent. and S. America; a weed of gardens, yards, and waste places in N. America, as in S B.C. (Langley and Agassiz, E of Vancouver; New Denver, about 35 mi N of Nelson), Alta. (Boivin 1966b), Sask., S Man. (N to Grand Beach, near the S end of L. Winnipeg), Ont. (N to the N shore of L. Huron and the Ottawa dist.), Que. (N to Rimouski, Rimouski Co.), N.B., P.E.I., and N.S. [*Adventina* Raf.; *G. aristulata* Bickn.; *G. parviflora* var. *hispida* DC.].

G. parviflora Cav.
A native of Mexico and S. America; introd. into N. America in similar habitats as those of *G. ciliata* (but much less common), as in B.C. (T.M.C. Taylor 1966b), S Ont. (N to Toronto; J.M. Macoun 1906), and SW Que. (N to Montreal; R. Campbell, Can. Rec. Sci. 6(6):342–51. 1895, and Raymond 1950b).

GNAPHALIUM L.   [8992]   Cudweed, Everlasting

1   Heads clustered (sometimes solitary) at the ends of the branches of a usually ample

terminal corymb or panicle of corymbs, the individual clusters not conspicuously leafy-bracted; phyllaries uniformly white to yellowish white or tan; pappus-bristles distinct, falling separately; lower leaves narrowly oblanceolate, the others linear to lanceolate or linear-oblong (sometimes slightly broader in *G. chilense*); stems commonly unbranched below the inflorescence, to over 7 dm tall.

2 Leaves adnate-clasping but their bases not decurrent down the stem; annuals or biennials.

   3 Inflorescence compact and often subcapitate (tending to consist of only 1 or a few dense clusters of heads); leaves copiously white-woolly on both sides, not at all glandular; (s B.C.) . . . . *G. chilense*

   3 Inflorescence open-corymbiform, the lower branches commonly elongating and branching near the end; leaves green and from glabrous to slightly glandular or slightly woolly above, white-woolly beneath; (?B.C.; Ont. to N.S.) . . . . . . *G. obtusifolium*

2 Leaves sessile and with decurrent lines running down the stem for an appreciable distance below their bases; heads generally numerous in small clusters, commonly forming a broad open inflorescence; (B.C. and Alta.).

   4 Perennial, the several stems from a short-lived taproot; leaves more or less tomentose but not at all glandular; phyllaries acute, woolly only at base or not at all . . . . *G. microcephalum*

   4 Annual or biennial, the stems simple or branched at base; leaves more or less glandular-hairy at least above, sometimes also somewhat tomentose.

      5 Phyllaries mostly sharply acute, generally yellowish or somewhat dingy; stem usually conspicuously glandular-hairy, becoming tomentose in the inflorescence (rarely to near the base, as well as glandular); (B.C.; Ont. to N.S.) . . . . *G. viscosum*

      5 Phyllaries mostly obtuse or broadly rounded at summit, tending to be pearly-white; stem usually thinly tomentose and only slightly or scarcely glandular-hairy . . . . [*G. californicum*]

1 Heads in terminal leafy-bracted clusters or leafy-bracted spikes (in depauperate individuals, sometimes solitary); mature stems diffusely branched from base.

  6 Heads in small capitate sessile clusters; involucres 3 or 4 mm high; pappus-bristles falling separately; copiously white-tomentose or -woolly annuals.

   7 Leaves to about 4 cm long, linear to narrowly oblanceolate; mature phyllaries usually discoloured (greenish or brownish) at tip, moderately woolly below; plant appressed-tomentose; (transcontinental; ?introd.) . . . . *G. uliginosum*

   7 Leaves 1 or 2 cm long, rather more broadly oblanceolate; phyllaries brown, usually with whitish tips, densely woolly toward base; plant loosely floccose; (s B.C. to s Sask.) . . . . *G. palustre*

  6 Heads in distinct spikes (in depauperate individuals, sometimes solitary or capitate); involucre at least 5 mm high; flowering stem simple or sparingly branched at base.

   8 Heads solitary and terminal or up to 5 in a spike; leaves linear, mostly crowded in basal tufts; flowering stem at most about 1 dm tall; pappus-bristles distinct, falling separately; dwarf tufted perennial; (Que., Labrador, and Nfld.) . . . . *G. supinum*

   8 Heads commonly more numerous in elongate spikes; leaves broader; stems commonly taller; pappus-bristles united at base and deciduous in a ring.

      9 Annual or biennial; basal spatulate leaves and oblanceolate or spatulate stem-leaves round-tipped, covered beneath with a white felt-like coat of enmeshed subappressed hairs; phyllaries brown to chestnut or purple; achenes papillate; (sw B.C.; s Ont.) . . . . *G. purpureum*

      9 Perennials with a stout caudex bearing tufts of acutish-tipped leaves; achenes sparsely strigose.

         10 Basal leaves linear-oblanceolate, 1-ribbed, at most about 1 cm broad; spike up to 4/5 of the height of the plant; phyllaries whitish or pale brown, with a dark spot near tip; (Ont. to Nfld. and N.S.) . . . . *G. sylvaticum*

         10 Basal leaves oblanceolate, 3-ribbed, to 3 cm broad; spike at most about half the height of the plant; phyllaries with broad dark margins and a narrow pale centre and base; (Que., Labrador, and Nfld.) . . . . . . *G. norvegicum*

[G. californicum DC.]
[The report of this species of the w U.S.A. (Oreg. to Calif.) from B.C. by J.M. Macoun (1896; Revelstoke and Ainsworth) is based upon G. viscosum, the relevant collections in CAN. (G. decurrens var. cal. (DC.) Gray).]

G. chilense Spreng.
/t/WW/ (Hs) Open, usually moist places, often in disturbed soil, from s B.C. (Vancouver Is.; Cascade; Yale; Manning Provincial Park, SE of Hope; Salmon Arm, about 50 mi E of Kamloops) and Mont. to s Calif. and Tex. [G. sprengelii H. & A.].

G. microcephalum Nutt.
/T/W/ (Hs) Dry open places, often sandy or rocky, from s B.C. (Vancouver Is. and adjacent islands and mainland N to Lytton; CAN; V) and sw Alta. (Waterton Lakes; Breitung 1957b) to s Calif. and Colo. [Incl. G. thermale Nels.].

G. norvegicum Gunn.
/aST/E/GEA/ (Hp). Damp humus slopes, grassy depressions, and ledges of Que. (coast of E Hudson Bay between ca. 56° and 60°N; Ungava Bay; mts. at 55°04'N, 67°12'W; Shickshock Mts. of the Gaspé Pen.), Labrador (between ca. 55°N and Ramah, 58°52'N), and NW Nfld.; w Greenland N to ca. 74°N, E Greenland N to 66°20'N; Iceland; Europe; NW Asia; introd. on Sakhalin Is. [G. sylvaticum of Labrador reports, not L.; G. sylv. vars. brachystachyum Ledeb. and fuscatum Wahl.]. MAPS: Hultén 1958: map 29, p. 49; Atlas of Canada 1957: map 5, sheet 38.
   The purported citation by Gmelin from "Russian America" (Alaska) noted by John Macoun (1884) is undoubtedly the same as the one noted by Hultén (1950) for G. sylvaticum, the "America rossica" report by K.F. von Ledebour (Flora Rossica pt. 2:610. E. Schweizerbart, Stuttgart. 1844; G. sylvaticum var. macrostachyum Ledeb.) being based upon a misinterpretation by Ledebour of Gmelin's "Nec in America septentrionali deficit."

G. obtusifolium L. Catfoot
/T/EE/ (T) Fields, thickets, and clearings from Ont. (N to Ottawa) to Que. (N to Berthierville, about 35 mi NE of Montreal), N.B., P.E.I., and N.S., s to Tex. and Fla.; reported from SE B.C. by Ulke (1935; Wilmer, about 65 mi SE of Golden), where probably introd. [Incl. var. praecox Fern. (G. polycephalum Michx.)].

G. palustre Nutt. Western Marsh Cudweed
/T/W/ (T) Moist places (often alkaline) and dried pool-beds from s B.C. (N to Kamloops), s Alta. (Milk River, Castor, Hand Hills, and Redcliffe; CAN), and s Sask. (N to Saskatoon; Breitung 1957a) to s Calif. and N.Mex.

G. purpureum L. Purple Cudweed
/T/X/ (T (Hs)) Dry, sandy or clayey, often disturbed soils from sw B.C. (Vancouver Is. and adjacent islands; CAN; V) to Kans., Ill., Ohio, s Ont. (Leamington, Essex Co., and Port Colborne, Welland Co.; CAN), N.Y., and New Eng., s to Calif., Tex., and Fla.; tropical America. [G. ustulatum Nutt.].

G. supinum L. Alpine Cudweed
/aST/E/GEA/ (Ch) Damp ravines and exposed rocks and gravels of Que. (Hudson Strait s to NE Hudson Bay at ca. 57°N; Ungava Bay watershed s to ca. 54°45'N; Tabletop Mt., Gaspé Pen.), N Labrador (s to ca. 54°N), N Nfld., Mt. Katahdin, Maine, and Mt. Washington, N.H.; w and E Greenland N to ca. 70°N; Iceland; Europe; w Asia. MAPS: Böcher 1954: fig. 28 (top; bottom left), p. 111; Hultén 1958: map 99, p. 119.

G. sylvaticum L.
/T/EE/EA/ (Hp) Fields, rocky slopes, borders of woods, and clearings from Ont. (collection in OAC from Algonquin Park, w of Pembroke; collections in TRT from the Nipissing and Muskoka districts E of Georgian Bay, L. Huron) to Que. (N to L. St. John, Anticosti Is., and the Gaspé Pen.), St-Pierre and Miquelon, Nfld., N.B., P.E.I., and N.S., s to N New Eng.; Iceland; Europe; w Asia.

(Concerning a purported citation from Alaska, see *G. norvegicum*). MAP: Hultén 1958: map 118, p. 137 (noting another total-area map by Saxer).

G. uliginosum L.   Low Cudweed
?Eurasian (considered by some auth., e.g., Fernald *in* Gray 1950, as native in N. America in spite of its very weedy nature): introd. in gardens, fields, and waste places in N. America, as in Alaska–Yukon (N to ca. 65°N), Dist. Mackenzie (N shore of Great Slave L.; W.J. Cody, Can. Field-Nat. 77(2):126. 1963), B.C., Alta. (N to L. Athabasca), Sask. (known only from Loon Lake, 54°02'N; Breitung 1957*a*), Man. (known only from Angusville, about 80 mi NW of Brandon), Ont. (N to the NW shore of L. Superior near the Minn. boundary), Que. (N to the Côte-Nord, Anticosti Is., and Gaspé Pen.), Nfld., N.B., P.E.I., and N.S., and in w Greenland at ca. 60° and 70°N. [*G. palustre sensu* Lowe 1943, not Nutt., the relevant above-noted Angusville collection in WIN]. MAP: Hultén 1968*b*:882.

G. viscosum HBK.
/T/X/   (Hs (T))   Meadows, pastures, borders of woods, and clearings from s B.C. (N to Salmon Arm and Revelstoke; not known from Alta.–Sask.–Man.) to Ont. (N to Nipigon, N shore of L. Superior), ·Que. (N to Ste-Anne-de-la-Pocatière, Kamouraska Co., and Bic, Rimouski Co.; MT), N.B., P.E.I., and N.S., s to Oreg., Mexico, Tenn., Pa., and New Eng. [*G. macounii* Greene; *G. decurrens* Ives, not L.].

GRINDELIA Willd.   [8833]   Gumweed, Tarweed, Resinweed

1   Heads discoid, usually not over 1.5 cm broad; leaves scarcely clasping, seldom over
    1 dm long and 1.5 cm broad; plant glabrous; biennial or short-lived perennial; (?B.C.) . . . .
    . . . . [*G. columbiana*]
1   Heads normally radiate, the rays yellow; upper stem-leaves mostly sessile and more or
    less clasping.
    2   Tips of at least the middle and lower phyllaries regularly reflexed; leaves glabrous;
        biennial or short-lived perennial; (s Dist. Mackenzie–B.C. to Man.; introd. in Ont. and
        Que.) . . . . *G. squarrosa*
    2   Tips of phyllaries loose or spreading but not regularly reflexed; leaves usually more or
        less villous; taprooted perennial, often with a caudex; (s ?Alaska–w B.C.) . . . . .*G. integrifolia*

[G. columbiana (Piper) Rydb.]
[The report of this species of the w U.S.A. (Wash., Oreg., and Idaho) from s B.C. by Eastham (1947; Grand Forks, near the U.S.A. boundary sw of Trail) requires confirmation. (*G. nana* ssp. *col.* Piper and var. *discoidea* (Nutt.) Gray; *G. disc.* Nutt., not H. & A.).]

G. integrifolia DC.
/t/W/   (Hs)   Salt marshes, rocky shores, and various inland habitats from w B.C. (Queen Charlotte Is.; Vancouver Is. and adjacent islands and mainland; concerning an early report of *G. stricta* from Alaska, *see* Hultén 1968*a*) to N Calif. [*G. collina* Henry (*G. stricta* var. *coll.* (Henry) Steyerm.); incl. vars. *aestivalis* and *autumnalis* Henry and *macrophylla* Greene; *G. aggregata* Steyerm.; *G. andersonii* Piper; *G. hendersonii* and *G. lanata* Greene; *G. oregana* Gray; *G. stricta* DC. and its var. *aestuarina* Steyerm.; *G. nana sensu* Carter and Newcombe 1921, not Nutt.; *Donia* ?*glutinosa sensu* Hooker 1834, not R. Br.].

G. squarrosa (Pursh) Dunal
/sT/WW/   (Hs)   Dry open places, the aggregate species from s Dist. Mackenzie (Wood Buffalo National Park at ca. 60°15'N; CAN) and B.C. (N to Chilcotin, Cariboo, Kamloops, and Glacier, Rogers Pass) to Alta. (N to Wood Buffalo National Park), Sask. ("Common in dried up potholes on prairie"; Breitung 1957*a*), and Man. (presumably native southwards; introd. N to The Pas, as also in Ont. and Que.), s to Calif., Tex., and Minn.
1   Heads discoid, lacking ray-ligules; leaves copiously serrulate; [*G. nuda* Wood; reported
    as introd. in s Ont. by Soper 1949] . . . . var. *nuda* (Wood) Gray
1   Heads radiate, the ray-ligules yellow.

2   Ray-ligules rarely more than 25; achenes typically with 1 or more short marginal knobs at summit; leaves entire to sharply toothed, the teeth not callous-tipped; perennial with a taproot and often a branched caudex; [var. *integerrima* (Rydb.) Steyerm. (*G. integerrima* Rydb.); *G. nana* Nutt. and its var. *integrifolia* Nutt.; B.C.: reported from the Dry Interior as far N as Hanceville, ca. 52°N, by Eastham 1947] .... var. *integrifolia* (Nutt.) Boivin

2   Ray-ligules usually more than 25; achenes typically lacking apical knobs; leaves regularly callous-serrulate to sometimes sharply toothed or entire; biennial or short-lived perennial.

3   Leaves entire or remotely serrulate, or the lower ones often coarsely and irregularly toothed or somewhat pinnatifid; mostly a short-lived perennial; [*G. perennis* Nels. and its f. *pseudopinnatifida* Löve and Bernard; apparently native from SW Dist. Mackenzie–B.C. to S Man. (also introd. in stockyards at The Pas); the plant introd. in Ont. (N to near Thunder Bay; reported from 22 counties by Herbert Groh, Can. Field-Nat. 43(5):106. 1929), Que. (near Montreal; Mont-St-Pierre, Gaspé Co.), and Nfld. (Rouleau 1956) is probably chiefly or wholly referable here] .... var. *quasiperennis* Lunell

3   Leaves closely and evenly serrulate or crenulate-serrulate.

4   Upper and middle leaves mostly linear-oblong or oblanceolate, commonly over 5 times as long as broad; rays to 14 mm long; [*G. serrulata* Rydb.; S Sask. (Ogema, about 60 mi S of Regina; Breitung 1957a), SE Man. (a collection in CAN from Morris, about 30 mi S of Winnipeg, has been placed here by Steyermark), and Ont. (introd.; Fernald *in* Gray 1950)] .... var. *serrulata* (Rydb.) Steyerm.

4   Upper and middle leaves mostly ovate or oblong, commonly not over 4 times as long as broad; rays rarely over 10 mm long; [*Donia* Pursh; reports from Canada are probably chiefly or wholly referable to the above taxa] .... [var. *squarrosa*]

GUTIERREZIA Lag.   [8835]

G. sarothrae (Pursh) Britt. & Rusby   Match-brush, Broom-Snakeroot
/T/WW/   (Ch)   Dry plains, prairies, and foothills (ascending to higher elevations southwards) from SE Wash. to Alta. (N to Calgary), Sask. (N to Humboldt, about 60 mi E of Saskatoon), and S Man. (N to Millwood, about 85 mi NW of Brandon), S to Calif., Mexico, and Kans. [*Solidago* Pursh; *G. diversifolia* Greene; *Brachyris* (*G.*) *euthamiae* Nutt.].

HAPLOPAPPUS (*Aplopappus*) Cass.   [8852]   Golden-weed

(Ref.: Hall 1928)
1   Plant shrubby and branching, not at all caespitose, to 6(9) dm tall, the branches brittle; leaves linear to narrowly oblanceolate, entire, to about 6 cm long and 4 mm broad, straight or twisted, glandular-glutinous; heads clustered at the ends of the branches but forming a racemose inflorescence in taller plants; ray-ligules mostly 1–5 (or none in some heads), 6–12 mm long; achenes elongate; (S B.C.) .... *H. bloomeri*

1   Plants herbaceous (sometimes with a short aerial woody caudex), often densely caespitose.

2   Ray-ligules none (or inconspicuous in *H. carthamoides*), the solitary or several heads discoid or apparently so.

3   Basal and lowermost stem-leaves more or less reduced and early deciduous, the others firm, mostly oblong-spatulate, spinulose-toothed, to about 4 cm long and 1 cm broad; involucres to 1 cm high, their phyllaries commonly merely green-tipped, the lower ones the smallest; disk-corollas to 8 mm long; achenes narrowly turbinate (top-shaped), to 3 mm long, densely white-hairy; (S Alta. to S Sask.) .... *H. nuttallii*

3   Basal leaves tufted, to 4 dm long and 4 cm broad, their entire to spiny-toothed oblanceolate blades tapering gradually to a long petiole; stem-leaves generally

more or less reduced and tending to become sessile up the stem; involucres 1.5–3 cm high, their lowest phyllaries often leaf-like; disk-corollas to 14 mm long; achenes elongate, glabrous; (S B.C.) .... *H. carthamoides*

2 Ray-ligules present, yellow and conspicuous.

    4 Leaves deeply 1–2-pinnatifid into narrow segments, to about 6 cm long; heads commonly several on each stem; phyllaries spinulose-tipped, in 4 or 5 unequal series; achenes at most 2.5 mm long; plant more or less greyish with a woolly pubescence; (S Alta. to S Man.) .... *H. spinulosus*

    4 Leaves entire or sharply serrulate; phyllaries not spinulose-tipped; achenes elongate; plants not greyish-woolly.

        5 Plants not densely caespitose and not at all mat-forming, the caudex simple or moderately branched, the flowering stems solitary to several, leafy or subscapose, often well over 2 dm tall; basal leaves oblanceolate to narrowly elliptic, subglabrous or more or less villous-tomentose.

            6 Heads characteristically several to rather numerous; involucres about 1 cm high, their conspicuously green-tipped phyllaries in 3 or 4 series of markedly unequal length; disk-corollas to 7 mm long; basal leaves usually sharply serrulate, to 2 dm long and 3.5 cm broad; stems to 5 dm tall; (S Dist. Mackenzie, Alta., and Sask.) .... *H. lanceolatus*

            6 Heads characteristically solitary (sometimes 1 or 2 smaller short-peduncled heads in the upper axils of robust plants); basal leaves to about 1.5 dm long.

                7 Involucres to nearly 1.5 cm high, their phyllaries either subequal or in unequal series, green throughout or prominently green-tipped; disk-corollas to 1 cm long; leaves entire, to 3 cm broad; stems to 4 dm tall .... [*H. integrifolius*]

                7 Involucres at most about 1 cm high, their phyllaries mostly subequal and generally green throughout; disk-corollas to 7 mm long; leaves usually sharply serrulate, to about 1.5 cm broad; stems to 3 dm tall; (?Sask.) .... [*H. uniflorus*]

        5 Plants more or less densely caespitose and generally mat-forming, with a much-branched caudex, numerous tufted, obscurely petioled basal leaves, and more or less numerous 1-headed stems rarely over 2 dm tall, the leaves all entire.

            8 Stems relatively leafy; leaves strongly glandular-puberulent, oblanceolate to spatulate or oblong, the basal ones the longest but the cauline ones fairly well developed; involucres to 11 mm high, their subequal phyllaries relatively loose and herbaceous or somewhat chartaceous; ray-ligules to 11 mm long; creeping rhizomes commonly present; (S B.C. and SW Alta.) .... *H. lyallii*

            8 Stems scapose or few-leaved; leaves seldom glandular, mostly crowded at the ends of the short woody caudical branches and forming a mat; phyllaries firm, green-tipped; creeping rhizomes wanting.

                9 Phyllaries in 2 nearly equal series; leaves narrowly linear (almost filiform), to about 2 cm long and less than 1 mm broad, hispid-ciliate on the margins, otherwise glabrous; plant less than 1 dm tall; (the Yukon) .... *H. macleanii*

                9 Phyllaries in 3 or 4 series; leaves narrowly oblanceolate, to about 1 dm long and 7 mm broad; plants to over 1.5 dm tall.

                    10 Phyllaries oval or oblong, rounded or very obtuse at tip, in 3 or 4 series of markedly unequal length; involucres to 13 mm high; (S Sask.) .... *H. armerioides*

                    10 Phyllaries lanceolate to ovate-lanceolate, acute to acuminate, less markedly unequal; involucres to 10 mm high .... [*H. acaulis*]

[*H. acaulis* (Nutt.) Gray]

[Reports of this species of the W U.S.A. (Idaho and Mont. to Calif. and Colo.) from Sask. by J.M.

Macoun (1894; taken up by Rydberg 1922, Fraser and Russell 1944, and Breitung 1957a) are based upon *H. armerioides,* most or all of the relevant collections in CAN, DAO, SCS, and SASK. (*Chrysopsis* Nutt.; *Stenotus* Nutt.; *S. (H.) falcatus* Rydb.; incl. the glabrous extreme, var. *glabratus* Eat. (*C. (S.) caespitosa* Nutt.)).]

H. armerioides (Nutt.) Gray
/T/WW/ (Hs) Dry hills and plains from Mont. and s Sask. (Cypress Hills, Eastend, Elbow, Wood Mountain, Lebret, Estevan, Short Creek, and Moose Mt., N to Whiteshore L. and Floral, both ca. 52°N; CAN; DAO; SCS; SASK; the report from Man. by Rydberg 1922, taken up by Lowe 1943, requires confirmation) to Ariz., N.Mex., and Nebr. [*Stenotus* Nutt.; as noted above, basis of reports of *H. acaulis* from Sask.].

H. bloomeri Gray
/t/W/ (N) Dry rocky slopes and open woods in the foothills and valleys up to moderate elevations from southernmost B.C. (Keremeos, about 20 mi SW of Penticton; V; reported from Westbridge, SE of Penticton, by Eastham 1947) to Calif. [*Chrysothamnus* Greene; *Ericameria* Macbr.].

H. carthamoides (Hook.) Gray
/t/W/ (Grt) Meadows and open hillsides at low to moderate elevations from southernmost B.C. (Keremeos, about 20 mi SW of Penticton; CAN; reported from Summerland, about 10 mi NW of Penticton, by Eastham 1947; reports from Alta. require confirmation) and Mont. to N Calif., Nev., and Wyo. [*Pyrrocoma* Hook.; *P. rigida* Rydb.].

[H. integrifolius Porter]
[The inclusion of B.C. and Sask. in the range of this species of the w U.S.A. (Idaho, Mont., and Wyo.) by Rydberg (1922) requires clarification. Breitung (1957a) notes that the Patience L., Sask., citation by Fraser and Russell (1944) refers to *H. lanceolatus* var. *vaseyi.* (*Pyrrocoma* Greene).]

H. lanceolatus (Hook.) T. & G.
/sT/WW/ (Hs) Moist alkaline meadows and open slopes (ranges of Canadian taxa outlined below), s to Calif., Utah, and w Nebr.

1 Heads rarely more than 4 or 5, in a narrow racemiform inflorescence (the lower peduncles relatively short); phyllaries abruptly acute; [*Pyrrocoma vaseyi* (Parry) Rydb.; *P. integrifolia* of Sask. reports, not Greene, according to Breitung 1957a; reported from Patience L., Sask., by Breitung] .... var. *vaseyi* Parry
1 Heads commonly numerous in a corymbiform or subpaniculate inflorescence; phyllaries acuminate.
  2 Plant rather copiously lanate, usually single-headed; [reported by W.J. Cody, Can. Field-Nat. 70(3):126. 1956, from near Fort Smith, s Dist. Mackenzie, at 60°03′N, the type locality, and from near Beaverlodge, Alta., 55°13′N] .... var. *sublanatus* Cody
  2 Plant glabrous or slightly villous, with several to many heads in a corymbiform inflorescence; [*Donia lanceolata* Hook., the type from between Carlton House, Sask., and Edmonton, Alta.; *Pyrrocoma* Greene; s Alta. (Milk River; MacLeod; Oyen; N of Empress) and s Sask. (Cypress Hills, Bracken, Swift Current, Mortlach, Whiteshore L., and Saskatoon; Breitung 1957a); the apparent report from Man. by John Macoun 1884, requires confirmation, no Manitoba citations being given by Hall 1928] ....
.... var. *lanceolatus*

H. lyallii Gray
/T/W/ (Hsr) Cliffs and talus slopes (often above timberline) from s B.C. (N to the Marble Range NW of Clinton; CAN; the type material was taken by Lyall along the B.C.–Wash. boundary. in the Cascade Mts.) and sw Alta. (N to the Banff dist.; CAN) to Oreg., Nev., and Colo. [*Stenotus* Howell; *Tonestus* Nels.].

H. macleanii Brandegee
/S/W/ (Ch) Dry rocky slopes up to about 3,000 ft in the Yukon (between ca. 61°30′ and 64°N;

type from near Dawson). [*Stenotus* Heller; *S. borealis* Rydb.]. MAPS: Hultén 1968*b*:854, and 1950: map 1099, p. 1673.

H. nuttallii T. & G.
/T/W/   (Ch (Hp))   Dry open places (often on alkaline clays) from Mont. to s Alta. (Lethbridge and the Red Deer R. valley; CAN) and s Sask. (Cypress Hills, Swift Current, Bracken, Mortlach, Elbow, and Moose Jaw; CAN; Breitung 1957a), s to Nev., Ariz., and N.Mex. [*Sideranthus (Machaeranthera) grindelioides* (Nutt.) Britt., not *H. gr.* DC.].

H. spinulosus (Pursh) DC.   Iron-plant
/T/WW/   (Hp (Ch))   Dry plains, prairies, and foothills from Alta. (N to Vermilion, 53°22'N; CAN), Sask. (N to Saskatoon; CAN), and sw Man. (N to Millwood, about 85 mi NW of Brandon) to Baja Calif., Mexico, Tex., Okla., and Minn. [*Amellus* Pursh; *Eriocarpum* Greene; *Sideranthus* ?Sweet; *S. (Diplopappus) pinnatifidus* Nutt.].

[*H. uniflorus* (Hook.) T. & G.]
[The type locality of this species of the w U.S.A. (N to Oreg., Idaho, and Mont.) was given by Hooker (1834; *Donia uniflora*) as "Plains of the Saskatchawan and Prairies of the Rocky Mountains." This is the apparent basis of the inclusion of Sask. in the range by Rydberg (1922) and later authors but Breitung excludes it from the flora of that province on the basis of the evidently too broad geographical designation of the type locality. (*Donia* Hook.; *Pyrrocoma* Greene).]

HELENIUM L.   [9305]   Sneezeweed

1   Disk yellow, to over 2 cm broad; ray-florets yellow throughout, fertile; denuded receptacle
    depressed-globose; (B.C. to Que.) .... *H. autumnale*
1   Disk purple or brownish purple, to about 1 cm broad; rays sometimes tinged with purple at
    base, the ray-florets sterile; denuded receptacle ovoid; leaves linear to lanceolate; (introd.
    in Ont. and sw Que.) .... *H. nudiflorum*

H. autumnale L.   Common Sneezeweed
/ST/X/   (Hs (Hp))   Rich thickets, meadows, and shores, the aggregate species from w Dist. Mackenzie (N to near Fort Simpson at ca. 62°50'N; CAN) and B.C.–Alta. to Sask. (N to Ile-à-la-Crosse, 55°27'N; DAO), Man. (N to Wekusko L., about 90 mi NE of The Pas), Ont. (N to the Ottawa dist.), Que. (N to Berthier-en-bas, about 25 mi NE of Quebec City), N.Y., and New Eng., s to Ariz., Tex., and Fla.
1   Stem to 1.5 m tall; larger leaves coarsely dentate or serrate, to 5.5 cm broad; rays to 2.5
    cm long and 12 mm broad; disk to 2.3 cm broad; plant commonly glabrate; [s Ont.
    (Tobermory, Bruce Co., where intermixed with var. *canaliculatum*); M.L. Fernald,
    Rhodora 45(540):490. 1943] .... var. *autumnale*
1   Stem usually lower; leaves to 4 cm broad, entire to shallowly (sometimes coarsely)
    toothed.
    2   Rays to 2.5 cm long; disk to 2 cm broad; plant commonly glabrate, to 1.2 m tall;
        [*H. grandiflorum* Nutt.; *H. macranthum* Rydb.; s Dist. Mackenzie–B.C. to Sask.] ....
        .... var. *grandiflorum* (Nutt.) T. & G.
    2   Rays to about 2 cm long.
        3   Leaves firm and subrigid, linear to lanceolate, entire or shallowly (rarely coarsely)
            toothed; rays commonly about 2 cm long; plant often glabrate; [var. *?fylesii*
            Boivin; *H. canaliculatum* Lam.; *H. ?pubescens* Ait.; Ont. and sw Que.] ....
            .... var. *canaliculatum* (Lam.) T. & G.
        3   Leaves membranaceous, lanceolate to elliptic, the larger ones relatively coarsely
            toothed; rays mostly about 1 cm long; plant commonly minutely puberulent; [*H.
            montanum* Nutt.; B.C. to w Ont.] .... var. *montanum* (Nutt.) Fern.

H. nudiflorum Nutt.
A native of the E U.S.A. (Kans. to N.C., s to Tex. and Ga.); rapidly spreading as a weed northwards, as in Ont. (Lake Superior Provincial Park, E shore of L. Superior; Southampton, Bruce Co.; Long

Point and near Simcoe, Norfolk Co.; Niagara Falls, Welland Co.; CAN; TRT) and sw Que. (Old Chelsea, about 10 mi NW of Hull; CAN). [*H. flexuosum* Raf.].

The above Simcoe, s Ont., station is noted by Groh (1947) as probably the colony from which the first Canadian collection was taken by W. Herriot in 1927, described by Landon (1960) as, "An isolated station of 100 acres... has persisted for 50 years. Is naturally confined but shows potential weed habits."

HELIANTHELLA T. & G.   [9212]

H. uniflora (Nutt.) T. & G.
/t/W/   (Hp)   Open woods and hillsides from s B.C. (valleys of the Dry Interior N to the Marble Mts., NW of Clinton, and near Princeton; CAN; V) and Mont. to Oreg., Nev., and N.Mex. [*Helianthus* Nutt.]. MAP: W.A. Weber, Am. Midl. Nat. 48(1): map 3, p. 16. 1952.

The B.C. plant is referable to var. *douglasii* (T. & G.) Weber (*H. doug.* T. & G.; phyllaries conspicuously hirsute-ciliate rather than only slightly so; disk to 2.5 cm broad rather than usually not over 2 cm; ray-ligules mostly 3 or 4 cm long rather than mostly 2 or 3 cm long).

HELIANTHUS L.   [9200]   Sunflower.   Soleil

1   Leaves at once sessile or very short-petioled (petioles at most about 5 mm long) and broadly rounded to truncate at base, triangular-lanceolate, shallowly but rather coarsely toothed, tapering nearly uniformly from near base to apex, triple-nerved, harsh above, commonly paler and pubescent beneath, all opposite, subhorizontally spreading; stem glabrous and often glaucous; disk yellow, to 1.5 cm broad; phyllaries lance-attenuate, in 2 or 3 loose subequal series; (Ont. and s Que.) .... *H. divaricatus*
1   Leaves not at once sessile or subsessile and broad-based (if broad-based, at least the lower leaves with winged or wingless petioles over 5 mm long).
  2   At least the upper leaves (below inflorescence) of the primary axis alternate.
    3   Receptacle flat or nearly so; disk brown to purplish brown; annuals with fibrous roots.
      4   Phyllaries oblong-lanceolate to broadly ovate, usually rather long-ciliate and with some long hairs on the back, chiefly ovate or ovate-oblong, abruptly contracted to the attenuate tip; central bracts of receptacle short-hairy but not bearded at summit; leaves ovate, dentate, to over 3 dm broad, the lower ones often cordate at base .... *H. annuus*
      4   Phyllaries lanceolate, barely ciliate, closely short-hispid on the back, tapering gradually to the attenuate tip; central bracts of receptacle conspicuously white-bearded at tip; leaves often narrower, entire or undulate-dentate, rarely over 1 dm broad and rarely cordate .... *H. petiolaris*
    3   Receptacle convex to low-conical; disk yellow; phyllaries lance-attenuate; leaves lanceolate to oblong-lanceolate; perennials with short or elongate rhizomes.
      5   Stems essentially glabrous at least below the inflorescence; leaves commonly triple-nerved at base, minutely pubescent beneath.
        6   Heads small, commonly several or numerous, their disks to about 1 cm broad; rays to about 1.5 cm long; leaves lanceolate, abruptly narrowed to base, taper-pointed, low-toothed except at base; (?Ont.) ....[*H. microcephalus*]
        6   Heads larger and usually fewer, the disk to over 2 cm broad; rays to about 3 cm long.
          7   Leaves lance-linear to lanceolate, very strongly scabrous above, entire or nearly so, cuneate at base, usually only the ones close to the inflorescence alternate; (transcontinental) .... *H. nuttallii*
          7   Leaves lanceolate to oblong-ovate, cuneate or rounded at base, only slightly or scarcely scabrous above, coarsely toothed; (introd. in Ont.) ......
            .... *H. grosseserratus*
    5   Stems scabrous-hispid or -hirsute at least above; disks to over 2.5 cm broad.
      8   Leaves ovate, coarsely toothed, triple-nerved for at least half their length, the blade to 1.5 dm broad, commonly nearly half as broad as long,

        minutely puberulent beneath, decurrent on the broadly winged upper half of the petiole; phyllaries about equalling the height of the disk; rhizomes freely tuberiferous; (introd.) . . . . *H. tuberosus*

    8   Leaves lanceolate to lance-oblong, commonly at least 3 times as long as broad, gradually tapering to the rather short petiole; rhizome short.

        9   Leaves lanceolate to lance-oblong, flat, commonly triple-nerved at base; stems generally spreading-hirsute; phyllaries conspicuously long-ciliate, not much surpassing the disk; rays to about 3 cm long; (Ont. to N.S.) . . . . *H. giganteus*

        9   Leaves lanceolate, often infolded (conduplicate) and with recurving tips, only the midrib prominent beneath; stems finely appressed-pubescent; phyllaries finely appressed-canescent, seldom ciliate, conspicuously surpassing the disk; rays to 4 cm long; (s B.C. to w Ont.; introd. eastwards) . . . . *H. maximiliani*

2   Leaves opposite or subopposite (or the upper leaves sometimes alternate); receptacle convex to low-conical.

   10 Disk reddish- to purplish-brown (colour of corolla-lobes); phyllaries firm and tightly appressed in several strongly unequal series, their tips neither attenuate nor foliaceous; leaves entire or the larger ones shallowly serrate.

     11 Stem subscapose, leafy only at or near the base (the upper leaves abruptly reduced), appressed-pubescent to glabrate above, often densely spreading-hirsute toward base; petioles of lower leaves slender, winged only at summit, to 1.5 dm long; disk at most about 1.5 cm broad; phyllaries short-acuminate; (introd. in s ?Ont.) . . . . [*H. occidentalis*]

     11 Stem leafy for about 3/4 of its length, scabrous, its upper leaves more gradually decreasing in size; petioles of lower leaves winged nearly to base, at most about 3 cm long; disk to 2 cm broad; phyllaries acute to blunt; (B.C. to w Ont.; introd. eastwards) . . . . *H. laetiflorus*

   10 Disk yellow.

     12 Stems clustered on the crown of a thickened, often somewhat turnip-shaped taproot, scabrous to subglabrous; leaves lanceolate or narrower, entire, triple-nerved, rough-hairy or scabrous, subsessile or shortly wing-petioled; phyllaries loose, lance-linear, acuminate or attenuate, generally conspicuously spreading-hirsute, especially marginally; (s ?B.C.) . . . . [*H. cusickii*]

     12 Stems from slender to stout creeping rhizomes (these often tuberous-thickened), not taprooted.

       13 Phyllaries closely appressed, lanceolate, acute; leaves up to 15 pairs, lanceolate to lance-ovate, long-acuminate, subentire to low-serrate; rays to 5 cm long; (var. *laetiflorus;* introd. in B.C., s Ont., s Que., and s Nfld.) . . . .
. . . . *H. laetiflorus*

       13 Phyllaries long-attenuate or more or less foliaceous at tip, at least the outer ones spreading.

         14 Leaves abruptly narrowed but broadly decurrent to base of petiole (thus appearing sessile), ovate-lanceolate to ovate, acute or acuminate, shallowly serrate or the upper ones entire, triple-nerved from somewhat below the middle, soft-tomentose beneath; phyllaries shorter than the disk; rays to 4 cm long . . . . [*H. doronicoides*]

         14 Leaves evidently petioled, triple-nerved from near base; phyllaries about equalling or surpassing the disk.

           15 Leaf-petioles mostly at least 1.5 cm long; leaves ovate, at least the larger ones coarsely toothed.

             16 Stem glabrous or nearly so; leaves relatively smooth and thin, only slightly decurrent on the petioles, green both sides; rhizomes slender, rarely bearing tubers; phyllaries attenuate, often much surpassing the disk; (Ont. to N.B.) . . . . . . *H. decapetalus*

             16 Stem scabrous-hispid; leaves thick and hard, scabrous above, rather densely velvety beneath with loose or spreading hairs,

broadly decurrent down the upper half of the petiole; rhizome
coarse and freely tuberiferous; phyllaries not much surpassing
the disk, their tips often recurving; (introd.) . . . . *H. tuberosus*
15 Leaf-petioles mostly less than 1.5 cm long; leaves entire or the
larger ones shallowly toothed.
   17 Leaves broadly lanceolate, relatively smooth and thin, nearly
equally green on both sides, their petioles barely winged at
base of blade; phyllaries lance-linear, attenuate, much sur-
passing the disk; (introd. in s ?Ont.) . . . . [*H. trachelifolius*]
   17 Leaves thick and hard, scabrous above, paler beneath,
narrowed to short winged petioles; phyllaries about equalling
the disk.
      18 Stem smooth and glaucous (or merely sparsely hispid
above); leaves broadly lanceolate to ovate, whitened
beneath with relatively soft pubescence, usually merely
abruptly narrowed to base and broadest somewhat below
the middle; (s Ont. to N.B.) . . . . *H. strumosus*
      18 Stem hirsute at least toward base; leaves triangular-
lanceolate, usually broadly rounded or subtruncate and
broadest near the base, hirsute beneath; (introd. in s Ont.) . . . . . .
. . . . *H. hirsutus*

H. annuus L.   Common Sunflower.   Tourne-soleil or Soleil
/T/WW/   (T)   Plains, bottomlands, and other rich soils of w N. America; cult. and spread to fields,
roadsides, and waste places eastwards and northwards but perhaps native in s B.C. (N to
Kamloops; CAN), s Alta. (Crowsnest L., Pincher Creek, Fort Macleod, Deer Creek, sw of Medicine
Hat, and Walsh; CAN), s Sask. ("Common in rich clay soils in the prairie region"; Breitung 1957a),
and s Man. (N to Stony Mountain, N of Winnipeg); introd. in cent. Alaska (Tanana Hot Springs, ca.
65°N), sw Dist. Mackenzie (near Fort Simpson at ca. 62°51'N; CAN), N Man. (Churchill; Beckett
1959), Ont. (N to the Ottawa dist.), Que. (N to Rimouski, Rimouski Co.; RIM), N.B. (Kent and St.
John counties; GH; NBM), P.E.I., and N.S.; introd. in Eurasia. [*H. macrocarpus* DC.; *H. aridus*
Rydb. in part; incl. *H. lenticularis* Dougl.]. MAP: Hultén 1968b:883.
   Forma *fallax* Boivin (ray-ligules orange rather than yellow) is known from Sask. (type from
Forget) and s Man. (Morden; DAO).

[*H. cusickii* Gray]
[The inclusion of B.C. in the range of this species of the w U.S.A. (Wash. and Idaho to Calif.) by
Rydberg (1922) is possibly based upon a collection in CAN from near Princeton, s B.C., where
taken by J.M. Macoun in 1905, this, however, referred by Porsild to *Helianthella uniflora*.]

H. decapetalus L.
/T/EE/   (Grh (Hpr))   Open woods, thickets, and streambanks from Nebr. to Minn., Ont. (N to
Casselman, near Ottawa; CAN; TRT), Que. (N to near Quebec City; MT), and N.B. (Boivin 1966b;
not known from P.E.I. or N.S.), s to Mo., Ky., and Ga. [*H. frondosus* L.].

H. divaricatus L.
/T/EE/   (Grh (Hpr))   Woods, thickets, and dry clearings from Ont. (N to Renfrew and Carleton
counties; reports from Sask. and Man. refer, at least in part, to the habitally very similar *Heliopsis
helianthoides*) to sw Que. (N to Pontiac and Gatineau counties) and Maine, s to Ark., Tenn., and
Ga.

[*H. doronicoides* Lam.]
[The tentative report of this species of the E U.S.A. (N to Minn. and N.J.) from Man. by Lowe (1943)
is probably based upon *H. tuberosus* var. *subcanescens,* an apparently relevant collection from
Rosewood in WIN. The report from the Kaministikwia Valley near Thunder Bay, Ont., is perhaps
referable to *H. tuberosus,* known from that locality.]

H. giganteus L.
/T/EE/ (Hp (Hpr)) Moist ground, rich thickets, and clearings from Ont. (N to Kapuskasing, ca. 49°20'N; CAN; reports from Alta.-Sask.-Man. require clarification, perhaps being largely based upon *H. nuttallii;* early collections by John Macoun in CAN from s B.C. require further study) to Que. (N to Cap-à-l'Aigle, Charlevoix Co., and the Gaspé Pen. at Price, Matane Co.; CAN), N.B. (Boivin 1966*b*), and N.S. (Yarmouth Co.; ACAD; not known from P.E.I.), s to an uncertain limit in the E U.S.A. largely through confusion with *H. nuttallii.*

H. grosseserratus Martens
A native of the E U.S.A. (N to N.Dak. and Ohio); cult. and spreading to rich thickets, roadsides, and waste places elsewhere, as in ?Sask. (the reports by Rydberg 1922 and 1932, require confirmation), ?Man. (the report from MacGregor by Shimek 1927, may be referable to *H. nuttallii*), and Ont. (N to Cochrane, 49°04'N).

H. hirsutus Raf.
A native of the E U.S.A. (N to Minn. and Pa.); introd. elsewhere, as in s Ont. (collection in MICH from near a grain elevator at Point Edward, Lambton Co.; collection in TRT from Toronto).

H. laetiflorus Pers.
/T/(X)/ (Grh) Dry prairies and plains from B.C. to w Ont., s to N.Mex., Mo., and Ind.; introd. along roadsides and railways elsewhere, as in the Yukon, Ont., Que., and the Maritime Provinces, and probably the more northern parts of the Prairie Provinces. MAPS and synonymy: *see* below.

1 Disk of head yellow (the disk-corollas with yellow lobes); leaves lanceolate to lance-ovate, long-acuminate, to 2.5 dm long, borne at up to 15 nodes; [*H. scaberrimus* Ell.; introd. in B.C. (Boivin 1966*b*), s Ont. (Lambton, Wentworth, and York counties), Que. (Valois, near Montreal; Lionel Cinq-Mars, Ann. ACFAS 18:80. 1952), and s Nfld. (Topsail Road; GH); MAP: Sarah Clevenger and C.B. Heiser, Jr., Rhodora 65(762): fig. 3 (U.S.A. native area), p. 123. 1963] . . . . var. *laetiflorus*
1 Disk reddish- to purplish-brown.
   2 Stem harshly scabrous, to 2.5 m tall, with up to 15 nodes; leaves oblong-lanceolate to lance-ovate, acuminate, to about 3 dm long; phyllaries lanceolate to narrowly ovate; [f. *rig.* (Cass.) Boivin; *Harpalium (Hel.) rig.* Cass.; *H. atrorubens sensu* Hooker 1833, not L.; B.C. (N to Dawson Creek, ca. 55°40'N) to Alta. (N to Peace River, 56°14'N), Sask. (collections in CAN from near Battleford and Indian Head appear referable here; not listed by Breitung 1957*a*), and Man. (N to The Pas, where probably introd.; apparently native southwards); introd. in Ont. (N to Peninsula, N shore of L. Superior), Que. (Gatineau and Laprairie counties), and P.E.I. (Charlottetown); MAP: Clevenger and Heiser, loc. cit., fig. 2, p. 123] . . . . var. *rigidus* (Cass.) Fern.
   2 Stem less harsh above, usually lower and with rarely more than 9 nodes; leaves subrhombic-lanceolate to -ovate, subacute or bluntish, to 1.5 dm long; phyllaries oblong or oblong-oval; [*H. sub.* Rydb.; the Yukon (Dawson, where perhaps introd.; CAN), Alta. (N to Dunvegan, 55°54'N; ?introd.), Sask. (N to Tisdale and Hudson Bay Junction), and Man. (N to Bield, s of Duck Mt.); introd. in Ont. (N to Schreiber and Timmins; perhaps native at Ingolf, near the s Man. boundary), Que. (N to Matane Co., Gaspé Pen.), N.B. (Sussex, Kings Co.), and N.S. (Port Williams, Kings Co.); MAPS: Clevenger and Heiser, loc. cit., fig. 1 (*H. sub.*), p. 123; Hultén 1968*b*:883] . . . .
. . . . var. *subrhomboideus* (Rydb.) Fern.

H. maximiliani Schrad.
/T/WW/ (Grh (Hpr)) Dry prairies and plains (often in disturbed or waste places) from s B.C. (Victoria, Vancouver Is., and Griffin L., near Kamloops; V) to s Alta. (Medicine Hat; CAN), Sask. (N to Tisdale, 52°51'N; CAN), Man. (N to Wekusko L., about 90 mi NE of The Pas; CAN), and w Ont. (probably native at Ingolf, near the s Man. boundary), s to Idaho, Colo., Tex., Ark., and Wisc.; introd. elsewhere, as in E Ont., Que. (Nominingue, Labelle Co.; Montreal dist.), and P.E.I. (a garden-escape at Southport, Queens Co.; ACAD). MAP (together with a discussion of the *H. nuttallii* complex): R.W. Long, Brittonia 18(1): fig. 3, p. 74. 1966.

Forma *pallidus* Schrad. (ray-ligules whitish rather than yellow) is known from SE Man. (near Otterburne, about 30 mi S of Winnipeg; Löve and Bernard 1959).

[H. microcephalus T. & G.]   Small Wood-Sunflower
[The Torrey and Gray report of a Goldie collection in Ont. noted by John Macoun (1884; *H. parviflorus*) requires clarification. (*H. parviflorus* Bernh., not HBK.).]

H. nuttallii T. & G.
/T/X/   (Grt)   Meadows and other moist or wet places at low to moderate elevations from B.C. (N to Dawson Creek, ca. 55°35'N) to Alta. (N to near Grande Prairie, 55°12'N), Sask. (N to Tisdale, 52°51'N), Man. (N to Benito, N of Duck Mt.), Ont. (N to SW James Bay at Fort Hope, 51°34'N), Que. (N to the Gaspé Pen.), Nfld., N.B., and N.S. (Boivin 1966*b*; not known from P.E.I.), S in the West to Calif., N.Mex., Okla., and Mo. MAPS and synonymy: *see* below.
1   Larger basal leaves mostly not over twice as long as broad; [*H. rydbergii* and *H. subtuberosus* Britt.; B.C. to Man.; MAP: R.W. Long, Brittonia 18(1): fig. 5, p. 76. 1966] ....
.... ssp. *rydbergii* (Britt.) Long
1   Larger basal leaves to 4 or 5 times as long as broad.
  2   Basal leaves with petioles at most about 5 mm long; [range of the species in Canada, the type from Turtle Mt., SW Man.; MAP: on the above-noted map by Long] ....
.... ssp. *canadensis* Long
  2   Basal leaves with petioles 1–3 cm long; [*H. fascicularis* Greene; *H. giganteus* var. *utahensis* Eat. (*H. utah.* (Eat.) Nels.); S Alta. to S Man.; MAP: R.W. Long, Brittonia 18(1): fig. 4, p. 75. 1966] .... ssp. *nuttallii*

[H. occidentalis Riddell]
[The report of this species of the E U.S.A. (N to Minn. and Ohio) from S Ont. by Dodge (1915; introd. along railways and in waste places in Lambton Co.; taken up by Soper 1949) requires confirmation.]

H. petiolaris Nutt.
/T/WW/   (T)   Dry prairies, plains, roadsides, and waste places, the native range uncertain because of the weedy nature of the species, but, tentatively, from Mont. to S Alta. (N to the Red Deer R.; CAN), S Sask. (Breitung 1957*a*), and S Man. (Aweme; Brandon; Glenboro) to Ariz. and Iowa; apparently introd. in S B.C. (John Macoun 1886) and Ont. (N to Cobalt, about 80 mi NE of Sudbury). [*H. aridus* Rydb. in part; *H. ?pumilus* Nutt., not L.].

H. strumosus L.
/T/EE/   (Grh (Hpr))   Open woods, thickets, and clearings from N.Dak. and Minn. to S Ont. (N to Northumberland and Hastings counties), Que. (N to Lotbinière Co.), and N.B. (Carleton and Northumberland counties; CAN; NBM; not known from P.E.I. or N.S.), S to Okla., Ark., Ala., and Ga. [Incl. var. *mollis* T. & G., not *H. mollis* Lam.].

[H. trachelifolius Mill.]
[The reports of this species of the E U.S.A. (N to Nebr. and Minn.) from S Ont. by Dodge (1915; Lambton Co.), Zenkert (1934; Turkey Point, Norfolk Co.), and Soper (1949) require clarification, perhaps being based upon the scarcely separable *H. strumosus.*]

H. tuberosus L.   Jerusalem Artichoke.   Topinambour
Perhaps native in the S U.S.A. and tropical America; a garden-escape or persisting from old plantings elsewhere, as in SE Sask. (Northgate, about 140 mi SE of Regina; A.J. Breitung, Am. Midl. Nat. 61(2):512. 1959), Man. (N to Dauphin, N of Riding Mt.; CAN), Ont. (N to near Thunder Bay), Que. (N to the Gaspé Pen. at Bonaventure), N.B. (John Macoun 1884), P.E.I. (Charlottetown; GH), and N.S. MAP: Sarah Clevenger and C.B. Heiser, Jr., Rhodora 65(762): fig. 4, p. 123. 1963.

Much of our material is referable to var. *subcanescens* Gray (leaves more generally soft-tomentose beneath than those of the typical form, all or nearly all opposite rather than the upper ones alternate).

HELIOPSIS Pers. [9157]

H. helianthoides (L.) Sweet   Ox-eye
/T/(X)/ (Hp (Grh)) Dry woodlands, thickets, prairies, and waste places from s B.C. (N to Revelstoke; CAN, ?introd.; not known from Alta.) to Sask. (between Moosomin, 50°07′N, and Tisdale, 52°51′N; CAN), Man. (N to Birtle, about 60 mi NW of Brandon), Ont. (N to near Thunder Bay), Que. (N to Cabano, Temiscouata Co.), and N.B. (St. John R. system; introd. at Summerside, P.E.I., and in Nfld.; the Que. plant may also be introd.), s to N.Mex., Kans., Mo., Ala., and Ga. [*Buphthalum* L.; *Helianthus (Heliopsis) laevis* L.]. MAPS (aggregate species): T.R. Fisher, Ohio J. Sci. 57(3): fig. 6, p. 188. 1957, and 58(2): fig. 1, p. 97. 1958.

Apart from the occurrence of the typical form in s Ont. (Ausable R., Lambton Co.; Boivin 1966*b*; also introd. at Summerside, P.E.I.), our material is referable to var. *scabra* (Dunal) Fern. (var. *occidentalis* (Fisher) Steyerm., at least in part; *H. scabra* Dunal; leaves harshly scabrous above rather than smooth or only slightly scabrous; achenes pubescent on the angles when young rather than glabrous throughout).

HIERACIUM L. [9607] Hawkweed. Épervière

1 Flowers white or creamy; heads usually numerous, on slender minutely stipitate-glandular peduncles in a compound corymbiform inflorescence; stellate hairs wanting; leaves oblanceolate or oblong, long-hirsute especially on the veins beneath, the lower ones on rather short long-hirsute petioles, commonly very shallowly and remotely toothed, the stem-leaves much reduced, sessile and entire; stem to about 8 dm tall, long-hirsute below, often glabrous above; (B.C. to sw Sask.) . . . . *H. albiflorum*
1 Flowers mostly yellow (orange-red in *H. aurantiacum*); stellate hairs often present on the involucre (sometimes hidden by the longer setae), often also on the stem and leaves.
  2 Stem soft, long-hirsute (commonly glabrous in *H. florentinum*), scapose or with 1 or 2 (rarely 3) bracts or small leaves near base; leaves lanceolate or oblanceolate, entire or minutely toothed; phyllaries subequal, with or without some small bractlets at base; (introd.).
    3 Rhizome short, nonstoloniferous; flowers yellow; scape to over 1 m tall.
      4 Leaves finely stellate-pubescent beneath; slender sterile branches ascending or spreading from among the basal leaves . . . . *H. praealtum*
      4 Leaves essentially glabrous or sparingly setose; sterile branches rarely developed . . . . *H. florentinum*
    3 Rhizome cord-like and elongate; plant becoming freely stoloniferous.
      5 Heads orange-red; involucre to 8 mm high, dark-villous with gland-tipped hairs; heads 5 to many, crowded; leaves long-setose on both sides or glabrate above; scape to about 7 dm tall . . . . *H. aurantiacum*
      5 Heads yellow.
        6 Heads commonly several or many, rather crowded; scape to over 1 m tall.
          7 Leaves glaucous, essentially glabrous above; plant freely and early stoloniferous . . . . *H. floribundum*
          7 Leaves green, abundantly setose on both surfaces; plant tardily stoloniferous . . . . *H. pratense*
        6 Heads often solitary (if up to 3 or 4, on finally elongate peduncles); scapes rarely over 5 dm tall.
          8 Leaves white-tomentose beneath at least when young, setose on both sides; scape filiform, commonly less than 2.5 dm tall, with usually only 1 well-developed head . . . . *H. pilosella*
          8 Leaves green and merely setose on both sides; heads commonly 2, 3, or 4.
            9 Involucre usually at least 1 cm high; scape to 4 or 5 dm tall; leaves to over 3 cm broad . . . . *H. flagellare*
            9 Involucre less than 1 cm high; scape smaller and more slender; leaves at most about 1.5 cm broad . . . . *H. auricula*

2 Stem firm, leafy or subscapose (scapose in *H. murorum* and *H. triste*); phyllaries either more or less imbricated or 1-rowed and with small bractlets at base; flowers yellow.

  10 Phyllaries 1-rowed but with a row of small bractlets at base; pappus-bristles all of one length; heads at most 2.5 cm broad; leaves subentire to shallowly undulate-dentate, the basal or lower cauline ones markedly larger than the progressively reduced (and often few) middle and upper ones.

    11 At least the lower bracts of the inflorescence foliaceous, the stem leafy nearly or quite to the top; (Ont. to N.S.).

      12 Stem slender, glabrous above, villous at base; panicle lax and open; heads with at most about 30 flowers, on long filiform glabrous or glandular-stipitate peduncles; leaves lanceolate, thin, glabrous, glaucous beneath . . . . *H. paniculatum*

      12 Stem coarse, densely pubescent at least above; corymb stiff and more compact; heads with 40 or more flowers, on stout tomentose (often dark-glandular) peduncles; leaves broader, thick, not glaucous, commonly pubescent at least beneath . . . . *H. scabrum*

    11 Inflorescence scaly-bracted, the stems leafy mostly below middle or only near base.

      13 Heads solitary or few on the finely stellate-pubescent to densely long-hairy stem, this usually less than 3 dm tall, scapose, the leaves all in a basal tuft or sometimes 1 (rarely 2) near the base of the stem; (B.C.–Alta.) . . . . *H. triste*

      13 Heads commonly more numerous, the inflorescence corymbiform or paniculiform; stems commonly taller and relatively leafy.

        14 Involucre (to 12 mm high), leaves, and stem shaggy-villous with a dense coat of white (becoming yellowish) hairs commonly about 4 or 5 mm long; leaves lanceolate or oblanceolate; (B.C.–Alta.) . . . . . *H. albertinum*

        14 Involucre less densely pubescent.

          15 Stem densely coated, at least below, with finally rust-coloured hairs 1 or 2 cm long; leaves oblanceolate to spatulate-oblong, long-hairy; involucre to 13 mm high, copiously stipitate-glandular to summit; (s Ont.) . . . . *H. longipilum*

          15 Stem lacking the above type of pubescence; involucres usually about 1 cm high.

            16 Leaves subglabrous (or merely inconspicuously stellate) and often glaucous above, sparsely or moderately setose below; involucres finely stellate and with few to many short, gland-tipped, mostly blackish bristles; (?B.C.–Alta.) . . . . [*H. scouleri*]

            16 At least some of the leaves distinctly long-setose above, scarcely glaucous (except sometimes in *H. venosum*).

              17 Leaves narrowly lanceolate to oblanceolate, usually acutish, sparsely to moderately setose (sometimes also stellate) beneath, to about 2.5 dm long (including the petiolar base), commonly extending in gradually reduced size to or slightly above the middle of the stem; involucres finely stellate and more or less stipitate-glandular, often copiously long-setose; (B.C.–Alta.) . . . . *H. cynoglossoides*

              17 Leaves broadly oblanceolate to elliptic, ovate, or obovate, usually obtuse or rounded at tip, mostly less than 1.5 dm long, chiefly basal or sub-basal (stem rarely leafy to middle); involucres glabrous or stipitate-glandular, obscurely or not at all stellate; (Ont.).

                18 Inflorescence open-corymbiform, the heads on long, filiform, glabrous or sparingly stipitate-glandular peduncles; leaves often purple-veined or mottled, more or less setose beneath, especially along the margins and midrib . . . . *H. venosum*

                18 Inflorescence subcylindric, the sparsely to densely stipitate-glandular peduncles relatively short; leaves

setose chiefly above, minutely stellate beneath ....
.... *H. gronovii*

10 Phyllaries in 3 or more distinctly unequal series; pappus-bristles of different lengths; heads usually over 2.5 cm broad; leaves often relatively coarsely toothed or shallowly lobed.

    19 Stem subscapose or with a few scattered leaves rapidly reduced in size upwards.

        20 Lowest leaves (elliptic or oval) rounded or cordate at base, their basal teeth divergent or reflexed; involucre densely stipitate-glandular; scape naked or with only 1 or 2 leaves; (introd.) .... *H. murorum*

        20 Lower leaves tapering to the petiole; stem with up to 12 scattered leaves.

            21 Involucre and pedicels glandular-stipitate, with or without a few long nonglandular hairs; leaves setose on both faces, often mottled with purple or bronze; (introd.) .... *H. vulgatum*

            21 Involucre and pedicels glandless or only minutely glandular, copiously long-pilose with nonglandular hairs; (Que. to Nfld. and N.S.) ... *H. robinsonii*

    19 Stem leafy nearly or quite to the inflorescence, the numerous leaves more gradually reduced in size.

        22 Leaves with revolute scabrous margins, linear to narrowly lanceolate, the upper ones narrowed or rounded at base; (transcontinental) .. *H. scabriusculum*

        22 Leaves flat, lanceolate to narrowly ovate, smoothish, the upper ones rounded or truncate to subcordate at base.

            23 Involucre glandless; phyllaries obtuse, the lowest ones spreading; panicle usually over 1/3 the total height of the plant; hairs of lower internodes and leaf-surfaces bulbous-based; (introd. in SW Que. and N.S.) .... *H. sabaudum*

            23 Involucre usually more or less stipitate-glandular; phyllaries attenuate, all appressed; panicle usually less than 1/3 the height of the plant; hairs (when present) slender; (transcontinental) .... *H. canadense*

H. albertinum Farr
/T/W/ (Hs) Dry open places in the foothills and up to moderate elevations in the mts. from SE B.C. (Trail to Cranbrook and Flathead; CAN) and SW Alta. (N to the type locality at Lake Louise, NW of Banff) to Oreg., Idaho, and Mont. [Intergrading through *H. absonum* Macbr. & Pays. with *H. cynoglossoides,* with which it should perhaps be merged].

H. albiflorum Hook.
/ST/W/ (Hs) Open woods and hillsides, mostly at moderate elevations, from SE Alaska (N to ca. 60°N) and SW Dist. Mackenzie (Mackenzie Mts. at ca. 62°30′N; CAN; not yet reported from the Yukon) through B.C. and Alta. (type from N of the Smoky R., N of 56°N) to SW Sask. (Cypress Hills; CAN; DAO; reported from Beausejour, NE of Winnipeg, Man., by Ernest Lepage, Nat. can. (Que.) 88(2):49. 1961, where probably introd.), S to Calif. and Colo. [*H. ?vancouverianum* Arv.-Touv.].

H. aurantiacum L.   Devil's Paint-brush, Orange Hawkweed.   Bouquets rouges
European; often a troublesome weed in fields and clearings in N. America, as in SE Alaska (Juneau), S B.C., Alta. (N to Edmonton), S Man. (Winnipeg), Ont. (N to SW James Bay at 51°16′N), Que. (N to the Gaspé Pen.), Nfld., N.B., P.E.I., and N.S. [*H. brunneocroceum* Pugsley].

A hybrid with *H. floribundum* (× *H. dorei* Lepage) is reported from the type locality, near Moore's Mills, Charlotte Co., N.B., and from Hunter River, Queens Co., P.E.I., by Ernest Lepage (Nat. can. (Que.) 94(5):618. 1967). A hybrid with *H. pilosella* (× *H. stoloniflorum* Waldst. & Kit.) is reported from Que. by Boivin (1966b; St-Foy, near Quebec City). This was later taken as the type of × *H. stoloniflorum* nm. cayouetteanum Lepage (loc. cit., 1967), who also notes a collection from St-Simon, Rimouski Co., E Que., as the type of nm. *laurentianum* Lepage.

H. auricula L.
European; apparently known in N. America only from near Kentville and Waterville, Kings Co., N.S. (ACAD; NSPM; Roland 1947; Fernald *in* Gray 1950).

H. canadense Michx.

/T/(X)/ (Hp) Thickets, rocky slopes, clearings, and disturbed ground, the aggregate species from N B.C. (Fernald *in* Gray 1950) to Alta. (N to L. Athabasca), Sask. (Cypress Hills and Meadow Lake, 54°08'N; Breitung 1957*a*; reports from Man. require confirmation), Ont. (N to w James Bay at ca. 53°N), Que. (N to SE Hudson Bay at 55°15'N and the Côte-Nord), S Labrador (N to the Hamilton R. basin), Nfld., N.B., P.E.I., and N.S., S to Wash. (?Oreg.), Mont., S.Dak., Mich., N.H., and Maine.

There has been much confusion between this species and *H. scabriusculum* and the above outline of the general range is tentative. The following more detailed outline of the distribution of various taxa in our area is based upon studies by Lepage. According to Ernest Lepage (Nat. can. (Que.) 88:45. 1961), *H. canadense* var. *hirtirameum* Fern. (type from Nfld.) is a hybrid between *H. canadense* and *H. scabrum* (× *H. fernaldii* Lepage; styles yellow). A hybrid between *H. kalmii* (*H. can.* var. *kalmii*) and *H. scabrum* has been named × *H. fassettii* Lepage; styles brown; SW Que.). Other hybrids reported by Lepage are: *H. canadense* × *H. (lachenalii) vulgatum* (× *H. grohii* Lepage, the type from Rivière-du-Loup, E Que.; also known from L. St. John); *H. canadense* × *H. scabriusculum* (× *H. dutillyanum* Lepage, the type from Pagwa, Ont.; also known in Ont. from Ingolf and Hare Is., L. Superior); and *H. kalmii* × *H. scabriusculum* (Ont.: Ingolf; near Thunder Bay).

1 Styles brown; phyllaries glabrous or stipitate-glandular; upper leaves rounded or cuneate at base (not triangular in outline).
   2 Stem robust, to about 1 cm thick at base; leaves irregularly toothed (some of the teeth longer than the others); inflorescence often much branched; [*H. fasc.* Pursh; *H. kalmii* var. *fasc.* (Pursh) Lepage; reported by Ernest Lepage, Nat. can. (Que.) 87(4):88. 1960, from Ont. (North Bay, L. Nipissing; Mt. McBean, N of L. Huron), Que. (Pontiac, Hull, Terrebonne, Berthier, Champlain, and Quebec counties), N.B. (St. Stephen), and N.S. (Halifax Co.)] . . . . var. *fasciculatum* (Pursh) Fern.
   2 Stem more slender; leaves entire, denticulate, or regularly dentate, the teeth of more uniform length.
      3 Leaves entire or obscurely denticulate; phyllaries dark brown to blackish, more or less stellate-hairy, the longest ones to over 1 cm long; [*H. kalmii* var. *sub.* Lepage, the type from the Nottaway R., Que.] . . . . var. *subintegrum* Lepage
      3 Leaves regularly denticulate or dentate; phyllaries yellowish green to dark green, often purple-tinged, generally less than 1 cm long; [*H. kalmii* L. and its var. *magnilacustre* Lepage; *H. eremocephalum* Gand.; *H. virgatum* Pursh; reported by Lepage, loc. cit., 1960, from S Ont. (French River Harbour), SW Que. (Maskinongé and Huntingdon counties), and N.S. (near Halifax)] . . . . var. *kalmii* (L.) Scoggan
1 Styles yellow; phyllaries stipitate-glandular; upper leaves truncate or subcordate at base (more or less triangular in outline).
   4 Leaves horizontally spreading, relatively long, bearing slender attenuate cilia along the margins and the midrib beneath, their teeth deltoid-apiculate; [Que.: Rivière-du-Loup, l'Islet, and Gaspé counties and Grosse-Isle, Magdalen Is.; Lepage, loc. cit., 1960] . . . . var. *divaricatum* Lepage
   4 Leaves mostly more or less strongly ascending, usually less strongly toothed, their hairs mostly stouter . . . . var. *canadense*
      5 Hairs of the stem, leaves, and pedicels to 3 mm long, intermixed with shorter hairs; [var. *hirtirameum sensu* Lepage, loc. cit., 1960, not Fern.; incl. var. *hirt.* f. *rufescens* Lepage, the hairs rust-coloured; S Ont. (var. *hirt.* f. *ruf.* reported from the Algoma dist. N of L. Huron by Lepage, loc. cit., 1960); f. *pilosius* reported from the type locality on a cliff near Eskasoni Brook, Cape Breton Co., N.S., by Ernest Lepage, Nat. can. (Que.) 88(2):45. 1961] . . . . f. *pilosius* Lepage
      5 Hairs of stem, leaves, and pedicels about 1 mm long; [var. *latifolium* T. & G.; *H. sabaudum (umbellatum) canadense* (Michx.) DC.; *H. macrophyllum* Pursh; *H. oxoacrum* Gand.; *H. prenanthoides* Hook., not Vill.; Ont. (N to the Attawapiskat R. at ca. 53°N), Que. (N to SE Hudson Bay at ca. 55°15'N and the Côte-Nord; type from L. Mistassini), Labrador (N to the Hamilton R. basin), Nfld., N.B., P.E.I., and N.S.] . . . . f. *canadense*

In addition to the above taxa, the very similar *H. laevigatum* Willd. of Eurasia (apparently differing chiefly from *H. canadense* in its distinctly wing-petioled lower leaves and its very narrow phyllaries) is reported (as *H. tridentatum* Fries) from Ont. by Lepage (loc. cit., 1960; Perth, Leeds, and

Grenville counties) and from sw Que. (Montreal) and N.B. (Northumberland, York, and St. John counties) by Ernest Lepage (Nat. can. (Que.) 85(4):81–82. 1958). Closely related Old World species or Greenland endemics reported from Greenland include, according to Joergensen, Soerensen, and Westergaard (1958), *H. acranthophorum, H. devoldii, H. eugenii, H. musartutense, H. nepiocratum,* and *H. stiptocaule* Omang, *H. inuloides* Tausch, *H. rigorosum* (Laest.) Almq., and *H. strictum* Fries. However, the genus is notorious for the "splitting" of species by European authors and most of these should probably be regarded as races or "microspecies".

H. cynoglossoides Arv.-Touv.
/T/W/ (Hs) Dry foothills up to moderate elevations in the mts. from s B.C. (valleys of the Dry Interior N to Lac la Hache, about 35 mi SE of Williams Lake; CAN; V) and s Alta. (N to Banff, E to the Cypress Hills) to Oreg., Utah, and Wyo. [*H. griseum* Rydb.].

H. flagellare Willd.
European; becoming a troublesome weed in E N. America, as in Que. (N to St-Simon, Rimouski Co.), N.B., P.E.I. (Rocky Point, Queens Co.; GH), and N.S.
   According to Ernest Lepage (Nat. can. (Que.) 94(5):611–15. 1967), this species is a hybrid between *H. caespitosum* Dum. (*H. pratense* of the present treatment) and *H. pilosella,* varying degrees of gene-exchange between the respective parents apparently producing the following recognizable nothomorphs:
1   Plant with an evident dominance of *H. pilosella* characters, the slender stem to 2(3) dm
    tall, branches, when present, arising along the lower half, the long superficial stolons
    sometimes producing new stems from their erect tips, the inflorescence consisting of
    rarely more than 1 or 2 heads; [Que. (St-Simon, Rimouski Co.), N.B. (Charlotte, York,
    Albert, and Carleton counties), and N.S. (Annapolis and Victoria counties)] . . . .
    . . . . nm. *cernuiforme* (Naegeli & Peter) Lepage
1   Plant intermediate or with an evident dominance of *H. caespitosum (pratense)*
    characters, the robust stem to 4 dm tall, bearing up to 6 heads.
    2   Branches rarely over 4 cm long, generally arising high on the stem.
        3   Involucre glandular, otherwise with few or no hairs; [not known from Canada] . . . .
            . . . . [nm. *flagellare*]
        3   Involucre distinctly hairy as well as glandular; [N.B.: St. Stephen, Charlotte Co.] . . . .
            . . . . .nm. *amauracron* (Missbach & Zahn) Lepage
    2   Branches generally long; stolons superficial or subterranean.
        4   Involucre glandular, otherwise with few or no hairs; [not known from Canada] . . . .
            . . . . [nm. *glatzense* (Naegeli & Peter) Lepage]
        4   Involucre abundantly hairy as well as glandular; [E Que.: St-Simon, Rimouski Co.,
            the type locality, and Cap-aux-Meules, Magdalen Is.] . . . . nm. *pilosius* Lepage

H. florentinum All.   King Devil.   Épervière des Florentins
European; a very aggressive weed in E N. America, as in Ont. (N to Sudbury and Ottawa), Que. (N to the Gaspé Pen.), Nfld. (St. Andrews; GH), N.B., and N.S. [*H. ?piloselloides* Vill.].

H. floribundum Wimm. & Grab.   Yellow Devil
European; introd. into fields and clearings in N. America, as in B.C. (Boivin 1966*b*), Ont. (N to Ottawa), Que. (N to Amos, 48°34′N), Nfld., N.B., P.E.I., and N.S.

H. gronovii L.
/t/EE/ (Hs) Dry open woods, thickets, and clearings from Kans. to Ill., Mich., s Ont. (Essex, Lambton, Lincoln, Middlesex, Waterloo, and Wentworth counties; CAN; OAC; TRT; there is also a collection in TRT from Algonquin Park, Renfrew Co., where probably introd.; the report from P.E.I. by McSwain and Bain 1891, requires confirmation), N.Y., and Mass., s to Tex. and Fla. [Incl. *H. marianum* Willd. and *H. pensylvanicum* Fries].

H. longipilum Torr.
/t/EE/ (Hs) Dry prairies and open sands from Minn. to s Ont. (Simcoe, Norfolk Co., and Sarnia, Lambton Co.; CAN; OAC; reported from Wentworth Co. by John Macoun 1884), s to E Tex. and La.

H. murorum L.   Golden Lungwort
European; locally introd. along roadsides and in fields, thickets, and open woods in N. America, as in SE Alaska (Wrangell; CAN, detd. Ernest Lepage; not listed by Hultén 1950 or 1968b), sw B.C. (Agassiz; Eastham 1947), Ont. (N to Arnprior, Renfrew Co.), Que. (N to St-Vallier, Bellechasse Co.; reported N to Cap-à-l'Aigle, Charlevoix Co., by R. Campbell, Can. J. Sci. 6(6):342–51. 1895), Nfld., N.B., and N.S.
Reports from Greenland are referred by Joergensen, Soerensen, and Westergaard (1958) to H. lividorubens Almq. and H. atratum Fries (incl. H. hyparcticum Almq. and H. stelechodes Omang), these, however, perhaps better treated as races or "microspecies"; (see note under H. canadense).

H. paniculatum L.
/T/EE/   (Hp)   Open woods and thickets from Ont. (N to Ottawa) to Que. (reported N to Chicoutimi, E of L. St. John, by A. Gagnon, G.W. Corrivault, and A. Morin, Ann. ACFAS 6:107. 1940), N.B. (Fredericton; Groh and Frankton 1949b; not known from P.E.I.), and N.S., s to Ala. and Ga.
Forma glandulosum Hoffm. (pedicels stipitate-glandular rather than glabrous) is common in our area. A hybrid with H. scabrum is reported from N.S. by M.L. Fernald (Rhodora 24(286):208. 1922; Bridgewater, Lunenburg Co.).

H. pilosella L.   Mouse-ear.   Oreille de souris
Eurasian; a very troublesome weed of sterile fields and pastures in N. America, as in sw B.C. (Vancouver Is. and Vancouver; Eastham 1947), Ont. (N to Ottawa), Que. (N to Charlevoix and Saguenay counties), St-Pierre and Miquelon, Nfld., N.B., P.E.I., and N.S.
Var. niveum Muell.-Aarg. (leaves permanently white-pannose beneath rather than becoming green) is known from N.S. (Canard, Kings Co.; ACAD).
H. alpinum L. (incl. H. angmagssalikense Omang; habitally similar to H. pilosella but larger-flowered) is a Eurasian species reported as native in Greenland (W Greenland N to 62°22'N, E Greenland N to 70°40'N; see Greenland map by Böcher 1938: fig. 109, p. 195). MAP: Hultén 1958: map 71, p. 91.

H. praealtum Gochnat   King Devil
European; locally abundant in grasslands and pastures in N. America, as in w B.C. (Terrace, near Prince Rupert; V), Ont. (N to the Nipissing dist.), Nfld. (Fernald in Gray 1950), N.B. (York and Sunbury counties; DAO, detd. Frankton), and P.E.I. (near Alberton, Prince Co.; DAO). [Incl. var. decipiens Koch].

H. pratense Tausch   King Devil.   Épervière des prés
European; a very aggressive weed in clearings and pastures in N. America, as in B.C. (Boivin 1966b), Ont. (N to Kapuskasing, 49°24'N), Que. (N to Anticosti Is. and the Gaspé Pen.), St-Pierre and Miquelon, Nfld., N.B., P.E.I., and N.S. [H. caespitosum Dum., the correct name through strict priority; H. cladanthum Arv.-Touv.].

H. robinsonii (Zahn) Fern.
/ST/E/   (Hs)   Ledges and clayey shores from Que. (E James Bay N to ca. 54°20'N; L. Mistassini; L. St, John; Bell R. at ca. 48°N; Mt-Tremblant; Rivière-du-Loup, Temiscouata Co.; Gaspé Pen.), SE Nfld., and N.S. (Seal Is., Yarmouth Co., and Inverness and Victoria counties, Cape Breton Is.; not known from N.B. or P.E.I.; reports from Labrador probably refer to the H. vulgatum complex) to Maine and N.H. [H. smolandicum ssp. rob. Zahn; H. smol. sensu Fernald 1925, not Almq.; H. ungavense Lepage]. MAP: Ernest Lepage, Nat. can. (Que.) 87(4): fig. 13, p. 104. 1960.

H. saubadum L.
European; collections in CAN from sw Que. (Montreal, where taken by Hincks in 1848) and N.S. (Chester Basin, Lunenburg Co., where taken by C.A. and U.F. Weatherby in 1941) have been placed here by Ernest Lepage. [Incl. H. vagum Jord.].

H. scabriusculum Schwein.
/ST/X/   (Hp)   Open woods, thickets, and sandy or rocky shores (ranges of Canadian taxa

outlined below), s to Oreg., Idaho, Colo., S.Dak., Mo., Ill., Mich., sw Que., and N.B. MAP and synonymy: *see* below.

1 Phyllaries bearing some stalked glands.
  2 Stem and leaves devoid of long hairs; [B.C. (N to Vanderhoof, ca. 54°N; type from 100-Mile House, about 75 mi NW of Kamloops) and s Alta. (Cypress Hills, Pincher Creek, Chief Mt., and Devil's Head L.; Ernest Lepage, Nat. can. (Que.) 87(3):71. 1960)] . . . . var. *saximontanum* Lepage
  2 At least the lower part of the stem and the lower leaves more or less pilose . . . .
    . . . . var. *scabrum* (Schwein.) Lepage
    3 Styles yellow; [type from Pigeon L., s of Edmonton, Alta.] . . . . . f. *xanthostylum* Lepage
    3 Styles brown; [*H. canadense* var. *scabrum* Schwein., not *H. scabrum* Michx.; according to Lepage, loc. cit., 1960, this taxon occurs in B.C. N to near Vanderhoof, ca. 54°N, and in Alta. N to Edmonton, reports from elsewhere in our area presumably referring to the other taxa] . . . . f. *scabrum*
1 Phyllaries nonglandular.
  4 Upper leaves and the upper part of the stem more or less hirsute.
    5 Upper part of stem densely hirsute; [type from Meadow Lake, 54°08′N, Sask.; Ernest Lepage, Nat. can. (Que.) 88 (2):43. 1961] . . . . var. *perhirsutum* Lepage
    5 Upper part of stem only moderately hirsute . . . . var. *columbianum* (Rydb.) Lepage
      6 Styles brown; [known only from Wash.] . . . . [f. *phaeostylum* Lepage]
      6 Styles yellow; [*H. col.* Rydb.; *H. canadense* var. *col.* (Rydb.) St. John; reported by Lepage, loc. cit., 1960, from B.C., Alta., Sask., and Man.] . . . . . . f. *columbianum*
  4 Stem and leaves lacking long hairs . . . . var. *scabriusculum*
    7 Styles yellow; [reported by Lepage, loc. cit., 1960, from Alta., Sask., Man., Ont., and sw Que. (type from St-Adolphe, Argenteuil Co.)] . . . . f. *chrysostylum* Lepage
    7 Styles brown; [*H. umbellatum* of auth., perhaps not L.; *H. manitobense* Gand.; *H. canadense sensu* Porsild 1943, and Raup 1947, not Michx.; *H. can.* f. *lepagei* Vict.; *H. can.* var. *angustifolium* T. & G. in major part; ?Alaska (Fernald *in* Gray 1950, as synonym of *H. umbellatum*) and the Yukon (N to ca. 64°N) to Great Bear L., Great Slave L., L. Athabasca (Alta. and Sask.), Man. (N to the Churchill R. at ca. 57°N), Ont. (N to w James Bay at 54°22′N; *see* James Bay map by Dutilly, Lepage, and Duman 1954: fig. 17, p. 121), Que. (N to E James Bay at ca. 52°15′N), N.B., and P.E.I. (Mt. Vernon, Queens Co.); MAP (aggregate species): Hultén 1968*b*:960] . . . . f. *scabriusculum*

H. scabrum Michx.
/T/EE/ (Hp) Dry open or wooded places from Minn. to Ont. (N to L. Nipigon, N of L. Superior), Que. (N to the E James Bay watershed at ca. 53°50′N, the Côte-Nord, and Gaspé Pen.; the report from Nfld. by Waghorne 1898, requires confirmation), N.B., P.E.I., and N.S., s to Mo., Tenn., and Ga.

The typical form has the lower stem-internodes (and the leaf-midribs beneath) densely villous with brownish hairs averaging at least 2 mm long. Var. *leucocaule* Fern. & St. John (stem densely white-tomentose to base, with an admixture of dark glands; leaves minutely glandular-pilose on both sides) is known from the type locality, Sable Is., N.S. Var. *tonsum* Fern. & St. John (stem glabrate below or minutely hispid; leaves glabrous on both sides or sparingly setose above; type from Grindstone Is., Magdalen Is.) occurs throughout the Canadian area.

[*H. scouleri* Hook.]
[Reports of this species of the w U.S.A. (Wash. and Mont. to Calif.) from B.C. require confirmation, most so-named B.C.–Alta. collections in CAN being referred to *H. cynoglossoides* by Boivin.]

H. triste Willd.
/ST/W/eA/ (Hs) Meadows and rocky places at low to fairly high elevations from the Aleutian Is. (type locality), Alaska–Yukon (N to ca. 64°N), and Great Bear L. through B.C.–Alta. to Calif. and N N.Mex.; the Andes of s S. America; Kamchatka. [Incl. var. *fulvum* Hult., with yellowish-brown indument, and var. *tristiforme* Zahn, intermediate between var. *triste* and var. *gracile*]. MAP: Hultén 1968*b*:958.

The typical form has relatively large heads, the involucres and upper part of the stem clothed with long greyish-black (exceptionally yellowish-brown) villous hairs, the peduncles not bearing stalked glands. According to Hultén's map, it is confined in N. America to Alaska–Yukon–Dist. Mackenzie. Also in this area but ranging southwards to the w U.S.A. is var. *gracile* (Hook.) Gray (*H. gracile* Hook., the type a Drummond collection "On the more elevated Rocky Mountains", probably in Alta.; *H. hookeri* Steud.; *H. ?arcticum* Froel.; incl. *H. gracile* vars. *alaskanum* Zahn, *detonsum* Gray, and *yukonense* Porsild), the heads averaging smaller, the involucres and upper part of the stem hirsute with short black hairs, with or without a few long villous hairs, the peduncles often bearing stalked glands. MAPS (*H. gracile*): Hultén 1968*b*:959; Porsild 1966: map 152, p. 85; Raup 1947: pl. 37.

H. venosum L.   Poor Robin's Plantain, Rattlesnake-weed
/T/EE/   (Hs)   Open woods and clearings from Ont. (N to the E shore of L. Superior at Pancake Bay, about 35 mi NW of Sault Ste. Marie; CAN; reports from Man. require confirmation; not known from Que. or the Atlantic Provinces) to s Maine, s to Mo., La., and Fla.

The typical form (at least some of the rosette-leaves remotely long-setose above) is reported from s Ont. by Gaiser and Moore (1966; Lambton Co.). Most of our material is referable to var. *nudicaule* (Michx.) Farw. (*H. gronovii* var. *nud.* Michx.; incl. var. *subacaulescens* T. & G.), the rosette-leaves, except sometimes the very lowest, glabrous. According to Gleason (1958), *H. venosum* is a variable species that apparently hybridizes with several others, including *H. gronovii*, *H. paniculatum,* and *H. scabrum.*

H. vulgatum Fries
European; widely introd. along roadsides and in fields and waste places in N. America, as in B.C. (Boivin 1966*b*), Ont. (N to Kapuskasing and Ottawa), Que. (N to SE Hudson Bay at ca. 56°10′N, the Côte-Nord, Anticosti Is., and Gaspé Pen.), Labrador (N to Ramah, 58°52′N), Nfld., N.B., P.E.I., and N.S.; w Greenland N to ca. 69°35′N, E Greenland N to 66°19′N. [*H. ?molle* and *H. ?pusillum sensu* Pursh 1814, and Hooker 1833, not Jacq. nor Willd., respectively].

This is a very polymorphic species, possibly native in E Canada and Greenland with the inclusion of the following scarcely separable taxa: *H. amitsokense* (Almq.) Dahlst., *H. dovrense* Fries, *H. groenlandicum* Arv.-Touv. (native on Anticosti Is., E Que., and in Labrador and Nfld. according to Fernald *in* Gray 1950), *H. ivigtutense* (Almq.) Omang, *H. lachenalii* Gmel. (incl. *H. argilaceum* and *H. cheriense* Jord., *H. irriguum* Fries, and *H. strumosum* (Linton) Ley), *H. maculatum* Smith, *H. plicatum* Lindeb., and *H. scholanderi* and *H. sylowii* Omang.

HYMENOPAPPUS L'Hér.   [9292]

H. filifolius Hook.
/T/WW/   (Hs)   Dry, often sandy or gravelly places in the plains and foothills from cent. Wash. to Mont., s Alta. (Belly R., Milk R., Whitemud R., and Cardston; CAN), and s Sask. (Dirt Hills, Missouri Coteau; CAN; reported by Breitung 1957*a*, from Rockglen, Ormiston, and Big Muddy Valley), s to Baja Calif., Tex., and N.Dak. MAP: combine the maps by B.L. Turner, Rhodora 58(692): fig. 28, p. 220, fig. 36, p. 226, and fig. 37, p. 231. 1956.

According to Turner's above-noted fig. 37, the Alta.–Sask. plant is var. *polycephalus* (Osterh.) Turner (*H. poly.* Osterh.; stem bearing up to 60 heads rather than at most 40, with at most 8 leaves rather than up to 12, the terminal segment of the rosette-leaves at most 3 cm long rather than up to 5 cm).

HYMENOXYS Cass.   [9304]

1   Leaves entire, essentially all basal, linear to broadly oblanceolate, to about 8 cm long and 1 cm broad; phyllaries subequal, free to base; head solitary on the scape, the rays to 2 cm long; (s Alta. and sw Sask.; s Ont.) .... *H. acaulis*
1   Leaves mostly basal but some cauline and alternate, pinnatifid to near the midrib into usually 3 narrowly linear divisions to about 4 cm long and 2 mm broad; phyllaries in 2

series of unequal length, those of the outer series united to about the middle; heads commonly 2 or 3, the rays to 1.5 cm long; plant puberulent or glabrate; (s Alta. and Sask.) . . . . *H. richardsonii*

H. acaulis (Pursh) Parker
/T/D/ (Hr (Ch)) Dry places at low to fairly high elevations from Idaho and Mont. to s Alta. (N to Lethbridge; CAN) and sw Sask. (Cypress Hills, where taken by John Macoun in 1894; CAN), s to s Calif., Tex., and Kans. (locally to Ohio); isolated in s Ont. (Georgian Bay and Manitoulin Is., L. Huron; CAN; TRT). [*Gaillardia* Pursh; *Actinea* Spreng.; *Actinella* Nutt.; *Tetraneuris* Greene; *T. septentrionalis* Rydb.].

The s Ont. plant is referable to var. *glabra* (Gray) Parker (*Actinella glabra* (Gray) Nutt.; *Tetraneuris (Actinea) herbacea* Greene; *T. simplex* Nels.; leaves green, sparingly appressed-silky, glabrate in age, rather than copiously and permanently appressed-silky).

H. richardsonii (Hook.) Cockerell
/T/W/ (Hs (Ch)) Dry plains and rocky hillsides from s Alta. (N to Lethbridge; CAN) and Sask. (N to near Carlton House, about 35 mi sw of Prince Albert) to Ariz. and w Tex. [*Picradenia rich.* Hook., the type from near Carlton House, Sask.; *Actinea* Ktze.; *Actinella* Nutt.; *H. macounii* (Cock.) Rydb.].

HYPOCHAERIS L.   [9572]   Cat's-ear

1   Essentially glabrous annual from a taproot; rays scarcely surpassing the involucre, only about twice as long as broad; central achenes slender-beaked, the outer ones usually beakless; leaves to about 1.5 dm long and 3.5 cm broad; (introd. in s B.C. and s Ont.) . . . . *H. glabra*
1   Hispid-leaved perennial from a fibrous-rooted caudex; rays distinctly surpassing the involucre and about 4 times as long as broad; achenes all with beaks shorter than or exceeding the length of the body; leaves to about 3.5 dm long and 7 cm broad; (introd., essentially transcontinental) . . . . *H. radicata*

H. glabra L.
Eurasian; introd. in disturbed or waste places in sw B.C. (Vancouver Is. and adjacent islands; CAN; V) and the Pacific states; reported from s Ont. by Stroud (1941; Wellington Co.).

H. radicata L.
Eurasian; introd. in lawns, pastures, fields, and waste places in N. America, as in sw B.C. (Vancouver Is. and adjacent islands and mainland E to Hope; CAN; V), Sask. (Scott, about 35 mi sw of North Battleford; Breitung 1957a), Ont. (N to Ottawa), Que. (N to the Gaspé Pen. at Gros-Morne and Bonaventure; CAN; GH), St-Pierre and Miquelon, Nfld. (St. John's; GH), N.B. (Fredericton; DAO), and N.S. (Yarmouth and Victoria counties; ACAD; DAO).

INULA L.   [9061]

1   Disk of heads rarely over 2 cm broad, the outer phyllaries linear; leaves more or less white-woolly beneath, entire or minutely toothed, rather narrowly lanceolate, less than 1 dm long, the cauline ones subcordate-clasping; (introd. in s Ont.) . . . . *I. britannica*
1   Disk to over 4 cm broad, the outer phyllaries broadly ovate; leaves softly tomentose-felted beneath, rather coarsely and doubly serrate, the elliptical basal ones to 4 dm long, the ovate cauline ones cordate-clasping; (introd. in B.C. and from Ont. to N.S.) . . . . *I. helenium*

I. britannica L.
Eurasian; apparently known in the wild state in N. America only from a single stand on the banks of the Etobicoke R. in Peel Co., s Ont., w of Toronto, where first taken by J.A. Simon in 1928 and well established, there also being collections in TRT taken in 1935, 1942, and 1950, and in CAN taken by the present writer in 1960.

I. helenium L.   Elecampane
Eurasian; introd. along roadsides and fencerows and in rich fields and clearings in N. America, as in SE B.C. (Cloverdale, E of Vancouver; V), Ont. (N to Ottawa), Que. (N to the S coast of the Gaspé Pen.), ?Nfld. (Boivin 1966*b*; not listed by Rouleau 1956), N.B. (Carleton and Kings counties), and N.S.

IVA L.   [9141]   Marsh-Elder, Sumpweed

1   Annual to over 2 m tall; heads crowded in naked spike-like racemes disposed in axillary and terminal panicles, the involucres 2 or 3 mm long; leaves long-petioled, ovate to rhombic (or the lowest cordate), coarsely and often doubly serrate, rough-hairy above, often finely silky-pubescent beneath; (introd., transcontinental) .... *I. xanthifolia*
1   Woody-based perennials commonly less than 5 dm tall; heads nodding in the axils of the upper bract-like leaves, solitary or in spike-like racemes.
　　2   Leaves essentially sessile, linear to narrowly oblong or narrowly ovate, entire, thick, pubescent or glabrate, at most about 3 cm long; heads 4 or 5 mm broad, the involucres 3 or 4 mm long; (S B.C. to S Man.) .... *I. axillaris*
　　2   Leaves (except the uppermost) distinctly petioled, lanceolate to oval or elliptic, sharply serrate, puberulent or strigose above, often glabrous beneath, somewhat fleshy, to about 1 dm long; heads 5 or 6 mm broad, the involucres to 5 mm long; (N.S.) ....
　　　　... *I. frutescens*

I. axillaris Pursh   Poverty-weed
/T/WW/   (Grh)   Dry (often alkaline or disturbed) places in the valleys, plains, and foothills from S B.C. (valleys of the Dry Interior N to the Nicola R., SW of Kamloops, and Vernon) to Alta. (N to Grande Prairie, 55°10′N), Sask. (N to near Carlton, about 35 mi SW of Prince Albert), and S Man. (N to Dauphin, N of Riding Mt.), S to Calif. and Okla. [Incl. var. *robustior* Hook.]. MAP: I.J. Bassett, G.A. Mulligan, and C. Frankton, Can. J. Bot. 40(9): fig. 1, p. 1245. 1962.

I. frutescens L.
/T/EE/   (Ch (N))   Saline marshes and shores from W N.S. (Yarmouth, Kings, and Hants counties, where considered native by Fernald *in* Gray 1950, but introd. by Roland 1947; a stand seen by the writer in 1957 at Avonport, Hants Co., appeared to be native) and the Atlantic and Gulf states to Fla. and Tex.
　　The N.S. plant is the scarcely distinct var. *oraria* (Bartlett) Fern. & Grisc. (*I. or.* Bartl.; plant to about 2 m tall rather than 3.5 m; leaves to 5 cm broad rather than 3 cm; heads to 6 mm broad rather than 5 mm; achenes averaging about 3 mm long rather than less than 2.5 mm).

I. xanthifolia Nutt.   Marsh-Elder.   Fausse herbe à poux
Native in the W U.S.A. (Wash. to N.Mex. and Tex.; abundantly introd. elsewhere along roadsides and streambanks and in waste places, as in B.C. (N to Dawson Creek, ca. 55°40′N), Alta. (N to Beaverlodge, 55°13′N), Sask. (N to Saskatoon), Man. (N to Churchill; Beckett 1959), Ont. (N to Thunder Bay and Ottawa), Que. (N to L. St. John), N.B., P.E.I., and N.S. [*Cyclachaena* Fresn.; *I. paniculata* Nutt.].

JAUMEA Pers.   [9262]

J. carnosa (Less.) Gray
/t/W/   (Hpr)   Tidal flats and coastal marshes from SW B.C. (Vancouver Is. and adjacent islands; CAN; V) to S Calif. [*Coinogyne* Less.].

KRIGIA Schreb.   [9560]   Dwarf Dandelion

1   Plant with a 1–3-leaved stem forking above and to about 7 dm tall, its oblong or oval clasping leaves mostly entire; radical leaves wing-petioled, often toothed and sometimes pinnatifid; involucre to about 1.5 cm high; achenes 15–20-ribbed; pappus consisting of numerous fragile bristles and short inconspicuous scales; (S Man. and S Ont.) .... *K. biflora*

1 Plant scapose, the several stems leafless or leafy only near base, to about 3 dm tall; leaves (at least the later ones) often pinnatifid; involucre to about 7 mm high; achenes 5-angled; pappus consisting of 5–7 broad scales alternating with scabrous bristles . . . .
. . . . *[K. virginica]*

K. biflora (Walt.) Blake
/T/(X)/ (Hs) Woodlands, meadows, and fields from SE Man. (Teulon, about 30 mi N of Winnipeg; CAN; reported from near L. Winnipeg by John Macoun 1884) to S Ont. (Essex and Lambton counties; CAN; OAC) and S New Eng., S to Ariz., N.Mex., Mo., and Ga. [*Hyoseris* Walt.; *H. (K.) amplexicaulis* Michx.; *Adopogon (Cynthia) virginicum* (L.) Ktze., not *Hyoseris virg.* L.].
The Teulon, Man., plant is referable to f. *glandulifera* Fern. (peduncles stipitate-glandular rather than glabrous).

[*K. virginica* (L.) Willd.]
[Reports of this species of the E U.S.A. (N to Wisc. and Maine) from Ont. by John Macoun (1884; noting a report by Gray) and Soper (1949) are based upon *K. biflora* (relevant collections in CAN). A collection in Herb. V from Bliss Landing, near Vancouver, SW B.C., has been placed here by J.W. Eastham but requires further study. (*Hyoseris* L.).]

LACTUCA L. [9596] Lettuce. Laitue

1 Achenes very flat, with a strong median nerve on each face and occasionally an additional pair of very obscure nerves; beak soft and filiform, about equalling or longer than the achene; pappus white; flowers yellow, sometimes drying purplish.
  2 Fruiting involucres at most 1.5 cm high; heads with commonly less than 20 flowers; achenes (including beak) usually less than 7 mm long; pappus mostly about 6 or 7 mm long; leaves neither prickly-toothed nor strongly glaucous; (Man. to N.S.; introd. westwards) . . . . *L. canadensis*
  2 Fruiting involucres at least 1.5 cm high; heads with up to 40 or more flowers; achenes (including beak) to 1 cm long; pappus to 12 mm long.
    3 Leaves strongly glaucous, sinuate or sinuate-pinnatifid, prickly-toothed on the margins and more or less prickly on the midrib beneath; (S Sask. and S Man.) . . . .
    . . . . *L. ludoviciana*
    3 Leaves green, scarcely prickly-toothed, pinnatisect, the lateral lobes oblong-obovate, commonly broadest above the base; (Ont. to N.S.) . . . . *L. hirsuta*
1 Achenes flat or moderately compressed, distinctly or prominently several-nerved on each face.
  4 Beak filiform, equalling or longer than the body of the achene (3 or 4 mm); pappus white; flowers pale yellow or greenish yellow, often drying purplish; leaves sagittate-clasping, typically pinnately lobed; annuals or biennials; (introd.).
    5 Leaves or their lobes entire or nearly so, not prickly-toothed.
      6 Stem-leaves ovate to orbicular, cordate-clasping, simple; inflorescence a dense corymbose panicle; flowers often streaked with violet . . . . *[L. sativa]*
      6 Stem-leaves linear-lanceolate, sagittate-clasping, simple or pinnatifid with a few distant narrow lobes, the blades commonly held vertically (with one edge up) and all more or less in one plane; inflorescence a narrow spike-like panicle with short erect branches; flowers often reddish beneath . . . . *L. saligna*
    5 Leaves and their lobes prickly-toothed.
      7 Ripe achenes olive-grey, short-bristly at apex; bracts with spreading sagittate auricles; stem-leaves sagittate-clasping, usually held vertically (often all in one plane), the lower ones often pinnatifid, with narrow distant lateral lobes . . . .
      . . . . *L. scariola*
      7 Ripe achenes blackish, more or less glabrous at apex; bracts with appressed rounded auricles; stem-leaves cordate-clasping, not held vertically, obovate-oblong and essentially unlobed or pinnatifid with broad lobes . . . . *[L. virosa]*
  4 Beak none or at most about 1 mm long; leaves rarely clasping.
    8 Pappus light-brown; achenes greyish brown, mottled; flowers typically bluish, up

to over 35 in a head; fruiting involucre to 14 mm high; leaves irregularly pinnatifid, sometimes runcinate, coarsely toothed; annual or biennial; (transcontinental) . . . . *L. biennis*

8   Pappus white.

    9   Flowers typically yellow, only 5 in a head; involucre at most about 11 mm high, consisting of 4 or 5 elongate phyllaries and very short calyculate ones at base; achenes dark red to blackish; leaves very thin, runcinate-lyrate, the large terminal angulate segment cordate at base; glabrous annual or biennial; (introd. in s B.C., Ont., and s Que.) . . . . *L. muralis*

    9   Flowers typically blue.

        10   Perennial from a deep-seated rhizome; flowers showy, blue or blue-purple, commonly 20 or more in a head; involucre to 2 cm high, the phyllaries in 3 or 4 rows; achenes short-beaked; leaves pale or glaucous, entire or the lower ones commonly more or less pinnatifid; (B.C. to James Bay) . . . *L. tatarica*

        10   Biennials with a basal rosette and a taproot; involucre to 1.5 cm high; flowers usually less numerous, blue or bluish; outer achenes often distinctly thick-beaked, the inner ones beakless; leaves often pinnatifid; (s Man. and s Ont.) . . . . *L. floridana*

L. biennis (Moench) Fern.

/sT/X/   (Hs)   Damp thickets and clearings from SE Alaska (N to ca. 60°N) and southernmost Yukon (near the B.C. boundary) to B.C., Alta. (N to Fort Saskatchewan), Sask. (N to Waskesiu Lake, ca. 54°N), Man. (N to Dawson Bay, N L. Winnipegosis), Ont. (N to Sandy L. at ca. 53°N, 93°W), Que. (N to SE James Bay at 51°29′N, L. Mistassini, and the Côte-Nord), Labrador (N to the Hamilton R. basin), Nfld., N.B., P.E.I., and N.S., s to Calif., Colo., and N.C. [*Sonchus* Moench; *S. (Mulgedium) leucophaeus* Willd., not *L. leucophaea* Sibth.; *L. multifida* Rydb.; *L. terrae-novae* Fern.; *L. spicata* of auth., not *S. spicatus* Lam.]. MAP: Hultén 1968*b*:952.

    Forma *integrifolia* (T. & G.) Fern. (leaves unlobed rather than irregularly pinnatifid, sometimes runcinate) is known from s Ont. (Puslinch, Wellington Co.; TRT) and sw Que.

L. canadensis L.

/T/EE/   (Hs)   Thickets and clearings, the aggregate species from E Man. (N to Lac du Bonnet, about 50 mi NE of Winnipeg) to Ont. (N to Nipigon, N shore of L. Superior), Que. (N to Cap-à-l'Aigle, Charlevoix Co., L. St. John, and the s Gaspé Pen.), N.B., P.E.I., and N.S., s to Tex. and Fla.; introd. westwards to s B.C. and Colo.

1   Most of the leaves unlobed; [var. *integrifolia* (Bigel.) Gray (*L. integrifolia* Bigel., not Nutt.); var. *montana* Britt.; *Sonchus ?pallidus* Willd.; Ont. to P.E.I. and N.S.; introd. in s B.C. (Vancouver Is.; Agassiz; Sicamous)] . . . . var. *canadensis*

1   Most of the leaves deeply lobed; [SE Man. to P.E.I. and N.S.; var. *longifolia* also introd. in s B.C.].

    2   Lobes of the leaves linear-falcate, usually entire; [*L. longifolia* Michx.; *L. elongata* Muhl.] . . . . var. *longifolia* (Michx.) Farw.

    2   Lobes of the leaves broadly falcate or obovate and obliquely truncate, entire or toothed . . . . var. *latifolia* Ktze.

L. floridana (L.) Gaertn.

/T/EE/   (Hs)   Moist woods, thickets, and clearings from Minn. to SE Man. (near Otterburne, about 30 mi s of Winnipeg; Löve and Bernard 1959), s Ont. (Essex and Kent counties; John Macoun 1884; Bernard Boivin, Rhodora 55(654):225. 1953; Core 1948), Ill., Ohio, N.Y., and Mass., s to Tex. and Fla. [*Sonchus* L.; *Mulgedium* DC.].

    Var. *villosa* (Jacq.) Cronq. (*L. villosa* Jacq.; leaves uncleft rather than lyrate or runcinate-pinnatifid; achenes all or nearly all beakless rather than the outer ones often distinctly thick-beaked) is known from s Ont. (Pelee Is., Essex Co.; Core 1948).

L. hirsuta Muhl.

/T/EE/   (Hs)   Dry open woods, thickets, and clearings from Ont. (N to Carleton Co.; TRT; not listed by Gillett 1958) to Que. (N to Oka, about 20 mi sw of Montreal), P.E.I. (Mt. Stewart, Queens Co.; GH; not known from N.B.), and N.S., s to Tex., La., and Va.

The report from s Man. by J.M. Macoun (1897; Killarney) is based upon *L. ludoviciana,* the relevant collection in CAN. The report N to Bic, Rimouski Co., E Que., by Scoggan (1950) requires confirmation. Our material is chiefly referable to the glabrous extreme, var. *sanguinea* (Bigel.) Fern. (*L. sang.* Bigel.). However, both it and the typical form are reported from s Ont. by Gaiser and Moore (1966; Lambton Co.).

L. ludoviciana (Nutt.) Riddell
/T/WW/ (Hs) Prairies, shores, and roadsides from Idaho and Mont. to s Sask. (Regina and Gainsborough; Breitung 1957a) and s Man. (Killarney, NE of Turtle Mt.; Grande Clarière, sw of Brandon; Aweme, SE of Brandon; St. Vital, near Winnipeg; CAN; WIN; the tentative report from Ont. by Boivin 1966b, requires clarification), s to Tex., Ark., Mo., Ill., and Wisc. [*Sonchus* Nutt.].

Some of our material may be referable to f. *campestris* (Greene) Fern. (*L. camp.* Greene; flowers blue or purple from the first rather than yellow, drying purplish; reported from SE Man. by Lowe 1943).

L. muralis (L.) Fresen.   Wall-Lettuce
Eurasian; locally introd. along roadsides and in waste places in N. America, as in sw B.C. (Vancouver Is. and adjacent islands and mainland E to Manning Provincial Park, about 30 mi SE of Hope; CAN; V), Ont. (Hamilton and Ottawa), and sw Que. (Rigaud and Montreal). [*Prenanthes* L.; *Mycelis* Reichenb.].

L. saligna L.   Willow-leaved Lettuce
European; introd. along roadsides and in waste places in N. America, as in s Ont. (Essex, Lambton, and York counties) and Que. (Rouleau 1947).

Forma *ruppiana* (Wallr.) Beck (leaves linear to lanceolate and entire rather than oblong and runcinate-pinnatifid) is known from sw Que. (Sherbrooke; CAN).   .

[*L. sativa* L.]   Garden Lettuce
[?Asiatic; cult. and occasionally found along roadsides and on waste heaps and garden-refuse in N. America but scarcely established, as in Alta. (Boivin 1966b) and Ont. (Hamilton and Ottawa).]

L. scariola L.   Prickly Lettuce
Eurasian; introd. along roadsides and in fields and waste places in N. America, as in B.C. (N to Kamloops; V), Alta. (Moss 1959), Sask. (Breitung 1957a), Man. (N to Winnipeg), Ont. (N to the Algoma dist. N of L. Huron and Ottawa), Que. (N to the Gaspé Pen.; Groh 1944b), N.B. (Moncton; ACAD; CAN), P.E.I. (Charlottetown; ACAD), and N.S. [*L. serriola* L., the original spelling, later corrected by Linnaeus].

Forma *integrifolia* (Bogenh.) Beck (var. *int.* Bogenh.; *L. integrata* (Gren. & Godr.) Nels.; *L. virosa sensu* Lowe 1943, not L., relevant collections in WIN and Herb. Man. Prov. Mus., Winnipeg; leaves unlobed or only the lowest ones pinnatifid rather than nearly all of the leaves pinnatifid) is represented by the above N.B. and P.E.I. (and some of the Man.) collections.

L. tatarica (L.) Meyer
/ST/(X)/EA/ (Gr) Moist meadows, prairies, thickets, and clearings from cent. Alaska (known only from ca. 65°N; not known from the Yukon), Great Bear L., Great Slave L., and B.C.–Alta. to Sask. (N to Prince Albert and Cumberland House), Man. (N to the N end of L. Winnipeg; introd. at Churchill), Ont. (N to the Fawn R. at ca. 55°N, 88°W, and sw James Bay), and Que. (known only from SE James Bay N to ca. 53°N; introd. in New Eng. and possibly in N.S., *Mulgedium pulchellum* reported from there by Lindsay 1878), s to Calif., N.Mex., Okla., Mo., Wisc., and Mich.; Eurasia. [*Sonchus* L.]. MAP: Hultén 1968b:952.

The N. American plant has been separated as the intergrading and barely distinguishable ssp. *pulchella* (Pursh) Stebbins (*Sonchus (Mulgedium; Lactuca) pulchellus* Pursh; *S. (M.)* ?*acuminatus* Willd.; *M. heterophyllum* Nutt.; *S. sibiricus sensu* Hooker 1833, not L.; stem less branched and averaging slightly taller than that of the typical form, bearing 10–30 entire or remotely and minutely denticulate leaves below the inflorescence rather than 3–12 conspicuously spinulose-denticulate leaves, the phyllaries usually more closely imbricate, their tips tending to be narrower and more attenuate; *see* G.L. Stebbins, Madroño 5(4):123–24. 1939).

[L. virosa L.]
[Eurasian; the tentative report from Tanana Hot Springs, cent. Alaska, by Hultén (1950) and the report from Man. by Lowe (1943) are based upon *L. scariola* f. *integrifolia,* relevant collections in CAN and WIN, respectively.]

LAPSANA L.   [9555]

L. communis L.   Nipplewort.   Herbe aux mamelles
Eurasian; introd. along roadsides and in waste places in N. America, as in SE Alaska (Juneau and Ketchikan; Hultén 1950), B.C. (N to Queen Charlotte Is. and Revelstoke), SE Man. (Winnipeg; Herb. Man. Prov. Mus., Winnipeg), Ont. (N to Sault Ste. Marie, North Bay, and Ottawa), Que. (N to L. St. John and Rimouski, Rimouski Co.), Nfld., N.B., P.E.I. (Charlottetown; GH), and N.S., and in w Greenland (Disko, ca. 70°N; CAN). MAP: Hultén 1968b:941.

[LAYIA H. & A.]   [9258]   Tidy-tips

[L. glandulosa (Hook.) H. & A.]
[The report of this species of the w U.S.A. (Wash. and Idaho to Baja Calif. and Ariz.) from B.C. by John Macoun (1886; noting a report by Gray, "Barren ground, British Columbia to California"), this taken up by Henry (1915), Rydberg (1922), and Hitchcock et al. (1955), requires confirmation. (*Blepharipappus* Hook.).]

LEONTODON L.   [9574]   Hawkbit

1   Scape commonly forking, scaly-bracted, to about 8 dm tall; heads erect before anthesis; involucre to 13 mm high, its phyllaries in several unequal series; achenes to 7.5 mm long; pappus consisting entirely of plumose chaffy-based bristles; leaves laciniate-toothed to pinnatifid, usually somewhat pubescent; (introd., transcontinental) . . . . *L. autumnalis*
1   Scape simple and usually naked, the solitary head nodding before anthesis; leaves subentire to rather shallowly pinnate-lobed.
  2   Pappus of some of the outer flowers reduced to a short laciniate crown (the achenes of these flowers commonly less scabrous than those of the inner flowers); achenes commonly less than 6 mm long; involucre to 11 mm high, its glabrous or hirsute subequal phyllaries subtended by a ring of minute bractlets; leaves hispid; scape filiform, to about 3.5 dm tall; (introd. in B.C.) . . . . *L. taraxacoides*
  2   Pappus similar in all flowers, consisting of an inner row of long plumose broad-based bristles and an outer row of much shorter barbed bristles; achenes to over 6 mm long; involucre to 1.5 cm high, its bristly-hispid phyllaries in 2 or 3 unequal series; leaves and scape bristly-hispid, the upwardly thickened scape to about 7 dm tall; (introd. in s Ont.) . . . . *L. hispidus*

L. autumnalis L.   Fall-Dandelion.   Liondent d'automne
Eurasian; introd. along roadsides and in fields in N. America (ranges of Canadian taxa outlined below). MAP and synonymy: *see* below.
1   Involucres and peduncles essentially glabrous; [*Apargia* Willd.; *Oporinia* Don; Alaska (Fairbanks; *see* Hultén 1968a); s B.C. (Vancouver Is. and adjacent islands; Langley Prairie, SE of Vancouver); Ont. (N to Timmins, 48°28′N), Que. (N to the Côte-Nord), Labrador (Hamilton R. basin), Nfld., N.B., P.E.I., and N.S.; sw Greenland; MAP (aggregate species): Hultén 1968b:942] . . . . var. *autumnalis*
1   Involucres and summits of peduncles densely spreading-pubescent with blackish hairs; [*Apargia pratensis* Link; E Que. (Gaspé Pen. at Port Daniel, Bonaventure Co.), Labrador (Hamilton R. basin), Nfld., N.B., P.E.I., and N.S.; Greenland] . . . . var. *pratensis* (Link) Koch

L. hispidus L.
Eurasian; locally introd. into fields and waste places in N. America, as in s Ont. (Cambridge (Galt), Waterloo Co., where taken by W. Herriot in 1902; CAN; OAC). [*Apargia* Willd.; *L. hastilis* var. *vulgaris* Koch].

L. taraxacoides (Vill.) Mérat
European; locally introd. into fields, pastures, and waste places in N. America, as in w B.C. (Queen Charlotte Is. and near Victoria, Vancouver Is., where first taken by James Fletcher in 1885; CAN; DAO); the report from s Ont. by F.H. Montgomery (Can. Field-Nat. 62(2):91. 1948; noting a 1910 report of *L. nudicaulis* from Cambridge (Galt), Waterloo Co., by W. Herriot) requires confirmation. [*Hyoseris* Vill.; *Thrincia (L.) leysseri* Wallr.; *Crepis (L.; Apargia) nudicaulis* of auth., not L.; *L. hirtus* of auth., not L.].

LIATRIS Schreb.   [8826]   Button-Snakeroot, Blazing-star

(Ref.: Gaiser 1946)
1   Pappus plumose, about 1 cm long; leaves linear or linear-oblanceolate, firm, punctate; stems to about 8 dm tall.
   2   Heads rarely if ever with more than 6 flowers, sessile or nearly so in a spike-like inflorescence; corolla at most 12 mm long, the tube pilose within; phyllaries with long acute or acuminate tips, commonly ciliate; leaves ciliate, otherwise glabrous; stems several from crowns of an elongate (sometimes corm-like) deep vertical subterranean trunk, glabrous; (s Alta. to s Man.) . . . . *L. punctata*
   2   Heads with up to 35 or more flowers, sessile or short-peduncled; corolla to 14 mm long, the inner surface of the lobes pubescent; coriaceous phyllaries rounded, mucronate, or abruptly acuminate at summit, they and the leaves sublustrous and eciliate; stems solitary from a roundish corm, glabrous or sparingly hirsute; (s Ont.) . . . .
. . . . *L. cylindracea*
1   Pappus merely finely barbed; phyllaries mostly round-tipped; stems usually solitary from corm-like rhizomes.
   3   Heads cylindric-campanulate, with less than 20 flowers, sessile in a usually dense spike; corolla glabrous within; leaves usually glabrous, linear to linear-lanceolate, the lowermost rarely over 2 cm broad; stem usually glabrous, to 2 m tall; (s Ont.) . . . . *L. spicata*
   3   Heads becoming hemispheric or subglobose, with up to 50 or more flowers, mostly distinctly peduncled (or sessile in *L. aspera*); stem usually more or less pubescent at least above.
      4   Heads becoming hemispheric (the tooth-fringed phyllaries remaining erect and loosely appressed), long-peduncled to occasionally subsessile, to 3 cm thick (or the terminal head often much larger); corolla-tube glabrous within; leaves ciliate, otherwise glabrous to densely scabrous-pubescent on both surfaces, the basal ones to 1.5 cm broad; stem pubescent at least above; (Alta. to Man.) . . . . . *L. ligulistylis*
      4   Heads becoming subglobose (the phyllaries soon puckered or squarrose-spreading), to 2.5 cm thick; corolla-tube pilose within at base; leaves eciliate.
         5   Phyllaries ciliate, with a very narrow or obsolete scarious summit, pubescent to merely scabrous; heads usually peduncled in an open raceme or occasionally a panicle; leaves scabrous to densely pubescent, the basal ones broadly obovate, to 5 cm broad; stem usually rather densely pubescent . . . .
. . . . [*L. scariosa*]
         5   Phyllaries with a broad scarious eciliate coloured summit, glabrous; heads sessile to short-peduncled in a spike or spicate raceme; leaves glabrous or merely scabrous, the basal ones linear-lanceolate, rarely over 2 cm broad; stem glabrous to puberulent or sparingly strigose; (s Ont.) . . . . *L. aspera*

L. aspera Michx.
/t/EE/   (Gst)   Dry, often sandy soil from N.Dak. to Wisc., Mich., s Ont. (Essex, Kent, and Lambton counties; CAN; GH; TRT; US), W.Va., and N.C., s to E Tex. and Fla. [*Lacinaria scabra* Greene]. MAP: L.H. Shinners, Am. Midl. Nat. 29: map 4, p. 30. 1943.

The s Ont. plant is referable to var. *intermedia* (Lunell) Gaiser (*Lacinaria scariosa* var. *int.* Lunell; *Li. scariosa sensu* John Macoun 1884, and Dodge 1914, not (L.) Willd., relevant collections in CAN; *L. ?squarrosa sensu* Macoun 1886, not (L.) Michx.; leaves smooth rather than scabrous; stem glabrous or sparingly pubescent at summit rather than at least the upper third puberulent or sparingly grey-strigose). Reports of the typical form from Sask. (as by Breitung 1957*a*) and Man. (as by Lowe 1943) are probably all based upon *L. ligulistylis,* relevant collections in CAN, DAO, and WIN. × *L. sphaeroidea* Michx. is apparently the most abundant phase of a hybrid-series involving *L. aspera* or its var. *intermedia* and one or more other *Liatris* species. Gaiser (1946) reports it from Essex and Lambton counties, s Ont. Its area is shown in a MAP by Shinners (loc. cit., map 5, p. 30). Another hybrid-complex between × *L. sphaeroidea* or *L. aspera* and *L. cylindracea* (× *L. gladewitzii* (Farw.) Shinners) is known from Kent and Lambton counties, s Ont.

L. cylindracea Michx.
/t/EE/ (Gst) Ledges and dry soils from Minn. to s Ont. (N to Manitoulin Is., N L. Huron; GH; N.C. Fassett, Rhodora 35(420):388. 1933), s to Mo., Ill., Ind., Ohio, and N.Y. [*L. squarrosa* var. *intermedia* DC.; *L. squarrosa sensu* Hooker 1833, not (L.) Michx.].
Forma *bartelii* Steyerm. (flowers white rather than rose-purple) is reported from s Ont. by Gaiser and Moore (1966; Lambton Co.).

L. ligulistylis (Nels.) Schum.
/T/WW/ (Gst) Dry to moist, often sterile soil from Alta. (N to Edmonton; CAN) to Sask. (N to N of Prince Albert; CAN), Man. (N to Cranberry Portage, about 20 mi SE of Flin Flon; WIN), and Wisc., s to N.Mex. and S.Dak. [*Lacinaria* Nels.; *L. aspera sensu* Breitung 1957*a*, and Lowe 1943, not Michx., relevant collections in CAN, DAO, and WIN]. MAP: L.H. Shinners, Am. Midl. Nat. 29: map 6, p. 30. 1943.
Forma *leucantha* Shinners (corolla white rather than rose-purple) is known from Sask. (Boivin 1966*b*) and s Man. (Griswold, about 20 mi w of Brandon; DAO). A hybrid with *L. squarrosa* var. *glabrata* (Rydb.) Gaiser (this not known from Canada), × *L. creditonensis* Gaiser, is reported from s Ont. by Gaiser (1946; type from her test-garden at Crediton, Huron Co., s Ont., where originating from parent plants brought in from the U.S.A.).

L. punctata Hook.
/T/WW/ (Gst) Dry prairies and plains, often in sandy soil, from Mont. to Alta. (N to Red Deer; CAN), Sask. (N to 15 mi w of Saskatoon; CAN; type, as first collection cited, a Drummond collection from "Plains of the Saskatchawan", either in Sask. or Alta.), and Man. (N to St. Lazare, about 75 mi NW of Brandon), s to Mexico and Tex. [*Lacinaria* Ktze.]. MAP: *Atlas of Canada* 1957: map 11, sheet 38.
Forma *albiflora* Sheldon (corolla white rather than rose-purple) is known from Sask. (Boivin 1966*b*) and s Man. (Souris Co.; DAO).

[*L. scariosa* (L.) Willd.]
[Reports of this species of the E U.S.A. (W.Va. and Pa. to S.C.) from Sask. and Man. by Hooker (1833) are probably referable to *L. ligulistylis.* Reports from s Ont. by John Macoun (1884) and Dodge (1914) are based upon *L. aspera* var. *intermedia,* relevant collections in CAN and WIN.]

L. spicata (L.) Willd.
/t/EE/ (Gst) Meadows and swampy places from Wisc., Mich., and Ohio to s Ont. (Essex and Lambton counties; CAN; GH; QUK; TRT; reported by Lionel Cinq-Mars et al., Nat. can. (Que.) 98(2):197. 1971, as introd. near an abandoned garden at Oka, near Montreal, Que.), Pa., and N.J., s to La. and Fla. [*Serratula* L.; *L. ?pycnostachya sensu* C. Rousseau, Nat. can. (Que.) 98(4):727. 1971, perhaps not Michx.].

LUINA Benth. [9403]

L. hypoleuca Benth.
/t/W/ (Hp) Ledges, crevices, and talus slopes from SW B.C. (Vancouver Is. and adjacent islands and mainland E to Manning Provincial Park, about 30 mi SE of Hope; CAN; V) to cent. Calif.

LYGODESMIA D. Don [9598] Skeletonweed

1 Branches spine-tipped, rigidly spreading; stems several from a taproot and branching caudex, bearing tufts of pale or brownish wool at the base; lower leaves to 3 cm long, the others reduced and scale-like; heads commonly short-peduncled and borne laterally on the branches; florets and longer phyllaries mostly 3–5; plant glabrous or minutely scabrous; (s ?B.C.) .... [*L. spinosa*]
1 Branches not spine-tipped, terminated by heads; stem solitary (but much-branched from near the base), lacking woolly tufts.
   2 Pappus light brown; florets and longer phyllaries usually 5; lower leaves at most about 1 dm long, those of the branches reduced and awl-like; perennial with a deep-seated rootstock; (s B.C. to s Man.) .... *L. juncea*
   2 Pappus white; florets and longer phyllaries usually 8 or 9; leaves to about 2 dm long, only the uppermost ones strongly reduced; annual; (s Alta. to sw Man.) .... *L. rostrata*

L. juncea (Pursh) Don
/T/WW/ (Gr) Dry prairies and plains, often in sandy soil, from SE B.C. (Similkameen R., SW of Princeton; Wardner, S of Cranbrook; CAN; V) to Alta. (N to Edmonton; CAN), Sask. (N to near Saskatoon; CAN), and s Man. (N to Rossburn, about 70 mi NW of Brandon), S to Ariz., N.Mex., and Ark. [*Prenanthes* Pursh].

L. rostrata Gray
/T/WW/ (T) Dry sandy prairies and plains from s Alta. (Moss 1959), s Sask. (Cypress Hills, Beverly, and Mortlach; CAN; Breitung 1957a), and sw Man. (N to St. Lazare, about 75 mi NW of Brandon) to Colo. and Kans.

[*L. spinosa* Nutt.]
[The inclusion of B.C. in the range of this species of the W U.S.A. (Mont. to Calif. and Ariz.) by Rydberg (1922; *Plei. spin.*) requires clarification. (*Pleiacanthus* Rydb.).]

MADIA Molina [9253] Tarweed

1 Leaves opposite (except often the uppermost reduced ones); (s B.C.).
   2 Leaves to over 1 dm long and 1 cm broad; involucres 4–6 mm high; ray-ligules to 1 cm long; receptacular bracts united into a cup about the sterile disk-florets; pappus consisting of several usually ciliate-fringed scales; heads on elongate bracted peduncles, the lateral peduncles often surpassing the main axis; biennial or short-lived perennial, commonly with a short rhizome, to about 7 dm tall; (Vancouver Is. and adjacent islands) .... *M. madioides*
   2 Leaves to about 2 cm long and 2.5 mm broad; involucres 2–4 mm long; ray-ligules minute; receptacular bracts united about the usually solitary fertile disk-floret; pappus none; heads in the terminal forks of the divaricately branching stem and in small cymose clusters; slender annual rarely to 2 dm tall; (s ?B.C.) .... [*M. minima*]
1 Leaves alternate (except often the lowermost ones); annuals.
   3 Involucres less than 5 mm high; ray-ligules very short and inconspicuous; receptacular bracts united about the solitary fertile disk-floret; heads mostly on filiform naked peduncles; leaves to 4 cm long and 2 mm broad; stems to 3 dm tall, branched above; (s B.C.) .... *M. exigua*
   3 Involucres at least 6 mm long; receptacular bracts distinct, each enveloping an achene; stems to over 8 dm tall.
      4 Involucres fusiform, to 9 mm high and 5 mm broad (when pressed); ray-ligules (0)1–3(5), inconspicuous, about 2 mm long; heads commonly clustered; leaves to 7 cm long and 5 mm broad; (s B.C. to sw Man.; introd. elsewhere) ...... *M. glomerata*
      4 Involucres ovoid or broadly urceolate, to 12 mm long and mostly 6–12 mm broad (when pressed); ray-ligules 5–13, 3–7 mm long.
         5 Plant rough-hairy, stipitate-glandular usually only above the middle; leaves to 11 cm long and 1 cm broad; heads not clustered; (s B.C.) .... *M. gracilis*

5   Plant rough-hairy and strongly stipitate-glandular, the glands extending nearly
or quite to the base of the stem; leaves to 18 cm long and 12 mm broad;
heads often clustered; (introd. in s B.C., Ont., Que., and Nfld.) .... *M. sativa*

M. exigua (Sm.) Gray
/t/W/   (T)   Dry grasslands and open woods from the plains and foothills up to moderate
elevations in the mts. from s B.C. (Vancouver Is. and adjacent islands and mainland E to Lower
Arrow L. and Trail; CAN; V) to N Baja Calif. and Nev. [*Sclerocarpus* Sm.; *Harpaecarpus* Gray; incl.
M. filipes Gray].

M. glomerata Hook.
/T/WW/   (T)   Dry open places from the valleys and foothills to moderate elevations in the mts.
from B.C. (N to Revelstoke) to Alta. (N to McLennan, 53°42′N), Sask. (N to Swift Current and
Regina), sw Man. (Portage la Prairie; DAO), and Minn., s to Calif., Ariz., and Colo.; introd. along
roadsides and in waste places elsewhere, as in s Alaska (N to ca. 62°N), the Yukon (near Dawson,
ca. 64°N; CAN), Ont. (Earlton, NE of Sudbury), and Que. (Longueuil, near Montreal; Trois-Pistoles,
Temiscouata Co.; Capucins, Matane Co., NW Gaspé Pen.). MAP: Hultén 1968*b*:885.

M. gracilis (Sm.) Keck
/t/W/   (T)   Dry open places (often along roadsides or in other disturbed areas) from the valleys to
moderate elevations in the mts. from s B.C. (Vancouver Is. and adjacent islands and mainland E to
Cascade, sw of Trail; CAN; V) to N Baja Calif. and Utah; Chile. [*Sclerocarpus* Sm.; *Madorella
(Madia) dissitiflora* Nutt.; *Madorella (Madia) racemosa* Nutt.].

M. madioides (Nutt.) Greene
/t/W/   (Hs)   Open woods from sw B.C. (Vancouver Is. and adjacent islands; CAN; V) to Calif.
[*Anisocarpus* Nutt.; *M. nuttallii* Gray].

[M. minima (Gray) Keck]
[Reports of this species of the w U.S.A. (Wash. and Idaho to Calif.) from B.C. by Henry (1915),
Rydberg (1922), and Hitchcock et al. (1955) and a so-named collection in Herb. V from Sooke,
Vancouver Is., require confirmation. (*Hemizonia* Gray; *Hemizonella* Gray; *Hemizonia (Hemizonella)
durandii* Gray).]

M. sativa Molina
Native in the w U.S.A. (Wash. to Calif.) and Chile; introd. along roadsides and in waste places
elsewhere, as in s B.C. (Vancouver Is. and mainland E to near Creston; CAN; V; possibly native in s
B.C.), Ont. (between Blind River and Spanish, N shore of L. Huron; OAC), sw Que. (Chambly and
Lotbinière counties; MT), and Nfld. (Rouleau 1956).
    Boivin (1966*b*) reports the species only from B.C. and the above reports from other areas in
Canada may prove referable to *M. glomerata*. Some of the B.C. material (and most or all of the
material from elsewhere in Canada, if correctly identified) is referable to var. *congesta* T. & G. (*M.
capitata* Nutt.; heads crowded in 1–few clusters at the ends of the stem and branches rather than
scattered or in scattered clusters).

MATRICARIA L.   [9339]   Wild Chamomile

1   Heads discoid; disk-corollas 4-lobed; receptacle conic, acute; achenes with 4 ribs on the
inner side, smooth on the back and between the ribs; pappus an obscure crown or none;
leaves 2–3-pinnatifid; bruised plant with fragrance of pineapple; (introd., transcontinental) ......
.... *M. matricarioides*
1   Heads with white rays; disk-corollas normally 5-lobed; leaves 2-pinnatifid.
    2   Receptacle conic, acute; achenes with 5 slender raised ribs on the inner side, smooth
on the back and between the ribs; rays at most 1 cm long; disk to about 1 cm broad;
bruised plant with fragrance of pineapple; (introd., transcontinental) .... *M. chamomilla*
    2   Receptacle convex to hemispheric, obtuse; achenes with 3 strong corky almost

wing-like ribs on the inner side, roughened on the back and between the ribs; plants nearly scentless.

3   Phyllaries with dark-brown to blackish margins; pappus-crown usually entire; heads 1 to several; rays at most 1 cm long; stem simple to rather abundantly branched, to about 3 dm tall; (transcontinental in arctic and subarctic regions) . . . . *M. ambigua*

3   Phyllaries with greenish to light-brown margins; pappus-crown distinctly dentate; heads several to numerous; (introd., transcontinental) . . . . *M. maritima*

M. ambigua (Ledeb.) Kryl.
/aS/X/GEA/ (Hp (Ch)) Moist sandy seashores (sometimes in grassy places near human habitations) from the coasts of Alaska–Yukon–Dist. Mackenzie–Dist. Keewatin to Banks Is., N Baffin Is., and northernmost Que. (Hudson Strait and Ungava Bay; reported from Labrador by Boivin 1966b), S in the West to SW Alaska (Nunivak Is. at ca. 60°N), farther eastwards S to NE Man. (S to York Factory, Hudson Bay, 57°N), N Ont. (coast of NW James Bay at ca. 55°N), James Bay (South Twin Is., ca. 53°N), and the coast of E James Bay, Que. (S to ca. 53°45′N); southernmost Greenland and w Greenland between ca. 73° and 74°30′N; Iceland; NE Europe; N Asia. [*Pyrethrum* Ledeb.; *Tripleurospermum* Löve & Löve; *T. phaeocephalum* (Rupr.) Pobed.; *Chrysanthemum (M.) grandiflorum* Hook.; *Chamomilla (M.) hookeri* Rydb.; *P. inodorum* var. *nanum* Hook.; *M. maritima* var. *nana* (Hook.) Boivin; *M. inodora* var. *phaeocephala* Rupr.]. MAPS: Hultén 1968b:890 (*Tri. phaeo.*); Porsild 1957: map 315, p. 200.

According to Q.O. Kay (Watsonia 7(3):130–41. 1969), this northern element of the *M. inodora* complex should more correctly be known as *Tripleurospermum phaeocephalum* (*see* synonymy).

M. chamomilla L.
Eurasian; introd. and locally abundant along roadsides and waste places in N. America, as in SW B.C. (mouth of the Fraser R.; V), Alta. (N to Fort Saskatchewan; CAN), Sask. (N to Crooked River, 52°51′N; Breitung 1957a), Man. (N to The Pas), Ont. (N to Cache L., Algonquin Provincial Park, Renfrew Co.; CAN), Nfld., N.B., and N.S.; Greenland.

According to Jan Toman and Frantisék Starý (Taxon 14(7):224–28. 1965), our plant should be known as *M. recutita* L., embracing the concept of *M. chamomilla* as described by Linnaeus in 1755, his 1753 description, published together with *M. recutita*, evidently applying to a different plant. Some of our Nfld. and N.B. material is referable to var. *coronata* (Gay) Coss. & Germ. (*M. cor.* Gay; pappus of achenes consisting of a distinct short crown, rather than obsolete).

M. maritima L.
Eurasian; introd. in fields and waste places in N. America, as in cent. Alaska (Fairbanks, ca. 65°N), Dist. Mackenzie (Fort Providence, w of Great Slave L. at ca. 61°20′N; W.J. Cody, Can. Field-Nat. 75(2):67. 1961; the report from Great Bear L. by John Macoun 1884, is probably referable to *M. ambigua*), B.C. (Vancouver Is. and North Pine, ca. 55°N), Alta. (N to Fort Saskatchewan), Sask. (N to Emma Lake, 53°34′N; Breitung 1957a), Man. (N to Ethelbert, E of Duck Mt.), Ont. (N to near Thunder Bay), Que. (N to the Côte-Nord and Gaspé Pen.), Labrador (Hamilton R. basin), Nfld., N.B., P.E.I., and N.S. MAP (aggregate species): Hultén 1968b:890 (*Trip. inod.*; incl. the range of the typical form).

1   Stem depressed, its branches horizontal or drooping; leaves to about 4 cm long, their fleshy linear segments to 5 mm long; heads to about 3 cm broad; ray-ligules to about 12 mm long; [*Chamomilla* Rydb.; *M. inodora* var. *salina* Bab.; locally introd. in waste ground near ports in Que. (York and Bonaventure, Gaspé Pen.; GH; RIM) and the E U.S.A.] . . . . var. *maritima*

1   Stem ascending or erect, its branches ascending; leaves larger, their nearly filiform segments to 2 cm long; heads 3 or 4 cm broad; ray-ligules to 2 cm long; [*Dibothrospermum agreste* Knaf; *M. (Chamomilla; Chrysanthemum; Pyrethrum; Tripleurospermum) inodora* L.; the common transcontinental form in N. America] . . . . var. *agrestis* (Knaf) Wilmott

M. matricarioides (Less.) Porter   Pineapple-weed
Eurasian (?Asiatic); introd. along roadsides and in fields and waste places in N. America and

rapidly spreading, as in the Aleutian Is., Alaska (N to ca. 69°N), the Yukon (N to ca. 65°30′N), Dist. Mackenzie (N to the Mackenzie R. Delta), B.C., Alta., Sask. (Breitung 1957a), Man. (N to Reindeer L. and Churchill), Ont. (N to Fort Severn, Hudson Bay, ca. 56°N), Que. (N to s Ungava Bay), Labrador (Boivin 1966b), Nfld., N.B., P.E.I., and N.S.; Greenland. [*Artemisia mat.* Less., the type from Unalaska, Aleutian Is.; *M. discoidea* DC.; *Santolina suaveolens* Pursh, not *M. suav.* L.]. MAP: Hultén 1968b:889.

MICROSERIS Don [9559]

1 Plants more or less caulescent, bearing leaves a short distance above the base of the stem, the leaves linear or broader and often laciniate; heads mostly solitary but sometimes several in robust individuals; involucres 1–2 cm high, calyculate (subtended by a series of much smaller bractlets); pappus consisting of up to 20 narrow scales tipped with a long, white, distinctly plumose, bristle-like awn; achenes to 8 mm long; glabrous or slightly scurfy perennial to 6 dm tall; (s B.C. and sw Alta.) . . . . *M. nutans*
1 Plants scapose, the leaves all in a basal rosette; head solitary; pappus-awns not plumose.
2 Pappus consisting of 5 lanceolate, glabrous or scabrous, chaffy scales tapering to an awn slightly to considerably longer than the body; involucres to 1.5 cm high, calyculate; achenes to 6 mm long; leaves linear and entire or more often with spreading linear lobes; glabrous or scurfy annual to 3.5 dm tall; (Vancouver Is.) . . . .
. . . . *M. bigelovii*
2 Pappus consisting of 10 or more members; involucres to 2.5 cm high, not calyculate, their phyllaries subequal or slightly imbricate; achenes to 1 cm long; leaves linear, long-acuminate, entire, their margins often crisped or wavy and minutely white-ciliate; scapes to about 3 dm tall.
3 Pappus consisting of 40–80 mixed capillary bristles and very slender attenuate scales; phyllaries often speckled; leaves to 2 cm broad, rarely over 15 times as long as broad; scape glabrous or more commonly villous-tomentose above; (B.C. to Man.) . . . . *M. cuspidata*
3 Pappus consisting of 10–30 slender, attenuate, bristle-like, white scales; phyllaries generally with a dark midrib and sometimes also finely speckled with blackish-purple; leaves mostly over 20 times as long as broad; scape glabrous or puberulent; (s B.C.) . . . . *M. troximoides*

M. bigelovii (Gray) Schultz-Bip.
/t/W/ (T) Open moist or grassy places from sw B.C. (several localities on s Vancouver Is.; CAN) to Calif. [*Calais* Gray]. MAP: K.L. Chambers, Contrib. Dudley Herb. 4(7): fig. 21a, p. 305. 1955.

M. cuspidata (Pursh) Schultz-Bip.
/T/WW/ (Hr) Dry open places, often sandy or gravelly, from Mont. to s Alta. (N to Banff; CAN; the tentative report from the Mackenzie R. Delta by Porsild 1943, requires clarification), s Sask. (Lebret and Lumsden, both SE of Moose Jaw; Breitung 1957a), and sw Man. (N to Fort Ellice, about 75 mi NW of Brandon, where taken by John Macoun in 1879; MTMG), s to Colo., Okla., Mo., and Wisc. [*Agoseris* Steud.; *Nothocalais* Greene; *Troximon* Pursh].

M. nutans (Geyer) Schultz-Bip.
/T/W/ (Hs) Chiefly in open, rather moist places at low to rather high elevations from s B.C. (N to the Marble Range, NW of Clinton; CAN) and sw Alta. (Waterton Lakes; Breitung 1957b) to Calif., Utah, and Colo. [*Scorzonella* Geyer; *Calais* and *Ptilophora* Gray; *Ptilocalais* Greene].

M. troximoides Gray
/t/W/ (Hr) Dry open places in the lowlands and foothills from southernmost B.C. (collections in CAN, detd. Calder and Porsild, from between L. Osoyoos and Midway, where taken by J.M. Macoun in 1905; collection in V from Penticton) and Mont. to N Utah. [*Nothocalais* Greene; *Scorzonella* Jeps.].

[MIKANIA Willd.] [8818]

[M. scandens (L.) Willd.] Climbing Hempweed
[The report of this species of the E U.S.A. (N.Y. and Maine to Tex. and Fla.) from s Ont. by John Macoun (1884; Malden, Essex Co.) requires confirmation, as, also, the T.J. Burgess report from Amherstburg noted by Soper (1962; *see* his s Ont. map 25, fig. 22, p. 36, indicating published reports). If the reports prove valid upon the location of the relevant voucher-specimens, the species is almost certainly extinct in s Ont., evidently not having been taken since that time. The report from Canada by A. Michaux (1803) also requires clarification, probably resulting from too loose an application of that name with respect to present political boundaries. (*Eupatorium* L.).]

ONOPORDUM L. [7113]

O. acanthium L. Scotch Thistle
Eurasian; introd. (perhaps sometimes a garden-escape) along roadsides and in fields and waste places in N. America, as in sw B.C. (Nanaimo, Vancouver Is., where taken by John Macoun in 1887 and 1908; CAN), Ont. (N to Bruce, Grey, Frontenac, and Lanark counties; the report from Winnipeg, Man., by Lowe 1943, is based upon *Cirsium drummondii*, the relevant collection in WIN), N.B., and ?N.S. (John Macoun 1884).

PETASITES Mill. [9381] Sweet Coltsfoot

1 Leaves very large (to over 1 m broad), cordate-rotund to round-reniform, sharply
   sinuate-toothed, becoming glabrate, on stout hollow petioles to 2 m long, these with
   dilated sheathing bases.
   2 Heads whitish; (introd. in s Ont.) .... *P. japonicus*
   2 Heads pale reddish-violet; (introd. in B.C.) .... *P. hybridus*
1 Leaves smaller, usually more persistently tomentose at least beneath; petioles shorter,
   their bases not conspicuously dilated.
   3 Leaves coarsely dentate, unlobed or with only 1 or 2 pairs of relatively shallow lobes
      toward base, floccose above, densely white-tomentose beneath; achenes about 3 mm
      long.
      4 Leaves ovate- or triangular-sagittate, to 2.5 dm long, unlobed and with rarely
         fewer than 20 teeth on each margin; fruiting heads to 2.5 cm long; (trans-
         continental) .... *P. sagittatus*
      4 Leaves triangular-cordate to reniform, smaller, evidently lobed and with mostly not
         more than 15 teeth on each margin; fruiting heads about 1.5 cm long; (western
         arctic, subarctic, and alpine regions) .... *P. frigidus*
   3 Leaves deeply lobed, green and essentially glabrous above, cordate-deltoid to
      reniform.
      5 Leaves white-tomentose beneath, cleft about half-way to the midrib; (trans-
         continental) .... *P. vitifolius*
      5 Leaves glabrous or only thinly tomentose beneath, mostly cleft more than 2/3 to
         the midrib.
         6 Leaves glabrous on both surfaces except for a slight ciliation on the veins
            beneath and on the margins; rhizome thickish; (the Yukon–Dist. Mackenzie) ....
            .... *P. arcticus*
         6 Leaves thinly tomentose beneath; rhizome slender and cord-like; (trans-
            continental) .... *P. palmatus*

P. arcticus A.E. Porsild
/aS/W/ (Grh) Known only from open or lightly wooded, clayey or shaly slopes along the coast of the Yukon (between Kay Point and King Point; CAN) and in NW Dist. Mackenzie (type from East Branch, Mackenzie R. Delta; CAN; *see* Porsild 1943:74). MAP: W.J. Cody, Nat. can. (Que.) 98(2): fig. 26, p. 155. 1971.

P. frigidus (L.) Fries
/aST/WW/EA/ (Grh) Wet tundra, moist woods, and shores from the Aleutian Is. and coasts of Alaska–Yukon–Dist. Mackenzie and w Dist. Keewatin to Banks Is., Prince Patrick Is., and Melville Is., s through the mts. of B.C. and sw Alta. (Waterton Lakes; Breitung 1957*b*) to the Wenatchee Mts. of Wash.; NE Europe; N Asia. [*Tussilago* L.; *Nardosmia* Hook.; *N. angulosa* Cass.; *P. alaskanus* Rydb.; *P. gracilis* Britt.]. MAPS: Hultén 1968*b*:913; Porsild 1957: map 321, p. 201; Raup 1947: pl. 36.

P. hybridus (L.) Gaertn., Mey., & Scherb. Butterbur
Eurasian; locally introd. into waste places of the E U.S.A. (Mass. to Pa.; Fernald *in* Gray 1950) and known in Canada from sw B.C. (Vancouver Is. and the lower Fraser Valley at New Westminster, Whannock, Steveston, Ladner, and Abbotsford; Groh 1947, *P. vulgaris*; Groh's report of it from Niagara Falls, Welland Co., s Ont., is based upon *P. japonicus,* the relevant collection in OAC). [*Tussilago* L.; *P. vulgaris* Desf.].
Concerning the B.C. plant, Eastham (1947) writes, "An Old World sp[ecies] introd. by the Japanese who use the leaf-stalks as a vegetable. Well-established and spreading by strong creeping rootstocks in the vicinity of former Japanese dwellings; gives indication of becoming a persistent weed."

P. japonicus (Sieb. & Zucc.) Schmidt Butterbur
Asiatic; apparently known in the wild state in N. America only from Niagara Falls and vicinity, Welland Co., s Ont. (OAC; TRT), where first taken by Beck in 1935. [*Nardosmia* Sieb. & Zucc.; *P. vulgaris sensu* Groh 1947, as to the s Ont. plant, and Soper 1949, not Desf.].
Concerning the s Ont. plant, Montgomery (1957) writes, "It has been known there for about 20 years and now covers a low, wet, wooded river flat to the extent of one-eighth of an acre."

P. palmatus (Ait.) Gray
/ST/X/eA/ (Grh) Moist woods and swampy places from the Yukon (N to ca. 68°N) and the Mackenzie R. Delta to Great Bear L., Great Slave L., L. Athabasca (Alta. and Sask.), s Dist. Keewatin, northernmost Man.–Ont., Que. (N to Ungava Bay), Labrador (N to Hebron, 58°12′N), Nfld., N.B., P.E.I., and N.S., s to Calif., Minn., Mich., and Mass.; E Asia (Hultén's map). [*Tussilago palmata* Ait., the type from Nfld.; *Nardosmia* Hook.; *N. (P.) hookeriana* and *speciosa* Nutt.; *P. frigidus* var. *palm.* (Ait.) Cronq.; *P. palm. (spec.)* var. *frigidus* Macoun, not *P. frigidus* (L.) Fries]. MAP: Hultén 1968*b*:914.

P. sagittatus (Banks) Gray
/aST/X/ (Grh) Meadows and bogs from Alaska (N to ca. 66°30′N), the Yukon (N to ca. 68°N), and the coast of Dist. Mackenzie to sw Dist. Keewatin, Man. (N to Churchill), and northernmost Ont.–Que.–Labrador, s to Wash., Idaho, Mont., Colo., Minn., Wisc., cent. Ont. (s to sw James Bay and the shore of L. Superior near Thunder Bay), cent. Que. (s to SE James Bay and the Knob Lake dist. at ca. 54°45′N), and Labrador (s to Makkovik, 55°05′N). [*Tussilago sagittata* Banks, the type from "Hudson's Bay"; *Nardosmia* Hook.; *P. dentatus* Blank.]. MAPS: Hultén 1968*b*:914; Porsild 1957: map 322, p. 201.

P. vitifolius Greene
/ST/X/ (Grh) Swampy ground and wet woods from Alaska (N to ca. 70°N), the Yukon (N to ca. 67°N), and the Mackenzie R. Delta to L. Athabasca, s Dist. Keewatin, northernmost Ont., Que. (N to Ungava Bay), and Labrador (N to Kangalaksiorvik, 59°25′N; GH), s to N Oreg., Alta. (s to Red Deer; CAN), Sask. (s to the Cypress Hills; DAO), s Man. (type from Emerson), cent. Ont. (s to near Thunder Bay), N Minn., and Que. (s to Anticosti Is. and the Gaspé Pen.). [Incl. *P. hyperboreus* Rydb., *P. nivalis* and *P. trigonophyllus* Greene, and *Nardosmia (P.) corymbosa* Hook.; *P. frigidus* var. *hyperboreoides* Hult.]. MAP (w area; *P. hyperboreus*): Hultén 1968*b*:913.
Concerning *P. hyperboreus,* Hultén (1968*b*) writes, "Frequently forms hybrid swarms with *P. frigidus.* Possibly a hybridogen species stabilized from the hybrid *P. frigidus* × *palmatus* in a period when the American and eastern Asiatic ranges of *P. palmatus* were confluent." The material reported from Alaska–Yukon as *P. frigidus* × *sagittatus* by Hultén (1950) is probably based upon

P. vitifolius, A.L. Bogle (Rhodora 70(784):533–51. 1968) presenting convincing evidence that this taxon has arisen through hybridization between *P. (frigidus* var.) *palmatus* and *P. sagittatus*.

PICRIS L.   [9575]   Ox-tongue

1   Phyllaries in 2 rows of about equal length, the outer ones lance-ovate to ovate, to 8 mm broad, the narrow inner ones thickened below; achenes with a slender fragile beak to about 4 mm long; pappus densely plumose; leaves entire, at least the upper ones and the phyllaries spinulose-tipped; plant bristly-hispid .... *P. echioides*
1   Phyllaries all narrow (less than 3 mm broad), in 3 or 4 series of unequal length; achenes beakless or nearly so; pappus sparingly plumose; leaves irregularly toothed; plant subglabrous to spreading-hispid .... *P. hieracioides*

P. echioides L.
Eurasian; locally introd. along roadsides and in fields and waste places in N. America, as in Alta. (Grande Prairie, Peace River dist., 55°10′N; Groh 1947), Sask. (Prince Albert; DAO), Ont. (Wellington, Welland, and Prince Edward counties), sw N.B. (St. Stephen, Charlotte Co., where taken by J. Vroom in 1882; NBM), and ?N.S. (Groh and Frankton 1949*b*).

P. hieracioides L.
Eurasian; locally introd. along roadsides and in fields and waste places in N. America, as in s B.C. (reported from near Victoria, Vancouver Is., by Groh 1947; reported from Wilmer, about 65 mi se of Golden, by Ulke 1935) and Ont. (N to Ottawa; Groh 1947); ssp. *kamtschatica* (Ledeb.) Hult. (*P. kamt.* Ledeb.) possibly native on Attu Is., w Aleutian Is. MAP: Hultén 1968*b*:943.

POLYMNIA L.   [9122]

P. canadensis L.   Leafcup
/t/EE/   (Hp)   Moist calcareous woods, ravines, and bases of cliffs from s Ont. (N to Huron, Halton, and Lincoln counties; *see* s Ont. map by Soper 1962: fig. 23, p. 38) to Vt., s to Okla., La., Tenn., and Ga.
     Forma *radiata* (Gray) Fassett (ray-ligules whitish, to 1 cm long, rather than minute or abortive) is known from Essex Co., s Ont. (East Sister Is. and Middle Sister Is. of the Erie Archipelago; Core 1948).

PRENANTHES L.   [9606]   Rattlesnake-root

1   Lower leaves tapering into winged petioles, the upper ones partly clasping; pappus creamy to light brown; flowers pink; heads ascending; leaves oval to oblanceolate; (B.C. to N.S.) .... *P. racemosa*
1   Lower leaves slender-petioled, with truncate to cordate-hastate bases, usually variously lobed or 3–5-cleft; flowers usually white or creamy; heads drooping or spreading.
    2   Involucres and pedicels commonly with at least a few long coarse hairs; principal phyllaries mostly 8, the short outer ones lanceolate; pappus creamy; leaves thickish, variously lobed and often pinnatifid; (s ?Ont.) .... [*P. serpentaria*]
    2   Involucres and pedicels glabrous or minutely pubescent; outer phyllaries somewhat broader in outline.
        3   Pappus reddish- or cinnamon-brown; flowers whitish, 8 or more in a head; primary phyllaries about 7, glabrous; (Sask. to E Que.) .... *P. alba*
        3   Pappus normally white to sordid or pale brown.
            4   Principal phyllaries 5, glabrous; flowers greenish white, 5 or 6 in a head; (Ont. to N.S.) .... *P. altissima*
            4   Principal phyllaries about 7 or 8 (rarely about 13); flowers 8 or more in a head.
                5   Branches of the inflorescence glabrous; involucres glabrous; lower leaves mostly deeply 3-parted (the divisions occasionally finely dissected); petioles essentially wingless; (Que. eastwards) .... *P. trifoliolata*

5 Branches of the inflorescence copiously soft-puberulent; involucres
glabrous or puberulent; leaves all merely rather remotely undulate-dentate,
on broadly winged petioles; (B.C. and Alta.) . . . . *P. alata*

P. alata (Hook.) Dietr.
/sT/W/ (Hp) Streambanks and other moist, often shaded places from the Aleutian Is. and s
Alaska (N to ca. 61°N) through w B.C. and w Alta. (Waterton Lakes and the Swan Hills, about 110
mi NW of Edmonton; the report from Carberry, sw Man., by R.M. Christy, J. Bot. 25:294. 1887,
requires clarification) to Wash. and Oreg. [*Nabalus alatus* Hook., the type locality given as "Fort
Vancouver and Observatory Inlet, on the North-West coast of America", Fort Vancouver being in
Wash., Observatory Inlet in w B.C. at ca. 55°N; *Sonchus (N.; Mulgedium; P.) hastatus* Less., not *P.
hastata* Thunb.; *P. lessingii* Hult.; *N. boottii sensu* Dawson 1875, not DC.]. MAP: Hultén 1968b:957.
Our Alta. material is referable to var. *sagittata* Gray (*P. (N.) sag.* (Gray) Nels.; inflorescence
narrow and spike-like, the branches all short rather than open-corymbiform, the main branches
relatively long).

P. alba L.   White Lettuce, Rattlesnake-root
/T/EE/ (Hs (Hsr)) Rich woods and thickets from Sask. (N to Battleford, 52°45'N; Breitung 1957a)
to Man. (N to the N end of L. Winnipegosis; CAN), Ont. (N to sw James Bay at 52°11'N; RIM), Que.
(N to L. Mistassini and Rivière-Ouelle, Kamouraska Co.; early reports from the Atlantic Provinces
are chiefly based upon *P. trifoliolata,* relevant collections in several herbaria), and New Eng., s to
S.Dak., Mo., Tenn., and Ga. [*Nabalus* Hook.].

P. altissima L.
/T/EE/ (Hs (Hsr)) Moist woods from Ont. (N to Ottawa; the report from s Man. by Lowe 1943,
taken up by Fernald *in* Gray 1950, is based upon *P. alba,* the relevant collection in WIN) to Que. (N
to Anticosti Is. and the Gaspé Pen.; the report from Nfld. by A.P. de Candolle, *Prodromus
systematis naturalis regni vegetabilis.* Treuttel et Würtz, Paris. Part 7, p. 241. 1838, is probably
referable to *P. trifoliolata*), N.B., P.E.I., (Malpeque, Prince Co.; CAN), and N.S., s to Tenn. and Ga.
[*Nabalus* Hook.; *N. (P.) ?cordatus* Willd.].
The typical form is essentially glabrous, the leaves coarsely toothed to deeply 3–5-parted. Forma
hispidula Fern. (leaves hispidulous beneath; stem villous) is known from Que. and N.S. Forma
integra Rousseau (at least the stem-leaves essentially entire and relatively narrow) is known from
the type locality, L. Wickenden, Anticosti Is., E Que.

P. racemosa Michx.
/sT/X/ (Hs) Moist thickets, meadows, and shores, the aggregate species from E B.C. (Pouce
Coupe in the Peace River dist. at ca. 55°45'N; V) to Alta. (N to Ma-Me-O Beach, ca. 54°N), Sask. (N
to Carlton, about 35 mi sw of Prince Albert), Man. (N to the Hayes R. about 40 mi sw of York
Factory), Ont. (N to w James Bay at ca. 53°N), Que. (N to E James Bay at 53°25'N, L. Mistassini,
and the Côte-Nord; not known from P.E.I.; reports from Nfld. require confirmation), N.B., and N.S.,
s to Mont., Colo., S.Dak., Mo., Ohio, and New Eng.
An apparent hybrid with *P. trifoliolata* (× *P. mainensis* Gray) is known from E Que. (near
Rivière-du-Loup, Temiscouata Co.), N.B. (St. John R. system), N.S. (Sandy Cove, Digby Co.;
NSPM), and N Maine.
1 Phyllaries at most 10; flowers commonly not more than 15 . . . . ssp. *racemosa*
    2 Leaves merely minutely dentate . . . . var. *racemosa*
        3 Flowers pink; [*Nabalus rac.* Hook.; range of the species] . . . . f. *racemosa*
        3 Flowers whitish; [known only from the type locality, Longueuil, near Montreal,
            Que.] . . . . f. *rollandii* Vict. & Rousseau
    2 Leaves more or less lyrate-pinnatifid; [s Ont.: Windsor, Essex Co.; Walpole Is.,
        Lambton Co.] . . . . var. *pinnatifida* Gray
1 Phyllaries 10 or more; flowers up to 25 or more; [reported by Arthur Cronquist, Rhodora
    50(590):30. 1948, from Alta. (type from Beaver Hill L.), ?Sask. (Cronquist citing an old
    Bourgeau collection), Man. (the cited Macoun collection from Morris), and Que. (the cited
    Marie-Victorin collection from L. St. John), and reported from James Bay (Ont. and Que.) .
    by Dutilly, Lepage, and Duman 1954 and 1958] . . . . ssp. *multiflora* Cronq.

[P. serpentaria L.]   Gall-of-the-earth
[The report of this species of the E U.S.A. (N to Ohio and Mass.) from S Ont. by Soper (1949) requires confirmation. Reports from Que. and the Atlantic Provinces are chiefly based upon *P. trifoliolata,* relevant collections in several herbaria. (*Nabalus* Hook.).]

P. trifoliolata (Cass.) Fern.   Gall-of-the-earth
/T/E/   (Hs (Hsr))   Thickets, clearings, and dry slopes from Que. (N to L. St. John, the Côte-Nord, Anticosti Is., and Gaspé Pen.) to S Labrador (Forteau, 51°28′N; GH), Nfld., N.B., P.E.I., and N.S., s to Tenn. and N.C.
1   Pappus cinnamon-brown; phyllaries lead-colour or blackish, the outer ones ovate-lanceolate to ovate; plant rarely over 7 dm tall; [*P. (Nabalus) nana* (Bigel.) Torr.; E Que. (Côte-Nord and Gaspé Pen.), S Labrador (Forteau, 51°28′N; GH), Nfld., and N.S.] . . . . var. *nana* (Bigel.) Fern.
1   Pappus straw-coloured or light brown; phyllaries green or purple-tinged, the lower ones lance-deltoid; plant to over 1.5 m tall . . . . var. *trifoliolata*
    2   Leaves marginally ciliate with reddish hairs to 1 mm long, lightly pubescent beneath; [E Que.: type from Mt-St-Pierre, Gaspé Pen.; Brion Is., Magdalen Is.] . . . . f. *ciliata* Vict. & Rousseau
    2   Leaves glabrous; [*Nabalus* Cass.; essentially the range of the species but less northern than var. *nana*] . . . . f. *trifoliolata*

PSILOCARPHUS Nutt.   [8965]

1   Receptacular bracts averaging about 3 mm long (up to 3.8 mm) at maturity; achenes to 1.7 mm long; leaves mostly linear-oblong, to about 3.5 cm long and 6 mm broad, up to 9 times as long as broad; plant more or less silky-tomentose, the pubescence moderately loose and not very dense; (Vancouver Is. and S Alta.) . . . . *P. elatior*
1   Receptacular bracts averaging about 2 mm long (at most 2.7 mm) at maturity; achenes to 1.2 mm long.
    2   Leaves linear to linear-lanceolate, to 2 cm long and 3 mm broad, up to 12 times as long as broad; achenes narrowly oblong or elliptic-oblong, broadest near the middle; tomentum usually fine, short and close or occasionally somewhat loose, silvery, generally persistent . . . . [*P. oregonus*]
    2   Leaves spatulate, oblanceolate, or oblong, to 1.5 cm long and 5 mm broad, mostly not over 6 times as long as broad; achenes broadly oblanceolate to narrowly obovate, broadest above the middle; tomentum generally thin and rather loose, often partly deciduous; (sw B.C.) . . . . *P. tenellus*

P. elatior Gray
/T/W/   (T)   Open moist places and dried beds of vernal pools from sw B.C. (S Vancouver Is.; several localities, first taken by John Macoun in 1887; CAN) and SE Alta. (Redcliff, near Medicine Hat; CAN) to Oreg. and Idaho. [*P. oregonus* var. *el.* Gray]. MAP: Arthur Cronquist, Res. Stud. Wash. State Univ. 18(2): map 2, p. 82. 1950 (the SE Alta. station should be indicated).

[*P. oregonus* Nutt.]
[The report of this species of the W U.S.A. (Wash. and Idaho to Calif.) from S Vancouver Is., B.C., by Carter and Newcombe (1921; taken up by Eastham 1947) is probably based upon *P. elatior.*]

P. tenellus Nutt.
/t/W/   (T)   Open moist places and dried beds of vernal pools from sw B.C. (S Vancouver Is.; several localities, first taken by John Macoun in 1887; CAN) to N Baja Calif. MAP: Arthur Cronquist, Res. Stud. Wash. State Univ. 18(2): map 4, p. 87. 1950.

RATIBIDA Raf.   [9178]   Prairie-Coneflower

1   Disk columnar, to 4 cm long, often longer than the yellow or partly or wholly purplish-brown rays; pappus an awn-tooth at the summit of the inner angle of the ciliate achene

(often also a smaller tooth on the outer angle); leaf-segments linear to lanceolate, essentially entire; perennial from a taproot and a short caudex; (B.C. to s Man.; introd. in s Ont.) .... *R. columnifera*
1   Disk ellipsoid, at most 2 cm long, much shorter than the uniformly pale-yellow rays; achenes smooth, lacking a pappus; leaf-segments lanceolate, entire to coarsely serrate; perennial from a stout woody rhizome or sometimes a short caudex; (s Ont.) ...... *R. pinnata*

R. columnifera (Nutt.) Wooton & Standl.
/T/WW/   (Hs)   Dry plains, prairies, and ravines from SE B.C. (Keremeos; Kootenay; between Cranbrook and Wardner; Fairmont Hot Springs) to s Alta. (N to Red Deer), Sask. (N to Saskatoon), and s Man. (N to Millwood, about 85 mi NW of Brandon; occasionally introd. eastwards, as in Ont.: Essex, Lambton, York, Lennox-Addington, and Carleton counties), s to Calif., Mexico, Tex., Ark., Mo., Ill., and Minn. [*Rudbeckia* Nutt.; *Lepachys* Macbr.; *Rud. columnaris* Pursh]. MAP: E.L. Richards, Rhodora 70(783): fig. 9, p. 389. 1968.
Forma *denudata* Boivin (ray-ligules wanting) is known from Medicine Hat, Alta., and the type locality, Val Marie, sw Sask. Forma *pulcherrima* (DC.) Fern. (the ray-ligules partly or wholly purplish-brown rather than uniformly yellow) is found essentially throughout the area.

R. pinnata (Vent.) Barnh.
/t/EE/   (Hs)   Prairies, thickets, and dry open woods from Nebr. to Minn., s Ont. (Essex, Kent, and Lambton counties; CAN; OAC), and N.Y., s to Okla., Ark., and Ga. [*Rudbeckia* Vent.; *Lepachys* T. & G.]. MAP: E.L. Richards, Rhodora 70(783): fig. 10, p. 390. 1968.

RUDBECKIA L.   [9178]   Coneflower

1   Leaves entire or only shallowly toothed, coarsely hirsute; disk purple or brown-purple, its corollas with spreading lobes; (introd.).
     2   Leaves narrowly lanceolate or oblanceolate, mostly 3-ribbed; pappus none; stigmas slender-subulate; receptacular chaff more or less hispid near the acute summit; (introd., transcontinental) .... *R. hirta*
     2   Leaves oval or ovate, 3–5-ribbed; pappus a short crown; stigmas short and blunt; receptacular chaff minutely pubescent near the blunt summit; (introd. in s Ont.) .... [*R. grandiflora*]
1   Leaves (at least the lower ones) deeply lobed; disk-corollas with ascending lobes; pappus a short crown; stigmas short and blunt.
     3   Leaves pinnately 5–7-cut or 3-lobed, nearly glabrous on both faces or more or less strigose or hirsute beneath; disk dull greenish-yellow or greyish; receptacular chaff viscid-puberulent near the blunt summit; stem glabrous; (Man. to N.S.) .... *R. laciniata*
     3   Leaves (at least the lower) 3-lobed or -parted, sometimes pinnately 5–7-parted, subglabrous or sparingly strigose; disk black-purple; receptacular chaff glabrous, abruptly mucronate; stem subglabrous or somewhat hirsute; (introd. in Ont. and sw Que.) .... *R. triloba*

[*R. grandiflora* (Sweet) DC.]
[Native in the E U.S.A. from Iowa and Mo. to Tex. and Okla.; introd in s Ont. (woods, Norfolk Co.; TRT), where, however, probably not established. (*Centrocarpha* Sweet).]

R. hirta L.   Black-eyed Susan.   Marguerite jaune
Apparently native in the Great Plains area of the cent. U.S.A. and in open woods and thickets of the E U.S.A.; introd. elsewhere, as in B.C. (N to Golden), Alta. (N to Chip L., about 60 mi w of Edmonton), Sask. (N to Hudson Bay Junction, 52°52′N), Man. (N to Cross Lake, NE of L. Winnipeg), Ont. (N to Cochrane, ca. 49°N), Que. (N to L. St. John and Anticosti Is.), Nfld., N.B., P.E.I., and N.S.
Our material is chiefly or wholly referable to var. *pulcherrima* Farw. (*R. lanceolata* Bisch.; *R. serotina* Nutt.; leaves entire or finely serrate rather than coarsely toothed, the blades of the basal ones commonly 4 or 5 (rather than about twice) as long as broad, the stem-leaves also relatively narrow; *see* R.E. Perdue, Jr., Rhodora 59(708):293–96. 1957). Forma *homochroma* Steyerm. (disk greenish yellow rather than dark purple or brown; ray-florets greenish yellow rather than orange) is

reported from Cap Jaseux, on the Saguenay R., Que., by R. Cayouette (Que. Minist. Agric. Serv. Rech. Enseignment Contrib. 107. 1970.).

R. laciniata L.
/T/X/ ((Hpr (Hs)) Streambanks and moist places from Mont. to s Man. (N to Dauphin, N of Riding Mt.; CAN; reports of *R. ampla* from Sask. by Rydberg 1922 and 1932, require confirmation), Ont. (N to an uncertain limit; introd. at Kapuskasing and Ottawa), Que. (N to the Montreal dist.; perhaps introd. N to the Quebec City dist.), and N.S. (apparently native at the edge of a thicket along the Black R., Kings Co., the type locality of the pubescent but probably completely intergrading extreme, var. *gaspereauensis* Fern.; a garden-escape at Barrington, Shelburne Co., as also near Grand Falls and St. John, N.B., and Inverness, Prince Co., P.E.I.), s to Ariz., Tex., and Fla. [Incl. var. *gaspereauensis* Fern. and *R. ampla* Nels.].

The "double-flowered" var. *hortensis* Bailey ("golden glow"; most or all of the disk-florets with long yellow rays, the head ligulate rather than radiate) is reported as a garden-escape in s Ont. by Gaiser and Moore (1966; Lambton Co.) and from Que., N.B., and ?P.E.I. by Boivin (1966*b*).

R. triloba L.   Brown-eyed Susan
Native in the E U.S.A. from Minn. to N.Y., s to Okla., Tenn., and Ga.; introd. or a garden-escape elsewhere, as in Ont. (N to Ottawa) and sw Que. (Chambly, near Montreal; MT).

SAUSSUREA L.   [9457]

1   Phyllaries in 3 or 4 series of markedly unequal length, the lowest ones ovate, the upper ones lanceolate; upper leaves rarely surpassing the corymbiform inflorescence.
  2   Lower leaves triangular-ovate to -cordate, sharply toothed, the blade to about 1.5 dm long and half as broad (upper leaves more lanceolate and becoming sessile), their lower surfaces sometimes persistently cobwebby-woolly; receptacle naked or more commonly with a few long bristly hairs toward the centre; coarse fibrous-rooted perennial, the usually several stems to over 1 m tall; (SE Alaska-B.C.) ...... *S. americana*
  2   Leaves all alike, from linear and entire to lanceolate or elliptic-lanceolate and prominently though remotely toothed, they and the involucres densely hairy to glabrate; receptacle bristly; stems to 4 or 5 dm tall, the plant perennial by a dark elongated rhizome; (western arctic and subarctic regions) .... *S. angustifolia*
1   Phyllaries subequal or in at most 2 or 3 series of less markedly unequal length, nearly all lance-acuminate.
  3   Receptacle naked or merely with a few bristly hairs toward centre; plants thinly arachnoid-villous especially about the inflorescence, sometimes becoming more or less glabrate; (Alaska and mts. of B.C. and Alta.) .... *S. nuda*
  3   Receptacle copiously beset with long hyaline bristly hairs.
    4   Plant long-villous with entangled, multicellular, viscid-glandular hairs, commonly not over 1 dm tall, the crowded heads often surpassed by the upper leaves; (Alaska–Yukon–w Dist. Mackenzie) .... *S. viscida*
    4   Plant copiously arachnoid-tomentose but not viscid, commonly over 2 dm tall, the upper leaves rarely surpassing the inflorescence .... [*S. tilesii*]

S. americana Eat.
/sT/W/   (Hp) Moist meadows and rocky slopes at moderate to high elevations from SE Alaska (N to ca. 60°N; probably also in s Yukon, Hultén's map indicating a station on the Yukon–B.C. boundary) through B.C. (collections in CAN, detd. Porsild and Cronquist, from the Haines Road at ca. 59°30′N and Columbia L., sw of Creston) to Oreg. and Idaho. MAP: Hultén 1968*b*:936.

S. angustifolia (Willd.) DC.
/aSs/WW/eA/   (Hs) Dry tundra and mts. up to high elevations from the coasts of Alaska–Yukon–Dist. Mackenzie (E to Coronation Gulf) to Dist. Keewatin (N to near the Arctic Circle), s to s Alaska–Yukon–Dist. Mackenzie–Dist. Keewatin; E Siberia. [*Serratula* Willd.; *Sau. monticola* and *S. ?multiflora* Richards., not *Sau. mult.* DC.]. MAPS Hultén 1968*b*:936; Porsild 1966: map 154 (solid dots only), p. 86; *Atlas of Canada* 1957: map 6, sheet 38.

The high-grown robust extreme may be distinguished as f. *ramosa* Jord. (type from Fairbanks, Alaska).

S. nuda Ledeb.
/ST/W/eA/ (Hs) Seashores, alpine meadows, and rocky slopes in the mts. (confined in N. America to Alaska–B.C.–Alta., the ranges of taxa outlined below); E Asia. MAP and synonymy: *see* below.
1 Heads generally long-peduncled, rarely surpassed by the leaves, these usually rather obscurely toothed; stems to 4 dm tall; [*S. alpina* vars. *ledebourii* (Herder) Gray (*S. led.* Herder) and *remotifolia* Hook. (*S. "remotiflora"* (Hook.) Rydb., a puzzling report of this from Sask. by Rydberg 1922); *S. led.* vars. *nuda* and *subsinuata* (Ledeb.) Herder (*S. sub.* Ledeb.); Alaska (N to ca. 69°N); MAP (aggregate species): Hultén 1968*b*:938] . . . .
. . . . var. *nuda*
1 Heads usually all crowded in a capitate cluster, often surpassed by the upper leaves; leaves generally more strongly toothed; stems commonly less than 2 dm tall; [*S. alpina* var. *densa* Hook. (*S. densa* (Hook.) Rydb.); mts. of S B.C. (Mt. Benson, Vancouver Is.; Paradise Mt., near Windermere; Mt. Assiniboine, near Field; Kicking Horse L.) and SW Alta. (Crowsnest Pass; Banff; Canmore, near Banff; L. Louise; Jasper); the type is a Drummond collection from "Elevated parts of the Rocky Mountains", presumably in Alta.]. *S. amara* (L.) DC. (*S. glomerata* Poir.; a Siberian species habitally similar to *S. nuda* but the inner phyllaries terminated by a broad scarious appendage rather than unappendaged) is reported as introd. at Debold, near Grande Prairie, Alta., by Groh (1944*b*), where, however, probably not established . . . . var. *densa* Hook.

[*S. tilesii* Ledeb.]
[This N Asiatic species extends as far eastwards as Bering Strait and Bering Is., W of the W Aleutian Is., but has not as yet been found in N. America. MAP: Hultén 1968*b*:938.]

S. viscida Hult.
/aSs/W/eA/ (Hs) Dry tundra and up to fairly high elevations in the mts. from the coasts of Alaska–Yukon to S-cent. Alaska, S Yukon, and w-cent. Dist. Mackenzie (between ca. 63°15′ and 65°N); NE Siberia. MAP: combine the maps by Hultén 1968*b*:937 (var. *visc.* and var. *yuk.*).
All our material except for a single collection from the W tip of the Seward Pen., Alaska, is referred by Hultén (1968*b*) to var. *yukonensis* (Porsild) Hult. (*S. angustifolia* var. *yuk.* Porsild, the type from Bolstead Creek, w-cent. Dist. Mackenzie; *S. densa sensu* Anderson Bakewell, Rhodora 45(536):316. 1943, not (Hook.) Rydb.; leaves sparingly floccose and viscid-pubescent rather than copiously so), this considered as possibly a hybrid between *S. angustifolia* and *S. viscida* by Hultén.

SENECIO L. [9411] Groundsel, Ragwort, Squaw-weed. Séneçon

(Ref.: Greenman 1916)
1 Stems more or less equally leafy throughout, the upper leaves only gradually reduced, no well-developed tuft of basal leaves present.
  2 Leaves entire or merely coarsely toothed to shallowly 1-pinnatifid; ray-ligules normally conspicuous.
    3 Stems commonly about 1 dm tall (or sometimes longer and sprawling), from a branching woody caudex surmounting a taproot; leaves thickish, spatulate or oblanceolate to obovate, irregularly few-toothed, to about 4 cm long and 2 cm broad, tapering to short wing-margined petioles; involucre to 12 mm high, with about 13 phyllaries; rays to 1 cm long; achenes puberulent; (mts. of S B.C. and SW Alta.) . . . . *S. fremontii*
    3 Stems taller; leaves mostly over 4 cm long; achenes glabrous.
      4 Heads very large (the disk to 4 cm broad, the involucre to about 1.5 cm high), usually not more than 5; rays to 2 cm long; leaves oblanceolate to obovate, subentire to coarsely toothed, fleshy, lustrous-green above, white-felted beneath, the lower ones early deciduous; stem more or less white-woolly;

perennial from a deep vertical rhizome; (sandy and gravelly Pacific and
Atlantic coasts) .... *S. pseudo-arnica*
4　Heads smaller (the disk less than 1.5 cm broad, the involucre and rays
commonly not over 1 cm long), several; leaves not white-felted beneath.
5　Plant pubescent at least in the inflorescence; leaves thickish, undulate or
rather irregularly dentate to shallowly pinnatifid, the radical ones petioled
and commonly early deciduous, all except sometimes the lowermost of the
stem-leaves sessile by a more or less clasping base; stems stout, soft and
easily flattened, rarely over 8 dm tall; annual or biennial; (transcontinental
in aquatic or marshy habitats) .... *S. congestus*
5　Plants nearly or quite glabrous; leaves thinner and shining, more regularly
and shallowly toothed, only the upper ones sessile and more or less
clasping; rhizomatous perennials.
6　Phyllaries not dark-tipped; heads several to many; leaves narrowly to
broadly triangular (commonly with nearly straight lateral margins),
broadly truncate to sagittate-cordate at base, rather coarsely toothed;
stems to over 1.5 m tall; (s Alaska, s Yukon, Dist. Mackenzie, and mts.
of B.C. and Alta.) .... *S. triangularis*
6　Phyllaries conspicuously dark-tipped; heads 3 or 4; leaves narrowly to
broadly lanceolate (with distinctly convex margins), gradually tapering
to a cuneate base, more shallowly toothed; stems to about 4 dm tall;
(mts. of s Yukon and N B.C.) .... *S. sheldonensis*
2　Leaves (at least the principal ones) deeply 1–3-pinnatifid.
7　Ray-ligules none or inconspicuous; annuals; (introd.).
8　Bractlets at base of involucre about half as long as the glandular-hispid
phyllaries; achenes glabrous; rays minute, recurving; whole plant heavily
glandular and strong-smelling .... *S. viscosus*
8　Bractlets minute; achenes short-strigose, particularly on the angles.
9　Bractlets and tips of glabrous phyllaries (about 20) blackish; rays none;
plant not viscid .... *S. vulgaris*
9　Bractlets and minutely pubescent phyllaries (about 12) not black-tipped;
rays minute, recurving; plant more or less glandular-viscid .... *S. sylvaticus*
7　Ray-ligules conspicuous.
10　Leaves deeply and rather uniformly 2–3-pinnatifid, cobwebby beneath when
young (soon glabrate); branches of inflorescence cobwebby; phyllaries
rhombic-oblanceolate, broadest above the middle; marginal achenes glabrous,
those of the disk minutely pubescent; stems tough, to about 12 dm tall; (introd.
in B.C. and from Ont. to Nfld. and N.S.) .... *S. jacobaea*
10　Leaves 1-pinnatifid into relatively broad lobes, the lobes themselves toothed to
incised; phyllaries mostly lanceolate, broadest at or below the middle; plants
glabrous or sparingly and minutely pubescent.
11　Principal leaves subpalmately divided to near the base into 3 sharply and
doubly serrate broad segments; achenes glabrous; (Aleutian Is.) ....
.... *S. cannabifolius*
11　Principal leaves pinnately lobed.
12　Rays shallowly 3-lobed at apex; achenes glabrous or sparingly and
minutely pubescent; perennial with an erect stem to over 8 dm tall;
(B.C. to w Ont.) .... *S. eremophilus*
12　Rays entire; achenes minutely pubescent on the ribs; annual (rarely
biennial or perennial) with a flexuous stem to about 3 dm tall,
decumbent at base; (introd. in N.B. and N.S.) .... *S. squalidus*
1　Stems abundantly leafy only toward base, the upper leaves greatly reduced; perennials.
13　Plants usually distinctly pubescent at anthesis; heads usually radiate and conspicuous
(often essentially discoid in *S. yukonensis* and *S. werneriaefolius* and atypically so in
other species).
14　Involucres copiously pilose or woolly; heads often solitary, at most 5 or 6,
relatively large, the involucres to 16 mm high, the disks to over 1.5 cm broad

(except in *S. yukonensis*); leaves entire to minutely callous-denticulate or shallowly undulate.

15 Phyllaries and summit of peduncle moderately pilose with multicellular hairs with purple cross-walls; achenes glabrous; heads often solitary; leaves rarely over 4 cm long (including the petiole); stems to about 2 dm tall; (western arctic and subarctic regions) . . . . *S. atropurpureus*

15 Phyllaries and summit of peduncle usually rather copiously floccose-lannate with whitish or yellowish entangled hairs; stems commonly taller.

16 Achenes strigose-hirsute, about 3 mm long; rays to 2.5 cm long; phyllaries white-lanate; heads 1–5; leaves floccose on both surfaces, to about 1.5 dm long (including petiole) and 2 cm broad; (Alaska to Dist. Mackenzie) . . . .
. . . . *S. lindstroemii*

16 Achenes glabrous; leaves commonly green and glabrate above, whitish-tomentose or floccose beneath.

17 Pubescence of involucre and summit of peduncle white; heads commonly solitary, large (the disk to 2.5 cm broad, the involucre to 16 mm high), the usually showy rays to 2 cm long; leaves to 2 dm long (including the petiole) and 3 cm broad; (mts. of s B.C. and s Alta.) . . . .
. . . . *S. megacephalus*

17 Pubescence of involucre and summit of peduncle distinctly yellowish; heads commonly 2 or more, somewhat smaller (the pale-yellow rays small or the head essentially discoid); leaves to about 1 dm long (including the petiole) and 1.5 cm broad; (Alaska–Yukon–w Dist. Mackenzie) . . . . *S. yukonensis*

14 Involucres glabrous or merely more or less puberulent (but usually then glabrate); heads commonly more numerous and smaller (involucre to about 1 cm high, disk to about 1 cm broad); achenes nearly or quite glabrous.

18 Pubescence loosely crisp-villous or cobwebby (sometimes very sparse by anthesis); leaves entire to irregularly dentate, to 2.5 dm long (including the petiole) and 6 cm broad; rays 6–15 mm long; stem stout, fibrous-rooted from a very short erect crown, to about 7 dm tall; (B.C. to Man.) . . . . *S. integerrimus*

18 Pubescence finer, more tomentose or floccose (or at first cobwebby in *S. elmeri* but thinly so or even wanting at anthesis).

19 Leaves callous-denticulate to sharply dentate (the teeth nearly horizontally divergent), narrowly to broadly oblanceolate, to about 2 dm long.

20 Involucre to 8 mm high, the conspicuously blackened tips of the phyllaries glabrous; stems to 5 dm tall, mostly solitary from a short rhizome, thinly tomentulose when young, generally subglabrate by anthesis; (var. *lugens*; mts. of B.C. and Alta., chiefly in wet alpine meadows) . . . . *S. integerrimus*

20 Involucre to 12 mm high, the often dark or blackish tips of the phyllaries minutely villous; stems to 3 dm tall, terminating the branches of a well-developed woody caudex, more or less cobwebby when young but thinly so or even glabrate at anthesis; (s B.C.) . . . . *S. elmeri*

19 Leaves entire to crenate or crenate-serrate (the teeth directed forward), or more or less lobed or pinnatifid.

21 Plants usually less than 1.5 dm tall, thinly tomentulose (becoming more or less glabrate), the several lax scapose stems terminating a branching woody caudex; leaves spatulate to elliptic or rotund-obovate, to 2.5 cm long and 1.5 cm broad; heads 1–6, commonly rather long-peduncled, radiate or discoid; (s ?B.C.) . . . . [*S. werneriaefolius*]

21 Plants otherwise.

22 Plant usually less than 4 dm tall, more or less strongly white-tomentose (often less so in age); leaves narrowly oblanceo-late to broadly elliptic or narrowly ovate, the blades less than 5 cm long; (B.C. to Man.) . . . . *S. canus*

22 Plant to 7 dm tall, the tomentum relatively thin and obscure, the leaves generally narrower; (s B.C.) .... *S. macounii*

13 Plants mostly glabrous from the first (if more or less tomentose when young, glabrate by anthesis except for sparse inconspicuous tomentum at the base of the stem and in the leaf-axils; *S. elmeri, S. integerrimus,* and *S. werneriaefolius* may key out here); achenes usually nearly or quite glabrous (copiously strigose in *S. tridenticulatus*).

23 Leaves entire or dentately toothed or lobed (the teeth or lobes nearly horizontally divergent).

24 Leaves coarsely dentate or lobed to rather deeply pinnatifid, their blades usually less than 1 dm long, commonly cordate-rotund to reniform and up to 1.5 times as broad as long; phyllaries not conspicuously black-tipped; rays usually relatively numerous.

25 Heads solitary; leaves to about 1.5 cm long and 2 cm broad, deeply 3–7-lobed to about 1/3 of their width; stems to about 2 dm tall; (Queen Charlotte Is., B.C.) .... *S. newcombei*

25 Heads commonly rather numerous; leaves to about 1 dm long, merely more or less deeply dentate; stems to about 8 dm tall; (Man. to Nfld. and N.S.) .... *S. aureus*

24 Leaves entire to sharply salient-dentate, the basal ones elliptic or oblanceo-late, cuneate-based (gradually tapering to a long winged petiole), the blade to over 2 dm long and about 7 cm broad, none of the leaves at all pinnatifid; phyllaries commonly distinctly black-tipped; rays few or wanting; (B.C. and Alta.).

26 Leaves generally entire (occasionally irregularly denticulate); plant more or less glaucous, the robust stem to about 2 m tall; (s B.C.) .... *S. hydrophilus*

26 Leaves generally dentate (rarely subentire); plants scarcely glaucous, to about 1 m tall; (s B.C. and sw Alta.) .... *S. foetidus*

23 Leaves crenately or serrately toothed or lobed (the teeth directed forward; species of the preceding contrasting lead 23 may key out here when the teeth are few and chiefly apical), at least the middle and lower stem-leaves commonly deeply pinnatifid.

27 Basal leaves mostly distinctly subcordate to cordate at base; inflorescence corymbiform; rays typically present; achenes glabrous.

28 Basal leaves lanceolate to narrowly ovate, commonly 2 or 3 times as long as broad, acutish at summit; rays generally pale yellow; (Que. to the Maritime Provinces) .... *S. robbinsii*

28 Basal leaves mostly broadly ovate to rotund or reniform, about as broad as or broader than long, obtusish or rounded at summit.

29 Rhizomes slender and horizontally creeping, to about 3 dm long, sending up scattered flowering stems or tufts of leaves; basal leaves ovate to rotund, deeply cordate, on slender petioles to 2.5 dm long; heads deep yellow, on peduncles to over 1 cm long; (Man. to Nfld. and N.S.) .... *S. aureus*

29 Rhizomes stout and relatively short, usually forking into a tuft of crowded crowns; basal leaves commonly oblong-ovate in outline, cuneate to somewhat cordate at base, on thickish petioles commonly not much over 1 dm long; heads pale yellow, on peduncles usually not over 6 or 7 mm long; (B.C. to Man.) .... *S. pseudaureus*

27 Basal leaves cuneate to rounded or truncate at base (rarely subcordate to reniform).

30 Achenes pubescent; inflorescence corymbiform; rays pale yellow; leaves narrowly oblanceolate in outline (some of the basal ones occasionally entire); stems several from a caespitose branching caudex; (sw Dist. Mackenzie to s Sask.–Man.) .... *S. tridenticulatus*

30 Achenes typically glabrous; stems often solitary.

31 Heads solitary (rarely 2); leaves thickish, the blades to about 2.5 cm

long and broad (or broader); stems to about 3 dm tall; (Alaska to w
Dist. Mackenzie and the mts. of B.C. and Alta.; E Que. and Nfld.) . . . .
. . . . *S. resedifolius*

31 Heads normally 2 or more in a usually corymbiform inflorescence;
stems often taller.
  32 Heads typically discoid, the rays wanting or inconspicuous; basal
    leaves elliptic to rotund or somewhat reniform; (essentially
    transcontinental).
    33 Phyllaries narrowly linear, at most 1 mm broad, green or with
      purple tips; disk pale yellow or lemon-colour; heads commonly
      more than 6; achenes drab or grey-brown; denuded receptacle
      jagged or fringed around the achene-pits; leaves relatively thin,
      the basal ones mostly elliptic to broadly ovate, cuneate to
      subtruncate at base, dentate or their bases often lacerate . . . .
      . . . . *S. indecorus*
    33 Phyllaries linear-oblong, to 2 mm broad, usually purplish at
      least toward the tips; disk deep orange to orange-red; heads
      rarely more than 5; achenes red to dark red-brown; denuded
      receptacle smooth; leaves firm and rather fleshy, the basal
      ones mostly ovate or obovate to cordate or reniform, coarsely
      dentate . . . . *S. pauciflorus*
  32 Heads typically radiate, the rays conspicuous.
    34 Leaves rather fleshy; phyllaries linear-lanceolate or -oblong, to
      2 mm broad; denuded receptacle jagged or fringed around the
      pits of the reddish-brown achenes; basal leaves obovate or
      cuneate-obovate; (Alaska–B.C. to Sask.) . . . . *S. streptanthifolius*
    34 Leaves relatively thin; phyllaries linear, at most 1 mm broad;
      denuded receptacle smooth.
      35 Petioles of basal leaves broadened upwardly and gradually
        merging with the obovate to subrotund, often cuneate-
        based crenate blade; phyllaries abruptly narrowed above
        middle; filiform elongate stolons early developed, these
        terminated by new rosettes; (s Ont.) . . . . *S. obovatus*
      35 Petioles slender to near summit and merging abruptly with
        the lanceolate to oblanceolate or obovate blade; phyllaries
        gradually tapering to apex; filiform stolons not developed;
        (transcontinental) . . . . *S. pauperculus*

S. atropurpureus (Ledeb.) Fedtsch.
/aSs/WW/A/ (Hsr) Moist tundra and in the mts. up to fairly high elevations, the aggregate
species from the coasts of Alaska–Yukon–Dist. Mackenzie to N Banks Is. and SE Victoria Is., S to S
Alaska–Yukon–w Dist. Mackenzie; Asia. MAPS and synonymy: *see* below.
1 Rootstock elongate and filiform; rosette-leaves wanting; phyllaries usually green (but their
  pubescence often purplish); [*Cineraria frigida* Richards., the type from "Barren grounds
  from Point Lake to the Arctic Sea"; *S. frigidus* (Richards.) Less. and its f. *schraderi*
  Greenm. and var. *ulmeri* Steffen; N. American range of the species; MAP: Hultén
  1968b:928] . . . . ssp. *frigidus* (Richards.) Hult.
1 Rootstock stout; basal rosette well developed; phyllaries purplish (as also the purplish
  indument).
  2 At least the basal leaves remotely but distinctly sharp-toothed . . . .
  . . . . ssp. *tomentosus* (Kjellm.) Hult.
    3 Involucre densely tomentose; [*Cineraria frigida* f. *tomentosa* Kjellm., the type from
      St. Lawrence Is., Alaska; *S. kjellmanii* Porsild; Alaska–Yukon–w Dist. Mackenzie;
      MAPS: Hultén 1950: map 1192b, p. 1682, and 1968b:928] . . . . var. *tomentosus*
    3 Involucre rather thinly tomentose; [*S. frigidus* var. *dent.* Gray; Alaska–SW Yukon;
      MAP: Hultén 1950: map 1192c, p. 1682] . . . . var. *dentatus* (Gray) Hult.

2 Leaves entire or nearly so; [*Cineraria atrop.* Ledeb.; *C. (S.) ?integrifolia sensu* Richardson 1823, and Hooker 1834, not L.; NW Alaska, reports from elsewhere in N. America referring to the above taxa; MAPS (the first two of the aggregate species): Porsild 1957: map 324, p. 201; *Atlas of Canada* 1957: map 6, sheet 38; Hultén 1968*b*:927] . . . . ssp. *atropurpureus*

S. aureus L.
/sT/EE/ (Hsr) Damp thickets and woods, meadows, swampy places, and shores (often in calcareous soils; ranges of Canadian taxa outlined below), s to S.Dak., Mo., Ark., Ala., and Fla.
1 Basal leaves mostly sharply toothed or shallowly lacerate, at least toward base . . . .
. . . . var. *aquilonius* Fern.
   2 Ray-ligules present; [Man. (Löve and Bernard 1959) to Ont. (N to Big Trout L. at ca. 53°45′N), Que. (N to E James Bay at 54°25′N and the Côte-Nord; type from below Tabletop Mt., Gaspé Co.), Nfld., N.B., P.E.I., and N.S.] . . . . f. *aquilonius*
   2 Ray-ligules wanting, the heads discoid; [*S. pseudaureus* f. *ecor.* Fern., the type from along the Madeleine R., Gaspé Co., E Que.] . . . . f. *ecoronatus* Fern.
1 Basal leaves blunt-toothed.
   3 Basal leaves oblong-oval, rounded to base or some of them barely subcordate; [*S. semicordatus* M. & B.; Ont. (Fernald *in* Gray 1950) to Que. (N to E James Bay at ca. 51°30′N, L. Mistassini, and Anticosti Is.), Nfld., N.B., and N.S.] . . . .
. . . . var. *semicordatus* (Mackenz. & Bush) Greenm.
   3 Basal leaves broadly ovate to rotund, distinctly cordate-based.
      4 Plant essentially glabrous from the first or only sparsely short-tomentose and becoming glabrate; involucres commonly not over 8 mm high; basal offshoots rarely purple, at most 5 mm thick.
         5 Stems slender, mostly solitary; basal leaves thinnish, less than 4 cm long; phyllaries at most 7 mm long; [*S. gracilis* Pursh; s Man. (Löve and Bernard 1959), s Ont., and Que. (N to the Nottaway R. at 51°18′N and L. Mistassini)] . . . .
. . . . var. *gracilis* (Pursh) Wood
         5 Stems stouter, often 2 or more; basal leaves firm, to 1.5 dm long; phyllaries to 9 mm long; [Ont. (N to s James Bay at 51°16′N) to Que. (N to E James Bay at 53°54′N, L. Mistassini, and the Côte-Nord), N.B., P.E.I., and N.S.] . . . .
. . . . var. *intercursus* Fern.
      4 Plant copiously long-floccose when young; involucres to 11 mm high; basal offshoots purplish, to 1 cm thick at anthesis; [U.S.A. only, reports from Canada referring to the above taxa] . . . . [var. *aureus*]

S. cannabifolius Less.
/sT/W/eA/ (Hp) Meadows on Attu Is., w Aleutian Is. (a report from Sitka, SE Alaska, is considered erroneous by Hultén 1950); E Asia. [*S. palmatus* (Pall.) Ledeb., not Less. nor La Peyr.]. MAP: Hultén 1968*b*:932.

S. canus Hook.   Silvery Groundsel
/T/WW/ (Hs) Dry, often rocky places from the plains and foothills to fairly high elevations in the mts. from B.C. (N to Lac la Hache, about 35 mi SE of Williams Lake) to Alta. (N to Banff; CAN), Sask. (N to N of Prince Albert; CAN), and Man. (N to Grand Rapids, near the NW end of L. Winnipeg), s to Calif., Colo., and Nebr. The type is a Drummond collection from "Banks of the Saskatchawan". [Incl. var. *acraeus* Greene; *S. purshianus* Nutt.]. MAP: G.L. Stebbins, Madroño 6(8): fig. 2, p. 247. 1942.
   The tentative report from Ont. by Boivin (1966*b*) is probably based upon that by F.H. Montgomery (Can. Field-Nat. 62(2):91. 1948; Cambridge (Galt), Waterloo Co.), the relevant collections in CAN, however, taken by W. Herriot in 1905, 1907, and 1920, being referred to *S. pauperculus* var. *thompsoniensis* by Bernard Boivin.

S. congestus (R. Br.) DC.   Marsh-Fleabane
/AST/X/EA/ (Hs) Fresh, saline, or alkaline marshes, shores, and margins of ponds, the aggregate species from the coasts of Alaska–Yukon–Dist. Mackenzie–Dist. Keewatin to Prince

Patrick Is., Melville Is., N Baffin Is., and northernmost Que., S to S B.C.–Alta.–Sask.–Man., N.Dak., Iowa, Wisc., Ont. (S to the N shore of L. Superior and SW James Bay), Que. (S to SE James Bay, L. Mistassini, and the Côte-Nord; not known from the Maritime Provinces), and S Labrador (N to Indian Harbour, 54°27'N; CAN); Eurasia. MAPS and synonymy: *see* below.

1 Inflorescence more or less villous-lanate.

    2 Corymb dense, copiously woolly with long multicellular hairs; stem rarely over 3 dm tall; [*Cineraria congesta* R. Br., the type from Melville Is.; *S. palustris* var. *congestus* (R. Br.) Hook. and its f. *polycricos* Polunin; transcontinental; MAPS (aggregate species): Porsild 1957: map 325, p. 201; Hultén 1968b:926] . . . . var. *congestus*

    2 Corymb more open and less villous; stem to about 8 dm tall; [*Cineraria palustris* L.; *S. palustris* (L.) Hook., not Velloso; *S. ?kalmii (Cineraria ?canadensis* L.) *sensu* Hooker 1834, not Nutt.; the common representative southwards] . . . . var. *palustris* (L.) Fern.

1 Inflorescence merely short-pubescent or with only a few long hairs, relatively open; [B.C. (Vancouver Is.), Alta. (N to Wood Buffalo National Park at 59°31'N), and Man. (N to Churchill)] . . . . var. *tonsus* Fern.

S. elmeri Piper

/T/W/ (Hs) Cliffs and talus slopes from SW B.C. (valleys of the Chilliwack, Skagit, Fraser, and Thompson rivers N to the Marble Range, NW of Clinton, and Tranquille L., near Kamloops; CAN; V) to Wash. [*S. crepidineus* Greene].

S. eremophilus Richards.

/T/WW/ (Hp) Moist or wet ground, open woods, and thickets from B.C. (N to Prince George and Dawson Creek, ca. 55°40'N; introd. along roadsides N to Muncho L. at ca. 58°N) to Alta. (native southwards; probably introd. N to Wood Buffalo National Park at 59°31'N), Sask. (N to 12 mi N of Prince Albert; CAN), Man. (N to The Pas; type from Cedar L., N of L. Winnipegosis at ca. 53°20'N), and W-cent. Ont. (Pigeon L., SW of Thunder Bay; F.K. Butters and E.C. Abbe, Rhodora 55(653):200. 1953), S to Ariz., N.Mex, and Nebr. MAP: Hultén 1968b:933.

    The report from SE-cent. Alaska by Hultén (1968b; Tok) is said by him to represent an introd. plant. The report N to Fort Franklin, Dist. Mackenzie, ca. 65°10'N, by Hooker (1834) requires confirmation but may also be based upon an introd. specimen, it being reported as evidently introd. in S Dist. Mackenzie by both W.J. Cody (Can. Field-Nat. 70(3):128. 1956; Salt River, 60°06'N, where taken along a roadside in an old burn) and J.W. Thieret (Can. Field-Nat. 75(3):120. 1961; disturbed soil along a roadside SW of Great Slave L.).

S. foetidus Howell

/T/W/ (Hs) Wet meadows in the foothills and mts. from S B.C. (Rossland, near the U.S.A. boundary SW of Trail; CAN) and SW Alta. (Waterton Lakes and Milk River Ridge; CAN; Breitung 1957b) to Oreg., Idaho, and Mont. [Incl. *S. hydrophiloides* Rydb.].

S. fremontii T. & G.

/T/W/ (Hs) Cliffs and talus slopes at high elevations from S B.C. (N to the Marble Range NW of Clinton and mts. near the Alta. boundary W of Banff; CAN) and SW Alta. (N to Banff; CAN) to Calif. and Colo. [Incl. the robust extreme, var. *blitoides* (Greene) Cronq.].

S. hydrophilus Nutt.

/T/W/ (Hs) Fresh or alkaline swampy places in the valleys and foothills from B.C. (mouth of the Dean (Salmon) R. SW of Bella Coola at ca. 52°10'N and Kootenay Flats, W of Creston; CAN) and Mont. to Calif., Colo., and S.Dak. [*S. pacificus* (Greene) Rydb.].

S. indecorus Greene

/ST/X/ (Hs) Damp ground and calcareous rocks and slopes from cent. Alaska–Yukon (N to ca. 65°N; *see* Hultén 1950: map 1198, p. 1683) to Great Bear L., Great Slave L., L. Athabasca (Alta. and Sask.), SW Dist. ?Keewatin (CAN), NE Man. (known only from Churchill), Ont. (N to the Shamattawa R. at 54°47'N and W James Bay N to 52°11'N), Que. (N to E James Bay at 52°37'N and the Koksoak R. S of Ungava Bay at 57°42'N), and Labrador (between ca. 54°30' and 58°15'N), S to N Calif., Idaho, Mont., N Mich., cent. Ont. (S to near Thunder Bay and Cochrane), and Que. (S to L. Mistassini, L. St. John, Bic, Rimouski Co., and the Gaspé Pen.). [*S. idahoensis* Rydb.; *S. pauciflorus* var. *fallax* Greenm. and f. *?ornatus* Boivin; *S. discoideus* of auth., not (Hook.) Britt.]. MAPS: Raup 1947: pl. 36; the NE Canadian limits are indicated in a map by Lepage 1966: map 20, p. 244.

S. integerrimus Nutt.
/aST/WW/ (Hs (Hsr)) Dryish to moist open places and woods at low to fairly high elevations, the aggregate species from the coasts of Alaska–Yukon–Dist. Mackenzie (E to Coronation Gulf) through B.C.–Alta. and s Sask.–Man. to Calif., Colo., Iowa, and Minn. MAPS and synonymy: *see* below.

1 Pubescence rather fine and of a tomentose nature (generally very sparse at anthesis); phyllaries very conspicuously black-tipped; stems from a short, thick, ascending or horizontal rhizome; [*S. lugens* Richards., the type from the Coppermine R., N Dist. Mackenzie; *S. imbricatus* Gardn.; Alaska–Yukon–Dist. Mackenzie and mts. of B.C. and sw Alta.; MAPS (*S. lugens*): Hultén 1968*b*:935; Raup 1947: pl. 36 (indicating a station in extreme sw Sask., this presumably based upon collections by the Macouns of var. *exaltatus* from Farewell Creek and the Cypress Hills, Sask., distributed as *S. lugens*] . . . . var. *lugens* (Richards.) Boivin
1 Pubescence loosely crisp or arachnoid-villous; stems from a very short erect crown.
  2 Phyllaries relatively broad and evidently and consistently black-tipped; peduncle of the terminal head consistently thickened and much shorter than the others; plant often rather persistently hairy.
    3 Rays bright yellow; basal leaf-blades generally oblanceolate to elliptic; [*S. exaltatus* Nutt.; *S. atriapiculatus* and *S. scribneri* Rydb.; *S. columbianus* Greene; *S. hookeri* T. & G.; B.C. (N to Spences Bridge), sw Alta. (N to Banff), and sw Sask. (Cypress Hills; Farewell Creek)] . . . . var. *exaltatus* (Nutt.) Cronq.
    3 Rays white or creamy; basal leaf-blades generally deltoid to subcordate; [*S. lugens* var. *ochroleucus* Gray; *S. leibergii* Greene; B.C. (headwaters of the Fraser R.; near Princeton; Cascade Mts.) and Alta. (Boivin 1966*b*)] . . . .
       . . . . var. *ochroleucus* (Gray) Cronq.
  2 Phyllaries relatively narrow, only minutely and irregularly black-tipped (if at all); peduncle of the terminal head often as long as the others; plant essentially glabrate at anthesis; [B.C. to Man. (N to The Narrows of L. Manitoba)] . . . . var. *integerrimus*

S. jacobaea L. Tansy-Ragwort, Stinking Willie
Eurasian; locally aggressive along roadsides and in fields and pastures in N. America, as in sw B.C. (Vancouver Is. and adjacent islands and mainland; CAN; V), Ont. (N to Ottawa), Que. (N to the E Gaspé Pen. at York; GH), St-Pierre and Miquelon, Nfld., N.B., P.E.I., and N.S. Reports from Man. require confirmation, perhaps being referable chiefly or wholly to *S. eremophilus*.

S. lindstroemii (Ostenf.) Porsild
/aSs/W/EA/ (Hs) Alpine meadows and slopes from the coasts of Alaska–Yukon and the Mackenzie R. Delta to sw Yukon, with an isolated area in the Beartooth Mts. of Mont. and Wyo.; Eurasia. [*S. integrifolius* var. *lind.* Ostenf.; *S. bivestitus* Cronq.; *S. denalii* Nels.; *S. fuscatus* of auth., perhaps not Hayek]. MAPS: Hultén 1968*b*:927 (*S. fusc.*); Porsild 1966: map 157, p. 86.

S. macounii Greene
/t/W/ (Hs) Dry open places and open woods from sw B.C. (Vancouver Is. and adjacent islands and ?mainland; CAN; type from Mt. Benson, Vancouver Is.; a collection in V from the mainland at Chilcotin, near Lillooet, requires confirmation, as do the reports of *S. fastigiatus* from Alta.–Sask. by J.M. Macoun 1896) to s Oreg. [*S. fastigiatus* var. *mac.* (Greene) Greenm.; *S. fast.* Nutt., not Schwein.].

S. megacephalus Nutt.
/T/W/ (Hs) Open rocky places at moderate to high elevations from SE B.C. (Manning Provincial Park, about 30 mi SE of Hope; Flathead, SE of Fernie; South Kootenay Pass, on the B.C.–Alta.

boundary) and sw Alta. (Waterton Lakes; CAN; reported from the Belly R. by John Macoun 1884) to Idaho and Mont. [Incl. *S. amplectens* Gray and its var. *taraxacoides* Gray (*S. tar.* (Gray) Greene)].

S. newcombei Greene
/T/W/ (Hs) Known only at low to moderate elevations in w B.C. (Queen Charlotte Is., where first taken by Newcombe in 1897, the type from Moresby Is.; CAN; DAO; *see* Calder and Taylor 1968:543–46).

Eric Hultén (Sven. Bot. Tidskr. 62(4):525. 1968), not having seen material of this taxon, suggests that it may be merely part of the very variable population of *S. resedifolius* but a comparison of collections in CAN indicates that the two taxa are fairly distinct.

S. obovatus Muhl.
/t/EE/ (Hsr) Calcareous cliffs and open or wooded slopes from Mich. to s Ont. (known only from the Bruce Pen., L. Huron, in Bruce and Grey counties; CAN; TRT) and s N.H., s to Ala. and S.C.

S. pauciflorus Pursh
/ST/X/ (Hs) Moist cliffs and subalpine to alpine meadows from Alaska (N to near the Arctic Circle) to the Yukon (N to ca. 65°N), Great Bear L., and Great Slave L., s through B.C. and sw Alta. (N to Jasper) to N Wash.–Idaho–Wyo. (and reputedly to Calif.), with scattered stations eastwards in NE ?Man. (a 1936 collection by Polunin from Churchill has been placed here by Fernald; Hultén's map also indicates a station presumably on the Nelson R. about 175 mi s of Churchill), Ont. (N shore of L. Superior near Thunder Bay; Greenman 1916), Que. (SE James Bay; Ungava Bay and its N watershed; Knob Lake dist. at ca. 55°N; Côte-Nord; Gaspé Pen.), Labrador (N to Komaktorvik Fjord, ca. 59°17′N; type, as first collection cited, from Labrador), and NW Nfld. [Var. *atropurpureus* Boivin; *S. aureus* vars. *borealis* Gray and *discoideus* Hook. (*S. disc.* (Hook.) Britt.) in part]. MAP: Hultén 1968b:930 (incl. *S. indecorus*).

S. pauperculus Michx.
/ST/X/ (Hs) Peats and wet rocks, ledges, and gravels (often calcareous), the aggregate species from Alaska–Yukon (N to ca. 69°N) to Great Bear L., Great Slave L., Alta. (N to Wood Buffalo National Park at 59°16′N), Sask. (N to L. Athabasca and Hasbala L. at ca. 60°N), Dist. ?Keewatin (Boivin 1966b), Man. (N to Churchill), northernmost Ont., Que. (N to Ungava Bay and the Côte-Nord), Labrador (N to Carol L., ca. 53°N, 67°W), Nfld., N.B. (St. John R. system), P.E.I. (Summerside, Prince Co.; D.S. Erskine 1960), and N.S., s to Oreg., N.Mex., Nebr., Ill., Ala., and Ga. MAPS and synonymy: *see* below.

1 Plant more or less persistently floccose-tomentose until anthesis or later; [*S. balsamitae (flavovirens)* var. *thomp.* Greenm.; *S. plattensis* Nutt.; *S. ?farriae* and *S. willingii* Greenm.; s Dist. Mackenzie–B.C. to s Ont. (Essex, Lambton, Waterloo, and Bruce counties] . . . . var. *thompsoniensis* (Greenm.) Boivin
1 Plant soon glabrate or essentially so.
  2 Phyllaries linear-oblong, their margins essentially parallel and tapering only above the middle . . . . var. *firmifolius* Greenm.
    3 Heads discoid, ray-ligules wanting; [*S. gaspensis* f. *ver.* Fern., the type from Anticosti Is., E Que.] . . . . f. *verecundus* Fern.
    3 Heads radiate; [*S. gaspensis* Greenm. and its vars. *firmifolius* (Greenm.) Fern. and *victorinii* Rousseau; E Que. (Côte-Nord; Anticosti Is.; type from Percé, Gaspé Co.) and w Nfld.] . . . . f. *firmifolius*
  2 Phyllaries narrower in outline, tapering gradually from near the base . . . . . var. *pauperculus*
    4 Heads discoid; [var. *balsamitae* f. *inchoatus* Fern.; Alaska, SE Dist. Keewatin, James Bay (Manawanan Is.), and E Que. (Cap-des-Rosiers, Gaspé Co.)] . . . .
     . . . . f. *inornatus* Fern.
    4 Heads radiate; [incl. vars. *balsamitae* (Muhl.) Fern. (*S. balsamitae* Muhl.) and *neoscoticus* Fern.; *S. flavovirens* and *S. tweedyi* Rydb.; *S. multnomensis* Greenm.; transcontinental; MAPS (aggregate species): Raup 1947: pl. 36; Hultén 1968b:931] . . . . f. *pauperculus*

S. pseudaureus Rydb.
/T/WW/ (Hs) Moist meadows, thickets, and woodlands from B.C. (N to Burns Lake, W of Prince George at ca. 54°N; Eastham 1947) to Alta. (N to Jasper; CAN), Sask. (N to Amisk L. at ca. 54°35′N; Breitung 1957a), and SW Man. (Carberry and MacGregor; CAN; reports from farther eastwards are probably chiefly referable to *S. aureus* var. *aquilonius,* for a comparison with which see M.L. Fernald, Rhodora 45(540):501–02. 1943), s to Calif. and N.Mex.

This taxon is scarcely separable from the *S. aureus* complex and if merged with that species, as presumably done by Boivin (1966b), would form part of a transcontinental complex.

S. pseudo-arnica Less.   Seabeach Groundsel.   Roi des champs
/ST/D (coastal)/eA/ (Hp) Sandy or gravelly seashores and upper beaches: Pacific coast from the Aleutian Is. and W Alaska (N to Cape Lisburne, ca. 69°N) to SW B.C. (s to Vancouver Is.); Atlantic coast from Labrador (N to Windy Tickle, 55°45′N) to E Que. (St. Lawrence R. estuary from St-Jean-Port-Joli, l'Islet Co., to the Côte-Nord, Anticosti Is., and Gaspé Pen.), Nfld., N.B. (Grand Manan, Charlotte Co.; GH), and N.S. (Sable Is. and Guysborough and Yarmouth counties; not known from P.E.I. or the U.S.A.); coast of NE Asia. [*Arnica maritima* L.; *A. doronicum* Pursh, not Jacq.]. MAPS: Hultén 1968b:933; Fernald 1925: map 28, p. 259, and 1918b:map 10, pl. 12.

Forma *rollandii* (Vict.) Fern. (*S. roll.* Vict.; ray-ligules wanting, the heads discoid) is known from E Que. (Ste-Flavie, NW Gaspé Pen., and the Mingan Is. of the Côte-Nord, the type locality of var. *roll.* Vict.).

S. resedifolius Less.
/aST/D/EA/ (Hs) Exposed cliffs, chiefly at subalpine to alpine elevations, from the E Aleutian Is. and coasts of Alaska (type from St. Lawrence Is.), the Yukon, and NW Dist. Mackenzie to Banks Is. (CAN), s through B.C. (Vancouver Is.; Crowsnest Pass, and North Kootenay Pass, the last two on the B.C.–Alta. boundary; CAN) and SW Alta. (Waterton Lakes; CAN) to Wash., Mont., and Wyo., with a disjunct area in E Que. (Shickshock Mts. of the Gaspé Pen.) and Nfld.; NE Europe; Asia. [*S. conterminus* and *S. hyperborealis* Greenm.; *Cineraria lyrata* Ledeb.; *S. ovinus* Greene; *S. fernaldii* f. *lingulatus* Fern.; *S. lyallii* Klatt, not Hook. f.; *S. subnudus (S. cymbalarioides* Buek, not Nutt.) of Canadian reports in part, perhaps not DC.]. MAPS: *Atlas of Canada* 1957: map 18, sheet 38; Fernald 1929: map 11, p. 1492, 1925: map 29, p. 259, and 1924: map 2, p. 560; combine the maps by Hultén 1968b:929 (*S. resed.* and *S. cont.*) and 930 (*S. hyper.*).

Forma *columbiensis* (Gray) Fern. (var. *col.* Gray 1884, the type locality given as "Mucklung River, British Columbia" but that river not listed in recent B.C. gazetteers; *S. hyperborealis* var. *col.* (Gray) Greenm.; *S. fernaldii* Greenm.; heads discoid, the rays obsolete or very short (not the radiate-headed form as stated by Fernald *in* Gray 1950)) occurs throughout the range.

S. robbinsii Oakes
/T/E/ (Hs) Peaty meadows, fields, and thickets from Que. (N to the Gaspé Pen. along the Ste-Anne-des-Monts R.; CAN; GH), N.B., P.E.I. (Queens and Kings counties; CAN; GH), and N.S. to Tenn. and N.C. [*S. aureus* var. *lanceolatus* Oakes].

S. sheldonensis A.E. Porsild
/Ss/W/ (Hs) Subalpine meadows of the Yukon (N to ca. 64°N; type from between Mt. Sheldon and Mt. Riddell) and northernmost B.C. (Dease L. region at ca. 58°30′N). See A.E. Porsild, Can. Field-Nat. 64(1):43–44. 1950, and 1951b:334–35. MAPS: Hultén 1968b:934; Porsild 1966: map 158, p. 86.

S. squalidus L.   Oxford Ragwort
European; apparently introd. in N. America only in N.B. (St. John, where taken by Warner in 1892; NBM) and N.S. (near Point Pleasant Park, Halifax; ACAD; DAO).

S. streptanthifolius Greene
/ST/WW/ (Hs) Moist or dryish open places and woodlands at low to moderate elevations from the Yukon (N to ca. 65°N; reported from Alaska by Boivin 1966b) to Great Bear L., Great Slave L., and L. Athabasca (Alta. and Sask.), s to Calif. and N.Mex. [*S. cymbalarioides* var. *strept.* (Greene) Greenm., var. *borealis* (T. & G.) Greenm. (*S. aureus* var. *bor.* T. & G.), and ssp. *moresbiensis*

Calder & Taylor; *S. crocatus* and *S. jonesii* Rydb.; *S. dileptifolius* and *S. mutabilis* Greene; *S. cymb.* Nutt., not Buek; *S. resedifolius sensu* Carter and Newcombe 1921, not Less.]. MAP (*S. cymb.;* not indicating the occurrence in Alaska): Hultén 1968*b*:931.

S. sylvaticus L.
Eurasian; introd. into waste places, clearings, rocky slopes, and open woods in N. America, as in sw B.C. (Vancouver Is. and adjacent islands and mainland E to Hope), Que. (N to the Gaspé Pen. at Gaspé Basin; CAN), Nfld., N.B., P.E.I., and N.S.

S. triangularis Hook.
/ST/W/ (Hp) Moist meadows and open places, mostly at moderate to high elevations, from Alaska (N to ca. 63°N), the Yukon (N to ca. 65°N), and sw Dist. Mackenzie (CAN) through B.C. and sw Alta. (N to Jasper; CAN; the inclusion of Sask. in the range by Rydberg 1922, requires confirmation) to Calif. and N.Mex. The type is a Drummond collection from "Moist Prairies among the Rocky Mountains", probably in Alta. [*S. prionophyllus* Greene; *S. saliens* Rydb.]. MAP: Hultén 1968*b*:934.

S. tridenticulatus Rydb.
/sT/WW/ (Hs) Calcareous outcrops and sandy or gravelly prairies from sw Dist. Mackenzie (Kakisa L., ca. 61°N, detd. Barkley; J.W. Thieret, Can. Field-Nat. 75(3):120. 1961), s Sask. (Old Wives L., sw of Moose Jaw; Greenman 1916), and sw Man. (N to Petrel, about 25 mi NE of Brandon) to N.Mex. and Tex. [*S. densus* Greene; *S. manitobensis* Greenm.; *S. mutabilis* Nels., not Greene].

S. viscosus L.   Sticky Groundsel
Eurasian; introd. in waste places, railway ballast, and about ports in N. America, as in B.C. (N to Prince Rupert and Terrace; Eastham 1947), s Man. (Winnipeg), Ont. (N to Grenville and Dundas counties), Que. (N to the Gaspé Pen.), Nfld., N.B., P.E.I., and N.S.

S. vulgaris L.   Common Groundsel
Eurasian; a weed of cult. and waste land in N. America, as in Alaska (N to ca. 68°N), the Yukon (N to ca. 64°N), Dist. Mackenzie (N to Great Slave L.) and all the provinces (in Ont., N to sw James Bay; in Labrador, N to Hopedale, 55°28′N); sw Greenland. MAP: Hultén 1968*b*:932.

[S. werneriaefolius Gray]
[The report of this species of the w U.S.A. (N to Idaho and Mont.) from SE B.C. by John Macoun (1886, as *S. petraeus*; w summit of North Kootenay Pass; taken up by Henry 1915, as *S. petrocallis*) and so-named collections in Herb. V from Penticton and Flathead, SE B.C., require confirmation. (*S. petraeus* and *S. saxosus* Klatt; *S. petrocallis* Greene).]

S. yukonensis A.E. Porsild
/aSs/W/ (Hs) Damp mossy tundra of Alaska–Yukon (N to the arctic coast; type from the upper Rose R., the Yukon) and w Dist. Mackenzie (at ca. 64°30′N). [*S. alaskanus* Hult.]. MAP: Hultén 1968*b*:926.

SILPHIUM L.   [9131]   Rosinweed

1   Stem nearly square in cross-section, essentially glabrous, leafy; leaves narrowly ovate to deltoid, scabrous, coarsely toothed, the upper ones united by a cup-like base around the stem, the petioles of the lower ones connate-clasping; involucre to 2.5 cm high; disk to 2.5 cm broad; (s Ont.; introd. in sw Que.) .... *S. perfoliatum*
1   Stem roundish or obscurely angled.
    2   Stem nearly naked except for a few scattered bracts, essentially glabrous, from a woody taproot; leaves narrowly to broadly ovate or elliptic, usually cordate at base, sharply toothed (occasionally pinnatifid), glabrous or scabrous; involucre to 2.5 cm high; disk to 2.5 cm broad; (s Ont.) .... *S. terebinthinaceum*
    2   Stem leafy.

3 Leaves lanceolate, subentire or irregularly toothed, mostly not more than 2 dm long, commonly in whorls of 3–5 (occasionally opposite or even alternate), usually scabrous above and more or less hirsute beneath; stem from a stout caudex, glabrous and glaucous; involucre less than 2 cm high; disk less than 2 cm broad; (s ?Ont.) .... [*S. trifoliatum*]

3 Leaves deeply pinnatifid or 2-pinnatifid, alternate, hirsute chiefly along the veins beneath, the lower ones to over 4 dm long; stem from a woody taproot, rough-bristly; involucre to over 4 cm high; disk to about 3 cm broad; (introd. in s Ont.) .... *S. laciniatum*

S. laciniatum L.   Compass-plant
Native in the ε U.S.A. (N to N.Dak. and Mich.); introd. along railway tracks in s Ont. (Maidstone, Essex Co., where taken by Howard in 1955; GH; OAC; *see* s Ont. map 27a by Soper 1962: fig. 24, p. 39).
The common name "compass-plant" derives from the tendency of the plant to present the edges of its vertical leaves north and south, probably an adaptation to maximum sun exposure.

S. perfoliatum L.   Cup-plant
/t/EE/   (Hs)   Rich woods, thickets, and prairies from S.Dak. to s Ont. (apparently native in Essex and Kent counties and possibly in the High Park region of Toronto; *see* s Ont. map 27b by Soper 1962: fig. 24, p. 39; reported from Ottawa by Groh 1946, where probably a garden-escape but not listed by Gillett 1958; introd. in sw Que. (Montreal; Groh 1946), Pa., and s New Eng.), s to Okla., Mo., Miss., and Ga.

S. terebinthinaceum Jacq.   Prairie-Dock
/t/EE/   (Hs)   Prairies and openings from Ind. and Ohio to s Ont. (apparently native in Essex, Lambton, Brant, and Haldimand counties; *see* s Ont. map 27c by Soper 1962: fig. 25, p. 40), s to Ala. and Ga.

[S. trifoliatum L.]
[The report of this species of the ε U.S.A. (N to Ind. and Pa.) from s Ont. by Dodge (1914; Amherstburg, Essex Co.) requires confirmation.]

SILYBUM Adans.   [9464]

S. marianum (L.) Gaertn.   Milk-Thistle
European; an occasional garden-escape in N. America, as in sw B.C. (Victoria and Nanaimo, Vancouver Is.; Herb. V; Henry 1915), Sask. (Eastend, about 75 mi sw of Swift Current; SCS), Ont. (N to Ottawa), sw Que. (near Montreal), N.B. (Gloucester, Victoria, and Kent counties), and N.S. (Halifax). [*Carduus* L.].

SOLIDAGO L.   [8849]   Goldenrod.   Verge d'Or

1 Inflorescence a compound, more or less flattish-topped corymb.
   2 Leaves not glandular-punctate, subentire or low-serrate above the middle, the lower ones long-petioled, persistent, much longer than the upper ones; anther-filaments freed below the summit of the corolla-tube; phyllaries more or less striate; heads distinctly peduncled.
      3 Leaves harsh, greyish with a usually dense short pubescence, the median ones mostly not over 6 times as long as broad; stem puberulent; mature achenes 10–15-ribbed; rays 8–14, well developed; plants of dry habitats; (Alta. to Ont.) .... *S. rigida*
      3 Leaves smooth, glabrous except for the minutely scabrous margins, the median ones usually more than 6 times as long as broad; stems glabrous or somewhat puberulent about the inflorescence; achenes with at most 7 nerves, glabrous or nearly so; rays about 6–9, rather short; plants of moist habitats.

4   Leaves flat, strictly 1-nerved, the lower ones oblanceolate to spatulate, obtusish at apex; achenes commonly 3–5-angled, scarcely nerved; (s Ont.) . . . . *S. ohioensis*

4   Leaves often longitudinally folded, tending to be 3-nerved, linear-lanceolate, their acute tips commonly divergent or recurved; mature achenes evidently 5–7-nerved; (SE Man. and Ont.) . . . . *S. riddellii*

2 Leaves more or less glandular-punctate, linear-lanceolate to lanceolate, mostly acute to attenuate at tip, entire or essentially so, fairly uniform in length, the lower ones soon deciduous; anther-filaments freed at the summit of the corolla-tube; phyllaries not striate; heads mostly sessile or subsessile in small clusters; rays commonly 8–25, small; achenes pubescent.

5   Leaves 1-nerved or sometimes with an additional pair of obscure lateral nerves; (s Ont. and N.S.) . . . . *S. tenuifolia*

5   Leaves distinctly 3-nerved (sometimes 5-nerved), seldom subtending reduced leaves or branches.

6   Inflorescence commonly elongate (flowering branches often arising from leaf-axils well down toward the middle of the stem); phyllaries mostly narrow and acute; (B.C. and ?Alta.) . . . . *S. occidentalis*

6   Inflorescence more compact (the flowering branches mostly arising from leaf-axils near the top of the stem, the outer branches sometimes overtopping the inner ones); phyllaries obtuse to acuminate; (transcontinental) . . . . *S. graminifolia*

1 Inflorescence axillary to thyrsoid or paniculate (if corymbiform, the heads racemose on the branchlets of the inflorescence).

7 Heads in small axillary clusters or in a terminal panicle or thyrse with the heads spirally arranged and not in strongly 1-sided racemes.

8   Heads large, the involucre usually at least 8 mm long; phyllaries thin; achenes glabrous; leaves chiefly elliptic to ovate, coarsely and sharply serrate, pinnate-veined, abruptly contracted to the petiole, the lower ones much longer than the middle and upper ones; (Ont. to Labrador, Nfld., and N.S.) . . . . *S. macrophylla*

8   Heads smaller, the involucre usually less than 8 mm long (occasionally 9 mm long in species with firm phyllaries).

9   Stem-leaves only gradually decreasing in size upwardly, the lower ones smaller or not much larger than the median ones and soon deciduous, on short winged petioles; separate basal rosettes wanting.

10 Mature achenes essentially glabrous; phyllaries obtuse or rounded, glutinous, yellowish; involucre mostly 3–5 mm high; leaves thick and firm, pinnate-veined, entire or the lower ones slightly toothed; (s Ont.) . . . . *S. speciosa*

10 Mature achenes pubescent.

11 Leaves pinnate-veined; heads in axillary clusters mostly surpassed by their subtending leaves; rays 3 or 4(5); (Ont. to N.S.).

12 Leaves ovate, abruptly narrowed to a short winged petiole; stem somewhat angled, zigzag, from a slender freely stoloniferous rhizome; involucre to 6 mm high . . . . *S. flexicaulis*

12 Leaves elongate-lanceolate to narrowly oblong, tapering to the subsessile base; stem terete, glaucous, from a stout, tardily stoloniferous rhizome; involucre less than 5 mm high . . . . *S. caesia*

11 Leaves more or less strongly triple-nerved from base; rays generally more than 6.

13 Plant ashy with minute close puberulence, to about 5 dm tall; leaves thickish and firm, entire; rays less than 10; (s Alta. to sw Man.) . . . . *S. mollis*

13 Plant not ashy-puberulent, to over 1.5 m tall; leaves relatively thin, sharp-serrate; rays usually more than 10; (E Que. and Nfld.) . . . . *S. calcicola*

9 Stem-leaves very unequal, the lower ones the largest and usually relatively the broadest; rosettes of large basal leaves from separate offshoots usually present.

 14 Phyllaries of at least the outer 2–4 series with strongly recurved green tips; involucre 5–9 mm high; achenes glabrous; principal leaves sharp-serrate; (Ont. to N.B.) .... *S. squarrosa*

 14 Phyllaries with ascending tips.

  15 Leaves more or less pubescent on one or both surfaces (except in *S. hispida* var. *tonsa*); mature achenes glabrous; involucre to 6 mm high.

   16 Rays cream-colour to nearly white; outer phyllaries with well-defined green tips; (Ont. to N.S.) .... *S. bicolor*

   16 Rays orange-yellow; phyllaries with less well-defined green tips; (Sask. to Nfld. and N.S.) .... *S. hispida*

  15 Leaves glabrous to minutely puberulent.

   17 Plant minutely ashy-puberulent throughout; achenes glabrous or occasionally sparingly hairy; involucre to 5 mm high, the phyllaries long-attenuate; inflorescence a compact thyrse or panicle of stiffly ascending racemes; (E Ont. to N.S.) .... *S. puberula*

   17 Plant glabrous to pilose (sometimes minutely puberulent in *S. spathulata*).

    18 Principal leaves elliptic to obovate or more or less rhombic, distinctly toothed; stem essentially glabrous; inflorescence an elongate, rarely branched, terminal thyrse of axillary clusters; rays 7 or 8; involucre to 7 mm high; achenes pubescent; (s Ont.) .... *S. sciaphila*

    18 Principal leaves somewhat narrower, entire or shallowly few-toothed .... GROUP 1

7 Heads in a panicle or thyrse, borne on the upper side of the branches in 1-sided racemes.

 19 Basal leaves much the largest, often forming rosettes, the upper stem-leaves much reduced; stems solitary or few .... GROUP 2 (p. 1604)

 19 Basal leaves not much larger than the stem-leaves and rarely forming rosettes; stems few or clustered .... GROUP 3 (p. 1605)

GROUP 1

1 Inflorescence corymbiform; heads few, on white-villous pedicels; rays commonly 12 or more; achenes pubescent; lower leaves densely ciliate with long soft curling hairs, at least toward base; (transcontinental) .... *S. multiradiata*

1 Inflorescence racemiform or thyrsiform (rarely corymbiform), its branches and pedicels glabrous to pilose but scarcely villous; lower leaves with glabrous or scabrous margins, rarely ciliate with curling hairs.

 2 Achenes glabrous or nearly so; rays 4–6; involucre 4–6 mm high; stem glabrous except in the inflorescence, this with strongly ascending branches.

  3 Stem-leaves rather remote; basal leaves to about 7 cm broad; plant of peaty or damp habitats; (Man. to Labrador, Nfld., and N.S.) .... *S. purshii*

  3 Stem-leaves 18 or more, often rather crowded; basal leaves to 1 dm broad; plants of usually dry habitats; (s Ont.) .... *S. speciosa*

 2 Achenes more or less pubescent (sometimes glabrous in *S. missouriensis*).

  4 Pedicels and branches of inflorescence glabrous or nearly so; involucre to 5 mm high; stem solitary; rhizome cord-like, bearing long slender stolons; (B.C. to w Ont.) .... *S. missouriensis*

  4 Pedicels and branches of inflorescence pubescent; involucre to 6 mm high; stems solitary or tufted, nonstoloniferous but often with basal rosettes of leaves; (B.C. to N.S.) .... *S. spathulata*

GROUP 2 (see p. 1603)

1 Stem and leaves minutely and closely ashy-puberulent; rays bright yellow; achenes pubescent; involucre to about 6 mm high.
  2 Leaves mostly sessile, distinctly 3-ribbed, coriaceous; panicle compact, its branches ascending; stem to about 5 dm tall, from a cord-like slender-stoloniferous base; (s Alta. to s Man.) . . . . *S. mollis*
  2 Leaves mostly more or less petioled, not prominently 3-ribbed, thinner, the upper ones usually with axillary tufts of reduced leaves; panicle-branches mostly spreading; stem to over 1 m tall, from a nonstoloniferous stoutish caudex; (B.C. to N.S.) . . . .
  . . . . *S. nemoralis*
1 Stem and leaves glabrous or pubescent but not closely ashy-puberulent; rays mostly deep yellow or orange-yellow.
  3 Pedicels and branches of inflorescence glabrous; stem and leaves normally glabrous or nearly so.
    4 Leaves fleshy, entire, smoothish, the bases of the lower ones distinctly clasping; involucre to 7 mm high, with narrow acutish soft phyllaries; pappus at least 3.5 mm long; disk-corollas at least 4 mm long; mature achenes over 2 mm long, pubescent; (saline coastal habitats in E Que. and the Atlantic Provinces) . . . .
    . . . . *S. sempervirens*
    4 Leaves scarcely fleshy, often toothed, mostly scabrous on the margins, not strongly clasping; involucre rarely over 5 mm high, its broader phyllaries mostly firm and obtuse; pappus not over 3.5 mm long; disk-corollas to 4 mm long; achenes rarely over 2 mm long.
      5 Heads with less than 15 flowers; phyllaries linear-oblong, relatively thin; plant of bogs and swamps; (SE Man. to the Atlantic Provinces) . . . . *S. uliginosa*
      5 Heads with at least 15 flowers; phyllaries broader and firmer; upper leaves often with axillary tufts of reduced leaves; plants of dry habitats.
        6 Rhizome cord-like, horizontal, abundantly slender-stoloniferous; leaves distinctly 3-ribbed, the basal ones entire or shallowly toothed; (B.C. to w Ont.) . . . . *S. missouriensis*
        6 Rhizome short, only occasionally slender-stoloniferous; leaves scarcely or not at all 3-ribbed, the basal ones more or less sharply serrate; achenes usually pubescent; (Man. to N.S.) . . . . *S. juncea*
  3 Pedicels and branches of inflorescence distinctly pubescent.
    7 Leaves fleshy, entire, smoothish, the bases of the lower ones distinctly clasping; achenes pubescent; involucre to 7 mm high; (saline coastal habitats in E Que. and the Atlantic Provinces) . . . . *S. sempervirens*
    7 Leaves not fleshy, at least the basal ones toothed (if distinctly clasping, their margins ciliate or scabrous); involucre to 5 mm high.
      8 Leaves loosely long-pilose on the principal veins beneath and usually sparsely so over the lower surface, unevenly coarse-serrate, the basal ones rarely in definite rosettes, elliptic-ovate to rhombic, acuminate; achenes pubescent; (s Ont. and ?N.S.) . . . . *S. ulmifolia*
      8 Leaves glabrous or at most scabrous-margined.
        9 Upper half of stem squarish in cross-section, the angles narrowly winged; leaves minutely but strongly papillate-scabrous above with forward-pointing hairs, their margins not ciliate; achenes minutely pubescent; (s Ont.) . . . . *S. patula*
        9 Stem not markedly square; leaves smooth or scabrous but not conspicuously papillate, their margins usually minutely ciliate; achenes essentially glabrous (sometimes pubescent in *S. juncea*).
          10 Lower stem-leaves clasping 1/2 to 3/4 around the stem, their thickish obscurely veiny blades elongate and tapering into the petioles; heads not more than 15-flowered; plants of bogs or marshes; (SE Man. to the Atlantic Provinces) . . . . *S. uliginosa*

10 Lower stem-leaves scarcely clasping; heads 15–20-flowered; plants of dryish habitats.

   11 Basal leaves rather abruptly contracted to the petiole; upper leaves closely serrate; panicle-branches copiously pilose; (s Ont.) . . . . *S. arguta*

   11 Basal leaves gradually tapering to the petiole; upper leaves entire or obscurely toothed; panicle-branches only sparsely hairy; (Man. to N.S.) . . . . *S. juncea*

GROUP 3 (see p. 1603)

1 Leaves not 3-ribbed (only the midrib strongly prominent beneath), at least the basal ones usually pinnate-veined (except in *S. odora*); achenes short-pubescent (sometimes subglabrous in *S. odora*).

  2 Leaves minutely translucent-punctate, all entire, not obviously pinnate-veined, glabrous except for the scabrous margins, anise-scented when bruised; achenes short-pubescent or subglabrous; involucre to 5 mm high . . . . [*S. odora*]

  2 Leaves not translucent-punctate, at least the lower ones distinctly pinnate-veined and more or less toothed; achenes usually permanently short-pubescent.

    3 Stems more or less hairy at least above the middle, solitary or in small clumps from creeping rhizomes; involucre to 5 mm high; (Ont. to the Atlantic Provinces) . . . .
. . . . *S. rugosa*

    3 Stems glabrous throughout or at least to near the inflorescence.

      4 Leaves loosely long-pilose on the principal veins beneath and usually sparsely so over the lower surface, unevenly coarse-serrate, the basal ones elliptic-ovate to rhombic, acuminate; stems from a branched caudex; involucre to 4.5 mm high; plant of dry habitats; (s Ont. and ?N.S.) . . . . *S. ulmifolia*

      4 Leaves glabrous except for the appressed-serrate margins, elliptic to oblanceolate; stems from creeping rhizomes; plant of wet to dryish habitats; involucre to 6 mm high; (sw N.S.) . . . . *S. elliottii*

1 Leaves 3-ribbed at least near base (the 2 lateral nerves prolonged parallel to the midrib).

  5 Stem densely pilose at summit below the inflorescence; (transcontinental) . . . .
. . . . *S. canadensis*

  5 Stem glabrous or only sparsely pilose at least below the inflorescence.

    6 Pedicels and panicle-branches glabrous; achenes glabrous or sparsely pubescent; leaves mostly subtending axillary fascicles; stems solitary, freely stoloniferous; (B.C. to w Ont.) . . . . *S. missouriensis*

    6 Pedicels and panicle-branches pilose; achenes short-pubescent; leaves mostly lacking axillary fascicles; stems commonly clustered, only tardily stoloniferous; (B.C. to N.S.) . . . . *S. gigantea*

S. arguta Ait.
/T/EE/ (Hs) Open woods, thickets, and clearings from s Ont. (N to Waterloo and Hastings counties) to Maine, s to Ill., Ala., and N.C.

S. bicolor L. White Goldenrod, Silverrod
/T/EE/ (Hs (Hsr)) Open woods and dry sterile soil from Mich. to Ont. (N to the Ottawa dist.), Que. (N to Magdalen Is.; CAN and GH, detd. Fernald), N.B., P.E.I., and N.S., s to Ark. and Ga. [*Aster* Nees; *S. virgaurea sensu* Lindsay 1878, not L., the relevant collection in NSPM].

    Although the whitish or creamy ray-ligules of this taxon contrast strongly in fresh material with the orange-yellow ones of *S. hispida*, these two species are very difficult to separate in the herbarium and reports of *S. bicolor* from elsewhere in Canada other than as indicated above require confirmation.

S. caesia L. Blue-stem Goldenrod
/T/EE/ (Hpr) Rich woods, thickets, and clearings from Wisc. to Ont. (N to Michipicoten, about 100 mi NW of Sault Ste. Marie, and Ottawa), Que. (N to Montreal; reported N to Cacouna,

Temiscouata Co., by D.P. Penhallow 1891), N.B. (St. John; NBM; not known from P.E.I.), and N.S. (Annapolis, Kings, and Halifax counties), s to Tex. and Fla.

A hybrid with *S. flexicaulis* is reported from sw Que. by Boivin (1966*b*). Forma *axillaris* (Pursh) House (*S. ax.* Pursh; heads in loose axillary clusters rather than forming a loosely paniculate leafy inflorescence) appears to be the common phase in our area.

S. calcicola Fern.
/T/E/ (Hpr) Rich woods and rocky or gravelly thickets of E Que. (Notre-Dame-du-Portage, Temiscouata Co., to the Gaspé Pen.), Nfld., and N New Eng. [*S. virgaurea* var. *calc.* Fern.; possibly of hybrid origin].

S. canadensis L. Canada Goldenrod. Bouquets jaunes
/ST/X/ (Hpr) Thickets, clearings, fields, meadows, and roadsides, the aggregate species from Alaska–Yukon–w Dist. Mackenzie (N to near the Arctic Circle) to B.C.–Alta., Sask. (N to near Prince Albert), Man. (N to Gillam, about 165 mi s of Churchill), Ont. (N to the Severn R. at ca. 55°30′N), Que. (N to SE James Bay at 52°37′N, L. Mistassini, and the Côte-Nord), Labrador (N to the Hamilton R. basin), Nfld., N.B., P.E.I., and N.S., s to Calif., N.Mex., Tex., and Fla.

A hybrid with *S. juncea* is tentatively reported from sw Que. by Boivin (1966*b*). A hybrid with *S. rugosa* is postulated by Malte for a collection in CAN from Indian Point, N.B., and this taxon is tentatively reported from P.E.I. by Boivin (1966*b*). × *S. erskinei* Boivin (*S. canadensis* × *S. sempervirens*) is known from the type locality, Wood Is., P.E.I. Collections in CAN and GH from Shelburne Co., N.S., are referred by M.L. Fernald (Rhodora 24(286):205. 1922) to a hybrid between *S. canadensis* and *S. uniligulata* (*S. uliginosa* var. *linoides* of the present treatment). MAP and synonymy: *see* below.

1 Involucre and disk-corollas mostly less than 3 mm long.
  2 Summit of stem minutely pubescent; [transcontinental].
    3 Leaves green, merely somewhat pilose on the nerves beneath; [*S. lepida sensu*
      Fernald, not DC.; *S. serotina* var. *?minor* Hook.] . . . . var. *canadensis*
    3 Leaves roughish with a greyish puberulence on both sides; [*S. gilvocanescens*
      and *S. lunellii* Rydb.; *S. dumetorum* Lunell; *S. pruinosa* Greene; *S. lepida* var.
      molina Fern.] . . . . var. *gilvocanescens* Rydb.
  2 Summit of stem and lower leaf-surfaces pilose; [E Que.: Percé, Gaspé Co.] . . . .
. . . . var. *hargeri* Fern.
1 Involucre and disk-corollas at least 3 mm long.
  4 Branches of the pyramidal panicle strongly divergent or recurved; heads slender, with
    less than 20 flowers; phyllaries markedly scarious-margined, the green midrib
    conspicuous; leaves cinereous-puberulent; [*S. scabra* Muhl., not Willd.; *S. altissima*
    L.; *S. hirsutissima* Mill.; w Ont. to N.B. and N.S.; reports of *S. altissima* from P.E.I. are
    thought by D.S. Erskine 1960, probably to refer to *S. rugosa*] . . . .
    . . . . var. *scabra* (Muhl.) T. & G.
  4 Branches of the panicle strongly ascending, their tips rarely recurved; heads broader,
    with usually more than 20 flowers; phyllaries subherbaceous, their midribs less
    conspicuous; leaves not cinereous-puberulent; [*S. serotina* var. *sal.* Piper; *S. sal.*
    (Piper) Rydb.; *S. elongata* Nutt.; *S. lepida* vars. *el.* (Nutt.) Fern. and *fallax* Fern.;
    transcontinental; MAP: Hultén 1968*b*:854] . . . . var. *salebrosa* (Piper) Jones

S. elliottii T. & G.
/T/E/ (Hpr) Swampy open ground and thickets from N.S. (Digby, Yarmouth, Shelburne, Queens, and Halifax counties; *see* N.S. map 439 by Roland 1947:578) and the Atlantic states s to Ga. [Incl. var. *ascendens* Fern.].

A hybrid with *S. rugosa* is reported from N.S. by M.L. Fernald (Rhodora 24(286):204. 1922; Belleville, Yarmouth Co.).

S. flexicaulis L.
/T/EE/ (Hpr) Rich woods and thickets from N.Dak. to Ont. (N to Ottawa), Que. (N to the Côte-Nord and Gaspé Pen.; not known from Anticosti Is.), N.B., P.E.I., and N.S. (the report from Nfld. by Bachelot de la Pylaie 1823, requires clarification), s to Kans., Iowa, Tenn., and N.C.

Forma *subincisa* Vict. & Rousseau (leaves moderately incised rather than merely divergently sharp-serrate, the teeth to 1 cm long) is known from the type locality, Mt-Royal, Montreal, Que. A purported hybrid with *S. macrophylla* is reported from N.S. by J.S. Erskine (Rhodora 55(649):19. 1953; Amethyst Cove, Kings Co.).

S. gigantea Ait.
/sT/X/ (Hpr) Meadows, damp thickets, and borders of woods from sw Dist. Mackenzie (near Fort Simpson, ca. 62°N; W.J. Cody, Can. Field-Nat. 75(2):66. 1961) and B.C. to Alta. (N to the Peace R. at ca. 59°N; John Macoun 1884; not listed by Raup 1935), Sask. (N to Prairie River, 52°52'N; CAN), Man. (N to Cross Lake, NE of L. Winnipeg), Ont. (N to the sw James Bay watershed at 51°15'N), Que. (N to Duparquet, W. Abitibi Co., ca. 48°30'N, and the Gaspé Pen.), N.B., P.E.I., and N.S., s to Oreg., N.Mex., Tex., and Ga. [*S. serotina* var. *gig.* (Ait.) Gray; incl. *S. serotina* f. *huntingdonensis* Beaudry and the broad-leaved extreme, var. *pitcheri* (Nutt.) Shinners (*S. pit.* Nutt.)].

Var. *serotina* (Ait.) Cronq. (var. *leiophylla* Fern.; var. *shinnersii* Beaudry; *S. ser.* Ait.; leaves glabrous on both sides rather than at least pilose on the veins beneath; mature achenes usually pubescent rather than often glabrous or subglabrous) occurs essentially throughout the range.

S. graminifolia (L.) Salisb.
/sT/X/ (Hpr) Damp to dryish shores, thickets, and meadows, the aggregate species from B.C.–Alta. to Great Slave L., Alta. (N to Wood Buffalo National Park at 58°37'N), Sask. (N to L. Athabasca), Man. (N to York Factory, Hudson Bay, 57°N), Ont. (N to the Fawn R. at ca. 55°30'N, 88°W), Que. (N to the Ekwan R. E of James Bay at 53°44'N and the Côte-Nord), Nfld., N.B., P.E.I., and N.S., s to N.Mex., S.Dak., Mo., Ky., and N.C.
1   Heads slender, at most about 20-flowered, the outer ovate acutish phyllaries abruptly
    replaced by linear-oblong, acute or acuminate inner ones; [*Euthamia media* and *E.*
    camporum Greene; reported by Raup 1936, from L. Athabasca, Alta. and Sask.] . . . .
    . . . . var. *media* (Greene) Harris
1   Heads usually more than 20-flowered, the outer ovate phyllaries gradually merging with
    the oblong, obtuse to merely acute inner ones.
    2   Leaves broadly lanceolate, obtusish or subacute, less than 1 dm long and mostly not
        more than 10 times as long as broad; plant essentially glabrous . . . .
        . . . . var. *major* (Michx.) Fern.
        3   Upper leaf-axils bulblet-bearing; [known only from the type locality on the Albany
            R. sw of James Bay, Ont.] . . . . f. *gemmans* Lepage
        3   Leaf-axils not bulblet-bearing; [vars. *grahamii* Rousseau, *septentrionalis* Fern.,
            and *tricostata* (Lunell) Harris; *S. lanceolata* var. *major* Michx.; the common form
            northwards] . . . . f. *major*
    2   Leaves narrowly lanceolate, attenuate at tip, to 1.5 dm long and 20 times as long as
        broad.
        4   At least the upper leaves more or less densely spreading-hairy; [*Euthamia nuttallii*
            Greene; s Man. (Löve and Bernard 1959) to Nfld. and N.S.] . . . .
            . . . . var. *nuttallii* (Greene) Fern.
        4   Leaves essentially glabrous except for the scabrous margins . . . . var. *graminifolia*
            5   Upper leaf-axils bulblet-bearing; [known only from the type locality on the
                Kenogami R., Ont.] . . . . f. *bulbifera* Lepage
            5   Leaf-axils not bulblet-bearing; [*Chrysocoma* L.; *Euthamia* Nutt.; *S. lanceolata*
                L.; Man. to Nfld. and N.S.; reported as introd. with cranberry plants in bogs at
                Ucluelet, Vancouver Is., B.C., by J.M. Macoun, Ottawa Naturalist 26(12):167.
                1913]. . . . . f. *graminifolia*

S. hispida Muhl.
/sT/EE/ (Hs (Hsr)) Dry or moist fields, shores, and rocky places (chiefly calcareous), the aggregate species from Sask. (N to Carswell L. at 58°35'N; G.W. Argus, Can. Field-Nat. 78(3):147. 1964) to Man. (N to the Nelson R. about 155 mi s of Churchill), Ont. (N to the Severn R. at ca. 55°45'N), Que. (N to E James Bay at ca. 53°50'N, L. Mistassini, and the Côte-Nord), Nfld., N.B., P.E.I., and N.S., s to Ark., Tenn., and Ga.

1 Stem and both surfaces of basal leaves copiously pilose.
  2 Stems ashy- or whitish-pilose; [incl. var. *disjuncta* Fern.; *S. bicolor* var. *concolor* T. & G.; Man. to Nfld. and N.S.) .... var. *hispida*
  2 Stems woolly-villous; (*S. lanata* Hook., the type from "Plains of the Saskatchewan"; Sask. (N to Carswell L. at 58°35′N) to Nfld.] .... var. *lanata* (Hook.) Fern.
1 Stem glabrous or short-pubescent; upper surface of basal leaves glabrous or sparingly short-pubescent.
  3 Leaves essentially glabrous beneath, the basal ones at most about 2.5 cm broad; [Ont. (N to w James Bay at 52°25′N) to Nfld. (type from Blomidon) and N.S.] ....
    .... var. *tonsa* Fern.
  3 Leaves sparingly hirtellous beneath, the basal ones averaging over 3 cm broad; [known from the type locality, Bonne Bay, w Nfld., and E Que. (Anticosti Is.; J. Rousseau 1950)] .... var. *arnoglossa* Fern.

S. juncea Ait.

/T/EE/ (Hs (Hsr)) Dry or moist open places and thickets from s Man. (N to Riding Mt.; the inclusion of Sask. in the range by Fernald *in* Gray 1950, requires clarification; not listed by Breitung 1957a) to Ont. (N to Renison, s of James Bay at ca. 51°N; Ilmari Hustich, Acta Geogr. 13(2):47. 1955), Que. (N to L. Timiskaming at ca. 47°30′N; the report from Anticosti Is. by Verrill 1865, requires confirmation, as also the report from Nfld. by M. Southcott, *Some Newfoundland Wild Flowers*. St. John's, Newfoundland. 1915), N.B., P.E.I., and N.S., s to Mo., Tenn., and Ga.

Forma *ramosa* (Porter & Britt.) Fern. (var. *ram.* P. & B.; panicle-branches erect, the individual racemes scarcely recurved at tip, rather than branches widely divergent, the racemes recurved at tip) is reported from N.B. by M.L. Fernald (Rhodora 38(450):208. 1936). Forma *scabrella* (T. & G.) Fern. (*S. arguta* var. *scab.* T. & G.; leaves scabrous rather than glabrous or merely marginally short-ciliate; panicle-branches often sparingly short-hairy rather than glabrous) occurs essentially throughout the range.

S. macrophylla Pursh

/ST/EE/ (Hsr) Damp woods and thickets from Ont. (N to Kapuskasing, 49°24′N) to Que. (N to s Ungava Bay and the Côte-Nord), Labrador (N to Saglek Bay, 58°29′N), Nfld., N.B., P.E.I., and N.S., s to N.Y. and Mass.

The tentative report from s Alta. by John Macoun (1884; Tail Creek, Red Deer R.) requires clarification. Forma *pseudomensalis* Beaudry (achenes appressed-hairy rather than glabrous) is known from the type locality in Fernald Pass, Mt. Logan of the Shickshock Mts., Gaspé Pen., E Que. The reduced northern and alpine extreme may be distinguished as var. *thyrsoidea* (Mey.) Fern. (*S. thyrsoidea* Mey., the type from Okak, Labrador, 57°33′N; *S. virgaurea sensu* Pursh 1814, not L.). Its f. *mensalis* (Fern.) Beaudry (*S. mensalis* Fern., the type from Tabletop Mt., Gaspé Pen., E Que.; achenes appressed-pilose rather than glabrous) is known from Que. (Shickshock Mts. of the Gaspé Pen. and Chimo, s Ungava Bay; J.R. Beaudry, Nat. can. (Que.) 91(6–7):195. 1964). Its f. *ramosissima* Lepage (plant bearing flowering branches nearly to the base of the stem) is known from the type locality, Fort George, James Bay, Que., ca. 54°N.

S. missouriensis Nutt.

/T/WW/ (Hsr) Dry prairies, sands, and gravels, the aggregate species from B.C. (N to Taylor Flats in the Peace River dist. at ca. 56°N) to Alta. (N to Dunvegan, 55°54′N), Sask. (N to Prince Albert), Man. (N to The Pas), and w Ont. (NW shore of L. Superior near Thunder Bay; also known from Lambton and Wentworth counties, s Ont., where perhaps introd., as also in ?Tenn. and N.J.), s to Ariz., Tex., and Mo.
1 Plants to about 9 dm tall; basal and lowermost stem-leaves mostly deciduous; inflorescence distinctly secund (the heads 1-sided on the branches); [*S. glaberrima* Martens; Alta. to w Ont.] .... var. *fasciculata* Holz.
1 Plants seldom over 5 dm tall; lowermost leaves mostly persistent, the middle and upper ones relatively few and reduced.
  2 Heads relatively large, the involucre mostly 4 or 5 mm high; inflorescence seldom at all secund; [*S. concinna* Nels.; s Alta.; collections from Field and Wilmer, B.C., are

also placed here by Eastham 1947, but Boivin 1967a, refers one (not stating which) to
S. multiradiata, the other collection not having been seen by him] .... var. *extraria* Gray
2   Heads mostly smaller, the involucre rarely as much as 5 mm high; inflorescence
tending to be somewhat secund; [var. *montana* Gray; B.C. (N to Taylor Flats, ca.
56°N) to Ont. (N to near Thunder Bay)] .... var. *missouriensis*

S. mollis Bartl.
/T/WW/ (Hpr) Dry plains, prairies, and sandy roadsides from W Mont. to S Alta. (clayey ditch at
Lethbridge, where perhaps introd.; CAN), Sask. (N to Saskatoon; CAN), and SW Man. (N to
Brandon; DAO), S to N.Mex., Okla., and Minn. [*S. nemoralis* var. *incana* (T. & G.) Gray (*S. inc.* T. &
G.)].

S. multiradiata Ait.
/aST/X/eA/ (Hsr) Meadows and rocky places at low to high elevations, the aggregate species
from the Aleutian Is. and coasts of Alaska–Yukon–Dist. Mackenzie to Victoria Is., Great Bear L.,
Great Slave L., S Baffin Is., and northernmost Ungava–Labrador (the type material being cult.
specimens originating from Labrador), S in the West to Calif. and N.Mex., farther eastwards S to
cent. Sask. (Ile-à-la-Crosse, 55°27'N, and Waskesiu Lake, 53°55'N; Breitung 1957a), Man. (S to
Cross Lake, NE of L. Winnipeg; CAN), N Ont. (coasts of Hudson Bay–James Bay), Que. (S to S
James Bay, Bic, Rimouski Co., and the N shore and Shickshock Mts. of the Gaspé Pen.), Nfld.,
N.B. (near Hillsborough, Albert Co.; P.R. Roberts, Rhodora 67(769):92. 1965; not known from
P.E.I.), and N.S. (St. Paul Is., Inverness Co., Cape Breton Is.; GH); Chukch Pen., NE Siberia. MAPS
and synonymy: *see* below.
1   Heads in open roundish-topped corymbiform clusters; phyllaries often subobtuse; lower
leaves rather consistently ciliate-margined; [*S. scop.* (Gray) Nels.; *S. ciliosa* Greene; *S.
corymbosa* Nutt.; the Yukon–Dist. Mackenzie–B.C. to Man.; MAP: Hultén 1968b:852] ....
.... var. *scopulorum* Gray
1   Heads in close compact clusters; phyllaries subacute to attenuate; lower leaves often
ciliate only toward base.
   2   Involucres commonly 4 or 5 mm high, with usually about 15 phyllaries; [N Dist.
Mackenzie (Eskimo L. basin); SW James Bay, Ont.; Gaspé Pen.; E Que. (Mt. Albert;
type from Cap-des-Rosiers)] .... var. *parviceps* Fern.
   2   Involucres commonly 6 or 7 mm high, with usually at least 20 phyllaries; [incl. var.
arctica (DC.) Fern. (*S. virgaurea* var. *arctica* DC.); transcontinental; type from
Labrador; MAPS: Porsild 1957: map 304, p. 198; Raup 1947: pl. 35; Hultén 1968b:852] .....
.... var. *multiradiata*

S. nemoralis Ait.
/sT/X/ (Hs (Hsr)) Dry prairies, sterile soils, and open woods from B.C. (N to Tête Jaune, about
55 mi W of Jasper, Alta.) to L. Athabasca (Alta. and Sask.), Man. (N to The Pas.), Ont. (N to
Schreiber, N shore of L. Superior), Que. (N to Ste-Anne-de-la-Pocatière, Kamouraska Co.; CAN;
reports from Anticosti Is. by John Macoun 1884, and Schmitt 1904, may be based upon *S. hispida,*
as also the report from Nfld. by Waghorne 1898), N.B., P.E.I., and N.S., S to Mont., Ariz., Tex., and
Ga.
    Our B.C.–Alta.–Sask. material (and most of our Man. material except from SE Man.; *see* Löve and
Bernard 1959:432) is referable to var. *decemflora* (DC.) Fern. (*S. dec.* DC.; *S. longipetiolata* Mack.
& Bush; *S. pulcherrima* Nels.; basal leaves relatively narrow, subentire or shallowly toothed rather
than distinctly crenate-serrate; heads all pedicelled rather than subsessile or pedicelled).

S. occidentalis (Nutt.) T. & G.
/T/WW/ (Hpr) Moist valleys and plains from S B.C. (Keremeos, Penticton, Salmon Arm,
Okanagan L., and Golden; CAN; reported from between Summerland and Osoyoos and from
Kinbasket L., about 60 mi N of Revelstoke, by Eastham 1947; reports from Alta. require con-
firmation) to Calif., N.Mex., and Nebr. [*Euthamia* Nutt.].

[*S. odora* Ait.]   Sweet Goldenrod
[The report of this species of the E U.S.A. (N to Mo. and N.H.) from N.S. by Lindsay (1878; near

Grand L., ?Halifax Co.) requires clarification. The tentative report from near Hamilton, s Ont., by John Macoun (1884; presumably taken up by Soper 1949) is probably referable to *S. missouriensis,* known from Hamilton (TRT), this supported by an 1892 Macoun collection in CAN from Sandwich, Essex Co., distributed as *S. odora* but proving to be *S. missouriensis.* Other collections in TRT from Windsor, Essex Co., and St. Thomas, Elgin Co., may also prove to be this latter species.]

S. ohioensis Riddell

/T/EE/ (Hs) Wet prairies, calcareous bogs, and sandy shores from Wisc. and Ill. to s Ont. (Lambton, Lincoln, Waterloo, Simcoe, and Bruce counties; CAN; OAC; TRT) and NW N.Y. [*Aster* Ktze.; *Oligoneuron* Jones; *S. houghtonii sensu* John Macoun 1884, at least in part, a relevant collection from Red Bay, Bruce Co., s Ont., in CAN].

A hybrid with *S. ptarmicoides (Aster ptarm.* of the present treatment) is reported from s Ont. by Boivin (1967a; × *S. krotkovii* Boivin, the type from Stokes Bay, Bruce Co.).

S. patula Muhl.

/t/EE/ (Hs (Hsr)) Meadows, wet woods, and ledges from Minn. to s Ont. (Lambton, Norfolk, Middlesex, Welland, Lincoln, Waterloo, Wellington, and York counties; CAN; GH; OAC; TRT) and Vt., s to La. and N.C.

S. puberula Nutt.

/T/E/ (Hsr) Dry or peaty sterile soils, sands, and rocky barrens from Que. (N to L. St. John, Tadoussac, Saguenay Co., and Magdalen Is.; CAN; GH; MT; a collection in TRT from Brockville, NE shore of L. Ontario, Ont., has also been placed here) to N.B., P.E.I., and N.S., s in the Atlantic and Gulf states to NW Fla. and Miss. [Incl. vars. *borealis* and *expansa* Vict.].

Forma *albiradiata* Schofield & Smith (ray-ligules white rather than yellow) is known from the type locality near Goat Is., Lunenburg Co., N.S.

S. purshii Porter

/ST/EE/ (Hsr) Peaty or damp places from Man. (N to Lac du Bonnet, about 50 mi NE of Winnipeg) to Ont. (N to the Fawn R. at ca. 54°30′N, 88°W), Que. (N to Ungava Bay and the Côte-Nord), Labrador (N to Makkovik, 55°05′N), Nfld., N.B., P.E.I., and N.S., s to Minn., Ind., Pa., and N.Y. [*S. chrysolepis* Fern.; *S. humilis* Pursh and its var. *abbei* Boivin; *S. uliginosa* of Canadian reports in part, not Nutt.].

A hybrid with *S. rugosa* is reported from sw Que. by J.R. Beaudry and D.L. Chabot (Can. J. Bot. 37(2):216. 1959; St-Adolphe, Argenteuil Co.).

S. riddellii Frank

/T/EE/ (Hs) Wet prairies, swamps, and ditches from SE Man. (near Otterburne, about 30 mi s of Winnipeg; Löve and Bernard 1959) to s Ont. (Essex, Kent, Lambton, and Bruce counties; CAN; TRT), s to Mo., Ohio, and Va. [*Oligoneuron* Rydb.].

A hybrid with *S. rigida* (× *S. maheuxii* Boivin) is reported from the type locality, Kleefeld, SE Man., by Boivin (1966b). See note under *Aster ptarmicoides* var. *lutescens.*

S. rigida L.

/T/(X)/ (Hs (Hsr)) Dry prairies, thickets, and open woods from Alta. (N to near the B.C. boundary at ca. 56°N; CAN) to Sask. (N to McKague, 52°37′N; CAN), Man. (N to Steeprock, about 100 mi N of Portage la Prairie; CAN), Ont. (near Thunder Bay; Essex, Elgin, Lambton, Middlesex, Brant, Waterloo, Renfrew, and York counties; CAN; TRT), and Mass., s to N.Mex., Tex., La., and Ga. [*Oligoneuron* Rydb.].

Apart from the s Ont. material, our plant is referable to var. *humilis* Porter (var. *canescens* (Rydb.) Breitung; *Oligoneuron (S.) canescens* Rydb.; *S. parvirigida* Beaudry; achenes with a few short loose hairs near the summit rather than completely glabrous; stem relatively short and slender). A hybrid between *S. parvirigida* and *S. riddellii* is reported from SE Man. by Löve and Bernard (1959; near Otterburne, about 30 mi s of Winnipeg).

S. rugosa Ait.

/T/EE/ (Hpr) Damp woods, thickets, and meadows, the aggregate species from Ont. (N to the E shore of L. Superior about 35 mi NW of Sault Ste. Marie) to Que. (N to SE James Bay, L. Mistassini, the Côte-Nord, and Gaspé Pen.), Nfld., N.B., P.E.I., and N.S., S to Tex. and Fla.

1 Stem merely scabrous-puberulent to short-hispid; leaves rounded at base, usually rather
   shallowly toothed, hispid beneath; [*S. aspera* Ait.; *S. asperata* Pursh; Ont. is included in
   the range given by Gleason 1958] .... var. *aspera* (Ait.) Fern.

1 Stem sordid-villous at least above; leaves tapering at base, commonly coarsely
   sharp-serrate, more or less villous beneath .... var. *rugosa*

  2 Lower panicle-branches often overtopped by the subtending leaves, these to over 1
     dm long; [*S. villosa* Pursh; Ont. to Nfld. and N.S.] .... f. *villosa* (Pursh) Beaudry

  2 Lower panicle-branches much surpassing the subtending leaves, these mostly less
     than 7 cm long; [Ont. to Nfld. and N.S.; reports from Man. require confirmation; MAP:
     R.H. Goodwin, Rhodora 39(459): fig. 1 (aggregate species; incomplete northwards),
     p. 23. 1937]. A purported hybrid with *S. uliginosa* (× *S. beaudryi* Boivin) is reported
     from the type locality, St-Adolphe-de-Howard, SW Que., by Boivin 1966*b*. × *S.*
     asperula Desf. is apparently a series of hybrids of various forms of *S. rugosa* and *S.*
     sempervirens. It is known from Que., N.S. (*see* map by Goodwin, loc. cit., fig. 3, p.
     25), and P.E.I. (Queens Co.; ACAD) .... f. *rugosa*

S. sciaphila Steele

/T/EE/ (Hsr) Ledges, cliffs, and sands from Minn. to Mich. and S Ont. (Boivin 1966*b*; Wasaga Beach, S Georgian Bay, L. Huron), S to Iowa and Ill.

S. sempervirens L.   Seaside Goldenrod

/T/E/ (Hsr) Saline or brackish (sometimes fresh) places near the coast from Que. (St. Lawrence R. estuary from Berthier-en-Bas, Montmagny Co., to Tadoussac, Saguenay Co., and the Gaspé Pen.; not known from Anticosti Is.) to Nfld., N.B., P.E.I., N.S., N.J., and Va. [*S. laevigata* and *S. viminea* Ait.]. MAP: R.H. Goodwin, Rhodora 39(459): fig. 2, p. 24. 1937.

    Forma *ochroleuca* Weatherby (ray-ligules very pale-yellow, almost white, rather than deep yellow) is known from the type locality, Parrsboro, Cumberland Co., N.S. A hybrid with *S. uliginosa* is reported from N.S. by Boivin (1966*b*; St. Paul Is., Inverness Co., Cape Breton Is.).

S. spathulata DC.

/ST/X/ (Hsr) Meadows and rocky places at low to alpine elevations, the aggregate species from Alaska–Yukon (N to ca. 68°N) and the Mackenzie R. Delta to Great Bear L., Great Slave L., L. Athabasca (Alta. and Sask.), Man. (N to the Nelson R. about 155 mi S of Churchill; CAN), Ont. (N to Schreiber, N shore of L. Superior), Que. (N to Anticosti Is.; CAN; MT), N.B., and N.S. (not known from P.E.I.), S to Calif., Ariz., N.Mex., Wisc., Mich., and Va. MAP and synonymy (together with a distinguishing key to two closely related, if not identical, "microspecies" of E Que.): *see* below.

1 Basal leaves tending to be acute or acutish; [Ont. eastwards] ....
   .... *S. spathulata* ssp. *randii* (Porter) Cronq.

  2 Heads large (the involucres to 8(9) mm high), numerous in an often branched and
     loose inflorescence; [*S. humilis* var. *gil.* Gray; *S. racemosa* var. *gil.* (Gray) Fern.; S
     Ont. and N.B.] .... var. *gillmanii* (Gray) Cronq.

  2 Heads mostly smaller (the involucre often not over 6 mm high).

    3 Basal leaves up to 20 times as long as broad, tending to be subentire;
      inflorescence tending to be relatively loose and raceme-like, the heads on
      pedicels to over 1 cm long; [*S. racemosa* Greene; Ont., Que., and N.B.] ....
      .... var. *racemosa* (Greene) Gl.

    3 Basal leaves mostly not more than 8 times as long as broad, tending to be sharply
      toothed; inflorescence tending to be relatively compact and thyrsoid, the heads on
      pedicels rarely over 4 mm long; [*S. randii* (Porter) Britt.; Ont., Que., and N.S.] ....
      .... var. *randii*

1 Basal leaves tending to be broadly obtuse or rounded at summit.

  4 Leaves pale green, thickish; involucres 4 or 5 mm long, their phyllaries with an

inconspicuous midrib; lobes of disk-corollas at most 1 mm long; pedicels to over 1 cm
long .... S. spathulata ssp. spathulata
5   Stems to 8 dm tall; basal leaves mostly oblanceolate; inflorescence relatively
    elongate and loose, the heads on pedicels to over 1 cm long; [S. ?chlorolepis
    Fern.; S. confertiflora DC., not Nutt.; S. decumbens var. oreophila (Rydb.) Fern.
    (S. oreophila Rydb.); S. glutinosa Nutt.; Alaska–B.C. to Man.; ?Ont.; E ?Que.;
    MAP (S. dec. var. oreo.): Hultén 1968b:853] .... var. neomexicana (Gray) Cronq.
5   Stems to about 1.5 dm tall; basal leaves mostly spatulate or obovate;
    inflorescence short and compact; [S. humilis (glutinosa) var. nana Pursh; S.
    decumbens Greene; the Yukon–B.C.–Alta.] .... var. nana (Gray) Cronq.
4   Leaves dark green, relatively thin; involucres over 5 mm high, their phyllaries with a
    conspicuous deep-green midrib and tip; lobes of disk-corollas to over 1 mm long;
    inflorescence compact, the heads on pedicels rarely over 4 mm long; [types from
    Anticosti Is., E Que.].
    6   Involucre to 8 mm long; inner phyllaries oblong, often over 1 mm broad; lobes of
        disk-corollas at least 1.3 mm long .... S. anticostensis Fern.
    6   Involucre at most about 6 mm long; inner phyllaries spatulate, rarely over 1 mm
        broad; lobes of disk-corollas about 1 mm long; [× S. raymondii Rousseau, an
        apparent hybrid between S. victorinii and S. (spathulata var.) racemosa, is
        reported from the type locality, Anticosti Is., E Que., by J. Rousseau 1950] ....
        .... S. victorinii Fern.

S. speciosa Nutt.
/t/EE/   (Hs)   Dry to moist prairies, thickets, and open woods, the aggregate species from Minn.
and Mich. (the report E to Sask. by Fernald in Gray 1950, requires clarification) to s Ont. (N to Bruce
and York counties), N.Y., and Mass., s to Tex. and Ga.
1   Stem-leaves relatively few; lower leaves narrow as in var. angustata but usually
    persistent; [S. jejunifolia Steele; S. uliginosa var. jej. (Steele) Boivin; S. klughii Fern.; s
    Ont., the type of S. klughii from Oliphant, Bruce Co.] .... var. jejunifolia (Steele) Cronq.
1   Stem-leaves numerous; inflorescence dense.
    2   Lower leaves oblanceolate to spatulate-oblong, mostly entire, generally deciduous;
        [var. rigidiuscula T. & G. (S. rig. (T. & G.) Porter); s Ont.: Lambton, Peel, and York
        counties] .... var. angustata T. & G.
    2   Lower leaves ovate to oblong or obovate, entire or serrate, mostly persistent;
        [S. conferta Mack.; E U.S.A. only, Canadian reports referring to the above taxa] ....
        .... [var. speciosa]

S. squarrosa Muhl.
/T/EE/   (Hs (Hsr))   Rich open woods, thickets, and clearings from Ont. (N to Constance Bay,
about 20 mi w of Ottawa) to Que. (N to L. Timiskaming at ca. 47°30′N and the sw Gaspé Pen. at
Matapédia) and N.B. (CAN; GH; not known from P.E.I. or N.S.), s to Ky. and N.C.

S. tenuifolia Pursh
/T/EE/   (Hpr)   Swampy ground and dry to wet sands, gravels, and peats, the aggregate species
from Mich., s ?Ont., and Ind. to w N.S. (E to Halifax Co.; see N.S. map 441 for S. galetorum by
Roland 1947:578), s to Va.
1   Primary stem-leaves usually subtending fascicles of reduced leaves or sterile branches;
    leaves thin, acuminate; disk-florets 5–7, fewer than the rays; [Euthamia Greene; s Ont.
    (Middle Sister Is. of the Erie Archipelago, Essex Co.; Core 1948) and w N.S.] ... var. tenuifolia
1   Primary stem-leaves seldom subtending reduced leaves or branches; leaves thickish,
    obtuse to short-acute; disk-florets usually 12 or more, at least as numerous as the rays;
    [S. galetorum (Greene) Friesner; N.S., the type from Salmon L., Yarmouth Co.] ....
    .... var. pycnocephala Fern.

S. uliginosa Nutt.
/sT/X/   (Hsr)   Moist to dryish thickets and acid swamps, bogs, and rocks (ranges of Canadian
taxa outlined below), s to Wisc., Ohio, N.Y., and N.H.

1 Stem to 1.5 dm tall, with up to 40 leaves; lower leaves to 8 cm broad; panicle to 4.5 dm long and 2.5 dm broad; [incl. var. *levipes* Fern.; *S. neglecta* T. & G.; *S. uniligulata* var. *negl.* (T. & G.) Fern.; s Ont. (Lambton Co.) and N.S.] . . . . var. *uliginosa*
1 Stem less than 1 m tall, commonly with not more than 20 leaves; lower leaves mostly not over 3 cm broad.
  2 Panicle relatively narrow, to 2.5 dm long and 1 dm thick; [*S. linoides* T. & G.; *S. neglecta* var. *lin.* (T. & G.) Gray; *S. uniligulata* (DC.) Porter; *S. humilis* var. *peracuta* Fern.; SE Man. (CAN; DAO) and Ont. (N to w James Bay at ca. 52°10′N) to Que. (N to s James Bay and the Côte-Nord), Labrador (N to the Hamilton R. basin), Nfld., N.B., P.E.I., and N.S.] . . . . var. *linoides* (T. & G.) Fern.
  2 Panicle corymbiform to broadly pyramidal, to about 1.5 dm thick; [*S. terrae-novae* T. & G., the type from Nfld.; *S. uniligulata* var. *ter.* (T. & G.) Fern.; E Que. (Magdalen Is.), Nfld., N.B., and N.S.] . . . . var. *terrae-novae* (T. & G.) Fern.

S. ulmifolia Muhl.
/t/EE/ (Hsr) Dry rocky woods and thickets from SE Minn. to s Ont. (Lincoln and Welland counties; CAN; TRT), Vt., and Mass. (the inclusion of N.S. in the range by Gleason 1958, requires clarification), s to Tex., Okla., Ark., and Ga.

SONCHUS L. [9595] Sow-Thistle. Laiteron

1 Perennials with deep vertical roots and extensively creeping horizontal rootstocks; heads to 5 cm broad, the flowers bright yellow to orange-yellow; involucre to over 2 cm long; achenes with at least 5 prominent longitudinal ribs, strongly rugose; (introd., transcontinental) . . . . *S. arvensis*
1 Annuals with taproots; heads less than 2.5 cm broad, pale yellow; involucre rarely over 12 mm long; (introd., transcontinental).
  2 Achenes 3-nerved on each side, otherwise smooth; leaves strongly spiny-toothed, their basal auricles rounded . . . . *S. asper*
  2 Achenes striate and transversely wrinkled, papillate; leaves with soft spiny teeth and acute basal auricles . . . . *S. oleraceus*

S. arvensis L. Field-Sow-Thistle
Eurasian; introd. along roadsides and in fields and waste places in N. America, as in s Alaska (N to ca. 59°N), sw Yukon (Whitehorse; CAN), sw Dist. Mackenzie, and all the provinces (in Man., N to Churchill; in Ont.–Que., N to James Bay).
    The following key includes characters used by Fernald *in* Gray (1950) to separate the nonglandular *S. arvensis* var. *glabrescens* from the nonglandular (but supposedly distinct) *S. uliginosus* Bieb. Following cytological studies, however, W. Shumovich and F.H. Montgomery (Can. J. Agric. Sci. 35(6):601–05. 1955) conclude that these two taxa are both tetraploids (2n = 18) and that the name *S. uliginosus* Bieb. should, through priority, be used to unite the nonglandular taxa as a species distinct from the hexaploid (2n = 27) *S. arvensis* var. *arvensis*, the glandular plant.
1 Involucres and peduncles stipitate-glandular.
  2 Gland-tipped hairs few; [type from a plant nursery at Ottawa, Ont.; denoted by Boivin 1967a, as a hybrid between *S. arvensis* and its var. *glabrescens*] . . . .
. . . . var. *shumovichii* Boivin
  2 Gland-tipped hairs numerous; [Alaska (Juneau, Hyder, Burroughs Bay, and Port Vita; Hultén 1950), sw Dist. Mackenzie (J.W. Thieret, Can. Field-Nat. 75(3):120. 1961), B.C. (Henry 1915), Alta. (Moss 1959), Sask. (Breitung 1957a), Man. (N to the Carrot R. N of The Pas), Ont. (N to w James Bay at ca. 53°N), Que. (N to E James Bay at ca. 51°30′N and the Côte-Nord), Nfld., N.B., P.E.I., and N.S.; MAP (aggregate species): Hultén 1968b:950] . . . . var. *arvensis*
1 Involucres and peduncles glabrous or with some small and obscure tufts of tomentum, but not stipitate-glandular.
  3 Involucres broadly campanulate to hemispherical; phyllaries uniformly deep green to lead-colour; [*S. ?uliginosus* Bieb.; essentially the range of the species but not

definitely known from Alaska–B.C. and in Man. extending N to Churchill] ....
.... var. *glabrescens* Guenth., Grab., & Wimm.
3   Involucres relatively slender, their pale phyllaries white-margined ...... *S. uliginosus* Bieb.

S. asper (L.) Hill   Spiny-leaved Sow-Thistle.   Chaudronnet
Eurasian; introd. along roadsides and in fields and waste places in N. America, the ranges of
Canadian taxa outlined below.
1   Stem and floral-axis stipitate-glandular with reddish glands; [probably throughout the
    range; known definitely from Man. (Bowsman River, N of Duck Mt.), Ont. (Kapuskasing),
    and Que. (Timiskaming)] .... f. *glandulosa* Beckh.
1   Plant not stipitate-glandular.
    2   Stem-leaves undivided; [Man. (near Otterburne, about 30 mi s of Winnipeg), Ont.
        (Kapuskasing), Que. (Charlevoix and Saguenay counties), N.B., and P.E.I.] ....
        .... f. *inermis* (Bisch.) Beck
    2   Stem-leaves more or less pinnatifid; [*S. oleraceus* var. *asper* L.; s Alaska–s
        Yukon–B.C.–Alta. to Sask. (N to Tisdale, 52°51′N), Man. (N to Swan R., N of Duck
        Mt.), Ont. (N to Kapuskasing, 49°24′N), Que. (N to the Côte-Nord), Labrador (Hamilton
        R. basin), Nfld., N.B., P.E.I., and N.S.; w Greenland N to ca. 70°N; MAP: Hultén
        1968b:951] .... f. *asper*

S. oleraceus L.   Common-Sow-Thistle, Milk-Thistle.   Laiteron potager
Eurasian; introd. in cult. fields and waste ground in N. America, as in cent. Alaska (Mendenhall;
Hultén 1950), sw Dist. Mackenzie (near Simpson at 62°51′N; Raup 1947), and all the provinces (in
Ont., N to w James Bay at ca. 53°N), in w Greenland N to ca. 70°N. [Var. *triangularis* Wallr.]. MAP:
Hultén 1968b:951.
    Forma *lacerus* (Willd.) Beck (leaf-lobes all narrow and subequal rather than the terminal half of
the leaf much larger than the lateral lobes) is known from P.E.I. (wharf at Charlottetown; ACAD)
and probably occurs throughout the range.

STEPHANOMERIA Nutt.   [9576]   Rush-Pink, Skeletonweed

1   Flowers mostly 10–21, the heads relatively large, with up to 8 principal phyllaries;
    achenes longitudinally ribbed and grooved; leaves entire or more often with distant salient
    sharp teeth or slender lobes; stem mostly simple but sometimes branched at the base, to
    3 dm tall, from deep-seated creeping roots; (B.C.) .... *S. lactucina*
1   Flowers mostly 5, the heads relatively small, with mostly 5 principal phyllaries; stems
    several from a taproot often surmounted by a stout branching caudex.
    2   Plants mostly not over about 2 dm tall, the principal leaves runcinate-pinnatifid;
        achenes more or less rugose-tuberculate and pitted as well as longitudinally grooved
        (rather than ribbed); (sw Alta. to sw Sask.) .... *S. runcinata*
    2   Plants to about 7 dm tall, with filiform or linear, entire or toothed leaves; achenes
        longitudinally ribbed and grooved, otherwise smooth or very nearly so; (s B.C.) ....
        .... *S. tenuifolia*

S. lactucina Gray
/t/W/   (Gr)   Dry slopes and pine forests in the foothills from B.C. (T.M.C. Taylor 1966b, the
locality or localities not given but presumably near the U.S.A. boundary in the Dry Interior) to Calif.
and Nev.

S. runcinata Nutt.
/T/WW/   (Grt)   Dry plains and foothills from Mont. to sw Alta. (along the St. Mary R., s of
Lethbridge, where taken by John Macoun in 1895; CAN; reported from near Milk River by J.M.
Macoun 1896) and sw Sask. (J.M. Macoun 1896; Wood Mountain, about 85 mi SE of Swift Current),
s to Colo. and NW Nebr. [*S. minor sensu* John Macoun 1884, not (Hook.) Nutt.].

S. tenuifolia (Torr.) Hall
/t/WW/   (Grt (Gr))   Dry rocky places from the plains to moderate elevations in the mts. from s

B.C. (valleys of the Dry Interior N to Clinton, about 60 mi NW of Kamloops, E to Keremeos, near the U.S.A. boundary SW of Penticton) and Mont. to Calif. and Tex. [*?Prenanthes* Torr.; *Ptiloria* Raf.; *Lygodesmia* Shinners; *L. (S.) minor* Hook.].

[TAGETES L.] [9311] Marigold

[T. patula L.] French Marigold
[This Mexican species (the genus not keyed out above but closely related to *Helenium*) is reported as introd. but not persisting in Que. by C. Rousseau (Nat. can. (Que.) 98:727. 1971; Ste-Foy, near Quebec City). It is a bushy-branched annual, the leaves pinnately divided into about 12 lanceolate to oblong serrate leaflets (the teeth tipped with a long weak awn), the long-peduncled heads solitary, about 3.5 cm broad, the numerous rays yellow with red markings.]

TANACETUM L. [9341] Tansy. Tanaisie

1 Heads numerous in a flattish-topped corymbose inflorescence, the disks at most 1 cm broad; leaves 2-pinnatifid, essentially glabrous; stem erect, glabrous; (introd., transcontinental) .... *T. vulgare*
1 Heads fewer, the disks over 1 cm broad; stem more or less villous or lanate, at least in youth.
   2 Heads commonly not more than 3, the disks to 2.5 cm broad; rays to 4 mm broad, protruding by as much as 5 mm; leaves essentially 2-pinnatifid, the ultimate segments comparatively broad; stem nearly erect, villous; (Alaska) .... *T. bipinnatum*
   2 Heads numbering up to 20 or more, the disks rarely over 2 cm broad; rays to 2.5 mm broad, protruding at most 3 mm; leaves mostly 3-pinnatifid; stems more or less decumbent at base.
      3 Divisions of leaves obtuse or rounded at apex, the ultimate ones oblong; heads commonly rather numerous (but seldom more than 20); rays deeply 3-lobed, not much surpassing the adjacent disk-florets; (?B.C.) .... [*T. douglasii*]
      3 Divisions of leaves acute, the ultimate ones lanceolate; heads usually fewer; rays obscurely lobed, markedly surpassing the adjacent disk-florets; (trans-
continental) .... *T. huronense*

T. bipinnatum (L.) Schulz-Bip.
/aSs/W/EA/ (Hsr) Sandy or peaty places in Alaska (N to ca. 71°N; the report from the Mackenzie R. Delta by Porsild 1951a, is apparently referable to *T. huronense*); NE Europe; N Asia. [*Chrysanthemum* L.; *Pyrethrum* Willd.; *T. (Artemisia) kotzebuense* Bess.]. MAP: Hultén 1968b:892 (*Chrys. bip.*).

[T. douglasii DC.]
[The inclusion of B.C. in the range of this species of the W U.S.A. (Wash. to Calif.) by Hitchcock et al. (1955) is referred by Calder and Taylor (1968) to *T. huronense* var. *huronense,* with which they merge it in synonymy. (*Chrysanthemum* Hult.).]

T. huronense Nutt.
/ST/(X)/ (Hsr) Peaty, sandy, or gravelly shores and slopes (chiefly calcareous), the aggregate species from Alaska–Yukon (N to ca. 67°30′N), the Mackenzie R. Delta, and W B.C. (Queen Charlotte Is. and Vancouver Is.) eastwards in scattered stations in Canada, Mich., and Maine as outlined below. MAP and synonymy: *see* below.
1 Heads commonly more than 6, the flowering stem to about 8 dm tall and with up to about 15 leaves; [*Chrysanthemum bipinnatum* ssp. *hur.* (Nutt.) Hult.; *T. pauciflorum sensu* Richardson 1823, and Hooker 1833, not DC. nor Fisch.; Alaska-Yukon-NW Dist. Mackenzie–W B.C.; shores of Lakes Huron, Michigan, and Superior (type from Mich.); MAP (aggregate species): Hultén 1968b:893 (*Chrys. bip.* var. *hur.*)] .... var. *huronense*
1 Heads commonly less than 6, the flowering stems rarely over 4 dm tall.
   2 Leaves to 3 dm long and about 1.5 dm broad, their acute primary segments comparatively remote and with rather remote ultimate segments; flowering stem to 4.5

dm tall, with up to 5 heads and 10 leaves; [incl. var. *monocephalum* Boivin; Mich., Ont. (N to S James Bay), E Que. (Kamouraska, Temiscouata, Matapédia, and Bonaventure counties), N.B. (Restigouche R. and St. John R. systems, the type from Woodstock), and Maine] .... var. *johannense* Fern.

2   Leaves at most about 1 dm long and 5 cm broad, their bluntish primary segments with rather crowded divisions.

   3   Flowering stem to over 3 dm tall, glabrous or sparingly pilose, with up to about 10 sparsely pilose leaves and up to 6 heads; [*Omalanthus (T.) camphoratus sensu* Hooker 1833, as to the York Factory, Man., plant, not Less.; N Alta. (L. Athabasca); NE Man. (York Factory region); Ont. (N to w Hudson Bay at ca. 56°40′N); Que. (E James Bay–Hudson Bay N to Hudson Strait; type from Anticosti Is.] .... var. *bifarium* Fern.

   3   Flowering stem at most about 2 dm tall, copiously lanate, with rarely more than 4 white-lanate leaves and 1 or 2 heads; [var. *floccosum* Raup; f. *lanatum* Rousseau; N Sask. (L. Athabasca); cent. Ont. (Winisk, 55°12′N); Que. (E James Bay; Ungava Bay; Anticosti Is.); Nfld. (type from Ingornachoix Bay)] ....
.... var. *terrae-novae* Fern.

T. vulgare L.   Common Tansy, Golden-buttons
Eurasian; introd. along roadsides and borders of woods and in fields and waste places in N. America, as in SE Alaska (Hyder, Douglas, and Sitka; Hultén 1950), sw Dist. Mackenzie (J.W. Thieret, Can. Field-Nat. 76(4):208. 1962), s B.C. (Vancouver Is.; Yale; Kootenay L.), Alta. (N to Edmonton), Sask. (N to Golburn, 52°46′N), Man. (N to Swan River, N of Duck Mt.), Ont. (N to Renison, s of James Bay at ca. 51°N), Que. (N to the Côte-Nord, Anticosti Is., and Gaspé Pen.), ?Labrador (Boivin 1966*b*), Nfld., N.B., P.E.I., and N.S. [*Chrysanthemum* Bernh.]. MAP (*Chrys. vulg.*): Hultén 1968*b*:891.

   Forma *crispum* (L.) Fern. (var. *crispum* L.; leaves deeply incised and crisped rather than merely cut-toothed) occurs nearly throughout our range.

TARAXACUM Zinn   [9592]   Dandelion.   Pisenlit

1   Flowers creamy white or pale yellow, often suffused with pink; achenes copiously tuberculate above the middle, commonly smoothish below; inner phyllaries not conspicuously corniculate-appendaged (horn-tipped); leaves subentire or shallowly and broadly few-lobed; (arctic regions) .... *T. hyparcticum*

1   Flowers sulphur- to orange-yellow.

   2   Mature achenes mostly tuberculate nearly to base, the tubercles crowded in the upper half, the surface of the achene lacking conspicuous flat areas between the tubercles.

      3   Achene-beak at most about 5 mm long; involucre blackish, less than 1.5 cm high, with rarely more than 12 phyllaries, these with callous or only slightly corniculate-appendaged tips; leaves subentire, dentate, or sinuate; scape less than 1 dm tall; (arctic–subarctic regions) .... *T. phymatocarpum*

      3   Achene-beak at least 6 mm long; involucre lighter in colour; leaves mostly dentate to runcinate; scapes commonly taller.

         4   Many phyllaries with a conspicuous corniculate appendage near the tip; outer phyllaries short and appressed or, if otherwise, broader than the inner ones; pappus creamy.

            5   Inner phyllaries at most about 2 cm long, their tips with short appendages or none; outer phyllaries tightly appressed, short, firm, with conspicuous white margins; achene-beak less than 1.5 cm long; (chiefly subarctic and alpine regions) .... *T. ceratophorum*

            5   Inner phyllaries to about 2.5 cm long, mostly with coarse appendages much surpassing their tips; outer phyllaries about 2/3 as long as the inner, finally loosely spreading or recurving, lacking conspicuous white margins; achene-beak to 17 mm long; (E Que. and w Nfld.) .... *T. laurentianum*

         4   Most or all phyllaries unappendaged near tip; outer phyllaries thin and

herbaceous (if strongly recurving, only slightly broader than the inner ones); pappus white.

    6   Leaves commonly broad at base and with mostly entire teeth and lobes; inner phyllaries during anthesis united only at base; achenes pale brown or reddish, the body to 4.5 mm long, the beak to 9 mm long; (E Que., S Labrador, and Nfld.) . . . . *T. ambigens*

    6   Leaves commonly slender-petioled, their lobes and teeth themselves sharply toothed; inner phyllaries during anthesis united up to 6 mm above base; achenes commonly paler, the body less than 4 mm long, the beak to 12 mm long; (E Que., Nfld., and N.B.) . . . . *T. latilobum*

2   Mature achenes tuberculate only above middle or, if occasionally below, the surface of the achene with broad flat areas between the remote tubercles.

    7   Many phyllaries with a conspicuous corniculate appendage near tip.

        8   Achenes red or reddish purple, the body to 3.5 mm long, the beak to 8 mm long; pappus creamy or sordid; rays sulphur-yellow; leaves slender-petioled, cleft nearly or quite to the midrib into long narrow lobes with intermediate shorter lobes; (introd. from S Dist. Mackenzie–B.C. to N.S.) . . . . *T. laevigatum*

        8   Achenes greyish, drab, olivaceous, or pale brown; rays commonly orange-yellow.

           9   Leaves deeply sinuate, the narrow lobes lacerate at base and with intermediate narrow lobes; achenes about 4 mm long, with a slender tip (pyramid) to 1.5 mm long below the beak, this to 13 mm long; pappus white; (E Que. and NW Nfld.) . . . . [*T. longii*]

           9   Leaves shallowly toothed or, if deeply lobed, the lobes chiefly entire; fruit with a stouter pyramid at least half as broad as long.

              10 Outer phyllaries straw-colour or whitish brown, lance-attenuate; achenes about 3.5 mm long, the beak to 11 mm long; rays orange-yellow; leaves broadly oblanceolate; (the Yukon–B.C. to S Baffin Is. and Que.) . . . . *T. dumetorum*

              10 Outer phyllaries herbaceous, greyish brown, ovate or short-lanceolate; achenes to 4.5 mm long, the beak less than 9 mm long; pappus creamy; rays pale yellow; leaves linear-oblanceolate; (transcontinental in arctic, subarctic, and alpine regions) . . . . *T. lacerum*

    7   Most or all phyllaries unappendaged near tip.

      11 Achene-body to about 4.5 mm long, attenuate into the pyramid (this to 2 mm long), the beak commonly 6 or 7 mm long; leaves shallowly and broadly few-lobed, tapering into long slender petioles; scapes rarely over 2 dm tall; (N Que. and N Labrador) . . . . [*T. torngatense*]

      11 Achene-body more or less abruptly contracted to the short pyramid; scapes commonly taller.

        12 Leaves mostly broad at base or with broadly winged petioles, the lobes commonly entire, the intermediate lobes few or none; achenes to 5 mm long, olivaceous or greyish brown; (chiefly arctic and subarctic regions) . . . . *T. lapponicum*

        12 Leaves usually narrowed to slender petiolar bases, at least the longer lobes toothed and with frequent intermediate smaller lobes or teeth; achenes at most 4 mm long, olivaceous or tawny; (introd., trans-continental) . . . . *T. officinale*

T. ambigens Fern.
/T/E/ (Hr) Calcareous ledges, meadows, and shores of E Que. (Côte-Nord and Gaspé Pen.; CAN; GH), S Labrador (Forteau, 51°28′N), and Nfld. (type from Port au Choix). [Incl. var. *fultius* Fern.].

T. ceratophorum (Ledeb.) DC.
/AST/X/GEA/ (Hr) Meadows, ledges, and cliffs (chiefly calcareous) from the Aleutian Is. and Alaska (N to the N coast) to the Yukon (N to ca. 63°N), Great Bear L., Great Slave L., N Sask. (L.

Athabasca and Hasbala L.), Man. (N to Churchill; not known from Ont., the report from Moosonee, s James Bay, by Dutilly and Lepage 1947, being referred by Dutilly, Lepage, and Duman 1954, to *T. scanicum,* merged with *T. laevigatum* in the present treatment), s Baffin Is., Que. (coasts of Hudson Bay and Ungava Bay; Bic, Rimouski Co.; Gaspé Pen.), and Nfld. (not known from the Maritime Provinces), s in the West through B.C.–Alta. to Calif. and N.Mex.; w Greenland N to ca. 70°N; northernmost Greenland; Spitsbergen; Eurasia. [*Leontodon* Ledeb.; incl. *T. aleuticum* Tatew. & Kitamura, *T. chamissonis* and *T. ovinum* Greene, *T. pellianum* Porsild, *T. ruberaceum* Hagl., and *T. brachyceras, T. hyperboreum, T. lateritium,* and *T. trigonolobum* Dahlst.; *T. officinale* var. *?glaucescens* Koch; *T. ?carthamopsis* M.P. Porsild; *T. montanum* Nutt. (*Leontodon monticola* Rydb.), not (Mey.) DC.]. MAPS: Hultén 1968*b*:945; Raup 1947: pl. 37.

T. dumetorum Greene

/aST/X/ (Hr) Meadows and calcareous ledges from s-cent. Yukon (CAN) and sw Dist. Mackenzie (Fort Smith and Fort Providence, ca. 61°20'N; W.J. Cody, Can. Field-Nat. 75(2):68. 1961) to B.C.–Alta., Sask. (N to Prince Albert; CAN; cited in synonymy under *T. ceratophorum* by Breitung 1957*a*), Man. (N to Cross Lake, NE of L. Winnipeg), Ont. (N to Nipigon and sw James Bay at ca. 53°N; *see* Ont.–Que. James Bay map by Dutilly, Lepage, and Duman 1954: fig. 18, p. 125), Que. (Hudson Strait; James Bay watershed s to the Harricanaw R. at ca. 50°N), and s Baffin Is., the s limits very uncertain through confusion with other species. [*Leontodon* Rydb.; incl. *T. russeolum* Dahlst.].

T. hyparcticum Dahlst.

/Aa/(X)/G/ (Hr) Grassy tundra (often near animal burrows or human habitations) from Banks Is. to Prince Patrick Is. and Ellesmere Is. (N to ca. 81°N), s to the coast of Dist. Mackenzie and Baffin Is. at ca. 67°N; northernmost Greenland s to ca. 77°30'N. MAPS: Hultén 1968*b*:948; Porsild 1957: map 329, p. 202; Savile 1961: map D, p. 928.

T. lacerum Greene

/aST/X/G/ (Hr) Meadows and moist places from the coasts of Alaska–Yukon–Dist. Mackenzie–Dist. Keewatin to Banks Is., Prince Patrick Is., N Baffin Is., and northernmost Ungava–Labrador, s in the mts. of the West to s B.C.–Alta. (type from the upper Liard R., N B.C.), farther eastwards s to N Sask. (L. Athabasca; included in the synonymy of *T. ceratophorum* by Breitung 1957*a*), Man. (s to Cross Lake, NE of L. Winnipeg), northernmost Ont. (known only from the Hudson Bay coast at ca. 56°45'N), Que. (s to SE James Bay, the Côte-Nord, Anticosti Is., and Gaspé Pen.; not known from the Maritime Provinces), and Nfld.; w Greenland N to ca. 73°N. [Incl. *T. arctogenum, T. canadense, T. ceratodon, T. groenlandicum, T. malteanum, T. pseudonorvegicum,* and *T. umbrinum* Dahlst., *T. arcticum* (Trautv.) Dahlst., *T. mutilum* Greene, and *T. ochraceum* Hagl.; *T. leptoceras* of Greenland reports, not Dahlst.]. MAPS: Hultén 1968*b*:946; Porsild 1957: map 327, p. 201; Raup 1947: pl. 37; Savile 1961: map K, p. 929.

T. laevigatum (Willd.) DC.   Red-seeded Dandelion

European; introd. in dry sterile soils in N. America, as in sw Dist. Mackenzie (Fort Simpson, ca. 62°N; W.J. Cody, Can. Field-Nat. 75(2):68. 1961), B.C. (N to Kamloops and Revelstoke), Alta. (Moss 1959), Sask. (N to Foam L. at 53°36'N; Breitung 1957*a*), Man. (N to Gillam, about 165 mi s of Churchill; DAO), Ont. (N to the N shore of L. Superior at Schreiber and Rossport, and Moosonee, s James Bay, ca. 51°20'N), Que. (N to Berthier-en-Bas, Montmagny Co.; Marcel Raymond, Ann. ACFAS 7:105. 1941), N.B., and N.S. [*Leontodon* Willd.; *T. (L.) erythrospermum* Andrz.; incl. *T. scanicum* Dahlst.]. MAP: Hultén 1968*b*:947 (the N. American range given for the section *Erythrosperma* applies here).

The report of the similarly red-fruited *T. eriophorum* Rydb. (native in the w U.S.A. from Wash. to Wyo.) from Alaska by Hitchcock et al. (1955) is probably referable to the European *T. scanicum* Dahlst., reported from Alaska by Hultén (1950; 1968*b*) but included in *T. laevigatum* in the present treatment. The report from sw Alta. by Breitung (1957*b*; Waterton Lakes) may also prove referable to *T. laevigatum,* distinguished from *T. eriophorum* as follows:

1 Inner phyllaries commonly corniculate-appendaged; leaves tending to be deeply incised most of their length; (introd. weedy species) .... *T. laevigatum*

1   Inner phyllaries seldom appendaged; leaves less dissected; [native montane species of
    the w U.S.A.] .... [*T. eriophorum* Rydb.]

T. lapponicum Kihlm.
/aST/(X)/GEA/ (Hr) Meadows, damp ledges, shores, and alpine slopes, the range very
uncertain through confusion with other species but tentatively from the Aleutian Is. to the N coast of
Alaska, cent. Yukon, the Mackenzie R. Delta, N Sask. (L. Athabasca; included in *T. ceratophorum*
by Breitung 1957a), Man. (N to Churchill, s perhaps to Norway House, off the NE end of L.
Winnipeg), N Ont. (coast of Hudson Bay s to Fort Severn, ca. 56°N), James Bay (South Twin Is., ca.
53°N), Que. (coasts of Hudson Bay–James Bay and Ungava Bay; Côte-Nord, Anticosti Is., and
Gaspé Pen.), Labrador (s to ca. 55°N), and Nfld.; s half of w and E Greenland; Eurasia. [Incl. *T.
alaskanum* Rydb., *T. dentifolium* Hagl., and *T. acidolepis, T. croceum, T. kamtschaticum, T.
maurostylum, T. purpuridens,* and *T. rhodolepis* Dahlst.; *T. officinale sensu* Fernald and
Sornborger 1899, at least as to the Ramah, Labrador, plant, the relevant collection in GH; *T.
(Leontodon) ?rupestre* Greene; *L. (T.) ?scopulorum* (Gray) Rydb.]. MAPS: Porsild 1957: map 328, p.
201; Raup 1947: pl. 37 (indicating a station in cent. B.C. at Mt. Selwyn, ca. 56°N); the maps by
Hultén 1968b, for *T. alaskanum* and *T. kamtschaticum* apply here for the w area.

T. latilobum DC.
/T/E/ (Hr) Rocky slopes and talus, often calcareous, from E Que. (N coast of the Gaspé Pen.;
the reports from B.C. by Henry 1915, and from Man. by Jackson et al. 1922, require clarification) to
Nfld. (type material taken by Bachelot de la Pylaie in 1823), N.B. (St. John and Grand Manan; CAN;
GH; NBM), and Maine. [*Leontodon* Britt.].

T. laurentianum Fern.
/T/E/ (Hr) Calcareous meadows, ledges, and shores of E Que. (Mingan Is. of the Côte-Nord
and Anticosti Is.; CAN; GH) and w Nfld. (type from Ha-Ha Bay). [Perhaps best merged with *T.
ceratophorum* or *T. dumetorum*].

[*T. longii* Fern.]
[Calcareous turfs and gravels of Que. (Chimo, s Ungava Bay; Grande-Rivière, Gaspé Pen.; DAO;
GH) and NW Nfld. (type from Ha-Ha Mt.). Perhaps best merged with *T. ceratophorum*.]

T. officinale Weber   Common Dandelion
European; a very common weed in lawns, fields, and waste places in N. America, as in Alaska (N to
ca. 65°N), the Yukon (N to ca. 62°N), Dist. Mackenzie (N to Great Slave L.), and all the provinces (in
Man., N to Churchill; in Labrador, N to the Hamilton R. basin). [*T. dens-leonis* Desf.; *Leontodon (T.)
taraxacum* L.; *L. (T.) vulgare* Lam.; incl. var. *palustre* (Sm.) Blytt. (*L. (T.) palustre* Sm.)]. MAP: Hultén
1968b:945.

T. phymatocarpum Vahl
/AS/X/GeA/ (Hr) Calcareous ledges and alpine slopes from the coasts of Alaska–Yukon–Dist.
Keewatin to northernmost Ellesmere Is., s to s Baffin Is.; w and E Greenland N to ca. 70°N; NE
Siberia. [Incl. *T. eurylepium* Dahlst. and *T. pumilum* Dahlst., not Gaud.]. MAPS: Hultén 1968b:948;
Fernald 1933: map 12, p. 122.

[*T. torngatense* Fern.]
[Known only from granitic cliffs of N Labrador, the type from Nachvak Bay, ca. 59°N. Perhaps best
merged with *T. lapponicum*.]

NOTE

In addition to many merged in synonymy under the species treated above, the following species
("?microspecies") of *Taraxacum* have been reported from Alaska–Canada–Greenland (the ranges
of most of the Alaskan ones mapped by Hultén 1950 and 1968b):
(1) Alaska–Yukon endemics: *T. atkaense* Tatew. & Kitamura, *T. carneocoloratum* Nels., and many
species described by Haglund (*T. andersonii, angulatum, arietinum, aureum, caligans, cal-*

lorhinorum, chlorostephum, chromocarpum, cinericolor, decorifolium, demissum, eyerdamii, fabbeanum, festivum, hypochoeropsis, kodiakense, leptoglossum, leptopholis, maurolepium, microceras, mitratum, multesimum, ochraceum, paralium, patagiatum, phalolepis, pribilofense, scotostigma, signatum, speirodon, and *sublacerum*).

(2) Alaska–Yukon endemic: *T. vagans* Haglund.

(3) the Yukon endemic: *T. latilimbatum* Haglund.

(4) the Yukon–B.C. endemic: *T. flavovirens* Haglund.

(5) Alaska–Dist. Mackenzie endemic: *T. integratum* Haglund.

(6) Alaska; Europe: *T. retroflexum* Lindb. f. (introd. in Alaska; collections in CAN and RIM from Temiscouata and Rimouski counties and the Gaspé Pen., E Que., have also been placed here by Haglund); *T. dahlstedtii* and *T. undulatum* Lindb. f., the former introd. in Alaska and E Que.

(7) Alaska; Asia: *T. glabrum* DC. (*T. kamtschaticum* Dahlst.; *T. lyratum* of auth. in part, not *Leontodon lyratus* Ledeb.); *T. collinum* DC.; *T. scanicum* Dahlst. (introd. in Alaska; also reported from w James Bay, Ont., by Dutilly, Lepage, and Duman 1954, and a collection in RIM from Cacouna, Temiscouata Co., E Que., has been referred to it by Haglund); *T. sibiricum* Dahlst.

(8) Greenland endemic: *T. amphiphron* Böcher.

(9) Greenland; Europe: *T. devians* and *T. naevosum* Dahlst.; *T. atroglaucum, T. cyclocentrum, T. dilutisquameum,* and *T. latispinulosum* Chr.

(10) Greenland; Eurasia: *T. campylodes* Haglund, *T. firmum* and *T. islandiciforme* Dahlst., *T. nivale* Lange, and *T. curvidens, T. davidssonii,* and *T. pleniflorum* Chr.

Collections from Quebec have been referred by Haglund to the following species: *T. disseminatum* Haglund (Bic, Rimouski Co.; RIM); *T. kjellmanii* Dahlst. (Cacouna, Temiscouata Co., and Bic, Rimouski Co.; RIM); *T. lingulatum* Markl. (Cacouna; RIM); *T. sublaeticolor* Dahlst. (Port Daniel, Gaspé Pen.; RIM); *T. tenebricans* Dahlst. (Cacouna and the Gaspé Pen.; CAN; DAO; RIM); *T. tumentilobum* Markl. (Nouvelle, Gaspé Pen.; DAO; RIM). *T. densifolium* Kihlm. is also reported from Quebec by Rouleau (1947).

TETRADYMIA DC. [9410]

T. canescens DC. Horse-brush
/t/W/ (N) Dry open plains and foothills from S B.C. (valleys of the Dry Interior N to Kamloops and Vernon; CAN; V) and Mont. to Calif. and N.Mex.

THELESPERMA Less. [9236] Green-thread

T. marginatum Rydb.
/T/W/ (Grh) Dry plains and foothills from S Alta. (Fort Macleod, w of Lethbridge, and Medicine Hat, the type locality, where taken by John Macoun in 1894; CAN; the inclusion of Sask. in the range by Rydberg 1922, requires confirmation; not listed by Breitung 1957a) through w Mont. to sw Wyo. [*T. ambiguum sensu* J.M. Macoun 1895, not Gray, the relevant collection in CAN].

TOWNSENDIA Hook. [8895]

(Ref.: Larsen 1927; Beaman 1957)
1 Plants annual, biennial, or short-lived perennial, from densely strigose or short-pubescent to subglabrate, the leafy stems from a crown surmounted by a basal tuft of persistent leaves, usually over 5 cm tall; basal leaves oblanceolate, long-petioled, the stem-leaves similar but much reduced.
  2 Rays lavender to blue or purplish, distinctly bluish when dried, to 2 cm long; disk commonly over 2 cm broad; involucre to about 1.5 cm high; (SE B.C. and sw Alta.) . . . .*T. parryi*
  2 Rays pinkish, not at all bluish on drying, to 12 mm long; disk 1 or 2 cm broad; involucre to 1 cm high . . . . [*T. florifer*]
1 Plants perennial, stemless and caespitose, usually not over 5 cm tall, the taproot surmounted by a stout branching caudex.
  3 Leaves (and involucre) conspicuously woolly-villous with long loose hairs, spatulate to obovate, to about 1.5 cm long and 4 mm broad; heads either sessile among the

leaves or solitary on naked scapes to 5 cm tall; ray-ligules lavender or bluish violet, usually 1 or 2 cm long; pappus readily deciduous .... [*T. spathulata*]
3   Leaves (and involucre) subglabrate or merely strigose; heads subsessile among the erect leaves; ray-ligules white or pinkish; pappus persistent.
    4   Involucres to 16 mm high, their larger phyllaries to 2.5 mm broad; rays to 18 mm long; disk-corollas to about 12 mm long; leaves oblanceolate, to 4 mm broad; (s B.C. to sw Man.) .... *T. exscapa*
    4   Involucres to 12 mm high, their phyllaries at most about 1 mm broad; rays mostly less than 12 mm long; disk-corollas mostly not over 8 mm long; leaves linear or nearly so, mostly only 1 or 2 mm broad .... [*T. mensana*]

T. exscapa (Richards.) Porter
/sT/WW/ (Hr (Ch)) Dry valleys, plains, and foothills from SE B.C. (Columbia Valley N to Invermere and Windermere; Eastham 1947; a remarkable station for *T. hookeri* reported on dry embankments at about 2,500 ft in sw Yukon by Porsild 1966) to Alta. (N to Jasper; CAN), Sask. (N to the type locality near Carlton, about 35 mi sw of Prince Albert), and sw Man. (N to Routledge and Brandon), s to Ariz., Tex., and Kans. [*Aster exscapa* Richards., the type a Richardson collection from Carlton House Fort, Sask.; *T. sericea* Hook.; incl. *T. hookeri* Beaman]. MAPS: Hultén 1968*b*:855 (*T. hook.*); Porsild 1966: map 159 (*T. hook.*), p. 86; *Atlas of Canada* 1957: map 11, sheet 38; Beaman 1957: map 12, p. 105; Larsen 1927: pl. 1, p. 6.

[*T. florifer* (Hook.) Gray]
[The report of this species of the w U.S.A. (Wash. and Mont. to Nev. and Utah) from sw Alta. by John Macoun 1884, was later referred by Macoun 1886, to *T. parryi* var. *alpina,* the relevant collections in CAN. (*Erigeron* Hook.; *Stenotus* T. & G.). The MAPS by Beaman (1957: map 13, p. 115) and Larsen (1927: pl. 1, p. 6) indicate no Canadian stations.]

[*T. mensana* Jones]
[The inclusion of Alta. in the range of this species of the w U.S.A. (Mont. and Idaho to Utah, Colo., and S.Dak.) by Hitchcock et al. (1955) appears to be based upon its citation in synonymy under *T. sericea* Hook. by Larsen (1927:30). Hooker (1834) cited two collections for his *T. sericea,* a Richardson collection from Carlton House, Sask., and a Drummond collection from "dry banks of the Saskatchewan and among the Rocky Mountains", tentatively including *Aster (T.) exscapa* in synonymy. The former collection is the type of *T. exscapa,* now accepted as a distinct species and interpreted by Beaman (1957:100) as including the Drummond Rocky Mountain plant selected by Larsen as the type of *T. sericea,* a later-published name. The MAP by Beaman (1957: map 8, p. 90) indicates no Canadian stations.]

T. parryi Eat.
/T/W/ (Hs) Open ground at moderate to rather high elevations from SE B.C. (Crowsnest Pass, on the B.C.–Alta. boundary; Beaman 1957) and sw Alta. (N to Scalp Creek, 51°43′N; CAN) to Oreg., Wyo., and Colo. [Incl. var. *alpina* Gray]. MAPS: Beaman 1957: map 11, p. 98; Larsen 1927: pl. 1, p. 6.

[*T. spathulata* Nutt.]
[The citation of this species of the w U.S.A. (known only from Wyo. according to Beaman 1957, but also ascribed to Alta. and Idaho by Hitchcock et al. 1955) from s Alta. by Larsen (1927; High River, 50°35′N, the 1884 G.M. Dawson collection in GH) is probably based upon *T. parryi* (another 1884 Dawson collection from the same locality in CAN, revised by Beaman). MAPS: Beaman 1957: map 15, p. 120 (no Canadian stations); Larsen 1927: pl. 1, p. 6 (the Alta. area should probably be deleted).]

TRAGOPOGON L. [9579] Goat's-beard. Salsifis

1   Flowers purple; pappus brownish; peduncles upwardly enlarged; phyllaries to 4 cm long; (introd.) .... *T. porrifolius*
1   Flowers yellow; pappus whitish; (introd.).

2 Peduncles slenderly cylindric, not enlarged in flower, scarcely so in fruit; corolla canary-yellow; involucre to about 3 cm long, the phyllaries commonly less than 2.5 cm long; achenes (including beak) at most 2.5 cm long; leaves with undulate-recurving tips . . . . *T. pratensis*

2 Peduncles evidently enlarged under the head; corolla sulphur-yellow; involucre and phyllaries longer; achenes (including beak) to about 3.5 cm long; leaves with straightish tips . . . . *T. dubius*

T. dubius Scop.   Goat's-beard

European; introd. along roadsides and in fields and clearings in N America, as in B.C. (N to Prince George, ca. 54°N), Alta. (N to Beaverlodge, 55°13′N), Sask. (Breitung 1957a), Man. (N to Lac du Bonnet, about 50 mi NE of Winnipeg), Ont. (N to Thunder Bay and Monteith, NE of Timmins at ca. 48°40′N), and Que. (N to Hull and Montreal). [*T. major* Jacq.].

A hybrid with *T. porrifolius* (× *T. mirus* Ownbey) is reported from S Ont. by Boivin (1966b; Port Colborne, Welland Co.), who also tentatively reports one with *T. pratensis* (× *T. crantzii* Dichlt) from S Ont.

T. porrifolius L.   Salsify, Oyster-plant, Vegetable-oyster

European; introd. or a garden-escape to roadsides and fields in N. America, as in S B.C. (N to Kamloops), Alta. (Waterton Lakes; CAN, detd. Porsild), SE Man. (near Otterburne, about 30 mi s of Winnipeg), Ont. (N to Schreiber, N shore of L. Superior; CAN), Que. (N to Ste-Anne-de-la-Pocatière, Kamouraska Co.; QSA), N.B. (Bass River, Kent Co.; NBM), and N.S. (Grand Pré, Kings Co.; Roland 1947).

Forma *montgomeryi* Boivin (ligules white rather than purple) is known from the type locality, Port Colborne, Welland Co., S Ont. A hybrid with *T. pratensis* (× *T. mirabilis* Rouy) is reported from S Ont. by Boivin (1966b).

T. pratensis L.   Goat's-beard

Eurasian; introd. along roadsides and in fields and waste places in N. America, as in B.C. (N to Prince George, ca. 54°N; Eastham 1947), Alta. (Boivin 1966b), Sask., Man. (N to Winnipeg), Ont. (N to Monteith, where growing with *T. dubius*), Que. (N to Rimouski, Rimouski Co.), N.B., P.E.I., and N.S.

Forma *roseomarginatus* Thell. (phyllaries pinkish or roseate rather than greenish white) is reported from Ont. by Gillett (1958; Ottawa dist.).

TUSSILAGO L.   [9380]

T. farfara L.   Coltsfoot.   Pas-d'âne

Eurasian; introd. along damp ledges, clays, and brooksides in N. America, as in SW B.C. (Vancouver Is.; Herb. V), Ont. (N to the Ottawa dist.), Que. (N to Anticosti Is. and the Gaspé Pen.), Nfld., N.B., P.E.I., and N.S.

VERNONIA Schreb.   [8751]   Ironweed

1 Principal phyllaries abruptly narrowed to prolonged filiform tips; pappus purplish; inflorescence loose and open; leaves more or less pubescent beneath; stem glabrous or thinly pubescent . . . . [*V. noveboracensis*]

1 Principal phyllaries rounded, obtuse, or short-cuspidate at summit.
  2 Leaves punctate beneath when dry, they, the stem, and the achenes glabrous; pappus purplish; inflorescence very dense; (s Sask.–Man.) . . . . *V. fasciculata*
  2 Leaves not punctate, more or less pubescent beneath; achenes usually more or less pubescent on the ribs; inflorescence loose and open; (s Ont.).
    3 Stem more or less pubescent or tomentose; leaves thinly to densely tomentose beneath (at least along the veins) with long crooked hairs; pappus usually tawny, sometimes purplish; (s ?Ont.) . . . . [*V. missurica*]
    3 Stem essentially glabrous; leaves thinly pubescent beneath with minute straight hairs; pappus purplish; (s Ont.) . . . . *V. altissima*

V. altissima Nutt.
/t/EE/ (Hp) Damp rich soil from Mo. to Ohio, s Ont. (Essex, Kent, Lambton, and Lincoln counties; CAN; TRT), and N.Y., s to La. and Ga. MAP: Cain 1944: fig. 41, p. 303.

Var. *taeniotricha* Blake (peduncles and veins of the lower leaf-surfaces bearing multicellular hairs with dark-purple cross-walls) is known from s Ont. (Bradley's Marsh, Dover Twp., Kent Co.; TRT).

V. fasciculata Michx.
/T/EE/ (Hp) Rich moist ground and prairies from SE Sask. (Weyburn, about 60 mi SE of Regina; Breitung 1957a) to s Man. (Morris and Otterburne; CAN; Löve and Bernard 1959), Minn., and Ohio, s to Tex., Okla., and Mo. MAPS: Cain 1944: fig. 39, p. 297, and fig. 41, p. 303.

Our material is referable chiefly or wholly to var. *corymbosa* (Schwein.) Schub. (*V. cor.* Schwein.; involucres to 9 mm high, their exposed phyllary-tips to 3 mm broad, rather than involucres to 8 mm high, their exposed phyllary-tips at most 2 mm broad, the stem averaging lower and the leaves broader than those of the typical form).

[V. missurica Raf.]
[This species of the E U.S.A. (N to Iowa and Ohio) is reported from Essex and Lambton counties, s Ont., by Dodge (1914; 1915) and collections in CAN, GH, and TRT from those counties and Kent Co. have been referred to it. However, as pointed out by Cain (1944:302–05), it intergrades so completely (through the hybrid-swarm that *V. illinoensis* Gl. is now thought to consist of) with *V. altissima* and *V. fasciculata* that its occurrence in Ont. requires confirmation (as also, indeed, its retention as a distinct species). MAP: Cain 1944: fig. 41, p. 303.]

[V. noveboracensis (L.) Michx.]
[The reports of this species of the E U.S.A. (N to Ohio, W.Va., N.Y., and Mass.) from s Ont. by John Macoun (1884) and Dodge (1914) are based upon *V. altissima,* relevant collections in CAN. (*Serratula* L.; *S. (V.) praealta* L.).]

[WYETHIA Nutt.] [9193]

1 Leaves and involucres glabrous, resinous-varnished, the basal leaves to about 6 dm long and 1.5 dm broad; heads commonly several, the central one the largest; rays to 5 cm long . . . . [*W. amplexicaulis*]
1 Leaves hirsute or strigose-hirsute, they and the sparsely hairy but conspicuously ciliate phyllaries not resinous-varnished; basal leaves to about 5 dm long and 1 dm broad; heads commonly solitary; rays to 3.5 cm long . . . . [*W. angustifolia*]

[W. amplexicaulis Nutt.] Mule's-ears
[This species of the w U.S.A. (N to Wash. and Mont.) is tentatively reported from the "Borders of British Columbia" by John Macoun (1886) on the authority of Gray (1884) and from Kootenay, SE B.C., by Henry (1915). No Canadian material has been seen, however, and its occurrence in B.C. requires confirmation.]

[W. angustifolia (DC.) Nutt.]
[A collection in the herbarium of Manning Provincial Park, SE of Hope, B.C., has been referred to this species of the w U.S.A. (Wash. to Calif.) but requires confirmation. (*Alarconia* DC.; *Helianthus hookerianus* DC.; *H. longifolius* Hook., not Pursh).]

[XANTHISMA DC.] [8837]

[X. texanum DC.] Star-of-Texas
[This Texan species is reported from s Ont. by J.K. Shields (Rhodora 56(665):103. 1954; along a sandy roadside in Townsend Twp., Norfolk Co.), where taken by M. Landon in 1937 but probably not established, no other collections, apparently, having been made since that date. (*Centauridium drummondii* T. & G.).]

XANTHIUM L.   [9148]   Cocklebur, Clotbur.   Lampourde

1   Leaves narrowly to broadly lanceolate, essentially entire, attenuàte to both ends, commonly subtended by 3-parted spines; fruiting bur beakless or with a single (rarely 2) inconspicuous beak; (introd.) .... *X. spinosum*
1   Leaves ovate to cordate-rotund, commonly lobed, not subtended by spines; fruiting bur with 2(3) strong, often hooked beaks.
　　2   Surface of bur glabrous or merely minutely pilose or glandular between the prickles.
　　　　3   Prickles stout, about 1 mm thick at base, strongly arching to the hooked tip, commonly not more than about 50 visible on each face of the reddish-brown bur ....
　　　　　　.. [*X. curvescens*]
　　　　3   Prickles slenderly linear-subulate or bristleform, scarcely thickened at base, straight or arching only at summit.
　　　　　　4   Prickles commonly not over 3 mm long, bristleform, mostly less than 50 visible on each face of the yellow-green bur; beaks 1 or 2 mm long .... *X. strumarium*
　　　　　　4   Prickles to 7 mm long, 100 or more visible on each face of the brownish bur; beaks to 6 mm long.
　　　　　　　　5   Body of mature bur lustrous and essentially glabrous; prickles smooth or only remotely glandular, their bases much narrower than the intervening spaces .... *X. chinense*
　　　　　　　　5   Body of mature bur dull or sublustrous, often pubescent; prickles mostly glandular-hispid below, their bases about as broad as the intervening spaces .... *X. pensylvanicum*
　　2   Surface of bur distinctly pilose or hispid.
　　　　6   Prickles and beaks of bur as thick as or thicker than the length of their superficial hairs; beaks to about 6 mm long.
　　　　　　7   Burs densely prickly, 200 or more slenderly subulate prickles visible on each face; beaks with inflexed or hooked tips .... *X. pensylvanicum*
　　　　　　7   Burs with usually less than 100 strongly arching and strongly hooked prickles visible on each face; beaks strongly incurving .... *X. orientale*
　　　　6   Prickles and beaks of bur much narrower than the length of their elongate basal hairs.
　　　　　　8   Beaks subulate, to 10 mm long, their bases at most 1/3 as thick as the length of the beak; mature burs warm brown to reddish brown; leaves dentate ....
　　　　　　　　.. *X. strumarium*
　　　　　　　　9   Burs at most about 3 cm long and 2 cm thick, the body less than 2 cm long and 1 cm thick; beaks at most 7 mm long and 2 mm thick at base; prickles mostly not over 7 mm long .... *X. italicum*
　　　　　　　　9   Burs to 4 cm long and thick, the body to 2.5 cm long and 2 cm thick; beaks to 10 mm long and 3.5 mm thick at base; prickles to over 10 mm long ....
　　　　　　　　　.. *X. oviforme*
　　　　　　8   Beaks at most about 6 mm long, their stout bases 2 or 3 mm thick; mature burs drab to pale brown; prickles mostly not over 5 mm long.
　　　　　　　　10 Leaves shallowly undulate; beaks soon strongly incurved, their tips finally approximate or crossing .... *X. echinatum*
　　　　　　　　10 Leaves prominently dentate, the deltoid teeth nearly as long as or longer than broad; beaks erect or nearly so, straightish or hooked at tip ....
　　　　　　　　　.. *X. strumarium*

NOTE

The present treatment of *Xanthium* in Canada must be regarded as tentative. According to Arthur Cronquist (Rhodora 47(564):402. 1945), "The determination of species of *Xanthium* has become a formidable task, undertaken by many botanists only when it becomes unavoidable and then with serious misgivings."

In the words of Wiegand (*in* Wiegand and Eames 1926:414, footnote), "I am now greatly in doubt as to the existence of more than one real species in the group represented by *X. chinense* Mill., *X.*

pennsylvanicum Wallr., *X. italicum* Mor., and other related forms. The foliage in these forms is practically identical, and the only differences of any moment are in the burs, which are indeed highly variable. Extreme forms of burs, however, are often found in the same colony, as though sporadically produced. A large suite of specimens is almost sure to show a nearly or quite unbroken series through the various forms. In every attempt to segregate the burs into species, so many transitional specimens have been found as to do unwarranted violence to any species concept. It is probably wise to treat all North American Xanthiums as one species except *X. spinosum* L. and possibly *X. strumarium* L. and *X. echinatum* Murr. *X. strumarium*, however, is scarcely distinct, and with more study may also be included. *X. echinatum* may be a real species, as it has a distinct coastal range and seems to behave as though genetically distinct. Provisionally, the oldest name, *X. orientale* L., is here taken for the group (when *X. strumarium* and *X. echinatum* are excluded).''

Hitchcock et al. (1955) accept *X. strumarium* as a distinct species, including *X. canadense, X. chinense, X. oviforme, X. pensylvanicum,* and *X. varians* in its synonymy. The only other species they list for the NW U.S.A. is the European *X. spinosum.*

M.L. Fernald (Rhodora 48(568):70–74. 1946) strongly criticizes the conservative treatments advocated by Wiegand and Cronquist, later (Fernald *in* Gray 1950) listing 10 native North American species (and 5 introduced from the Old World). Doris Löve and Pierre Dansereau (Can. J. Bot. 37(2):173–208. 1959) accept a total of 23 native and introduced species for North America, their MAP, fig. 4, p. 186, indicating the distribution of these in the major political units.

X. chinense Mill.
/T/X/ (T) Moist ground, roadsides, and cult. or waste land from s B.C. (Dry Interior between Keremeos and Osoyoos, s of Penticton; CAN; perhaps actually referable to *X. strumarium* (*X. canadense* being reported from Penticton by Henry 1915, this referred to *X. italicum* by Eastham 1947); not definitely known from Alta.–Sask.; reported from Man. by Lowe 1943, and indicated for s Man. in the above-noted MAP by Löve and Dansereau), s Ont.–sw Que. (Löve and Dansereau, map), Vt., and Mass., s to Calif., Tex., and Fla.; (not in Asia, *chinense* a misnomer, the type actually from Mexico). [Incl. *X. americanum* Walt. and *X. pungens* Wallr.].

[X. curvescens Millsp. & Sherff]
[Known only from the shores of L. Champlain, Vt., according to Fernald *in* Gray (1950) but reported from Que. by Rouleau (1947: presumably the shores of L. Champlain in Missisquoi Co.). According to Fernald, it is probably a local hybrid between *X. chinense* and *X. orientale*.]

X. echinatum Murr.   Sea-Burdock
/T/X/ (T) Although Fernald *in* Gray (1950) reports this species only from "Beaches, dune-hollows and borders of saline marshes along the coast, N.S. to Va.", the above-noted map by Löve and Dansereau indicates a range from Idaho and Mont. to s Sask., s Man., s Ont., sw Que., and Maine, s to Utah, Colo., Mo., Mich., and N.C.; introd. in Europe. [*X. canadense* var. *echinatum* (Murr.) Gray].

X. italicum Moretti
/T/X/E/ (T) Low grounds, streambanks, and cult. or waste land from s B.C. (N to Vernon; CAN), s ?Alta.–Sask. (included in *X. strumarium* by both Moss 1959, and Breitung 1957a), s Man. (N to near Killarney and Winnipeg; CAN), ?Ont. (Fernald *in* Gray 1950; not indicated on the map by Löve and Dansereau), Que., N.B. (Fredericton; ACAD; DAO), and P.E.I. (Charlottetown; MT; not known from N.S.) to Calif., Mexico, Tex., and Fla.; W.I.; S. America; s Europe.

X. orientale L.
Eurasian; introd. along shores and in waste places in N. America, as in sw Que. (shores of the St. Lawrence R. around Montreal; shores of the Richelieu R.; shores of L. Champlain in Missisquoi Co.) and Vt.

X. oviforme Wallr.
Apparently native in the w U.S.A. (but included in *X. strumarium* by Hitchcock et al. 1955), the above-noted map by Löve and Dansereau indicating its occurrence in Wash. and Oreg.; introd. elsewhere, as in sw Que. (Montreal dist.; GH; MT) and from Mich. to Vt. and Pa.

X. pensylvanicum Wallr.
/T/X/ (T) Moist ground and cult. or waste land from Oreg. to N.Dak., Ont. (N to Russell Co.; TRT), sw Que. (Fernald *in* Gray 1950), and Mass., s to s Calif., Tex., and Fla.

X. spinosum L. Spiny Cocklebur
European (*see* M.L. Fernald, Rhodora 48(568):74. 1946, concerning the improbability of "Neolithic Bulgarians" coming to South America for the plant!); waste places in N. America, as in sw B.C. (ballast at Nanaimo, Vancouver Is., where taken by John Macoun in 1887; CAN), SE Sask. (Steelman, about 45 mi NE of Estevan; Breitung, 1957a), s Ont. (Middlesex, Waterloo, Wentworth, and York counties; CAN; OAC; TRT), Que. (Rouleau 1947), and N.B. (ballast at St. John, where taken by G.U. Hay in 1877; ACAD).

X. strumarium L.
/T/X/ (T) Moist ground, shores, and waste or cult. land, the aggregate species from B.C. (Lulu Is.; V) to Alta. (*X. commune* reported N to McMurray, 56°44′N, by Raup 1936), Sask. (N to Saskatoon; CAN), Man. (Otterburne; Löve and Bernard 1959), Ont. (N to Ottawa), Que. (N to Hull and Montreal; reports from the Atlantic Provinces require confirmation), and Mass., s to Calif., N.Dak., and Pa.; introd. in Eurasia.
1 Bur straight-beaked, usually yellowish green, merely puberulent, less than 2 cm long . . . .
. . . . var. *strumarium*
1 Bur incurved-beaked, usually yellowish brown or brownish.
　2 Burs usually over 2 cm long (to 3.5 cm), the surface between the prickles often stipitate-glandular, the lower part of the prickles conspicuously spreading-hirsute with viscid hairs; [*X. canadense* Mill.; incl. *X. varians* Greene and *X. commune*, *X. glanduliferum,* and *X. macounii* Britt.] . . . . var. *canadense* (Mill.) T. & G.
　2 Burs commonly less than 2 cm long, the surface between the prickles gland-dotted or slightly glandular-puberulent to subglabrous, the prickles hirsute; [*X. macrocarpum* var. *glab.* DC.; *X. glab.* (DC.) Britt.] . . . . var. *glabratum* (DC.) Cronq.

Index to Latin Names of Families, Genera and Species in the Systematic Section

Roman type is used for maintained genera and species, for excluded genera and species [treated in the text in square brackets], and for maintained family names (which appear in CAPITALS).

Italic type is used for synonyms and incidental references to maintained species.

Part 2: pages 91 to 545
Part 3: pages 547 to 1115
Part 4: pages 1117 to 1626.

Abies, 178
  alba, 183
  amabilis, 179
  americana, 181, 188
  balsamea, 179
  balsamifera, 179
  canadensis, 183, 188
  denticulata, 183
  douglasii, 186
  engelmannii, 182
  grandis, 179
  heterophylla, 188
  hookeriana, 188
  lasiocarpa, 179
  mariana, 183
  menziesii, 186
  mertensiana, 188
  mucronata, 186
  nigra, 183
  pattoniana, 188
  rubra, 183
  subalpina, 179
  taxifolia, 186
Abietia
  douglasii, 186
Abronia, 661
  acutalata, 661
  latifolia, 661
  micrantha, 661
  umbellata, 661
Abutilon, 1088
  abutilon, 1088
  avicennae, 1088
  theophrasti, 1088
Acalypha, 1054
  digynea, 1054
  rhomboidea, 1054
  virginica, 1054
ACANTHACEAE, 1399
Acer, 1073
  barbatum, 1074, 1077
  canadense, 1075
  circinnatum, 1074
  coccineum, 1076
  dasycarpum, 1076
  douglasii, 1074
  fraxinifolium, 1075
  ginnala, 1074

  glabrum, 1074
  grandidentatum, 1077
  interior, 1075
  macounii, 1074
  macrophyllum, 1074
  montanum, 1077
  negundo, 1075
  nigrum, **1075**, 1077
  pensylvanicum, 1075
  platanoides, 1075
  pseudo-platanus, 1076
  regelii, 1077
  rubrum, 1076
  saccharinum, **1076**, 1077
  saccharophorum, 1075, 1077
  saccharum, 1075, **1076**
  spicatum, 1077
  striatum, 1075
  subserratum, 1074
  tataricum, 1074
ACERACEAE, 1073
Acerates
  hirtella, 1250
  longifolia, 1251
  viridiflora, 1252
Achillea, 1457
  alpicola, 1458
  arenicola, 1458
  asplenifolia, 1458
  borealis, 1458
  dentifera, 1458
  filipendulina, 1458
  lanulosa, 1458
  ligustica, 1458
  megacephala, 1458
  millefolium, 1458
  multiflora, 1459
  multiplex, 1459
  nigrescens, 1458
  occidentalis, 1458
  pannonica, 1458
  ptarmica, **1458**, 1459
  setacea, 1458
  sibirica, 1459
  subalpina, 1458
  tomentosa, 1458
Achlys, 760
  triphylla, 760
Acinos
  arvensis, 1318
  thymoides, 1318
Acmispon
  americanus, 1007
Acnida
  altissima, 660
  cannabina, 659
  ruscocarpa, 659, 660
  tamariscina, 660
  tuberculata, 660
Acomastylis
  calthifolia, 921
  humilis, 921, 923
  rossii, 923

Aconitum, 718
  bicolor, 719
  chamissonianum, 719
  columbianum, 719
  delphinifolium, 719
  fischeri, 719
  insigne, 719
  lycoctonum, 719
  maximum, 719
  napellus, 719
  paradoxum, 719
  semigaleatum, 719
  septentrionale, 719
  variegatum, 719, **720**
Aconogonum
  phytolaccaefolium, 631
Acorus, 453
  calamus, 453
Acrolasia
  albicaulis, 1117
  ctenophora, 1117
  dispersa, 1118
  gracilis, 1117
Acroptilon
  picris, 1518
Acroschizocarpus
  kolianus, 846
Acrostichum
  alpinum, 171
  areolatum, 173
  ilvense, 172
  platyneuros, 154
  thelypteris, 170
Actaea, 720
  alba, 720
  americana, 720
  arguta, 720
  asplenifolia, 720
  brachypetala, 720
  brachypoda, 720
  caudata, 720
  eburnea, 720
  × ludoviciana, 720
  neglecta, 720
  pachypoda, 720
  palmata, 759
  racemosa, 728
  rubra, 720
  spicata, 720
Actinea
  acaulis, 1571
  herbacea, 1571
  richardsonii, 1571
Actinella
  acaulis, 1571
  glabra, 1571
  lanata, 1547
  richardsonii, 1571
Actinomeris, 1459
  alternifolia, 1459
  squarrosa, 1459
Adenarium
  maritimum, 680

1627

Index